# Construction Materials Reference Book

Fully updated to reflect the latest materials and their applications, this second edition of the *Construction Materials Reference Book* remains the definitive reference source for professionals involved in the conception, design and specification stages of a construction project. The theory and practical aspects of each material are covered in detail, with an emphasis on properties and appropriate use, enabling a deeper understanding of each material and greater confidence in their application.

Containing 38 chapters written by subject specialists, a wide range of construction materials are covered, from traditional materials such as stone through masonry and steel to advanced plastics and composites.

With diagrams, reference tables, chemical and mathematic formulae, and summaries of the appropriate regulations throughout, this is the most authoritative construction materials guide available. This edition features extra material on environmental issues, whole life costing, and sustainability, as well as the health and safety aspects of both use and installation.

**David Doran** graduated in 1950 with an external BSc (Eng). He then undertook National Service with the Royal Engineers in Malaya from 1953 to 1955. He is a Fellow of ICE and IStructE, a former member of Council of IStructE and Chairman of five Task Groups. From 1965 to 1985 he was Chief Engineer with George Wimpey plc, where he was Manager of the Civil & Structural Design Department with a staff of 300, and Head of QA for Wimpey Construction. Upon leaving Wimpey, David became an independent consultant and is retained by GBGeotechnics, Cambridge. He has been involved in several important publications since 1988.

**Bob Cather** graduated in Materials Science from the University of Bath. He initially worked with The Marley Tile Co. Ltd. on product development and subsequently spent more than 30 years with Ove Arup & Partners. Bob has considerable expertise in the selection, design and performance of materials in the built environment. This expertise has enabled his specialist contribution to projects and for clients worldwide, covering all areas of construction.

# Construction Materials Reference Book

## Second Edition

Edited by David Doran
and Bob Cather

LONDON AND NEW YORK

First edition published 1992
by Butterworth Heinemann

This edition published 2014
by Routledge
2 Park Square, Milton Park, Abingdon, Oxon, OX14 4RN

Simultaneously published in the USA and Canada
by Routledge
711 Third Avenue, New York, NY 10017

*Routledge is an imprint of the Taylor & Francis Group, an informa business*



*British Library Cataloguing in Publication Data*
A catalogue record for this book is available from the British Library

*Library of Congress Cataloging-in-Publication Data*
Construction materials reference book / edited by David Doran, Bob Cather. — Second edition.
pages cm
Includes bibliographical references and index.
1. Building materials—Handbooks, manuals, etc. I. Doran, David. II. Cather, Bob.
TA403.4.C66 2013
624.1'8—dc23

2012044271

ISBN13: 978-0-7506-6376-2 (hbk)
ISBN13: 978-0-08-094038-0 (ebk)

Typeset in Helvetica and Times New Roman
by Keystroke, Station Road, Codsall, Wolverhampton

Printed and bound in Great Britain by
CPI Group (UK) Ltd, Croydon, CR0 4YY

# Contents

# List of Contributors

**Godfrey Arnold** BSc CEng MIChemE
Godfrey Arnold Associates

**Ben Bowsher** BSc MIM MIQA
Executive Director UKCARES

**James Broughton** BEng(Hons) PhD
Departmental Head of Joining Technology Research Centre, Oxford Brookes University

**Chris Brown**
Formerly of Arup

**John Bull** Eur Ing BSc PhD DSc CEng FICE FIStructE FIHT FIWSc
Head of Civil Engineering, Brunel University

**Michael Bussell** BSc(Eng)
Retired Consultant, formerly Arup

**Geoffrey B. Card** Eur Ing BSc(Eng) PhD FICE
GB Card & Partners

**Bob Cather** BSc CEng FIMMM
Consultant, formerly Director of Arup Materials Consulting

**Vince Coveney** PhD
University of the West of England, Bristol

**David Doran** FCGI BSc(Eng) DIC CEng FICE FIStructE
Consultant, formerly Chief Engineer with Wimpey plc

**Geoff Edgell** BSc PhD CEng MICE FIM
Technical Director, CERAM Building Technology

**Charles Fentiman** MSc PhD
Director, Shire Green Roof Substrates

**Geoffrey Griffiths** FICE FIAT
Associate Director UK Infrastructure, Arup

**R. Harris**
Wimpey Environmental

**Barry Haseltine** MBE FCGI FREng DIC BSc(Eng) FICE FIStructE
Consultant

**Nick Hay** BSc
Project Manager, Copper Development Agency

**Shaun A. Hurley** BSc PhD MRSC
Consultant, formerly with Taylor Woodrow Technology Centre

**Allan R. Hutchinson** BSc PhD CEng MICE FIMMM FIMI
Head of Sustainable Vehicle Engineering Centre, Oxford Brookes University

**Stephan Jefferis** MA MEng MSc PhD CEng CEnv FICE CGeol FGS FRSA
Director Environmental Geotechnics Ltd; Visiting Professor, Department of Engineering Science, University of Oxford

**Robert Langridge** BSc MIAT
Asphalt Quality Advisory Service

**Andrew Lawrence** MA(Contab) CEng MICE MIStructE
Associate Director, Arup

**Ian Liddell** CBE FREng MA DIC CEng MICE
Consultant, formerly Buro Happold

**Arthur Lyons** MA MSc PhD Dip Arch Hon LRSA
De Montfort University

**Bryan Marsh** BSc PhD CEng MICE FICT FCS
Associate, Arup

**Stuart Moy** BSc PhD CEng FICE
Southampton University, School of Civil Engineering and the Environment

**Peter Olley**
Director, C. Olley & Sons Ltd

**David K. Peacock** CEng FIMMM
Titanium Marketing & Advisory Services

**Roger Plank** PhD BSc(Eng) CEng FIStructE MICE
Professor of Civil Engineering, University of Sheffield

**C. D. Pomeroy** DSc, CPhys, FinstP, FACI, FSS
Formerly of the British Cement Association

**Edwin Stokes** BEng CEng MIMMM
Stokes Consulting

**S. R. Tan** MSc CEng FIMMM
INEOS Vinyls UK Ltd

**David Thompsett** MA(Cantab)
Independent Consultant, Technical Advisor to British Plastics Federation, Formerly Jablite Ltd

**Robert Viles** CChem MRSC
Chief Technologist, Fosroc International

**Tony Wall** PhD
Consultant to the Zinc Information Centre

**David Wilson** MBE BSc PhD
Consultant, International Lead Association

**John Woods** Cert Ed AIP RP FIOR
Technical Officer, Lead Sheet Association

**Tim Yates** BTech BA PhD
Technical Director, BRE Ltd

# 1 Introduction

**David Doran** FCGI BSc(Eng) DIC CEng FICE FIStructE
Consultant, formerly Chief Engineer with Wimpey plc

**Bob Cather** BSc CEng FIMMM
Consultant, formerly Director of Arup Materials Consulting

## Choices in materials

To design and construct successfully with materials, it is important to develop an understanding of their inherent properties, their method of manufacture and the constraints and local conditions imposed on them by incorporation into a particular construction. This philosophy is central to the successful adoption of and design with materials in the real world.

For some, adoption of a particular material may be a means to an end to create a final product. For others, there may be more interest in understanding the inner nature of materials and how this understanding might help to create better or more interesting design and construction solutions. Different solutions, new and established, may not all require the same level of materials knowledge, but there will be few that would not benefit from some enhancement of understanding.

The range of properties of interest for a material to be used can be wide, and there is a strong impulse to increase this range.

In mechanical properties, strength and stiffness are frequent requirements. Strength considerations may extend to strength in compression, tension and bending. Strain behaviour under imposed loadings, static or dynamic, for short-term or sustained periods in cracked or uncracked elements begins to demonstrate the complexity of understanding that might be required. Other facets of behaviour that might be important to understand include thermal properties (thermal conductivity, thermal expansion and specific heat capacity), acoustic behaviour, optical characteristics (reflection and transmission), and electrical and magnetic responses.

Increasingly over recent decades a fuller understanding of materials and products under fire conditions has become essential. The importance of this understanding has been driven by people-safety and by commercial-loss considerations. Fire behaviour has wider parameters than some simple concept of 'burnability' or 'inflammability', requiring understanding of ignition characteristics, flame spread, heat release, strength loss, smoke and potential toxic fume emission. These issues, although based in materials and product contexts, will frequently be assessed in relation to whole-building and building occupant behaviour.

Another aspect of the performance of materials that has achieved higher profile relates to the retention of properties with time – the 'durability'. It is reasonably straightforward to determine the mechanical, physical or other properties at the beginning of utilisation, but how much of the properties will be lost and at what rate upon exposure to agencies such as rot, corrosion, UV radiation, freezing and thawing, insect attack or biological growths? In many situations it is the retained properties at the end of life that may be of greater importance. Of course, the material alone does not determine the lifetime performance; there is likely to be substantial influence from the mode and method of construction.

The desire to use materials better may derive from a desire to avoid known and recurring 'problems'. It may be driven by desires to do something differently – to build higher, to span further, to achieve longer life, perhaps with reduced aftercare or maybe just to use a smaller quantity of materials.

Using less material might be seen as one of the key routes to address the widespread concerns on the environmental impact of civilisations and industries. These concerns may be directed at 'global warming', carbon dioxide emissions, resource depletion, toxic wastes and emissions and numerous other targets. There appear to be many different routes to considering these issues and, at present, no universal description of need and solution. The extent to which changes to materials alone can provide adequate solutions, and to which change to living needs to be adopted, cannot at present be simply resolved.

It is clear that the proper understanding of material composition, properties and behaviour can form an essential role in reducing the impact on the environment of the use of construction materials. Much is already in place – the use of industrial wastes from iron making and power generation as a component of cement and concrete, the recycling of metallic components, use of alternative fuel sources to heat material production kilns, and the development of higher efficiency insulants are some prominent examples.

The 'natural world' offers many opportunities to develop construction materials solutions that not only reduce direct impacts of use but ought to be long-term options with more renewable supplies of materials. Such options may take us beyond wider and beneficial use of timber, straw, cork and sheep's wool to sources for resins, fibres, adhesives, rubbers, polymers, etc. for wide application. Beyond this we can look to the natural world for inspiration on a completely different range of practical solutions – the science of biomimetics.

Over the history of the utilisation of materials, the selection process and detail of use has been by a process that somewhat parallels 'natural selection' (also perhaps known as 'trial and error'). We use things, and those that work we use again; those that don't work we discard. Increasingly technology developments have given us the capabilities – should we choose – to understand the chemistry and microstructure of material, generally headed 'materials science'. Thus we now have greater ability to understand how existing materials work and perhaps, where beneficial, to design specific new materials for specific applications.

In parallel to the developments in the technology of materials there are considerable changes in the technologies of knowledge acquisition, dissemination and uptake. Together these knowledge technologies form an important basis for enabling the proper selection and utilisation of materials in construction applications. This *Construction Materials Reference Book* is one substantial component of the knowledge resource in materials.

There are many potential sources and kinds of information built up for materials in construction. This information and experience can be delivered by different mechanisms: word of mouth, research dissemination, dedicated and general electronic systems and printed word. The publication of this second edition of the *Construction Materials Reference Book* recognises the continuing value of a printed-word compilation of an up-to-date, authoritative and expert record of the best knowledge on materials and their application to construction.

This reference book is the sequel to the successful first edition published in 1992. It has been thoroughly updated and refreshed by the use of a mixture of new authors and some who contributed to the first edition. There is much new content in this edition.

It was neither desirable nor practical to supply authors with a straitjacket of guidelines from which to develop their themes and content. The editors wanted to achieve a record of maximum materials expertise and knowledge but in a format that carried across the whole book. The following list was made available to each contributor to use, as appropriate, as a broad framework for each chapter.

- Introduction & general description
- Sources
- Manufacturing process
- Chemical composition
- Physical properties
- Dimensional stability
- Durability
- Use and abuse
- Proprietary brands
- Hazards in use
- Coatings (paints and anti-corrosion)
- Performance in fire
- Sustainability and recycling
- References and bibliography

The perceived readership for this book is twofold:

(a) professionals (architects, engineers, surveyors, builders and materials specialists), particularly those who are contemplating using a material for the first time
(b) students or laypersons seeking an introduction to construction materials.

Such a book could not have been produced without support and encouragement. The editors would like to thank Alex Hollingsworth (who originally commissioned the book), Brian Guerin, Mike Cash, Mike Travers, Lan Te, Jodi Cusack, Liz Burton and, of course, the authors for bringing this project to a successful conclusion.

# 2 Aluminium

**John Bull** Eur Ing BSc PhD DSc CEng FICE FIStructE FIHT FIWSc
Head of Civil Engineering, Brunel University

## Contents

## 2.1 Introduction

Aluminium is the second most used metal throughout the world. Small quantities of aluminium were isolated in 1827, but it was after 1886, with the invention of the modern smelting process, that aluminium started to develop its potential.

Aluminium and its alloys are widely used in the construction industry, for example as shown in Figure 2.1. Aluminium has excellent corrosion resistance, good conduction of electricity and is strengthened by alloying, cold working or strain hardening. Aluminium and its alloys do not lose ductility or become brittle at low temperatures.

Aluminium and its alloys have a major advantage in that they can be extruded into a wide range of profiles, as shown in Figure 2.2. Further, aluminium can be repeatedly recycled and each recycling requires only 5% of the energy used for manufacturing new aluminium. In countries such as Brazil and Japan, the recycling rate for aluminium cans is over 90%.

**Figure 2.1** The aluminium-clad dome of the San Gioacchino Church in Rome, in service since 1897

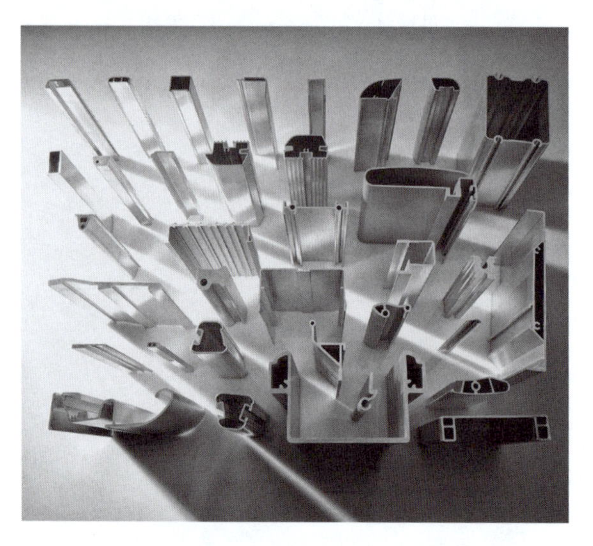

**Figure 2.2** A selection of extruded aluminium sections/profiles

## 2.2 A description of aluminium's properties

An aluminium alloy is an alloy in which aluminium predominates, by mass, over each of the other elements, provided it does not conform to the definition of aluminium. The definition of aluminium is a metal with a minimum content of 99.0% by mass of aluminium, with a content by mass of any other element, within the following limits:

- a total content of iron and silicon not greater than 1.0%
- a content of any other element not greater than 0.10% except for copper, which can have a content of up to 0.20% provided that neither the chromium nor the manganese content exceeds 0.05%.

Aluminium is malleable and easily worked by common manufacturing and shaping processes, and is ductile. Aluminium and almost all of its alloys are highly resistant to corrosion, are excellent conductors of electricity and are non-magnetic, non-combustible and non-toxic.

## 2.3 Sources of aluminium

Bauxite, the principal ore source for new aluminium, is composed primarily of one or more aluminium hydroxide compounds and is mined using open-cast methods as shown in Figures 2.3 and 2.4. About 85% of bauxite mined is used to produce the alumina, aluminium oxide, using the Bayer chemical process from which is produced aluminium using the Hall-Héroult electrolytic process. In 2011 more than 220 million tonnes (Mt) of bauxite was mined.[1] The major locations of extraction are Australia (67 Mt), China (46 Mt), Brazil (28 Mt), India (20 Mt), Guinea (18 Mt), Jamaica (10 Mt), Russia (6 Mt), Kazakhstan (5 Mt) and Suriname (5 Mt).

In 2011 the world primary aluminium (aluminium tapped from electrolytic cells or pots during the electrolytic reduction of metallurgical aluminium oxide) production[2] was estimated at 25.6 Mt, with north America contributing 5.0 Mt, east and central Europe 4.3 Mt, west Europe 4.0 Mt, GAC/Gulf region 3.5 Mt, Asia 2.5 Mt, Oceania 2.3 Mt, South America 2.2 Mt and Africa 1.9 Mt. Aluminium primary production in the UK in 2010 was 0.186 Mt.[1, 3]

**Figure 2.3** Bauxite mining in progress

**Figure 2.4** Bauxite mine shown in Figure 2.3 following reinstatement

**Figure 2.5** A selection of the 98% of aluminium scrap from the demolished Wembley Stadium being recovered for recycling

## 2.4 Manufacturing aluminium

The refining of the mined bauxite ore is completed in two stages: first the Bayer process, which obtains alumina from the bauxite ore, and second, the Hall-Héroult process that turns the alumina into aluminium. Four tonnes of bauxite makes 2 t of alumina, which makes 1 t of aluminium.[4, 5]

### 2.4.1 The Bayer process[4]

Bauxite, mined in the form of granules, is digested, depending on the property of the ore, at 140 to 240°C and pressures up to 3.5 MPa, with caustic soda to dissolve the aluminium, leaving the iron, silicon and titanium compounds undissolved. The residues are filtered and washed to leave liquor that contains only aluminium in the caustic solution. The aluminium is precipitated out as a hydrate, filtered, washed and calcined to produce the alumina, aluminium oxide. The excess caustic is removed from the residues and reused. The residue, known as red mud, is returned to the mining areas being restored.

### 2.4.2 The Hall-Héroult process[4]

In a primary smelter, aluminium is produced by an electrolytic process. The finely powdered alumina is dissolved in a molten bath of cryolite at about 950°C. The cryolite forms the electrolyte of the cell. The consumable anode and the permanent cathode are both made from carbon. The cells, known as pots, run at low voltage but very high amperage, typically 200,000 in the UK. Primary aluminium smelters use 14 kW h of electricity to produce 1 kg of aluminium. Consequently some 60% of the western world's primary aluminium is produced using 'clean' electricity such as hydroelectric power or where other forms of energy are low cost and plentiful, such as Bahrain and Dubai.

The molten aluminium produced at the cathode is periodically siphoned from the pot and sent to be cast into ingots for remelting, extrusion billets or rolling slabs. Generally, the aluminium produced is of 99.7% purity or better. More usually the molten aluminium from the cells is transferred to a holding furnace. There it is alloyed with a variety of elements such as iron, silicon, magnesium and copper. The alloy is then cast into an extrusion billet or a rolling slab using a semi-continuous process known as direct chill casting. These products can be sent directly to the extrusion presses and rolling mills for fabrication into semi-finish products, such as extrusions, sheet, plate and foil.

### 2.4.3 The recycling of aluminium and its alloys[6, 7]

Aluminium and its alloys are recycled, converting aluminium scrap into deoxidiser for the steel industry, foundry ingots and master alloys. Good-quality scrap is recycled for the production of extruded and rolled products. Scrap from new unused aluminium and aluminium alloys, being within the control of the aluminium industry, has a recycling rate of almost 100%. For aluminium previously sold to end consumers in transport, packaging, engineering, building, etc. as shown in Figure 2.5, some 73% is being returned to the aluminium industry to be recycled. The aluminium industry is working hard to increase this percentage. For example, aluminium used in building and construction has a recycling rate of 92% to 98%, depending on the type of building. Recycling rates of aluminium used in transport applications are over 95%.

Quality control is strictly exercised to ensure that the alloys conform to their specifications and compositional control. During the melting and casting operation the metal is degassed and cleaned using inert gas and filtration to remove hydrogen and other impurities.

### 2.4.4 Product form

The usual aluminium and aluminium alloy product forms are: forgings and forging stock, wire and drawing stock (electrical, welding, mechanical), drawn products, extruded products, foil, finstock (foil for heat exchanger applications), sheet, strip and plate, can stock and closures, slugs, electro-welded tube and alloys for foodstuff application.

EN573-4 gives tables for the form of products available for wrought aluminium and aluminium alloys for major fields of application. The tables give applications and product forms for each of the eight alloy groups for each alloy designation. Supplementary information is given concerning whether the mechanical properties are specified in the corresponding European Standards.

### 2.4.5 The fabrication process

#### *2.4.5.1 Casting*[8]

Externally, aluminium castings are found in aircraft, buses, cars, ships, spacecraft, trains and almost all vehicular transport where light weight reduces fuel requirements. Castings can be anodised to provide highly corrosion-resistant and coloured surfaces to enhance the appearance of building structures. Internally, aluminium castings are used for computers, cooking utensils, furniture, refrigerators, tables, washing machines and other lightweight high-technology equipment. Of aluminium castings, 60% are used in transport applications, 15% for domestic and office equipment, 6% for general engineering products and 5% in the building and construction industry.

The aluminium casting process may be by sand, pressure die, permanent mould, plaster or investment casting.

- Sand casting is a versatile and low-cost process, but is not as dimensionally accurate as, nor does it have the surface finish quality of, other casting processes. It does have the advantage of flexibility of numbers of castings produced.
- Pressure die-casting is the predominant casting method and is used for large-quantity production of small parts weighing up to 5 kg.
- High-pressure die casting is made by injecting molten aluminium alloy into a metal mould under high pressure. Rapid injection and solidification, under this high pressure, combine to produce a dense, fine-grained surface structure, with excellent wear and fatigue properties, close tolerances and good surface finishes.
- Die-castings cannot be easily welded or heat treated due to entrapped gases, but techniques such as vacuum die-casting can reduce entrapped gases. Heat treating the aluminium alloy die castings improves dimensional and metallurgical stability.
- In low-pressure die-casting, molten metal is introduced into metal moulds at pressures up to 170 kPa. The process is highly automated and thinner thicknesses can be cast than by permanent mould castings.
- Permanent mould castings, with a maximum weight of 10 kg, are used for high production runs. The tooling costs are high, but lower than for pressure die-casting. Destructible cores and complex cavities can be used. The castings have good dimensional tolerances and excellent mechanical properties which can be enhanced by heat treatment.
- In shell mould casting, a mould is made of resin-bonded sand, in the form of a shell from 10 mm to 20 mm thick. The castings have finer surface finishes and greater dimensional accuracy than sand casting. Equipment and production costs are relatively high, with the size and complexity of the castings being limited.
- In plaster casting, moulds are made of plaster and have high reproducibility, fine detail, close tolerances and a good surface finish. Although the cost of the basic equipment is low, the operating costs are high.
- Investment casting, used for precision engineered parts, uses refractory moulds formed over expendable wax or thermoplastic patterns. The molten metal is cast into the fired mould and produces components requiring almost no further machining. The process produces thin walls, good tolerances and fine surface finishes.
- In centrifugal casting the shape and size of the castings are limited and the cost is high. However, the integrity of the castings comes closer to that of wrought products and equates well with permanent mould castings.

#### *2.4.5.2 Rolling*[9]

Hot and cold rolling is used to reduce ingot thickness from up to 600 mm to as low as 0.05 mm and is divided into three main products as follows:

- plate – a flat material over 6 mm thick, used mainly as structural components
- sheet – a flat cold-rolled material, between 0.2 mm and 6 mm thick, used mainly in the construction and transport industries
- foil – a cold-rolled material less than 0.2 mm thick, used mainly for packaging.

In rolling, rectangular cast aluminium slabs, weighing up to 20 t, are heated to 500°C and then passed repeatedly through a rolling mill until either the required plate thickness is attained or the metal is thin enough to be coiled for further rolling. When in coil form, aluminium is fed through a series of cold rolling mills which successively reduce the metal thickness and recoil it after each rolling pass until the required thickness is obtained. Annealing may be required between passes, depending on the required final temper.

Rolling replaces the coarse cast structure with a stronger and more ductile material. The subsequent degrees of strength and ductility are functions of the amount of rolling, the rolling temperature, the alloy composition and the use of annealing

#### *2.4.5.3 The extrusion process*[10]

Aluminium can be extruded into any complex tight-tolerance shape with virtually no further machining being required. The cost of tooling is low when compared with the rolling, casting, forging and moulding of other metals. Some extruded shapes are unattainable by other processes. Aluminium can be placed where strength is needed in the extrusion, giving optimal structural efficiency.

Aluminium extrusions are produced by heating aluminium billets to 500°C and extruding or forcing the hot aluminium through a steel die. As the extruded section emerges it is cooled and cut to the desired length. Heat treatment is then used to optimise the material's mechanical properties. Various finishes such as natural silver or colour anodised film, a full range of colours in polyester powder coatings and electro-phoretic white/bronze acrylic paint can be applied for protection and improved appearance.

#### *2.4.5.4 Superplastic forming*[11]

Superplastic forming is a hot stretching process where aluminium sheet is forced over or into a one-piece die by the application of air pressure. Elongations of up to 10 times are possible. Only a limited number of superplastic alloys are at present available.

#### *2.4.5.5 Aluminium tube*

Aluminium and aluminium alloy tubes have uniform wall thickness and are normally round, square or hexagonal in cross-section.

*2.4.5.5.1 Welded tube* A continuous strip of aluminium or aluminium alloy is roll-formed into a tube shape and welded to form a complete tube. The complete tube may have the excess weld metal removed to obtain a smoother finish and may be subsequently rolled to form other cross-sectional shapes.

*2.4.5.5.2 Drawn tube* Drawn tube is a high-quality seamless product that can be produced with a thin wall section from both heat-treatable and non-heat-treatable alloys. A hollow cross-section billet is produced by extrusion over a mandrel. The size of the bloom is reduced by drawing it through a die, which determines the outside diameter of the tube. The inner diameter is defined by a plug.

*2.4.5.5.3 Extruded tube* Extrusion allows aluminium sections to be produced in an almost unlimited range of shapes. The fact that the cost of the die is very low makes the production of non-standard shapes an economic proposition. The 6000 series alloys are the most extrudable, followed by the 7000 series and then the 2000 series. The four-digit numerical system describing aluminium is explained in section 2.6.1.1.1.

#### *2.4.5.6 Forging*[12]

In forging, machining, joints and welds are eliminated and a fully wrought structure with improved shock and fatigue resistance, a high strength to weight ratio, good surface finish and the elimination of porosity is produced. Precision forgings are used for highly stressed parts in the construction, aerospace and automotive industries.

The forging of aluminium, usually involving heat-treatable alloys, is carried out in a similar way to other metals. Blanks are cut from extruded stock or an ingot and before forging preheated to between 400°C and 500°C, depending on the alloy. In hand forgings, the blank is hot worked between flat dies, using a pneumatic hammer or a press. Further machining produces the final component. In the non-heat-treatable alloys, where the mechanical properties depend on the degree of cold working, cold forging is possible. Simple components may be pressed or stamped directly from extruded stock. Die forgings use shaped dies, giving a high degree of dimensional consistency and reducing the machining to the finished form, and have good mechanical properties and structural integrity.

#### *2.4.5.7 Aluminium powder and paste*[13]

Commercially available particulate aluminium falls into three main categories.

- Atomised aluminium powder used in powder metallurgy for rapid solidification technology, metal matrix composites and mechanical alloying. Combined with powder extrusion, rolling and forgings allow the creation of new alloys and forms of material.
- Aluminium flake powder is used to make 'lightweight' concrete, as hydrogen gas bubbles form by reaction with the very alkaline cement paste, producing porous concrete.
- Aluminium paste is used in anti-corrosion paints and reflective roof coatings with bitumen, and as a coloured paint pigment.

## 2.5 The use of aluminium and aluminium alloys

### 2.5.1 Aluminium and aluminium alloys in the building, construction and offshore industries

Aluminium and aluminium alloys have a wide range of applications including: architectural hardware, conservatories, curtain walling, doors, exterior cladding, glazing, greenhouses, heating, ladders, partitions, prefabricated buildings, rain water goods, motorway sign gantries, roofing, scaffolding, shop fronts, signs, structural glazing, ventilating ducting and windows. Also structurally, as roof members, space frame constructions and whole structural sections such as offshore platform helidecks. In addition, aluminium offers the designer a immeasurable range of extruded profiles and decorative finishes such as anodising, coatings using powder, wet spray or electrophoretic techniques.

### 2.5.2 Rolled aluminium and aluminium alloys[9,11]

Rolled aluminium and aluminium alloys are used in many industries, including:

- *aircraft* – cladding, fitments and structural members
- *aerospace* – cladding, satellites and space laboratory structures
- *building industry* – cladding, guttering, insulation and roofing

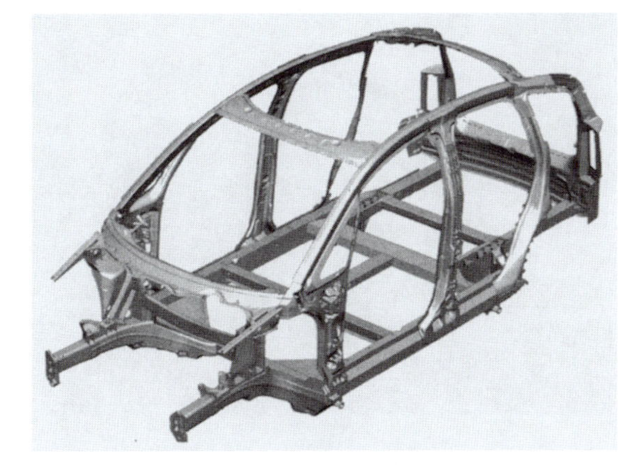

**Figure 2.6** Aluminium space frame of a motor vehicle

- *chemical industry* – chemical carriers, process plant and vessels
- *electrical industry* – busbars, cable sheathing, transformer windings and switchgear
- *food industry* – handling and processing equipment
- *general engineering* – cladding, heat exchangers and panelling
- *packaging* – beer barrels, bottle caps, cans, containers and wrapping
- *printing* – lithographic plates
- *rail industry* – coach panelling, freight wagons, structures and tankers
- *shipping* – hulls, interior fitments and superstructures
- *transport* – buses, lighting columns, radiators, tankers, tippers, traffic signs, trim and truck bodies.

Figures 2.6, 2.7, 2.8 and 2.9 show typical uses of structural aluminium.

## 2.6 Types of aluminium and aluminium alloys

Aluminium and its alloys fall into two categories:

- heat-treatable, where the alloy can be strengthened by suitable thermal treatment
- non–heat-treatable, where the alloy cannot be substantially strengthened by thermal treatment.

### 2.6.1 Classification of aluminium and aluminium alloys

#### *2.6.1.1 The numerical designations*

BS EN 573-1 gives the European designations of aluminium and aluminium alloys, as illustrated in the following example.

> *EN AW-5154A*. EN shows it is a European designation listed in a European Code. EN is followed by a space. A represents aluminium and W represents a wrought product. After the W the hyphen is followed by the international designation consisting of four digits, representing the chemical composition and if required, a letter identifying a national variation; this designation is attributed by the Aluminium Association via an international registration procedure.[14]

*2.6.1.1.1 The four-digit numerical system* An alloying element is an element intentionally added for the purpose of giving the metal certain special properties and for which minimum and maximum limits are specified.

**Figure 2.7** The Investec Media Centre at Lord's Cricket Ground, London. Designed in aluminium by Future Systems and fabricated in the Pendennis shipyard

**Figure 2.8** The aluminium Thames Water Tower, London

**Figure 2.9** Aluminium helicopter deck on a platform in the North Sea

- 1xxx (1 000 series): aluminium of 99% purity and greater. The second digit indicates alloy modifications in the impurity limits of alloying elements, with 0 indicating unalloyed aluminium having natural impurity limits. The last two of the four digits indicate to the nearest 0.01% the minimum aluminium percentage.
- The 2xxx to 8xxx series are grouped by the major alloying elements as follows:
  - 2xxx (2000 series) copper
  - 3xxx (3000 series) manganese
  - 4xxx (4000 series) silicon
  - 5xxx (5000 series) magnesium
  - 6xxx (6000 series) magnesium and silicon
  - 7xxx (7000 series) zinc
  - 8xxx (8000 series) other elements
  - 9xxx (9000 series) unused series.

The alloy designations in the 2xxx to 8xxx series are determined by the alloying elements ($Mg_2Si$ for the 6xxx alloys) present in the greatest mean percentages. If the greatest mean percentage is common to more than one alloying element, the choice of the group will be in the order of the group sequence Cu, Mn, Si, Mg, $Mg_2Si$, Zn or others. The second digit in the alloy designation indicates the original alloy and alloy modifications, with 0 indicating the original alloy. The last two of the four digits have no special significance and serve only to identify the different aluminium alloys in the group.

## 2.6.2 Temper designations

Temper designations given in BS EN 515 are for all forms of wrought aluminium and aluminium alloys and for continuously cast aluminium and aluminium drawing stock and strip intended to be wrought. The definitions used are as follows.

- Ageing is precipitation from supersaturated solid solution resulting in a change of properties of the alloy, usually occurring slowly at room temperature (natural ageing) or more rapidly at elevated temperatures (artificial ageing).
- Annealing is a thermal treatment to soften the metal by removing strain-hardening or by coalescing precipitates from solid solution.
- Cold working is plastic deformation of the metal at temperatures and rates such that strain hardening occurs.
- Solution heat-treatment consists of heating the metal to a suitable temperature, holding the metal at that temperature long enough to allow the constituents to enter into solid solution and then cooling rapidly to hold the constituents in solution.
- Strain-hardening modifies the metal structure by cold working, producing an increase in strength and hardness but a loss of ductility.

### *2.6.2.1 Basic temper designations*

The basic temper designation consists of letters and follows the hyphen after the alloy designation.

- *F – As fabricated.* F applies to the products of the shaping process in which no special control over thermal conditions or strain hardening is employed. There are no mechanical property limits for this temper.
- *O – Annealed.* Applies to products that are annealed to obtain the lowest strength temper. The O may be followed by a digit other than 0.
- *H – Strain-hardened.* H applies to products subjected to cold work after annealing or hot forming, or to a combination of cold work and partial annealing or stabilizing, to secure specified mechanical properties. There are always at least two

digits after H, the first giving the type of thermal processing and the second the degree of strain hardening. Any third digit identifies special processing techniques.

- *W – Solution heat-treatment.* W describes an unstable temper and applies only to alloys that spontaneously age at room temperature after solution heat treatment. The designation is specific only when the period of natural ageing is indicated; e.g. W ½ h.
- *T – Thermally treated to produce stable tempers other than F, O or H.* T applies to products that are thermally treated, with or without supplementary strain-hardening, to produce stable tempers. The one or more digits following T indicate the specific sequence of treatments.

*2.6.2.2 Subdivisions of the basic temper designation*

*2.6.2.2.1 Subdivisions of the O temper designation*

- *O1 – High temperature annealed and slow cooled.* O1 applies to wrought products that are thermally treated at approximately the same time and temperature required for solution heat treatment and slow cooled to room temperature, to accentuate ultrasonic response and/or provide dimensional stability. It is applicable to products that are to be machined prior to solution heat treatment by the user. Mechanical property limits are not specified.
- *O2 – Thermo-mechanical processed.* O2 applies to a special thermo-mechanical treatment to enhance formability and is applicable to products that are to be super-plastically formed prior to solution heat treatment by the user.
- *O3 – Homogenized.* O3 applies to continuously cast drawing stock or strip subjected to a high-temperature soaking treatment to eliminate or reduce segregations, thus improving subsequent formability and/or response to solution heat treatment.

*2.6.2.2.2 Subdivisions of H temper designations*

**2.6.2.2.2.1 First digit after H** The first digit after H indicates the specific combination of basic operations.

- *H1x – Strain hardened only.* H1x applies to products that are strain hardened to obtain the desired strength without supplementary thermal treatment.
- *H2x – Strain hardened and partially annealed.* H2x applies to products that are strain hardened more than the desired final amount and then reduced in strength to the desired level by partial annealing. For alloys that age-soften at room temperature, the H2x tempers have the same ultimate tensile strength as the corresponding H3x tempers. For other alloys, the H2x tempers have the same minimum ultimate tensile strength as the corresponding H1x tempers and slightly higher elongation.
- *H3x – Strain hardened and stabilised.* H3x applies to products that are strain-hardened and whose mechanical properties are stabilised either by low temperature thermal treatment or as a result of heat introduced during fabrication. Stabilisation usually improves ductility. This designation is applicable only to those alloys that, unless stabilised, gradually age-soften at room temperature.
- *H4x – Strain hardened and lacquered or painted.* H4x applies to products that are strain hardened and that may be subjected to some partial annealing during the thermal curing which follows the painting of lacquering operation.

**2.6.2.2.2.2 Second digit after H** The second digit following H indicates the final degree of strain hardening, as identified by the minimum value of the ultimate tensile strength. Tempers between O and Hx8 are designated by numerals 1 to 7.

- 8 has been assigned to the hardest tempers normally produced.
- 4 designates tempers whose ultimate tensile strength is midway between that of the O and the Hx8 tempers.
- 2 designates tempers whose ultimate tensile strength is midway between that of the O and the Hx4 tempers.
- 6 designates tempers whose ultimate tensile strength is midway between that of the Hx4 and the Hx8 tempers.
- 1, 3, 5 and 7 designate tempers intermediate between those defined above.
- 9 designates tempers whose minimum ultimate tensile strength exceeds that of the Hx8 tempers by at least 10 MPa.

**2.6.2.2.2.3 Third digit after H** When used, the third digit indicates a variation of a two-digit temper. It is used when the degree of control of temper and/or the mechanical properties differ from, but are close to, that for the two-digit H temper designation to which it is added, or when some other characteristic is significantly affected, e.g. Hx11, H112, H116, Hxx4, Hxx5. Other digits after H may be used to identify other variations of a subdivision.

*2.6.2.2.3 Subdivisions of T*

**2.6.2.2.3.1 First digit after T** The first digit following T identifies the specific sequences of basic treatments, with numerals 1 to 9 being assigned as follows.

- *T1* – Cooled from an elevated temperature shaping process and naturally aged to a substantially stable condition.
- *T2* – Cooled from an elevated temperature shaping process, cold worked and naturally aged to a substantially stable condition.
- *T3* – Solution heat treated, cold worked and naturally aged to a substantially stable condition.
- *T4* – Solution heat treated and naturally aged to a substantially stable condition.
- *T5* – Cooled from an elevated temperature shaping process and then artificially aged.
- *T6* – Solution heat treated and then artificially aged.
- *T7* – Solution heat treated and over-aged/stabilised.
- *T8* – Solution heat treated, cold worked and then artificially aged.
- *T9* – Solution heat treated, artificially aged and then cold worked.

**2.6.2.2.3.2 Additional digits after T** One or more digits may be added to designations T1 to T9 to indicate solution heat treatment and or precipitation treatment, the amount of cold work after the solution heat treatment or the stress relieving operation, as follows.

- *Stress relieved by stretching:* Tx51, Txx51, Tx510, Txx510, Tx511, Txx511.
- *Stress relieved by compression:* Tx52, Txx52.
- *Stress relieved by combined stretching and compression:* Tx54, Txx54.

The 51, 52 and 53 digits may be added to the designation W to indicate unstable solution heat treatment and stress-relieved tempers.

- *T42* – Solution heat treated from annealed or F temper and naturally aged to a substantially stable condition.
- *T62* – Solution heat treated from annealed or F temper and artificially aged.

The variations of the T7 tempers are designations that apply to products which are artificially over-aged to improve a property such as stress corrosion resistance, fracture toughness, exfoliation corrosion resistance, or to obtain a good compromise between the abovementioned properties and the tensile strength, e.g. T79, T76, T74 and T73.

There may be a second numeral after T to indicate a temper extension.

- 1 as a second digit after T may be used to indicate a solution heat treatment at lower than standard temperature, a limited rate of quenching, a limited and controlled amount of cold work or an artificial ageing in under-ageing conditions.
- 1 and 3 to 9 as a second digit after T3, T8 or T9 indicates increasing amounts of cold working after solution heat treatment or after artificial ageing.
- 1 and 3 to 5 as a second digit after T5 or T6 indicates different degrees of under-ageing.
- 6 as a second digit after T5 or T6 indicates a level of mechanical properties, respectively higher than T5 or T6, obtained through the special control of the process.

## 2.7 Chemical compositions and mechanical properties

### 2.7.1 The master alloys[15]

Aluminium master alloys are concentrated alloys of an element pre-dissolved in aluminium. They comprise aluminium with metals such as boron, chromium, copper, iron, manganese, silicon, strontium, titanium and zirconium, and adjust the composition of aluminium alloys and control the final properties. Grain refiners containing titanium and boron are added to aluminium to enhance the metallurgical structure, increasing the casting speed and improving surface quality.

### 2.7.2 Chemical composition

The chemical composition and form of wrought products is given in BS EN 573-2 and is intended to supplement the four-figure designations given in BS EN 573-1. The designations of aluminium and aluminium alloys are based on chemical symbols followed by numbers indicating the purity of the aluminium or nominal content based on the chemical composition limits given in BE EN 573-3.

If the chemical symbol based designation is used, then there will be the prefix EN, followed by a blank, then the letter A representing aluminium and the letter W identifying wrought products or ingots to be wrought. W is separated from the following designation by a hyphen as follows: EN AW-5052, or EN AW-5052 [Al Mg2.5] or for exceptional use EN AW-Al Mg2.5.

To ensure consistency with other national and international standards and in particular with ISO 209-1, whose code of designation is based on the same principles:

- where the composition of an alloy is strictly identical to the composition of an alloy registered by ISO, the ISO designation shall be used
- where the composition of an alloy does not correspond to the composition of any alloy in ISO 209-1, a new designation for the alloy will be created.

#### *2.7.2.1 Coded designation of wrought aluminium and aluminium alloys*

Designations for unalloyed aluminium consist of the international chemical symbol for aluminum (Al) followed by the percentage purity, e.g.: EN AW-1199 [Al 99.99], EN AW-1070A [Al 99.7]. Al is separated by a blank space from the percentage purity. If a low-content element is added to the unalloyed aluminium, the symbol of the element is added without a space after the percentage purity: EN AW-1100 [Al 99.0Cu].

For aluminium alloys with several added alloying elements, they are arranged in order of decreasing nominal content, with numbers expressing the mass percentage content of the considered elements, e.g.: EN AW-6061 [Al Mg1SiCu], EN AW-2014 [Al Cu4SiMg]. If these contents are equal, the alloying elements, restricted to four, are arranged in alphabetical order of symbols, e.g.: EN AW-2011 [Al Cu6BiPb].

To distinguish between two alloys of similar composition, additional designations are used, given in decreasing priority, e.g.: EN AW-5251 [Al Mg2], EN AW-5052 [Al Mg2,5], EN AW-6063 [Al Mg0,7Si].

In certain alloys, the base metal is of high purity and it is necessary to give the specified content in full to two decimal places, e.g.: EN AW-5305 [Al 99.85Mg1].

BS EN 573-3 specifies, in percentage by mass, the chemical composition of aluminium and aluminium alloys. The numerical and the chemical symbols alloy designations are given for each of the eight series of aluminium and aluminium alloys.

### 2.7.3 Mechanical properties

For aluminium and aluminium alloys:

- BS EN485-2 gives the mechanical properties sheet, strip and plate
- BS EN754-2 gives the mechanical properties for cold drawn rod/bar and tube
- BS EN755-2 gives the mechanical properties for extruded rod/bar, tube and profiles
- BS EN1592-2 gives the mechanical properties for HF seam-welded tubes.

#### *2.7.3.1 Hardness*

There is no simple relationship between hardness and tensile strength for aluminium and aluminium alloys. However, there are portable hardness testers with graduations that relate to aluminium and are used for surveillance checking and distinguishing between stocks of different alloys, or stocks in different heat tempers.

## 2.8 Physical properties

### 2.8.1 Material constants for normal temperature design

Table 2.1 gives the typical range of material constants for aluminium and its alloys. Table 2.2 gives the values used in

**Table 2.1** Range of values of material constants for aluminium and its alloys[9]

| *Property* | *Range of values* |
|---|---|
| Coefficient of linear expansion | $16 \times 10^{-6}$ to $24 \times 10^{-6}$ per °C |
| Electrical conductivity (% of the International Annealed Copper Standard (IACS)) | 20% to 63.8% |
| Electrical resistivity | 2.7 to 8.62 μΩ cm |
| Melting range | 475 to 770°C |
| Modulus of elasticity | 69,000 to 88,000 N/mm$^2$ |
| Poisson's ratio | 0.3 to 0.35 |
| Proof stress (0.2 per cent) | 60 to 520 N/mm$^2$ |
| Tensile strength | 55 to 580 N/mm$^2$ |
| Thermal conductivity | 117 to 244 W/m °C |
| Unit mass | 2650 to 2840 kg/m$^3$ |

**Table 2.2** Values of material constants used in EN 1999-1-1 structural design calculations

| *Property* | *Value* |
|---|---|
| Coefficient of linear thermal expansion | $\alpha = 23 \times 10^{-6}$ per °C |
| Modulus of elasticity | $E = 70{,}000$ N/mm$^2$ |
| Poisson's ratio | $\nu = 0.3$ |
| Shear modulus | $G = 27{,}000$ N/mm$^2$ |
| Unit mass | $\rho_{al} = 2700$ kg/m$^3$ |

structural design calculations for aluminium alloys at 20°C, for normal temperature design and for aluminium and its alloys covered by EN1999-1-1. For service temperatures between 80°C and 100°C a reduction in strength must be taken into account. However, the reduction in strength is recoverable for temperatures between 80°C and 100°C when the temperature is reducing to normal temperature design.

In general both the tensile strength and elongation of aluminium and its alloys are greater at sub-zero temperatures than at normal temperature design. None of the aluminium alloys suffer from brittleness at low temperatures and there is no transition point below which brittle fracture occurs.[11]

### 2.8.2 Material constants for elevated temperatures

*2.8.2.1 General*

The values of the material constants at elevated temperatures associated with fire are given below in accordance with EN1999-1-2. The effectiveness of fire protection materials for aluminium is performed by test, but at present there is no European Standard for testing such materials for aluminium.

*2.8.2.2 Proof stress*

For thermal exposure of up to 2 h, the 0.2% proof stress $f_o$ is multiplied by a strength reduction factor $k_{o,\theta}$ whose value depends on temperature, alloy and temper. For temperatures up to 100°C $k_{o,\theta}$ has a value of between 0.90 and 1.00; at 250°C, $k_{o,\theta}$ has a value of between 0.23 and 0.82; at 350°C, $k_{o,\theta}$ has a value of between 0.06 and 0.39; and at 550°C, $k_{o,\theta} = 0$.

*2.8.2.3 Modulus of elasticity*

The modulus of elasticity $E_{al,\theta}$ after 2 h of exposure to elevated temperature changes from 70,000 N/mm$^2$ at 20°C to 67,900 N/mm$^2$ at 100°C, to 54,600 N/mm$^2$ at 250°C, to 37,800 N/mm$^2$ at 350°C, to 0 N/mm$^2$ at 550°C.

*2.8.2.4 Unit mass*

The unit mass is independent of temperature and remains at $\rho_{al} = 2700$ kg/m$^3$.

*2.8.2.5 Thermal elongation (coefficient of thermal expansion)*

The relative thermal elongation $\Delta l/l$ for $0°C < \theta_{al} < 500°C$ is calculated using:

$$\Delta l/l = (0.1 \times 10^{-7} \times \theta^2_{al}) + (22.5 \times 10^{-6} \times \theta_{al}) - (4.5 \times 10^{-4}) \quad (2.1)$$

where $l$ is the length at 20°C and $\Delta l$ is the temperature-induced elongation.

*2.8.2.6 Specific heat*

The specific heat $c_{al}$ for $0°C < \theta_{al} < 500°C$ is calculated using:

$$c_{al} = [(0.41 \times \theta_{al}) + 903] \text{ J/kg °C}. \quad (2.2)$$

*2.8.2.7 Thermal conductivity*

The thermal conductivity $\lambda_{al}$ for $0°C < \theta_{al} < 500°C$ depends on the alloy. For alloys in the 3xxx and 6xxx series:

$$\lambda_{al} = [(0.07 \times \theta_{al}) + 190] \text{ (W/m °C)} \quad (2.3)$$

For alloys in the 5xxx and 7xxx series:

$$\lambda_{al} = [(0.1 \times \theta_{al}) + 140] \text{ (W/m °C)} \quad (2.4)$$

## 2.9 Special properties

### 2.9.1 Creep

For normal temperature design, creep is not considered. However, creep is greater the higher the temperature. There are a number of aluminium alloys with satisfactory creep performance at 200°C to 250°C, but most aluminium alloys show increasing creep at 300°C and higher.[11]

### 2.9.2 Fatigue strength

Due to the much lower mass of aluminium and its alloys than that of steel, the ratio of variable actions to permanent actions is high and the fatigue design of aluminium is more critical than that for steel.

The generalised form of the fatigue strength curves, stress range to number of cycles, is given in tables of classified detail categories and fatigue strength curves in EN1999-1-3. Unclassified details are assessed by reference to published data or by fatigue testing.

The limit state of fatigue uses one of the three following methods.

- *Safe life design:* based on the calculations of damage during the structure's design life using standard lower bound endurance data and an upper bound estimate of the fatigue loading. The method provides a conservative estimate of fatigue life and does not require in-service inspection.
- *Damage-tolerant design:* based on monitoring fatigue crack growth using an inspection programme applied throughout the life of the structure.
- Design assisted by testing: only used when the necessary data is not available from standards or other reliable sources.

### 2.9.3 Fire

The thermal conductivity and specific heat of aluminium are four times and twice respectively that of steel. Heat is conducted away from hot spots faster in aluminium, which extends serviceability time but raises temperature in other parts of the structure. However, more heat is required to bring the same mass of aluminium to a specific temperature, e.g. welding.

Under normal fire conditions, aluminium and its alloys, like many other construction metals, are considered non-combustible and obtain the highest possible classification against fire penetration and spread of flame.

### 2.9.4 Thermite sparking[16]

The thermite reaction is an exothermic reduction of iron oxide (rust) by aluminium particles in an intimately mixed smear. Once it is initiated considerable heat builds up, generating a spark that can ignite an explosive gas mixture (e.g. methane–air) if present. The thermite spark occurs either at the initial impact or when a previously deposited aluminium smear is subsequently struck either by steel, or aluminium in the presence of rust. Thermite sparking is eliminated by anodising the aluminium or painting or galvanising the steel surfaces.

### 2.9.5 Corrosion of aluminium[17]

Corrosion is deterioration by chemical or electrochemical reaction with the environment. Aluminium is corrosion-resistant in many environments due to the inert film of aluminium oxide that forms on its surface.

#### *2.9.5.1 Galvanic corrosion (bi-metallic corrosion)*

Aluminium, when in electrical contact with more noble metals such as copper, lead, nickel or steel, will begin to corrode if water is present. When in contact with stainless steel, aluminium corrosion is negligible provided the protective oxide film on the stainless steel remains in place. Galvanic corrosion can be eliminated by inserting an insulating material between the metals. Galvanic corrosion can be put to good use to provide cathodic protection. To reduce steel corrosion, aluminium bars are connected to the steel as anodes and the aluminium corrodes.

#### *2.9.5.2 Crevice corrosion*

Crevice corrosion in aluminium occurs when an aqueous solution is trapped in a crevice or at a joint. The rate of corrosion is generally low as the corrosion products restrict the entrance to the crevice; however, the corrosion products can cause distortion of thin sections.

#### *2.9.5.3 Pitting*

Pitting is localised corrosion, usually occurring on an externally exposed aluminium surface immersed in a moist environment. If the pitted surface dries out, the pitting will stop.

#### *2.9.5.4 Intergranular corrosion*

Intergranular corrosion, where a path is corroded preferentially along the grain boundaries of the metal, is unusual and generally only occurs with specific alloys or environments. The mechanism is electrochemical and dependent on the formation of local cells at the grain boundaries. The degree of susceptibility of an alloy to intergranular attack can vary depending on its microstructure, which is a result of the metallurgical history and thermal treatment. Heat treatments are used to diminish the intergranular attack.

#### *2.9.5.5 Exfoliation corrosion*

Exfoliation corrosion, which is unusual, is a specific type of selective attack that proceeds along grain boundaries which run parallel to the surface of the metal, forcing the layers apart causing the metal to swell. Metal flakes may be pushed up and even peel from the metal surface. It is most common in the heat-treatable Al–Mg–Cu and Al–Zn–Mg–Cu alloys and is associated with a marked directionality of the grain structure. If the grain structure has no directionality, exfoliation corrosion does not usually occur.

#### *2.9.5.6 Stress corrosion cracking*

For this unusual form of corrosion to occur, the three following requirements must be satisfied.

- The alloy must be susceptible due to its composition and microstructure (5xxx series alloys with Mg > 3.5%).
- The alloy must be in a corrosive environment (over 70°C for long periods of time).
- The alloy must be under tensile stress.

In the absence of any one of the above requirements, stress corrosion cracking does not occur.

## 2.10 Durability and protection[17]

### 2.10.1 General[17]

The corrosion resistance of aluminium and its alloys is due to the inert self-sealing protective oxide film that forms on its surface on exposure to oxygen. The oxide film is stable within the pH range of about 3 to 9.

In mild environments no protection is needed for the majority of alloys.

In moderate industrial conditions darkening and roughening of the surface will take place. As the atmosphere becomes more aggressive, surface discolouration and roughening increase, as do the white powdery surface oxides. Added protection may be necessary.

In costal and marine environments the surface will roughen and acquire a grey, stone-like appearance, with some alloys requiring protection. The rate of corrosion decreases rapidly with time as the oxide film builds, but in a few cases, e.g. exposure to caustic soda, the corrosion rate increases linearly. Special precautions may be necessary if the aluminium, containing copper as a major alloying element, is immersed in sea water.

Aluminium subject to water runoff from fresh concrete will stain or corrode, and protection of the aluminium is required

In a mild rural environment, aluminium and its alloys have little initial loss of reflectivity for up to three years followed by almost no change for up to 80 years. Tropical environments are in general no more harmful than temperate environments, although certain alloys are affected by long exposure to high ambient temperatures.

### 2.10.2 Alloy durability[17]

The 1xxx series has low strength but excellent corrosion resistance.

The 2xxx series alloys are very strong and are used for structural purposes. However, their corrosion resistance is not high and they are significantly affected in heavily polluted industrial or marine environments. They require protection in aggressive environments.

The 3xxx series alloys are stronger than the 1xxx series and have good corrosion resistance but may need lacquering or painting in heavily polluted industrial or marine environments.

The 4xxx series has similar properties to the 3xxx series but is not often used.

The 5xxx series alloys have superior mechanical properties to the 1xxx, 3xxx and 4xxx series and better corrosion resistance than the 6xxx series. The 5xxx series can be used in marine conditions where total immersion in sea water is required.

The 6xxx series alloys are the most common aluminium alloys, being widely used in architecture, engineering and transport. The good corrosion resistance of these alloys means that they can be used in marine and industrial environments.

The 7xxx series are high-strength alloys, having reduced corrosion resistance. Protection in moderate or severe industrial, marine and urban environments is required to prevent, for example, stress corrosion cracking.

### 2.10.3 Protection

EN1999-1-1 categorises aluminium alloys into durability ratings, A, B and C in descending order of durability. These ratings determine the need and degree of required protection. Under normal atmospheric conditions alloys listed in EN1999-1-1 do not require protective treatment. However, during execution, measures should be taken to ensure that no corrosion develops and that all parts are well ventilated and drained.

EN1090-3 lists the protection methods, which include:

- anodic oxidation
- surface preparation
- coatings, pre-treatment, base coat, final coat
- coatings with bitumen or bituminous combinations
- repair coatings
- passivation.

#### *2.10.3.1 Overall protection*

The most common method of protection is anodising, which is an electrolytic process and produces a dense, chemically stable protective aluminium oxide film that is an integral part of the underlying aluminium. Most films produced by anodising are translucent and show the silvery look of the aluminium, but various colours are achievable varying the anodising process. Other methods include chemical conversion coatings and various paint finishes. Sacrificial anodes, e.g. zinc, can be used to protect aluminium when used in the marine environment.[17]

The required protective treatment will be described in the project specification and will consider the corrosion mechanism, exposure condition, material thickness and alloy durability. The treatments range from no protection to protection normally required except in special cases, protection that depends on the special conditions of the structure, and immersion in sea water not recommended. EN 1090-3 and EN508-2 give further details.

The protection of the internal voids of hollow sections must be considered but may not be required if the internal void can be sealed.

#### *2.10.3.2 Aluminium in contact with aluminium and other metals*

Due to electrochemical attack on the aluminium, aluminium contact surfaces in crevices, contact with certain metals or washings and joints of aluminium to aluminium or to other metals and contact surfaces in bolted, high-strength friction grip bolted, riveted or welded joints require protection as given in Table D2 of EN 1999-1-1 in addition to that required by Table D1. Details of the corrosion protection procedures are given in EN 1090-3 and in EN 508-2.

Where pre-painted or protected components are assembled, additional sealing of the contact surfaces is defined in the project specification with consideration given to the expected life of the structure, the exposure and the protection quality of the pre-protected components.

Where the metals being joined to aluminium are aluminium, painted steel, stainless steel and zinc-coated steel with the bolt or rivet material being aluminium, stainless steel or zinc coated steel, protection is related to three types of environment:

- *atmospheric:* rural dry unpolluted, rural mild; industrial urban moderate, industrial urban severe
- *marine:* non-industrial, industrial moderate, industrial severe
- *immersed:* fresh water, sea water.

For the metal being joined there are five increasing levels of procedures that can be applied: 0, 0/X, X, Y and Z. Procedure 0 is no treatment through to procedure Z where procedures shall be established by agreement with the parties involved.

For bolts and rivets, the treatment is specified in increasing procedures 0, 1 and 2. Procedure 0 says no treatment through to procedure 2, which gives four requirements.

There are further treatments a and z. Procedure a considers painting where dirt may be entrapped or moisture retained. Procedure z considers additional protection of zinc-coated structural parts as a whole.

#### *2.10.3.3 Aluminium in contact with concrete, masonry or plaster*

Aluminium in contact with dense compact concrete, masonry or plaster in a dry unpolluted or mild environment should be coated with a bituminous paint, or a coating providing the same protection. In an industrial or marine environment at least two coats of heavy-duty bituminous paint should be used. The surface of the contacting material should be similarly painted. Submerged contact between aluminium and such materials should be separated by suitable mastic or a heavy duty damp-course layer. When embedded in concrete the protection should extend at least 75 mm above the concrete surface.

Lightweight concrete and similar products need consideration if water or rising damp can cause a steady supply of aggressive alkali from the cement.

#### *2.10.3.4 Aluminium in contact with timber*

Timber in contact with aluminium in an industrial, damp or marine environment should be primed and painted, as some of the more acid timbers can cause corrosion. Oak, chestnut and western red cedar, unless well seasoned, can be harmful to aluminium.

Timber preservatives accepted as safe with aluminium are: creosote, zinc napthenates, zinc carboxylates and formulations containing non-ionic organic biocides.

Copper naphthenate, fixated CC-, CCA- and CCB-preservatives, formulations containing boron compounds or quaternary ammonium compounds are only used in dry situations where the aluminium surface has a substantial application of sealant.

Preservatives such as non-fixing inorganic formulations containing water-soluble copper or zinc compounds, and also formulations containing acid and alkaline ingredients ($pH < 5$ and $pH > 8$) should not be used.

#### *2.10.3.5 Aluminium in contact with soil*

Aluminium in contact with soil should be protected with at least two coats of bituminous paint, hot bitumen or plasticised coal tar pitch. Additionally, wrapping tapes may be used.

#### *2.10.3.6 Aluminium immersed in water*

Aluminium immersed in fresh, sea or contaminated water, such as shown in Figure 2.10, should be of durability rating A with fastenings made from aluminium or corrosion-resisting steel, or be welded.

#### *2.10.3.7 Aluminium in contact with chemicals used in the building industry*

Chemicals such as fungicides and mould repellents that contain compounds based on copper, lead, mercury or tin will, under wet or damp conditions, cause aluminium to corrode. Some cleaning materials ($pH < 5$ and $pH > 8$) should not be used as they can affect the surface of the aluminium and quick and adequate water rinsing will be required.

**Figure 2.10** High-speed catamaran ferry with aluminium hull and superstructure

*2.10.3.8 Aluminium in contact with building industry insulating materials*

Glass fibre, polyurethane and various insulating products need testing for compatibility with aluminium if they are to be used under damp and/or saline conditions. If compatibility becomes a concern, a sealant should be applied to the aluminium.

## 2.11 Materials selection

### 2.11.1 General

The choice of a suitable aluminium alloy depends on its availability, durability, formability, physical properties, required form, strength, weldability, etc. Table 2.3 gives the structural aluminium alloys listed in EN 1999-1-1.

### 2.11.2 Heat-treatable wrought alloys

*2.11.2.1 Alloys EN AW-6082 and EN AW-6061*

EN AW-6082 is a widely used heat-treatable alloy and often the principal structural alloy for welded and non-welded applications. It has good to excellent strength, durability rating B, excellent weldability, is good to fair for decorative anodising, and is available in sheet, strip, plate, extruded products such as bar, rod, profile and tube plus cold drawn tube and forgings. It is a common alloy for extrusions, plate and sheet from stock. It is increasingly used in the marine environment and is normally used in the fully heat-treated condition EN AW-6082-T6.

EN AW 6061 is a widely used heat-treatable alloy for welded and non-welded applications. It has good to fair strength, durability rating B, excellent weldability and is good to fair for decorative anodising. It is available in extruded products such as bar, rod, tube, profile and cold drawn tube and is normally used in the fully heat-treated condition EN AW-6061-T6.

EN AW-6082 and EN AW-6061 have high strength after heat treatment, good formability in the T4 temper and good machining properties. There is a loss of strength in the HAZ; however, post-weld natural ageing can recover some of the strength. If it is used in extrusions they are restricted to thicker, less intricate shapes than with other 6xxx series alloys.

*2.11.2.2 Alloy EN AW-6005A*

EN AW-6005A is recommended for structural applications and is available only in the extruded form of bar, rod, tube and profile but can be extruded into more complex shapes than EN AW-6082 or EN AW-6061, especially for thin-walled hollow shapes. It has good strength, durability rating B, excellent weldability by both TIG and MIG, and is good to fair for decorative anodising. The loss of strength in the HAZ is similar to EN AW-6082 and EN AW-6061. Its machining properties are similar to EN AW-6082 but the corrosion resistance of the welded and unwelded components is similar or better than for EN AW-6082.

*2.11.2.3 Alloys EN AW-6060, EN AW-6063 and EN AW-6106*

These alloys are recommended for structural applications. EN AW-6060 and EN AW-6063 are available in extruded form of bar, rod, tube and profile and in cold drawn tube. EN AW-6106 is available in extruded profile only. All three have good to fair strength, durability rating B and excellent weldability. EN AW-6060 is excellent for decorative anodising, while EN AW-6063 and EN AW-6106 are excellent to good for decorative anodising. They are used if appearance is a priority over strength and are particularly suited to anodising and similar finishing processes. They are readily weldable by both MIG and TIG processes and lose strength in the HAZ.

*2.11.2.4 Alloy EN AW-7020*

EN AW-7020 is recommended for welded and non-welded structural applications. It is available in sheet, strip and plate, in extruded products such as bar, rod, tube and profile and cold drawn tube. It has excellent strength, durability rating C, excellent weldability, is good to fair for decorative anodising but is not as easy to produce in complicated extrusions as the 6xxx alloys. It is normally used in the fully heat-treated EN AW-7020 T6 condition

**Table 2.3** The structural alloys given in EN 1999-1-1

| *3xxx series alloys* | *5xxx series alloys* | *6xxx series alloys* | *7xxx series alloys* | *8xxx series alloys* |
|---|---|---|---|---|
| EN AW-3004 | EN AW-5005 | EN AW-6060 | EN AW-7020 | EN AW-8011A |
| EN AW-3005 | EN AW-5049 | EN AW-6061 | | |
| EN AW-3103 | EN AW-5052 | EN AW-6063 | | |
| | EN AW-5083 | EN AW-6005A | | |
| | EN AW-5454 | EN AW-6106 | | |
| | EN AW-5754 | EN AW-6082 | | |

and has a better post-weld strength than 6xxx series alloys. EN AW-7020 is sensitive to environmental conditions. If heat treatment is not applied after welding, the need for protection of the HAZ must be checked. If EN AW-7020 T6 is subject to cold working it may become susceptible to stress corrosion cracking, consequently close collaboration between the designer and the manufacturer on the alloy's intended use and likely service conditions is essential. This alloy is widely used in mainland Europe, but based on the experience of the 7xxx military alloys, where they are stressed to the limit, UK use is much more pessimistic due to their perceived proneness to stress corrosion cracking and sensitivity to grain direction.

### 2.11.3 Non-heat-treatable wrought alloys

*2.11.3.1 General*

In the 5xxx wrought non-heat-treatable alloys, EN AW-5049, EN AW-5052, EN AW-5454, EN AW-5754 and EN AW-5083 are recommended for structural applications, have durability rating A, excellent weldability and are excellent to good for decorative anodising. Other non-heat-treatable alloys considered for less stressed structural applications are EN AW-3004, EN AW-3005, EN AW-3103 and EN AW-5005; these have durability rating A, excellent weldability and are excellent for decorative anodising, except for EN AW-3103 which is good for decorative anodising.

*2.11.3.2 Alloys EN AW-5049, EN AW-5052, EN AW-5454 and EN AW-5754*

These alloys are suitable for welded or mechanically joined structural parts subject to moderate stress. They are ductile in the annealed condition and lose ductility rapidly with cold forming. They have very good resistance to corrosion attack, especially in a marine atmosphere. They are suitable only for simple extruded shapes and can be easily machined in the harder tempers.

EN AW-5049 is available as sheet, strip and plate and has good to fair strength.

EN AW-5052 is available in sheet, strip and plate, extruded bar and rod, simple sections of extruded tube and profile and cold drawn tube and has good to fair strength.

EN AW-5454 is available in sheet, strip and plate, extruded bar and rod, simple sections of extruded tube and profile and has good to fair strength.

EN AW-5754 is available in sheet, strip and plate, extruded bar and rod, simple sections of extruded tube and profile, cold drawn tube and forgings and has good to fair strength. It is the strongest 5xxx series alloy, offering practical immunity to intergranular corrosion and stress corrosion.

*2.11.3.3 Alloy EN AW-5083*

EN AW-5083 is the strongest structural non-heat-treatable alloy in general commercial use, having good to excellent strength, excellent weldability, very good corrosion resistance, and is good to excellent for decorative anodising. It is ductile in the soft condition, has good forming properties, but loses ductility with cold forming and may then become hard with low ductility. It is available in sheet, strip and plate, extruded bar and rod, simple sections of extruded tube and profile, cold drawn tube and forgings.

EN AW-5083 may in all tempers (Hx), especially in H32 and H34 tempers, be susceptible to intergranular corrosion, which can develop into stress corrosion cracking under sustained loading, but special tempers such as H116 have been developed to minimise this effect. EN AW-5083 is not recommended for use where it is subjected to further heavy cold working and/or service temperatures above 65°C, where EN AW-5754 should be used instead.

EN AW-5083 is easily welded by both MIG and TIG, but if strain-hardened materials are welded the properties in the HAZ revert to the annealed value. Due to the high magnesium content, it is particularly hard to extrude, but it has good machining qualities in all tempers.

*2.11.3.4 Alloys EN AW-3004, EN AW-3005, EN AW-3103 and EN AW-5005*

EN AW-3004, EN AW-3005, EN AW-3103 and EN AW-5005 are available in sheet, strip and plate, and have poor to fair strength, excellent weldability and durability rating A. They are harder than 'commercially pure' aluminium and have high ductility. EN AW-3004, EN AW-3005 and EN AW-5005 are excellent for decorative anodising while EN AW-3103 is good for decorative anodising. EN AW-3103 and EN AW-5005 are also available in extruded bar, rod, tube and profile and cold drawn tube.

*2.11.3.5 Alloy EN AW-8011A*

EN AW-8011A is used increasingly in the building industry for facades due to its advantages in fabrication. It is available in sheet, strip and plate, has poor to fair strength, durability rating B, good weldability and is poor to fair for decorative anodising.

### 2.11.4 Cast products

*2.11.4.1 General*

Six foundry alloys are recommended for structural application and all have good weldability: four heat-treatable alloys (EN AC-42100, EN AC-42200, EN AC-43000 and EN AC-43300) and two non-heat-treatable alloys (EN AC-44200 and EN AC-51300).

*2.11.4.2 Heat-treatable alloys EN AC-42100, EN AC-42200, EN AC-43000 and EN AC-43300*

These alloys respond to heat treatment and are suitable for chill or permanent mould casting. EN AC-43300 is also suitable for sand casting. They are not usually used for pressure die casting except by using advanced casting methods. They have good weldability and a resistance to corrosion B. The highest strength is achieved using EN AC-42200-T6 but this has lower ductility than EN AC-42100.

EN AC-42100 has good castability, fair to good strength but is poor for decorative anodising.

EN AC-42200 has good castability, good strength but is poor for decorative anodising.

EN AC-43300 has excellent castability, good strength but is not recommended for decorative anodising.

EN AC-43000 has excellent to good castability, poor strength and is not recommended for decorative anodising.

*2.11.4.3 Non-heat-treatable casting alloys EN AC-44200 and EN AC-51300*

Both alloys are suitable for sand, chill or permanent mould casting, but are not recommended for pressure die casting. EN AC-44200 has excellent castability, good weldability, poor strength, a durability rating of B and is not recommended for decorative anodising. EN AC-51300 has fair castability, good weldability, poor strength, durability rating A and is excellent for decorative anodising.

## 2.12 Fabrication and construction

Aluminium and its alloys can be easily shaped by the usual industrial metalworking processes including casting, extrusion, forging and rolling.[5]

Aluminium fabrication requires special measures, and equipment must be of good quality and well polished. A dedicated segregated fabrication workshop is required, as are approved welders and staff trained in the handling of aluminium. Aluminium when heated shows no colour change and when molten must not come into contact with water as it may explode. The welding of aluminium requires special consideration due to its tough, adherent oxide film and high thermal conductivity. Welding should be in draught-free, dry conditions in an area separated from other materials.

### 2.12.1 Cutting[18]

Aluminium is cut using the same types of tools as used for steel: e.g. machining, notching, routing, sawing and shearing. With aluminium sawing is much faster, but the size and shape of the teeth and the speed of cutting are different. Flame cutting is not recommended as it alters the mechanical properties of the alloy and produces an uneven edge. High-pressure water jets and lasers are used for precision cutting.

### 2.12.2 Drilling and punching

Drilling in aluminium is easier and quicker than in steel and can be carried out at higher speeds. The drilled hole is larger than in steel and excessive heat can be generated, so coolant is required. When punching, a drill or reamer is used to obtain the final hole size.

### 2.12.3 Bending and forming

Similar equipment is used for the bending and forming of aluminium as is used for steel. Minimum bend radii for a range of alloys, tempers and thicknesses are recommended and springback is greater than for steel. Pure aluminium and non-heat-treatable alloys can be easily formed; however, inter-stage annealing during some forming operations may be needed for the higher alloyed materials. Heat-treatable alloys can be formed in the annealed condition or immediately after solution treatment and some cold work is possible after room-temperature ageing. After full heat treatment, forming is difficult, especially with the stronger alloys.[11]

### 2.12.4 Machining

Cast and wrought aluminium alloys can be easily machined at high speeds. Cutting fluids, which are removed after machining, are used to increase lubrication and reduce overheating. For wrought alloys machining is improved by cold work and heat treatment, and all alloys are most easily machined in their hardest temper.[11]

### 2.12.5 Bolting and screwing

Bolted connections may be made from aluminium, stainless steel or steel with a protective layer, e.g. cadmium or zinc, and should be stronger and more corrosion-resistant than the alloys being joined. Aluminium and stainless steel bolts are the preferred solution while the threads of aluminium bolts should be lubricated. Steel bolts with a protective layer have a limited life before the protective layer ceases to exist.[18]

### 2.12.6 Welding

#### *2.12.6.1 Quality requirements*

Welders and welding operators must be suitably qualified. The quality levels required are specified by the designer and for EN1999-1-1 should be according to EN 1090-3. The manufacturer must demonstrate compliance with the required quality levels. The specification of the level of quality requirements depends on EN ISO 3834-1 as follows:

- extent and significance of safety critical products
- complexity of manufacture
- range of products manufactured
- range of different materials used
- extent to which metallurgical problems may occur
- extent to which manufacturing imperfections affect product performance.

The quality requirements of EN 1090-3 relate to four execution classes as follows.

- *Execution class I* – EN ISO 3834: Part 4: Elementary quality requirements.
- *Execution class II* – EN ISO 3834: Part 3: Standard quality requirements.
- *Execution classes III and IV* – EN ISO 3834: Part 2: Comprehensive quality requirements.

A welding plan must be drawn up for execution classes II, III and IV. If no execution class is specified, class II applies.

The execution class is selected based on the following four conditions, although in practice only the first three apply:

- consequence of a structural failure, e.g. high, medium or low
- type of loading, e.g. fatigue, or predominantly static (high or low tensile stress)
- level of design action effect as compared to the resistance of the cross-section
- technology and procedures to be used.

#### *2.12.6.2 Arc welding*

For general engineering fabrication, two arc welding gas-shielded processes are used: the MIG (metal inert gas using direct current) process for heavier construction and the TIG (tungsten inert gas using alternating current) process for thinner materials. The inert gases shield the arc and the weld pool from air and prevent oxidation. Arc welding is used on the 1xxx, 3xxx, 5xxx, 6xxx and the weaker 7xxx series alloys, but not for the 2xxx and the stronger 7xxx alloys. The effect of arc welding in the HAZ can be a strength loss of up to 72%, although it is usually within the range of 0% to 40%.[18] Table 3.6 of EN1999-1-1 gives alloy combinations that may be welded together.

Normally with MIG and TIG welding, no preheating of the metal is necessary, but preheating up to 50°C is used to prevent weld defects, perhaps when the metal is cold and there is condensation. For large multi-pass welds, the interpass temperatures are restricted to reduce the size and severity of the HAZ softening. The electrode/filler wire to be used depends on the alloys being welded and is specified by the designer.

#### *2.12.6.3 Friction stir welding*

In the solid phase friction stir welding process the plates to be joined are butted firmly together and a rotating tool is pushed into the plates and moved along the line of the weld. Frictional heating causes the plates to become plastic and they are welded together.

Friction stir welds have better fatigue strength and require less preparation than arc welding.

*2.12.6.4 Other forms of welding*

Other forms of welding are available:

- stud, spot and plasma arc welding
- spot, seam and flash resistance welding
- pressure and explosive solid phase welding where other metals such as steel can be joined to aluminium.

### 2.12.7 Adhesive bonding

Adhesive bonded joints can be used with aluminium and all its alloys, but this requires an expert technique and should be used with care. The joints do not weaken the aluminium and have good fatigue performance plus good joint stiffness.[18] As tension loads perpendicular to the plane of the adhesive can cause joint peeling, the joint is designed to transmit only shear forces.

Adhesive bonding is not be used for main structural joints unless testing has established its validity including environmental and fatigue effects. Adhesive bonding is used for plate stiffener combinations and other secondary stressed conditions with loads being carried over as large a bonded area as possible.

The surfaces to be joined must be pre-treated, usually by chemical conversion and anodising. Adhesives such as single or two-part modified epoxies, modified acrylics and one- or two-part polyurethanes are usually used and have characteristic shear strengths between 20 and 35 MPa. Higher characteristic shear strengths may be used if validated by tests to ISO 11003.

### 2.12.8 Finishes[19]

Anodising and organic coatings are much more widely used surface finishes than chemical and mechanical finishes and plating.

Anodising increases the thickness of the chemically stable protective translucent aluminium oxide film that naturally occurs on aluminium. The thick oxide film does not flake or peel off, but contains capillary pores where a variety of colours can be introduced and sealed in. Highly reflective finishes are available. A range of thicknesses of the anodic film are used, e.g. exterior applications have thicker films than interior applications. Thick, abrasion-resistant films are also produced. Anodised aluminium, which can be cleaned using water or mild soap or detergent, is used for aircraft parts, balustrading, curtain walling, electronic components, furniture, handles, kitchen trim, lighting systems, nameplates, windows and yacht masts.

The usual methods of applying paint to aluminium are electrophoretic, powder coating and wet spraying. The electrophoretic technique gives sound uniform coatings, but the number of colours available is limited. Polyester powder coatings provide sound durable coatings in a wide range of colours. Wet spraying is usually non-factory applied and must not contain materials that corrode aluminium such as copper, mercury, tin or lead.

### 2.12.9 Handling and storage of aluminium and its alloys[11]

As aluminium is a soft material compared with steel, it needs careful handling and protection from surface damage, e.g. aluminium should not be dragged but lifted clear and then moved. Plate, sheet, sections and tubes should be stored vertically with an adequate surrounding air flow.

Aluminium should be stored in dry conditions such that condensation does not occur, and separately from other metals, building materials, some timbers and timber preservatives.

## 2.13 Standards

In 1975 the Commission of the European Community decided on an action programme in the field of construction that led to an initiative to establish a set of harmonised technical rules, the Eurocode programme, for the design of construction works. There are 10 Structural Eurocodes. The Eurocodes relevant to this chapter are EN 1999 Eurocode 9 Design of aluminium structures and its supporting Eurocodes.

The Eurocode standards provide common structural design rules for everyday use for the design of whole structures and component products. There are also National Annexes which contain information on those parameters left open for national choice, known as Nationally Determined Parameters. It must be clearly stated and agreed which National Annex is to be used.

When using the UK edition of EN 1999 it is necessary to refer to the two published documents PD 6702-1[20] and PD 6705-3.[21]

## 2.14 Acknowledgements

The immense help and support of the Aluminium Federation, especially Tom Siddle, is gratefully acknowledged.

## 2.15 References

1 US Geological Survey (2012), *Minerals information, bauxite and alumina statistics and information, mineral commodities summaries*, Reston, VA.
2 The International Aluminium Institute (2007), *Form 150, Primary aluminium production*, London, UK, date of issue 20 March 2012.
3 ALFED (2012), *Annual report of the Aluminium Federation for the year 2011*, Aluminium Federation, Birmingham, UK.
4 ALFED (2004), *Fact Sheet 2 – Primary aluminium production*, Aluminium Federation, Birmingham, UK.
5 ALFED (2004), *Fact Sheet 1 – Aluminium the metal*, Aluminium Federation, Birmingham, UK.
6 ALFED (2004), *Fact Sheet 9 – Aluminium wrought remelt*, Aluminium Federation, Birmingham, UK.
7 ALFED (2004), *Fact Sheet 8 – Aluminium and recycling*, Aluminium Federation, Birmingham, UK.
8 ALFED (2004), *Fact Sheet 5 – Aluminium casting*, Aluminium Federation, Birmingham, UK.
9 European Aluminium Association, *Rolled products*, www.alueurope.eu/rolled-products.
10 European Aluminium Association, *Extruded products*, www.alueurope.eu/extruded-products.
11 ALFED (2009), *The properties of aluminium and its alloys*, Aluminium Federation, Birmingham, UK.
12 ALFED (2004), *Fact Sheet 6 – Aluminium forging*, Aluminium Federation, Birmingham, UK.
13 ALFED (2004), *Fact Sheet 7 – Aluminium powder and paste*, Aluminium Federation, Birmingham, UK.
14 The Aluminium Association (2009), *International alloy designations and chemical composition limits for wrought aluminium and wrought aluminium alloys*, The Aluminium Association, Washington, DC.
15 ALFED (2004), *Fact Sheet 10 – Master alloys*, Aluminium Federation, Birmingham, UK.
16 Wimpy Offshore and Alcan Offshore (1990), *Aluminium design guide: basic guide on the use of aluminium in the offshore industry*, Wimpy Offshore, London, UK; Alcan Offshore, Gerrards Cross, UK.

17 ALFED (2004), *Fact Sheet 19 – Aluminium and corrosion*, Aluminium Federation, Birmingham, UK.
18 Dwight, J. (1999), *Aluminium design and construction*, E&FN Spon, London, UK.
19 ALFED (2004), *Fact Sheet 11 – Aluminium finishing*, Aluminium Federation, Birmingham, UK.
20 PD 6702-1, *Structural use of aluminium – Part 1: Recommendations for the design of aluminium structures to BS EN 1999*. BSI, London, UK.
21 PD 6705-3, *Structural use of steel and aluminium—Part 3: Recommendations for the execution of aluminium structures to BS EN 1090-3*. BSI, London, UK.

### 2.15.1 Standards

EN ISO 3834-1: 2005. *Quality requirements for fusion welding of metallic materials – Part 1: Criteria for the selection of the appropriate level of quality requirements*, London: BSI.
EN ISO 3834-2: 2005. *Quality requirements for fusion welding of metallic materials – Part 2: Comprehensive quality requirements*, London: BSI.
EN ISO 3834-3: 2005. *Quality requirements for fusion welding of metallic materials – Part 3: Standard quality requirements*, London: BSI.
EN ISO 3834-4: 2005. *Quality requirements for fusion welding of metallic materials – Part 4: Elementary quality requirements*, London: BSI.
EN ISO 14554-1: 2001. *Quality requirements for welding – Resistance welding of metallic materials – Part 1: Comprehensive quality requirements*, London: BSI.
EN ISO 14554-2: 2001. *Quality requirements for welding – Resistance welding of metallic materials – Part 2: Elementary quality requirements*, London: BSI.
EN12258-1: 1998. *Aluminium and aluminium alloys – Terms and definitions – Part 1 General terms*, London: BSI.
ISO209-1: 1989. *Wrought aluminium and aluminium alloys – Chemical composition and forms of product – Part 1: Chemical composition*, London: BSI.

### 2.15.2 Structural design standards

EN1090-1. *Execution of steel structures and aluminium structures – Part 1: General technical delivery conditions for structural steel and aluminium components*, London: BSI.
EN1090-3. *Execution of steel structures and aluminium structures – Part 3: Technical rules for the execution of aluminium structures*, BSI London: BSI.
EN1990. *Basis of structural design*, London: BSI.
EN1991. *Actions on structures*, London: BSI.
EN1991-1-2. *Actions on structures Part 1–2: Actions on structures exposed to fire*, London: BSI.
EN1993. *Design of steel structures*, London: BSI.
EN1995-1-1: 2004. *Design of timber structures – Part 1-1: Common rules and rules for buildings*, London: BSI.
EN1999-1-1. *Design of aluminium structures – Part 1-1: General structural rules*, London: BSI.
EN1999-1-2. *Design of aluminium structures – Part 1-2: Structural fire design*, London: BSI.
EN1999-1-3. *Design of aluminium structures – Part 1-3: Structures susceptible to fatigue*, London: BSI.
EN1999-1-4. *Design of aluminium structures – Part 1-4: Cold formed structural sheeting*, London: BSI.
EN1999-1-5. *Design of aluminium structures – Part 1-5: Shell structures*, London: BSI.

### 2.15.3 Chemical composition, form and temper definition of wrought products standards

EN 508-2: 2000.*Roofing products from metal sheet – Specifications for self supporting products of steel, aluminium or stainless steel sheet – Part 2: Aluminium*, London: BSI.
BS EN 515: 1993. *Aluminium and aluminium alloys – Wrought products – Temper designations*, London: BSI.
BS EN 573-1: 2004. *Aluminium and aluminium alloys – Chemical composition and form of wrought products, Part 1: Numerical designation system*, London: BSI.
BS EN 573-2: 1995. *Aluminium and aluminium alloys – Chemical composition and form of wrought products, Part 2: Chemical symbol based designation system*, London: BSI.
BS EN 573-3: 2003. *Aluminium and aluminium alloys – Chemical composition and form of wrought products, Part 3: Chemical composition*, London: BSI.
BS EN 573-4: 2004. *Aluminium and aluminium alloys – Chemical composition and form of wrought products, Part 4: Forms of products*, London: BSI.
EN1396: 1997. *Aluminium and aluminium alloys – Coil coated sheet and strip for general applications – Specifications*, London: BSI.
EN10002-1: 2001. *Tensile testing of metallic materials – Part 1: Method of test at ambient temperature*, London: BSI.

### 2.15.4 Technical delivery conditions standards

EN485-1: 1994. *Aluminium and aluminium alloys – Sheet, strip and plate – Part 1: Technical conditions for inspection and delivery*, London: BSI.
EN586-1: 1998. *Aluminium and aluminium alloys – Forgings – Part 2: Technical conditions for inspection and delivery*, London: BSI.
EN754-1:1997. *Aluminium and aluminium alloys – Cold drawn rod/bar and tube – Part 1: Technical conditions for inspection and delivery*, London: BSI.
EN755-1: 1997. *Aluminium and aluminium alloys – Extruded rod/bar, tube and profiles – Part 1: Technical conditions for inspection and delivery*, London: BSI.
EN1592-1: 1998. *Aluminium and aluminium alloys – HF seam welded tubes — Part 1: Technical conditions for inspection and delivery*, London: BSI.
EN12020 -1: 2001. *Aluminium and aluminium alloys – Extruded precision profiles in alloys EN AW-6060 and EN AW-6063 – Part 1: Technical conditions for inspection and delivery*, London: BSI.

### 2.15.5 Dimensions and mechanical properties standards

EN485-2: 2004. *Aluminium and aluminium alloys — Sheet, strip and plate, Part 2: Mechanical properties*, London: BSI.
EN 85-3: 2003. *Aluminium and aluminium alloys – Sheet, strip and plate, Part 3: Tolerances on shape and dimensions for hot rolled products*, London: BSI.
EN485-4: 1994. *Aluminium and aluminium alloys – Sheet, strip and plate, Part 4: Tolerances on shape and dimensions for cold rolled products*, London: BSI.
EN508-2: 2000. *Roofing products from metal sheet – Specifications for self supporting products of steel, aluminium or stainless steel – Part 2: Aluminium*, London: BSI.
EN586-2: 1994. *Aluminium and aluminium alloys – Forgings – Part 2: Mechanical properties and additional property requirements*, London: BSI.
EN586-3: 2001. *Aluminium and aluminium alloys – Forgings – Part 3: Tolerances on dimensions and form*, London: BSI.
EN754-2: 1997. *Aluminium and aluminium alloys – Cold drawn rod/bar and tube – Part 2: Mechanical properties*, London: BSI.
EN754-3: 1996. *Aluminium and aluminium alloys – Cold drawn rod/bar and tube — Part 3: Round bars, tolerances on dimension and form*, London: BSI.
EN754-4: 1996. *Aluminium and aluminium alloys – Cold drawn rod/bar and tube — Part 4: Square bars, tolerances on dimension and form*, London: BSI.
EN754-5: 1996. *Aluminium and aluminium alloys – Cold drawn rod/bar and tube – Part 5: Rectangular bars, tolerances on dimension and form*, London: BSI.
EN754-6: 1995. *Aluminium and aluminium alloys – Cold drawn rod/bar and tube — Part 6: Hexagonal bars, tolerances on dimension and form*, London: BSI.
EN754-7: 1998. *Aluminium and aluminium alloys – Cold drawn rod/bar and tube – Part 7: Seamless tubes, tolerances on dimension and form*, London: BSI.
EN754-8: 1998. *Aluminium and aluminium alloys – Cold drawn rod/bar and tube — Part 8: Porthole tubes, tolerances on dimension and form*, London: BSI.
EN755-2: 1997. *Aluminium and aluminium alloys – Extruded rod/bar, tube and profiles – Part 2: Mechanical properties*, London: BSI.

EN755-3: 1996. *Aluminium and aluminium alloys – Extruded rod/bar, tube and profiles – Part 3: Round bars, tolerances on dimension and form*, London: BSI.
EN755-4: 1996. *Aluminium and aluminium alloys – Extruded rod/bar, tube and profiles — Part 4: Square bars, tolerances on dimension and form*, London: BSI.
EN755-5: 1996. *Aluminium and aluminium alloys – Extruded rod/bar, tube and profiles — Part 5: Rectangular bars, tolerances on dimension and form*, London: BSI.
EN755-6: 1996. *Aluminium and aluminium alloys – Extruded rod/bar, tube and profiles — Part 6: Hexagonal bars, tolerances on dimension and form*, London: BSI.
EN755-7: 1998. *Aluminium and aluminium alloys – Extruded rod/bar, tube and profiles — Part 7: Seamless tubes, tolerances on dimension and form*, London: BSI.
EN755-8: 1998. *Aluminium and aluminium alloys – Extruded rod/bar, tube and profiles — Part 8: Porthole tubes, tolerances on dimension and form*, London: BSI.
EN755-9: 2001. *Aluminium and aluminium alloys – Extruded rod/bar, tube and profiles — Part 9: Profiles, tolerances on dimension and form*, London: BSI.
EN1592-2: 1997. *Aluminium and aluminium alloys – HF seam welded tubes – Part 2: Mechanical properties*, London: BSI.
EN1592-3: 1998. *Aluminium and aluminium alloys – HF seam welded tubes – Part 3: Tolerances on dimensions and shape of circular tubes*, London: BSI.
EN1592-4: 1998. *Aluminium and aluminium alloys – HF seam welded tubes – Part 4: Tolerances on dimensions and form for square, rectangular and shaped tubes*, London: BSI.
EN12020-2: 2001. *Aluminium and aluminium alloys – Extruded precision profiles in alloys EN AW-6060 and EN AW-6063 – Part 2: Tolerances on dimensions and form*, London: BSI.

### 2.15.6 Welding standards

EN439: 1994. *Welding consumables – Shielding gases for arc welding and cutting*, London: BSI.
EN970: 1997. *Non destructive examination of welds – Visual examination*, London: BSI.
EN1011-1: 1998. *Welding – Fusion welding of metallic materials – Part 1: General*, London: BSI.
EN1011-4: 2000. *Welding – Requirements for fusion welding of metallic materials – Part 4: Aluminium and aluminium alloys*, London: BSI.
EN1418: 1998. *Welding personnel. Approval testing of welding operators for fusion welding and resistance weld setters for fully mechanised and automatic welding of metallic materials*, London: BSI.
EN30042: 1994. ISO 10042:1992. *Arc welded joints in aluminium and its weldable alloys – Guidance on quality levels for imperfections*, London: BSI.
EN ISO 9606-2: 2004. *Qualification tests of welders – Fusion welding – Part 2: Aluminium and aluminium alloys*, London: BSI.
EN ISO 15614: 2. 2005. *Specification and approval of welding procedures for metallic materials – Part 1: Welding procedure tests for the arc welding of aluminium and its alloys*, London: BSI.
ISO18273: 2004. *Welding consumables – Wire electrodes, wires and rods for arc welding of aluminium and aluminium alloys. Classification*, London: BSI.

### 2.15.7 Adhesive standards

BS ISO 11003: 2001. Adhesives – Determination of shear behaviour of structural adhesives, London: BSI.

## 2.16 Bibliography

ALFED (2008), *Fact Sheet 1 – Aluminium in transport*, Birmingham, UK: Aluminium Federation.
ALFED (2008), *Fact Sheet 8 – Aluminium in building and construction*, Birmingham, UK: Aluminium Federation.
ALFED (2008), *Fact Sheet 15 – Aluminium in packaging*, Birmingham, UK: Aluminium Federation.
The Aluminium Association (2009), *International alloy designations and chemical composition limits for wrought aluminium and wrought aluminium alloys*, Washington, DC: The Aluminium Association.
Australian Aluminium Council, www.aluminium.org.au.
EAA, *Aluminium for future generations*, Brussels, Belgium: European Aluminium Association (EAA), www.aluminium.org.
International Aluminium Institute, www.world-aluminium.org
Mazzolani, F. M. (1995), *Aluminium alloy structures*, second edition, London, UK: Pitman.
Muller, U. (2011), *Introduction to structural aluminium design*, Caithness, UK: Whittles Publishing.
Spence, W. P. (1998), *Construction materials, methods and techniques*, New York, NY: Delmar Publishers.
Tindall, P., & Hoglund, T. (2012), *Designers' guide to Eurocode 9: Design of Aluminium Structures*, London, UK: ICE.

# 3 Copper

**Nick Hay** BSc
Project Manager, Copper Development Agency

## Contents

## 3.1 Introduction

It is not known whether copper or gold was the first metal used by mankind, but copper was certainly the first metal used for the production of tools and utensils. Copper artefacts have been dated to over 9000 years old. The earliest use was most probably nodules of metallic or 'native' copper, found on or near the surface of the earth, but it is known that copper was being smelted by the sixth century BC. The Copper Age continued until, fortuitously, ore deposits were used that contained both copper and arsenic. The arsenical bronze alloy produced could be worked into cutting tools with a longer life of the cutting edge, presumably at the expense of the life of the smelters! Other ores, containing copper and tin, produced a tougher tin bronze. Eventually, early man was able to smelt both copper and tin and then alloy them to produce tin bronzes remarkably similar in composition to those used today. The Bronze Age continued until man had learned not only how to smelt iron but also how to harden it to produce tools and weapons which, in their turn, were superior to bronze.

Copper is a very ductile metal, with comparatively good strength and toughness. This combination of ductility (that is, a large uniform plastic extension which makes it readily formable), strength and toughness (really its ability to avoid brittle failure) is characteristic of elements with the face-centred-cubic structure. These properties vary little over a very wide temperature range, from cryogenic temperature to near the melting point. By contrast, metals with the other common crystal structures, body-centred cubic and hexagonal close packed, show a ductile to brittle change at low temperatures. Copper is non-magnetic and has a higher electrical and thermal conductivity than any of the common metals other than silver. It can be machined, soldered, brazed and welded. The pure metal has an attractive colour, but a wide range of colours can be produced by alloying, or the surface appearance can be changed by plating with other metals, applying organic coatings or chemical treatments. The workability is markedly dependent on the concentrations of impurity elements present. In particular, these impurities may melt at relatively low temperatures and form grain boundary films that markedly lower the cold and hot workability. Copper is routinely manufactured to 99.99% purity, which allows very rapid drawing into wire. Copper and its alloys are also readily cast into shapes.

Pure copper has a higher resistance to oxidation at ambient temperature than iron, aluminium or any of the other common engineering metals. When oxidised, it forms a black outer scale of cupric oxide (CuO). Between this and the metal surface a thin reddish layer of cuprous oxide ($Cu_2O$) is formed, which adheres firmly to the metal surface and restricts the passage of oxygen to continue the oxidation reaction.

Copper is well known for its anti-bacterial properties and the copper content of brasses has a beneficial effect on restricting the growth of micro-organisms. Tests at the University of Southampton have shown the superiority of copper alloys compared to stainless steels in controlling harmful micro-organisms.[1] These tests strongly indicate that the use of copper alloys in applications including door knobs, push plates, fittings, fixtures and work surfaces would considerably mitigate MRSA in hospitals and reduce the risk of cross-contamination between staff and patients in critical care areas. It has also been shown that copper alloys are effective in controlling *E. coli OH157* and *Listeria monocytogenes*, both of which cause serious food poisoning.

Copper consumption can be attributed to five market sectors, of which construction consumes about 38% of copper take-off (including 13% for electrical installations within buildings), the remaining four sectors being electrical and electronics, transport, industrial equipment and consumer durables. It is estimated that only 12% of known copper reserves have been mined throughout history. Most copper is extracted from open cast mines which can be found in all the continents. Local environmental impact of mining is strictly controlled, and refining is carried out close to the main sources of the ore. Typical applications for copper and copper alloys are given in Table 3.1.

**Table 3.1** Typical applications for copper and copper alloys

| *Application* | *Material* | *Designation* |
|---|---|---|
| Electrical | Oxygen-free copper | CW008A |
| Pipework | Phosphorus deoxidised copper | CW024A |
| Roofing | Phosphorus deoxidised copper | CW024A |
| Cladding, | Phosphorus deoxidised copper | CW024A |
| façade, | Brass | CW502L |
| flashings | Aluminium bronze | CW503L |
| | Phosphor bronze | CuAl5 |
| | | CW450K |
| Gutters, downpipes | Phosphorus deoxidised copper | CW024A |
| Window frames, doors, shop fronts | Brass | CW620N |
| | | CW450K |
| | | CW501L |
| | | CW502L |
| | | CW505L |
| | | CW509L |
| | | CW508L |
| | Phosphor bronze | CW451K |
| | Nickel silver | CW401J |
| Hand-rails, balustrades | Brass | CW720R |
| | | CW501L |
| | | CW505L |
| | | CW620N |
| Door furniture | Phosphorus deoxidised copper | CW024A |
| | Brass | CW501L |
| | | CW502L |
| | | CW505L |
| | | CW509L |
| | Phosphor bronze | CW451K |
| | Nickel silver | CW401J |
| Damp-proof courses | Phosphorus deoxidised copper | CW024A |
| Ties, tension straps, hangers, | Aluminium bronze | CW307G |
| | | CC333G |
| brackets | Phosphor bronze | CW451K |
| Load-bearing fixings | Aluminium bronze | CW307G |
| | | CC333G |

## 3.2 Sources and production

Copper is found in the earth's crust and the oceans, although the amount in the latter is not yet commercially viable. The upper 10 kilometres of the crust is thought to contain an average of about 33 ppm of copper. For commercial exploitation, copper deposits generally need to be in excess of 0.5% copper, and preferably over 2%.

Over 160 copper minerals are known, of diverse appearance and colour. Most of these are very rare, and fewer than a dozen are at all common. Of these the most brilliantly coloured are the bright green banded malachite, bornite, which is iridescent, giving rise to its alternate name of peacock ore; and chalcopyrite, a mixed sulphide of copper and iron. The most common ores are copper sulphide, copper-iron sulphide and copper oxide.

Commercially exploited deposits of copper ores are found in many parts of the world, frequently associated with mountain building processes. Deposits occur at many locations in the

Americas, mainly in the United States and Chile, and in areas of the North American plains such as Michigan, Ontario, Quebec and Manitoba. In Africa the largest deposits are found in Zambia and Zaire, but copper is also mined in several other locations in Central and Southern Africa.

Modern methods of extraction allow economic leaching and electrowinning (copper ions plating out of solution onto a cathode) of copper from low-grade ores; extraction techniques are continually being refined and developed to achieve the most efficient removal of copper from a wide variety of ores.

The oxidised ores, consisting of the silicates, carbonates and sulphates, are treated by several methods, all involving some form of leaching of the crushed ore with sulphuric acid to produce impure solutions of copper sulphate. The copper-loaded solvent is then stripped of its copper by contacting with a strong sulphuric acid, producing a concentrated copper sulphate solution which is used as an electrolyte for electrowinning by the deposition of metallic copper on copper cathodes.

Sulphide ores are first mechanically crushed and ground so that nearly all copper mineral particles are freed from the rock or 'gangue'. Flotation by the injection of air and violent agitation is carried out with the pulverised ore held in suspension in water, to which surface-active agents have been added. The sulphide minerals are continuously drawn off from the surface and dewatered to produce copper sulphide concentrate, which is further treated in one of two ways.

Controlled roasting is a process in which the sulphides are burnt in air to give a product of copper sulphates and oxides suitable for acid leaching, as described earlier. The other method, matte smelting, is the most important for the extraction of copper from sulphides. There are several methods of smelting mattes (copper-iron sulphide and oxide slag) by the melting of concentrate at about 1200°C. The molten matte is then turned into 'blister-copper' by oxidation in a furnace. Finally, anodes (in what is known as tough pitch copper) for electrolytic refining are produced in a furnace in which sulphur is burned off with air blown through tuyeres, after which excess oxygen is removed.

## 3.3 Designations and properties[2]

The development of European Standards has led to a new numbering system for alloys. The system is a six-character, alpha-numeric series which, for copper-based materials, begins with C. The second character indicates the product form (see Table 3.2). The third, fourth and fifth characters are digits that allow precise identification of the alloy. The sixth character indicates the alloy group (see Table 3.3).

Thus, for a wrought 60/40 brass the material designation is now CW509L, formerly CZ109. Annex 3.1 at the end of this chapter gives designations and typical mechanical properties for some commonly used copper alloys.

**Table 3.2** Numbering system for copper alloys – product form

| *Character* | *Product form* |
|---|---|
| B | Ingot for re-melting |
| C | Cast products |
| F | Brazing/welding filler metals |
| M | Master alloys |
| R | Refined unwrought copper |
| S | Scrap form |
| W | Wrought products |
| X | Non-standardised materials |

**Table 3.3** Numbering system for copper alloys – precise identification

| *Character* | *Alloy group* | |
|---|---|---|
| A or B | Copper | 000–099 |
| C or D | Low alloyed copper (less than 5% alloying elements) | 100–199 |
| E or F | Miscellaneous copper alloys (5% or more alloying elements) | 200–299 |
| G | Copper-aluminium alloys | 300–349 |
| H | Copper-nickel alloys | 350–399 |
| J | Copper-nickel–zinc alloys | 400–449 |
| K | Copper-tin alloys | 450–499 |
| L or M | Copper-zinc alloys, binary | 500–599 |
| N or P | Copper-zinc–lead alloys | 600–699 |
| R or S | Copper-zinc alloys, complex | 700–799 |

## 3.4 Grades of copper

Most copper used for electrical purposes is 'high conductivity copper' made from material that has originally been electrolytically refined to high purity before being melted and cast without the addition of alloying elements or impurities. A little oxygen is all that is present to ensure that conductivity remains high, now around 101.5% compared with the standard set by the IEC in 1913.

- High conductivity (HC) electrolytically refined copper (sometimes known as tough pitch copper), with a nominal conductivity of 100% IACS (International Annealed Copper Standard), is used for most electrical applications such as busbars, cables and windings. High-conductivity copper is very readily worked hot and cold. It has excellent ductility, which means that it can be easily drawn to fine wire sizes and it is available in all fabricated forms.
- Oxygen-free high-conductivity copper is produced by casting in a controlled atmosphere and is used where freedom from the possibility of embrittlement is required.
- A special grade of oxygen-free high-conductivity copper (certified grade) with low residual volatile impurities is used for high vacuum electronic applications such as transmitter valves, wave guide tubes, linear accelerators and glass to metal seals.
- Deoxidised copper (usually deoxidised with phosphorus) is a material that can readily be brazed or welded without fear of embrittlement. It is used for the manufacture of tube for fresh water, hot water cylinders, pressure vessels and roofing and cladding sheet.
- Free-machining copper – an addition of 0.5% sulphur or tellurium raises the machinability rating of copper from 20% (based on 100% for free-cutting brass) to 90%. Applications for these free-machining grades include electrical components, gas-welding nozzles and torch tips and soldering iron tips.

A wide variety of high-conductivity copper alloys are available for special purposes, of which the following three are the most common.

- Copper-silver (0.01 to 0.14% Ag) has better creep resistance, up to 250°C, than copper itself and is therefore used in the manufacture of conducting and catenary wires, commutators, alternators and motors, where the capacity to resist temperature and stress is essential.
- Copper-cadmium alloys, with about 1% cadmium, are used for their wear-resistant properties for some heavy-duty catenary wires, which are familiar as the overhead electrical conductor wire seen on electric railway systems. These alloys have the best combination of strength and electrical

conductivity, but their use is declining due to health concerns over the use of cadmium.

- For electrical applications such as resistance welding electrodes, where service is at high temperature under heavy stress, a copper-chromium (up to 1% Cr) alloy is often employed. This is heat treatable to give good room-temperature mechanical properties which are maintained well as the operating temperature rises (400°C continuous rating is possible) and it retains conductivity of around 80% IACS. The addition of up to 0.2% zirconium confers even better elevated-temperature fatigue resistance.

As mentioned briefly above, deoxidised copper is used for the other major area of application of the coppers in building, the principal uses being for central heating systems, pipe for gas and water supply, gutters and downpipes and sheet for roofing and cladding. Copper's high resistance to corrosion has led it to become the principal material for plumbing systems in the UK. The strength of copper means that it can be used in low- and high-pressure systems. It is a material that can be used for many different pipework applications: hot and cold water, natural gas and fire sprinkler systems. The ability of copper to form a protective and aesthetically pleasing surface, or patina, by weathering has encouraged its use for roofing and cladding of buildings over many centuries. The roofscapes of many capital cities across the world are punctuated by 'green' copper domes and roofs. Architects utilise the durability, low maintenance, ductility and aesthetic qualities of copper on small and large, private and public, ordinary and iconic buildings. It is now possible to obtain 'brown' pre-oxidised and 'green' pre-patinated copper sheets for architectural applications.

Copper has a melting point of 1083°C and an annealing temperature range of 200°C to 650°C. The high melting point of copper means that it is not subject to creep (plastic flow). Brasses typically have a melting range from 800°C to 970°C, with annealing temperature ranges of 450°C to 680°C. When copper and copper alloys are subjected to sustained heat, perhaps as a result of a fire, the material will maintain its integrity although the strength and other mechanical properties will be reduced to those of the annealed condition. Subsequent use of a component that has been subjected to elevated temperatures should take account of the annealed condition mechanical properties.

## 3.5 Copper alloys

### 3.5.1 Brass[3]

Brasses are copper alloys in which the main alloying element is zinc. The generic term 'brass' covers a wide range of materials suitable for many different types of application. Good corrosion resistance, machinability, formability and thermal and electrical conductivity are properties characteristic of all the brasses, together with ductility retained above and below ambient temperatures, good spark-resistance and low magnetic permeability.

Copper-zinc alloys with up to about 36% zinc have a single-phase metallurgical structure, the $\alpha$ phase, a solid solution of zinc in copper. Cap copper (up to 5% Zn) is used for ammunition percussion caps. The gilding metals (10% to 20% Zn) are used for architectural metalwork, papermaking, jewellery strip and applications requiring suitability for brazing and enamelling. The cartridge brasses (30% Zn) have the maximum ductility of the copper–zinc range and are used for deep drawing. Common brass, containing 36% zinc, is the most usual composition used for brass sheet.

Brasses with more than about 37% zinc have a binary metallurgical structure (two-phase) and are known as $\alpha$–$\beta$ alloys. The $\beta$ phase is easily deformable when hot, and these alloys lend themselves more readily to hot-forming techniques than almost any other alloy used in engineering. Such compositions, all derived from Muntz metal, with about 40% zinc, allow the production of complex machinable high-strength shapes at low material cost.

Other elements are added to the brasses to produce alloys with improved properties, e.g. strength, ductility, corrosion resistance, machinability. Free-machining brass (containing 39% Zn and 3% Pb) has for decades been the standard alloy against which the machinability of other metals and alloys has been judged. The lead is present as fine particles that help chip forming of the swarf so that it can clear away from the tool tip.

Brasses may be regarded as medium-strength engineering materials with proof strengths in the range 250–500 N/mm$^2$ (see Annex 3.1). In this respect they are comparable to low-alloy constructional steels, some stainless steels and aluminium alloys in the series 2xxx, 7xxx.

The addition of manganese, iron and aluminium produces high-tensile brasses (sometimes incorrectly known as manganese bronzes) which, with enhanced strength and resistance to wear, impact and abrasion, are used for architectural and heavy-duty engineering applications, e.g. auto transmissions, mining equipment, gas valves, boat fixings.

The brasses have good resistance to atmospheric corrosion, forming a thin adherent oxide or tarnish layer in a dry atmosphere and the green patina in moist atmosphere. They are resistant to alkalis and organic acids, but in some potable waters and in seawater, brass may suffer from a form of attack known as dezincification (see Section 3.6.4). Brasses are also susceptible to another form of attack described as stress corrosion cracking (see Section 3.6.6).The lower copper content of the duplex brasses means that they are often preferred for the manufacture of die castings, since the good plasticity reduces the risk of hot tearing in the mould. $\alpha$ brasses and duplex brasses are listed in Tables 3.4 and 3.5.

Brass alloys exhibit a range of colours: the red/golden of the gilding metal, the yellow colour characteristic of the dual-phase alloys, the brownish hue of manganese bronze and the white of nickel silver. Additionally, the alloys are readily electroplated for surface appearance with chromium or silver.

Lead is seen as a hazardous substance and, because of this, its use is being limited. The Restriction of the use of Certain Hazardous Substances (RoHS) in Electrical and Electronic

**Table 3.4** $\alpha$ brasses

| *Alloy designation* | *ISO symbol* | *Description* | *Uses* |
|---|---|---|---|
| CW501L<br>CW502L<br>CW503L | CuZn10<br>CuZn15<br>CuZn20 | Gilding metals | Architectural applications such as cladding, ornamental grills and shopfronts, condenser tubes, marine hardware. The name arises from the attractive red to golden colour. |
| CW505L | CuZn30 | Cartridge brass | Cartridge cases, hence the name; grillwork and building furniture such as hinges, locks and plumbing fittings. |
| CW508L | CuZn37 | Common brass | General-purpose alloy suitable for forming. |
| CW706R | CuZn28Sn1As | Admiralty brass | Salt-water applications, particularly for mixtures of fresh and salt water. |

**Table 3.5** Duplex brasses

| *Alloy designation* | *ISO symbol* | *Description* | *Uses* |
|---|---|---|---|
| CW509L | CuZn40 | Muntz metal | Hot stamping of plumber's fittings. They are used architecturally as panel sheets and also as structural members, such as end-plates for condensers. |
| CW720R | CuZn40Mn1Pb1 | Manganese brass (bronze) | Window frames. Chocolate brown in colour. |
| CW712R | CuZn36Sn1Pb | Naval brass | First developed for marine applications. It is used for tubeplates for condensers and heat exchangers and for hardware submerged in seawater. |
| CW409J | CuNi18Zn20 | Nickel silver | Tableware, telecommunications components. White in appearance. |

Equipment Regulations 2005[4] allows a lead content of up to 4% by weight in copper alloys, although it should be noted that leaded brasses contain a maximum of 3% lead.

### 3.5.2 Phosphor bronze

Binary alloys of copper and tin are called bronzes and can contain up to 12% tin. An increase in hardness and strength is gained at minimal cost by the addition of phosphorus to a level of around 0.25% to make phosphor bronzes. Tin contents range from 4% up to 8% in wrought materials, or higher if the alloy is used as cast. Typical applications for phosphor bronzes are window frames, door furniture and cladding. Alloys containing the higher tin level are particularly suitable for severe operating conditions. Possessing high corrosion resistance, excellent tensile and fatigue strength, superior wear resistance and bearing/frictional properties, this type of alloy finds application for heavy-duty bearings, bushes and gears, high-performance engine components, high-strength switch parts, thrust washers, slides, pistons and many others. Leaded phosphor bronzes can be produced that machine almost as easily as free-cutting brass.

Tin oxidises more readily than copper, and tin oxide has a higher density than the metallic alloy, so any tin oxide formed while the metal is molten remains entrapped within the metal after it has solidified, markedly reducing the ductility. Either zinc or phosphorus can be added to deoxidise the melt, but phosphorus is the more effective. Wrought copper-tin alloys containing about 0.03% residual phosphorus are called phosphor bronzes (designation CW450K–CW453K). These contain a maximum of 8.5% tin to produce a single-phase ($\alpha$) alloy but, with the higher tin contents in this range, it may be necessary to anneal the as-cast alloy in order to homogenise the microstructure and remove any traces of $\beta$ phase prior to cold working. The single-phase alloys are readily fabricated into sheet, bar and rod by either hot or cold working, but the usual restrictions with copper alloys on impurities such as lead, which produce low-melting-point phases, apply if the alloy is to be hot worked. Lead additions to improve machinability are only tolerated in alloys that are shaped by cold working.

Alloys containing more than 8.5% tin can be hot worked, but are not commonly used in the wrought condition because of the brittleness of the intermetallic phases. These compositions are more commonly used in the cast form.

Gunmetal, originally used for casting bronze cannon, is the generic name for a series of alloys that contain 2–11% tin and up to 6% nickel to improve the strength and hardness. These alloys contain zinc to act both as a deoxidiser and as an alloy element to replace part of the more expensive tin and copper and to improve the casting properties. The zinc content ranges from 2% to 9% and increases as the tin content is decreased. Up to 5% lead may be added to increase machinability. These compositions are used for the production of bronze statues and also for pumps and valves. Bronze bearings are normally cast with about 20% tin and up to 2% each of nickel and zinc. The lead content ranges from 8% to 23%, depending on the plasticity required for bedding-in the bearing. Bells are often cast in a 20% tin bronze and higher tin contents, up to 40%, have a long history of application for mirror surfaces.

### 3.5.3 Aluminium bronze[5]

A range of copper alloys contain up to 12.5% aluminium and frequently other elements such as nickel, iron, manganese and silicon. Varying the proportions of these results in a range of strong, tough alloys with excellent resistance to corrosion and wear that are ideal for a wide variety of demanding engineering applications, e.g. restraint and load-bearing fixings, pump impellers and casings and valves for gas industries (see Table 3.6).

**Table 3.6** Aluminium bronzes

| *Alloy designation* | *ISO symbol* | *Uses* |
|---|---|---|
| CW303G | CuAl8Fe3 | When cold worked, these alloys work harden very rapidly and are difficult to deform, so they are usually supplied in the hot-worked condition as plate, sheet, strip, seamless tube, bar, forgings, which are given only moderate cold deformation to form the required shapes. They are not normally formed by extrusion because the abrasive oxide coating covering a relatively strong substrate results in fairly rapid wear of the extrusion dies. |
| CW307G | CuAl10Ni5Fe4 | High-strength alloys for use in aggressive media when wear resistance and good impact strength are required. Used for ties, hangers and tension straps. |
| CC331G<br>CC333G | CuAl10Fe2–C | For highly stressed components in corrosive environments where high wear and shock loads may be encountered. |
| | CuAl10Fe5Ni5–C | Pumps, bearings, tools, bushings, housings and load-bearing fixings. |

Their strength (tensile strength 600–760 N/mm$^2$; see Annex 3.1), and in many respects their corrosion resistance, is better than most stainless steels, especially in aggressive marine environments. They are available both as high-integrity castings in weights up to many tonnes and in the usual wrought forms such as plate, forgings, extrusions and as welding wire. They are weldable for fabrication of large components (see Section 3.7).

The aluminium bronzes possess the excellent natural corrosion resistance of all copper alloys enhanced by the protective film of alumina (aluminium oxide) formed very rapidly under normal operating conditions. If damaged, this film is self-healing, which means that the alloys can be used in service conditions when abrasion can be expected.

They have an attractive golden colour and have good resistance to tarnish, due to the formation of a transparent alumina surface layer which also confers good wear resistance. In general, the alloys retain their strength well at elevated temperatures, oxidise at a low rate and have good creep and high-temperature-fatigue properties. One of the attributes of copper alloys is their non-sparking behaviour, which makes them suitable for use in potentially combustible and explosive environments, e.g. mining. The high hardness attainable with aluminium bronzes facilitates their use in place of steel for applications in these conditions.

Probably the most important attribute of the aluminium bronzes is their outstanding corrosion resistance to a wide range of environments including strong, non-oxidising acids, organic acids, chlorides and sulphur dioxide. They have very high resistance to conditions that in most other metals would give rise to pitting or crevice corrosion, but should not be used in contact with strong alkalis, which can dissolve the alumina surface film. These alloys are not immune to stress corrosion cracking (see Section 3.6.6), but are very much less susceptible than the brasses to this type of damage.

Aluminium bronzes show very good resistance to high-velocity erosion (impingement attack) and to cavitation. The former occurs when high-velocity fluids, often containing suspended solids, impinge on the metal surface. If the protective surface film is removed in this zone, the underlying metal can corrode rapidly. But the alumina surface film is hard and firmly adherent, so it is not easily removed. If the film is damaged, the protective layer is rapidly reformed. Cavitation is the result more of mechanical damage than of corrosion. In highly turbulent water, vapour bubbles may form at the metal surface where the fluid flow produces low-pressure pockets. These bubbles then collapse on the metal surface where the fluid pressure is higher, producing shock waves (hammering) which can detach the protective film and fragments of the underlying metal by the propagation of fatigue cracks. The high strength of some of the alloys and the tenacity of the surface film make them strongly resistant to this form of damage. For the same reasons, these alloys have high resistance to corrosion fatigue.

Aluminium bronzes are widely used for pump casings and impellers, valves, in fresh and saline water systems, particularly where high-velocity fluid flows are encountered, and for piping and condensers for high-pressure steam. Reinforcing bars and clamps for the repair of old buildings are now frequently made from aluminium bronze instead of iron, to avoid the swelling that occurs from the formation of rust. Cast and wrought forms are used for load-bearing plates and expansion joints in bridges and buildings. Architectural uses include cladding, where the golden yellow colour is sometimes selected as a cheaper alternative to gold alloy coatings.

The alumina surface coating on the aluminium bronzes can create problems in joining, similar to those encountered in joining of pure aluminium and its alloys, but compounded by the higher thermal conductivity of the copper-based alloys. The most successful joining method, producing high-strength joints compatible with the strength of the alloys, is fusion welding by gas shielded arc processes, using an inert gas (usually argon or helium) to prevent oxidation of the aluminium in the molten pool of weld metal.

### 3.5.4 Copper-nickel

Like the aluminium bronzes, the copper-nickel alloys (previously referred to as cupronickels) have came into their own during the past century. Following the 1914–1918 war, the British Navy required condenser tubes with improved resistance to failure when handling seawater in harbours, estuaries and other polluted waters. This led to the development of copper-nickel-iron alloys in the 1930s (see Table 3.7).

The two most used alloys of copper and nickel contain around 10% and 30% nickel with other elements. The addition of nickel to copper improves strength and corrosion resistance but good ductility is retained. Excellent resistance to corrosion attack by marine environments is combined with the resistance of copper to biofouling.

In the past few decades high-strength copper-nickels have been developed, with proof strength up to 820 N/mm$^2$. These alloys, strengthened by the presence of minute nickel–aluminium precipitates in the microstructure, are selected for bolting (with resistance to hydrogen embrittlement) and other highly loaded components on naval vessels and offshore oil and gas structures. They show anti-galling properties, i.e. nuts can be tightened and untightened without sticking. Their tensile strength is equal to that of steel bolts. High-strength cast copper–nickel, containing chromium, is used for pumps and valves in naval vessels.

### 3.5.5 Copper-beryllium

The addition of beryllium to copper gives an alloy capable of being heat-treated and cold worked to provide exceptionally good mechanical properties at room and elevated temperatures. For many years these alloys have been used extensively for demanding applications such as springs, contacts, heavy-duty engineering and electrical components and moulds for plastics and glass production.

Because of their high strength and hardness they are used for the manufacture of non-sparking tools for use in hazardous environments. There are two basic types of alloy – one contains nearly 2% beryllium with some nickel and/or cobalt; the other contains about 0.5% beryllium and 2% cobalt.

**Table 3.7** Copper–nickels

| *Alloy designation* | *ISO symbol* | *Uses* |
|---|---|---|
| CW352H | CuNi10Fe1Mn | Seawater piping aboard ships and offshore oil and gas production platforms. The alloy is also employed in sheet form to sheathe the hulls of ships and clad the legs of offshore platforms. |
| CW354H | CuNi30Mn1Fe | Developed initially to provide an even better material for condenser tubing. Other applications for these copper–nickels include plant construction for desalination by distillation, vehicle hydraulic systems and the production of coinage. |

## 3.6 Corrosion mechanisms

Copper is usually classed as a noble metal, with high resistance to corrosion due to an adherent oxide layer. It is inert in potable water and has good resistance both to seawater and to deaerated, non-oxidising acids. It is less resistant to alkalis (which may be leached out of concrete), to ammonia and to sulphur compounds. Correct materials selection and system and component design will reduce, if not eliminate, the potential for corrosion.

### 3.6.1 Galvanic corrosion (bi-metallic corrosion)

When different metals or alloys are in contact with one another in an electrolyte (seawater, fresh water, rain, dew, condensation) they affect one another's resistance to corrosion. Usually one – the more 'noble' – will cause some degree of accelerated attack (galvanic corrosion) on the other and will itself receive a corresponding degree of protection. Table 3.8 shows a number of common metals and alloys in their order of nobility in sea water and may be used to give some indication of the possible galvanic corrosion effects of coupling dissimilar metals. In general, the further the two metals are apart in the electrochemical series, the greater the effect will be.

**Table 3.8** Galvanic series for common metals and alloys

| *Corroded end: anodic - least noble* |
|---|
| Magnesium |
| Magnesium alloys |
| Zinc |
| Aluminium-magnesium alloys |
| 99% aluminium |
| Cadmium |
| Aluminium-copper alloys |
| Aluminium-copper-magnesium (Duralumin) |
| Steel or iron |
| Cast iron |
| Chrome-iron (active) |
| Ni-resist |
| 18/8 chromium-nickel austenitic steel (active) |
| 18/8 Mo steel (active) |
| Hastelloy C |
| Lead–tin solders |
| Lead |
| Tin |
| Nickel (active) |
| Inconel (active) |
| Hastelloy A |
| Hastelloy B |
| Brasses |
| Copper |
| Bronzes |
| Copper-nickel alloys |
| Monel |
| Silver solder |
| Nickel (passive) |
| Inconel (passive) |
| Chrome-iron (passive) |
| 18/8 chromium-nickel austenitic steel (passive) |
| 18/8 Mo chromium-nickel austenitic steel (passive) |
| Silver |
| Graphite |
| Gold |
| Platinum |
| *Protected end: cathodic – most noble* |

Among the brasses themselves there are small differences of electrochemical potential, those of highest copper content being more noble. In particular, the $\alpha$ brasses are somewhat more noble than the $\beta$ brasses; this shows itself in the tendency for the beta phase in $\alpha$-$\beta$ brasses to suffer preferential attack, but the difference between the two is not great.

The extent to which galvanic corrosion takes place depends not only on the difference between the two metals in the galvanic series but also on their relative areas, exposed to the electrolyte, sufficiently close to one another for significant corrosion currents to flow through the electrolyte between them. If the effective area of more noble (cathodic) metal greatly exceeds that of the other metal, galvanic attack on the less noble metal may be severe but, if the area of cathodic metal is smaller than that of the other metal, the effect will be negligible. For example, stainless steel trim in a brass valve is quite acceptable but brass bolts on a stainless steel structure would certainly not be.

It is possible to prevent galvanic corrosion by electrically insulating the more noble and less noble metals from one another. A difficulty arises because the metallic connection between the two members of the galvanic couple does not have to be by direct contact between them. Whenever steps are taken to insulate the two members of a potential galvanic couple from one another, it is important to check, before they are brought in contact with the water or other electrolyte for which they are to be used, that the desired absence of electrical continuity between them has been achieved.

As an alternative to insulating the two members of a couple from one another, one or both of them can be isolated from the electrolyte by coating or painting. In some cases the anodic member needs to be painted or coated to protect it from corrosion that would take place even in the absence of the galvanic effect.

### 3.6.2 Crevice corrosion

Crevice corrosion occurs when two components, such as flanges, are in close contact but there is a thin film of water between them. In these circumstances the free access of oxygen is limited, creating conditions where corrosion can occur within the crevice. Correct joint design can eliminate the possibility of crevice corrosion.

### 3.6.3 Pitting corrosion

Pitting produces very localised attack, often in approximately hemispherical form beneath a small adherent mound of green corrosion product. If this mound is carefully removed, crystals of red cuprous oxide can usually be seen in the cavity. Pitting corrosion can lead to failure of a component by way of pinholes that perforate the material.

Pitting corrosion is not a serious problem with brasses. $\alpha$ brasses inhibited against dezincification can, however, suffer pitting under some circumstances. Service in slow-flowing sulphide-polluted sea water is most likely to produce pitting in aluminium brass, but this alloy sometimes develops pitting corrosion in fresh water service.

Pitting can, in certain circumstances, be a problem for copper pipework.[6] There are three forms of pitting and they result from specific chemistries.

1 Carbon film pitting, sometimes referred to as 'Type 1', causes corrosion of cold water pipes carrying hard deep well waters. Modern manufacturing techniques have all but eliminated this form of corrosion as carbon films are removed from tube bores.
2 Hot soft water pitting corrosion, sometimes referred to as 'Type 2', is extremely rare in the UK: it seldom causes failure in less than about 10 years. It occurs in hot water pipes in some soft water areas, specifically if the operating temperature is above 60°C.

3 Excessive use of flux used in the soldering of copper tube and fittings for water services, resulting in flux runs within the bore of the tube, may cause corrosion and should be avoided.

### 3.6.4 Selective leaching

The best known example of selective leaching in copper alloys is 'dezincification', the leaching of zinc from the brass under certain conditions that leaves behind a mass of porous copper. The attack is most severe in stagnant water containing high concentrations of oxygen and carbon dioxide. The surface of the alloy dissolves in the solution and copper is redeposited on the surface in a sponge-like form, while bulky zinc corrosion products accumulate in the liquid. Dezincification is not usually a problem with zinc contents up to about 20% but becomes progressively worse as the zinc content is increased beyond this level. Alloys containing the β phase are particularly prone to this form of attack, since the β phase is attacked preferentially. Attack can be prevented in single-phase α alloys by the addition of up to 0.1% arsenic and a specific heat treatment, and arsenical brass should be specified for all aqueous applications. However, two-phase (duplex) and β brasses are not protected by arsenic additions and these alloys should not be used in contact with water where dezincification may occur.

### 3.6.5 Erosion corrosion

Copper and copper alloys owe their long-term corrosion resistance to the protective effect of thin, adherent films of corrosion products, which form during the early life of the component and form a barrier between the metal surface and its corrosive environment. Water flow conditions that produce high water velocities at the protected metal surface can generate shear forces sufficient to cause local removal of the protective corrosion product film, exposing bare metal to corrosion, and to sweep away the fresh corrosion products resulting from this exposure before they can form a new protective layer. Such conditions are obviously associated with high average water velocities, but arise particularly where excessively turbulent flow gives rise to local water velocity much higher than the average flow rate. The severe local attack that results is commonly termed impingement attack or, more accurately, since it is the result of corrosion of the metal combined with erosion of the corrosion product film, erosion corrosion.

Metal that has suffered erosion corrosion exhibits a smooth water-swept surface usually without corrosion products. Localised attack, often associated with local turbulence, immediately downstream of an obstruction gives individual water-swept pits, undercut on the upstream side and often horseshoe-shaped with the open end of the horseshoe pointing downstream. More widespread attack produces a broad smooth surface in which small horseshoe-shaped features are often visible.

### 3.6.6 Stress corrosion cracking

Stress corrosion cracking (SCC) or 'season cracking' occurs only in the simultaneous presence of a sufficiently high tensile stress and a specific corrosive environment. For copper and copper alloys the environment involved is usually one containing ammonia or closely related substances such as amines, but atmospheres containing between 0.05% and 0.5% of sulphur dioxide or nitrites by volume can also cause stress corrosion cracking, as can mercury and mercurous nitrate.

Stress corrosion cracking is usually localised and, if ammonia has been involved, may be accompanied by black staining of the surrounding surface. The fracture surface of the crack may be stained or bright, according to whether the crack propagated slowly or rapidly. The cracks run roughly perpendicular to the direction of the tensile stress involved. For example, drawn tube that has not been stress relief annealed has a built-in circumferential hoop-stress; consequently exposure to an ammoniacal environment is liable to cause longitudinal cracking. Stress corrosion cracking in pipes that have been cold bent without a subsequent stress relief anneal occurs typically along the neutral axis of the bend. Stress corrosion cracking due to operating stresses is transverse to the axis of the applied stress. A stress relief anneal at about 300°C will reduce internal stress, without affecting the hardness, and will thus eliminate the potential for SCC.

## 3.7 Joining[7]

Since copper does not form an oxide coat spontaneously on exposure to air, it is relatively easy to clean the surface and join together by soldering or brazing techniques before an oxide has reformed (see Table 3.9). Copper surfaces are readily coated with molten tin, requiring only a flux such as ammonium chloride or zinc chloride to prevent reoxidation. The fluxes are corrosive to copper, however, and must be washed or wiped off the surface after treatment. Solder alloys are specified in EN 29453; while the standard includes alloys with lead content, it is recognised that lead bearing solders are being phased out in a number of applications because of health concerns over the lead, particularly in terms of recycling products at end-of-life. Tin–silver solders are now used for soldering joints in potable water systems, as it is not now permitted to use leaded solders in drinking water installations. It is worth noting that there are a number of flame-free joining options for copper pipework, including compression jointing, press fittings and push-fit fittings.

Copper can be brazed with a copper alloy containing up to 50% zinc to lower the melting point. Copper-zinc alloys have slightly inferior corrosion resistance compared with pure copper, which may result in corrosion at the joint in arduous service conditions, and some zinc may be lost by volatilisation if the brazing temperature is too high. Silver may be added to the brazing alloy to depress the melting point further, without adverse effect on the corrosion resistance. Tough-pitch coppers are usually brazed in a furnace in a reducing atmosphere to avoid the risk of hydrogen embrittlement.

The very high thermal conductivity of copper creates problems in fusion welding as preheating of the joint region and high rates of heat input are required to obtain satisfactory welds. Flame welding is best avoided with the un-deoxidised, tough-pitch grades because of the risk of reaction between hydrogen and the copper oxide.

Copper-rich materials are used extensively as filler rods in the welding of other high-melting-point metals. Phosphorus deoxidised tough-pitch copper, sometimes with up to 3% silicon and/or manganese, is popular, or with up to 3% zinc for autogenous welding.

## 3.8 Recycling and sustainability

Copper and copper alloy scrap can be recycled relatively cheaply, with low power consumption. The amount of energy required to recycle copper is 75–92% lower than that needed to convert copper ores to metal.[8] The recycling of copper and its alloys plays an important part in the economics of production and has been undertaken since the copper industry began. The cost of the raw material can be significantly reduced if an alloy can be made with recycled material. If the scrap is pure copper, a high-quality product can be made from it. Similarly, if scrap is kept segregated, and consists only of one alloy composition, it is easier to remelt to a good quality product conforming to standard.

**Table 3.9** Suitability of copper alloys to jointing processes

| | *ISO designation* | *Soldering* | *Brazing* | *Bronze welding* | *Oxy-acetylene welding* | *Gas shielded arc welding* |
|---|---|---|---|---|---|---|
| *Coppers* | | | | | | |
| CW004A | Cu-ETP | 1 | 2 | 4 | 4 | 3 |
| CW005A | Cu-FRHC | 1 | 2 | 4 | 4 | 3 |
| CW006A | Cu-FRTP | 1 | 2 | 4 | 4 | 3 |
| CW024A | Cu-DHP | 1 | 1 | 2 | 2 | 1 |
| CW008A | Cu-OF | 1 | 2 | 3 | 4 | 4 |
| *Low-alloyed coppers* | | | | | | |
| CW118C | CuTeP | 1 | 2 | 4 | 4 | 4 |
| – | CuCd | 1 | 2 | 4 | 4 | 4 |
| CW100C | CuBe1.7 | 2 | 2 | 4 | 4 | 3 |
| *Phosphor bronzes* | | | | | | |
| CW450K | CuSn4 | 1 | 1 | 4 | 3 | 2 |
| CW451K | CuSn5 | 1 | 1 | 4 | 3 | 2 |
| CW452K | CuSn6 | 1 | 1 | 4 | 3 | 2 |
| CW453K | CuSn8 | 1 | 1 | 4 | 3 | 2 |
| *Aluminium bronzes* | | | | | | |
| CW303G | CuAl8Fe3 | 2 | 3 | 4 | 4 | 2 |
| CW307G | CuAl10Ni5Fe4 | 2 | 3 | 4 | 4 | 2 |
| *Copper-nickels* | | | | | | |
| CW352H | CuNi10Fe1Mn | 1 | 2 | 4 | 4 | 1 |
| CW354H | CuNi30Mn1Fe | 1 | 1 | 4 | 2 | 1 |
| *Brasses* | | | | | | |
| CW501L | CuZn10 | 1 | 1 | 4 | 2 | 2 |
| CW502L | CuZn15 | 1 | 1 | 4 | 2 | 2 |
| CW503L | CuZn20 | 1 | 1 | 4 | 2 | 2 |
| CW505L | CuZn30 | 1 | 1 | 4 | 2 | 3 |
| CW508L | CuZn37 | 1 | 1 | 4 | 2 | 3 |
| CW509L | CuZn40 | 1 | 2 | 4 | 2 | 2 |
| *Leaded brasses* | | | | | | |
| CW614N | CuZn39 Pb3 | 1 | 2 | 4 | 4 | 4 |
| *Special brasses* | | | | | | |
| CW702R | CuZn20A12As | 2 | 2 | 4 | 3 | 3 |
| CW706R | CuZn28Sn1As | 1 | 1 | 4 | 2 | 3 |
| CW602N | CuZn36Pb2As | 1 | 2 | 4 | 4 | 4 |

1, excellent; 2, good; 3, fair; 4, not recommended.

When costing the production of a component, allowance can be made for recovery of money through the reclamation and sale of clean process scrap. There are many forms and sources of scrap that may be utilised. Components that are beyond their useful life are a valuable source. In addition, scrap from manufacturing processes such as trimming, fettling (of castings) and machining is most useful.

Good-quality high-conductivity copper scrap, for example, generated during power cable drawing, can be recycled by simple remelting. Where contamination has occurred, it is normally remelted and cast to anode shape so that it can be electrolytically refined. In cases where scrap is contaminated with certain elements such as tin and lead – as a result of being tinned or soldered – it is more economic to take advantage of this contamination to produce a copper alloy (gunmetal or bronze) which needs these additions as alloying elements.

Where copper and copper alloy scrap is very contaminated and unsuitable for melting, it can be recycled by other means either to recover the copper as the metal or to give some of the many copper compounds essential for use in industry and agriculture.

Copper is one of the most commonly recycled materials in the world. Recycling reduces the need to exploit natural resources and also reduces environmental pollution.

Sustainable development is that which meets the needs of the present without compromising the ability of future generations to meet their own needs (World Commission on Environment and Development, 1987).[9] From this definition the component parts of sustainable development have been established under the headings of:

(a) environmental protection
(b) economic growth
(c) social considerations.

### 3.8.1 Environmental protection

- Copper is a finite resource; it is mined from ore deposits formed in the ground millions of years ago. However, very little copper is used up, since it can be endlessly recycled without loss of properties; it is conserved for future generations.
- When copper is mined and refined to 99.9% purity, gases such as $SO_2$ and dust are released. Although these are collected by the metal producers to protect the environment, with recycling there are virtually no emissions.
- Recycling copper (and other materials) reduces landfill costs. Copper forms a very small percentage of the materials found dumped on landfill sites; it is too valuable to throw away.
- Recycling uses 10% of the energy that would be used to mine and produce the same copper. So recycling helps to conserve the world's supply of fossil fuels.
- Because of its superior conductivity, copper is the material of choice for transferring electric power. In electric motors and other components, more copper improves efficiency by reducing wasteful heat loss. This means less energy demand

per unit of output, which means fewer greenhouse-gas emissions which are associated with climate change.

### 3.8.2 Economic growth

- Copper is essential to technology, enabling peak performance from advanced microprocessors and other miniature components that drive the digital economy of today and tomorrow.
- Recycled copper is worth up to 90% of the cost of original copper. It is impossible to detect the difference between new and recycled copper, both of which are used to produce high-value, long-lasting products by companies who require no subsidies. Clearly recycling provides an economic benefit.

### 3.8.3 Social considerations

- Copper is well known for its anti-bacterial properties. A strong case has been presented for the value of copper and copper alloys to help control the incidence of infection from cross-contamination due to dangerous foodborne and hospital-borne pathogens, such as *E. coli* O157, *Campylobacter*, *Listeria monocytogenes*, *Salmonella* and the difficult-to-treat Meticillin-resistant *Staphylococcus aureus* (MRSA). This case is based on copper's intrinsic ability to quickly inactivate these dangerous microbes at both refrigerated temperature (4°C) and room temperature (20°C).
- For centuries copper pipes and vessels have been used to convey clean drinking water. Copper limits the growth of water-borne pathogens such as *Legionella pneumophila*.
- Copper is also a micro-nutrient vital for all forms of plant and animal life and thus serves a critical function in agriculture to ensure soil fertility and productive yield of vital foodstuffs.
- Modern social and business life would be impossible without electricity, available instantly at the point of use. Electricity conducted by copper encounters much less resistance compared to any other commonly used metal; its electrical conductivity is 60% higher than aluminium. Copper is a key material for generation, distribution and use of electricity and is found in cables, generators, motors and transformers. Because of its electrical conductivity, copper is a key material in energy efficiency and renewable energy generation. All household electrical appliances, electronic and telecommunication devices also contain significant quantities of copper.

### 3.8.4 Embodied energy

The embodied energy of a material (in J/kg) is a measure of the total energy consumed during every phase of the life cycle of a product, from cradle to grave, so for copper it includes energy used during mining and extraction, manufacture, disassembly and final recycling. The long life and recyclability of copper products has a positive impact on their embodied energy.

The copper industry has developed up-to-date life-cycle data for its tube, sheet and wire products.[10] The information has been prepared in co-operation with recognised life cycle practitioners, using international methodologies (ISO standards), leading software (GaBi), and proprietary production data collected from across the copper industry. For more information visit www.copper-life-cycle.de.

## 3.9 Health and environment

Copper is a natural element within the earth's crust which has been incorporated into living organisms throughout the evolutionary process. It is an essential nutrient required by all higher life forms. Copper is required as part of a balanced diet. It is especially important for pregnant women, the developing foetus and newborn babies. A typical recommended daily requirement is 1–2 mg for adults and 0.5–1 mg for children.[11] Nature is well adapted to making the best use of copper, protecting itself from any negative effects. This applies at the most basic levels right up to the most complex metabolic functions of the human body. It also holds true with long-term effects of man's use of copper on buildings.

With copper roofing small amounts of copper are carried off the roof in the rainwater. Through natural processes of binding to organic matter, adsorption to particles and precipitation, the copper runoff finally comes to rest in a mineral state as part of the earth's natural background of copper material. More than 99% of the copper ions released from a roof will bind with organic and inorganic matter in the soil and will not therefore be bio-available.[12]

## 3.10 References


1 J. O. Noyce, H. Michels and C. W. Keevil (2006). Potential use of copper surfaces to reduce survival of epidemic meticillin-resistant *Staphyloccus aureus* in the healthcare environment. *Journal of Hospital Infection*, *62*: 289–297.
2 Pub 120, *Copper and copper alloys – Compositions, applications and properties*. Copper Development Association, Hemel Hempstead, UK, 2004.
3 Pub 117, *The brasses, properties and applications*. Copper Development Association, Hemel Hempstead, UK, 2005.
4 RoHS Directive. *The Restriction of the Use of Certain Hazardous Substances in Electrical and Electronic Equipment Regulations 2005*. No. 2748. European Council and Parliament, Brussels, Belgium.
5 H. Meigh (2000), *Cast and wrought aluminium bronzes – Properties, processes and structure*. IOM Communications, Grantham, UK.
6 P. McIntyre and A. D. Mercer (Eds) (1993). *Corrosion and related aspects of materials for potable water supplies*. The Institute of Materials, London.
7 Pub 98 (1994), *Joining of copper and copper alloys*. Copper Development Association, Hemel Hempstead, UK.
8 *Copper fact sheet – Copper recycling* (2005). International Copper Association, New York, NY.
9 World Commission on Environment and Development (1987). *Our common future*, Oxford University Press, Oxford, UK.
10 S. Gößling-Reisemann, L. Tikana, H. Sievers and A. Klassert. *Life cycle methodology for copper – Allocation and recycling approaches*. COM2007 Conference, Metsoc, Toronto, Canada.
11 E. M. Ward, C. L. Keen and H. L. McArdle (2003). *A review: The impact of copper on human health*. International Copper Association, New York, NY.
12 *Copper fact sheet – Copper and roof run-off*, International Copper Association, New York, NY, 2005.

## Annex 3.1 Material designations and typical mechanical properties for some commonly used copper and copper alloys

**Table 3A.1** Commonly used copper and copper alloys

| *Old BS designation* | *EN designation* | *ISO symbol* |
|---|---|---|
| *Copper* | | |
| C101 | CW004A | Cu-ETP |
| C102 | CW005A | Cu-FRHC |
| C103 | CW008A | Cu-OF |
| C104 | CW006A | Cu-FRTP |
| C105 | – | Cu-AS |
| C106 | CW024A | Cu-DHP |
| C107 | – | Cu-AsP |
| C108 | – | |
| C109 | CW118C | CuTeP |
| C111 | CW114C | CuSP |
| *Aluminium bronze* | | |
| CA102 | – | |
| CA104 | CW307G | CuAl10Ni5Fe4 |
| CA106 | CW303G | CuAl8Fe3 |
| AB1 | CC331G | CuAl10Fe2-C |
| AB2 | CC333G | CuAl10Fe5Ni5-C |
| *Beryllium copper* | | |
| CB101 | CW100C | CuBe1.7 |
| *Brass* | | |
| CZ101 | CW501L | CuZn10 |
| CZ102 | CW502L | CuZn15 |
| CZ103 | CW503L | CuZn20 |
| CZ106 | CW505L | CuZn30 |
| CZ108 | CW508L | CuZn37 |
| CZ109 | CW509L | CuZn40 |
| CZ110 | CW702R | CuZn20Al2As |
| CZ111 | CW706R | CuZn28Sn1As |
| CZ112 | CW712R | CuZn36Sn1Pb |
| CZ136 | CW720R | CuZn40Mn1Pb1 |
| DZR1 | CC752S | CuZn35PbAl-C |
| HTB1 | CC765S | CuZn35Mn2Al1Fe1-C |
| *Gunmetals and leaded bronzes* | | |
| LG2 | CC491K | CuSn5Zn5Pb5-C |
| LB2 | CC495K | CuSn10Pb10-C |
| *Phosphor bronze* | | |
| PB101 | CW450K | CuSn4 |
| PB102 | CW451K | CuSn5 |
| PB103 | CW452K | CuSn6 |
| PB104 | CW453K | CuSn8 |
| PB1 | CC481K | CuSn11P-C |
| PB2 | CC483K | CuSn12-C |
| *Copper-nickel* | | |
| CN102 | CW352H | CuNi10Fe1Mn |
| CN107 | CW354H | CuNi30Mn1Fe |

A full listing of copper and copper alloy designations and compositions can be found in CR13388 – Copper and copper alloys – Compendium of compositions and products (BSI).

Table 3A.2 gives typical mechanical properties for wrought copper alloys. For more precise information reference should be made to the product standard.

**Table 3A.2** Mechanical properties for wrought copper alloys

| *EN designation* | *Symbol* | *0.2% Proof strength (N/mm²)* | *Tensile strength (N/mm²)* | *Elongation (%)* | *Hardness (HV)* | *Nominal min. conductivity (% IACS)* |
|---|---|---|---|---|---|---|
| CW004A | Cu-ETP | 50–340 | 200–400 | 50–5 | 40–120 | 100 |
| CW005A | Cu-FRHC | 50–340 | 200–400 | 50–5 | 40–120 | 100 |
| CW008A | Cu-OF | 50–340 | 200–400 | 50–5 | 40–120 | 100 |
| CW006A | Cu-FRTP | 50–340 | 200–400 | 50–5 | 40–120 | |
| CW024A | Cu-DHP | 50–340 | 200–400 | 50–5 | 40–120 | |
| CW118C | CuTeP | 200–320 | 250–360 | 7–2 | 90–110 | 90 |
| CW114C | CuSP | 200–320 | 250–360 | 7–2 | 90–110 | 93 |
| CW307G | CuAl10Ni5Fe4 | 400–530 | 600–760 | 15–5 | 170–220 | |
| CW303G | CuAl8Fe3 | 180–210 | 460–500 | 30 | 125–135 | |
| CW100C | CuBe1.7 | 200–1100 | 410–1300 | 35–3 | 100–400 | 30 |
| CW501L | CuZn10 | 120–560 | 240–600 | 45–2 | 60–165 | |
| CW502L | CuZn15 | 120–590 | 260–630 | 50–2 | 65–170 | |
| CW503L | CuZn20 | 120–590 | 260–630 | 50–2 | 65–170 | |
| CW505L | CuZn30 | 130–810 | 300–830 | 55–1 | 65–200 | |
| CW508L | CuZn37 | 130–800 | 280–820 | 50–1 | 65–190 | |
| CW509L | CuZn40 | 200–420 | 340–500 | 45–2 | 90–150 | |
| CW702R | CuZn20Al2As | 140–380 | 340–540 | 60–20 | 80–160 | |
| CW706R | CuZn28Sn1As | 110–410 | 320–460 | 60–20 | 80–160 | |
| CW712R | CuZn36Sn1Pb | 160–360 | 340–480 | 30–10 | 90–150 | |
| CW720R | CuZn40Mn1Pb1 | 160–350 | 350–550 | 20–10 | 100–170 | |
| CW450K | CuSn4 | 140–850 | 320–950 | 60–1 | 75–230 | |
| CW451K | CuSn5 | 140–850 | 320–950 | 60–1 | 75–230 | |
| CW452K | CuSn6 | 140–950 | 340–1000 | 60–1 | 80–250 | |
| CW453K | CuSn8 | 170–1000 | 390–1100 | 60–1 | 85–270 | |

# 4 Ferrous Metals
## An Overview

**Michael Bussell** BSc(Eng)
Formerly Ove Arup & Partners

### Contents

## 4.1 Introduction

Iron was and still is the metallic element most widely used in construction. Over time, it has been employed in three basic forms. Today, many types of *steel* are in use throughout the world, it having supplanted *wrought iron* (no longer made); *cast iron* is still employed in some new work, although on a much smaller scale than in the past. However, cast and wrought iron are to be found in many older buildings and other structures whose maintenance, conservation, and re-use has come to form a major part of the construction industry's workload in recent decades. A technical understanding of these forms of iron and their past use is an important asset when dealing with existing construction, just as a similar understanding of steel and the present-day forms of cast iron is needed for new construction.

This chapter offers a brief historical and technical overview of cast and wrought iron and steel in construction. It begins with a comparison of their key characteristics. This is followed by guidance on identifying and distinguishing the three forms, intended to assist when dealing with existing construction.

The making of all iron and steel begins with iron ore which, since post-medieval times, has been smelted in a furnace to produce liquid *pig iron*. This, and the early methods of large-scale steel-making, are described here. The subsequent processing of pig iron to make cast iron and wrought iron is described in the following Chapters 5 and 6, while modern-day steel-making is described in Chapter 7

This overview continues with a chronological summary of notable constructional uses of cast iron, wrought iron, and steel, including use in reinforced and prestressed concrete. It concludes with notes on durability, protection against corrosion, performance in fire, and issues of sustainability and recycling. These topics are addressed more fully in the succeeding chapters on the specific ferrous metals.

## 4.2 Key characteristics of cast and wrought iron and steel

Table 4.1 summarises and compares the key characteristics of the ferrous metals, from which it will be seen that cast and wrought iron and steel each have distinctive properties. This is especially so when 'historic' cast iron – grey cast iron, brittle and weak in tension – is contrasted with wrought iron and steel, both ductile materials (although nowadays spheroidal graphite cast iron is also available, with reasonable ductility).

Figure 4.1 illustrates and compares the stress–strain behaviour of typical examples of grey cast iron, wrought iron, and steel.

## 4.3 Distinguishing between cast and wrought iron and steel

With their differing properties, it is clearly important to be able to recognise the three forms of ferrous metals, particularly in structural use. Appearance will often suffice to identify the metal; Table 4.2 notes significant visual characteristics of the metals, while Table 4.3 lists original connection methods commonly used to connect structural elements and also other functional or decorative items such as gates and railings.

Dating (see Figure 4.3 on page 40) will often be helpful in distinguishing between wrought iron and steel – earlier than about 1880 is likely to be wrought iron; after 1900 is likely to be steel. If necessary, a small sample may be removed from a lightly stressed area for metallographic examination.

**Table 4.1** Key characteristics of cast and wrought iron and steel*

| *Material* | *Grey cast iron* | *Wrought iron* | *Steel* | *Spheroidal graphite cast iron* |
|---|---|---|---|---|
| Typical carbon content (%) | 2–5 | 0–0.1 | 0.05–1.5 | 3–4 |
| Form of carbon | Graphite flakes within iron | Trace | Alloyed with iron | Nodules within iron |
| Tensile strength (N mm²)† | 75–300 | 250–550§ | 400–650§ | 350–900 |
| Compressive strength (N mm²)† | 550–800 | 250–550§ | 400–600§ | 350–900 |
| Yield strength (N mm²)† | No defined yield | 180–300§ | 230–500§ | 220–600 (0.2% proof stress) |
| Elongation at failure† | 0.25–0.75% | 4–30%# | 15–30%# | 2–22%# |
| Elastic modulus (N mm²)† | 85–100 (at low stress) | 180–200 | 205–210 | 170 |
| Fracture mode | Brittle | Ductile‡ | Ductile | Ductile |
| Typical failure surface | Crystalline | Usually fibrous‡ | Usually amorphous | Amorphous |
| Method of forming sections/shapes | Casting when molten | Hot-rolling or forging | Hot-rolling, forging, or casting | Casting when molten |
| Weldability | Very difficult | Forge-weldable when heated; fillet welding unwise | Oxy-acetylene, later electric arc | Difficult |
| Corrosion resistance | As-cast faces good; otherwise similar to steel | Slightly better than steel | Fair – often requires protection | Similar to grey cast iron |

* Physical properties are heavily influenced, in particular, by the amount of carbon present and by the manufacturing and working process – see Chapters 5, 6 and 7. † Numerical values are *typical*; for grey cast iron and wrought iron (and early steel) the ranges reflect variability in manufacture, whereas for steel and spheroidal graphite cast iron they indicate the strengths obtainable by present-day specifications. § Higher strengths obtainable by cold-working, or by heat treatment or addition of trace elements (steel only), e.g. for wire or prestressing tendons. # Elongation at failure decreases with increasing strength. ‡ Some poor-quality or overworked wrought iron may fail in brittle mode with some crystalline fracture.

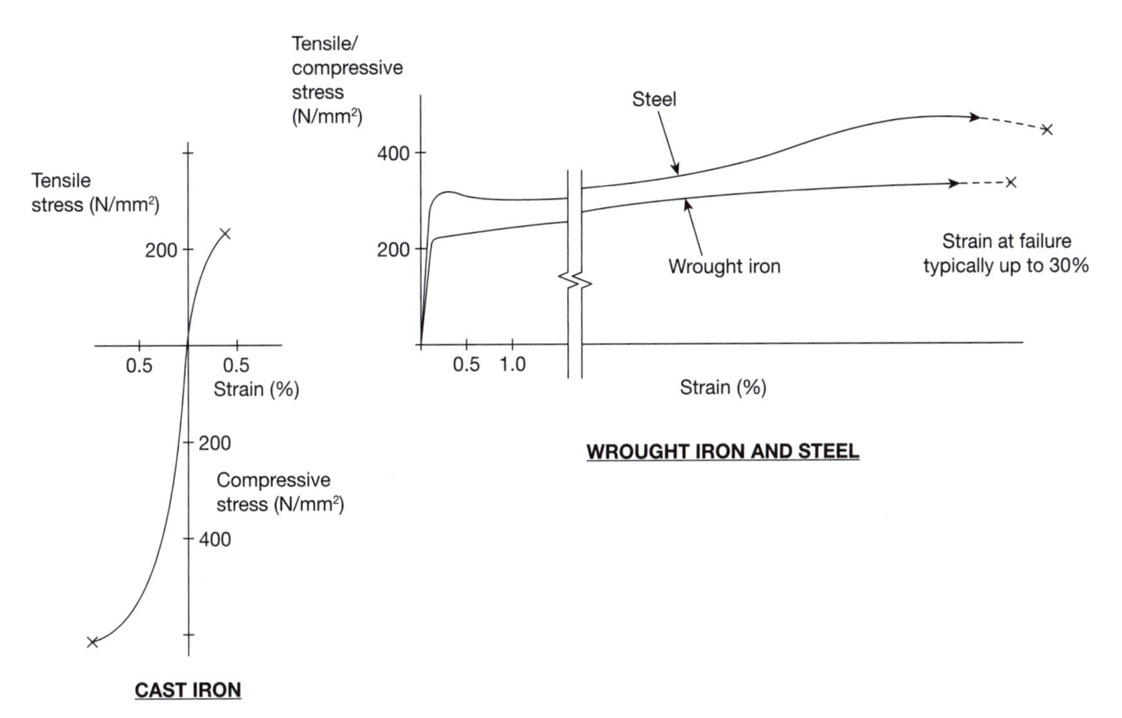


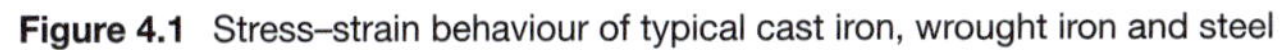

**Figure 4.1** Stress–strain behaviour of typical cast iron, wrought iron and steel

**Table 4.2** Visual characteristics as an aid to identification of the metal

| *Visual characteristic* | *Grey cast iron* | *Wrought iron* | *Steel* | *Spheroidal graphite cast iron* |
|---|---|---|---|---|
| Surface appearance (uncorroded) | Gritty or rough from mould; possible blowholes and/or continuous small 'ribs' from joint between mould halves | Smooth; when corroded, may appear delaminated into layers (flat surfaces) or wedges (rods) | Similar to wrought iron | Similar to grey cast iron when cast, usually then fettled to give smooth surface |
| Element profile | Typically thick; often ornate or complex profile varying along length; beams often with varying flange width and/or web depth; columns of fluted solid or round hollow section, occasionally H-section; also pipes, gutters, etc. and decorative elements | Typically ⌊, I, ⊥, or small I and [ sections; round bars; rolled sections or plates with angles, riveted together into girders or columns | Similar to wrought iron but often larger I and [ sections; hollow round or rectangular tubes; occasionally solid round steel columns; riveted or (later) welded compound sections | Similar to grey cast iron |
| Fracture surface | Crystalline, bright or grey | Fibrous (except in fatigue, from impact, or when overworked – crystalline) | Amorphous (except in fatigue or from impact when overworked – crystalline) | Amorphous |

**Table 4.3** Original connection methods used between structural elements

| *Connection method* | *Grey cast iron* | *Wrought iron* | *Steel* | *Spheroidal graphite cast iron* |
|---|---|---|---|---|
| Direct bearing | ✓ | ✓ | ✓ | ✓ |
| Plain bolts | ✓ | ✓ | ✓ | ✓ |
| Forge-welding | ✗ | ✓ | ✓ | ✗ |
| Rivets | ✗ | ✓ | ✓ (until 1960s) | ✗ |
| Oxy-acetylene welding | ✗ | ✗ | ✓ (early 20th C only) | ✗ |
| Electric arc welding | ✗ | ✗ | ✓ (from 1950s) | ✓ |
| High-strength friction grip bolts* | ✗ | ✗ | ✓ (from 1950s) | Possible |

* Godfrey (1965) provides a pithy explanation of why high-strength friction grip (HSFG) bolts soon found favour: 'despite the fact that H.S.F.G. bolts were more expensive than rivets, it was more economical to use them than rivets because fewer men were required for installation, the work could proceed much more quickly and they never worked loose under any conditions of service. In addition, there were the further advantages that there was less noise and no risk of fire' (both of which were very definitely inherent hazards in the riveting process).

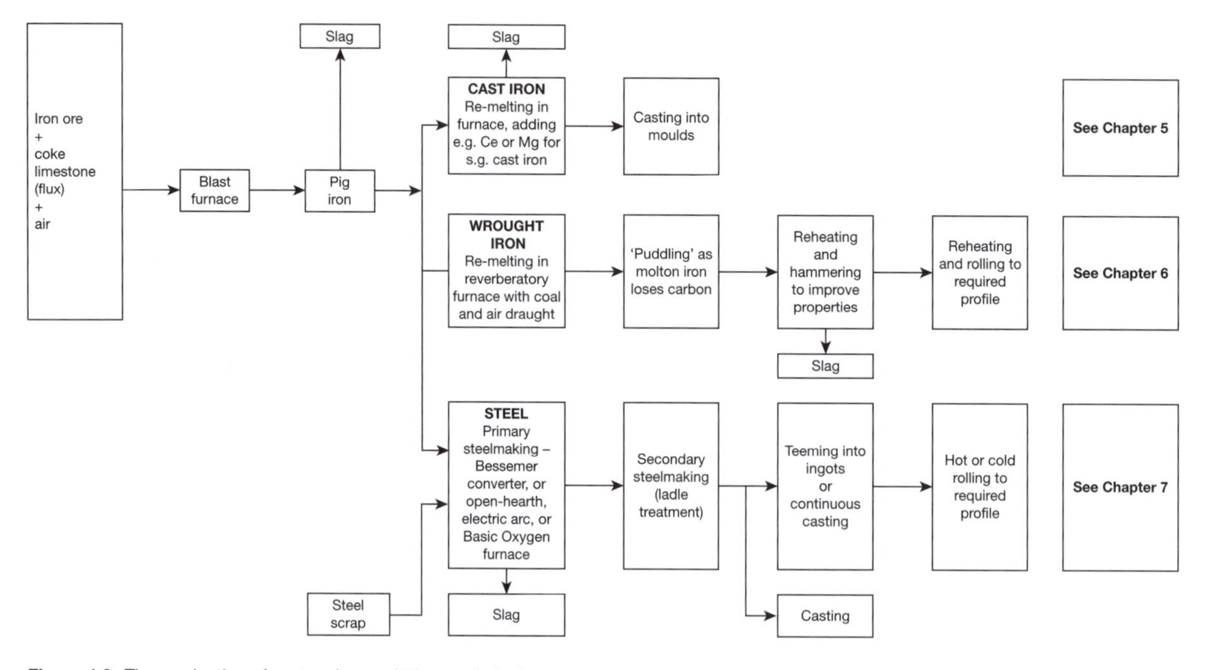


**Figure 4.2** The production of cast and wrought iron and steel

## 4.4 Making iron and steel

The processes and materials involved in the making of iron and steel are summarised in Figure 4.2. This shows the basic extraction of iron from ore by smelting in a blast furnace – a process common to all three forms of ferrous metal – and guides the reader to chapters in which the subsequent manufacturing processes specific to each of these forms are described.

### 4.4.1 Iron ores

Iron has a strong affinity for oxygen. This is evident when iron or steel, exposed to a damp atmosphere, oxidises to produce rust. Accessible deposits of iron – ores – are commonly therefore found as forms of iron oxide, or combined with carbon dioxide as carbonate.

The earliest iron ore to be exploited was bog-ore, a brown sesquioxide of iron ($Fe_2O_3$) formed by bacterial action in marshy ground. Hematite, a form of the sesquioxide found in both red and brown forms, occurs widely on a much larger scale. In the UK its discovery led to the setting-up of local ironworks, sometimes in areas that are not nowadays thought of as industrialised, such as the Forest of Dean and Cumbria. Magnetite, the magnetic black oxide of iron ($Fe_3O_4$), has a slightly higher iron content. In Sweden, in particular, its abundant deposits were exploited to establish a highly respected iron-making tradition.

Ferrous carbonate ($FeCO_3$) is to be found in strata of the Carboniferous and Jurassic periods, often known as ironstones. In the former it is often associated with coal, whose presence reduces the cost of smelting the ore; the Blackband ironstone of Lowland Scotland was a major economic ore. A large deposit of the later Jurassic ironstones ran in a belt from Lincolnshire south-west as far as Oxfordshire.

Today, it is cheaper to import ore from abroad than to win it from less accessible native deposits.

### 4.4.2 Early iron-making

In early days, iron was obtained by heating the ore with wood, whose carbon combined with the oxygen in the ore as gaseous carbon monoxide or dioxide to leave a small 'bloom' of solid iron, although with impurities remaining. This was a slow process and produced iron only in small quantities. Regrettably, the first call on the new material was as weaponry, while its earliest use in construction was limited to components where its strength and toughness were essential, such as hinges and locks. The iron, almost pure but containing residual impurities or 'slag', could be forged or hammered to shape, or in the archaic term 'wrought', when heated. Hence it took the name *wrought iron*.

In Europe, the introduction of the blast furnace, probably *c.* 1400 AD in Belgium (but long known in China) transformed the scale and speed of making of iron. Within a masonry furnace were stacked iron ore and charcoal, which was made from timber. Once the charcoal was ignited, a forced draught of air, the 'blast', was blown through the charge, accelerating combustion and raising the temperature. The blast was produced by water-powered bellows, so furnaces were typically sited by a stream or river in wooded country (which provided the charcoal). The furnace sides were lined with a suitable refractory material such as fireclay or dolomite, which could stand up to the sustained high temperatures generated during smelting. The charcoal provided both fuel and also the elemental carbon to combine with, and hence remove, the oxygen from the ore. More carbon combined with the iron to produce a molten alloy with typically 4–4¼% carbon content at about 1150°C – the lowest melting point for an iron–carbon compound, lower than the melting point of pure iron, 1535°C.

Impurities rose to the surface of the molten iron as a viscous slag and could be drawn off. The liquid iron could then be 'tapped', or run out of the furnace. Commonly it was poured onto the ground into formed channels, off which branched numerous depressions into which the molten iron ran and then solidified. The channels came to be known as 'sows' and the depressions as 'pigs', giving rise to the term *pig iron* for the newly made iron.

It was soon found that adding suitable crushed stone (typically limestone) to the furnace helped the process; it acted as a 'flux', combining with impurities in the ore to produce a light fusible 'slag' that floated on the molten iron and could be drawn off, so that the useful iron and the waste materials were separated. A further advance was to 'roast' or calcine the ore as-dug, before smelting, to burn off moisture and some impurities. This also drove off carbon dioxide from iron carbonate ores; the overall result was to reduce the amount of useless material charged into the furnace.

### 4.4.3 The early industrial production of iron

Large-scale production of pig iron began in the eighteenth century. An important breakthrough was the successful application of coal – or rather its derivative, coke – as the fuel and source of carbon for smelting iron ore. Coal had been tried, but as nearly all coal contains sulphur this caused difficulties. The sulphur combined with the molten iron, and if wrought iron was made from this it was 'red-short' – liable to break up if worked when heated. Coking the coal by heating it in a closed chamber drove off the sulphur derivatives and other volatile material (which in due course found use as 'coal-gas' for lighting and heating, the forerunner of today's natural gas). Abraham Darby successfully smelted iron ore with coke in 1709 at the significantly named hamlet of Coalbrookdale in Shropshire. The abundance and ready availability of coal led to its widespread use as fuel in preference to charcoal, whose production used much timber and had therefore basically tied earlier iron-smelting to areas where timber was abundant, such as the Ashdown Forest in Kent and Sussex, and the Scottish Highlands.

An additional advantage of smelting with coke was that it was both stronger and more porous than charcoal; this assisted the stacking of larger charges in the furnace with improved air-flow. In turn, the furnaces became larger. Once 'blown in' – first charged and fired up – the furnace would remain in continuous use until deterioration of the lining demanded re-lining, with tapping of iron and slag being alternated with re-charging of the ore, coke, and flux.

Coal and iron ore (and the limestone needed to flux the smelting process) were often found together in Carboniferous strata in areas such as the West Midlands, South Wales, and Lowland Scotland. These areas subsequently underwent rapid industrial development in the later 18th century as the demand for iron increased. Steam power replaced water power for providing the blast furnace's air draught, so that an ironworks could be sited where the coal and ore were to be found, being no longer reliant on a stream or river for power. Iron was needed to make these steam engines and those draining mines; for wagon wheels and rails for the many mineral tramways and early railways carrying coal and metal ores; and for many other industrial applications.

In 1828 James Neilson patented his 'hot-blast' process by which air was pre-heated before being blown into the blast furnace. This shortened the smelting time, and improved efficiency as it eliminated the need for calcining, reduced fuel consumption, and even made possible the direct use of coal as fuel. On the other hand, silica and phosphorus were not eliminated as thoroughly as in the traditional 'cold-blast' process, which made the production of poor-quality iron more likely unless greater care was taken.

The essence of the blast furnace was now defined; although many technical refinements were introduced subsequently, accompanied by increased size and efficiency, the modern furnace today operates on unchanged principles. Iron-making in the UK, as elsewhere, is now carried on in a few highly mechanised large-scale ironworks close to waterways capable of delivering imported ores, whereas until relatively recently it was practised in many smaller labour-intensive ironworks, usually located on or near native iron ore deposits with ready access to coal.

The large-scale manufacture of wrought iron was developed by Henry Cort, who in 1783 used a reverberatory furnace to blow heated air from a coal fire over pig iron, melting it and with the oxygen in the air combining with carbon in the pig iron. The iron had to be stirred or 'puddled' to expose all of it to the air; as the carbon content dropped, the melting point of the iron increased and it became a spongy solidifying ball. This could then be refined by reheating and hammering, before being rolled or otherwise formed into useful shapes. Wrought iron has always been shaped in solid form, whereas cast iron – as its name indicates – is shaped while still molten.

### 4.4.4 Early steel-making

Steel had been made from early times by heating iron in a sealed clay vessel with charcoal, so that the carbon in the charcoal slowly diffused into the iron. This produced tough steel, which could be further treated by heating and quenching in water, and sharpened to take a hard edge (unlike wrought iron). The process was called cementation as it hardened the metal, and the iron–carbon compound formed in the steel was known as cementite. In the 1740s Benjamin Huntsman, in Sheffield, refined the method by re-melting this steel in heat-resistant clay crucibles. This allowed impurities to rise to the surface to be skimmed off, while the carbon spread more evenly through the melt, resulting in a more consistent alloy. As with wrought iron, the making of such 'crucible' steel was a small-scale, laborious, and time-consuming process until the mid-19th century, although the name of Sheffield as the centre of its production became synonymous with the production of high-quality steel generally, and of steel tools and cutlery in particular.

It was Henry Bessemer, an inventor, who first devised a means of converting pig iron into steel on a large scale. His 'converter', details of which were published in 1856, looked and functioned not unlike a concrete-mixer. The empty converter was tipped on its side to receive a charge of molten pig iron. It was then tilted back to an upright position, and cold air was blown *through* (not *over*) the iron. Atmospheric oxygen combined with carbon in the iron and burnt off as carbon monoxide in a vigorous exothermic reaction that threw up streamers of slag and some iron, and also removed other volatile products. When the reaction ceased, it left iron with a low carbon content – effectively modern mild steel – which, by tilting the converter on its bearings, could be poured out and cast into ingots for subsequent shaping.

In his experiments, Bessemer had by chance used pig iron made from a hematite ore with a low phosphorus content. Whereas more than a trace of sulphur makes wrought iron and indeed steel 'red-short', as described above, phosphorus makes them 'cold-short' – brittle and liable to fracture at ordinary temperatures. It was two decades before reliable steel could be made from high-phosphorus ores, using a limestone or calcined dolomite (magnesium carbonate) lining to the converter rather than the more customary siliceous or 'acid' lining. The chemically 'basic' lining reacted with the phosphorus, and the resulting compounds were removed as part of the slag. This discovery, by Sidney Gilchrist Thomas and his cousin Percy Gilchrist, allowed use of the phosphatic iron ores widely found in Britain and elsewhere to produce reliable steel. Steel made in this way was known as 'basic' Bessemer steel or – particularly on the European continent – as Thomas steel, in contrast to the 'acid' Bessemer steel made from a low-phosphate ore with an acid lining to the converter.

Meanwhile, in the mid-1860s the open-hearth steel furnace had been developed by Siemens and the Martin brothers. This made use of a pre-heated air blast to improve efficiency, could handle scrap metal as well as new pig iron, and took considerably longer to produce steel than the Bessemer process – beneficially allowing the process to be more closely controlled. (This was in contrast to the potentially variable quality between one small 'puddled' wrought iron ball and the next, each produced on a small-scale batch basis, and to the rapid and hence largely uncontrollable formation of steel in the Bessemer converter.)

Once available in quantity as a reliable material, steel proved a strong competitor with wrought iron. It was roughly half as strong again and was tougher, although less malleable to forge and work. It was initially in great demand for rails, boilers, and ships' hulls, all of which benefited from its superior properties. The weight of a ship's hull in wrought iron, for example, could notionally be reduced by around one-third if steel were used instead, with a consequent profitable increase in freight tonnage capacity. Its use in construction, however, was slower to develop.

By 1900 the UK production of steel had eclipsed that of wrought iron (5 million tons as against 1 million tons), reversing the position of 30 years earlier (0.25 million tons of steel in 1870 compared with 3 million tons of wrought iron).

In 1913, Harry Brearley – again in Sheffield – found that adding chromium to steel inhibited the normal corrosion process. Nickel has also been added since; nowadays the resulting *stainless steel* finds wide constructional use, including as reinforcement (see Chapter 8).

Today, most steel is made in electric arc or induction furnaces, or by the more recent and now largely dominant Basic Oxygen Steel (BOS) process; these are described in Chapter 7.

## 4.5 Constructional uses of cast and wrought iron and steel

In medieval and post-medieval times, iron was expensive, and so its use in construction was broadly limited to where either its functional or aesthetic merits could be exploited. As examples, wrought iron made strong hinges and locks, while cast iron's durability and its ability to take up a moulded decorative profile were of value, both functionally and visually, in ornamental fire-backs and in grave-slabs for the wealthy.

The dramatic increase in both the supply of and the demand for iron generated by the Industrial Revolution saw its much wider application, particularly for structural use. Cast iron was first to be used structurally in the late 18th century, to be followed in the mid-19th century by wrought iron; this was challenged as that century drew to a close, and was then supplanted, by steel. In the same period, steel began to be used in combination with concrete in an innovative 'composite' construction material – reinforced concrete. This was further developed in the mid-20th century into prestressed concrete, with tendons of higher-strength steel.

Figure 4.3 shows time-lines for structural use of the three ferrous metals in the UK. Solid lines represent the main periods of use, while broken lines indicate periods of limited use.

Today, cast iron – with its ability to be formed in almost any required shape – is still often employed for functional and decorative uses, although for structural use the grey cast iron of the 19th century has been replaced by the more recently developed form, spheroidal graphite cast iron with its superior physical properties, notably in terms of strength and ductility (see Chapter 5). Its occasional structural use is indicated by the short broken line at the present-day end of the time-line for cast iron in Figure 4.3.

### 4.5.1 Cast iron

Cast iron was applied to structural purposes from the late 18th century. The 1779 Iron Bridge across the River Severn at Coalbrookdale survives; its 30 m arch span made it the first large cast iron bridge (see Figures 5.3 and 5.8).

The development of factories, textile mills, warehouses and the like with their high-value contents, and the associated fire hazards, spurred the use of so-called 'fireproof' construction – the term at the time meaning incombustible, rather than resistant to the effects of fire. Initial tentative use of structural cast iron was limited to columns, but soon beams too were being formed in cast iron rather than timber. A structural bay of a typical early 19th-century textile mill, re-erected at Manchester University, is shown in Figure 5.1.

Many textile mills and other large structures were built with cast iron beams until the 1860s and beyond, despite the availability by then of wrought iron. Non-structural uses of cast iron were manifold, from railings and rainwater goods and pipework to window-frames and boot-scrapers.

Although no longer used structurally, grey cast iron remains a significant material for non-structural functional and decorative roles in construction today. Unlike grey cast iron, the more recently introduced spheroidal graphite cast iron is ductile, and although more expensive it has found application where this property is significant, for example in utilities covers.

The properties and uses of cast iron are considered more fully in Chapter 5.

### 4.5.2 Wrought iron

Wrought iron began to be adopted for structures in the early 19th century, particularly for cables, tie-rods, and the like, in which forms it found use in suspension bridges and tie-members in roofs, a notable surviving example being the 1826 suspension bridge over the Conwy estuary in North Wales designed by Thomas Telford (Figure 6.5).

Large-scale use of structural wrought iron began in the late 1840s with the introduction of rolled angles, tees, and small I-sections, and the construction in the same period of large, riveted long-span railway bridges such as the Britannia Bridge over the Menai Strait on the London–Holyhead route, sadly destroyed by fire in 1970 (Figure 6.6). This use continued through the remainder of the century, for example in major railway stations such as St Pancras in London (Figure 6.8), and in bridges such as the Royal Albert Bridge carrying the mainline railway between Devon and Cornwall (Figure 6.9), although steel gradually took the place of wrought iron in structures as it became more readily available from the 1880s.

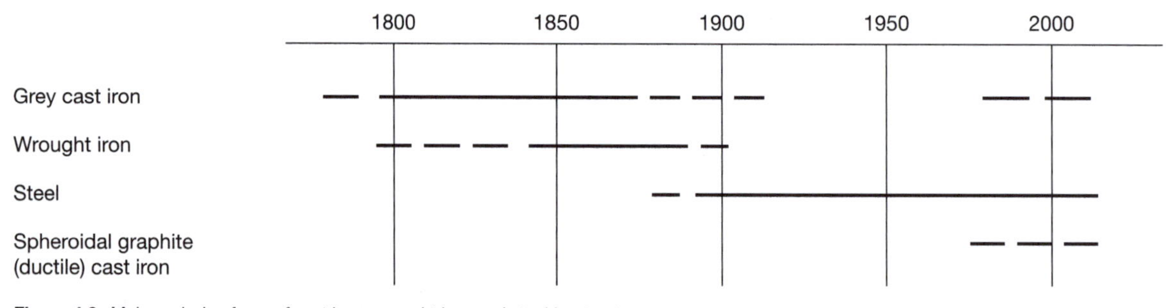


**Figure 4.3** Main periods of use of cast iron, wrought iron and steel in structures

**Figure 4.4** The Forth Railway Bridge

Non-structural use of wrought iron included railings, gates, screens, and indeed most articles that were subsequently made with steel.

The properties and uses of wrought iron are considered more fully in Chapter 6.

### 4.5.3 Steel

Henry Bessemer devised his 'converter' to make steel from pig iron in the 1850s, but technical problems with the process and the demand from the railway companies and others for this strong and tough material meant its the use in buildings and bridges did not really begin until the 1880s, after the Board of Trade had sanctioned its use in bridges in 1877. An early application was also one of the largest – the Forth Railway Bridge in Scotland, built in 1884–90 and incorporating some 50,000 tons of open-hearth steel connected by 6½ million rivets (Figure 4.4).

The subsequent development of more powerful rolling mills meant that larger steel I-sections could now be produced economically, often being substituted for riveted built-up girders with their greater labour content and slower delivery. However, riveted steel construction continued to be widely used for a variety of elements such as roof trusses, bridges, and building frames.

The wider use of steel framing in building construction was greatly advanced by the passing of the London County Council (General Powers) Act 1909. This, for the first time in the UK, laid down conditions of use, and prescribed loadings and permissible stresses to be used in the design of structural steel (and indeed cast and wrought iron too, a little late in the day!). No less important was its recognition that, when loads were carried by the steel frame, the walls on column lines needed only to be thick enough to be weatherproof (if external) or stable (if internal). Previously such walls had had to be sized by empirical rules as if loadbearing, resulting in thick and uneconomical masonry, particularly on the lower floors of taller buildings.

Subsequently steel has become by far the commonest ferrous metal used in construction. Hot-rolled high tensile steel was introduced in the 1930s, at the same time as electric-arc welding began to challenge riveting for connecting steel elements.

Steel was the first material to have its structural design codified in a British Standard with the publication of BS 449 in 1932, just as the proliferation by the various steel-makers of differing thicknesses and weights for I-sections of the same nominal dimensions had been a trigger to the formation of the Engineering Standards Committee, precursor of the British Standards Institution, in the early 1900s.

The properties and uses of steel are considered more fully in Chapter 7.

### 4.5.4 Reinforcement and prestressing tendons

Concrete (see Chapter 14) has been used since Roman times, essentially as an artificial stone replacing that natural material which has always been more expensive. It shares with stone, brickwork, and other forms of masonry a good resistance to compressive loads and is thus ideal for walls, columns, and arches. But it is brittle and weak in tension, and hence is unsuited for use in beams and slabs where it could break without warning under load.

Early experiments with reinforcing of concrete took place in the UK, continental Europe and the USA from the 1850s. But as with steel (although for different reasons, possibly including scepticism of its worth), it was near the end of the century before reinforced concrete started to be used in earnest for construction. François Hennebique, an energetic French entrepreneur, took out the first of a series of patents in 1892, for the use of reinforced concrete. His patent included the use of round mild steel bars and also flat steel strips. The restrictions of such patents meant that other early reinforced concrete systems had to use different steel section profiles, including bars with indentations or surface protuberances, cold-worked square twisted bars, square bars with 'wings' rolled on then slit and bent up, and bars in the form of a Cross of Lorraine (‡).

As patents expired, profiles became fewer and were standardised, based on the round mild steel bar. Early reinforcement with higher strength than mild steel was achieved by cold-working the steel; it was not until the 1930s that hot-rolled alloys with higher strength were being produced. These were ribbed to provide enhanced anchorage and bond to the surrounding concrete, just as the twisting of square or other profiled bars increased these properties as well as enhancing their strengths by cold-working. Welded fabric, commonly called 'mesh', offered useful prefabricated sheets, reducing steel-fixing labour on site.

Eugène Freyssinet was a pioneer in the development of *prestressed concrete* in the late 1920s. The idea of prestressing was to precompress the concrete, to overcome its tensile weakness and thereby make more effective use of the overall concrete section to resist bending effects in beams and slabs. Higher-strength steel was needed so that adequate precompression was maintained over time as the concrete inevitably underwent contraction due to drying shrinkage and also 'creep' (long-term shortening under sustained compression).

The various forms of steel used in prestressed concrete were called 'tendons' rather than reinforcement, as they were actually applying pre-loading to the concrete rather than reinforcing it. Types of tendon currently in use, as well as reinforcement for concrete, are described in Chapter 14.

Sutherland et al. (2001) review the history and technical development of reinforced and prestressed concrete construction, and include information on many earlier reinforcing systems.

## 4.6 Durability

With iron's strong affinity with oxygen, it is unsurprising that all ferrous metals are vulnerable to corrosion, evidenced as brown, red or black rust. Consequently, painting or other treatment is needed to protect iron or steel in exposed conditions.

Cast iron taken from the mould has relatively good corrosion resistance, attributed to the rapid 'chilling' of the molten iron as it makes contact with the mould material, which results in a hard, durable surface layer. However, if the surface is removed, for example by shot-blasting, then the freshly-exposed iron will rust as if it were steel.

Wrought iron has been found by experience in the long term to be slightly superior to steel in resisting corrosion; this is considered to be due to the beneficial presence of the silicate slag. Neither metal is as resistant as cast iron taken from the mould.

Particular concerns for corrosion include:

- exposure to aggressive environments such as sea-water and industrial atmospheres
- junctions between exposed and embedded metalwork, such as the bases of railings and columns on masonry plinths, where water can enter immediately beyond the limit of the paintwork
- surfaces and spaces where water can pond rather than drain away
- overlapping elements, such as in riveted or bolted lattice girders, where water can be trapped.

Severe and costly problems have been encountered on occasions in buildings where a structural steel frame has been embedded in external masonry walls with insufficient protection. Water penetration, particularly through neglected joints in the masonry, can cause rusting of the steel which in turn can both weaken the structure and lead to cracking and dislodging of masonry units (Gibbs, 2000).

Corrosion of steel reinforcement can be a serious consideration in severe exposure conditions such as structures in a marine environment or road bridges where de-icing salts might be used in cold weather. Protection against corrosion can be provided by appropriate concrete mix design and adequate thickness of concrete cover. Stainless steel reinforcement is an alternative, more costly to provide but with superior long-term performance.

Accelerated low-water corrosion is a relatively recently identified form of corrosion that can affect marine structures such as sheet-piled walls (Breakell et al., 2005).

Perhaps surprisingly, many species of timber can induce corrosion in iron and steel. This is due to volatile organic acid in the timber. Oak and Douglas fir are two species that are particularly aggressive. BRE Digest 301 gives advice on the problem and on suitable protective measures.

## 4.7 Coatings for corrosion protection

Traditional protection against corrosion has been by painting. Some of the materials customarily used, such as red lead, are now regarded as toxic and should not be used for new work or repainting (although use with care may be the preferred solution for heritage structures, applied with appropriate health and safety measures). Existing paint schemes may contain such materials, and should be sampled and chemically analysed before its removal is contemplated. It can otherwise present a health hazard both to those carrying out the work and those who may be affected by airborne dust, and also risks contaminating waterways.

Galvanising is an alternative method of providing corrosion protection, or it can augment a paint scheme in exposed locations. Modern formulations such as epoxy-pitch coatings can provide good protection with extended periods between maintenance, which is very useful for structures to which access for such work is difficult and costly, such as seaside piers.

BS5493 has since 1977 provided informed guidance on coatings for corrosion protection. It has been partially replaced by several standards; its status at the time of writing is noted on the BSI website as "current, obsolescent, superseded". The new standards include BS EN 12500 and 12501, respectively covering the likelihood of corrosion in atmospheric environments and soil, and BS12944 (in eight parts) on protective paint systems for protection of steel structures. Many of the recommendations in these standards for new structures may be applied to existing structures also.

Corrosion and corrosion protection of steel are discussed further in Chapter 7.

## 4.8 Performance in fire

Unprotected iron and steel is vulnerable to fire, with progressive loss of strength occurring as temperatures rise above 300–400°C. Wrought iron and steel beams will sag and columns can buckle, which can trigger a more widespread collapse. Cast iron has often been found to survive more successfully. This may well be because it was commonly used in industrial structures, such as mills, in which the cast iron beams were embedded in arched masonry floors that provided a massive 'heat-sink', slowing the temperature rise in the iron.

Building regulations will prescribe the required period of fire protection for iron and steel structures. This can be achieved by applying fire protection such as boarding, sprayed coatings, or intumescent coatings which – although applied as a thin coating – will swell up in a fire to provide thermally insulating protection to the metalwork. A recent development has been in the field of 'fire engineering', in which the behaviour of a fire can be modelled mathematically to show whether applied protection is necessary, or can actually be omitted.

## 4.9 Sustainability and recycling

Iron and steel can be recycled very easily, and indeed scrap provides much of the raw material in the modern steelworks. In terms of sustainability, the very best course for an iron or steel structure is for it to remain in use, involving minimal input of additional energy. Elegant but redundant cast ironwork may be more valuable if sold as architectural salvage rather than as scrap, while wrought iron can be offered for re-rolling by museums or specialist ironworkers (see Chapter 6).

## 4.10 References

*BRE Digest 301:1985. Corrosion of metals by wood.* Watford, UK: Building Research Establishment.

Breakell, J. E., Foster, K., Siegwart, M., et al. (2005). *Managing accelerated low-water corrosion.* London, UK: Construction Industry Research and Information Association.

*BS 5493:1977. Code of practice for protective coating of iron and steel structures against corrosion.* London, UK: British Standards Institution.

*BS EN 12500:2000. Protection of metallic materials against corrosion. Corrosion likelihood in atmospheric environment. Classification, determination and estimation of corrosivity of atmospheric environments.* London, UK: British Standards Institution.

*BS EN 12501-1:2003. Protection of metallic materials against corrosion. Corrosion likelihood in soil. General.* London, UK: British Standards Institution.

*BS EN 12501-2:2003. Protection of metallic materials against corrosion. Corrosion likelihood in soil. Low alloyed and non alloyed ferrous materials.* London, UK: British Standards Institution.

*BS EN ISO 12944:1998-2007. Paints and varnishes. Corrosion protection of steel structures by protective paint systems* (in eight parts). London, UK: British Standards Institution.

Gibbs, P. (2000). *Technical Advice Note 20: Corrosion in masonry clad early 20th century steel framed buildings.* Edinburgh, UK: Historic Scotland.

Godfrey, G. B. (1965). *High strength friction grip bolts.* London, UK: British Constructional Steelwork Association (Publication No 26).

Sutherland, J., Humm, D., and Chrimes, M. (Eds) (2001). *Historic concrete: Background to appraisal.* London, UK: Thomas Telford.

## 4.11 Bibliography

Bussell, M. (1997). *Appraisal of existing iron and steel structures.* London, UK: Steel Construction Institute.

Carr, J. C. and Taplin, W. (1962). *History of the British steel industry.* Oxford, UK: Blackwell.

Gale, W. K. V. (1967). *The British iron and steel industry: A technical history.* Newton Abbot, UK: David & Charles.

Gale, W. K. V. (1969). *Iron and steel.* London, UK: Longmans, Green.

Skelton, H. J. (1891). *Economics of iron and steel: Being an attempt to make clear the best every-day practice in the heavy iron and steel trades, to those whose province it is to deal with material after it is made.* London, UK: Biggs & Co.

Skelton, H. J. (1924). *Economics of iron and steel: Being an exposition of every-day practice in the heavy iron and steel trades.* London, UK: Stevens & Sons and H. J. Skelton & Co. Ltd.

Sutherland, R. J. M. (Ed.) (1997). *Structural iron, 1750–1850: Studies in the history of civil engineering, Volume 9.* Farnham, UK: Ashgate.

Thorne, R. (Ed.) (2000). *Structural iron and steel, 1850–1900: Studies in the history of civil engineering, Volume 10.* Farnham, UK: Ashgate.

Tylecote, R. F. (1992). *A history of metallurgy* (2nd edition). London, UK: Institute of Materials.

# 5 Cast Iron

**Michael Bussell** BSc(Eng)
Formerly Ove Arup & Partners

## Contents

## 5.1 Introduction

Cast iron is a material that has figured significantly in past use structurally and for other applications, in which its function is often complemented by a decorative treatment readily achieved in the casting process. It also remains a useful metal for present-day construction. So it is of interest both to those involved in the maintenance, conservation and re-use of existing construction, and to those engaged in the design and execution of new work.

Cast iron typically contains about 2–5% of carbon, which has the effect of lowering the melting point of the alloy well below that of pure iron (1535°C). This makes it possible to form castings of almost any required profile by pouring molten metal into a mould. In contrast, the other basic ferrous metals – wrought iron (no longer made), and steel – are shaped, typically by rolling or forging, from the hot solid.

Cast iron can be made with significantly different properties, largely determined by the way in which the carbon is present in the casting. This in turn is influenced by the cooling regime and any subsequent heat treatment during the casting process, and the presence of other elements in small quantities.

The commonest types to be found in construction today are grey cast iron, in which the carbon is largely in the form of graphite flakes, and ductile cast iron (developed after the Second World War), in which the carbon is in the form of spheres or nodules of graphite, giving rise to its alternative names of spheroidal graphite (often abbreviated to s.g.) or nodular cast iron. The different ways in which the carbon is held affect the material's properties: grey cast iron is weak in tension and fails in a brittle way, whereas ductile cast iron is more akin to steel in strength and malleability.

Since the first edition of this book was published in 1992, there have been notable changes affecting both the appraisal of existing structural cast iron and the use of new cast iron. A number of publications arising from research work at Manchester University led by Swailes (referenced below) have provided valuable guidance on the behaviour and use of cast iron in structures – a use that was common practice throughout the nineteenth century but lapsed, so that for most of the 20th century the material was barely known to and even less understood by most practising engineers. The findings from this work have reinforced confidence in the assessment of cast iron bridges, guidance on which was published in 1984 and is currently available in Highways Agency Standard BD21 (Highways Agency et al., 2001) and similar standards from other bodies with large bridge estates. This assessment approach has subsequently been extended to be applicable to building structures also (Bussell, 1997); it is explained more fully in Section 5.3.4.10 below.

For material specifications of new cast iron, the British Standards current when the first edition of this book appeared have been replaced by European Standards, available in the United Kingdom as BS EN standards. In particular, BS EN 1561 covers grey cast iron and BS EN 1563 covers s.g. cast iron.

Relevant to those interested in both existing and new uses of cast iron, the two former leading trade organisations serving the iron-founding industry – the British Cast Iron Research Association and the British Foundry Association – have merged with other bodies to become, respectively, Castings Technology International and the Cast Metals Federation.

## 5.2 General description

### 5.2.1 Types of cast iron

Four types of cast iron are in common engineering use; these and their properties and uses are summarised in Table 5.1. Of these, the white and malleable forms have found little use in the construction industry, and so are not considered further here. The term 'malleable iron' was unfortunately sometimes used in the past as a synonym for wrought iron, but its true nature – a treated cast iron – should be recognisable by inspection, aided by material testing if necessary.

Attention is focused on 'historic' – essentially grey – cast iron; modern grey cast iron; and ductile (s.g.) cast iron.

'Historic' grey cast iron will be considered here in detail. It is of particular relevance for civil and structural engineers dealing with the appraisal, repair and adaptation of existing structures; they need to be aware of its properties, and to recognise that it was

**Table 5.1** Types of cast iron

| *Type of cast iron* | *Microstructure* | *Physical properties* | *Uses* | *Notes* |
|---|---|---|---|---|
| Grey (or flake graphite) | Graphite in flake form in an iron matrix; flakes form discontinuities | Strong in compression; relatively weak in tension; good resistance to corrosion; limited ductility Easily machined and cut; very large castings practicable | Main form of cast iron used in construction, for columns, beams, decorative panels, etc., as well as rainwater goods, pipes, tanks and machinery | Historic cast iron nearly always grey iron Still used today for pipes, pipe fittings, manhole covers, etc. |
| White | No free graphite; carbon combined with iron as hard carbide (cementite) Low equivalent carbon; low silicon content | Very hard and very brittle; machined by grinding only | Surfaces needing high resistance to abrasion | Virtually irrelevant to construction industry |
| Malleable | Made by prolonged heat treatment of white iron castings, when carbide is transformed into graphite in nodular form | Very strong in tension as well as compression; good ductility | Hinges, catches, step-irons and similar castings of limited size; decorative panels of fragile design | Generally superseded by ductile iron, with similar properties but no need for heat treatment after casting |
| Ductile (or nodular or spheroidal graphite) | Addition of magnesium or cerium results in graphite being deposited in nodular form | Very strong in tension as well as compression; good ductility | Successor to malleable iron, with wide engineering applications | To date, limited but some notable uses in construction |

not made to the standards and quality control that are customary today. Following this, and to avoid textual repetition, the distinguishing properties and other aspects of modern grey and ductile cast iron are described only where these differ significantly from those of historic grey cast iron.

The terms 'grey' and 'white' describe respectively the dull greyish and the brighter, lustrous appearance of the freshly fractured surfaces of these forms of cast iron, reflecting the presence or absence of free graphite.

#### *5.2.1.1 'Historic' cast iron*

This term describes cast iron as was much used structurally between about 1780 and 1880 for columns, beams, arches, brackets, and the like, but equally it found wide application over a lengthier period for other functional and decorative components in construction (Sheppard, 1945; Gloag and Bridgwater, 1948; Lister, 1960). Much survives in use today. It is mainly grey cast iron.

The most significant physical properties of this type of iron are: its much greater strength in compression than in tension; its brittle mode of failure; and its non-linear behaviour under tensile load. These result in behaviour not shared by ductile cast iron or the other common ferrous metals – wrought iron and steel – which are considered in subsequent chapters.

#### *5.2.1.2 Modern grey cast iron*

Modern grey cast iron is essentially the same as historic grey iron, but is generally of a more consistent and higher quality, produced to comply with materials standards. Although it has uses in construction, it is no longer used structurally apart from the occasional replica casting.

Past and current uses of grey cast iron range from domestic applications in rainwater goods to the municipal side of civil engineering, including pipes and their fittings, bollards, manhole covers, gratings and other applications where simple robustness is needed rather than a calculable structural performance.

#### *5.2.1.3 Ductile (spheroidal graphite) cast iron*

Ductile cast iron has found limited (although usually 'high-profile') use in construction to date; this might well continue. It is a relatively modern material, introduced after the Second World War, and like modern grey cast iron its quality can be specified by reference to materials standards. Its most important properties are its ductility and its similar strength in both tension and compression, with strength and stiffness both being similar to those of rolled steel; yet it can be formed into complex shapes in moulds with the same freedom as can grey cast iron.

### 5.2.2 Identification of cast iron

Its name points up the fact that cast iron is formed into shape in a molten state, which in turn implies that it was poured into a mould that predetermined its shape. This in itself often affords sufficient evidence for cast iron to be distinguished from rolled or forged iron or steel. A continuous single profile that changes cross-section along its length, for example the beams shown in Figure 5.1, could never have been rolled through a mill or forged. Similarly, ornate detailing points to a casting. The moulding process (see below) can result in mould lines or 'pips' along the junction of the mould sections, again a diagnostic feature.

## 5.3 Historic cast iron

Cast iron used in construction until recently was almost exclusively grey cast iron, with nearly all the carbon present as 'flakes' of free graphite. Some white cast iron was used in construction, only as a starting-point towards the manufacture of small components of malleable cast iron produced by prolonged

**Figure 5.1** A bay of cast iron columns and beams from a textile mill, re-erected on campus at Manchester University

heat treatment of white iron castings. This process, as noted in Table 5.1, was quite widely used for decorative ironwork of a delicate character that would be subject to accidental damage. Nowadays ductile cast iron has inherent properties very similar to those achieved in malleable castings and is simpler to produce.

### 5.3.1 Sources

As today, cast iron was made from pig iron produced in a blast furnace by the smelting of iron ore, as described in Chapter 4.

### 5.3.2 Manufacturing process

In 1794 John Wilkinson patented the cupola furnace, smaller and simpler than the blast furnace. In this, pig iron could be re-melted and refined before being poured into moulds to make the required castings, which therefore no longer had to be formed at the ironworks where the iron had been made. Many small foundries sprang up in cities and towns to supply local needs for cast iron products.

The making of a casting began with the preparation of a pattern. This was the three-dimensional 'master', typically of wood or metal, on whose accuracy a successful casting depended. However, since molten iron shrinks by about 1% as it cools and solidifies, the pattern had to be made correspondingly slightly over-size. (In Imperial measurement the shrinkage allowance was usually 1/8 in. per foot, close to 1%.) A further consideration was that the pattern should not incorporate sharp re-entrant corners, abrupt or substantial changes in thickness, or other features that could cause cracking or weakness in the cooling iron. So it was the norm to 'sweep' sharp corners into an internally radiused or chamfered profile (this incidentally could further aid recognition of a casting). Pattern-making was – and remains – a skilled trade in which practical experience matters.

The casting was made in a moulding box, typically split into upper and lower halves. For successful casting, both the design and the use of the mould again required skill. The upper half of the box (the 'cope') was inverted, the pattern was laid into it, and sand or loam was then packed around it to form the required profile. (Some sands or loams were sufficiently cohesive to hold their shape, while others had to be baked to form a mould. Nowadays the material is often resin-bonded to ensure rigidity.) The same process was repeated with the lower half of the box (the 'drag'), the pattern was removed, and the two halves of the box were then pegged together. Holes were provided in the top half of the mould to receive the incoming molten metal, and also to allow the free escape of air displaced by the molten iron, which would otherwise burst up through the moulding material.

Solid castings were generally avoided, not only because they used more metal but because thick sections were slow to cool and solidify inwards from the mould face, which could result in serious weaknesses in the casting. Hollow profiles were formed – for example in circular-section columns and rainwater downpipes – by inserting a sand or loam core formed around a wrought iron bar. Such elements were typically cast horizontally in moulds on the foundry floor, so it was important to ensure that the hollow core was held in place at intervals along the mould by 'chaplets' – small metal spacers placed between the core and the outer mould faces, to ensure that the core did not literally float upwards when the denser molten iron entered the mould. This could result in a gross misalignment of the internal surface of the casting, which could 'banana' over its length, affecting strength.

Meanwhile, as the mould was being prepared, the pig iron was being re-melted in a small furnace. This was often done to remove some impurities remaining from the smelting process – Hatfield (1912, p. 5) vividly describes pig iron as a 'complex and variable mass of free iron and free carbon, carbides, double carbides, sulphides, silicides, and phosphides'. Furthermore, it was common to 'mix' iron from different sources, based on experience that showed that this enhanced the particular properties required for the castings being made. Swailes (1996, pp. 30–31) quotes from an 1859 paper on cast iron girders, which reported that a Manchester ironworks making I-beams would use equal quantities of five different irons for each charge: two different Staffordshire irons, one from Yorkshire, one from Scotland, with the fifth being 'scrap'. The resulting beams were cast on their sides, and were noted as being "a very good example of workmanship".

When the molten iron had reached the correct temperature – often judged by the foundryman's skilled eye looking through a spy-hole in the furnace – the slag-hole in the furnace would be breached to allow slag to be drained off the top of the molten iron, before the plug in the tap-hole was broken out. Now iron could be tapped and poured into crucibles on trolleys, to be carried to the moulds into which it would be ladled. Figure 5.2 shows iron being tapped in a small foundry some years ago.

Larger castings, such as those for the main ribs of the 1779 Iron Bridge (Figure 5.3), might be of a size to warrant air-casting, in which the profile was laid out in sand or loam on the floor and the molten iron was poured into the resulting 'gutter', with a top face open to the atmosphere. Such castings may often be identified by the more 'blotchy' texture of the top face, distorted by gas bubbles bursting in the pasty cooling iron which was not held to shape by a mould face.

Controlled slow cooling was essential in the making of sound castings. Slow cooling ensured that the carbon in the iron was deposited out of the iron as flake graphite, rather than left in combination as cementite (iron carbide), which resulted in white cast iron.

Some defects in casting were hard to avoid, particularly from gas bubbles that resulted in blow-holes (analogous to honeycombing in concrete). Less scrupulous or less competent ironfounders might mask surface defects with a pasty formulation typically including beeswax or resin, iron filings, and an aggressive liquid that caused rapid rusting of the filings, which then swelled up to fill the void. 'Beaumont egg' was one such potion, used to fill blow-holes and other defects in the cast iron pillars of the ill-fated Tay Railway Bridge.

Another concern was inherent in castings of any but the simplest shape. For example, in a cast I-beam, the iron at the

**Figure 5.2** Tapping molten iron from the furnace prior to casting

Figure 5.3 Curved air-cast iron ribs of the 1779 Iron Bridge over the River Severn

junction of the web and flange is gradually cooling and solidifying from the mould faces inwards, while the remaining core of semi-molten iron in the junction is being 'pulled' in three directions at once. The actual result can indeed be a 'spongy' matrix rather than dense iron, but recent studies (Swailes, 1996) have shown that the effect of this on strength is not necessarily significant. Likewise, some over-break of the mould material into the iron does not have a major effect on strength (Figure 5.4).

Potentially more serious – and not often visible – was a 'cold-shut', where molten iron flowing into the mould space met iron that had already started to solidify. This might have been a small spatter of iron thrown ahead of the flow, or a larger cooling surface. The risk was that the iron would not bond together as an entity, so that the junction could act effectively as a crack. Skilful mould design and care in pouring the iron were the best safeguards against such problems, and so they remain today.

Flaws can however occur in historic structures that are not discernible until shown up by fracture.

Modern ironfounding makes use of more advanced technology, but in essence follows the same process.

### 5.3.3 Chemical composition

Grey cast iron contains between 2% and 5% of carbon; in commercial grey cast iron it is more typically 3–4%. In molten iron the carbon is chemically combined in the form of cementite, iron carbide ($Fe_3C$). When the melt is allowed to cool slowly, the carbon precipitates out of the melt as flakes of graphite. White iron, in contrast, is produced when the iron is cooled quickly; here the carbon remains chemically combined with the iron as hard cementite.

Figure 5.4 Fractured surface of cast iron beam after test to failure, showing dark patches of moulding sand in the grey cast iron matrix

The microstructure of a typical grey cast iron with its flakes of graphite is shown in Figure 5.5 (a), to be compared with that of a malleable cast iron formed after prolonged heat treatment of white iron Figure 5.5 (b) and that of a modern ductile cast iron in Figure 5.5 (c).

Silicon (Si) and phosphorus (P) are present in small quantities in all commercial cast irons, and both elements influence the properties of the material. It is the equivalent carbon content that matters rather than the total carbon content when considering grey cast irons. This equivalent is defined as:

$$\text{Equivalent carbon content (\%)} = \text{total carbon content (\%)} + \frac{\text{Si (\%)} + \text{P (\%)}}{3}$$

The higher the equivalent carbon content, the lower will be the tensile strength. Grey cast iron typically has a higher free carbon content, and so is generally weaker than white cast iron.

Silicon promotes the precipitation of graphite as the iron solidifies, although more than 1–2% results in a substantial loss of strength. Phosphorus assists the casting process by increasing the fluidity of the iron, but makes for brittle iron.

In general, both metallurgical understanding and the ability to control the amount of such elements in molten iron were less advanced when historical grey cast iron was being made than they are now, and therefore the strength of the resulting castings

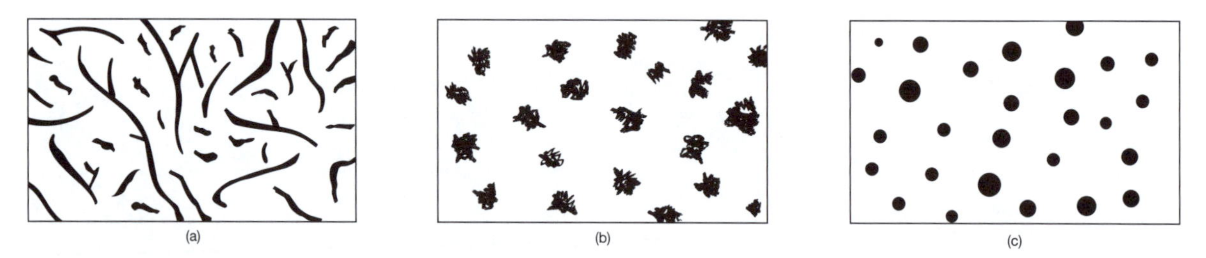


Figure 5.5 Microstructures of (a) grey cast iron with graphite flakes; (b) malleable cast iron with graphite nodules formed during heat treatment; (c) modern ductile cast iron with spheroidal graphite nodules induced chemically

and their other properties were likely to be more variable. That said, the more reputable foundries were well capable of producing good and consistent cast iron.

## 5.3.4 Physical properties

Grey cast iron has a granular texture which is evident when it has been fractured (Figure 5.4), in contrast to the fibrous texture of typical wrought iron (see Figure 6.1).

Throughout the period of its practical use, the strength of historic grey cast iron was described in Imperial units, sometimes in lb in.$^{-2}$ but more commonly in ton in.$^{-2}$. Conversion to the present-day SI unit of stress (1 N mm$^{-2}$, also written as 1 Pa) is given by 1 ton in.$^{-2}$ ≡ 2240 lb in.$^{-2}$ ≡ 15.44 N mm$^{-2}$.

### *5.3.4.1 Density*

BD21 (Highways Agency et al., 2001) gives a figure for historic cast iron density as 7200 kg m$^{-3}$.

### *5.3.4.2 Tensile strength*

The tensile strength of grey cast iron was recognised to be the defining property of the material for structural use, as it was typically only one-quarter to one-sixth of the strength in compression. Equally, the failure mode in tension was brittle and abrupt, without the yielding and extension before fracture that characterises both wrought iron and steel.

This difference in tensile and compressive strength, and the brittle failure mode in tension, can both be explained by considering the microstructure shown in Figure 5.5(a). The flakes of graphite have little or no tensile strength, but they are thin. Thus in compression they have almost no effect on the behaviour of the iron matrix. However, when the material is subjected to tension, the graphite offers no tensile resistance, and so the cross-sectional area of material available to resist the tensile force is less than that subject to compression. In addition, as this force increases, the flakes open out, with the matrix of the cast iron becoming progressively more distorted as failure approaches.

A very useful way of visualising this is to think of the grey cast iron as a slotted plate, with thin slots representing the embedded graphite flakes, having no tensile strength. Many of these will be at right angles, or nearly so, to a tensile stress applied from any direction, as shown diagrammatically in Figure 5.6. (Gilbert, 1957; Swailes, 1996). As the tensile stress increases, the slots distort from being thin and flat into a 'lozenge' shape. The tapered ends of the slots are also points of stress concentration in the iron, so that failure is finally triggered by a fracture running from a slot and spreading across the entire section – a brittle fracture.

A further consequence of this 'slotted plate' behaviour is that grey cast iron exhibits non-linear behaviour under load from the outset, as discussed below in Sections 5.3.4.5 and 5.3.4.7.

Many tensile tests were carried out on grey cast iron during much of the nineteenth century. A selection of results of tests,

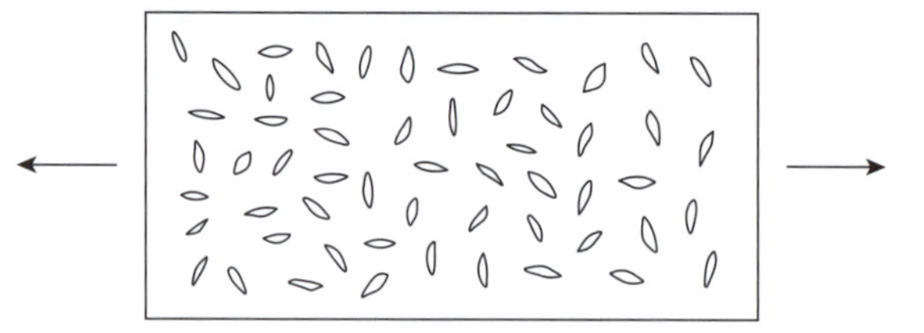

Figure 5.6 Slotted plate analogy for the effect of graphite flakes in grey cast iron

published between 1815 and 1887 and collected in Unwin (1910), is summarised in Table 5.2. (Cross-sectional areas of bars and stresses have been converted to SI units; in this table, as results were originally reported to three significant figures, bar areas are given to the nearest mm$^2$ and stresses to the nearest N mm$^{-2}$, rather than being rounded to the nearest 10 N mm$^{-2}$.)

It is clear from Table 5.2 that tensile strengths vary quite widely. One or two results are just below 75 N mm$^{-2}$ (about 5 ton in.$^{-2}$); most are higher, but are seldom above 150 N mm$^{-2}$ (almost 10 ton in.$^{-2}$) before the 1850s. Eaton Hodgkinson, who was responsible for most of the tests reported from this period, took great care to ensure that he was testing purely in tension, although his specimens were not of the form used today. Other test results of the same period show similar strengths.

After 1850 tensile strengths seemed to increase quite markedly, but it is possible that by that time testing was most common with high-grade iron such as that used for cannons. Highest strengths approach 300 N mm$^{-2}$ (nearly 20 ton in.$^{-2}$), but the variability remains wide.

More recent tensile strength tests on specimens cut from mid-nineteenth-century structures, using modern techniques, are not easy to find. Quite often bending tests have been carried out on both full-size beams and cut-out small specimens, without related direct tension tests. However, tensile tests on 20 specimens, mainly from quite large beams, in four separate studies between 1934 and 1987 gave results as follows (flexural strengths from these tests are included in Table 5.4):

- lowest – 104 N mm$^{-2}$ (6.7 ton in.$^{-2}$)
- highest – 243 N mm$^{-2}$ (15.7 ton in.$^{-2}$)
- mean – 187 N mm$^{-2}$ (12.2 ton in.$^{-2}$).

A more recent study of the strength of cast iron beams included tensile tests on eight 12–16 mm diameter specimens cut from the beams; strengths ranged from 121 to 154 N mm$^{-2}$ (Swailes and Parmenter, 1997).

A significant characteristic of grey cast iron is that it is stronger in flexural tension than in direct tension, which is relevant as present-day appraisal of historic cast iron for tensile capacity is almost invariably focused on beams, and sometimes columns subject to bending, rather than in direct tension. Flexural strength is considered in Section 5.3.4.5.

**Table 5.2** Tensile strength of grey cast iron (from Unwin, 1910, quoted also in Hatfield, 1912)

| *Experimenter and date of report* | *No. of tests* | *Cross-section of bars* (mm$^2$) | *Tenacity* (*ultimate tensile strength*) (N mm$^{-2}$) | |
|---|---|---|---|---|
| | | | *Range* | *Mean* |
| Minard and Desormes, 1815 | 13 | 148–323 | 79–140 | 111 |
| Hodgkinson and Fairbairn, 1837 | – | 645–2580 | 93–151 | 114 |
| Hodgkinson and Fairbairn, 1849 | 81 | 1935–2903 | 76–162 | 105 |
| Wade, 1856 | 6 | – | 317* | 211¶ |
| | 4 | – | – | 141† |
| Woolwich, 1858 | 53 | – | 65–236 | 161 |
| Turner, 1885‡ | – | 645 | 73–242 | – |
| Rosebank Foundry, 1887 | 23 | – | 100–281 | 236§ |
| Unwin, 1910 | 6 | 484 | – | 212 |
| | 3 | 645 | 230–267 | 242 |

*Highest result obtained. ¶Selected as good iron. †Selected as bad iron. ‡Special series of experimental test bars, with varying proportions of silicon. §Mean of 10 best specimens.

#### *5.3.4.3 Factors affecting tensile strength*

Several factors are widely recognised to affect the tensile strength of grey cast iron.

Carbon content, more properly equivalent carbon content, has already been identified as a key factor in Section 5.3.3; the higher this is, the lower will be the tensile strength.

Size, and the thickness of the casting, are both significant influences on strength. A thin section will cool more rapidly, and with a finer dispersion of graphite, than a thick one, and its matrix hence is likely to have a higher strength. This is acknowledged in the current British Standard for grey cast iron BS EN 1561 (see Section 5.4.2); tensile specimens cut from larger or thicker castings will typically give lower strengths than those from smaller test specimens cast from the same iron.

A further factor is that the surface of a casting cools quite quickly in contact with the mould material, and can take up a somewhat stronger 'chilled' hard texture, contributing relatively more to the strength of a thinner section. Knowledgeable iron-founders however argued that castings should be left for generous periods in the mould to cool slowly, without risking thermal shocks that could arise from too-early stripping from the mould.

Finally, as is now well understood, a thick casting is more likely to contain a flaw that could trigger a brittle fracture than is a thinner section. This is discussed more fully in Section 5.3.4.5.

#### *5.3.4.4 Compressive strength*

Grey cast iron has a good compressive strength; tests in the nineteenth century typically gave strengths for stocky specimens of the order of 50–70 ton in.$^{-2}$ (770–1080 N mm$^{-2}$) or better. Applied to the basic sizing of columns to carry load, this could indicate that quite slender sections would suffice, although in practice both buckling and eccentricity of loading would need to be considered, and these factors would point to use of a larger section.

#### *5.3.4.5 Flexural strength*

The bending strength of cast iron has been a source of confusion and debate ever since the early nineteenth century. The material's brittleness, its difference in strength between compression and tension, its non-linear behaviour under tensile loads, and the effect of casting size on strength are all factors that make grey cast iron a more challenging structural material to comprehend than today's commonplace materials such as steel and reinforced concrete.

Sutherland (1979–80 and 1985) was notable in focusing concern on the bending strength of grey cast iron, and considered at some length the development of its understanding in the nineteenth century in the first, 1992, edition of this reference book. Swailes (1996) reviewed the design, manufacture and reliability of historic grey cast iron beams, while Swailes and Parmenter (1997) reported results of tests on beams making use of modern techniques of instrumentation for recording test data.

Simple modulus-of-rupture tests on large grey cast iron beams show much wider variations in apparent strength with size of section than can be explained as solely due to variations in tensile strength. Table 5.3 illustrates these variations, as demonstrated by both contemporary and more recent tests on cast iron beams. It can be seen that on an elastic (modulus-of-rupture) basis, a 25 mm × 25 mm bar of grey iron could be up to 2½–3 times as strong in bending than was a large I-beam of similar material. Also, smaller specimens cut from large beams had a modulus of rupture 2–3 times that of the beams from which they were cut.

The standard test for quality of castings in the nineteenth century was for a small square bar, typically 1 inch (25 mm) square, to be laid horizontally on supports 4 feet (1.22 m) apart. The iron was subject to a central point load until it deflected 0.75 in. Anderson (1884, p. 35) notes that 'a good mixture, properly cast, will sustain a load of 620 lbs [2.76 k N) with a deflection of half an inch'. Although not taken to failure, this represents a *proof stress* of about 310 N mm$^{-2}$, comparable to test results for such bars shown in Table 5.3. While it is clear that this test would overestimate the flexural strength of full-size structural beams, it did nevertheless define a benchmark to be achieved, and also allowed comparison of different grades and mixes of cast iron.

Thomas Tredgold (1822) reported the results of bending tests on small bars. He derived a formula for the capacity of cast iron beams based on what he saw as the limit of elastic behaviour (in fact it was based on the modulus of rupture). He provided constants to adjust his formula for different types of loading and beam sections. Effectively he was advocating a design method for grey cast iron beams based on a safe tensile stress strength of 106 N mm$^{-2}$ with an implied factor of safety of between 2.6 and 3.8, whereas for larger beams this stress was dangerously near the upper limit. Given the engineering knowledge of the time, much of Tredgold's reasoning was sound; he had no reason to doubt the validity of extrapolation from small bending tests. What was less sound was his dismissal of ultimate tensile strengths of 112 N mm$^{-2}$ found earlier by Captain Sam Brown, and of 130–135 N mm$^{-2}$ as found by John Rennie.

Basing his thinking on the assumption of equal strength in tension and compression, Tredgold advocated cast iron I-beams with equal top and bottom flanges; such beams were widely used in bridges and major buildings until at least as late as the 1840s. In many cases, perhaps almost universally in major structures,

**Table 5.3** Results of flexural tests on different sizes of grey cast iron beam and on small specimens cut from beams tested

| *Source(s)* | *Shape of beams and depth (mm)* | *Results of flexural tests on beams or bars* | | | *Results of flexural tests on specimens from beams tested (25 mm × 25 mm, except as noted*)* | | | *Ratio X/Y* |
|---|---|---|---|---|---|---|---|---|
| | | *No. of tests* | *Modulus of rupture (N mm$^{-2}$)* | | *No. of tests* | *Modulus of rupture (N mm$^{-2}$)* | | |
| | | | *Range* | *Mean (X)* | | *Range* | *Mean (Y)* | |
| Sutherland, 1979–80 and 1985 (a)§ | Square, 25 | Many | 280–400 | 340 | – | – | – | – |
| | Rect., 76 (25 wide) | 1(?) | | 258 | – | – | – | – |
| | Rect., 76 (51 wide) | 1(?) | | 250 | – | – | – | – |
| | Square, 76 | 1(?) | | 200 | – | – | – | – |
| As above (b)§ | Small I or ⊥, 102–178 | 7 | 143–203 | 175 | – | – | – | – |
| BRE, 1984 | ⊥, 333 | 2 | 156–210 | 183 | – | – | – | – |
| Gough, 1934 | I, 298 | 6 | 138–190 | 156 | 12 | 394–464 | 427 | 2.7 |
| Cooper, 1987 | I, 360 | 1 | – | 122 | 5 | 218–250 | 237 | 1.9 |
| Chettoe et al., 1944 | I, 298–634 | 12 | 116–135 | 122 | 9 | 252–390 | 332 | 2.7 |
| BRS, 1968 (a)† | I, 460–675 | 6 | 101–118 | 108 | – | – | – | –. |
| As above (b)† | I, 460 | 2 | 158–159 | 158.5 | – | – | – | – |
| Swailes and Parmenter, 1997 (a) | I, 692 | 5 | 91–113 | 99 | 48 * | 197–274 | – | 2.0–2.8 |
| As above (b) | I, 762 | 5 | 99–134 | 117 | – | – | – | – |
| As above (c) | T, 762 (inverted) | 7 | 78–118 | 95 | – | – | – | – |

§Review of mainly nineteenth-century tests. †Beams in set (b) noted as superior to those in set (a). *Cut specimens 25–50 mm deep, 12.5–25 mm wide.

each beam was proof-loaded, the proof loads together with full dimensions quite frequently surviving on drawings of the period. Not surprisingly, some of these proof loads were very high.

Eaton Hodgkinson carried out an extensive series of tests on cast iron, the results of which partly corrected Tredgold's misconception (Hodgkinson, 1831). He showed that grey cast iron was actually much stronger in compression than in tension. He also deduced that in beams the neutral axis must rise towards the compression face, due to the non-linear behaviour in tension, so that the amount of material in tension increased with a corresponding increase in load-bearing capacity.

Having shown that the grey cast iron of his time was about six times as strong in compression as in tension and that the ultimate tensile strength of that iron was about 100–115 N mm$^{-2}$, Eaton Hodgkinson evolved his 'ideal section' of beam *with a bottom flange of up to six times the area of the top flange*. In practice this proportion was generally found to be impractical, and in most 'Hodgkinson' beams the proportion is nearer to 3:1 or 4:1. These beam profiles were widely used from the 1830s onwards, mainly in industrial and engineering structures, although Tredgold's thinking can be found in civic and domestic building construction until at least 1850.

For practical design, Hodgkinson also put forward a very simple formula for the ultimate bending strength of cast iron beams. It was extensively used throughout the nineteenth century and even quoted in handbooks after 1900. Compared with the strengths of beams found from tests, the formula broadly 'straddles' the real strengths, although it can overestimate the strength of some larger beams by 40% or more.

The non-linear stress–strain relationship in tension produces a shift of the neutral axis towards the compression face, and also results in a convex stress block; both contribute to the increased flexural strength compared with the axial tensile strength (see Section 5.3.4.7 below), although neither effect is enough in itself to produce the overall increase.

Little further work was done on the structural performance of cast iron after Hodgkinson's impressive achievements. This was not just because the problems appeared to be solved, but moreover that by the 1850s not only was cast iron beginning to get a bad name as an unreliable material (as a result of several notable failures), but it was also being overtaken by wrought iron, which was becoming widely and cheaply available and offered a more ductile and arguably more 'reliable' material.

The science of brittle materials would seem to offer explanations both for the lower tensile strength of large cast iron sections in direct and flexural tension and for the greater strength in flexure than in axial tension, both being influenced by the probability of the presence of flaws. Timoshenko (1953, pp. 360–361) describes the work of Weibull in 1939, which proposed formulae relating ultimate strength to size. The strength of larger sections was predicted to reduce, due to the increasing probability of a flaw being present as the volume of material increases. Similarly, for a given axial tensile stress, the entire cross-section is at risk from the failure-triggering effects of a flaw anywhere in it; whereas in bending the same tensile stress occurs only at the extreme fibre, with the stress decreasing towards the neutral axis. So the presence of a flaw in the body of a section in bending should trigger failure, in general, only at a higher flexural stress compared with the axial stress case.

Weibull's work is cited in Swailes and Parmenter (1997), with the proposal that it should be studied and applied to the structural behaviour of cast iron in particular. With the growing interest in the structural performance of grey cast iron generated by the need to appraise older structures, particularly for alteration or adaptive re-use, it is to be hoped that the next edition of this book may be able to report useful work to this end. However, the variability of grey cast iron arising from its manufacture must be borne in mind, and (as with wrought iron, for the same reason) a conservative approach to strength assessment should prudently be taken.

As a cautionary endorsement of this, the case of a cast iron beam failure in 2002 is summarised here; it has been reported elsewhere more fully (SCOSS, 2003, and references cited therein). The beam was one of a number that supported roof terraces over ground floor flats in a row of mid-nineteenth-century houses in London. It failed at midspan, fortunately when the flat below was unoccupied. The terrace construction was of brick arches topped with a levelling bed of weak lime concrete carrying several overlapping layers of thin tiles (the original 'waterproofing'), above which were screeding and asphalt from later waterproofing works. The calculated flexural tensile stress in the beam with the original finishes was in the region of 100 N mm$^{-2}$, i.e. at a level where failure would not be surprising. (It is probable that the beams had not been 'engineered', but instead had been supplied by the builder and sized by the iron-founder.) Subsequent addition of terrace finishes had increased this stress; also, there had been corrosion of the beam by water trapped within the enclosing concrete.

It is known that cast iron beams can 'creep' to failure under long-term high flexural stresses, and this collapse is largely attributable to overstressing from the outset, aggravated by corrosion effects. A similar collapse of a cast iron beam supporting a roof terrace occurred at Somerset House in London in 1869, some 35 years after construction, happily again without loss of life or injury (Swailes, 2003).

#### *5.3.4.6 Shear strength*

Because of the necessarily thick webs needed to ensure a good flow of molten iron for casting beams in grey iron, shear is not likely to be a problem when appraising existing construction. It is not even mentioned in most early publications on cast iron; Twelvetrees (1900) notes that shear strength is not less than tensile strength.

#### *5.3.4.7 Stress–strain relationship and elastic modulus*

Figure 5.7 shows typical stress–strain relationships in both tension and compression for grey cast iron. It can be seen that not only is the iron much weaker in tension than in compression, but it is also less stiff and, furthermore, the stress–strain curve is non-linear in tension almost from the start. This results from the distortion of the graphite flakes 'slots' as tensile stress increases. It means that the elastic modulus at the onset of loading (and probably within the normal working stress range) can be taken as the tangent modulus, whereas as tensile stress increases the (lower) secant modulus is more appropriate. In compression the 'closure' of the graphite 'slots' is less significant, so that the compressive stress–strain relationship is more nearly linear.

So any value given for the elastic modulus of grey cast iron needs to be qualified. Twelvetrees (1900) quotes a range of 65–95 kN mm$^{-2}$, whereas Gilbert (1957) gives a range of 85–145 kN mm$^{-2}$ for six cast irons with varying carbon and silicon content. Cast iron beams are usually of substantial cross-section for their loadings, so that a typical value in the range 85–100 kN mm$^{-2}$ should be accurate enough for deflection calculations on structural cast iron within the working stress range.

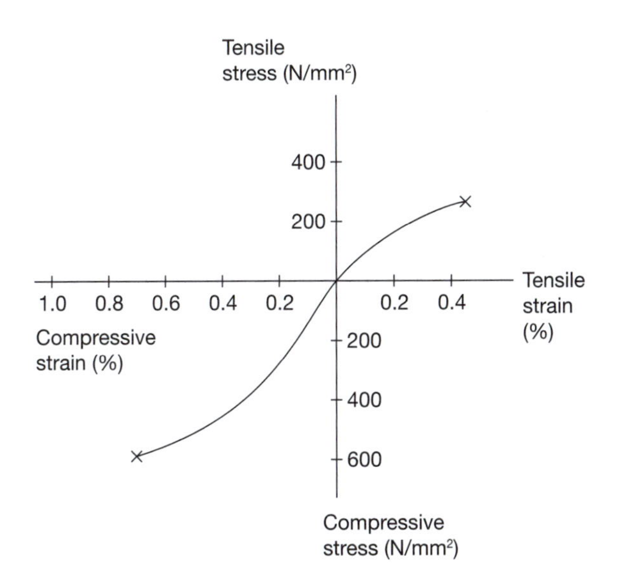


**Figure 5.7** Typical stress–strain relationship for grey cast iron

A significant consequence of the non-linear and lower elastic modulus in tension is that, as stresses increase, *the neutral axis moves towards the compression face*. Laboratory testing has shown that this shift can be as much as 12% at failure (Gilbert, 1957), although it is much lower at working stress. The shift mobilises a slightly increased area of iron in tension. This augments the increased bending resistance arising from the 'bellied' tensile stress block, which is 'swollen' compared to the triangular stress block of a linear-elastic material. These are further factors in the disparity of axial and flexural tensile strength discussed above.

The elongation of grey cast iron at failure is low, generally 0.25–0.75%, which may be compared with values of up to 30% for wrought iron and steel. This low overall elongation to failure is responsible for grey cast iron's reputation as a brittle material. However, if brittle materials are defined strictly as those that fail at or near the end of their linear elastic range with little or no non-elastic yielding, then grey cast iron is not brittle. In tension its non-elastic strain at failure is typically four times the elastic strain – but the total strain is still very small.

#### *5.3.4.8 Impact resistance*

There was no standard test for impact resistance in the nineteenth century to afford comparison between grey cast iron and wrought iron or steel. Few would claim that the resistance of grey cast iron to impact is very good. Furthermore, the resistance reduces gradually with falling temperature and with increasing phosphorus content.

Cast iron columns are vulnerable to impact from powered vehicles, for example fork-lift trucks working in a warehouse. The Health & Safety Executive has published an information sheet highlighting the risks and suggesting possible mitigating measures (HSE, 1999).

#### *5.3.4.9 Fatigue*

On the credit side it can be said that grey cast iron has a low sensitivity to fatigue, and also has good damping qualities. The endurance limit is typically at least one-third of the tensile strength (Angus, 1976). The material has low notch-sensitivity, due to the 'notches' already present from the graphite flakes in the iron matrix.

#### *5.3.4.10 Present-day structural appraisal of grey cast iron*

It must be made clear at the outset that grey cast iron always was designed, *and always should be assessed*, on a permissible stress basis under working or service loads. As a brittle material, grey cast iron has virtually no ductility, and will fracture before it can develop any perceptible plastic distortion. In this respect it differs from wrought iron and steel, which may be assessed on the basis of yield strength and factored loads, in the same way that steel is designed today.

Structural cast iron in the nineteenth century, like wrought iron, was designed by the understanding of the day, related to the ultimate tensile strength of the material. Factors of safety appeared to be generous – as late as 1900, Twelvetrees (1900) was suggesting 5–7 for the dead load on beams and columns, and 8–10 for live load. However, if design was based on the tensile strength of small specimens, the reduced strength of the actual full-size castings might well be one-third or less of the small specimen strength, so the actual factors of safety would be lower.

Beams for major structures were often specified to be proof-loaded. Records may even exist of required proof loads, and perhaps also of the actual loads applied. Such records are worth searching for as one of the most useful aids to appraisal.

The first British prescriptive code for the structural design of buildings was the 1909 London County (General Powers) Act (HMSO, 1909) – by which time, in reality, cast iron had virtually been abandoned for new structural use! The Act limited the permissible tensile stress in cast iron (including tensile bending stress) to 23 N $mm^{-2}$ (1.5 ton $in.^{-2}$) and compressive stress to 124 N $mm^{-2}$ (8 ton $in.^{-2}$). Columns were to be designed with reduced stresses that took account of slenderness effects.

In many existing structures, analysis of cast iron beams is likely to show tensile stresses in excess of this 1909 figure of 23 N $mm^{-2}$, although the structure is showing no signs of distress. It may also be noted that mid-twentieth-century guidance on appraisal of railway bridges – which often needed to be checked for heavier general or 'one-off' loads – frequently adopted a permissible tensile stress of 2.5–3 ton $in.^{-2}$ (39–46 N $mm^{-2}$).

This was taken into account in the 1984 edition of BD21, *Assessment of Highway Bridges and Structures*, and remains in the current edition (Highways Agency et al., 2001). The 1984 edition was drafted with input from British Railways, London Transport and British Waterways, among others, so that – despite recent privatisation and dissociation of some transport bodies – this approach should be applicable to any cast iron bridge. The Steel Construction Institute's guide to the appraisal of existing iron and steel structures (Bussell, 1997) draws on the guidance originally published in BD21, which may be extended to the appraisal of building structures. (These are usually subject to much less dynamic stressing than railway bridges.)

Both BD21 and the SCI guide give permissible stresses for cast iron beams that vary with the proportion of live to dead load according to published graphs, themselves based on equations. The maximum permissible stresses are given in Table 5.4, but it should be noted that the highest levels of these only apply to dead loads or favourable combinations of dead and live load. The equations should be used with the appropriate values of dead and live load. These permissible stresses are noted in BD21 (p. 3/4) to provide "a reasonable assurance" against fatigue failure.

For highway bridges, BD21 also gives advice on how the section modulus of beams may be notionally increased to allow for composite action with fill.

Assessment of columns is based on the Gordon-Rankine formula; a maximum permissible stress of 154 N $mm^{-2}$ for a very squat column section is reduced to take account of slenderness. For hollow circular columns, it is advisable to check the wall thickness by drilling – non-percussively, to avoid the risk of shattering the iron – at three points equally spaced around the circumference, before calculating load-bearing capacity; the internal column core might well have been displaced during casting, so a thickness based on a single drilled hole could be misleading.

If an assessment using these permissible stresses shows that an apparently sound cast iron structure is overstressed, helpful factors such as end fixity of beams, composite action and load-sharing are worth considering. Good performance over time, the type of loading, and the consequence of a possible failure are all factors to be considered before a final judgement is made. The guide to the general appraisal of existing structures (Institution of Structural Engineers, 1996) and the SCI guide (Bussell, 1997) both offer advice on approaches to appraisal.

More recently, Swailes and de Retana (2004) have published a useful review of the understanding of cast iron column strength, while Brooks et al. (2008) have described the analytical studies and the extensive testing programme undertaken to justify the cast iron columns supporting the train deck and platforms of St Pancras Station, including tests to failure. Both studies address what can be the significant issue of column flexure, which the Gordon-Rankine formula does not allow for.

### 5.3.5 Dimensional stability

The coefficient of linear expansion of grey cast iron for temperatures between 0 and 200°C is typically 11 × $10^{-6}$ per °C. Grey cast iron can undergo creep and permanent set under load, particularly in tension, but this is not generally significant at the service stress levels in structures.

### 5.3.6 Durability

Experience in structures over a century or more indicates that grey cast iron as-cast has better corrosion resistance than wrought iron, and considerably greater resistance than steel. This may be largely due to the thin layer of slightly 'chilled' hard white iron that forms on the surface of the casting as molten iron comes into contact with the mould material, producing a hard and semi-vitreous surface. In most situations it will last for decades or possibly even centuries with nothing more serious happening than the formation of a thin surface coating of rust. However, this cannot wholly be relied upon, and serious corrosion can occur locally where water lodges.

Cast iron is often painted for a pleasing appearance; such treatment protects the surface, but also serves to indicate corrosion when it starts to break down.

Opinions differ on whether existing cast iron should be blast-cleaned prior to repainting. On one hand, such treatment will remove the hard surface layer, after which the exposed cast iron faces have little better durability than steel. On the other hand, failure to remove surface corrosion products will risk poor adhesion and early breakdown of any corrosion protection scheme. A further consideration is the practicality of removing surface coatings (which may contain toxic material such as lead) from existing construction while in situ; this will call for great care to avoid the risk of contamination and pollution.

Extensive use was made in the nineteenth century of grey cast iron for marine structures, such as the piles and columns of seaside piers and the caissons of railway bridges. These have often survived with surprisingly little corrosion damage (although careful and regular inspection is recommended so that any damage can be remedied before it becomes serious). In tidal waters the ironwork at bed level may, however, suffer abrasion damage from shingle. In brackish waters and damp soil, cast iron may suffer anaerobic corrosion known as *graphitisation*, in which much of the iron is leached out by an electrolytic action between

**Table 5.4** Limits of permissible stresses for cast iron beams in BD21 and SCI guide

| *Stress* | *Dead load only or most favourable combinations of permanent and live load* | *All live load, no dead load* |
|---|---|---|
| Compression | 154 N $mm^{-2}$ (10 ton $in.^{-2}$) | 81.3 N $mm^{-2}$ (5.3 ton $in.^{-2}$) |
| Tension | 46 N $mm^{-2}$ (3 ton $in.^{-2}$) | 24.6 N $mm^{-2}$ (1.6 ton $in.^{-2}$) |
| Shear | 46 N $mm^{-2}$ (3 ton $in.^{-2}$) | 24.6 N $mm^{-2}$ (1.6 ton $in.^{-2}$) |

the different phases of the cast iron matrix to leave a spongy weak residue of graphite, insoluble salts, and finely divided iron enclosed by a black and blistered surface. Replacement is often the only practical remedy.

### 5.3.7 Use and abuse

The celebrated Iron Bridge over the River Severn at Coalbrookdale, Shropshire, completed in 1779 (Figures 5.3 and 5.8), was one of the earliest examples of structural cast iron in Britain. It was soon to be followed by other cast iron arch bridges and some church and domestic architecture, and then the textile mills of Bage, Strutt and others in the 1790s and early 1800s. Most notable of these mills was that completed in 1797 at Ditherington, on the outskirts of Shrewsbury (a few miles from the Iron Bridge). This is the first multi-storey building to have the internal framing of beams and columns entirely in iron (Figure 5.9).

In addition, cast iron was being used for pipes, in increasing quantities for machinery, and in items such as railings, quite apart from cannons and cannon balls. After 1800, and particularly after 1820, the structural use of cast iron grew out of all proportion, reaching a climax in the early 1840s.

Even though it was weaker in tension than in compression, the tensile strength and the stiffness of grey cast iron exceeded those of timber, and thus the spanning capacity of beams was immediately increased. In addition, the incombustibility of the material led to extensive use of 'fireproof' construction, first in the textile mills and later in public buildings of all types. Single iron castings of up to 1 m or more in structural depth and about 20 m long were used for bridge beams or in major industrial structures.

One advantage of the casting process was that the section profile could be varied along the length; for example the profile of a beam web could be curved along its top ('hog-backed') or on its underside ('fish-bellied'), as shown in Figure 5.1 and also in Figure 5.11; the increased depth at midspan provided enhanced bending resistance where it was needed. And structural elements could be of ornate profile, so that columns could be formed with Classical or florid ornamentation, while whole façades in iron could serve both structurally and as the weatherproof building envelope (for example the Gardner warehouse in Jamaica Street, Glasgow; Figure 5.10).

All-cast-iron structures with columns, main and secondary beams in the material were quite common around 1830–1850, as in the now-demolished Great Quadrangle Store of the late 1820s in the former Sheerness Naval Dockyard in Kent (Figure 5.11). The fish-bellied secondary beams neatly provided a grid of level bearings for the stone flags of this entirely incombustible structure.

Figure 5.12 is from the surviving 1858–60 Boat Store, also in Sheerness Dockyard, which has a pioneering multi-storey frame – reliant for its stability on the frame itself, not on enclosing masonry walls. The photograph shows the junction of a cast iron column, main beams of riveted wrought iron, and secondary ones of I-section cast iron. Care is needed in analysing such structures to make sure that the materials are correctly identified, and that the way they are connected together is fully understood.

Connections between cast iron members were typically made by direct bearing, both between consecutive lifts of columns (often with spigots to provide vertical continuity of iron) and where beams lodged onto the enlarged or flared head-plate of a column. In the latter, restraint against movement was typically provided by

**Figure 5.8** General view of the 1779 Iron Bridge over the Severn in Shropshire

**Figure 5.9** Ditherington Mill, Shrewsbury of 1797, the first multi-storey iron-framed building

bolts linking the ends of adjacent beams, or by wrought iron rings that were heated and then fitted onto 'ears' cast onto the beam faces, cooling to grip the ends of the beams tightly across the column (as in the re-erected frame shown in Figure 5.1).

Sutherland (1997) and Thorne (2000) have edited two volumes of informative key papers that describe notable developments and uses of structural cast iron in the periods 1750–1850 and 1850–1900.

By 1850 cast iron was being eclipsed by riveted wrought iron (see Chapter 6), especially for beams, and – although cast iron columns and brackets continued into the present century – by 1920 the use of the material in structures had all but ceased. One or two well-publicised failures of cast iron beams, such as that at Radcliffe's Mill in Oldham in 1844 and the Dee Bridge in 1847, did much to destroy faith in cast iron as a spanning structural material. Nevertheless, in other fields the use of grey cast iron has continued in the construction industry where 'chunky' robustness combined with good resistance to corrosion are more important than structural capacity; manhole covers, gratings and bollards are typical.

Abuse of cast iron columns by vehicle impact has already been mentioned; another problem for such columns can arise when hollow circular columns, used also as down-pipes for internal valley gutters in larger buildings, become blocked. If the trapped water then freezes in icy weather, the column can be shattered by the expansive action as ice forms from water. As always, regular inspection and necessary maintenance can avert serious consequences.

**Figure 5.10** The façade of the 1855–6 Gardner's warehouse in Glasgow, with cast iron panels forming both the structure and the envelope

### 5.3.8 Proprietary brands

In the nineteenth century, ironfounders developed their own 'mixes' for particular applications, and would market these under

**Figure 5.11** Cast iron columns, hog-backed main beams, and fish-bellied secondary beams in the 1820s Sheerness Dockyard Great Quadrangle Store, with stone-flagged flooring

**Figure 5.12** Column–beam junction in the Boat Store at Sheerness Dockyard

their own name – itself offered as a guarantee of quality – and a reference number, such as Butterley No. 2 (Swailes, 1996).

Several architectural and ornamental ironfounders became major producers, with companies such as Walter Macfarlane of Glasgow producing illustrated catalogues, sometimes in colour and works of art in themselves, listing and displaying literally thousands of different designs. Catalogues often ran to hundreds of pages, with engravings of standard profiles for columns, pipes, guttering, roof ridges, terminals and other details, railings, gates, panels, gratings, straight and spiral staircases, balconies, and many other products. This may have been before the days of British Standards, but there were clearly off-the-shelf standard designs to be had, as well as bespoke castings when required.

### 5.3.9 Hazards in use

Mention has already been made of the risk of grey cast iron cracking or shattering if struck by a vehicle or other impact. Care is needed too if drilling into what is suspected or known to be structural cast iron – a non-percussive drill setting must be used, to guard against the risk of the iron shattering.

### 5.3.10 Coatings for corrosion protection

Grey cast iron can be readily painted to provide both corrosion protection and a pleasant appearance. When painting partially exposed elements, such as 'barley-sugar' window mullions set into a stone window-sill, it is important to ensure that the gap between iron and stone is effectively sealed, so that water cannot drain into the gap and allow slow local corrosion.

### 5.3.11 Performance in fire

The loss of strength of grey cast iron becomes very marked above 400°C. Nevertheless, in actual fires cast iron columns have frequently survived while the wrought iron or steel structures that they supported have collapsed; this may well be due to the columns not being close enough to the seat of the fire to heat up sufficiently. Also, when cast iron has cracked in fires this has often been due to the thermal expansion and displacement of other elements of the structure, rather than the direct effect of fire on the cast iron.

A helpful study of the effect of fire on cast iron in structures was published in 1984 (Barnfield and Porter, 1984); a subsequent English Heritage review (Porter et al., 1998) drew attention to approaches based on fire engineering, in which for example the embedment of iron beams in a brick-arched floor offers a heat-sink effect, much enhancing the potential survival of the beams in fire.

Columns are by their nature more exposed, and so are more likely to need applied protection. Intumescent coatings can provide unobtrusive protection; these are applied as a thin film that swells up in the event of fire to provide an insulating layer to the iron, slowing down the effects of fire. They are particularly useful for the sympathetic protection of historic structures with decorative structural ironwork, where the coating is not so thick as to mask fine detail.

There is a view that cold water from a fire brigade's hoses may shatter hot cast iron after a fire by thermal shock, but the evidence for this appears to be largely apocryphal.

### 5.3.12 Sustainability and recycling

As with wrought iron, the most sustainable treatment of existing cast iron is continued use with proper care and maintenance.

#### *5.3.12.1 Retention in place*

Sound cast iron needs little maintenance. Unless appraisal has shown that a cast iron structural element is manifestly inadequate for future service, it may well be appropriate to retain the cast iron rather than replace or strengthen it, for example with steel, which may well require more expensive maintenance in the future. It has to be said that this calls for the exercise of engineering judgement, which could usefully be informed by some form of testing to provide supporting evidence of adequacy.

If strengthening of a historic structure is judged to be necessary, conservation principles favour the addition of supplementary structure to augment the cast iron elements, rather than their replacement – for example, by placing a steel column alongside an existing cast iron column. This approach retains the original historic fabric.

#### *5.3.12.2 Recycling*

Redundant cast iron is readily recycled as scrap metal for use in making iron or steel. However, there is a market for both structural and decorative cast iron components in sound condition in the field of architectural salvage. Sale for re-use may well yield more than scrap value, and also ensures the survival of good ironwork, albeit it may well be in a new location and function.

#### *5.3.12.3 Maintenance*

As already noted, grey cast iron has good corrosion resistance and, provided it is not exposed to unduly aggressively damp or chemical atmospheres and it is regularly painted, it should have a long life with minimal maintenance needs.

#### *5.3.12.4 Repairs*

At the most basic 'cheap and cheerful' level, there is a long tradition of repairs to cast iron structures by simple plating with wrought iron or steel plates or angles bolted across cracks or fractures, such as where the projecting flange of an I-section cast iron column has been snapped off by vehicle impact. Such repairs are arguably more for 'comfort' than for function, as the load in the ironwork will have redistributed the instant that the flange snapped off. Nevertheless, there may be a good case for leaving such repairs alone; at worst they do no harm.

Welding is often practical for repairs to non-structural grey cast iron, but needs to be carried out with care. It is best confined to castings of limited size that can be removed, taken to a works experienced in such repairs, pre-heated, and welded using the most appropriate electrodes. If in doubt, advice should be sought from the Welding Institute, Castings Technology International, or a specialist firm with demonstrable expertise in such work.

Where repairs to cracked grey cast iron structures need to be carried out *in situ*, 'metal stitching' is often suitable. This is a cold repair method, i.e. it does not involve welding or brazing. Lines of holes are drilled in a predetermined pattern at right angles to the crack, and ductile steel 'keys' or 'locks' are fitted into these lines to tie the metal together either side of the crack. Holes are then drilled along the line of the crack and tapped to receive tapered screws that fill the crack. Grinding-down and repainting can achieve a virtually invisible mend. Widely used in mechanical engineering, the technique has also found some applications in structural repairs; it is offered by a number of specialist firms, many of which can be found by an Internet search for 'metal stitching'.

A fuller description of metal stitching, and of other techniques for the repair of both structural and non-structural cast iron, is offered in Swailes (2006) and Wallis and Bussell (2008). Historic Scotland issues several useful guides to repair of non-structural iron, downloadable from its website.

A relatively new technique for strengthening structures is the application of externally bonded fibre-reinforced composites. This has been applied to a number of cast iron bridges, often to provide an adequate margin of safety against increased vehicle loading. Two studies provide useful information on the technique, and review experience to date (Cadei et al., 2004; Moy, 2001). Interest will focus on durability and long-term performance.

## 5.4 Modern grey cast iron

### 5.4.1 Manufacturing process

The properties of modern grey cast iron are fundamentally the same as those of its historic predecessor. The carbon is in the same flake graphite form, and there is a similar difference in strength and stiffness in tension and compression.

Likewise, the basic method of manufacture is unchanged from the days of historic cast iron, although the ways in which the material is specified and its quality is assured have changed with the advent of increasing national, and now international, standardisation.

In the nineteenth century, informed engineers specified their cast iron by the foundry's name and the particular iron 'mix' (e.g. Low Moor No. 1), knowing that this would deliver castings with the required properties. Today, in contrast, British (now European) Standards specify irons with a range of strength and other properties.

As with all construction products, but with cast iron in particular, specification is not enough to achieve quality; users are well advised to ensure that a prospective ironfounder is capable of producing the required castings.

Large modern foundries can produce castings in quantity to a high standard by industrialised methods. However, there are also small foundries working on more of a craft level, equally capable of doing very satisfactory work. Such small firms may give better value and be more adaptable than the bigger organisations, particularly for decorative ironwork or the small-scale replacement of castings that are not critical structurally.

At the same time, casting design must take account of the practicalities of the manufacturing process. For example, the avoidance of cracks during cooling depends very much on the design of the casting. Sharp internal angles should be avoided, and changes in wall thickness should be gradual, or – ideally – avoided altogether if this is possible, to minimise different rates of cooling within the casting. Equally, smooth transitions and curved junctions ease the flow of the metal during pouring, and thus reduce the danger of hidden blow-holes. Casting design is a skilled craft, and engineers and architects should wherever possible discuss the details of the casting with an experienced ironfounder, preferably the one who will be responsible for the actual casting, before finally fixing the shape and section thickness.

A Steel Construction Institute design guide offers useful advice on design, procurement, inspection and testing of modern-day castings (Baddoo, 1996). Engineers, and others not familiar with the cast iron industry, would do well to consult the Cast Metals Federation (incorporating the British Foundry Association) for advice on the selection of foundries for particular casting work. For more specific technical advice, Castings Technology International (formerly the British Cast Iron Research Association) may be a more appropriate source of information.

### 5.4.2 Physical properties of modern grey cast iron

The appropriate British Standard for grey cast iron is now BS EN 1561:1997, which has replaced BS 1452:1990. The Standard specifically excludes any requirements on methods of manufacture and chemical composition, but defines six grades of grey cast iron, designated either by the tensile strength of test bars machined from separately cast samples or by Brinell hardness tests on the casting. There is a linearly proportional empirical relationship between Brinell hardness and tensile strength, although hardness readings have a 'scatter' of about 20% either side of the mean line. If grey cast iron is to be used structurally it will always be preferable to specify by strength designation.

For each of the six grades, the Standard gives a mandatory value of tensile strength to be achieved from tests on separately cast samples of an as-cast diameter of 30 mm. It also specifies a mandatory tensile strength for cast-on samples of varying thickness (lower than the strength of the separately cast samples), and an anticipated (but not mandatory) tensile strength in the casting itself, again related to thickness. As with historic grey cast iron, tensile strength reduces as casting thickness increases. An Annex, classed as 'informative', lists other physical properties for all except the weakest grade; these properties are given for guidance, and are not criteria for acceptance or rejection.

Table 5.5 gives the main structural properties of grey cast iron as set out in the Standard. A dash (–) indicates that no figure is given in the Standard for the particular property of that grade of iron.

The Standard makes reference to the extensive work on the engineering properties of grey cast iron carried out by Gilbert of the former British Cast Iron Research Association, now Castings Technology International (Gilbert, 1977).

Further informative guidance in Tables A.1 and A.2 of the Standard includes:

- elongation at failure for all designations: in range 0.3–0.8%
- indicative values for fatigue strength and fracture toughness
- bending fatigue strength in the range (0.35–0.5) × tensile strength of separately cast test pieces
- density: 7100 kg $m^{-3}$ for EN-GJL-150, increasing linearly to 7300 kg $m^{-3}$ for EN-GJL-350
- Poisson's ratio for all designations: 0.26
- coefficient of linear expansion for all designations: $10.0 \times 10^{-6}$ per °C at temperatures between –100°C and +20°C, and $11.7 \times 10^{-6}$ per °C between 20°C and 200°C.

**Table 5.5** Main structural properties of grey cast iron (from Tables 1 and A.1, BS EN 1561:1997)

| *Designation symbol* | *Mandatory tensile strength of separately cast sample ($N\ mm^{-2}$)* | *0.1% proof stress: tension ($N\ mm^{-2}$)* | *Bending strength ($N\ mm^{-2}$)* | *Compressive strength ($N\ mm^{-2}$)* | *0.1% yield point: compression ($N\ mm^{-2}$)* | *Shear and torsional strength ($N\ mm^{-2}$)* | *Elastic modulus ($kN\ mm^{-2}$)* |
|---|---|---|---|---|---|---|---|
| EN-GJL-100 | 100–200 | – | – | – | – | – | – |
| EN-GJL-150 | 150–250 | 98–165 | 250 | 600 | 195 | 170 | 78–103 |
| EN-GJL-200 | 200–300 | 130–195 | 290 | 720 | 260 | 230 | 88–113 |
| EN-GJL-250 | 250–350 | 165–228 | 340 | 840 | 325 | 290 | 103–118 |
| EN-GJL-300 | 300–400 | 195–260 | 390 | 960 | 390 | 345 | 108–137 |
| EN-GJL-350 | 350–450 | 228–285 | 490 | 1080 | 455 | 400 | 123–143 |

Comparing the tensile strength of the grey cast iron designations in BS EN 1561:1997 with the typical strengths of historic grey cast iron, the large increase is very noticeable. Today, the strength of many engineering castings may exceed that of wrought iron, and may even approach that of steel. However, grey cast iron still has a very low elongation to failure and so is essentially a 'brittle' material, admittedly with good damping qualities but with low or at best uncertain resistance to impact.

Furthermore, tensile strength decreases with increasing thickness of the material and depends on the shape and proportions of the casting. For example, the Standard indicates that, for cast iron of designation EN-GJL-200, the anticipated strength of a casting of wall thickness of 10–20 mm is 180 N $mm^{-2}$; this drops to 155 N $mm^{-2}$ for a wall thickness of 20–40 mm.

### 5.4.3 Use of grey cast iron in construction today

Grey cast iron remains today in wide use in many parts of the construction industry. A notable area is that of rainwater goods, including gutters, hoppers, down-pipes and gullies. A modern manufacturing innovation has been the introduction of centrifugal casting for pipes and the like, in which the spinning mould generates the hollow cylindrical form without the need for an internal core. Modern patterns are often made of stiff plastic, and sometimes left in place to be melted out by the liquid iron as it enters the mould.

The modern technique of 'concasting', or continuous casting, is applied to steel-making but can also be used to produce castings of quite large – and solid – cross-section. Molten iron is extruded from a die and water-cooled. The technique has been considered for making replacement columns and piles for the restoration of a historic seaside pier, where the solid cross-section would provide a durable structural element with a considerably longer life than the traditional hollow cylindrical section.

This is an example of how, after decades of neglect, interest in grey cast iron for new use in structural work has revived to a limited extent, mainly for its appearance and flexibility of form rather than its physical properties. It is certainly suitable for columns in areas protected from accidental damage and where fire regulations are not paramount; it is also now used for brackets, bracing members and – to a very limited extent – for beams. In such situations, if tensile and bending strengths are important in relation to design stresses, proof-load testing of individual elements would be desirable, just as it was in the nineteenth century.

However, it remains questionable whether grey cast iron is a material to specify today for structures. A more recent, and in almost every respect superior, structural cast iron is now available. This is ductile or spheroidal graphite iron, the properties of which are discussed in the following section.

## 5.5 Ductile (spheroidal graphite) cast iron

### 5.5.1 History and manufacturing process

Malleable iron, already introduced in Table 5.1, is manufactured by the prolonged heat treatment of white iron followed by slow cooling. If the casting is heated in contact with iron oxide (such as hematite ore), much of the flake graphite diffuses into the oxide while the remainder is left as small clusters in a strong matrix. Alternatively, heating the casting in sand or a similarly inert material leads to the free carbon remaining in the casting, the flakes changing to become more numerous clusters (Figure 5.5(b)). Without the flake graphite to break up the structure, these malleable castings have greater tensile strength and ductility and better resistance to impact than does grey cast iron; indeed malleable castings behave more like steel.

Malleable cast iron was used widely in the second half of the nineteenth century, but because of the need to heat completed castings it was found more suitable for small components than for structural elements such as beams and columns. The heat treatment also made it appreciably more expensive than grey cast iron.

Although malleable castings are still made today and are covered by a current British Standard (BS EN 1562:1997), the material is being superseded to an increasing extent in the construction industry by spheroidal graphite cast iron, also known as ductile or nodular cast iron. In 1946 H. Morrogh of the former British Cast Iron Research Association discovered that a similar microstructure to that of malleable iron could be produced during casting by adding the rare earth element cerium to the molten iron; a similar discovery was made at about the same time in the USA by the addition of magnesium. A typical microstructure of ductile iron produced in this way is shown in Figure 5.5(c). The characteristic 'spheroids' or 'nodules' of graphite are clearly visible; these eliminate the adverse effects of the graphite flakes in grey cast iron, described in Section 5.3.4, so that the casting is essentially now ductile rather than brittle.

### 5.5.2 Physical properties of ductile cast iron

The appropriate British Standard for ductile cast iron is now BS EN 1563:1997, which has replaced BS 2789:1985. As for present-day grey cast iron to BS EN 1561, the Standard specifically excludes any requirements on methods of manufacture and chemical composition, but defines nine grades of ductile cast iron, designated by the tensile strength of test bars machined from separately cast samples, its 0.2% proof stress, and the percentage elongation at failure. A typical designation such as EN-GJS-400-18 indicates a ductile cast iron with a tensile strength of 400 N $mm^{-2}$ and minimum elongation of 18%.

As an alternative, the Standard specifies a mandatory tensile strength, 0.2% proof stress, and elongation for test pieces machined from cast-on samples of varying thickness. For casting thickness up to 30 mm the values required are the same as for the separately cast samples, reducing modestly with increasing casting thickness. The designation in the case of the EN-GJS-400-18 grade quoted above would then be EN-GJS-400-18U.

An Annex, classed as 'informative', lists other physical properties for all except the weakest grade; these properties are given for guidance, and are not criteria for acceptance or rejection.

Table 5.6 gives the main structural properties of ductile cast iron as set out in the Standard, when tensile strength and elongation are determined from separately cast samples. A dash (–) indicates that no figure is given in the Standard for the particular property of that grade of iron.

For the two lowest grades, 350 and 400, an impact resistance may also be specified to be achieved at either room temperature (around 23°C), or down to temperatures as low as –40°C; this requirement is specified by adding -RT or -LT to the designation. (Hardness requirements may also be specified with the agreement of the supplier.)

Further guidance in Table B.1 of the Standard includes:

- indicative values for fatigue strength and fracture toughness
- un-notched fatigue limit decreasing from (0.5–0.4 and lower) × tensile strength with increasing strength
- density: 7100 kg $m^{-3}$ for strengths up to 500 N $mm^{-2}$, 7200 kg $m^{-3}$ for higher strengths
- Poisson's ratio for all designations: 0.275
- coefficient of linear expansion for all designations: $12.5 \times 10^{-6}$ per °C.

For most structural uses, an ultimate strength of 350 N $mm^{-2}$ or higher in tension is very reasonable, especially if combined with good elongation (although it should be noted that elongation at failure reduces with increasing strength). Furthermore, ductile iron at the lower end of the strength table can have a defined level of resistance to impact, something lacking in grey cast iron. Thus it is now possible to combine the strength and ductility of wrought

**Table 5.6** Main structural properties of ductile cast iron (from Tables 1 and B.1, BS EN 1563:1997)

| *Designation symbol* | *Tensile strength ($N\ mm^{-2}$)* | *0.2% proof stress ($N\ mm^{-2}$)* | *Elongation (%)* | *Compress-ive strength ($N\ mm^{-2}$)* | *Shear and torsional strength ($N\ mm^{-2}$)* | *Elastic modulus ($kN\ mm^{-2}$)* |
|---|---|---|---|---|---|---|
| EN-GJS-350-22 | 350 | 220 | 22 | – | 315 | 169 |
| EN-GJS-400-18 | 400 | 240–250 | 18 | 700 | 360 | 169 |
| EN-GJS-400-15 | 400 | 250 | 15 | – | – | – |
| EN-GJS-450-10 | 450 | 310 | 10 | 700 | 405 | 169 |
| EN-GJS-500-7 | 500 | 320 | 7 | 800 | 450 | 169 |
| EN-GJS-600-3 | 600 | 370 | 3 | 870 | 540 | 174 |
| EN-GJS-700-2 | 700 | 420 | 2 | 1000 | 630 | 176 |
| EN-GJS-800-2 | 800 | 480 | 2 | 1150 | 720 | 176 |
| EN-GJS-900-2 | 900 | 600 | 2 | – | 810 | 176 |

iron and structural steel with the freedom of form that has always been a special feature of grey cast iron.

### 5.5.3 Use of ductile cast iron

With a tensile strength almost as high as that of steel and a good resistance to impact, ductile iron has technical advantages over virtually all other types of cast iron, which the construction industry has been surprisingly slow to recognise. However, it is now being used on an increasing scale both in mechanical engineering and in the construction industry. Pipes, tunnel-lining segments, heavy-duty covers and street furniture are increasingly being made of ductile iron in place of grey cast iron.

The material has found relatively little use in structures as yet. The joints in the British Steel Corporation's 'Nodus' space frame system are an exception, while there have been a number of one-off structural applications, for example at the Renault Centre, Swindon (although on a very small scale) and for actual roof trusses at the Menil Gallery in Houston, Texas (Figure 5.13); even there the actual castings of which the trusses are made are only about 1 m long.

The main rival to ductile cast iron is cast steel, with which even higher tensile strengths are possible. The Steel Construction Institute design guide already mentioned in the context of modern grey castings provides useful guidance on the properties and uses of ductile cast iron and cast steel (Baddoo, 1996). Case studies of

**Figure 5.13** Ductile cast iron roof trusses for the Menil Gallery, Houston (courtesy of Arup)

the use of both materials are described. The absence of a code of practice for structural design of such castings is noted, with the suggestion that use be made of a British Cast Iron Research Association publication (Gilbert, 1986). A more recent publication giving extensive information on properties is a data handbook for ductile cast iron (Castings Development Centre, 1997).

Ductile iron is more expensive than grey cast iron, but is typically cheaper than cast steel. However, material cost may be less than one-third of the total cost of the castings, when the costs of pattern-making and moulding are taken into account. A long run of a standardised casting will have lower unit costs than short runs of castings with even slight variations in profile, while complex profiles will be more difficult to cast successfully. So, repetition and simplicity are keys to the cost-effective use of any castings.

Ductile cast iron has better fluidity and is dimensionally more stable in the transition from the liquid to the solid state than is cast steel. It is thus easier to achieve finer detailing and more accurate castings with ductile iron. Its tensile strength is less than that of most cast steels, but it can readily be made more than adequate for most structural uses. As far as elongation to failure and impact resistance are concerned, ductile iron is generally comparable with the lower grades of cast steel; it is of course greatly superior in these respects to grey cast iron.

One advantage of cast steel over ductile iron is that it can be welded more easily and with greater certainty. Opinions differ on the advisability of welding ductile iron. The SCI guide advises (p. 30) that fusion welding can result in a hard and brittle heat-affected zone, and recommends that welding be avoided wherever possible, relying instead on bolted connections where necessary. However, even highly stressed castings have been welded successfully, especially in mechanical engineering where repetition makes research into techniques and closely controlled welding procedures more worthwhile than is usual with structures. Heat treatment after welding may be advisable, but this is much simpler with small components than with major structural sections.

There are many structural situations, as with columns or arch ribs for instance, in which the economy of ductile iron relative to cast steel, its freedom of form, and its robustness could in future be used to great advantage without any need for welding. Grey cast iron continued to be used for columns for several decades in the later nineteenth and even early twentieth centuries after it had been superseded by wrought iron and steel for beams and trusses. With ductile iron such use could be revived, but this time the columns would have a resistance to impact or to unintended bending moments that grey cast iron cannot offer. Jointing could be by socketing and bolting as in the nineteenth century, and the scope for slenderness and elegance of form would be large. With structural members used essentially in compression, even some purely positional welding could be tolerated without worry.

There are also other structural situations where ductile iron could have a place. In the minds of some structural engineers, *all* cast iron is brittle and a material to be avoided. This is a prejudice that needs to be refuted!

Assuming that there might be a rebirth of structural cast iron in the form of ductile iron, there will be a need for a high level of quality assurance and quality control, not just of the iron as a material but of the finished casting. Such a need arises with any 'new' material. It arose with precast concrete, which in general is being produced to a far higher level of quality today than that of four or five decades ago when the material first came into vogue. Perhaps a similar future awaits ductile cast iron?

## 5.6 Acknowledgements

The first acknowledgement must be to the author of this chapter in the first edition of this reference book, James (R. J. M.) Sutherland, whose knowledge and enthusiasm have inspired many – including the present writer – to learn more about historic cast iron. Revising this chapter for this new edition has necessarily required coverage of developments since the first edition, including the research on historic cast iron largely led by Swailes, and the replacement of British by European Standards for the specification of present-day grey and ductile cast iron. Nevertheless, much of what James Sutherland wrote nearly two decades ago has stood the test of time, and is retained here. He has also kindly reviewed and made helpful comments on the revisions to his original text.

In the first edition of this book, acknowledgements were made to the British Cast Iron Research Association (now Castings Technology International), the British Foundry Association (now the Cast Metals Federation), Glynwed Foundries, and Ballantine Bo'ness Iron Co. Ltd for information used in parts of this chapter, and also to John Thornton and Paul Craddock of Ove Arup & Partners, and J. A. Allen; G. N. J. Gilbert was noted as having been especially helpful. For this revised chapter in the second edition, further thanks are due to Alison Ford of the Library and Information Service at Castings Technology International.

## 5.7 References

Anderson, Sir J. (1884). *The strength of materials and structures*. London, UK: Longmans, Green.

Angus, H. T. (1976). *Cast iron: Physical and engineering properties* (2nd edition). London, UK: Butterworth.

Baddoo, N. R. (1996). *Castings in construction*. Ascot, UK: Steel Construction Institute.

Barnfield, J. R. and Porter, A. M. (1984). Historic buildings and fire: Fire performance of cast-iron structural elements. *The Structural Engineer*, *62A*(12), 373–380.

BRE (1984). *Loading tests on cast iron beams*. Watford, UK: Building Research Establishment (private report).

Brooks, I., Browne, A., Gration, D. A. and McNulty, A. (2008). Refurbishment of St Pancras – justification of cast iron columns. *The Structural Engineer*, *86*(111), 28–39.

BRS (1968). *Static strength of cast iron girders*. Watford, UK: Building Research Station Private Report.

*BS EN 1561:1997. Founding. Grey cast irons*. London, UK: British Standards Institution.

*BS EN 1562:1997. Founding. Malleable cast irons*. London, UK: British Standards Institution.

*BS EN 1563:1997. Founding. Spheroidal graphite cast iron*. London, UK: British Standards Institution.

Bussell, M. (1997). *Appraisal of existing iron and steel structures*. Ascot, UK: Steel Construction Institute.

Cadei, J. M. C., Stratford, T. J., Hollaway, L. C. and Duckett, W. G. (2004). *Strengthening metallic structures using externally bonded fibre-reinforced polymers*. London, UK: Construction Industry Research and Information Association.

Castings Development Centre. (1997). *Data Handbook for Ductile Cast Irons*. Sheffield, UK: CDC.

Chettoe, C. S., Davey, N. and Mitchell, G. R. (1944). The strength of cast iron girder bridges. *Journal of the Institution of Civil Engineers*, *22*(8), 243–307.

Cooper, P. E. (1987). *The behaviour of cast iron beams in bending*. Leeds University, UK (MEng thesis).

Fairbairn, W. (1864). *On the application of cast and wrought iron to building purposes*. London, UK: Longmans, Green (3rd edition).

Gilbert, G. N. J. (1957). *Factors relating to the stress/strain properties of cast iron: Research Report No. 459*. Birmingham, UK: British Cast Iron Research Association.

Gilbert, G. N. J. (1977). *Engineering data on grey cast irons – SI units*. Birmingham, UK: British Cast Iron Research Association.

Gilbert, G. N. J. (1986). *Engineering data on nodular cast irons: SI units*. Birmingham, UK: British Cast Iron Research Association.

Gloag, J. and Bridgwater, D. (1948). *A history of cast iron in architecture*. London, UK: Allen & Unwin.

Gough, H. J. (1934). Tests on cast iron girders removed from the British Museum. *Institution of Civil Engineers Selected Engineering Papers No. 161*. London, UK: ICE.

Hatfield, W. H. (1912). *Cast iron in the light of recent research*. London, UK: Charles Griffin.

Highways Agency, Scottish Executive Development Department, National Assembly for Wales, and Department for Regional Development. (2001). *The assessment of highway bridges and structures, Standard BD 21*. London, UK: The Stationery Office, and freely available to download from www.standardsforhighways.co.uk

HMSO. (1909). *London County Council (General Powers) Act 1909*. London, UK: His Majesty's Stationery Office.

Hodgkinson, E. (1831). Theoretical and experimental researches to establish the strength and best form of iron beams. *Memoirs of the Literary and Philosophical Society Manchester, 2nd Series, 5*, 407–554.

HSE. (1999). *Cast iron columns in buildings: The dangers of collapse from powered vehicle collision: HSE Information Sheet MISC 157*. Bootle, UK: Health & Safety Executive.

Institution of Structural Engineers (1996). *The appraisal of existing structures* (2nd edition; 3rd edition in preparation). London, UK: IStructE.

Lister, R. (1960). *Decorative cast ironwork in Great Britain*. London, UK: Bell.

Moy, S. S. J. (Ed.). (2001). *FRP composites: Life extension and strengthening of metallic structures: ICE design and practice guide*. London, UK: Thomas Telford.

Porter, A., Wood, C., Fidler, J. and McCaig, I. (1998). The behaviour of structural cast iron in fire: A review of previous studies and new guidance on achieving a balance between improvements in fire protection and the conservation of historic structures. *English Heritage Research Transactions, 1*, 11–20.

SCOSS. (2003). *Failure of cast iron beams: Safety Advisory Note SC/02/88*. Wirral, UK: Standing Committee on Structural Safety.

Sheppard, R. (1945). *Cast iron in building*. London, UK: Allen & Unwin.

Sutherland, R. J. M. (1979–80). Thomas Tredgold. Part 3: Cast iron. *Transactions of the Newcomen Society, LI*, 71–82.

Sutherland, R. J. M. (1985). Recognition and appraisal of ferrous metals. In *Symposium on Building Appraisal, Maintenance and Preservation*. University of Bath, UK.

Sutherland, R. J. M. (Ed.). (1997). *Structural iron, 1750–1850: Studies in the history of civil engineering, Volume 9*. Farnham, UK: Ashgate.

Swailes, T. (1996). 19th century cast-iron beams: Their design, manufacture and reliability. *Proceedings of the Instution of Civil Engineers, Civil Engineering, 114*, 25–35.

Swailes, T. (2003). 19th century 'fireproof' buildings, their strength and robustness. *The Structural Engineer, 81*(19), 27–34.

Swailes, T. (2006). *Guide for practitioners 5: Scottish iron structures*. Edinburgh, UK: Historic Scotland.

Swailes, T. and de Retana, E. A. F. (2004) The strength of cast iron columns and the research work of Eaton Hodgkinson (1789–1861). *The Structural Engineer, 82*(2), 18–23.

Swailes, T. and Parmenter, M. (1997). Full-scale loading tests on cast-iron beams and an investigation of size effects. In *Structural assessment* (K. S. Virdi, F. K. Garas, J. L. Clarke and G. S. T. Armer (eds.)), pp. 259–268. London, UK: E. & F. N. Spon.

Thorne, R. (Ed.). (2000). *Structural iron and steel, 1850–1900: Studies in the history of civil engineering, Volume 10*. Farnham, UK: Ashgate.

Timoshenko, S. P. (1953). *History of strength of materials*. London, UK: McGraw-Hill.

Tredgold, T. (1822). *A practical essay on the strength of cast iron*. London, UK: J. Taylor.

Twelvetrees, W. N. (1900). *Structural iron and steel*. London, UK: Fourdrinier.

Unwin, W. C. (1910). *The testing of materials of construction*. London, UK: Longmans.

Wallis, G. and Bussell, M. (2008). Cast iron, wrought iron and steel. In *Materials and Skills for Historic Building Conservation*, M. Forsyth (Ed.), pp. 123–159. Oxford, UK: Blackwell.

## 5.8 Bibliography

Many of the books and other publications cited as references contain numerous references and/or a bibliography. The following short list is on specific topics.

*BS EN 1560:1997. Founding. Designation system for cast iron. Material symbols and material numbers*. London, UK: British Standards Institution.

Castings Development Centre. (1997). *Data handbook for grey irons*. Sheffield, UK: CDC.

Gilbert, G. N. J. (1983). *Engineering data on malleable cast irons – SI units*. Birmingham, UK: British Cast Iron Research Association.

Swailes, T. and Marsh, J. (1998). *Structural appraisal of iron-framed textile mills: ICE design and practice guide*. London, UK: Thomas Telford.

## 5.9 Sources of advice on cast iron

The British Cast Iron Research Association was acknowledged in the first edition of this book as a source of information on cast iron. This organisation merged with Castings Technology International (formerly the Steel Castings Research and Trade Association) in 1996; after a brief period as the Castings Development Centre and also as Cast Metals Development, the enlarged trade association has been known since 2001 as Castings Technology International (www.castingstechnology.com).

The British Foundry Association, also acknowledged in the first edition, has likewise merged, in its case with the British Metal Casting Association and the British Investment Casting Association. Since 2001 the enlarged trade association has operated under the name of the Cast Metals Federation (www.castmetalsfederation.com).

Historic Scotland (www.historic-scotland.gov.uk) publishes guidance, some free, on the care of ironwork; much of this is generally applicable elsewhere.

The Scottish Ironwork Foundation (www.scottishironwork.org) offers technical advice on architectural ironwork; much of this too is generally applicable elsewhere.

The Welding Institute (www.twi.co.uk) is an authority on welding and welding procedures.

# 6 Wrought Iron

**Michael Bussell** BSc(Eng)
Formerly Ove Arup & Partners

## Contents

## 6.1 Introduction

Of all the manufactured materials covered in this book, wrought iron is arguably unique in that it is no longer made in the United Kingdom nor, in any significant quantities, anywhere else in the world. However, its inclusion here is merited as the material was used in many nineteenth-century buildings and other structures, while decorative wrought ironwork is to be found in construction dating from Norman times until very recently. Those engaged in the maintenance, conservation, and re-use of existing construction should therefore be alert to the possible presence of wrought iron, and have an understanding of its characteristics, in order that it may be looked after and continue to serve a useful life in the future.

Wrought iron is a versatile material, which can be readily hammered into a required shape when heated. As such, it was used in a huge variety of artefacts, both functional and decorative. Gates, railings, balustrades, hinges, locks, window-catches, inn-signs, weather-vanes, tie-rods, screws, nails, nuts and bolts are but a few of the many common constructional elements that were for centuries made in wrought iron at the hands of a blacksmith (Lister, 1970; Webber, 1971). Although production later took place also in iron-fabricators' workshops, this was essentially always a small-scale craft process.

The introduction of the 'puddling' furnace by Henry Cort in 1783 made possible the economical production of wrought iron on a scale that made it one of the great industries of the nineteenth century, with coal taking the place of charcoal to fuel the process. Even so, the nature of the puddling process was such that wrought iron was always made only in small batches. This has an importance influence on the structural properties of wrought iron, as discussed below.

From about 1850, wrought iron gradually took over from grey cast iron as the high-performance structural material. It continued as such until towards the end of the nineteenth century when, in turn, it was superseded by mild steel, a material of greater all-round strength capable of being formed industrially and economically into much larger structural sections. Its use gradually declined; the last wrought iron made in the United Kingdom was in 1973.

There is considerable scope for reworking of salvaged wrought iron. This is an important issue in conservation practice, as it will allow like-for-like repair of a material that is no longer being made; it is discussed more fully in Section 6.14.

When the first edition of this book was published in 1992, the author of this chapter – James Sutherland – pointed out that virtually no research had been carried out on wrought iron in the twentieth century. At that time, practical interest in the performance of existing structures was growing, spurred both by the economic need for their continued effective use (adapted for new functions as necessary), and by popular interest in the built heritage. This has generated renewed interest in the structural behaviour of wrought iron (and indeed cast iron), to which Sutherland has made a notable contribution. This revised chapter retains much of his work, while embracing more recent studies.

## 6.2 General description

Wrought iron is a relatively simple material, comprising almost pure iron with a very low carbon content, incorporating strings of slag aligned by its rolling during manufacture. Its structural properties are effectively the same in tension, compression and bending, and they are generally more predictable than those of cast iron. It is a ductile material and typically has a 'fibrous' failure surface in tension, unlike grey cast iron which fails in a brittle manner with a crystalline or granular surface (compare Figures 5.4 and 6.1).

**Figure 6.1** Typical fibrous surface of wrought iron plate exposed by cutting and then bending (courtesy of Thomas Swailes, University of Manchester)

Compared with cast iron, wrought iron is typically two to three times as strong in tension. Its strength is not much below that of mild steel, by which it was superseded, and which it broadly resembles in appearance, although it is less tough than steel when subject to impact or fatigue loading. Wrought iron is more readily worked than steel when heated; like steel it could be shaped by hot-rolling to form plates, bars, rods, wire, and structural sections such as angles, tees, and I-beams of modest size. These were typically connected by rivets or bolts. Large engineering sections could be hot-forged. Smaller functional and decorative components were shaped and hammered while hot.

Experience over time has shown that wrought iron resists corrosion somewhat better than does steel; this appears at least in part to be due to the presence of silicate slag incorporated in the iron as it is made.

The repair of wrought iron – particularly by welding – needs care, but is practicable. The traditional jointing method of riveting is nowadays a rare skill. It, and bolting, offer alternative connection methods.

### 6.2.1 Identification of wrought iron

In structural use, wrought iron can be visually distinguished from cast iron by the differing characteristics set out in Table 4.2 in Chapter 4.

Distinguishing between wrought iron and steel is less easy, as the two are generally similar in appearance. However, if the metalwork date is known then anything earlier than *c.*1880 can be identified as wrought iron with reasonable confidence, while anything later than *c.*1900 is steel. In the intervening two decades both materials were in use structurally; although, as pointed out in Section 6.6.13.1 below, for a structural assessment the two materials may be assumed to be of effectively equal yield strength.

If identification is essential, then a small sample can be carefully cut out of a lightly stressed area and removed. A hand lens applied to a polished and etched surface may distinguish wrought iron by the distinctive fine lines of slag strung out in the direction of rolling; steel is more isotropic and usually has a crystalline lustre. A spark test carried out on a clean surface with specialist equipment or observed by a skilled eye should be able to distinguish wrought iron (dull red spark) from steel (bright white). It may then be necessary to carry out chemical analysis, particularly if welding is proposed for repair or strengthening, or if there is concern about impact resistance (see Section 6.6.11).

## 6.3 Sources

The overview of iron and steel manufacture in Chapter 4 explained that early wrought iron was made in a direct reduction process by heating iron ore with timber or charcoal to produce iron with some impurities. The introduction in Europe of the blast furnace *c.* 1400 AD made it possible to smelt the ore into 'pigs' of iron with a significant carbon content, from which wrought iron was subsequently made by re-melting the pigs and removing the carbon. This, and the subsequent working of the iron into practical elements, is described immediately below.

## 6.4 Manufacturing process

The making of wrought iron was originally carried out in a 'finery', where pig iron was re-melted. Air was then blown over it, whose oxygen combined with the carbon in the molten iron as carbon monoxide and burnt off, leaving a spongy mass of almost pure iron (becoming solid as its carbon content dropped), together with impurities forming a slag.

In 1783, Henry Cort greatly improved the production of wrought iron by designing a 'reverberatory' furnace making use of coal. This was burnt in one chamber, and the flames and hot air were drawn into an adjacent chamber in which was the pig iron. A shaped roof reflected the heat downwards, melting the iron but preventing the sulphurous coal gases from tainting it by contact. To ensure that oxygen in the hot air combined with carbon in the iron, the molten metal was stirred by hand or 'puddled', using long bars. As the carbon content was reduced, the melting point of the iron increased, so that it gradually solidified to form spongy balls of iron and slag (known in the ironworking trade as 'cinder') – in a vivid image, the thickening decarburised mass was described as 'coming to nature'.

Joseph Hall found in 1816 that adding iron waste from slag to the puddling furnace accelerated the production of almost pure iron, as oxygen from iron oxides in the waste reacted energetically with and burnt off the carbon in the molten pig iron. This became known as 'pig boiling' or 'wet puddling', the latter name contrasting the process with the slower and more laborious 'dry' puddling of Cort's method.

The white-hot spongy ball of iron and slag taken from the puddling furnace weighed about 50 kg – about as much as could be man-handled. (The hard labour, the heat, the noise and the 12-hour working shifts common in ironworks combined to render the wrought iron-maker's job arguably as tough as any in heavy industry.) Held by large tongs, the ball was put under a heavy 'shingling' hammer, a form of trip-hammer, or later the direct-acting steam hammer invented by James Nasmyth in 1839 to deliver more powerful and more frequent blows. This hammering expelled much of the slag, some of which was molten, and resulted in a roughly cuboidal 'bloom'. It compacted the iron and began to transform the remaining slag from the globular or lump form, in which it had accumulated in the puddling furnace, into the threads or strands that are distinctive features of wrought iron.

The bloom was immediately hot-rolled into a bar, known as 'muck bar', a term that recognised that as yet the iron was not saleable. Next it had to be sheared into smaller bars, stacked in a pile, and reheated prior to re-rolling. This cycle was repeated up to four times. The first re-rolling produced 'crown bar' (usually marked with a crown), also known as 'merchant bar'. As its second name implied, it was of saleable quality and sold for general use. Successive re-rolling produced what was commonly classed in turn as Best, Best Best, and Best Best Best wrought iron (*sic*). It was generally accepted that re-rolling up to about six times produced progressive improvements in strength and consistency of quality, with a finer grain structure in the iron; but further re-rolling achieved little improvement, and could actually produce an inferior iron.

**Figure 6.2** Microstructure of a typical wrought iron, with threads or strand of slag aligned in direction of rolling

The rolling process oriented both the iron matrix and the remaining slag, so that these can be seen to lie in layers parallel to the direction of rolling (Figure 6.2).

It is intuitively clear that the properties of the wrought iron would differ along and across the grain, and many tests have shown this to be the case (see Sections 6.6.2 and 6.6.3 below). This was taken into account in the way that the sheared bars were stacked ready for re-rolling. So, for rods, wire and chains, the bars would all be aligned parallel to the direction of rolling. This was known as a plate pile. Often, particularly for structural I-sections and channels, some bars would be placed on edge on either side in a box pile, so that the re-rolling would improve the integrity of the iron top-to-bottom. In contrast, for boiler plates and other applications subject to two-dimensional in-plane stress, the bars were commonly stacked in alternate layers at right angles – a cross pile – with the aim of achieving similar strengths in both stressed directions.

In the same year that he had invented the puddling furnace, 1783, Henry Cort also devised a grooved rolling mill that could hot-roll the iron into useful shapes such as angles, tees and rods, rather than having to hand-forge them as was previously the case.

Forge-welding together heated pieces of wrought iron and resulted in sections larger could be made from a single bloom. This was important, particularly for structural applications. However, wrought iron products were never made on anything like the scale of either cast iron or later steel, both of which could be poured by the ton compared with the hand-held spongy ball of raw iron that came out of a puddling furnace. A telling illustration of the small-scale capacity of the individual puddling furnace is the fact that in north-east England in 1873 there were 1702 puddling furnaces – 54 in Northumberland, 508 in Cleveland, and 1140 in County Durham. Between them they produced 614,000 tons (624,000 tonnes) of wrought iron (Carr and Taplin, 1962, pp. 83–84).

This statistic reinforces the image of wrought iron as a 'small-batch' product, with the potential for variability that this inevitably implies in terms of materials, labour skill, and subsequent processing into the end-product. A calculation based on a suggestion by B. L. Hurst estimates the annual production capacity of a typical puddling furnace. Each two-hour 'heat' yielded four 50 kg balls of wrought iron from one 'charge'. So, in 12-hour shifts working seven days a week, about 8 tonnes of wrought iron was produced, i.e. around 420 tonnes a year. Annual production was some 2,300,000 tonnes in 1889, in the last great days of wrought iron manufacture. This implies that there were about 5500 puddling furnaces producing some 46,000,000 balls of wrought iron each year!

The manufacture of wrought iron changed little until its demise. The need to puddle the iron and remove it from the furnace, both tasks being done by hand, limited the size of a charge for puddling to about 200–250 kg, with each resulting ball of iron and slag weighing no more than about 50 kg. Larger sections had to be made by forge-welding two or more pieces of iron together, a process aided by the slag, which acted as a flux.

## 6.5 Chemical composition

The metallurgy of wrought iron is essentially simple. It is virtually pure iron in the form of ferrite, with usually less than 0.1% of carbon (compared with 0.1–1.8% for steel and 1.8–5.0% for cast iron). In the puddling process – particularly in Hall's 'wet-puddling' with its vigorous reaction – other elements such as silica, manganese, and the potentially embrittling phosphorus combined with oxygen and other impurities to form slag, much but not all of which was expelled by the shingling process and by subsequent reheating and re-rolling. (In practice it was impossible to remove all the slag, and indeed absolutely pure iron is soft and not really suitable for structural use.)

Residual silicon and phosphorus in the slag helped to make the iron more malleable when worked, and contributed to the ability of wrought iron to be forge-welded together when hot. The presence of the silicate slag also enhanced the corrosion resistance of the wrought iron. On the debit side, too much sulphur in the iron made it 'red-short', and likely to crumble or break up when worked hot, while too much phosphorus or silicon could make it 'cold-short' and brittle at room temperatures.

## 6.6 Physical properties

In contrast to the granular texture of cast iron, wrought iron has a fibrous texture, somewhat like timber, but this is evident only when the material has been torn, seriously corroded, or cleaned and polished for examination. A typical fibrous failure of wrought iron is shown in Figure 6.1, which may be compared with the crystalline failure surface of cast iron in Figure 5.4). (A 'crystalline' fracture of wrought iron can however occur under certain conditions; see Sections 6.6.10 and 6.6.12 below.)

Throughout the period of its practical use, the strength of wrought iron was described in Imperial units, sometimes as lb in.$^{-2}$ but more commonly in ton in.$^{-2}$. Conversion of the latter to the present-day SI unit of stress (1 N mm$^{-2}$, also written as 1 Pa) is given by 1 ton in.$^{-2}$ ≡ 2240 lb. in.$^{-2}$ ≡ 15.44 N mm$^{-2}$.

### 6.6.1 Density

BD21 (Highways Agency et al., 2001) gives wrought iron density as 7700 kg m$^{-3}$, slightly below the corresponding value for steel of 7850 kg m$^{-3}$.

### 6.6.2 Tensile strength

Many tensile tests were made on wrought iron, particularly when wrought iron was in use structurally from the 1820s to *c.*1900, but often only the ultimate strength was recorded without reference to final elongation, elastic limit or yield strength, or the elastic (Young's) modulus. Inadequate recording, and the repetition of results without giving full credits, makes it difficult either to compare these tests or to analyse them statistically. A selection of ultimate tensile strength test figures for bars, flats, angles, and the like is quoted in Table 6.1 from publications that drew together results from a number of sources, while Table 6.2 gives a selection of results for plates. These results may usefully be compared with some generalized values for strength put forward at different times and shown in Table 6.3, while Table 6.4 presents some results from more recent tests.

The wide 'spread' or range of ultimate strengths is immediately apparent, from 260 to 560 N mm$^{-2}$ (about 17–36 ton in.$^{-2}$). It might initially be thought that the strength of wrought iron would have increased as production techniques improved during the nineteenth century, but this does not appear to be the case. Some of the earliest test results show high tensile strengths for wrought iron, almost as high as those one would expect for mild steel today. Work-hardening may have contributed to these higher strengths.

What is more certain is that, with improved techniques, the strength of wrought iron became more *consistent* over time; the fairly narrow range of strength values given in the withdrawn British Standard BS 51:1939 (see Table 6.3) is probably more representative of typical wrought iron. This trend towards consistency parallels the experience of steel-making in the twentieth century. Nevertheless, because of the small-batch manufacturing process, the potential for variability of strength *within an individual structural element* must again be stressed.

*Proof loading* of important components as a means of quality assurance was quite common in the early and mid-nineteenth century. In 1821, Telford was proving the links for the Menai Suspension Bridge to a stress of 10.75 ton in.$^{-2}$ (166 N mm$^{-2}$), based on an ultimate strength of 27 ton in.$^{-2}$ (417 N mm$^{-2}$). This proof stress was comfortably below the yield found by Telford (221–329 N mm$^{-2}$, Table 6.5, as reported by Barlow), but more than double his working stress of 5 ton in$^{-2}$. Proof testing seemed to decline after the 1850s as confidence in calculation grew.

**Table 6.1** Typical test results for ultimate tensile strength of wrought iron bars, flats, angles, etc.

| *Author and date of publication* | *No. of tests* | *Ultimate tensile strength* | | *Source of results* |
|---|---|---|---|---|
| | | *Range (N mm$^{-2}$)* | *Mean (N mm$^{-2}$)* | |
| Barlow, 1837 | 9 | 420–491 | 448 | Telford / Brunton tests |
| | 7 | 407–482 | 453 | Samuel Brown tests |
| | 10 | 387–526 | 464 | Marc Brunel tests (best iron) |
| | 10 | 464–557 | 495 | Marc Brunel tests (best best iron) |
| | 6 | 417–510 | 479 | Marc Brunel tests (best iron) |
| Humber, 1870 | 6 | 285–386 | 346 | Ship straps, angles, etc. |
| | 4 | 283–303 | 291 | Very soft Swedish iron |
| | 188 | 308–476 | 397 | Bars; Kirkaldy tests |
| | 72 | 261–440 | 378 | Angle irons; Kirkaldy tests |
| Unwin, 1910 | 12 | 348–475 | 387 | Steel Committee tests |

**Table 6.2** Typical test results for ultimate tensile strength of wrought iron plate

| *Author and date of publication* | *No. of tests* | *Along or across grain* | *Ultimate tensile strength* | | *Plate thickness and relative strength across grain* |
|---|---|---|---|---|---|
| | | | *Range (N mm$^{-2}$)* | *Mean (N mm$^{-2}$)* | |
| Fairbairn, 1869 | 11 | Along | 312–407 | 348 | 6–7 mm; strength similar along and across grain |
| | 8 | Across | 286–425 | 356 | |
| Clark, 1850 | 14 | Along | 278–340 | 304 | 12–17 mm thick; strength on average 14% lower across grain |
| | 2 | Across | 258–262 | 260 | |
| Humber, 1870 | 16 | Along | 299–404 | 355 | Strength on average 8.2% lower across grain |
| | 16 | Across | 285–380 | 326 | |
| Unwin, 1910 | 51 | Along | 336–443 | 377 | Tests from Bohme, Berlin, 1884; strength on average 14% lower across grain |
| | 13 | Across | 307–360 | 325 | |

**Table 6.3** Figures for the ultimate tensile strength of wrought iron as published or specified at various times

| *Author and date* | *Element type and ultimate tensile strength (N mm$^{-2}$)* | | | | |
|---|---|---|---|---|---|
| | *Bars* | *Angles, tees, channels, etc.* | *Plates* | | |
| | | | *Along grain* | *Across grain* | *Cross-grain strength reduction* |
| Clark, 1850 | 371 | 371 | 309 | – | – |
| Humber, 1870 | 325–387 | – | 278–340 | – | – |
| Matheson, 1873 | 309–371 | 309–371 | 278–340 | – | – |
| Skelton, 1891 | 340–371 | 324–340 | 309–340 | 262–278 | 15–18% |
| Twelvetrees, 1900 | Flat: 324–371<br>Round: 293–340 | 309–355 | 278–340 | – | – |
| BS 51:1939,* amended 1948 (withdrawn) | 309–386 | 309–386 | 309–371 | 263–324 | 12–15% |

*Strength varies slightly with thickness of element: figures given cover full range.

**Table 6.4** Recent test results for the ultimate tensile strength of wrought iron

| *Author and date of publication* | *No. of tests* | *Ultimate tensile strength* | | *Source of samples, and notes* |
|---|---|---|---|---|
| | | *Range (N mm$^{-2}$)* | *Mean (N mm$^{-2}$)* | |
| Cullimore, 1967 | 10 | 278–352 | 322 | Flat tie bars from Chepstow railway bridge |
| Sandberg, 1986 (private comm.) | 4 | 327–340 | 336 | All from the same flat tie member |
| Morgan and Hooper, 1992 | 6 | 281–339 | 307 | Elements from *SS Great Britain*, including one plate, tested along grain |
| | 5 | 109–221 | 172 | As above, structural elements, across grain |
| | 1 | 225 | 225 | As above, plate, across grain |
| Morgan, 1999 | 10* | 234–417 | 368 | Bolts from 1883 bridge* |
| Moy et al., 2009 | 33 | 295–365 | † | Plate from girder bridge, along grain |
| | 27 | 260–301 | † | As above, plate, across grain |

*One specimen fractured at slag inclusion in shoulder, not in gauge length, and its result is not included here. †Range shows mean values from tests on several groups of specimens; overall mean value not reported.

### 6.6.3 Tensile strength of wrought iron plates

The rolling of wrought iron creates laminations, and the material will therefore exhibit anisotropic properties. The threads of slag have little if any effect on the strength of the wrought iron in the direction of rolling, but can reduce the strength across, i.e. perpendicular to the grain, as indicated in Tables 6.2–6.4.

In the nineteenth century it was suggested that the strength of wrought iron across the grain could be only two-thirds to three-quarters of the strength along the grain, and this cautious approach is reflected by advice in recent structural appraisal guides (Institution of Structural Engineers, 1996; Bussell, 1997). However, iron-makers were aware of this, so that for elements stressed both along and across the grain such as boiler plates, or the buckle plates used as decking in rail and road bridges, account could be taken of this strength variation by stacking the iron in a cross-pile pattern during re-rolling. This would result in some reduction of strength along the grain, compensated for by a corresponding enhancement of strength across it.

**Table 6.5** Typical test results for yield strength and ductility of wrought iron bars, flats, angles, etc.

| *Author and date of publication* | *No. of tests* | *Yield stress* | | *Elongation at failure (%)* | *Notes* |
|---|---|---|---|---|---|
| | | *Range (N mm$^{-2}$)* | *Mean (N mm$^{-2}$)* | | |
| Barlow, 1837 | 6 | 221–329 | 272 | – | Telford/Brunton tests |
| | 7 | – | – | 0.5–12.1 (mean 5.1) | Samuel Brown tests |
| | 10 | 278–417 | 340 | – | Marc Brunel tests (Best iron) |
| | 10 | 340–433 | 371 | – | Marc Brunel tests (Best Best iron) |
| | 3 | 155–186 | 170 | – | Peter Barlow (not to destruction) |
| Humber, 1870 | 4 | 166–185 | 171 | – | Very soft Swedish iron |
| Unwin, 1910 | 12 | 170–216 | 196 | – | Steel Committee tests |
| Cullimore, 1967 | 10 | 195–258 | 223 | 8–39† | Flat tie bars from Chepstow railway bridge |
| Sandberg, 1986 (private comm.) | 4 | 227–252 | 239 | – | All from the same flat tie member |
| Morgan and Hooper, 1992 | 5 | 192–230 | 211 | 13.4–34.0 | From elements of *SS Great Britain* |
| Morgan, 1999 | 10 | 216–265 | 241 | 1.9–12.4 | Bolts from 1883 bridge* |

*One specimen fractured at slag inclusion in shoulder, not in gauge length, and its result is not included here. †39% elongation obtained on specimen after machining off surface corrosion pitting.

Appraising engineers should be aware of this possible difference and, in this connection, look out for signs of lamination or splitting.

### 6.6.4 Compressive strength

Lower ultimate strengths in compression than tension are stated as fact in some early writings, and this was often repeated in later handbooks. This view was largely but not wholly dispelled by the research for the Britannia and Conway tubular bridges in the 1840s, which isolated the problem of buckling rather than 'squashing' in compression. Twelvetrees (1900, p. 60) acknowledged this: 'wrought iron, used under compressive stress in structures, seldom fails from insufficiency of strength, but rather from want of proper stiffening. The strength of wrought iron to resist compression is approximately the same as its tensile strength.'

Today, as with steel, it is generally agreed that, for practical purposes, strengths in tension and compression are the same. The last British Standard for wrought iron (BS 51:1939) specified tensile strength alone, with no mention of compression.

### 6.6.5 Flexural strength

In flexure, the tensile and compressive strengths of wrought iron – an essentially ductile material – can be taken to be the same as these strengths when axially loaded.

### 6.6.6 Shear strength

Shear strength of wrought iron has been given little attention compared with tensile strength.

For rivets, Clark (1850) quoted results of experiments made in the 1840s showing an average shear strength of 370 N mm$^{-2}$ in single shear and 340 N mm$^{-2}$ in double shear for rivet iron with a tensile strength of 340 N mm$^{-2}$. He concluded that the shear strength and the tensile strength are the same.

Later writers such as Warren (1894) and Unwin (1910) gave shear strengths of 67–77% of the tensile strength for different irons, but did not describe the test methods.

### 6.6.7 Variation of strength of wrought iron with size of sections

Most wrought iron structures were formed out of flats, plate, angles, T-sections, channels and joists, all of modest thickness (up to 25 mm and generally less). In this range the variation of strength with thickness appears to be small. Wrought iron tie-rods are seldom more than 50–60 mm in diameter, and again with these the tensile strength of the actual material seems to be sensibly constant.

However, in the case of some large forgings, the tensile strength has been shown to be appreciably reduced, even down to about 100 N mm$^{-2}$ as pointed out in a paper by Malet (1858–1859) and discussed by Unwin (1910). Appraising engineers would do well to look particularly closely at both real strength and stress levels in the comparatively rare situations where large forgings or thick tie-bars are used in wrought iron structures.

### 6.6.8 Stress–strain relationship and elastic modulus

The stress–strain relationship for a typical wrought iron is similar to that for mild steel (Figure 6.3). This shows a distinct flattening at yield and then an increase in slope, as one expects for steel. As with steel, work-hardened wrought iron is likely to exhibit a higher yield point but with a lesser elongation at failure.

Values for Young's modulus for wrought iron from some nineteenth century sources cited in this chapter range from 160 to 220 kN mm$^{-2}$, with a mean of around 195 kN mm$^{-2}$. More recent test values are in the range 173–199 kN mm$^{-2}$ (Cullimore, 1967) and 190–255 kN mm$^{-2}$ in tension, 153–180 kN mm$^{-2}$ in compression (Moy et al., 2009). The UK bridge assessment guide BD21 gives a value of 200 kN mm$^{-2}$ (Highways Agency et al., 2001).

### 6.6.9 Elastic limit, yield strength and ductility

In the nineteenth century, many of the tests on wrought iron were carried out to establish ultimate tensile strength; structural design was then typically based on dividing this value by a factor of safety – typically 4 – and using the result as a *permissible stress*.

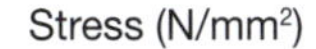


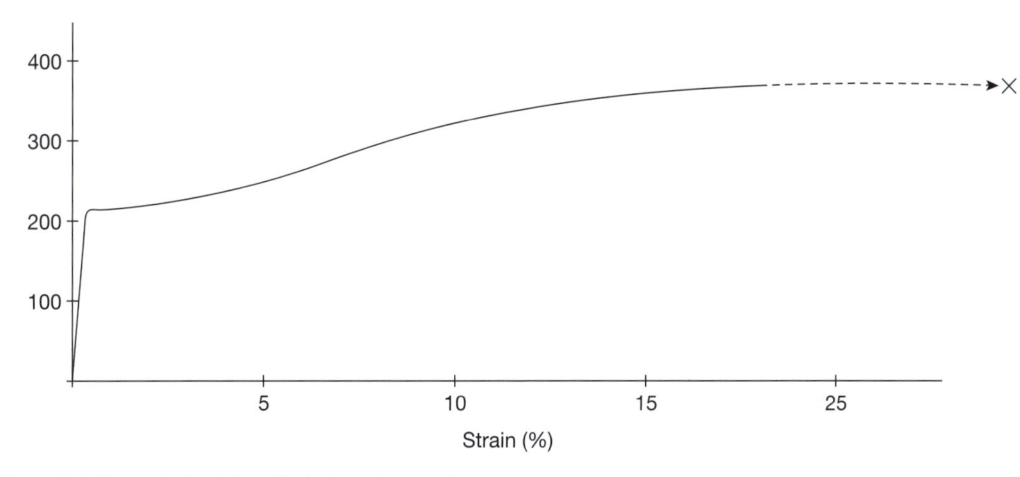


**Figure 6.3** Stress–strain relationship for typical wrought iron

There was relatively little interest at the time in establishing the elastic limit, or the yield strength on which present-day structural design of steel is based. (Recent work – Moy et al., 2009 – has been based on a 0.2% proof stress to represent yield strength, which in the past was seldom clearly defined.)

Despite this, it was widely recognised that wrought iron needed to be ductile, both to be readily hot-worked and forge-welded and to give reliable trustworthy service; so that, for example, an overloaded beam would sag – unlike a beam of brittle cast iron, which would give no evidence of distress before fracture. An informed commentator, writing in 1891, rightly argued that good ductility was to be preferred to a high ultimate strength: 'iron that shows a high breaking strain [meaning failure stress], may be practically valueless for general smithing purposes' (Skelton, 1891, p. 139).

Tables 6.5 and 6.6 show some reported test figures for the yield strength of structural rolled sections and plates, while Table 6.7 shows generalized values for strength put forward at different times.

As in the case of the elastic limit, there are few published values for elongation at failure and even fewer plots of stress against strain. Nevertheless, from the figures given in BS 51 as summarized in Table 6.7, it can be seen that an elongation at failure of between 10% and 35% was expected for linear sections

**Table 6.6** Typical test results for yield strength and ductility of wrought iron plate

| *Author and date of publication* | *No. of tests* | *Strength along or across grain* | *Yield stress* | | *Elongation at failure (%)* | *Notes* |
|---|---|---|---|---|---|---|
| | | | *Range (N mm$^{-2}$)* | *Mean (N mm$^{-2}$)* | | |
| Clark, 1850 | 12 | Along | – | – | 0.6–12.5 | Plate 12–17 mm thick |
| Unwin, 1910 | 51 | Along | 185–281 | 240 | 6.4–30.9 | Tests from Bohme, Berlin, |
| | 13 | Across | 200–237 | 215 | 1.7–16.9 | 1884; yield across grain 10% less than along it |
| Moy et al., 2009 | 33 | Along | 216–277 | † | 4.4–18.4 | Plate from girder bridge, along grain |
| | 27 | Across | 223–285 | † | 1.4–6.3 | As above, plate, across grain |

†Range shows mean values from tests on several groups of specimens; overall mean value not reported; values are measured 0.2% proof stress.

**Table 6.7** Figures for the yield strength and ductility of wrought iron as published or specified at various times

| *Author and date* | *Yield stress (N mm$^{-2}$)* | *Elongation at failure (%)* | *Notes* |
|---|---|---|---|
| Clark, 1850 | 185 | – | 'Useful' strength noted as 185 N mm$^{-2}$ |
| Matheson, 1873 | 155–170 | – | 'Permanent set' likely at 155–170 |
| Warren, 1894 | 185 | – | |
| BS 51:1939, amended 1948 | Not given | 10–35* | Bars, flats, angles, tees and channels |
| (withdrawn) | | 4–14* | Plates along the grain |
| | | 2–5* | Plates across the grain |
| BD21, 2001 | 220 | Not given | Value to be used in assessment |

*Depending on grade of wrought iron.

**Table 6.8** Effect of cold working on the properties of wrought iron

| *Element type* | *Treatment* | *Ultimate tensile strength ($N\ mm^{-2}$)* | *Elastic limit ($N\ mm^{-2}$)* | *Elongation to failure* (%) | *Source* |
|---|---|---|---|---|---|
| Bar | Hot rolled to 32 mm diameter | 402 | – | 20.3 | Unwin, 1910 (from Fairbairn) |
| | Subsequently cold rolled down to 25 mm diameter | 594 | – | 8.0 | |
| Plate | Hot rolled to 8 mm thick | 367 | 224 | 15.0 | Unwin, 1910 (from Considère) |
| | Subsequently cold rolled down to 7.1 mm thick | 460 | 408 | 7.0 | |
| Wire | Hot rolled | 283–449; mean 366 | – | – | Rankine, 1872 |
| | After cold working | 490–787; mean 635 | – | – | |

such as bars, rods, angles and channels, but possibly only 4–14% for plate along the grain and barely half those values across the grain. But, comparing these figures with an ultimate elongation of 0.75% or less for cast iron, one can see how welcome the ductility of wrought iron was to engineers when the material came into general use for bridges in the late 1840s.

Test results showed effectively identical yield stresses in tension and compression.

### 6.6.10 Effect of cold working on strength and ductility

As one might expect, cold working – whether cold rolling, cold hammering or wire drawing – has quite a marked effect on the properties of wrought iron, as with steel. Table 6.8 clearly shows the effect of cold working: it increases ultimate tensile strength and elastic limit, but reduces the elongation to failure and effectively makes the material more brittle. Wire drawing has a particularly marked effect on the strength of wrought iron, as can be seen from figures in Table 6.8, as quoted by Rankine. Strength may be more than doubled, but it is to be expected that ductility is very much reduced.

Cutting by shearing and forming holes by punching could have the same effect locally as cold rolling or hammering, i.e. the iron around the hole might become more brittle (Unwin, 1910). For this reason, some engineers insisted on drilled holes rather than punched ones. Others felt that the effect was too local to be significant.

The effects of work hardening, both beneficial and adverse, could of course be nullified by subsequent annealing.

Excessive working without adequate reheating might well be the cause of the brittle fracture observed during a laboratory test on a wrought iron beam joist (Figure 6.4).

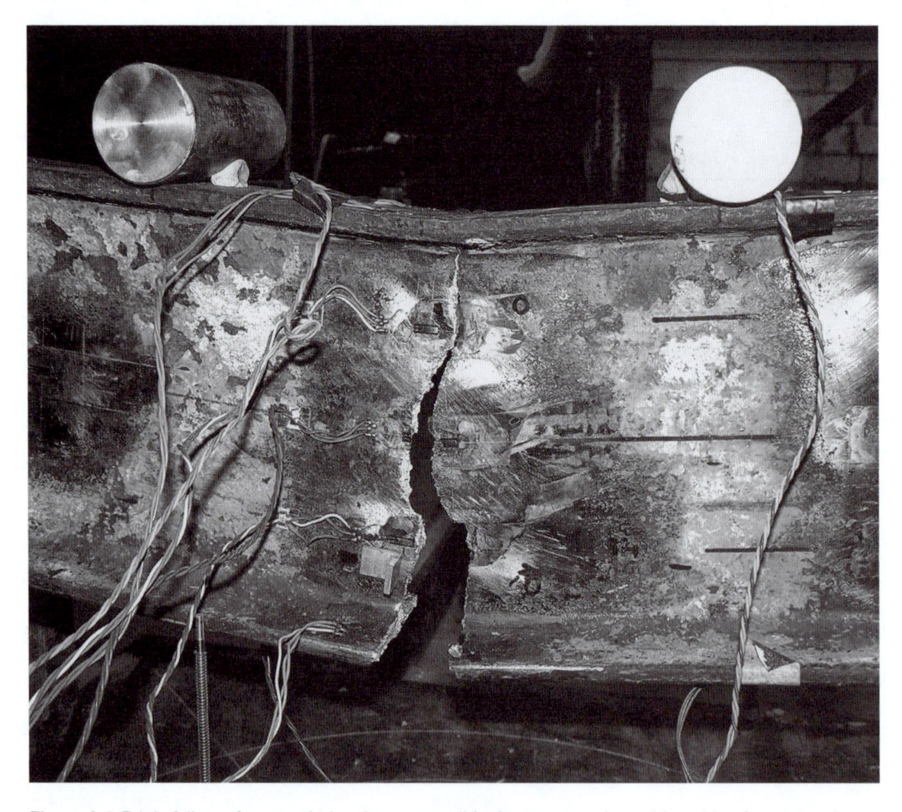

**Figure 6.4** Brittle failure of a wrought iron beam, possibly due to excessive cold-working (courtesy of Matthew O'Sullivan and Thomas Swailes, University of Manchester)

### 6.6.11 Impact resistance

Although the impact resistance of wrought iron was commonly argued to be good – and certainly vastly better than that of grey cast iron – some recent studies indicate that it can lack toughness. Limited tests by Sutherland, the original author of this chapter (unreported), have shown that wrought iron strained nearly to fracture will then break under a modest impact. A study of wrought iron from the 1843 *SS Great Britain*, now in Bristol and undoubtedly a structure, showed a poor impact toughness at room temperature, with an absorbed energy value as low as 4–6 J at room temperature, and a ductile–brittle transition temperature of about 50°C (Morgan and Hooper, 1992).

Moy et al. (2009) report along-grain Charpy values of 12–16 J and cross-grain values of 5–6 J for the flange plate of a riveted plate girder bridge.

A review of historical test data on wrought iron suggests that a high phosphorus content – an upper limit of 0.16% is suggested – will result in a drop in ductility and impact resistance (O'Sullivan and Swailes, 2008).

### 6.6.12 'Crystallisation' and fatigue

Past writers on wrought iron, notably in the nineteenth century, have alluded to the 'crystallisation' or 'embrittlement' of wrought iron over time, with its fracture being crystalline or granular, as shown in Figure 6.4 – more like a grey cast iron fracture (Figure 5.4), rather than fibrous as with typical wrought iron (Figure 6.1). Numerous papers were published on this behaviour, especially in relation to brittle failures of railway axles, whose fracture in traffic could have very serious consequences.

Today, expert opinion is inclined to the view that such obvious crystallinity existed unseen from the time of manufacture, and that in normal conditions of temperature and use the phenomenon of 'going crystalline' does not exist. If investigated today the failures in moving parts would probably be attributed to fatigue.

Railway axles for example were often made with sharp re-entrant 'shoulders' where the wheel hubs were fitted, a detail that nowadays is recognised as vulnerable to fatigue failure under the constant reversals of stress that the axles experienced in service. Rounded shoulders were found to give improved performance. Equally, the iron itself could have been over-worked or cold-worked during manufacture. The resulting material microstructure would then be more sensitive to repeated loading than adjacent ones where the structure was normally fibrous. Rankine (1872) went a long way towards dismissing the theory of progressive crystallization.

The classic work on fatigue was carried out by Wöhler and by Bauschinger, both on wrought iron and steel; their work, and that of other researchers, is described in Timoshenko (1953). More recent investigations have been mainly limited to steel, but in recent decades two interesting investigations have been carried out, one on iron from Brunel's dismantled Chepstow rail bridge and the other on his Clifton Suspension Bridge at Bristol (Cullimore, 1967; Cullimore and Mason, 1988). Both are generally reassuring in relation to the conditions to which these structures were subjected, but it could be misleading to generalise from these cases here; readers with a particular interest should refer to the published papers.

The likelihood of fatigue being a factor to consider in the appraisal of existing wrought iron structures is discussed briefly in Section 6.6.13.4.

### 6.6.13 Present-day structural appraisal of wrought iron

Structural wrought iron in the nineteenth century was designed by the understanding of the day, related to the ultimate tensile strength of the material as established by testing. Typically a factor of safety of 4 was applied to ultimate strength to derive a working or 'permissible' stress of 4–6 ton in.$^{-2}$ (62–93 N mm$^{-2}$). For railway bridges a working stress of 5 ton in.$^{-2}$ (77 N mm$^{-2}$) was laid down by the Board of Trade in 1859. This figure became almost standard in the nineteenth century in most fields, and was carried through into the 1909 London County (General Powers) Act (the so-called 'Steel Frame Act', HMSO, 1909). This was the first British prescriptive code for the structural design of buildings. The corresponding permissible stress for steel was given in this Act as 7.5 ton in.$^{-2}$ (116 N mm$^{-2}$), i.e. 50% higher than that for wrought iron.

Viewed from today, this was a 'safe' approach, because the yield stress of wrought iron is typically at least one-half of its ultimate tensile strength, so that a structure would have to be overloaded by at least 100% before showing signs of yield in the form of sagging or other distortion.

Nowadays, design practice for a ductile structural material such as steel is based on its yield strength rather than its breaking strength, while improvements in the understanding of structural behaviour and what makes for an adequate factor of safety have seen this factor reduced from 4 to, typically, 1.8–2.

It can be seen that calculations based on a working stress approach to an existing wrought iron structure may suggest that the structure is inadequate, either for a proposed new use or indeed for its continued present use. Applying these results could lead to possible extensive and costly repairs, or even a proposal for demolition and replacement. This will involve the engineer's client in expense, may mean the loss or substantial alteration of a historic structure, and is certainly not the most sustainable approach (see Section 6.14 below). All these considerations encourage appraisal of the structure based on the yield strength of wrought iron using current factors of safety, which will almost invariably result in an increased predicted strength.

This approach was adopted in the 1984 edition of BD21, *Assessment of Highway Bridges and Structures*, and remains in the current edition (Highways Agency et al., 2001). The 1984 edition was drafted with input from British Railways, London Transport and British Waterways, among others, so that – despite recent privatisation and dissociation of some transport bodies – this approach should be applicable to any wrought iron bridge. A current and widely used general guide to structural appraisal (Institution of Structural Engineers, 1996) endorses the approach, while the Steel Construction Institute's guide to the appraisal of existing iron and steel structures (Bussell, 1997) draws on the guidance originally published in BD21, which may be extended to the appraisal of building structures.

#### *6.6.13.1 Yield strength of wrought iron*

The Standards and guides just mentioned above adopt a characteristic yield stress of 220 N mm$^{-2}$ for wrought iron assessment, with a partial materials factor of 1.2.

The yield-based approach to assessment brings with it what may be the somewhat surprising position that, certainly for an initial assessment, it does not matter whether the material is wrought iron or steel. The yield strength and Young's modulus for the two materials are given in BD21 as shown in Table 6.9.

**Table 6.9** Yield strength and elastic modulus of wrought iron and pre-1955 steel

| *Material* | *Yield strength (N mm$^{-2}$)* | *Elastic modulus E (kN mm$^{-2}$)* |
|---|---|---|
| Wrought iron | 220 | 200 |
| Steel (pre-1955)* | 230 | 205 |

*Expected yield strengths for steel after 1955 are tabulated in BD21.

Given the difficulty of visually distinguishing the two materials, particularly when dating cannot be used to make a definite identification, this is a very useful convergence of properties. It may eliminate the need for material testing too (see below).

Applying the partial factors of 1.4 for dead and 1.6 for live load commonly used in building structural design today, and taking the mean of these, the *equivalent permissible (or working) stress* in wrought iron is, typically:

$$220 \div 1.2 \div 1.5 = 122 \text{ N mm}^{-2}$$

This is nearly 60% higher than the nineteenth century safe stress of 77 N mm$^{-2}$, and is still based on quite conservative partial factors of safety. The partial factors for dead and live load given in BD21 are generally slightly lower than the factors of 1.4 and 1.6 used in the comparison above, which indicates that a slightly higher equivalent working stress could be justified for appraisals. The ISE report on appraisal also discusses use of reduced partial factors of safety where conditions are known; similarly the SCI guide suggests that the partial dead load factor may be reduced from 1.4 to 1.2 when the actual dead-weight of construction is verified (Institution of Structural Engineers, 1996; Bussell, 1997).

(It is interesting to note in passing that the permissible stress for wrought iron given in the Department of Transport's earlier assessment code BE4: 1967 was 8.4 ton in.$^{-2}$ or 130 N mm$^{-2}$; this compares quite closely with the 7.9 ton in.$^{-2}$ or 122 N mm$^{-2}$ just derived above from BD21, its successor.)

BD21 guidance based on yield strength, introduced in 1984, appears to have proved acceptable for bridge assessment over more than two decades. For buildings and other structures, adoption of a yield-based assessment approach was slower. This is probably due in no small part to the values for permissible stresses inscribed in the 1909 London Act referred to above, which had appeared at a time when probably no wrought iron was being used in new structural work. No later figures for wrought iron were published having similar authority, and it is understandable that District Surveyors in London, and other building control officers, were reluctant either to agree higher permissible stresses or to consider a yield-based approach. When the SCI guide was being prepared – intended in particular to apply to buildings – it was clearly opportune to consider adopting the approach already set out in BD21. It seemed reasonable to assume that the criteria for safety in road or rail bridges, with large and varying live loads, were likewise adequate for buildings.

A draft of the SCI guide was submitted to the then Department of the Environment for review in relation to Building Regulations, and received no adverse comment. As a consequence, the guide as published offers the yield-based approach of BD21, although there it is suggested as a second-stage approach, to be considered if a first-stage working stress appraisal based on the 1909 London Act stresses does not immediately justify the structure.

The characteristic strength for wrought iron of 220 N mm$^{-2}$, originally adopted in BD21 in 1984 and later in the SCI guide, was based on over 500 tests analysed by the then British Railways Board. Most of these test results were collected by the Swiss across a number of European countries. The results of the analysis were:

- mean yield stress – 263 N mm$^{-2}$
- standard deviation – 24 N mm$^{-2}$
- characteristic yield stress – 221 N mm$^{-2}$ (95% confidence)
- lowest individual result – 184 N mm$^{-2}$.

Comparing these values with the limited number of yield figures given in Tables 6.5 and 6.6 shows a generally good measure of agreement.

The recommended yield strength of 220 N mm$^{-2}$ is a 'blanket' figure for all wrought iron. Adopting this value, and also the elastic modulus value of 200,000 kN mm$^{-2}$ from BD21 (itself close to the value of 205 000 kN mm$^{-2}$ given for steel; see Table 6.9), formulae in current codes of practice for steel may be applied to the assessment of wrought iron structures.

#### *6.6.13.2 Rivet strength*

Rivet iron was recognised as a material needing good ductility so that it would fill rivet holes when driven hot. Although rivet iron was sometimes stronger than the iron in the elements it connected, there is a good case for adopting the general figure of 220 N mm$^{-2}$. On the positive side, the ductility of rivets means that rivet groups can generally be assessed for strength on a plastic basis.

#### *6.6.13.3 Strength testing of materials*

BD21 plays down the value of testing wrought iron for strength in favour of adopting the 'prescribed' yield strength of 220 N mm$^{-2}$. Annex C warns that sampling of material from a particular structure will not necessarily give a more reliable estimate of strength than adopting this value. This is partly due to the inherent variability of wrought iron strength, resulting from the way in which it was made and worked, and partly due to statistical considerations that apply to establishing a characteristic strength for the structure based on a relatively small number of samples.

Work by Moy et al. (2009) confirmed the wide variability in material properties from samples taken from even a single flange plate of a girder bridge. In such circumstances, materials testing is unlikely to be cost-effective, and might indeed merely create uncertainty.

This advice reflects the 'small-batch' nature of wrought iron manufacture, as discussed above. Nevertheless, BD21 does warn that the iron should be examined carefully for laminations, defects and cracks, as well as corrosion. If the iron does not appear to be sound it suggests testing, and gives guidance on this. Clearly, individual appraising engineers need to exercise judgement on such matters, and seek specialist advice where necessary. And, as noted in Section 6.6.11, mechanical and chemical testing (in particular for phosphorus content) may be called for when impact resistance is a significant issue.

#### *6.6.13.4 Fatigue*

BD21 gives no positive guidance on fatigue in wrought iron bridges because of the problem of assessing past stress history. It makes reference to current bridge codes when fatigue endurance calculations are considered necessary.

In most buildings the problem of fatigue simply does not arise. Fluctuating load stresses are usually very low. Even with road bridges, the variations in stress due to vehicle loading may well be insignificant in most cases in relation to the fatigue life of the wrought iron, but appraising engineers would do well to check the possible levels of repeated change of stress as a matter of routine and consider these in relation to Cullimore's two papers (1967 and 1988) and relevant departmental standards. A recent paper on fatigue of riveted bridge girders – although for steel rather than wrought iron – offers a modified approach to fatigue assessment that may lead to enhanced life predictions (Xie et al., 2001).

However, care is needed when dealing with wrought iron that has been subject to high stresses for a long period. What has been described as 'strain age embrittlement' led to the brittle failure of such a highly stressed wrought iron beam in 1984 which injured two workmen (Bland, 1984).

## 6.7 Dimensional stability

The coefficient of linear expansion of wrought iron is of the same order as that of cast iron and steel, at $12 \times 10^{-6}$ per °C at normal working temperatures.

## 6.8 Durability

There seems to be general acceptance, based on experiences over centuries, that wrought iron has a somewhat better basic corrosion resistance than mild steel, although inferior to that of cast iron. The scientific basis for this performance does not appear to have been formally established, but it is accepted as fact by many who deal with the material on conservation projects. The durability of charcoal-smelted iron in particular may be expected to be good, as it contains no sulphur; there is evidence for this in the survival of large wrought iron obelisks and beams from medieval and earlier times in India.

Equally, in 'industrial' wrought iron made by the puddling process the original worked surface may retain millscale that provides a protective barrier, while the slag threads reduce the penetration of corrosion into the iron. Steel used in repairs to nineteenth-century wrought iron may be found half a century later to have rusted more than the original metal, so there is a good case on grounds of durability for keeping steel out of wrought iron structures wherever this is practicable, if necessary by using salvaged wrought iron (see Section 6.14).

Despite this, wrought iron is broadly as vulnerable to corrosion as steel when in aggressive conditions, and when water can lie, for example in the narrow gaps between joined sections in riveted structures where crevice corrosion can develop. The resulting rust expansion can pry the sections apart and fracture rivets. Victorian seaside piers with riveted below-deck lattice girders are a structural type that is at particular risk of crevice corrosion aggravated by salt-laden spray. Regular inspections are needed to detect damage and remedy it before expensive repairs become inevitable.

## 6.9 Use and abuse

### 6.9.1 Structural use

Early use of structural wrought iron from *c.*1800 was based on the simplest rolled sections – rods and flats used as the chains of suspension bridges, and tie-rods and fixings used in combination with timber or cast iron. The Conwy suspension road bridge in North Wales, designed by Thomas Telford and opened in 1826, is a good example of a confident use of wrought iron using simple components (Figure 6.5). Large-scale manufacture of rolled plate went almost in parallel, and was followed by bulb angles, tees, channels and joists.

It may seem curious that Henry Cort's remarkable improvements in the manufacture and forming of wrought iron in the 1780s – his puddling process, and his grooved rolling-mill – did not have much impact on the construction industry until at least the late 1830s, but there probably seemed little incentive to use a more expensive material to replace an established cheaper one, namely cast iron, which was apparently performing well.

It was the unprecedented traffic loadings of the railways, and the increasingly 'bad press' applied to cast iron following some notable collapses of cast iron structures in the middle 1840s, that did most for the wrought iron industry in Britain.

The real structural breakthrough came after the very successful completion of the long-span Conway and Britannia tubular railway bridges in North Wales in 1848 and 1850. They were not the first wrought iron girder bridges to be built, but it was the research studies for their design by Stephenson, Fairbairn and Hodgkinson between 1845 and 1847 that, more than anything else, established the use of wrought iron in construction. The tubes were irreparably damaged by fire in 1970, and were replaced by steel arches that

**Figure 6.5** Conwy Bridge, North Wales, by Thomas Telford (1826)

now carry both railway lines and a roadway. However, a small section of one tube has been re-erected close to the bridge, which allows a sense of its scale to be gained (Figure 6.6).

From the late 1850s until towards the end of the nineteenth century, riveted wrought iron was the 'advanced' structural material. Figures 6.7–6.9 show three examples of the application of wrought iron in large-scale structures in mid-century. With its superior tensile strength and ductility, the material found wide use in bridges and as beams, girders, and roof structures in buildings. Roof structures could be surprisingly delicate, as for example in Figure 6.7. Wrought iron however was little used for columns, for which cast iron with its superior compressive strength remained the common choice. Cast iron also continued in use for arches, brackets and similar small components where tensile stresses were very low or did not exist.

The availability of cheap steel from the 1880s led in turn to the decline of wrought iron as a structural material. However, with their similar nature and properties, the transition from wrought iron to steel was more evolutionary than the more visible change from cast iron elements (typically a single, sometimes complex, casting) to rolled wrought iron (typically made up of plates, angles, and sometimes rolled sections all riveted together).

Sutherland (1997) and Thorne (2000) have edited two volumes of authoritative and informative papers describing notable uses of structural wrought iron from 1750–1850 and 1850–1900, the latter volume also covering the advent of structural steel.

## 6.9.2 Connections

Traditional methods of joining wrought iron were forge-welding and riveting.

Forge-welding was essential to the making of wrought iron, being needed to fuse blooms into larger sections as described in Section 6.4. It was also used in fabrication, for example to forge a threaded end onto a tie-rod. Hot-working was also much used to shape both structural and decorative wrought iron; it was of course an integral part of the blacksmith's work, as well as that of now-extinct specialist crafts such as nail-making and chain-making.

**Figure 6.6** The surviving section of the huge tubular girders of the 1850 Britannia Railway Bridge over the Menai Strait, re-erected close to the replacement steel arch bridge after the tubes were irreparably damaged by fire in 1970

**Figure 6.7** Huddersfield Station roof of 1850 – delicate wrought iron trussing

**Figure 6.8** The 73 m span roof of St Pancras Station, completed in 1868 and recently restored to accommodate London's international passenger rail terminal

**Figure 6.9** The 1859 Royal Albert Bridge across the River Tamar west of Plymouth; two self-anchored 'suspension arches', each spanning 139 m, still carrying mainline rail traffic to and from Cornwall

Once wrought iron sections had been shaped, they were most commonly joined by riveting. Bolted connections might be adopted to suit the conditions on a particular site; adjustable wedged ends to tie-rods were also common. Making a riveted connection involved hammering to form a second head onto a red-hot headed shank placed in a hole in the parts to be joined. Riveting was carried out both in the iron fabricator's workshop and on site.

### 6.9.3 Ironmongery and decorative work

Wrought iron, being readily workable when heated, was a very versatile material and, as noted in the Introduction above, found wide use in many fields of construction in addition to its structural applications (Lister, 1970). The reputation of the material for decorative work in particular is attested by the term 'wrought iron' being often applied to present-day work such as gates and railings that have actually been made in steel!

## 6.10 Proprietary brands and British Standards

In the period when wrought iron was in widest use, from the late eighteenth century to *c.*1900, quality was defined by the ironmaker's own name – and reputation – and the trade-wide accepted grading of iron depending on how often it had been re-rolled, as described above. Among the best-quality wrought iron available was that made at the Lowmoor and Farnley works in Yorkshire, which were styled 'Best-Yorkshire', with excellent ductility and above-average toughness. As such it was valued for safety-critical applications such as railway couplings and chains. This quality did not come cheaply – Skelton (1924, p. 143) notes that for this wrought iron 'the market price is about double the cost of the best Staffordshire bar iron' – the latter itself regarded as a superior product.

Major users of wrought iron laid down their own requirements, typically defining ultimate tensile strength, elongation or contraction at failure, and bend tests both hot and cold. Such bodies included the Admiralty, the War Office, the Indian Government, and Lloyds.

The British Standard Report No. 51 of 1913, later to became BS 51, began the process of rationalising the grading of wrought iron away from repeated use of the term 'best', although the initial classification still acknowledged the superiority of Yorkshire wrought iron, with the grading, in descending order of quality, being:

- Best-Yorkshire – 'highest possible grade of iron'
- Grade A – 'iron for chains, cables and best smithy work'
- Grades B and C – 'iron for general purposes' (quoted in Skelton, 1924, p. 195).

BS 51 retained the A, B and C grading in its 1939 edition and the 1948 revision.

## 6.11 Hazards in use

Wrought iron is in itself not a hazardous material. However, toxic materials such as lead might well be present in paint coatings (see below).

Health and safety authorities today would doubtless look askance at some of the processes involved with wrought iron. Puddling molten iron and handling white-hot balls of spongy iron weighing about 50 kg with tongs were hot, strenuous and hazardous activities. Equally the process of riveting – so widely used for connections in both the fabricating shop and on site – would probably be unacceptable if suggested as a new technique today. Throwing, catching, and placing red-hot rivets was a skilled (and dangerous) task, while closing them involved noise and impact, aggravated when the pneumatic riveting tool replaced the sledge-hammer, greatly increasing the frequency of blows. It is perhaps fortunate that such techniques are no longer practised!

## 6.12 Coatings for corrosion protection

Despite its somewhat superior durability compared with steel, wrought iron will nevertheless corrode at a greater or lesser rate, depending on the severity of exposure conditions, and so throughout its use in construction it has generally received protective treatment. Provided that this has been well maintained, future treatment may be based on rubbing-down and application of further coatings selected for suitability and compatibility. In broad terms, its treatment can be similar to that applied to steel.

It might be tempting to begin by deciding on a complete strip-down of existing coatings by abrasive blasting back to bare metal. There are persuasive arguments to be considered before this approach is adopted, as follows.

- Removal of surface millscale and bonded rust on the wrought iron will reduce the inherent corrosion resistance of the material.
- Current conservation thinking is that the multi-layered coatings on a typical element are part of the history of the fabric, which can repay study and therefore should not be arbitrarily destroyed.
- The existing coatings may well contain lead, chromium, or other toxic materials; if so, their removal in situ will require thorough health and safety measures to avoid contamination or pollution by these products, which could include encapsulation of the working areas, measures to collect all loose material as it is removed from the ironwork, and protective clothing and health monitoring of the work force.

**Table 6.10** Strength of wrought iron and steel with increase in temperature

| *Temperature (°C)* | *Material strength as percentage of strength at 20°C (=100%)* | | |
|---|---|---|---|
| | *Wrought iron (Roelker)* | *Wrought iron (Franklin Institute)* | *Bessemer steel (Roelker)* |
| 400 | 72 | 83 | 62 |
| 600 | 22 | 15 | 27 |

This is not to argue against any treatment of existing coatings, which if damaged or missing will necessitate preparation before a new coating scheme can be applied. It may be appropriate to remove material where excess paint has obscured crisp detail, and to expose and clean areas where corrosion has occurred. Crevice corrosion is a particular concern, when water has become trapped in the thin spaces between adjacent elements – for example at the intersection of lattice girder diagonals, connected by rivets, or in the overlaps and junctions between ornamental ironwork in gates and railings. Here it may well be necessary to use a combination of mechanical, chemical and heat treatments to remove paint and corrosion product.

## 6.13 Performance in fire

The behaviour of wrought iron in fire is similar to that of structural steel, with its strength dropping sharply above about 400°C. Twelvetrees (1900) reproduced figures from late-nineteenth-century tests comparing the strength of wrought iron and steel as a percentage of their strengths at 20°C, shown in Table 6.10.

In general, fire protection measures suitable for structural steel are equally applicable to wrought iron.

## 6.14 Sustainability and recycling

### 6.14.1 Retention in place

The most sustainable approach to wrought iron in existing construction must be to maintain it properly so that it continues to give satisfactory performance. This approach satisfies the key considerations of economy, appearance, and the retention of what might well be historic fabric in a material that is no longer made.

Inspection, assessment, and treatment of wrought iron are all best carried out by those with previous experience of the material and its characteristics, although it is hoped that this chapter will provide a useful introduction for those unfamiliar with wrought iron. An informed approach to strength assessment, for example, should give a structure the best chance of being shown to be adequate, or of requiring only modest repair or strengthening.

### 6.14.2 Recycling

On occasions where the existing wrought iron structure has to be removed, or is judged to be beyond remedy – for example due to excessive and widespread corrosion, or a gross increase in loading requirements – then it could be offered to a commercial organisation or museum that can rework it (some are listed at the end of this chapter). Readers engaged in the demolition or alteration of wrought iron structures could assist this process by offering unwanted material, particularly in larger thicknesses and sections, for recycling. This will help to ensure that the finite amount of surviving wrought iron is retained in circulation for as long as possible. It might be noted too that the price offered for wrought iron for re-use is likely to be higher than its scrap price.

### 6.14.3 Maintenance

For exposed wrought iron the principal concern will be to maintain existing paint coatings and to ensure that water is excluded or shed as rapidly as possible.

As with all structures, regular inspection is the first essential, so that necessary maintenance can be identified and carried out before serious deterioration occurs.

### 6.14.4 Repairs to wrought iron

Repairs to structural wrought iron are likely to involve either the replacement of locally lost or corroded metalwork or the addition of elements to augment the capacity of the existing structure. As noted in Section 6.8, experience has shown that introducing steel into wrought iron structures may lead to accelerated corrosion of the repair material. Wherever possible, wrought iron should be used in repairs.

Removal of corroded or damaged rivets or bolts, if necessary, may be carried out by several methods. Drilling-out does require careful location of the drill, centred over the shank, to minimise damage to the parent metal. Burning-off of the heads requires skill to avoid damage to adjacent metal, followed by drilling-out or pressing-out of the residual shank. At least one specialist firm can offer a 'rivet-buster', a pneumatic tool that shears off the rivet head.

The traditional methods used for joining wrought iron sections, forge-welding and riveting are virtually extinct and in any event may not be applicable when repairs are being carried out to corroded or broken elements. Forge-welding requires the components to be heated in a hearth before they can be hammered together to form a reliable join, and this is clearly not practical for anything other than small sections that can be removed and replaced without difficulty. For smaller sections, riveting is likely to result in a conspicuous and unsightly repair, and is a scarce (but not entirely vanished) skill.

The standard methods of joining steel are by bolting and electric-arc welding, and these can be applied to the repair of wrought iron.

Bolting is a straightforward process and is particularly applicable to the repair and strengthening of structural wrought iron. (However, as with riveting, it may well be an inelegant solution for small sections.) The dome-headed tension-controlled bolt – a variant of the high-strength friction grip bolt – offers, like riveting, the assurance of a connection that will not loosen in service. For historically important wrought ironwork, the dome-headed bolt also results in a better visual match to the original rivets than is offered by the conventional bolt-head, on one face at least.

Electric-arc welding can be applied to wrought iron, but this does require a careful choice of suitable electrodes and suitably skilled welders. Two special problems with the welding of wrought iron, as opposed to steel, are the laminar nature of the wrought iron and the presence of the threads of slag. The laminar structure militates against the effectiveness of fillet welding onto a wrought iron surface when significant stresses are to be applied. Partial- or full-penetration butt welds across a section perpendicular to the grain are more promising because, with adequate preparation, the weld metal should link the iron laminae longitudinally across the whole cross-section. However, because the slag melts at a temperature below the fusion temperature of the iron, there is a danger during the process that the slag can flow across the weld face and result in a weakened connection. Advice may usefully be sought from a specialist body such as the Welding Institute, and trials may well repay the time and cost before the repairs proper are undertaken.

A fuller description of these and other techniques for the repair of both structural and non-structural cast iron is offered in Swailes (2006) and Wallis and Bussell (2008). For decorative wrought iron, Historic Scotland issues several useful guides to repair of non-structural iron, downloadable from its website.

A technique that offers promise for reducing the repair burden on structures subject to aggressive exposure conditions is that of cathodic protection, in which the electrolytic action causing corrosion is effectively suppressed by inducing an electric current in the metalwork. Two methods are available. Impressed current cathodic protection involves applying a relatively low voltage (20–30 V) to the metal, in conjunction with an anode, typically of titanium mesh or rods. This method should be applicable to the frames of building structures and other wrought iron elements such as ornamental gates, where electrical continuity can be achieved. Necessarily this method requires the permanent application of an electrical current. The alternative method, sacrificial cathodic protection, involves the provision of a suitable anode such as a zinc block which – as the name implies – corrodes preferentially, avoiding corrosion of the ironwork being protected. Again, electrical continuity is essential. This approach is well suited to piers and other marine structures. It does of course require the periodic replacement of the zinc as it is lost through corrosion.

Gibbs (2000) provides an introduction to the subject, particularly applicable to buildings, while the Corrosion Prevention Association offers broader guidance.

A relatively new technique for strengthening structures is the application of externally bonded fibre-reinforced polymers (FRP). Its use on wrought iron has to date been constrained by concern over the possible risk of delamination of the wrought iron, which also discourages the use of fillet welding in repair work. It is perhaps significant that a major study of metallic structure strengthening by FRP identified no example applied to wrought iron (Cadei et al., 2004, p. 180).

The repair of wrought iron – whether it be structural, otherwise functional, or decorative – is a fairly specialised field with relatively few practitioners. Periodical publications such as the *Building Conservation Directory* contain listings of specialists. Reputable firms should have no hesitation in describing past work and providing contact with previous clients.

## 6.15 Acknowledgements

As in the preceding chapter on cast iron, the first acknowledgement must be to the author of this chapter in the first edition of this reference book, James (R. J. M.) Sutherland, whose knowledge and enthusiasm have inspired many – including the present writer – to learn more about historic wrought iron. Revising this chapter for this new edition has necessarily required coverage of developments since the first edition. These have mainly concerned the appraisal of structural wrought iron, the relatively few but valuable recently published studies of the material's engineering properties, and growing experience in the conservation and repair of historic wrought iron structures. Nevertheless, much of what James Sutherland wrote nearly two decades ago has stood the test of time, and is retained here. He has also kindly reviewed and made helpful comments on the revisions to his original text.

Thanks are due to Lawrance (B. L.) Hurst for his thought-provoking calculation, quoted in Section 6.4, from which the annual production of individual 'balls' of wrought iron, each effectively 'hand-made', is estimated at about 46,000,000 at the peak of UK production in the late 1880s. It is little wonder that the properties of the material are found to be variable!

## 6.16 References

Barlow, P. (1837). *A treatise on the strength of timber, cast iron, malleable iron and other materials*. London, UK: John Weale.

Bland, R. (1984). Brittle failure of wrought-iron beams. *The Structural Engineer*, *62A*(10), 327.

*BS 51:1939 Wrought iron for general engineering purposes*. London, UK: British Standards Institution (amended 1948; since withdrawn).

*Building Conservation Directory* (annual). Salisbury, UK: Cathedral Communications.

Bussell, M. (1997). *Appraisal of existing iron and steel structures*. Ascot, UK: Steel Construction Institute.

Cadei, J. M. C., Stratford, T. J., Hollaway, L. C. and Duckett, W. G. (2004). *Strengthening metallic structures using externally bonded fibre-reinforced polymers*. London, UK: Construction Industry Research and Information Association.

Carr, J. C. and Taplin, W. (1962). *History of the British steel industry*. Oxford, UK: Blackwell.

Clark, E. (1850). *The Britannia and Conway tubular bridges*. London, UK: Day and Son.

Cullimore, M. S. G. (1967). The fatigue strength of wrought iron after weathering in service. *The Structural Engineer*, *45*(5), 193–199.

Cullimore, M. S. G. and Mason, P. J. (1988). Fatigue and fracture investigation carried out on the Clifton Suspension Bridge. *Proceedings of the Institution of Civil Engineers, Part 1*, *84*, 303–329.

Fairbairn, W. (1869). *Iron: Its history, properties and processes of manufacture*. Edinburgh, UK: A. & C. Black.

Gibbs, P. (2000). *Technical Advice Note 20: Corrosion in masonry clad early 20th century steel framed buildings*. Edinburgh, UK: Historic Scotland.

Highways Agency, Scottish Executive Development Department, National Assembly for Wales, and Department for Regional Development. (2001). *The assessment of highway bridges and structures, Standard BD21*. London, UK: The Stationery Office, and freely available to download from www.standardsforhighways.co.uk.

HMSO. (1909). *London County Council (General Powers) Act 1909*. London, UK: His Majesty's Stationery Office.

Humber, W. (1870). *Cast and wrought iron bridge construction*. London, UK:Lockwood.

Institution of Structural Engineers. (1996). *The appraisal of existing structures*. London, UK: IStructE (2nd edition; 3rd edition in preparation).

Lister, R. (1970). *Decorative wrought ironwork in Great Britain*. Newton Abbott, UK: David & Charles.

Malet, R. (1858–1859). On the coefficients $T_e$ and $T_r$ of elasticity and of rupture of wrought iron. *Proceedings of the Institution of Civil Engineers*, *XVIII*, 296–330.

Matheson, E. (1873). *Works in iron: Bridge and roof structures*. London, UK: E. & F. N. Spon.

Morgan, J. (1999). The strength of Victorian wrought iron. *Proceedings of the Institution of Civil Engineers, Structures & Buildings*, *134*(Nov.), 295–300.

Morgan, J. E. and Hooper, J. M. (1992). SS Great Britain – The wrought iron ship. *Metals and Materials*, *8*(12), 655–659.

Moy, S. S. J., Clarke, H. W. J. and Bright, S. R. (2009). The engineering properties of Victorian structural wrought iron. *Proceedings of the Institution of Civil Engineers*, *162*, 1–10.

O'Sullivan, M. and Swailes, T. (2008). A study of historical test data for better informed assessment of wrought-iron structures. In *Proceedings of the VIth International Conference on Structural Analysis of Historical Construction* (D. D'Ayala and E. Fodde (Eds)), Vol. 1, pp. 207–215. London, UK: Taylor & Francis.

Rankine, W. J. M. (1872). *A Manual of civil engineering*. London, UK: C. Griffin.

Skelton, H. J. (1891). *Economics of iron and steel: Being an attempt to make clear the best every-day practice in the heavy iron and steel trades, to those whose province it is to deal with material after it is made*. London, UK: Biggs & Co.

Skelton, H. J. (1924). *Economics of iron and steel: Being an exposition of every-day practice in the heavy iron and steel trades*. London, UK: Stevens & Sons and H. J. Skelton & Co..

Sutherland, R. J. M. (Ed.). (1997). *Structural iron, 1750–1850: Studies in the history of civil engineering, Volume 9*. Farnham, UK: Ashgate.

Swailes, T. (2006). *Guide for practitioners 5: Scottish iron structures*. Edinburgh, UK: Historic Scotland.
Thorne, R. (ed.) (2000). *Structural iron and steel, 1850–1900: Studies in the history of civil engineering, Volume 10*. Farnham, UK: Ashgate.
Timoshenko, S. P. (1953). *History of strength of materials*. London, UK: McGraw-Hill.
Twelvetrees, W. N. (1900). *Structural iron and steel*. London, UK: Fourdrinier.
Unwin, W. C. (1910). *The testing of materials of construction*. London, UK: Longmans.
Wallis, G. and Bussell, M. (2008). Cast iron, wrought iron and steel. In *Materials and skills for historic building conservation*, M. Forsyth (Ed.), pp. 123–159. Oxford, UK: Blackwell Publishing.
Warren, W. H. (1894). *Engineering construction in iron, steel and timber*. London, UK: Longman.
Webber, R. (1971). *The village blacksmith*. Newton Abbott, UK: David & Charles.
Xie, M., Bessant, G. T., Chapman, J. C., and Hobbs, R. E. (2001). Fatigue of riveted bridge girders. *The Structural Engineer*, *79*(9), 27–36.

## 6.17 Bibliography

Benson, R. (1967). Large forgings of wrought iron. *Transactions of the Newcomen Society*, *40*, 1–14.
Gale, W. K. V. (1964). Wrought iron: A valediction. *Transactions of the Newcomen Society*, *36*, 1–11.
Gale, W. K. V. (1965). The rolling of iron. *Transactions of the Newcomen Society*, *37*, 35–46.
Kirkaldy, D. (1863). *Results of an experimental enquiry into the tensile strength and other properties of various kinds of wrought-iron and steel*. Author.
Mott, R. A. (1979). Dry and wet puddling. *Transactions of the Newcomen Society*, *49*, 153–158.
*Report of the commissioners appointed to inquire into the application of iron to railway structures, 1849*. London, UK: Her Majesty's Stationery Office.
Sutherland, R. J. M. (1963–1964). The introduction of structural wrought iron. *Transactions of the Newcomen Society*, *36*, 67–84. (Also in (1997) *Structural iron, 1850–1950: Studies in the history of civil engineering, Volume 9*. (R. J. M. Sutherland, Ed.), pp. 271–291. Farnham, UK: Ashgate.)

## 6.18 Sources of advice on wrought iron

Historic Scotland (www.historic-scotland.gov.uk) publishes guidance on the care of ironwork, both free and to purchase.
The Corrosion Prevention Association (www.corrosionprevention.org.uk) publishes advice on corrosion prevention methods.
The Scottish Ironwork Foundation, Stirling (www.scottishironwork.org) offers architectural and technical advice on ironwork, much of which is generally applicable elsewhere.
The Real Wrought Iron Company (www.realwroughtiron.com), with its associated company Chris Topp & Co. (www.christopp.co.uk), provides information on wrought iron, its characteristics, and its care and repair.
The Welding Institute (www.twi.co.uk) is an independent organisation advising on welding and welding procedures.

## 6.19 Some commercial organisations working with salvaged wrought iron

Two commercial organisations (among others) that use wrought iron in repair and replacement work, and will be interested in acquiring salvaged wrought iron for future use, are:

- The Real Wrought Iron Company (www.realwroughtiron.com) – which also has a hot rolling mill
- Dorothea Restorations Ltd (www.dorothearestorations.com).

## 6.20 Museums having particular connections with wrought iron

Two museums that periodically arrange working demonstrations with wrought iron are:

- The Black Country Living Museum (Dudley, West Midlands, www.bclm.co.uk) – holds machinery and material from this area, once a major centre for prolific iron-making and iron-working, with numerous specialist crafts such as the making of chains and nails.
- The Ironbridge Gorge Museum (Telford, Shropshire, www.ironbridge.co.uk) – this museum has two puddling furnaces, a steam-hammer for shingling, and rolling mills, all from Thomas Walmsley's Atlas Ironworks in Bolton, Lancashire; the closure of this works in 1973 marked the end of wrought iron-making in the United Kingdom.

# 7 Steel

**Roger Plank** PhD BSc(Eng) CEng FIStructE MICE
Professor of Civil Engineering, University of Sheffield

## Contents

## 7.1 General description

Steel is an alloy of iron (Fe) with carbon (C) and various other elements, some of them being unavoidable impurities while others are added deliberately. The properties of the material are significantly affected not only by this chemical composition but also by thermal and mechanical treatments. Processes that improve one property may adversely affect others so compromises must be struck, and it is important that an appropriate grade of steel be specified depending on its intended use. A wide range of steels is therefore available to suit a variety of applications, and steels for construction form a subset of this range. *Smithell's Metals Reference Book* (Gale & Totemeier, 2004) provides comprehensive information on all steels.

Steel has a number of properties that make it suitable for use in construction: it has a high strength to weight ratio, good stiffness, and is ductile, tough, and resilient. It can also be readily formed into useful shapes, and when a product is no longer required, the material is easily recycled with no loss of quality.

## 7.2 The manufacturing process

The production of steel components for use in construction involves a number of distinct stages including primary and secondary steelmaking, casting, forming, and finishing.

### 7.2.1 Primary steelmaking

Two alternative routes are used for large-scale commercial steel production, namely the basic oxygen steelmaking (BOS) process and the electric arc furnace (EAF). The BOS route is a two-stage process, and accounts for most steel production. It is often referred to as the 'integrated steelmaking' route, involving first ironmaking and then conversion of the iron to steel. The basic raw material is iron ore, which takes the form of a variety of oxides of iron. Liquid iron, sometimes referred to as 'hot metal', is produced by heating the iron ore with coke and lime in a blast furnace. This has a carbon content of about 4–4.5% by weight and contains other impurities, including sulphur and phosphorus. As a result the material is brittle and unsuitable for most engineering applications. The BOS steelmaking process therefore involves a second stage to refine the iron and in particular to reduce the carbon content, typically to less than 1.5%, by blowing oxygen through the metal in a converter. In order to control production temperatures, the converter is first charged with a significant quantity of scrap steel, typically about 25% of the total charge, before the liquid iron is added. The steel at this stage generally contains between 0.03% and 0.07% of carbon, which is insufficient for many applications. Carbon must therefore often be *added* to the steel again during secondary steelmaking.

The alternative electric arc furnace route uses mainly scrap iron and steel as its feedstock, although iron, typically in the form of pellets, is used for critical steel grades to control the level of impurities. The scrap steel charge is carefully controlled according to chemical composition (e.g. low alloy, stainless), level of impurities (e.g. sulphur, phosphorus and copper), size and shape, and variability The 100% scrap charge makes the process more vulnerable to injurious tramp elements such as copper, nickel and tin which are difficult to remove because they are more stable than iron. To control these it is important to identify the sources of the scrap used in the production and to use a quality of scrap consistent with the required properties of the steel.

The cold charge is melted by electric arcs generated by graphite electrodes. Ultra high power (UHP) furnaces have been developed for mass steel production, creating the so-called mini-mills which can produce up to 200 tonnes of liquid crude steel in less than two hours.

Electric arc furnaces account for about 30% of steel production, and this is limited by the supply of acceptable scrap.

### 7.2.2 Secondary steelmaking

The liquid steel contains a small volume of impurities present in the feedstock, notably sulphur. It is further refined and its composition adjusted in subsequent processes referred to as secondary steelmaking. In this process, some elements are added and others removed in order to achieve the required specification, and the temperature, internal quality and inclusion content have to be carefully controlled. A variety of processes may be used, depending on the type of steel being produced. These include stirring, injection, and degassing. Ladle stirring is essential to homogenise composition and temperature, and assist the removal of inclusions. It is achieved by bubbling argon through the liquid steel or by electromagnetic stirring (EMS) and helps sweep out the unwanted oxygen, nitrogen and hydrogen, reducing the need for deoxidising additions. Alloying elements are added by injection as powder or wire, while vacuum degassing holds the melt under reduced pressure, removing residual gases that would otherwise yield a very porous product.

The secondary steelmaking process deoxidises the melt by the addition of manganese and silicon. These react with dissolved oxygen to form insoluble particles of oxides, suppressing the formation of gas bubbles of carbon monoxide and carbon dioxide. The deoxidising additions are usually made in the form of ferro-alloys which also contain carbon. This limits the amount of deoxidiser which can be added without changing the carbon content unacceptably. In such circumstances, further deoxidation may be achieved by adding aluminium or other elements having a high affinity for oxygen. These elements also help control grain size during hot rolling. Deoxidised steel is referred to as 'killed' or 'semi-killed'.

Manganese also reacts with the small amount of sulphur remaining in the steel, forming manganese sulphide. This prevents lenses of iron sulphide from forming between the iron crystals, which would create processing problems and have a deleterious effect on the properties of the steel, including embrittlement.

Carbon and manganese have a great influence on the properties of structural steels. In general carbon improves strength and manganese gives low-temperature toughness, and it may be necessary to control carefully the content of these elements to give the grade of steel required. In recent years so-called micro-alloyed steels or HSLA (high strength low alloy) steels have been developed. These steels are normalised or controlled rolled carbon–manganese steels which have been 'adjusted' by micro-alloying to give higher strength and toughness, combined with ease of welding. Small additions of aluminium, vanadium, niobium or other elements are used to help control grain size. Sometimes about 0.5% molybdenum is added to refine the microstructure.

### 7.2.3 Casting

After secondary steelmaking, the liquid steel is allowed to cool and solidify in the casting process. Traditional ingot casting has largely been replaced by continuous casting, in which the molten steel is formed directly into shapes that are more readily transformed into the final forms needed for practical applications. In this process the molten steel is delivered via a tundish into a water-cooled mould. This forms a thin solid shell around the molten metal, enabling it to be passed through a set of water-cooled rolls that gradually shape the steel into a solid form. This emerges as a 'strand' which can have a wide range of cross-sectional shapes depending on the final form intended. For example, the steel would be in the form of 'slabs' for subsequent rolling into plate and strip, 'blooms' for sections such as beams and columns, and 'billets' for products such as rod and wire. Processes have also been developed for continuously casting steel directly into thin strip (< 3 mm thick). As the strand emerges from the continuous casting process it is cut into suitable lengths for subsequent shaping.

### 7.2.4 Shaping steel into standard forms

The product that emerges from the continuous casting process has to be formed into usable forms; this is typically done by heating the steel to soften it, then physically shaping it into the required form. Both the thermal and mechanical treatments associated with this will influence the microstructure of the steel and hence its properties.

For construction products rolling is the most common method used for shaping and is particularly suitable for large-scale production of simple constant cross-sections, such as beams, columns, plate and sheet. It can be done hot or, for thinner sections, cold.

### 7.2.5 Hot rolling

Most structural sections for construction are formed by hot rolling in a range of standard sizes. The steel is preheated to 1200–1300°C and then passed through sets of rollers which gradually change the profile into the required section. The temperature of the steel falls as the rolling process continues, and reheating may be necessary. In the process known as controlled rolling, finishing temperatures are carefully controlled and are generally around 850°C. This gives a very fine grain size, which is associated with a good combination of strength and toughness.

Hot rolled sections are produced in a wide range of shapes and sizes (Figure 7.1). These are typically identified by section type (e.g. universal beam or universal column), serial size and mass per unit length. Each serial size corresponds to a different set of roller sizes, while changes in mass per unit length are achieved by increasing the thickness of the flanges during rolling. A typical designation therefore might appear as: 457 × 152 × 46 UB.

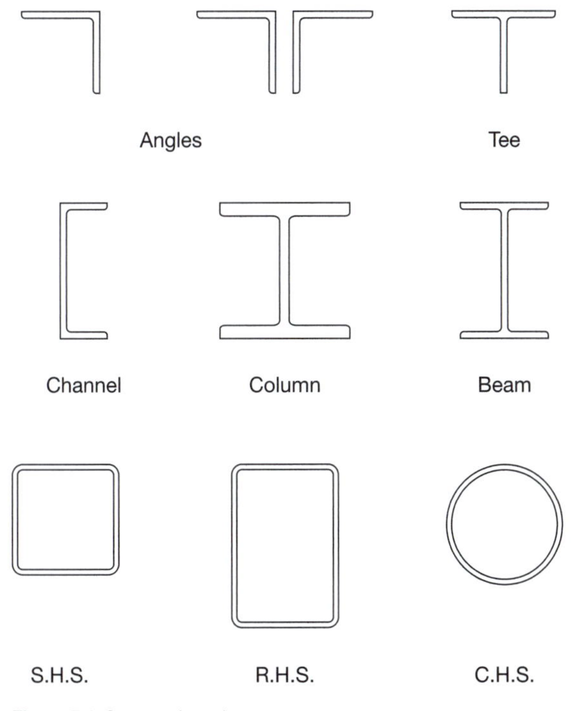


**Figure 7.1** Structural sections

This indicates that the section is a universal beam (UB) with a serial size 457 × 152 and has a mass per unit length of 46 kg/m. A number of different sections are available within the same serial size, each with a different weight per unit length, largely resulting from different flange thicknesses which are achieved by varying the roller details during production. The specific example quoted here is in accordance with British Standard sections. In other countries different conventions are used to identify specific sections.

Although modern rolling mills can achieve a relatively high degree of accuracy and consistency, it is essential to have some degree of rolling tolerances. This is more critical for hot-rolled products because of physical wear during rolling and inevitable temperature variations. Although these tolerances generally have very little effect when the products are used in normal fabrication for structural applications, they should be considered when detailing structures that need a close fit to more precise components such as sophisticated cladding or glazing systems. Site adjustments can often be allowed for, for instance by using slotted connections, and good co-ordination with the steel fabricator and other parties at the design stage should avert possible problems.

### 7.2.6 Cold forming

Cold-formed sections are produced by bending and shaping flat sheet steel, normally galvanised and 0.4–6 mm thick, at ambient temperatures. Components are manufactured using two different methods: roll-forming or brake pressing. Roll forming is a continuous process that shapes flat sheet or coiled, flat strip into continuous profiles or sections. The press brake process is used for small production runs by pressing individual sheets into the required form.

### 7.2.7 Tubular sections

Structural hollow sections can be produced by the seamless method or, more commonly, by welding. Seamless tube is made by spinning a solid cylindrical ingot in rotating rolls and forcing it around a mandrel which determines the internal diameter. Thick-walled tubes with good strength can be produced in this way.

The majority of hollow sections for structural applications are produced by cold rolling in which relatively thin strip is passed through a series of rollers that gradually fold it to form a closed section. The process is completed by a continuously welded longitudinal seam.

Typical section sizes and wall thicknesses for tubular sections are:

- circular sections from 20 mm to 500 mm diameter with thickness from 3 mm to 16 mm
- square and rectangular sections from 40 mm to 500 mm breadth/depth with thickness from 2 mm to 20 mm.

### 7.2.8 Cable and rod production

The production of rods and bars, such as those used in tension structures, uses a hot rolling process similar to that for structural sections. They are generally manufactured in very high strength, fine grain steel with a maximum carbon equivalent of 0.55% ensuring good weldability. Arc welding can therefore normally be carried out using standard techniques and low hydrogen rods. Typical diameters range from 10 to 100 mm with standard lengths up to about 12 m. Threaded ends to facilitate connections and tensioning are best formed by rolling.

Smaller diameter components such as wires are typically made by drawing a rolled rod through a tapered hole. This results in work hardening of the steel which increases the strength but reduces ductility and toughness.

**Table 7.1** Classification of steels according to carbon content

| | *Ultra low carbon steel* | *Low carbon steel* | *Medium carbon steel* | *High carbon steel* |
|---|---|---|---|---|
| Carbon properties | < 0.01% C<br>• High strength<br>• High toughness<br>• Good formability | < 0.25% C<br>• Reasonable strength<br>• High ductility<br>• Excellent fabrication | 0.25–0.7 % C<br>• High strength<br>• High toughness | 0.7–1.3 % C<br>• Very hard<br>• Low toughness |
| Treatment | Quenched | Annealed, normalised | Quenched and tempered | Quenched and tempered |
| Applications | Sheet materials for automobiles, pipelines, etc. | Structural steels, sheet materials white goods, etc. | Shafts, gears, connecting rods, rails, etc. | Springs, dies, cutting tools, etc. |

Cables are made up of a number of individual cold drawn steel wires, which are spun together and then assembled in a particular pattern. For higher capacities the wires are not circular in cross-section but are profiled to achieve better nesting in the final cable build-up. Their high strength makes them ideal for use in tension structures such as suspension bridges, as guys or stays to main structural members and for other purposes, such as supporting lifts. They are also widely used in prestressed concrete construction.

### 7.2.9 Castings

Some steel components are cast directly into shape, and the production process is not significantly different from that of the traditional process of casting steel into ingots. Cast steel is used for complex shapes that may be too expensive or impractical to fabricate. Examples include nodal joints and special fixings. Cast steel is more expensive and less accurate than cast iron, but is stronger and can be welded. Sand moulding is commonly used where relatively small numbers of castings are required, while the lost wax process, which is more expensive to set up but produces higher quality castings, is more suited to large production runs requiring a high standard of finish.

Solidification shrinkage is significant, but good process design can avoid porosity and other defects in the finished casting. The castings are usually heat treated to relieve any residual stresses caused by uneven cooling and to refine the grain structure of the material. In general, acceptable yield strengths and tensile strengths are obtained, but special attention may be required if toughness is important.

### 7.2.10 Cut and welded sections

While the vastly greater part of fabricated steel in construction is based on standard hot rolled sections such as straight beams and columns, non-standard steel sections can be fabricated where architectural or structural solutions dictate. Modified sections include tapered beams, plate girders, castellated beams and curved sections. Such shapes might be used for aesthetic reasons or where the largest suitable standard section is structurally inadequate.

Variations can be made by cutting and welding standard cross-sections or from plate, using normal fabrication techniques. Curved lengths of section have become popular and can be produced by specialist fabricators using a variety of bending techniques.

## 7.3 Chemical composition

The chemical composition of steel consists principally of iron and carbon but normally with other elements added in small quantities to achieve the required properties (alloy steels). Carbon is the single most important alloying element, influencing a range of properties. Although its percentage content by weight may seem small, because of its low density compared with iron, the content by volume is quite high and must therefore be carefully controlled.

Steels are often broadly classified according to carbon content, as illustrated in Table 7.1, which includes a brief description of the general characteristics of each type. Structural steels, such as those used for rolled sections, contain less than 0.25% carbon. Non-structural steels, such as strip for cold rolling to form cladding panels, have even lower carbon contents. As a result they have slightly reduced strength but increased ductility.

The lowest carbon content that can be achieved easily on a large scale is about 0.04%. This content is characteristic of sheet or strip steels intended to be shaped by extensive cold deformation, as in deep drawing. Carbon contents of more than 0.25% are used in the wider range of general engineering steels. These steels are usually quenched and tempered for use in a range of mechanical engineering applications, and may also be used for high strength bolts in some structural applications.

After carbon, manganese is the next most important alloying element and is added in amounts up to about 1.5%. It is important because it fixes the sulphur in the steel as manganese sulphide, deoxidises the steel, thereby reducing porosity, and increases the tensile strength. Manganese also brings other benefits associated with the grain structure of the steel, improving strength and low temperature performance. However, too much manganese can be harmful because it promotes the formation of brittle microstructures. A maximum manganese content is therefore generally specified, depending on the grade of steel.

The inclusion of small quantities of a wide range of other alloying elements can further affect the properties of steel, and their content needs to be carefully controlled. For example, up to 0.15% phosphorus by weight can harden steels, enabling accurate machining, but excessive quantities are injurious and cause embrittlement. Niobium, nickel and chromium improve strength, while molybdenum is often added to improve impact values. Hardness is increased by the addition of chromium and vanadium, while chromium and nickel are used in significant proportions in the manufacture of stainless steel.

In recent years so-called micro-alloyed steels or HSLA (high strength low alloy) steels have been the developed. These are normalised or controlled rolled carbon–manganese steels that have been adjusted by micro-alloying to give higher strength and toughness, combined with ease of welding. This is achieved through small additions of aluminium, vanadium, niobium or other elements to control grain size. Molybdenum may also be added to further refine the microstructure.

## 7.4 Microstructure

Most engineering properties of steel – strength, ductility, toughness – depend on its crystalline structure, grain size and other metallurgical characteristics. These are referred to as the microstructure of the steel and depend on its chemical composition and temperature–deformation history, including the effects of welding. Some mechanical properties of steels are independent of their microstructures, notably the elastic modulus which is constant regardless of the type or grade of steel. Control of microstructures to produce a very low inclusion content and a fine grain size has led to significant improvements in the available combinations of properties and is therefore a very important aspect of steel production. By altering the how the steel is heated and cooled during its production in relation to the mechanical processes to which it is subjected, the grain size can be modified.

The microstructure of steel can be seen by taking a thin section of the material and preparing it by polishing and etching for examination under a microscope. This can reveal the phase composition and other features such as non-metallic inclusions, principally of slag or manganese sulphide, entrained during production. These become elongated as a result of the rolling process, and although they do not affect strength they can reduce ductility and toughness. They also create an anisotropic material with significantly better properties in the rolling direction than through the thickness. This isn't because the crystals become elongated in the direction of rolling and in doing so develop greater strength along their length. In addition, any impurities are similarly elongated and represent zones of weakness that are much more pronounced through the thickness of the specimen.

Improvements in secondary steelmaking processes have led to reduced inclusion content, with sulphur levels of 0.01% or less, and better shape control of inclusions. This gives better ductility and toughness, and greatly improved through-thickness properties. The latter is particularly important for thick welded sections, such as the nodes of offshore structures.

### 7.4.1 Iron–carbon phases

Steel is a two-phase material, which means that atomic structure can take two basic forms. These are ferrite or α-iron, and austenite or γ-iron, which can dissolve considerably more carbon. These atomic structures depend not only on the chemical composition of the steel, but will modify as the material is heated and deformed.

The internal structure of ferrite, which is the most stable form of iron at room temperature, is composed of iron atoms in a regular three-dimensional pattern which is described as body-centred cubic (Figure 7.2). This structure changes to face-centred cubic (Figure 7.3) when heated, forming austenite. The change occurs at a particular temperature depending on carbon content – 910°C for pure iron and about 720°C for structural steel. The face-centred cubic (fcc) arrangement of atoms occupies slightly less volume than the body-centred cubic (bcc) structure. Consequently, when steel is heated, it expands as normal until the change temperature is reached, at which point there is a step contraction of about 0.5% in volume. The change is reversible if the steel is cooled slowly.

Ferrite can dissolve considerably less carbon (0.021% by weight at 910°C) than austenite (up to 2.03% by weight at 1154°C). So as austenite cools through 910°C, it changes to ferrite and in doing so releases the excess of carbon. This can lead to the formation of iron carbide or cementite, which is brittle material. The resulting steel has a microstructure that is a mixture of ferrite and cementite crystals, often arranged in a layered pattern referred to as pearlite because of its pearl-like appearance.

If the austenite is rapidly cooled past a critical temperature of 723°C, the resulting ferrite–cementite mixture takes on a finer, non-lamellar structure called bainite, which is similar in constitution to pearlite but is generally stronger and more ductile.

When the austenite is very rapidly cooled – a process known as quenching – the carbon does not have time to be released and can become locked within the BCC structure, forming martensite. Because the carbon atoms are too big to be accommodated within the interstices of the atomic structure, the basic form is distorted into a body-centred tetragonal (BCT) structure. The ability of the steel to form martensite, which is hard but brittle, is called the 'hardenability'. In construction this is important in the context of welding in particular.

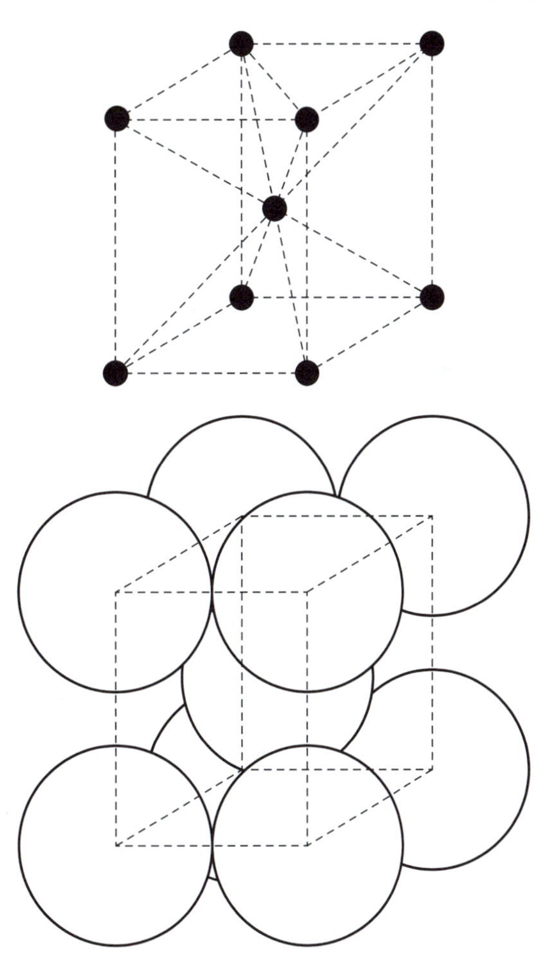

**Figure 7.2** Body-centred cubic crystal: unit cell

These changes can be used to manipulate the properties of the steel. For example, quenching and tempering is a two-stage process in which the initial rapid cooling leads to the formation of martensite; this is followed by controlled heating to transform some of the brittle martensite into bainite.

The influence of temperature and carbon content on the microstructure of steel is represented in the iron–carbon phase diagram (Figure 7.4), which is divided into areas of stable microstructures at particular compositions and temperatures. The diagram is characterised by the following areas.

- *Ferrite*. Almost pure iron (less than 0.02% carbon by weight), with the bcc crystal structure. It is stable at all temperatures below 720–910°C.
- *Cementite*. Iron carbide, $Fe_3C$, which contains about 6.67% carbon by weight.
- *Pearlite*. A fine lamellar mixture of about 90% ferrite and 10% cementite. The overall carbon content is 0.8% by weight and it is stable at all temperatures below about 720°C.

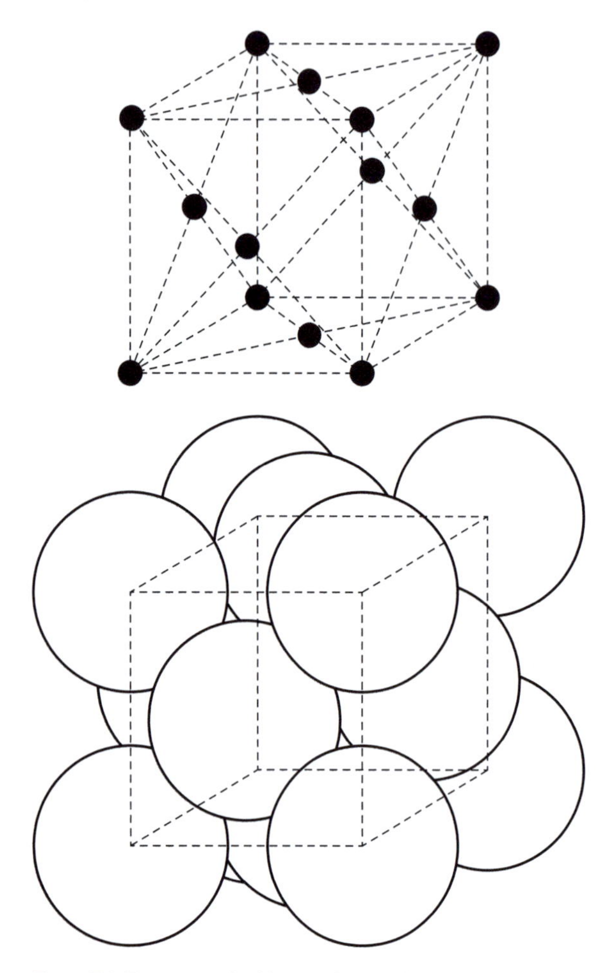

Figure 7.3 Face-centred cubic crystal: unit cell

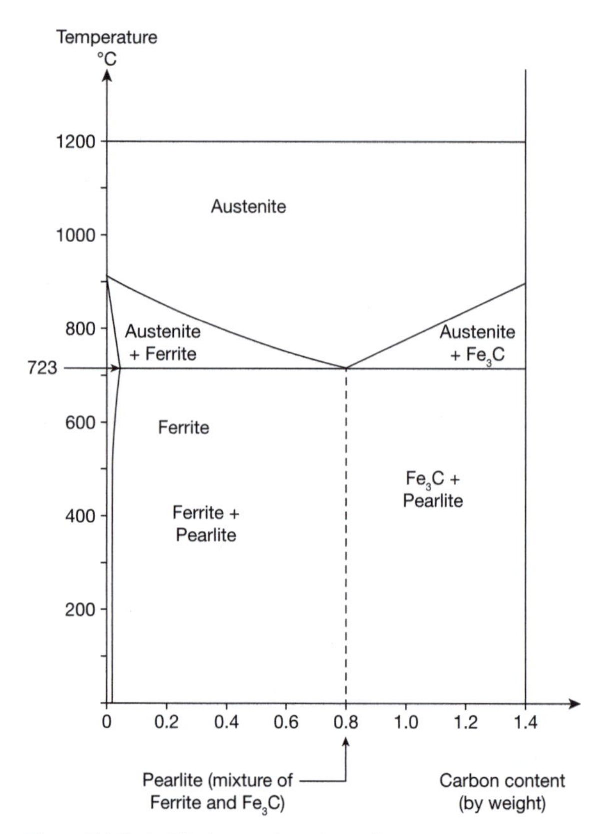


Figure 7.4 Part of the iron–carbon phase diagram

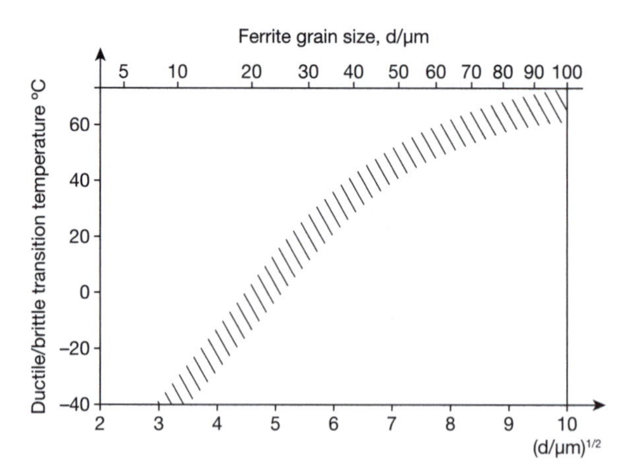


Figure 7.5 Effect of ferrite grain size on the ductile/brittle transition temperature of mild steel (0.11%)

- *Austenite.* Steel with the fcc structure, containing up to approximately 2% carbon by weight. It is stable at temperatures above 720–910°C.
- *Bainite and martensite.* Products found in steels cooled rapidly from temperatures in the range of austenite stability, or in more slowly cooled alloy steels.

The microstructure and therefore the properties of steel are clearly highly dependent on the carbon content and the heating history. The grain size of the ferrite is also crucially important in several respects. Most notably a finer grain size improves strength – both yield strength and ultimate tensile strength – and toughness, with a reduction in the ductile/brittle transition temperature (Figure 7.5).

### 7.4.2 Normalising and hot rolling of steels

Normalising is a heat-treatment process in which the steel is heated to about 850°C and allowed to cool naturally in still air. The main purposes are to relieve any residual stresses caused by rolling etc. and to control the ferrite grain size. At the high temperatures applied, the steel structure consists of grains of austenite. On cooling, ferrite grains grow from the austenite grain boundaries and, finally, pearlite is formed. Steels containing small amounts of aluminium or, better, titanium or niobium inhibit austenite grain growth, thereby producing fine-grained ferrite.

The rolling processes for producing structural steel sections affect the microstructure in a similar way to normalising. The early stages of rolling are carried out at temperatures well within the austenite range, where the steel is soft and easily deformed. The deformation breaks down the microstructure but the poly-crystalline structure of the austenite is continuously reformed by a process known as recrystallisation. The amount and rate of deformation and the rolling temperatures affect the grain size, with heavy deformations at low temperatures resulting in small grains. The deformation suffered by the material breaks up the coarse as-cast grain structure but, at these high temperatures, the atoms within the material can diffuse rapidly, which allows the deformed grains to recrystallise and reform the equiaxial polycrystalline structure of the austenite.

Heavy deformation at low temperatures in the austenite range gives finer recrystallised grains. If the rolling is finished at a temperature just above the ferrite + austenite region of the equilibrium diagram and the section is allowed to cool in air, an ordinary normalised microstructure having moderately fine-grained ferrite results. Modern controlled rolling techniques aim to do this, possibly rolling at lower temperatures to give even finer grains.

Low carbon steels in thin section sizes may be rolled at ambient temperatures. This breaks up the ferrite–pearlite structures and causes work hardening: recrystallisation cannot occur at these temperatures. The properties change with deformation.

Other methods of cold working, such as wire drawing, can impart larger deformations. With suitable metallurgical adjustment, high carbon steels can be rendered ductile enough to allow extensive deformation. Very large amounts of work hardening can give very high-strength wires for ropes and cables.

### 7.4.3 Rapidly cooled steels

Very rapid cooling, for example by quenching in cold water, causes the formation of martensite instead of ferrite and pearlite. Martensite is extremely strong but very brittle. Less rapid cooling results in the formation of bainite. Structural steels do not normally experience such rapid cooling. However, the process of welding can give rise to localised conditions that allow similar metallurgical transformations to take place.

### 7.4.4 The effects of welding on steel

Welding is a process of forming attachments between components which involves locally raising the temperature of the steel to melting point. The temperature of adjacent areas is also raised to well within the austenitic range. The area affected by the weld in this way is referred to as the heat affected zone (HAZ). As the weld cools further metallurgical changes associated with the phase diagram develop, and these clearly depend on the rate of cooling. In addition if the weld is unprotected, the molten metal in the weld pool can readily absorb atmospheric gases – oxygen and nitrogen – and this can cause porosity and brittleness in the solidified weld metal. Welding processes therefore generally include techniques to avoid such gas absorption. The high temperature changes also result in residual stresses locked into the steel, which can cause problems if the weld has significant defects.

The weld generally cools rapidly because of the heat sink provided by the components being connected. This would normally lead to the formation of low temperature transformation structures, such as bainite and martensite.

The microstructure of the HAZ is influenced by the chemical composition of the parent metal, which influences determines its hardenability (its tendency to form martensite), the heat input during welding, which affects the grain size, and the subsequent cooling rate. Generally, a high heat input rate leads to a coarser HAZ microstructure. The cooling rate is influenced by the thickness of the connected pieces and their temperature; preheating can be used to reduce the temperature gradient and hence slow down the cooling.

### 7.4.5 Carbon equivalent

The total alloy content of steel affects its hardness and this can be undesirable for welding. It is therefore important to limit the total content, particularly for structural steels which must be weldable. Some alloying elements have a greater influence than others, and this makes it difficult to compare the performance of different alloys made elements. Methods of measuring combinations of different alloying elements have therefore been developed, the most common being the equivalent carbon content. This is calculated using a formula such as:

$$\text{CEV} = \%\text{C} + \frac{\%\text{Mn}}{6} + \frac{\%\text{Cr} + \%\text{Mo} + \%\text{V}}{5} + \frac{\%\text{Ni} + \%\text{Cu}}{15}$$

The carbon equivalent value thus calculated enables the alloy to be compared with plain carbon steels, and hence provides a relative measure of weldability and hardness. As a general rule, hardening in the HAZ can be avoided by using a steel with a carbon equivalent less than about 0.42 depending on the welding process details; this effectively limits the maximum alloy content of structural steels.

### 7.4.6 Cold cracking

Martensite produced in the weld zone as a result of single pass welding is untempered and therefore hard and brittle. Together with volume changes associated with thermal effects and transformations in microstructure, this can produce local stresses big enough to crack the martensite. This is referred to as cold cracking, and can be aggravated if the weld has absorbed hydrogen, for example from moisture in the atmosphere or welding electrodes. Hydrogen dissolved in the weld metal diffuses to the hard HAZ, where it initiates cracks at sites of stress concentration. This diffusion can lead to cracking which may not occur until some days after the welding is completed. This is referred to as hydrogen cracking. Hard HAZs of low ductility are less able to cope with this problem than are softer and more ductile materials.

As cooling rate is important, welding thick plates is more likely to give rise to hydrogen cracking than for thin plates. To reduce the danger, the cooling rate, particularly between 800°C and 500°C, can be reduced by preheating thick plates before welding. This also helps by encouraging diffusion of hydrogen from the weld zone so that less remains after the weld has cooled.

Increasing the heat input rate of the welding process (where possible) is beneficial since it results in a slower cooling rate after welding and therefore a lower likelihood of hardening in the HAZ. For the same reason, there is less risk of hydrogen cracking when welding thin plates and sections, since the cooling rate in the HAZ is lower.

Limiting the presence of hydrogen by avoiding damp, rust and grease, by using controlled hydrogen electrodes and low hydrogen welding processes, for example metal inert gas or submerged arc welding, will also help to avoid cracking.

### 7.4.7 Weld metal solidification cracking

When a weld cools, the molten weld pool solidifies from the edges to the centre, and in the process any impurities such as sulphur and phosphorus are pushed ahead of the advancing boundary (Figure 7.6). This finally results in a concentration at the centre of the weld, forming a potential line of weakness which can crack when subject to transverse shrinkage strains. Weld metals with a low sulphur and phosphorus content are available for most structural steels. In addition to good process control and joint design, this should avert the problem.

### 7.4.8 Lamellar tearing

If significant residual welding stresses develop across the thickness of any of the components, they can readily lead to cracking because the through-thickness ductility of the plate is

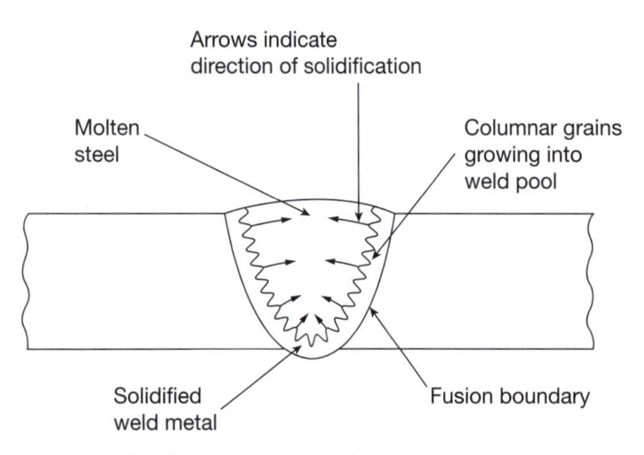


**Figure 7.6** Solidification of weld metal

very low due to the presence of planar inclusions such as sulphates and silicates. These form as the steel cools after casting, and are deformed into discs parallel to the surface during rolling. Lamellar tearing can be best avoided by joint design to avoid residual stresses in the through-thickness direction, and specifying steel with low inclusion content.

## 7.5 Stainless steels

Although stainless steels are not widely used for major structural components in construction, they have been used in some high-profile applications and this is increasing. However, their principal applications are in cladding, secondary components and fixings where the characteristics of corrosion resistance and appearance can provide a very cost-effective solution. The properties of stainless steel are achieved by the addition of significant amounts of chromium (>11%); other alloying elements may also be added. For example, nickel is added to enhance corrosion resistance and formability and small amounts of molybdenum may be introduced to improve resistance to pitting and crevice corrosion. When exposed to the atmosphere, the chromium reacts with oxygen to form chromium (III) oxide ($Cr_2O_3$), creating a dense, stable oxide film on the surface, and acting as a barrier to further corrosion. This protective film is extremely thin but is impervious, and readily re-forms if it becomes damaged.

There are three principal types of stainless steel: martensitic, ferritic, and austenitic.

- *Martensitic stainless steels* typically contain about 13% chromium and more than 0.1% carbon. They may be heat treated by quenching and tempering to give martensitic microstructures which are hard and strong but of poor toughness. Common applications include cutlery and razorblades; they are not widely used in construction.
- *Ferritic stainless steels* contain very little carbon (< 0.08%) with chromium contents ranging from 13–17%, and no nickel. They do not undergo the phase changes of plain carbon steels because only the ferrite phase (bcc crystal structure) is stable at temperatures below melting. Metallurgical control is confined to control of composition and of grain size through processing schedules. They have lower corrosion resistance, strength and ductility than austenitic grades, and are difficult to weld.
- *Austenitic stainless steels*, probably the most common, contain a minimum of 8% nickel, stabilising the austenite (fcc) crystal structure at all temperatures below melting. Consequently they are generally stronger and more ductile than ferritic steels, but also more expensive. Carbon contents must be maintained as low as practicable, typically below 0.1%. Metallurgical control is through composition and grain size.

Duplex stainless steels have a mixed austenitic/ferritic microstructure and are based on 22–23% chromium and 4–5% nickel additions. Some grades have better corrosion resistance than the standard austenitic stainless steels due to the higher content of chromium and presence of molybdenum and nitrogen. Duplex stainless steels are also stronger than austenitic steels and are readily weldable.

Precipitation hardening steels combine the best features of austenitic and martensitic steels. They can be heat treated to give a wide range of strengths, together with excellent toughness. They are mainly used in the aerospace industry, but some grades have been used in construction, for example for special heavy-duty connections in buildings.

Standards for grades and designations of stainless steel are set out in BS EN 10088: 1995, Stainless Steels. It includes specifications for the chemical compositions of particular grades of stainless steel, reference data on physical properties such as density, modulus of elasticity and thermal conductivity, and standards for chemical compositions and surface finishes for specific product types.

A stainless steel with a typical composition of 18% chromium and 10% nickel is commonly referred to as 18/10 stainless. Similarly 18/0 and 18/8 refer to chromium contents of 18% and nickel of 0% and 10% respectively. Particular grades of stainless steel are more formally identified by a more precise designation, according to a standard convention. For example, material printed with '1.4401' indicates the following:

| 1. | 44 | 01 |
|---|---|---|
| Denotes steel | Denotes one group of stainless steels | Individual grade identification |

The principal grades of stainless steel used in architecture are grades 304 and 316, and a fuller list of common names and designations is given in Table 7.2.

The protective skin that forms on the surface of stainless steel can be impaired by physical damage, although in practice this is unlikely because of the density of the film. Chemical damage is possible from the atmosphere or the underlying metal. Chlorides are particularly aggressive, and resistance to pitting in marine and other environments is improved by including up to 5% molybdenum in the steel composition. The underlying metal can also damage the protective skin of austenitic stainless steels through grain-boundary attack in which small quantities of residual carbon react with the chromium to form chromium carbide, limiting the amount of chromium available to form the protective skin. The problem is best avoided by carefully controlling the steel composition to ensure that carbon contents are as low as possible and that there is sufficient chromium. Niobium or titanium may be also added to tie up the carbon, releasing the chromium to contribute fully to its corrosion-resisting function.

## 7.6 Weathering steels

So-called weathering steels are plain carbon or low-alloy steels containing up to about 2% total of copper, chromium, nickel and phosphorus. When exposed to an external atmosphere these steels build up a dense layer of protective oxides on their surface in a similar manner to stainless steel. In most environments, the coating is sufficiently effective to protect the steel from further corrosion. However, the composition of the film is very different from that of stainless steel, with a distinctive 'controlled' rusty brown appearance, and there are a number of potential problems. The rust-like layer develops slowly, gradually providing more protection. In order to achieve a reasonably uniform colour, it is critical that surface has an even exposure. The surfaces must not allow water to stand on any areas, as the film forms only under

**Table 7.2** Principal grades of stainless steels

| *Family* | *BS EN 10088 designation* | *Popular name** | *Weight (%)* | | | | | *Attributes* |
|---|---|---|---|---|---|---|---|---|
| | | | *Cr* | *Ni* | *Mo* | *N* | *Others* | |
| Austenitic | 1.4301 | 304 | 18 | 9 | – | – | – | Good range of corrosion resisting and fabrication properties; readily available in a variety of forms, e.g. sheet, tube, fasteners, fixings; 1.4401 (316) has better pitting corrosion resistance than 1.4301 (304). Low carbon (L) grades should be specified where extensive welding of heavy sections is required. |
| | 1.4307 | 304L | 18 | 9 | – | – | Low C | |
| | 1.4401 | 316 | 17 | 12 | 2 | – | – | |
| | 1.4404 | 316L | 17 | 12 | 2 | – | Low C | |
| | 1.4541† | 321 | 18 | 10 | – | – | Ti | |
| Duplex | 1.4362 | 2304 | 23 | 4 | – | 0.1 | – | Higher strength and wear resistance than standard austenitic grades with good resistance to stress corrosion cracking. Grade 1.4462 (2205) has better corrosion resistance than 1.4362 (2304). |
| | 1.4462 | 2205 | 22 | 5 | 3 | 0.15 | – | |

*The popular name originates from the (now partly superseded) British Standards and AISI system. †Titanium is added to stabilise carbon and improve corrosion performance in the heat affected zones of welds. However, except for very heavy section construction, the use of titanium stabilised austenitic steels has largely been superseded by low carbon grades.

conditions of wetting and drying. It is also important to avoid crevices, which can lead to very rapid local corrosion, and ensure that rust run-off does not cause discolouration of adjacent surfaces. The film is much less robust than stainless steel, and even gentle abrasion can both cause damage to the protective layer and stain anything that brushes against it. Care should also be taken to ensure that the film is not damaged by chemical attack, from either the environment or the underlying metal. Chlorides are particularly aggressive in this respect, and the use of weathering steels in marine environments is not advised.

## 7.7 Physical properties of steels

For construction, the most important mechanical properties of steel under non-repeating loading are elastic modulus and yield stress. Fatigue strength is important in conditions where cyclical loading conditions exist, but these are unusual in construction. Poisson's ratio, ultimate strength and hardness are also relevant. The required combination of strength and toughness may be achieved by controlling alloying content, level of impurities, and physical treatment.

Properties such as elastic modulus, bulk density, specific heat, and coefficient of thermal expansion, which do not depend significantly on microstructure, are referred to as structure-insensitive. However, most mechanical properties including yield strength, ductility and toughness or fracture strength are structure-sensitive – that is, they are strongly influenced by the microstructure which in turn depends on the working and heat treatment processes.

### 7.7.1 Stress–strain curve for simple tension specimen

The mechanical properties of steel are commonly determined by tensile tests conducted on long and narrow specimens to give a uniaxial state of stress. Load ($F$) and elongation ($\Delta L$) are measured during testing. The nominal or engineering stress $\sigma$ ignores changes in cross-sectional area and is simply:

$$\sigma = F/A$$

while the nominal or engineering strain $\varepsilon$ is the ratio of the change in length ($L - L_0$) to the original length $L_0$:

$$\varepsilon = (L - L_0) / L_0 = \Delta L / L_0$$

The stress–strain curve can be divided into three regions: elastic, plastic and strain hardening. In the elastic range strain is proportional to stress, and relatively little deformation takes place. Most steels for construction show a distinct yield point marking the change from elastic to plastic behaviour (Figure 7.7). This yielding is characterised by the upper yield stress at the start of yielding, and the lower yield stress, but in most practical applications these details are ignored and a single yield point is assumed. In the plastic range, deformations increase rapidly for relatively small increases in load, while in the strain hardening region, deformation requires further increases in stress because of the interaction between dislocations.

As the specimen extends, its cross-section reduces. Initially this thinning is spread throughout the length, but in the strain hardening region it becomes localised, resulting in necking (Figure 7.8). The stress at which this develops is the ultimate tensile stress. Beyond this point, the load necessary to maintain elongation decreases although the local stress at the neck continues to increase. Finally the specimen fractures, and the percentage reduction of area at this point is a measure of the steel's ductility.

Stresses may exist in more than one direction, and this can significantly influence the onset of yielding. However, in construction such conditions are not common. The effects of multi-axial loading,

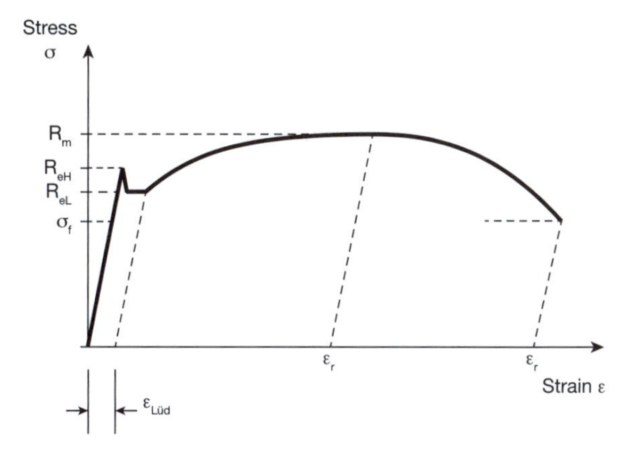


**Figure 7.7** Engineering stress–strain curve of a metal, e.g. mild steel, with a yield point

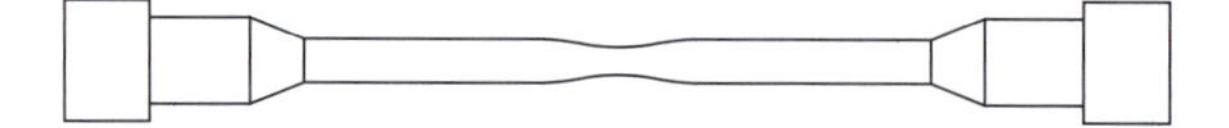

**Figure 7.8** Tension coupon at point of failure

temperature and strain rate may be important in some applications but are not normally considered for construction.

### 7.7.2 Toughness

Toughness is a measure of the resistance to fracture – materials with a low toughness are brittle in nature, while those with high toughness are ductile. In most circumstances steel behaves in a ductile manner with appreciable deformation prior to fracture. However, defects such as cracks or sharp notches which act as stress raisers, or localised areas where the mechanical properties have been compromised, such as the heat affected zone of a weld, can increase the risk of brittle failure, particularly at low temperatures.

Most steels exhibit tough ductile behaviour at normal temperatures, but at low temperatures can undergo a dramatic transition to brittle fracture. The temperature at which this change takes place is referred to as the transition temperature.

Hardening treatments to increase strength usually reduce ductility and increase the tendency towards brittle behaviour. In low-temperature environments steel alloys, typically with an increased manganese–carbon ratio and reduced nickel content, must be carefully specified, and in particular it is important that a material with a transition temperature well below the expected minimum is specified.

### 7.7.3 Hardness

Hardness is a measure of resistance to deformation and is normally measured by forcing a loaded indenter into the surface of the steel. Unlike tensile strength, for example, it is a relative measure and not a fundamental, absolute property of the steel. A number of different testing methods are used, and these can be static or dynamic. Dynamic testing is quicker but less precise than static testing. The Brinell test uses a hardened steel ball, the Vickers test uses a square-based diamond pyramid, and the Rockwell test uses a diamond cone indenter. After removing the load, the size of the indentation is measured. These are all static methods of testing and compared to dynamic testing, only the plastic deformation is taken into account. However, they are favoured in industry and research because of the consistency of test results.

## 7.8 Corrosion

In appropriate conditions, steel will combine with moisture, for example due to rainfall or condensation, and oxygen to form iron oxides in a process of corrosion. Other metals, including stainless steel, may produce environmentally stable oxide layers when similarly exposed, affording protection to the underlying metal.

Corrosion can affect not only appearance, but also structural performance. It may therefore be necessary to protect the steel from moisture and/or oxygen to prevent this rusting. There are a variety of systems and techniques for this and the exact specification will depend on the particular application and degree of exposure.

### 7.8.1 The process of corrosion

Corrosion of iron, including steel, is generally described as an electrochemical process, which can be described by the following overall equation:

$$4Fe + 3O_2 + 2H_2O = 2Fe_2O_3$$

The process actually occurs in two stages, starting at anodic areas on the surface where ferrous ions go into solution. Electrons are released from the anode and move through the metallic structure to the adjacent cathodic sites on the surface, where they combine with oxygen and water to form hydroxyl ions. These react with the ferrous ions from the anode to produce ferrous hydroxide which itself is further oxidised in air to produce hydrated ferric oxide, otherwise known as red rust.

Every metal has a different potential to corrode, and this is represented by the galvanic series. Noble metals such as gold and silver have a high potential value and are very resistant to corrosion. Some other metals such as aluminium have low potential values but produce stable oxide layers. The potential value of iron is higher than aluminium, chromium or zinc, but because the oxide layer is unstable, it is more prone to corrosion.

### 7.8.2 Bimetallic corrosion

When two dissimilar metals are in contact with an electrolyte – generally a conductive solution – an electric current passes between them and the metal with the lower potential value (the anode) corrodes. Some metals, such as nickel and copper, therefore promote the corrosion of steel, whereas others such as zinc corrode preferentially, thereby protecting the steel. The rate of bimetallic corrosion increases with the relative separation of the two metals in the galvanic series. It also depends on the nature of the electrolyte and the contact area.

Bimetallic corrosion is most serious for immersed or buried structures. In less aggressive environments such as stainless steel brick support angles attached to mild steel structural sections, the effect is minimal and no special precautions are required. However, in other circumstances such as more complex cladding systems, bimetallic corrosion must be prevented. For these higher risk cases, gaskets, sleeves, and similar electrically insulating devices should be used, or the complete assembly protected, for example by painting.

### 7.8.3 Rates of corrosion

The principal factors that affect corrosion rates of steel in air are time of wetness and atmospheric pollution – sulphates and chlorides in particular increase corrosion rates, reacting with the steel to produce soluble, corrosive salts. Furthermore, actual corrosion rates can vary markedly due to the effects of micro climate such as sheltering and prevailing winds. Corrosion rates cannot therefore be generalised, but only broadly classified as shown in Table 7.3.

Because corrosion depends on contact between the steel and moisture, the most common form of protection is to apply an impervious coating to the steel. An alternative is to use an alloy of steel that is resistant to corrosion – stainless steel is the most common example. In all cases the protection system should be considered in relation to the degree of exposure and the likely rate of corrosion. Good design and detailing can help to reduce the duration of wet contact with the steel, and thereby reduce the effective degree of exposure.

Under special circumstances, other methods of protection may be used such as cathodic protection or corrosion inhibitors. Cathodic protection is a method of corrosion protection in which a sacrificial element such as zinc is connected to the steel to form the anode of an electrochemical cell, the steel surface forming the cathode. The anode corrodes in preference to the steel, and eventually must be replaced. It is not commonly used in the construction industry except in some offshore applications.

Steel tubes containing water are generally protected using corrosion inhibitors which are added to the water in small concentrations. Some paints also include inhibitors such as zinc phosphate.

**Table 7.3** Atmospheric corrosion classification (BS EN ISO 12944 Part 2)

| *Corrosion category and risk* | *Loss of thickness of steel (μm)* | *Examples of typical environments in a temperate climate* | |
|---|---|---|---|
| | | *Exterior* | *Interior* |
| C1 Very low | < 1.3 | – | Heated buildings with clean atmospheres, e.g. offices, shops, schools, hotels |
| C2 Low | 1.3–25 | Atmospheres with low levels of pollution. Mostly rural areas | Unheated buildings where condensation may occur, e.g. sports halls |
| C3 Medium | 25–50 | Urban and industrial atmospheres, moderate sulphur dioxide pollution. Coastal areas with low salinity. | Production rooms with high humidity and some air pollution, e.g. laundries, breweries |
| C4 High | 50–80 | Industrial areas and coastal areas with moderate salinity | Chemical plans, swimming pools, coastal shipyards |
| C5-I Very high (industrial) | 80–200 | Industrial areas with high humidity and aggressive atmosphere | Buildings or areas with almost permanent condensation and high pollution |
| C5-M Very high (marine) | 80–200 | Coastal and offshore areas with high salinity | Buildings or areas with almost permanent condensation and with high pollution |

The thickness loss values are after the first year of exposure. Losses may reduce in subsequent years. In coastal areas in hot, humid zones, the thickness loss can exceed the limits of category C5-M. Special precautions must therefore be taken when selecting protective paint systems for structures in such areas.

## 7.9 Coatings

In a stable, dry, internal environment, unpainted steel will last indefinitely with no significant corrosion, but in exposed conditions, steel must be protected unless an allowance is made for corrosion loss (corrosion rates in various environments can be reasonably predicted). Various coating systems may be used for surface protection, depending on the degree of exposure.

### 7.9.1 Surface preparation

Surface preparation is very important to ensure good adhesion of applied coatings and to remove contaminants already on the surface that can cause corrosion under the coating. This involves the removal of surface dirt and grease which form during fabrication, and mill scale – a complex oxide that forms on the surface of the steel during the hot rolling process. Mill scale is unstable and vulnerable to cracking, allowing moisture to penetrate and attack the steel beneath.

Various methods of surface preparation and grades of cleanliness are specified in standards. These refer to the surface appearance of the steel after hand or power tool cleaning, abrasive blast cleaning, flame cleaning, or acid pickling. Surface cleaning by hand tools such as scrapers and wire brushes is relatively ineffective. Power tools offer a slight improvement but neither of these methods is used unless abrasive blasting is not possible. Flame cleaning is neither economic nor fully effective.

Abrasive blast cleaning is the most important and widely used method for cleaning mill-scaled and rusted surfaces. Abrasive blasting involves firing steel shot or grit onto the surface of the steel at high speed, either by compressed air or centrifugal impeller. The abrasives are recycled with separator screens to remove fine particles. This process can effectively remove all mill scale and rust.

Acid pickling, in which the steel is immersed in a bath of mild acids to dissolve the mill scale and rust without appreciably attacking the steel, is normally used as a preparation for hot dip galvanising.

Surface cleaning standards (BS EN/ISO 8501-1:2007) are generally specified in relation to different levels of visual cleanliness, and cleaned surfaces can be compared with a reference photograph to check compliance. The standard descriptors are:

- Sa 3 – blast cleaning to visually clean steel
- Sa 2½ – very thorough blast cleaning
- Sa 2 – thorough blast cleaning
- Sa 1 – light blast cleaning.

### 7.9.2 Paint coatings

Painting is the most common method of protecting steelwork from corrosion. Paint coatings should provide an effective barrier to water and air, and have good adhesion to the substrate. Some coatings may also contain corrosion-inhibiting chemicals providing additional protection to that of a simple barrier.

Paints are made by mixing and blending three components – binder, pigment and solvent.

- Pigments are finely ground inorganic or organic powders that provide colour, opacity, film cohesion and sometimes corrosion inhibition.
- The binder is the component that forms the film, and is traditionally based on resins or oils. However, inorganic compounds, synthetic polymers and rubbers are becoming more widely used.
- Solvents are used to dissolve the binder and to facilitate application at the paint. Solvents are usually water or organic liquids such as petroleum or alcohol. However, concerns about environmental protection have led to a significant move away from organic solvents, and water-based systems are much more widely used.

Paint is applied to the steel surface, producing a film that may be as little as 25 μm thick when dry. The degree of corrosion protection afforded is proportional to the dry film thickness.

The range of coatings on the market is enormous, although the number of generic types is limited. Primers are usually classified according to the main corrosion inhibitive pigments used in their formulation. Examples include zinc phosphate primers and metallic zinc primers. Each of these pigments can be incorporated into a range of binder resins, giving, for example, zinc phosphate alkyd primers, and zinc phosphate epoxy primers. Intermediate barrier coats and finishing coats are usually classified according to their binders – for example epoxies, vinyls, and urethanes.

A traditional, rather primitive, coating system consisted of a primer, undercoat and top coats. The primer wets the surface of the steel and provides good adhesion for subsequent coats. Intermediate coats are used to build the total thickness of the system. Generally the thicker the coating, the longer the life, but performance can also be improved by introducing flake pigments to increase effective path length through coating. Examples include micaceous iron oxide (MIO) or glass flake. The finishing coats are primarily cosmetic and determine the final appearance in terms of gloss, colour, etc.

Successive coats may be of the same generic types, but variations are possible. However, it is generally advisable to obtain all materials from the same manufacturer, and to take specialist advice.

Prefabrication primers, also referred to as blast primers, shop primers, temporary primers, or holding primers, are used on structural steelwork, immediately after blast cleaning, to maintain the surface in a rust-free condition until final painting can be undertaken. They are mainly applied to steel plates and sections before fabrication.

### 7.9.3 Metallic coatings

Metallic coatings generally provide protection in two ways. Firstly they form a barrier in the same way as paint, and secondly they act in a sacrificial manner, the metal corroding in preference to the steel, on the same basis as bi-metallic corrosion.

There are four principal methods of applying metal coating to steel: hot-dip galvanising, thermal spraying, electroplating and sherardising, although the last two are generally used only for small items such as fittings and fasteners. The protection afforded depends mainly on the coating metal and its thickness rather than the method of application.

The most common method is hot-dip galvanising in which the steel, after pickling in inhibited hydrochloric acid, is immersed first in a fluxing agent and then in molten zinc at a temperature of about 450°C. At this temperature, the steel reacts with the molten zinc to form a series of zinc/iron alloys on the surface. The thickness of the coating is influenced by various factors including the size and thickness of the workpiece and the surface preparation of the steel. Thicker components and abraded surfaces produce relatively thick coatings. For many applications, hot-dip galvanising is used without further treatment. However, the steel may be overpainted to provide extra protection, or for decoration. This combination is referred to as a 'duplex' system.

An alternative method of applying a metallic coating is spraying using molten zinc or aluminium. The metal, in powder or wire form, is fed through a heated nozzle, and molten globules are blown onto the steel surface by a compressed air jet. In contrast to hot-dip galvanising, the coating consists simply of a layer of the metal, and no alloying occurs at the surface of the steel. Spray coating can be carried out on site as well as in the fabrication shop and there is no limit on the size of workpiece. However, it is more expensive than hot-dip galvanising; in addition the resulting sprayed coating is more porous than a galvanised layer and hence normally needs sealing by separate syste

## 7.10 Performance in fire

Steel is incombustible and this property is one of the principal reasons why iron and subsequently steel were introduced as main structural components in the nineteenth century. However, like other structural materials its properties change with increasing temperatures. Most importantly both strength and stiffness reduce as the steel is heated, but changes in thermal properties, principally the specific heat, are also important because they affect the development of temperatures. Most of these changes are gradual, but the chemical transformation from ferrite–pearlite to austentite at about 700°C has a marked effect on thermal properties in particular.

### 7.10.1 Mechanical properties

Material softening can be represented in the form of a progressive degradation of the stress-strain curves as the temperature of the steel increases, with a reduction in both elastic modulus and 'yield' stress (Figure 7.9). These form the basis of simple strength

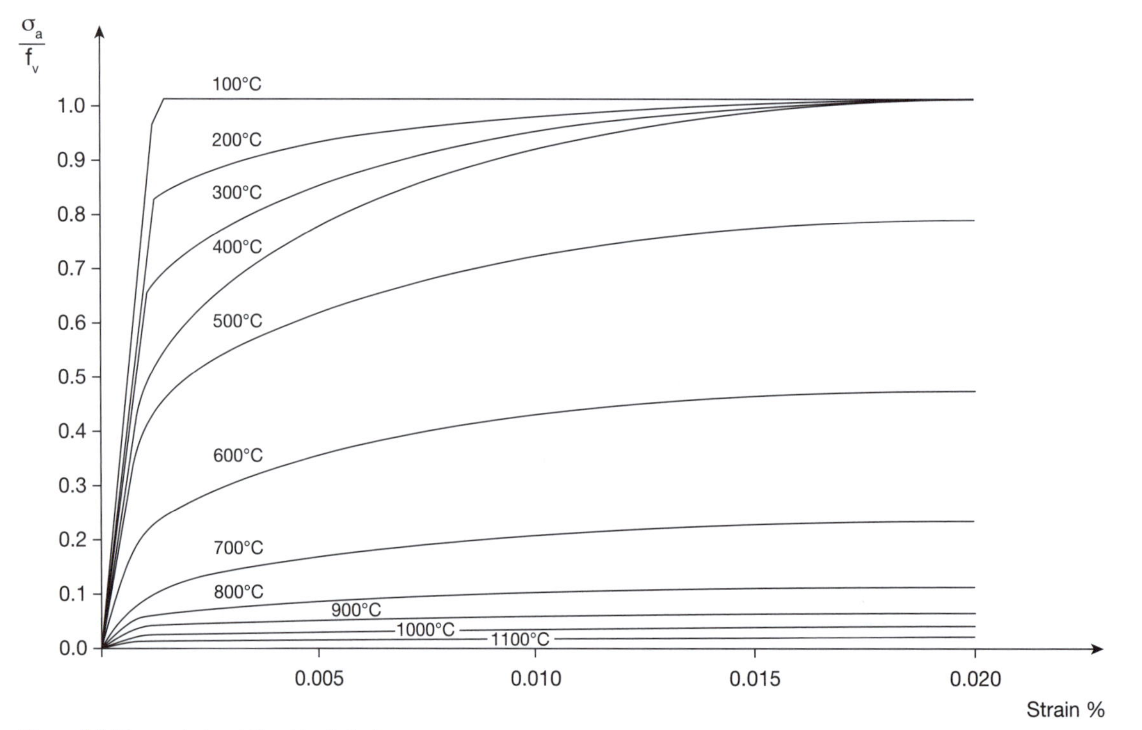


**Figure 7.9** Stress–strain relationship of steel

reduction data used in some design codes, although the definition of a 'yield' point at high temperatures is somewhat arbitrary and is effectively based on a strain limit. Although steel does not melt until about 1500°C, its strength has reduced almost to half at about 550°C, and only 23% of the ambient-temperature strength remains at 700°C. Some producers have developed special steels by adding alloys (e.g. with molybdenum) that allow greater retention of yield and ultimate tensile stress up to temperatures of about 800°C, but these have not enjoyed widespread use.

### 7.10.2 Thermal properties of steel at high temperature

The rate of heat transfer in any material depends on three properties: thermal conductivity ($k$), specific heat ($c$) and density ($\rho$). The thermal conductivity of steel reduces from 54 W m K to half this value by 800°C, but with no further decrease at higher temperatures. Its specific heat increases from 450 J kg K at 20°C to 700 J kg K at about 600°C, with a very dramatic change in the range 700–800°C associated with the phase change from pearlite to austentite at 700°C.

The density of steel is approximately 7850 kg/m$^3$, decreasing slightly with increasing temperature.

### 7.10.3 Steelwork fire resistance

The temperatures in a building fire can reach well in excess of 1000°C. This can significantly change the structural behaviour of the steelwork, firstly by softening of the material and secondly by restrained expansion. For example, with steel beams supporting a concrete slab on the upper flange the differential thermal expansion caused by shielding of the top flange, and the heat-sink function of the concrete slab, cause a 'thermal bowing' towards the fire in the lower range of temperatures.

This leads to changes in structural behaviour that interact with other parts of the building so that the net structural response can be very different from that at ambient temperature.

The fire resistance of a structural is measured by the length of time it can survive in a standard fire without exceeding specified performance limits. The standard fire defines the atmosphere temperature in relation to time as shown in Figure 7.10. One simplified interpretation of this fire curve is that temperatures rapidly exceed 550°C, and at this level the strength of steel has reduced to about 60% of its ambient temperature strength, effectively eliminating the factor of safety against collapse.

As a result, most steel-framed buildings are required to be protected so that the steel temperature remains below the critical value for the specified fire resistance period. This is typically achieved by encasing the steelwork with fire protection in the form of an insulating material.

### 7.10.4 Protected steel

The steel may be fire protected by active or passive means. Active systems such as sprinklers act directly to suppress the fire. Passive fire protection takes the form of insulating materials applied to the steel to maintain its temperature at a sufficiently low level for the duration of the fire. Materials may be non-reactive, such as boards and sprays, or reactive, such as intumescent coatings.

Boards and sprays are typically based on vermiculite or mineral fibres, and their performance is related to thickness. A wide range of commercial products is available. Other forms of non-reactive protection include fibre blankets and traditional materials such as concrete and masonry.

Intumescent coatings are paint-like substances that are inert at normal temperatures but swell dramatically on heating to form an insulating layer of charred material. They have traditionally been used for aesthetic reasons, but are becoming very popular for unexposed steelwork, mainly because off-site application can significantly reduce site construction time.

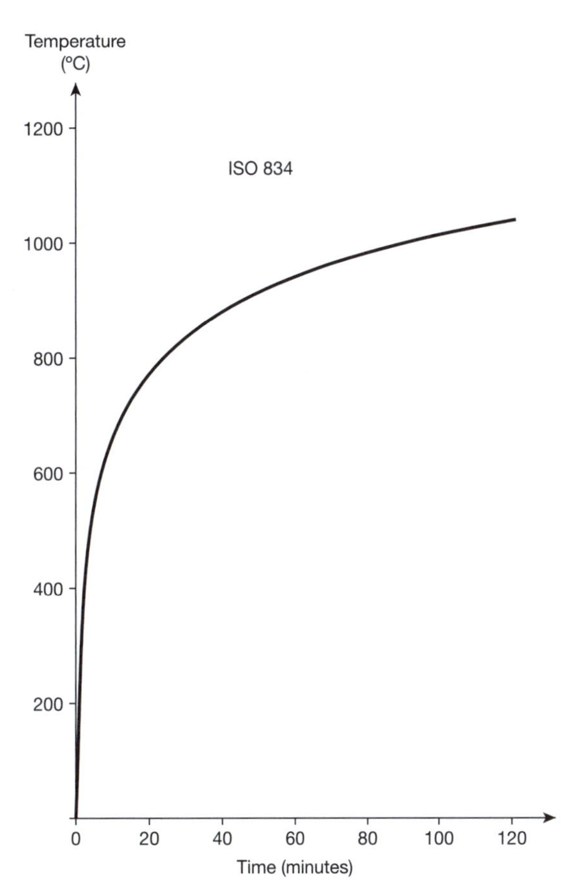


**Figure 7.10** Standard time–temperature curve in fire test

### 7.10.5 The heating of steel

The rate at which the temperature of a steel element increases depends on the size and shape of the cross-section. A stocky section will heat more slowly than a slender one, and this is related to its section factor, defined as the ratio of the exposed surface area to the volume of the member per unit length, $A/V$ – the higher the section factor, the faster the rate of heating. This can be important in assessing the performance of unprotected steel, but it is also used to determine applied fire protection requirements.

For a given fire resistance period the required thickness of fire protection depends on the section factor – the higher the section factor, the greater is the required thickness. Fire protection manufacturers therefore present design data for the required protection thickness in relation to section factor and fire resistance period.

### 7.10.6 New approaches to structural fire engineering design of steel structures

The cost of passive applied fire protection is significant, and it has been recognised that traditional assumptions about structural behaviour in fire are oversimplified. In particular, the following points have been recognised as potentially important.

- Steel temperatures are generally lower than those of the atmosphere, and parts of the section shielded from direct exposure to the fire may be considerably cooler.
- 'Yield' is not clearly defined at high temperatures.
- Fire is an accidental limit state, with lower design loads and less severe failure criteria.

- Secondary structural actions may become significant in fire conditions.

Extensive research has led to some significant improvements in the design of steel structures for fire, providing a more rational, consistent and cost-effective approach. These generally involve sophisticated modelling of the overall structural response, taking account of high temperature material properties, and provide a realistic indication of how a structure will behave in fire.

## 7.11 Sustainability

Sustainable development involves constructing in such a way that we maximise social and economic benefit while minimising environmental impacts associated with consumption of resources and pollution. In recent years, problems associated with climate change in particular have pushed care for the environment into the mainstream of public consciousness. For construction the issues are associated with use of materials, energy consumption, longevity and waste, and these are best considered through a life cycle analysis that accounts for impacts at all stages of construction, use and disposal.

The so-called 'greenhouse effect' is associated with changes in the 'transparency' of the atmosphere, decreasing the amount of thermal energy re-radiated from the earth's surface, while leaving the incoming radiation from the sun unaffected. The principal issues for the use of steel in construction are how the material is produced, and how it can contribute to more sustainable construction generally. Reducing the emission of certain gases, particularly carbon dioxide, will help to slow down climate change. The main source of carbon dioxide is the burning of fossil fuels, so the improvements in energy efficiency associated with modern steel production are important.

Recycling of scrap steel is also a significant factor. This offers a number of advantages compared with processing iron ore, with significant reductions in energy for production, extraction of raw materials, and waste. At the end of their useful life, steel products can be easily separated from other waste material by magnetic processes, and there is a well-established international infrastructure for recycling. This provides a reliable source for new steel production, with no loss of quality. As a result, steel is the most recycled of all materials, and almost half of the current world production is based on scrap material. Current data suggests that 94% of construction steel is recovered, with 10% at present being re-used and 84% recycled. The production of steel also generates waste by-products in the form of slags. Fortunately these have their own value and are generally used as the basis for other building products such as cement replacements and lightweight aggregates. These provide a further beneficial contribution to the environment, by reducing carbon dioxide emissions associated with the production of the replaced materials and reducing the need for new raw materials.

Efficient use of resources in production is important, but it is also necessary to consider the construction process and the operational stage of a building. The impact of the construction process on the local environment can be significant, even though it is temporary. The main issues are to avoid unnecessary disturbance – for example noise, dust, traffic – while at the same time using the minimum of resources including raw materials and energy, and creating a minimum of waste. Social issues such as using a local workforce and making improvements to the neighbourhood are also important. Because steel lends itself to prefabrication, site activity can generally be completed more quickly than other forms of construction, and waste is significantly reduced (waste arising during fabrication, for example from off-cuts, is invariably recycled). Also because it is a dry method of construction, it is easier to maintain a clean environment.

Because the life span of a building is typically 50 to 60 years, operational aspects of the building almost invariably have the most significant impacts on its sustainability. An integrated approach is therefore necessary for sustainable construction to be achieved, and minimising operational energy is a key issue. This may involve the use of natural ventilation and cooling to reduce the need for air conditioning, a well sealed and insulated envelope to minimise heat losses, and good use of natural lighting. The use of steel is consistent with all of these strategies.

Another factor is the adaptability of a building to the changing needs of the owners and users. Clearly, extending the useful life of buildings can reduce environmental impacts, and the key is to provide flexible spaces that can be used in a variety of ways. The most important aspect is to create column-free areas, which results in long beam spans. Fortunately steel has excellent strength and this can be readily achieved.

However adaptable a building may be, it is inevitable that any building will eventually have to be demolished or deconstructed, and clearly it is beneficial to recycle, or preferably reuse, as much material as possible. This is facilitated by reversible connections and dry, prefabricated construction, which are inherent features of steel construction.

## 7.12 References

*BS EN/ISO 8501-1:2007: Preparation of steel substrates before application of paints and related products – visual assessment of surface cleanliness. Part 1: Rust grades and preparation grades of uncoated steel substrates and of steel substrates after overall removal of previous coatings.* London: BSI, 2007

Gale, W. F. and Totemeier, T. C. (Eds). (2004). *Smithells metals reference book*, 8th ed. Oxford, UK: Elsevier Butterworth-Heineman,

## 7.13 Bibliography

Bates, W. (1984). *Historical structural steelwork handbook.* London, UK: BCSA.

BCSA. (2005). *Galvanising structural steelwork: An approach to the management of liquid metal assisted cracking.* London, UK: BCSA.

*BS 5950-1:2000. Standard use of steelwork in building: Code of practice for design in simple and continuous construction: hot rolled sections.* London, UK: BSI.

Corus. (2004). *Weathering steel.* Scunthorpe, UK: Tata/Corus.

Doran, D. K., Douglas, J. and Pratley, R. (2009). *Refurbishment and repair in construction.*.Caithness, UK: Whittles.

IStructE. (2008). *Manual for the design steelwork building structures to Eurocode 3.* London, UK: IStructE.

Joanides, F. and Weller, A. (2002). *Structural steel design to BS 5950:pt1.* London, UK: Thomas Telford.

SCI (2003). *Steel designers' manual*, 6th ed. Oxford, UK: Blackwell Science.

# 8 Steel Reinforcement for Reinforced Concrete

**Ben Bowsher** BSc MIM MIQA
Executive Director UKCARES

**Contents**

## 8.1 Introduction

This chapter is divided into three main sections.

1 Carbon steel bars for the reinforcement of concrete (BS4449: 2005).
2 Welded steel fabric for the reinforcement of concrete (BS4483: 2005).
3 Prestressing steel wire and strand (BS5896:1980) and bars (BS4486:1988) for the reinforcement of concrete.

In referring to the contents of this chapter, it should be borne in mind that standards are subject to revision from time to time and when a revised version is published it supersedes the previous version. The above British Standards have been written to support the European Standard for reinforcing steel, EN10080, compliance with which enables the manufacturer to apply the CE marking as required.

The term 'steel reinforcement' is used here to embrace bars, wire fabric and prestressing steels, although it is not universally accepted as including prestressing steels.

## 8.2 Bars and welded fabric

### 8.2.1 Terminology

Steel bars for use in reinforced-concrete construction are termed 'reinforcing bars', sometimes shortened to 'rebars'. With the increased use of automatic cutting and bending equipment by reinforcement fabricators, the smaller sizes, up to 16 mm, are commonly manufactured from bar supplied by the steel mill to the fabricator in coiled form. Coils are then decoiled, straightened and cut or cut and bent into standard or specified lengths and cut and bent shapes for delivery to the construction site. Bars, in coil form, may be manufactured by either hot or cold processing. In general terms, bars resulting from hot rolling will comply with the ductility grades BS4449 Grade 500A, 500B or 500C, while those resulting from cold processing will comply with ductility grade 500A.

Welded fabric is the correct term for bars pre-welded into sheet form. Fabric comprises a mesh of bars welded into sheets. Only material complying with BS4449 can be used, in any of the three ductility grades specified, but the welded product must comply with BS 4483.

Stainless-steel bars are also available to BS6744:2001, 'Austenitic stainless bars for the reinforcement of concrete' – with properties similar to bars suitable to BS4449.

### 8.2.2 Methods of manufacture

*8.2.2.2 Melting and casting*

BS4449 leaves the steel-making process to the discretion of the steel producer. In the UK and most commonly in the rest of Europe, the electric arc furnace (EAF) melting process is used. This process normally uses 100% scrap metal as the raw material. This is charged into the furnace and heat is applied by means of electrical discharge from carbon electrodes, thus melting the scrap. While little refining of steel occurs in the EAF furnace, subsequent ladle treatment is often employed.

Based on cast analysis and product analysis, the maximum values of specific alloying elements in the steel given in Table 8.1 are allowed in BS4449.

An EAF furnace generally produces 0.5 to 1.0 million tonnes per annum, making it ideally suited to smaller scale steel-making operations typically used for the manufacturing of reinforcing steel. The recycling principle employed by this process clearly provides a significant environmental benefit.

Molten steel is cast into 'billets', which are then further processed by hot rolling in straight bars or bars in coil form.

*8.2.2.3 Hot rolling bars*

All reinforcing steels go through a hot rolling operation in order to consolidate the product. In the hot rolling process, the cast billet is reheated to a temperature of 1100–1200°C, and then progressively rolled to smaller diameters to reduce its cross-section. At the end of the rolling mill, the product is sheared to the required length, bundled and labelled for despatch. All material is identified to the original cast of steel for the purposes of traceability.

The rib profile is rolled onto the steel in the last 'stand' of the hot rolling operation.

There are currently two common methods for achieving the required mechanical properties in hot rolled bar: in-line heat treatment and the use of micro-alloying additions. In-line heat treatment is sometimes referred to as quench and self tempering (QST). In this process, high-pressure water sprays are directed onto the surface of the bar as it exits the rolling mill. The short duration of the quench transforms only the surface of the bar to a hard metallurgical phase, while leaving the centre of the bar untransformed. After leaving the quench, the core cools slowly, transforming to a softer, tougher metallurgical phase. The heat diffusing out from the core 'tempers' the hard phase at the surface. The result is a relatively soft, ductile core, with a strong surface layer. The alternative of micro-alloying uses small amounts (< 0.15%) of certain alloying elements (normally vanadium) in order to achieve the required mechanical properties.

In other cases, the required mechanical properties are achieved after rolling, by applying further work to the hot rolled coil, usually by cold stretching. In this case, the shape is produced in the hot rolling operation, and the stretch produces around 3–4% reduction in cross-sectional area. The stretch is normally applied by putting the coil through a series of bending rolls, after which the product is layer wound onto spools. Layer winding enables improved processing by automatic bending machines, improving efficiency and providing better consistency of product.

Bars produced by any of the above methods perform similarly in concrete and are interchangeable.

*8.2.2.4 Cold rolling of bars*

For the production of some reinforcing steel bars, particularly in smaller sizes, the method commonly used is cold rolling. The

**Table 8.1** Chemical composition

| | *C* | *S* | *P* | *N* | *Cu* | *Carbon equivalent value (CEV)** |
|---|---|---|---|---|---|---|
| Cast analysis | 0.22 | 0.05 | 0.05 | 0.12 | 0.8 | 0.50 |
| Product analysis | 0.24 | 0.055 | 0.055 | 0.014 | 0.85 | 0.52 |

*For an explanation of CEV, see Section 8.3.5. See also notes in Table 2 of BS4449 for allowed deviations to the overall chemical composition, and Chapter 7 of this book.

**Table 8.2** Standard fabric types available from stock

| *Fabric reference* | *Longitudinal wires* | | | *Cross wires* | | | *Mass (kg m$^{-2}$)* |
|---|---|---|---|---|---|---|---|
| | *Nominal wire size* | *Pitch (mm)* | *Area (mm$^2$ m$^{-1}$)* | *Nominal wire size* | *Pitch (mm)* | *Area (mm$^2$ m$^{-1}$)* | |
| *Square mesh* | | | | | | | |
| A393 | 10 | 200 | 393 | 10 | 200 | 393 | 6.16 |
| A252 | 8 | 200 | 252 | 8 | 200 | 252 | 3.95 |
| A193 | 7 | 200 | 193 | 7 | 200 | 193 | 3.02 |
| A142 | 6 | 200 | 142 | 6 | 200 | 142 | 2.22 |
| *Structural mesh* | | | | | | | |
| B1131 | 12 | 100 | 1131 | 8 | 200 | 252 | 10.9 |
| B785 | 10 | 100 | 785 | 8 | 200 | 252 | 8.14 |
| B503 | 8 | 100 | 503 | 8 | 200 | 252 | 5.93 |
| B385 | 7 | 100 | 385 | 7 | 200 | 193 | 4.53 |
| B283 | 6 | 100 | 283 | 7 | 200 | 193 | 3.73 |
| *Long mesh* | | | | | | | |
| C785 | 10 | 100 | 785 | 6 | 400 | 70.8 | 6.72 |
| C636 | 9 | 100 | 636 | 6 | 400 | 70.8 | 5.55 |
| C503 | 8 | 100 | 503 | 5 | 400 | 49 | 4.51 |
| C385 | 7 | 100 | 385 | 5 | 400 | 49 | 3.58 |
| C283 | 6 | 100 | 283 | 5 | 400 | 49 | 2.78 |
| *Wrapping mesh* | | | | | | | |
| D98 | 5 | 200 | 98 | 5 | 200 | 98 | 1.54 |
| D49 | 2.5 | 100 | 49 | 2.5 | 100 | 49 | 0.77 |

Stock sheet size: length 4.8 m, width 2.4 m, area 11.52 m$^2$.

feedstock material for this process is a hot rolled, round section rod. Cold rolling is generally used for producing bars to BS4449 Grade 500A for the manufacture of welded fabric, but which can also be used for cutting and bending applications.

These bars undergo a process of deformation in which the rod is passed through a series of rolls. The material is forced into the gap between the rolls, and so is compressed. As in the hot rolling process, the rolls have grooves machined in them, such that three rows of ribs are formed on the cold rolled wire. Cold rolled wire has lower ductility than hot rolled steels, but has the advantages of good section control and coil presentation.

#### *8.2.2.5 De-coiling*

All coil products have to be de-coiled in some process before they can be used. Sometimes this is done as part of the processing on an automatic link-bending machine. Other reinforcement fabricators have de-coiling machines for producing straight lengths for further processing.

De-coiling processes are generally of two types: 'roller' straightening and 'spinner' straightening. It is important to realise that all de-coiling processes will have some effect on the final material properties. In some cases, the effect will be marginal. If badly performed, de-coiling can have a significant detrimental effect.

#### *8.2.2.6 Fabric manufacture*

Fabric produced to BS4483 comprises a square or rectangular mesh of bars that are welded together at some or all of the cross-over points in a shear-resistant manner. In practice, electric-resistance welding machines are used and for the standard ranges of fabric every intersection is welded (Table 8.2).

The most commonly produced sizes of bar in fabric are 5, 6, 7, 8, 9, and 10 mm, although fabric can be made with other sizes to suit different requirements. The standard fabric types available from stock are listed in Table 8.3. Other types of fabric are manufactured to specification as defined in BS4483.

**Table 8.3** Tensile properties

| *Standard* | *Ductility class* | *Min. yield stress (MPa)* | *Max. yield stress (MPa)* | *Tensile/ yield ratio* | *Elongation $A_{gt}$ (%)* |
|---|---|---|---|---|---|
| BS4449 | 500A | 500 | 650 | 1.05 | 2.5 |
| BS4449 | 500B | 500 | 650 | 1.08 | 5.0 |
| BS4449 | 500C | 500 | 650 | 1.15 | 7.5 |

## 8.3 Properties

### 8.3.1 Mechanical

Carbon steel bars according to BS4449 are used exclusively in both cut and cut and bent reinforcement to BS8666 and welded steel fabric to BS4483. BS4449 includes three ductility grades, any of which may be used to satisfy the above standards. The tensile properties of BS4449 are as follows given in Table 8.3.

### 8.3.2 Geometry/Bond

All bars to BS4449 are now ribbed, with a defined rib area requirement that ensures, via rib measurement or bond test, that the steel will bond satisfactorily with the concrete. Each ductility class has a defined rib pattern for ease of identification.

### 8.3.3 Sizes and lengths

It should be noted that the word 'size' is used in preference to 'diameter'. 'Size' implies 'nominal size'; that is, the diameter of a circle with an area equal to the effective cross-sectional areas of the bar. With a ribbed bar it is not possible to measure a simple dimension that equals the size of the bar, because the circle circumscribing the ribs will have a diameter greater than the nominal size.

Table 8.4 The nominal cross-sectional area of various bar sizes

| *Size (mm)* | *Cross-sectional area ($mm^2$)* | *Mass per metre (kg)* |
|---|---|---|
| 6 | 28.3 | 0.222 |
| 7 | 38.5 | 0.302 |
| 8 | 50.3 | 0.395 |
| 9 | 63.6 | 0.499 |
| 10 | 78.5 | 0.617 |
| 12 | 113 | 0.888 |
| 16 | 201 | 1.58 |
| 20 | 314 | 2.47 |
| 25 | 491 | 3.85 |
| 32 | 804 | 6.31 |
| 40 | 1257 | 9.86 |
| 50 | 1963 | 15.4 |

The cross-sectional areas derived from the mass per metre run using an assumed density of 0.00785 kg $mm^{-2}$ per metre run are given in Table 8.4. Grade 500 bars are commonly available and stocked in sizes 8, 10, 12, 16, 20, 25, 32, 40 and 50mm. Sizes 6, 7and 9 mm are commonly used in welded fabrics. Sizes 6, 7 and 9 mm are preferred diameters for the manufacture of welded fabric to BS4483 only.

The standard length of bars cut at the steel mill and available for suppliers is 12 m. The maximum length of bar obtainable by special arrangement and with adequate notice is 18 m. Longer lengths can be produced by a 'splicing' process with the use of mechanical couplers.

### 8.3.4 Tolerances

*8.3.4.1 Bars*

The tolerance applicable to cross-sectional area, expressed as the tolerance on mass per metre, shall not be more than ±4.5% for sizes greater than 8 mm and ±6.0% for sizes less than or equal to 8 mm. The permitted tolerance on length on bars supplied from a steel mill is +100/–0 mm, although other tolerances may be agreed at the time of the order. BS8666 specifies the tolerance on cut and bent bars. The tolerance on cut lengths is ±25 mm. For a bending dimension up to and including 1000 mm, a tolerance of ±5 mm is permitted. For bending dimensions over 1000 mm and up to and including 2000 mm, discrepancies of +5 mm and –10 mm are acceptable. Over 2000 mm the limits are +5 mm and –25 mm.

*8.3.4.2 Fabric*

The mass of an individual sheet of fabric shall be derived from a combination of the specified sheet dimensions, nominal bar sizes and specified pitches (centres) of bars in the sheet. The pitch of longitudinal and transverse bars shall not be less than 50 mm. The permitted deviations for welded fabrics are as follows.

- Length and width: ±25 mm or ±0.5%, whichever is the greater.
- Bar pitch: ±10mm or ±5%, whichever is the greater.
- Overhangs: to be agreed at the time of the order.

### 8.3.5 Weldability

All bars complying with BS4449 are defined as weldable. The ability of reinforcing steel to produce satisfactory welds will depend on a number of factors including the suitability of the welding procedure and the competence of the welder. For the steel itself, it is common to define weldability in terms of a carbon equivalent value (CEV). The maximum CEV as measured on steel bars, is specified in BS4449 as 0.52. The CEV) is calculated from the sum of the following percentage values:

$$C + \frac{Mn}{6} + \frac{Cr + Mo + V}{5} + \frac{Ni + Cu}{15}$$

BS7123, Specification for Metal Arc Welding of Steel for Concrete Reinforcement, specifies pre-heat temperatures for welds as a function of bar size, which are different if the CEV is above or below 0.42%. Alternatively, weld procedure tests may be used to demonstrate that a certain welding procedure can produce an acceptable weld at a certain CEV level. This is considered a protection against the formation of hard, brittle structures adjacent to the weld, which might be susceptible to fractures unless suitably controlled.

### 8.3.6 Bendability

*8.3.6.1 Test bending*

Bend performance is covered in BS4449 by means of a rebend test. The diameter of former (or mandrel) around which a bar must be bent without fracture or the formation of cracks is four times the diameter for bars equal to or less than 16 mm and seven times the diameter for sizes greater than 16 mm.

For reinforcing fabric after welding, the bars between welds must pass the rebend test specified in BS 4449.

*8.3.6.2 Bending in practice*

BS8666:2005 covers the bending dimensions and scheduling of steel reinforcement for concrete. Because all designers, detailers or fabricators need to be familiar with the whole of BS8666, it is not reproduced here. However, the minimum radii of former specified for practical use are given in Table 8.5. For practical bending of fabric, the radius of former used must be at least 12, 16, 20, 24 and 32 mm for wire size, 6, 8, 10, 12 and 16 mm, respectively.

An important limitation for the transport of bent bars is that the shape must be contained within an imaginary rectangle with the shorter sides not exceeding 2750 mm.

### 8.3.7 Product testing and certification

BS4449 gives recommendations for third-party certification of product conformity. This includes the determination by manufacturers of their statistical performance in the achievement of specified properties. Annex B of BS4449 describes the requirements for batch-acceptance testing, where third-party product certification does not exist. CARES is the independent product

Table 8.5 The minimum radii of former specified in BS8666 for practical use

| *Size (mm)* | *Radius (mm)* |
|---|---|
| 6 | 12 |
| 8 | 16 |
| 10 | 20 |
| 12 | 24 |
| 16 | 32 |
| 20 | 70 |
| 25 | 87 |
| 32 | 112 |
| 40 | 140 |
| 50 | 175 |

certification body in the UK which is accredited to provide such product certification.

### 8.3.8 Corrosion-resistant reinforcing steels

In normal atmospheric exposure conditions, with adequate concrete cover and appropriate detailing, corrosion of reinforcing steel in concrete is not a problem; the concrete provides a protective environment for the reinforcing steel. In certain aggressive environments, or where inadequate cover is provided, corrosion of reinforcing steels can occur, with effects that have been well documented. The problems of maintenance and replacement of corrosion-affected structures are thought to account for a multi-billion dollar problem worldwide.

#### *8.3.8.1 Corrosion of reinforcing steel in concrete*

Carbon reinforcing steel is generally passive in terms of corrosion behaviour at the pH levels normally found within concrete (pH = 13). However, corrosion can be accelerated by a number of different chemical processes in certain environmental conditions. The most commonly cited causes of corrosion of reinforcing steel in concrete are:

(a) *Carbonation of the concrete due to atmospheric attack.* The absorption of water and carbon dioxide from the atmosphere into concrete can cause a lowering of the pH, due to the formation of carbonic acid. Once the pH is sufficiently reduced, corrosion of the reinforcement can occur.
(b) *Ingress of chloride ions into the concrete.* When chloride ions reach a certain threshold level at the steel surface, passivity is broken down, and corrosion of the reinforcing steel proceeds.

Carbonation is normally controlled by the correct provision of concrete cover.

There are three widely accepted solutions to the prevention of corrosion of steel in concrete:

- galvanising
- epoxy coating
- stainless steel.

Where comparisons are being made between them, areas such as coating integrity, the role of concrete cover, the severity of the environment, the development of bond strength and relative economics must be considered. both galvanising and epoxy coating involve a coating onto the steel surface, and as such the corrosion performance of the steel depends on the nature and integrity of the coat and the performance of the coating process. Stainless steel is type of alloy steel which in itself is corrosion-resistant.

#### *8.3.8.2 Galvanising*

Galvanising, of which the most common method is by immersing reinforcement in a bath of molten zinc at around 450°C, provides a metallurgically bonded zinc coating over the surface of the reinforcement, either before or after bending. Any repairs required can be made by the application of zinc-rich paint. This coating can withstand a higher concentration of chloride ions than black bar before the onset of corrosion, while galvanized reinforcement can remain passive at lower pH values, offering an improved resistance to carbonation.

#### *8.3.8.3 Epoxy coating*

Epoxy coating may be applied by the dipping of cut or cut and bent bar into a fluidised bed of epoxy powder. Most commonly however it is applied to stock bar lengths, prior to cutting and bending, by an electrostatic spray process.

The epoxy coat acts as a barrier against the environment. Performance against both ingress of chloride ions and carbonation can be markedly improved over that of black bar, provided that the coating and the coating process are properly controlled. It is recognised however that for this to be achieved, the quality of the surface of the steel including preparation before coating, the performance of the epoxy powder, the coating process itself and equally importantly the quality of handling during cutting and bending, and delivery to and on site must all be carefully controlled.

The epoxy coat does not adversely affect the mechanical properties of the reinforcing steel and, provided that the coating thickness is not excessive, the bond strengths achieved can be acceptable.

#### *8.3.8.4 Stainless reinforcing steel*

Stainless steels are generally defined as those having a minimum of 12% chromium present as an alloying constituent. The presence of chromium results in a thin layer of stable chromium oxide forming on the surface of the steel. The oxide layer is passive, and highly resistant to atmospheric corrosion. Moreover, the oxide layer is instantaneously self-healing in oxidising conditions, so that cracks, defects or surface damage do not affect corrosion resistance. Stainless steels retain passivity in concrete at low pH levels, and high chloride concentrations, hence their use in structures at risk of chloride-induced corrosion.

Within the broad definition of stainless steels, many different types are available. BS EN 10088-3 lists more than 60 available alloy compositions. The different steels are grouped by type, according to their metallurgical structure: austenitic, ferritic, martensitic or duplex. Most stainless reinforcing steels are austenitic, with some duplex steels also being used; ferritic and martensitic steels are not normally used for reinforcing applications. BS6744:2001 includes six stainless steel designations from BS EN 10088-3, of varying chemical analysis and corrosion resistance characteristics (see Table 8.6).

BS6744:2001 specifies three grades with different mechanical properties. The term 'grade' is often confused with alloy type. In BS6744, 'grade' refers only to the mechanical properties, and

**Table 8.6** Stainless steel bar chemical composition (cast analysis) % by mass (from BS6744:2001)

| *BS EN 10088-1 steel designation number* | *C (max)* | *Si (max)* | *Mn (max)* | *S (max)* | *Cr (min/max)* | *Ni (min/max)* | *Mo (min/max)* | *P (max)* | *N (max)* |
|---|---|---|---|---|---|---|---|---|---|
| 1.4301[a] | 0.07 | 1.0 | 2.0 | 0.03 | 17.0/19.5 | 8.0/10.5 | – | 0.045 | ≤ 0.11 |
| 1.4436[a] | 0.05 | 1.0 | 2.0 | 0.015 | 16.5/18.5 | 10.5/13.0 | 2.5/3.0 | 0.045 | ≤ 0.11 |
| 1.4429 | 0.03 | 1.0 | 2.0 | 0.015 | 16.5/18.5 | 11.0/14.0 | 2.5/3.0 | 0.045 | 0.12/0.22 |
| 1.4462 | 0.03 | 1.0 | 2.0 | 0.015 | 21.0/23.0 | 4.5/6.5 | 2.5/3.5 | 0.035 | 0.10/0.22 |
| 1.4501[b] | 0.03 | 1.0 | 1.0 | 0.015 | 24.0/26.0 | 6.0/8.0 | 3.0/4.0 | 0.035 | 0.20/0.30 |
| 1.4529[b] | 0.02 | 0.50 | 1.0 | 0.010 | 19.0/21.0 | 24.0/26.0 | 6.0/7.0 | 0.03 | 0.15/0.25 |

[a]Nitrogen content of these steels may be increased to 0.22% max. [b]These designations are available on request but are only required for special applications.

**Table 8.7** Tensile properties of BS6744:2001 grades

| *Grade* | *0.2% Proof Stress, $R_{p0.2}$ (MPa)* | *Stress ratio, $R_m/R_{p0.2}$ (min) (%)* | *Elongation at fracture, $A_5$ (min)* | *Total elongation at maximum force, $A_{gt}$ (min) (%)* |
|---|---|---|---|---|
| 200 | 200 | 1.10 | 22 | 5 |
| 500 | 500 | 1.10 | 14 | 5 |
| 650 | 650 | 1.10 | 14 | 5 |

not to the chemical composition, which is determined by the steel type (designation). The three mechanical grades may be supplied in any of the steel designations. The three grades specified are given in Table 8.7.

It should be noted that the mechanical properties for both Grade 500 and Grade 650 would meet all of the mechanical requirements of BS4449:1997 Grade 460B. These grades can therefore be substituted for normal carbon steel, without significant design modification.

The most significant barrier to the use of stainless reinforcing steel is the cost. For this reason, stainless steel is normally used selectively in those parts of the structure most at risk of chloride ingress. The Highways Agency Design Guide BA 84/02 and BS6744:2001 both provide guidance on the appropriate specification of stainless reinforcement for particular exposure conditions. Table 8.8 summarises the guidance in BS6744.

### 8.3.9 Fibres for reinforcing concrete

Steel fibres may be used as an alternative in those applications where nominal reinforcement has traditionally been provided. These might include ground-bearing floor slabs, sprayed concrete for ground support, composite floors on metal decking and precast concrete tunnel linings.

Steel fibres can provide the concrete matrix with some post-cracking load capacity. For applications such as industrial floor slabs, design methodologies are now used that take advantage of this property. These methods may also be used for bar or fabric-reinforced sections, as long as the reinforcement is placed accurately. The correct use of spacers for reinforcement will ensure that this is the case.

Savings in the cost of supplying and fixing the conventional reinforcement that is replaced may offset the extra cost of adding fibres to the fresh concrete. The fibres, which should be uniformly distributed throughout the concrete mix so that the desired properties are achieved, may avoid any problems caused by the difficulties associated with the accurate location of conventional steel.

### 8.3.10 Welding of reinforcing steels

In the UK, designers, engineers and contractors are increasingly seeing the benefits of specifying factory-made pre-assembled welded fabrications to improve on-site productivity. There are now a number of appropriately CARES certificated reinforcement fabricators offering this service, and the use of pre-assembled welded fabrications is increasing significantly. Under controlled conditions away from the construction site:

- welding can provide efficient joining of reinforcement robust enough to survive transportation, lifting and installation
- when required, load-bearing joints can be produced to specified strength levels.

Pre-assembled welded fabrications can be manufactured into a range of shapes and sizes to suit applications such as pile reinforcement, beam and column cages, shear head reinforcement, diaphragm walls and 'carpet' reinforcement. The welding processes used for pre-assembled welded reinforcement fabrications are:

- gas shielded metal arc welding (metal inert gas (MIG) and metal active gas (MAG))
- manual metal arc welding (MMA), also known as 'stick' welding
- electrical resistance spot welding.

All of these may be used in appropriate applications. Control of welding processes is vital, since the combination of heat input and rapid cooling rates has the potential to produce hard, brittle structures, susceptible to fracture. The susceptibility to cracking is further increased if significant hydrogen absorption occurs in the welding process, which may cause hydrogen cracking in the zone adjacent to the weld, commonly known as the heat affected zone (HAZ). Improperly made welds introduce defects into the reinforcement, such as cracks, pores and other stress raisers which also impair mechanical properties, and can act as initiation sites for failure. These effects can lead to degradation of material properties, whether the welds are locational (tack) welds or

**Table 8.8** Guidance on the use of stainless steel reinforcement for different service conditions

| *Grade in BS EN 10088-1* | *Service condition* | | | |
|---|---|---|---|---|
| | For structures or components either with a long design life, or that are inaccessible for future maintenance | For structures or components exposed to chloride contamination with no relaxation in durability design (e.g. concrete cover, quality or water proofing treatment requirements) | Reinforcement bridging joints, or penetrating the concrete surface, and also subject to chloride contamination (e.g. dowel bars or holding down bolts) | Structures subject to chloride contamination where reductions in normal durability requirements are proposed (e.g. reduced cover, concrete quality or omission of water proofing treatment) |
| 1.4301 | 1 | 1 | 5 | 3 |
| 1.4436 | 2 | 2 | 1 | 1 |
| 1.4429 | 2 | 2 | 1 | 1 |
| 1.4462 | 2 | 2 | 1 | 1 |
| 1.4529 | 4 | 4 | 4 | 4 |
| 1.4501 | 4 | 4 | 4 | 4 |

1. Appropriate choice for corrosion resistance and cost. 2. Over-specification of corrosion resistance for the application. 3. May be suitable in some instances; specialist advice should be obtained. 4. Grade suitable for specialist applications, which should only be specified after consultation with corrosion specialists. 5. Unsuitable for the application.

structural welds. For this reason, all welding of reinforcing steels requires strict adherence to proven welding procedures and an acceptable level of welder competence.

Reinforcing steels used in the UK today are produced by a range of processes and with varying chemical analyses. It is therefore vital this be recognised and that proper control of the welding process be maintained if the negative effects of welding on the performance of the reinforcing steel are to be avoided. Welds must be fit for their intended purpose, whether they are load-bearing structural welds or non-load-bearing tack welds.

Welding of reinforcing steel may be conducted away from the shelter of a factory environment, although it should be appreciated that many of the controls described above are much more difficult to maintain on a construction site, with inherent risk to weld integrity and often site safety.

### 8.3.11 Mechanical splicing of reinforcing steels

The basic principle of splicing, which is common in reinforcing steel construction, is to lay two bars parallel to one another, over a certain length, and wrap them together with tying wire. The load present in the first bar is transmitted to the concrete by the bonding between steel and concrete, which then transmits the load to the second bar. Since this load transference is indirect, the efficiency of this joint depends on many factors, including the properties of the concrete. This complexity leads in turn to complex regulations for lapping, and design engineers and site engineers must be familiar with these design requirements if the job is to be done properly.

To improve in this area, the construction industry has developed the mechanical coupler as an alternative. There are many different types of coupler, but the most common are based on a threaded bar and coupler, a swaged coupler or a combination of the two. By creating an end-to-end bar connection, a continuous load path is created from one bar to another that is independent of the condition and quality of the concrete. Furthermore, the splice is easy to test relative to the overlapping method, which, if ever tested, would require a test in concrete. This would be expensive and time-consuming and, as a result, is never done in practice.

A number of proprietary splicing systems are sold, with various claims of performance and practical benefits. The most common of these are as follows.

- *Tapered thread*. The ends of the rebar are cut square and a tapered thread is cut onto the bar, using a set of dies and a custom-made threading machine, to suit the taper thread of the coupler. The coupler is assembled using a torque wrench, which should be calibrated for the purpose. A benefit of this system is that the bars are easily and correctly centred in the coupler, ensuring ease of application.
- *Parallel thread*. After cutting square, the ends of the bar are enlarged, or 'upset', by cold forging, such that the core diameter of the bar is increased to a predetermined diameter. A parallel thread is then cut onto the enlarged end. Using this technique, the effective diameter of the threaded bar is equal to the bar diameter, thereby ensuring failures within the bar and not the coupled joint. No torque wrench is required for assembly.
- *Swaged or thread/swage combination*. Swaging of a coupler, by applying radial pressure to the bar/coupler assembly and resulting in a pressure-sealed splice, is less common in today's market. This may be due to a very slow installation rate. To overcome this, there is a coupler that employs a combination of swaging and parallel threading to ensure a full-strength joint with flexibility of assembly. Sleeves, which are swaged onto the bar ends, are connected by means of a high-performance threaded stud, thus ensuring a full-strength joint.

The application of the coupler should be considered before settling on the use of a particular type. Broadly, applications include standard structural concrete (e.g. BS8110), bridge applications (e.g. BS5400) and nuclear applications (e.g. BNFL). Any certification used to support the suitability of particular couplers should be specific to these applications.

### 8.3.12 The behaviour of reinforcing steel when exposed to heat caused by a fire

Steels lose strength when exposed to increasing temperature, and this will therefore apply to reinforcing steel embedded in a concrete structure subject to elevated temperatures in a fire situation.

Typically, the static load on a reinforcing steel element within a concrete structure is designed to be around 55% of the room-temperature yield strength of the steel. This corresponds to the fact that, typically, the yield strength of reinforcing steel at 550°C is approximately 55% of the room temperature yield strength. 550°C is therefore commonly referred to as the 'critical temperature'. Therefore structures that are typically loaded will just remain stable at 550°C. This temperature will thus vary depending on the static loading conditions.

Protection of steel reinforcement is also a critical factor in any such considerations, and the performance of the reinforcing steel should account for the behaviour of concrete as a cover to the reinforcement. Any design approach should consider issues such as concrete type, the extent of concrete cover, and the degradation of concrete with time in a fire situation including spalling of concrete.

## 8.4 Prestressing steels

### 8.4.1 Terminology

Prestressing steels (or tendons) are produced in the form of wire and strand to BS5896:1980 or of bar to BS4486:1988.

### 8.4.2 Methods of manufacture

Steel for prestressing materials may be made by any process except the air and mixed air/oxygen bottom blown processes. The cast analysis must not show more than 0.040% sulphur or 0.040% phosphorus.

Wire made to BS5896:1980 is made by cold drawing a high carbon content steel rod through a series of reducing dies. The wire may be crimped mechanically or have the surface indented to improve bond when used in short units.

To enhance the mechanical properties further, the drawn wire is run through a mechanical straightening process followed by a stress-relieving low-temperature heat treatment or through a controlled tension and low-temperature heat-treatment process ('stabilising'). The latter process gives low relaxation properties. The wire is then rewound into large diameter coils (usually in 200–500 kg weights) from which it pays out straight. In this form it is suitable for making parallel bundles for post-tensioned applications.

Seven-wire strand made to BS5896:1980 is made by spinning six cold-drawn wires in helical form round a slightly larger straight central wire. The strand is then stress-relieved or 'stabilised', the latter imparting low relaxation behaviour.

Die-drawn strand is made by drawing a seven-wire strand through a die under controlled tension and temperature for low relaxation. The drawing process also compacts the wires, giving a larger steel area in the same circumscribing circle as the normal strand. Coils of strand are usually supplied in weights of 2–3 tonnes, for ease of handling on site.

Prestressing bars to BS4486:1988 are made by hot rolling low-alloy steels under controlled conditions and then stretching them in a controlled manner at about 90% of their characteristic

**Table 8.9** The most commonly* used sizes of wire, strand and bar

| *Type* | *Nominal diameter (mm)* | *Specified characteristic breaking load (kN)* | *Nominal area ($mm^2$)* | *Approximate length ($m\ kg^{-1}$)* |
|---|---|---|---|---|
| *Wire* | 7.0 | 60.4 | 38.5 | 3.31 |
| *Strand* | | | | |
| 'Standard' | 12.5 | 164 | 93 | 1.37 |
| | 15.2 | 232 | 139 | 0.92 |
| 'Super' | 12.9 | 186 | 100 | 1.27 |
| | 15.7 | 265 | 150 | 0.85 |
| 'Drawn' | 12.7 | 209 | 112 | 0.12 |
| | 15.2 | 300 | 165 | 0.77 |
| | 18.0 | 380 | 223 | 0.57 |
| *Bar type/strength* | | | | |
| Smooth or deformed 1030 | 26.5 | 568 | 552 | 0.230 |
| $N\ mm^{-2}$ | 32 | 830 | 804 | 0.158 |
| | 36 | 1048 | 1018 | 0.125 |
| | 40 | 1300 | 1257 | 0.101 |

*Products outside British Standard specifications often used: strand 15.4 mm/250 kN and 15.5 mm/261 kN. Refer to manufacturers' publications.

strength. Bars may be plain or deformed with a maximum length of 18 m. The prepared ends are threaded by cold rolling and bars can be joined end to end by couplers.

### 8.4.3 Sizes and properties

The most commonly used sizes of wire, strand and bar are given in Table 8.9. Other less used sizes of wire and strand are included in BS5896.

### 8.4.4 Mechanical properties

For prestressing steels it is customary for the manufacturer to supply a test certificate defining the size, breaking load, relaxation type and load-extension properties (the latter covering at least elongation at maximum load), 0.1% proof load or load at 1% extension and modulus of elasticity.

The requirements for relaxation and typical examples are given in Tables 8.10 and 8.11.

### 8.4.5 Quality assurance

UK suppliers of prestressing steels operate the CARES third-party product certification scheme in addition to high levels of inspection and testing.

### 8.4.6 Site precautions

Prestressing steels require careful handling and storage, and surface contamination and mechanical or heat damage must be avoided at all stages. The steel should be unloaded under cover and stored in clean, dry buildings free from the effects of wind-blown or soil-contaminated deleterious matter. Otherwise, storage should be off the ground by the use of impervious supports and the material should be covered with waterproof sheeting with air-circulation access below the underside of the steel. For long-term storage sealed polythene bags or tight wrapping should be used. A bag of silica gel or vapour-phase inhibitor should be included inside the bag or wrapping. Steel in ungrounted ducts also requires protection.

Prestressing steels should be unloaded carefully, avoiding shock loading and distortion of coils. Local damage should be avoided at all times and kinking of wire and strand in coil avoided. Bar threads should be protected.

Cutting should only be done by mechanical cutters on wire, and abrasive discs on strand. Flame cutting should only be used with the greatest care and supervision and stray heating of adjacent tendons must be avoided.

**Table 8.10** Maximum 1000 h relaxation* at 20°C

| | *Initial load/breaking load (%)* | | |
|---|---|---|---|
| | *60* | *70* | *80* |
| *Bar* | 3.5 | 6.0 | |
| *Strand and straight wire* | | | |
| Relaxation class 1 | 4.5 | 8.0 | 12.0 |
| Relaxation class 2 | 1.0 | 2.5 | 4.5 |

*From BS5896:1980 and BS4486:1988.

**Table 8.11** Typical 1000 h relaxation at 20°C for strand

| *Initial load/ breaking load (%)* | *Relaxation 1 (%)* | *Relaxation 2 (%)* |
|---|---|---|
| 60 | 3.4 | 1.00 |
| 65 | 4.4 | 1.04 |
| 70 | 5.6 | 1.10 |
| 75 | 7.6 | 1.60 |

# 9 Lead

**John Woods** Cert Ed AIP RP FIOR
Technical Officer, Lead Sheet Association

**David Wilson** MBE BSc PhD
Consultant, International Lead Association

Updated from the first edition text by

**Bob Cather** Bsc CEng FIMMM
Consultant, formerly Director of Arup Materials Consulting

**David Doran** FCGI BSc(Eng) DIC CEng FICE FIstructE
Consultant, formerly Chief Engineer with Wimpey plc

## Contents

## 9.1 Overview

Lead has been known and used for thousands of years, for a variety of reasons. Although some uses have disappeared with time as a result of technological change or substitution, for example, other new applications have been developed and in consequence the production and use of the metal have grown – and continue to grow – slowly but steadily. In 2004 about 6.7 million tonnes of lead was used throughout the world. Of this only about 3 million tonnes was produced by mining new ores, the major portion (3.7 million tonnes) being recycled from lead products that had reached the end of their useful lives. The proportion of lead produced by recycling is growing strongly and mine production is actually declining, making lead highly sustainable in terms of resource depletion.

Pure metallic lead is a relatively soft, malleable and ductile material that facilitates moulding into shapes and profiles. By contrast this *softness* can permit flow and distortion at barely above ambient temperature even under self weight. Pure lead has poor fatigue resistance and a high coefficient of thermal expansion that can lead to fatigue failure. The properties of lead in the above conditions can be beneficially modified by the use of alloying additions as described in greater detail below. Lead also possesses good vibration damping properties.

Pure lead has good corrosion resistance to sulphuric, chromic and phosphoric acids, forming insoluble salts on the metal surface that prevent further attack. In acid solutions, it is cathodic to iron and the iron is attacked preferentially. In alkaline solutions the polarity is reversed; lead is then anodic to iron and protects the iron from attack. Lead has good corrosion resistance in hard water, where the calcium and magnesium salts form compact, insoluble carbonate films on the surface, as evidenced by the lead sheeting in the Roman baths at Bath, which is still in good condition nearly 2000 years after its installation. In soft water condensates these salts are not present, so no protective film is formed and the lead slowly corrodes. The rate of corrosion then increases in proportion to the amounts of oxygen and carbon dioxide dissolved in the water but with possible consequences for both the lead and the water. Lead has reasonably good resistance to seawater, rural, industrial and marine atmospheres.

Care is required, however, to ensure that potential corrosive conditions are recognised and appropriate measures are taken to prevent attack from bimetallic couples, crevice corrosion, differential aeration, etc. Buried lead pipes can be attacked by a combination of incompletely cured cement and penetration of road de-icing salts. Even timber can produce corrosive conditions, because tannic and acetic acids, which attack lead, can be released from damp wood.

The most common construction uses of metallic lead are for cladding, roofing and flashings. In alloy form, it is a constituent of the common solders. The low solubility of lead in other metals is used to advantage in the production of free machining alloys. Lead has a high surface tension and very low solubility in copper and steel, for example, so lead additions to these metals exist as small globules disseminated through the solid metal. This property is put to good use during machining, since the turnings break off into small segments whenever the cutting tool encounters one of these soft lead particles.

Lead oxides are an important constituent of crystal glass, cathode ray tube glass (for radiation shielding) and ceramic glazes. Chemical compounds of lead are used as stabilisers in PVC. Lead compounds have also been widely used in the past as paint additives and as anti-knock additives for petrol, but their use has largely been phased out for health and environmental reasons. Lead is potentially harmful if absorbed by living organisms. Precautions should be taken to avoid inhalation of lead dust and fumes.

## 9.2 Sources and production

Lead is most commonly extracted from the earth as the sulphide (PbS), often in association with zinc sulphide (ZnS), but lead carbonate and lead sulphate ores are also recovered on occasions. The major suppliers are Australia, China, the USA and Peru. The sulphide ores are concentrated by crushing and flotation and then sintered to agglomerate and remove sulphur. The oxidised sinter cake is most commonly reduced in a lead blast furnace, in which the zinc may be removed by volatilisation, although there are also a number of more modern direct smelting processes. The molten lead is cooled to just above its melting point to separate the less soluble impurities which are skimmed off the surface. The metal is then reheated to about 750°C to oxidise out impurities such as arsenic, antimony and tin. Many lead ores have valuable silver and gold contents and zinc may be added at this stage to precipitate $Au_2Zn_3$ and $Ag_2Zn_3$, which form a crust on the top of the lead as the temperature is lowered. Any remaining zinc can be removed by vacuum distillation. Alternatively, the metal may be refined electrolytically – a treatment that is effective in removing bismuth impurities – but electrolytic refining of lead is very uncommon.

## 9.3 Metallic lead

Pure lead is very soft and creeps readily under small loads at ambient temperature. Consequently, the pure metal is only used where corrosion resistance or specific properties are required. The purest form in the European Standard EN 12659: 1999 is designated PB990 and contains a minimum of 99.99% lead by weight. Lower grades down to 99.94% are allowed to contain higher levels of several impurities, notably bismuth which is particularly difficult to separate from lead. Additions of various elements such as antimony, arsenic and tellurium can be used to refine the grain size and increase the strength of lead. Small additions of copper and tin can be used to improve corrosion resistance. For the grids in lead-acid batteries the former practice of adding antimony has been replaced almost universally by additions of calcium and tin, which reduce the amount of gassing and permit the batteries to be sealed. Hard lead containing up to 9% antimony is now uncommon. All grades are readily deformed by rolling to produce lead sheet, generally of thickness 0.5–2.0 mm, although foils as thin as 0.02 mm can be produced.

## 9.4 Lead sheet

Three methods of producing lead sheet are used in the UK: traditional sand cast, rolled lead (formerly known as milled) and the machine cast process. However, only rolled lead sheet can be manufactured to meet the requirements of BS EN 12588:1999 for lead and lead alloys – Rolled Lead Sheet for Building Purposes. The standard strictly limits the thickness tolerances of the sheet and the small amounts of other metals included in the chemical composition of the metal to control thermal fatigue and creep without affecting malleability.

When exposed to the atmosphere lead forms a surface protective patina or film which gives good resistance to rural, industrial and marine environments.

With relatively low installation costs and minimal maintenance, lead sheet is known to provide outstanding durability even in severe weather conditions. The metal can also be recycled at the end of its service life and nearly all of the lead produced in the UK is manufactured from reclaimed sources (e.g. lead sheet pipe and car batteries).

Lead sheet is very soft and can be worked and dressed to shape at ambient temperatures. Thermal expansion is relatively high, so it is customary to limit the area of the sheets and to provide only loose interlocking joints in order to minimise distortion caused by variations in summer to winter temperatures. BS6915:2001 covers the installation of rolled lead sheet and gives guidance on the types of expansion joints and maximum areas of lead sheet that can be considered for different applications, so the risk of fatigue cracking and creep is minimised.

Perhaps one drawback to the use of lead for roofing is the weight of the metal, as 1 mm thick sheet weighs approximately 12.5 kg per m$^2$, thus requiring stronger structural support than with some lighter cladding materials. However, while lead melts at 327.4°C it is incombustible and resistant to fire, and lead roofs can achieve suitable AA fire rating when tested to BS475-3 and laid on T & G boarding or concrete. The use of lead in roofs is however, a specialist area and perhaps has not yet been developed to its full potential.

Lead sheet is used also for simple flashing and damp-proof courses but should be protected by a layer of bitumen when it is in contact with fresh concrete to minimise corrosion. Other applications include shower trays, sink tops, and linings for bins storing containers of strong acids, where again the corrosion-resistance properties are an advantage.

The high density of lead is an advantage for shielding against X-rays and gamma rays. The high internal damping capacity, coupled with the high density and low stiffness, also makes lead sheet an excellent material for use as a sound barrier.

## 9.5 Other uses of lead

The low melting point, high ductility, high corrosion resistance and durability of lead are attributes that have resulted in a wide range of applications of the metal, from lead shot for shotgun ammunition to leaded windows. Electroplated or hot-dipped lead coatings were once commonly applied to engineering components but this use has now virtually disappeared. Extruded lead sheaths can be applied to underground and underwater power cables to prevent water ingress: although they are still used, the market is much smaller than in the past.

When alloyed with tin, lead forms solders with low melting points. The binary eutectic (38% lead, 62% tin) has a melting point of 183°C and has been used extensively in a wide variety of applications, of which electrical/electronic solders are the only significant use remaining, and even this is declining in the face of environmentally based pressures to substitute. Quaternary alloys with tin, cadmium and bismuth can have melting points as low as 70°C, making them useful for heat safety devices such as sprinkler systems.

Lead has a good resistance to seawater, forming a compact adherent corrosion film that makes the alloy ideal for the cathodic protection of steel structures in seawater. Lead blocks, or alloys of lead with silver, tin or antimony, are connected by a conductor to the steel and an impressed DC current is applied to reverse the normal corrosion polarity so that the lead behaves anodically to the steel.

The largest use of lead by far is in lead-acid batteries, in which it is used both in the metallic form and as lead compounds. Batteries consist of electrically conducting metallic grids onto which is pressed a paste comprising mainly lead oxide and powdered lead. Using sulphuric acid as an electrolyte, the battery can be charged and discharged repeatedly, with the lead oxide converting to lead sulphate on the positive electrode during discharge, and the lead converting to lead sulphate on the negative. On recharging, the reverse reactions occur. Traditionally the metal grids were alloyed with antimony, but in modern batteries calcium and tin are the most common alloying elements.

There are several other important uses of lead compounds, although two traditional applications – lead paint pigments and tetraethyllead antiknock additives for petrol – have virtually disappeared. Lead oxides are still a critical constituent of special glasses, in particular crystal glass and cathode ray tube glass in which it prevents the escape of X-rays. Lead compounds are also important components of ceramic glazes and of certain plastics (notably PVC), in which they act as 'stabilisers' to prevent degradation under the influence of heat and ultraviolet light.

A higher strength sheet can be produced by dipping a low carbon steel sheet in a bath of molten lead with a chloride flux, held at a temperature of about 350°C, to produce a coating thickness normally in the range 5–20 μm. The appropriate standard for these hot-dip coatings is BS6582-17. Lead does not readily wet and adhere to steel, so tin is usually added to the bath to form a lead–tin layer on the surface of the steel, to which the coating adheres more firmly. The tin content can range from 2% to 25% but for Terne coatings (Terne meaning dull, from the French), the tin content is usually in the range 3–15%. The lead coating serves as a lubricant in bending and forming operations and the coated sheet is readily joined by soldering or spot welding. The thickness of the lead coating can be controlled by variation of the bath temperature, time of immersion of the sheet in the bath and by rolling or vibrating the sheet as it emerges from the bath. The coated sheet is mainly used for roofs, fire doors, lining tanks in contact with petroleum products and for housings for electrical switchgear, fuse boxes, etc. in rich environments.

Lead is also widely used to isolate moving machinery and prevent transmission of mechanical vibrations to the floor. Very noisy equipment may be entirely encased in a lead-lined cubicle. Thicker lead sheet is used under foundations to insulate buildings from extraneous vibrations. One interesting development being explored in Japan is the use of lead as an earthquake damper. Instead of lead sheet, the building slab is isolated from the ground by a rubber pad which supports the entire weight of the building. Movements in the rubber pad due to terrestrial displacements are damped by columns of lead fixed between the base of the building and the earth.

## 9.6 Bibliography

BSEN 12588:1999, *Lead and lead alloys – Rolled lead sheet for building purposes.*

BS6915:2001, *Design and construction of fully supported lead sheet roof and wall coverings – Code of practice.*

*Rolled Lead Sheet* – The complete manual published by the Lead Sheet Association, Tonbridge, UK, 2003.

# 10 Titanium

**David K. Peacock** CEng FIMMM
Titanium Marketing & Advisory Services

## Contents

## 10.1 Introduction

Titanium is the fourth most abundant metallic element in the earth's crust. Discovered in the late eighteenth century, it was not until late in the 1940s that the metal was available commercially. Since that time titanium's high strength to weight ratio has guaranteed a leading position in aero engines and airframes, and the metal's outstanding resistance to corrosion in a wide range of aggressive media has dramatically extended the performance and life of chemical, petrochemical, desalination installations, offshore equipment and power generation plant. Titanium is the ninth industrial metal. Its position in this 'league' belies its importance in key, major applications worldwide.

## 10.2 Sources and production

The two principal titanium ores, rutile ($TiO_2$) and ilmenite (a mixture of titanium and iron oxides) both occur in massive deposits. Rutile is readily extracted from beach sands and considerable deposits of these in Western Australia have from the first supplied a significant part of total demand. Less than 10% of refined rutile is used for metal production, the balance being used primarily for pigments. The extraction of the metal from its oxide by the Kroll process is a lengthy, batch, chemical process involving chlorination of the rutile in the presence of coke to produce titanium tetrachloride; the reduction of the tetrachloride in a reactor with magnesium to produce titanium 'sponge' (a porous mass contaminated with chlorides and metallic magnesium); cleaning, crushing and blending of the sponge with alloying additions to use directly for the production of castings (rammed graphite or investment cast); or to produce an electrode for consumable vacuum arc melting (VAR); and finally melting to produce an ingot for processing by traditional forging, rolling, extrusion, drawing, etc. to provide a comprehensive range of semi-finished or 'mill' products. Alternatives to VAR currently include electron beam or plasma cold hearth re-melting. The development of reduced energy extraction processes continues, most notably with the FFC electrolytic deoxidation process, which not only consumes less energy but is also a more environmentally friendly process overall.

Considerable use is made of titanium and titanium alloy revert materials and scrap in the remelting processes. Bulky solids are welded together to form part or all of the VAR electrode, or used as loose feed to the cold hearth furnaces. Light scrap and turnings are normally compacted before use in either process. The ability to recycle titanium ensures maximum recovery of every form of reverted material and scrap and overall reduction of energy input to sustain metal supply. The sustained value of life-expired titanium parts and systems should always be taken into account in life cycle cost considerations. The probability that titanium process plant and other equipment will remain both clean and free of corrosion means that re-use of the whole plant or of elements of the plant, e.g. condenser tubing, may be considered, and may offer further economies in major equipment production and procurement cycles.

## 10.3 Why use titanium?

In all fields of engineering, designers, fabricators and end users now consider titanium for an ever-widening range of applications. The metal and its alloys, once seen as 'exotic', suffer less today from outdated and misguided notions about high cost, poor availability, and difficulties in fabrication. Without these past prejudices, engineers are ready to realise the benefits that titanium offers in applications as widely differing as aero engines, offshore platform pipework, automotive exhaust systems and springs, implants for the human body and ultra-lightweight roofing. Properly specified and designed, there are few other materials that can approach, economically or technically, the performance provided by titanium and its alloys in critical applications.

Titanium is as strong as steel, yet 45% lighter. Titanium alloys will work continuously at temperatures up to 600°C, resisting creep and oxidation, and down to liquid nitrogen temperatures without loss of toughness. Titanium will survive indefinitely without corrosion in seawater, and most chloride environments. The wide range of available titanium alloys enables designers to select materials and forms closely tailored to the needs of the application. The versatility of two basic compositions is such however that they continue to satisfy the majority of applications, and this level of common use remains a major factor in the cost-effective production, procurement and application of titanium.

The price per kilo of titanium is no guide to the cost of a properly designed component or piece of equipment. First cost is in any event only one part of the full cost equation. Maintenance, downtime and replacement costs, which may be a very significant element in plant designed for long and reliable service life, are another. Additional costs of energy associated with operating unnecessarily heavy or thermally inefficient equipment may be a third penalty on life cycle costs. Titanium is frequently specified for its ability to cut costs through reliable and efficient performance; for example, it is installed in steam turbine condensers and service water piping applications with a 40 year performance guarantee against corrosion failure. One manufacturer offers a 100 year no corrosion warranty on its metal supplied for architectural applications.

## 10.4 Titanium and its alloys

Titanium and its alloys may be characterised by their microstructures, and there are three main types of alloy:

- *alpha* – a close packed hexagonal crystal lattice
- *beta* – a body centred cubic lattice
- *alpha–beta* – a mix of the two.

Commercially pure titanium with oxygen as the principal 'alloy' is an alpha alloy. Certain nominally alpha alloys are additionally described as 'lean alpha' or 'near alpha' when only a relatively minor amount of beta phase is present in the structure.

Alpha–beta alloys are by far the most common type with the composition. Heat treatment potential varies. Ti-6Al-4V in this group is the workhorse alloy of the industry.

Beta alloys, which can be heat treated so as to introduce alpha phase by ageing after solution treatment, are more correctly designated 'metastable beta alloys'. Relatively few fully stable beta alloy compositions are available, with as yet no substantial commercial applications.

Other minor phases and intermetallic compounds can be created in certain alloys, and these may be specifically beneficial or specifically harmful to performance. Suppliers of the alloys concerned have detailed information on the generation or avoidance of such phases.

The principal attributes of titanium alloys of prime importance to the design engineer are:

- outstanding corrosion resistance
- superior strength-to-weight ratios
- low density and modulus
- high fatigue strength and fracture toughness
- excellent erosion resistance
- high heat transfer capability
- low thermal expansion coefficient.

Titanium alloys are also used because of their:

- excellent properties at cryogenic temperatures
- high temperature creep resistance

**Table 10.1** Frequently used titanium alloys

| *Alloy* | *Attributes/Applications* |
|---|---|
| Commercially pure (CP) titanium Grades 1, 2, 3 and 4 | Corrosion resistance with ease of fabrication and welding, strength increases with grade number |
| CP Ti with added palladium (Pd) Grades 7, 11, 16 and 17 | Pd provides enhanced corrosion resistance properties; otherwise as Gr1 (11, 17) and Gr 2 (7, 16) |
| CP Ti with added ruthenium (Ru) Grades 26 and 27 | Replacing Grades 7, 11, 16, 17 at lower cost properties otherwise as Gr1 (27) and Gr 2 (26) |
| Titanium alloy Ti-0.8Ni-0.3Mo Grade 12 | Higher strength than CP but with enhanced corrosion resistance similar to Pd and Ru grades |
| Titanium alloy Ti-6Al-4V Grade 5 | Workhorse general-purpose high-strength alloy |
| Ti-6Al-4V ELI Grades 23 and 29 | Variants of Grade 5 for specialised applications |
| Titanium alloy Ti-3Al-2.5V Gr. 9 (with Pd Gr.18 with Ru Gr. 28) | 'Half 6-4' more easily fabricated, medium-strength alloy. Gr.18 and 28 with enhanced corrosion resistance for specialised applications. |
| Alloy Ti-3Al-8V-6Cr-4Zr-4Mo Grade 19 | High strength beta alloy with well-established industrial applications including bolting |
| Alloy Ti-5Al-1Sn-1Zr-1V-0.8Mo Grade 32 | 'Navy' alloy fabricable and weldable for high-strength marine and offshore structural applications |

- non-magnetic character
- fire resistance
- short radioactive half life
- attractive appearance of natural and coloured surfaces.

The range of titanium alloys available ensures that almost every design and engineering requirement can now be satisfied. A convenient and widely used system for identification of the various grades of commercially pure (cp) titanium and titanium alloys used for both high-strength and corrosion-resisting applications is provided by the ASTM specifications. The full range of alloys extends from high ductility commercially pure titanium used in plate heat exchangers (where extreme formability is essential) to fully heat treatable alloys with strength above 1500 MPa. Corrosion-resistant alloys are capable of withstanding attack in the most aggressive sour oil and gas environments or geothermal brines at temperatures above 250°C. High-strength oxidation- and creep-resistant alloys see service in aero-engines at temperatures up to 600°C.

The two compositions that currently dominate applications are commercially pure titanium (ASTM Grade 2), selected for basic corrosion resistance with strength in the range 350–450 MPa, and the high-strength (900–1100 MPa) titanium alloy Ti-6Al-4V, (ASTM Grade 5).

Table 10.1 lists the alloys most frequently used in a wide range of applications and that are most commonly available from manufacturers and stockists. Grade numbers refer to the ASTM specifications. In addition, a number of Russian alloys are now being marketed and sold in the West. Details and specifications are available from the suppliers. Specifications include:

- VT1-0 Commercially Pure – similar to ASTM Grade 2
- VT6 – similar to ASTM Grade 5.

Properties in Table 10.2 relate to the class of alloys covered. The various forms of titanium and its alloys are covered by the following ASTM specifications:

- ASTM B 265 – sheet, strip and plate
- ASTM B 338 – seamless and welded tube
- ASTM B 348 – bars and billets
- ASTM B 363 – seamless and welded fittings
- ASTM B 367 – castings

**Table 10.2** Properties of titanium and its alloys

| *Description* | | *Commercially pure titanium (CP)* | *Medium strength non-heat treatable (Ti-6A1-4V)* | *Highest strength alloys* |
|---|---|---|---|---|
| *Alloy type* | | *Alpha* | *Alpha–Beta* | *Beta* |
| *Mechnical properties* | | | | |
| 0.20% Proof stress | MPa | 345–480 | 830–1000 | 1100–1400 |
| UTS | MPa | 480–620 | 970–1100 | 1200–1500 |
| Elongation | % | 20–25 | 10–15 | 6–12 |
| Hardness | HV | 160–220 | 300–400 | 360–450 |
| Tensile modulus | GPa | 103 | 114 | 69–110 |
| Torsion modulus | GPa | 44.8 | 42 | 38–45 |
| *Physical properties* | | | | |
| Density | g/cm$^3$ | 4.51 | 4.42 | 4.81–4.93 |
| Thermal expansion (0–220°C) | $10^{-6/0}$C | 8.9 | 8.3 | 7.2–9.5 |
| Thermal conductivity at room temp | W/m k | 22 | 6.7 | 6.3–7.6 |
| specific heat at room temperature | J/Kg/°C | 515 | 565 | 490–524 |

- ASTM B 381 – forgings
- ASTM B 861 – seamless pipe
- ASTM B 862 – welded pipe
- ASTM B 863 – wire.

Welding consumables to matching compositions are covered by AWS specification A5.16.

## 10.5 Titanium design codes

Each country tends to have its own regulations regarding the design, manufacturing and testing of pressure vessels and other equipment for both critical and general duties. These codes rarely contain detailed information on materials. Titanium is such a case and this has led to situations where titanium designs are based on codes originally produced for steel or other alloys, with consequent poor utilisation of the titanium and a more expensive product than could be achieved by the use of titanium property-based standards agreed between the buyer and the manufacturer. Supplements detailing titanium properties are available with a limited number of codes.

## 10.6 Corrosion resistance

Commercially pure (CP) titanium is one of the most corrosion-resistant engineering metals, and provides outstanding performance in a wide range of aggressive environments. Titanium is immune to corrosive attack by saltwater or marine atmospheres. It also exhibits exceptional resistance to a broad range of acids, alkalis, halogens, natural waters, corrosive gases, reducing atmospheres and organic media.

Titanium develops a thin, tenacious and highly protective surface oxide film. The surface oxide of titanium will, if scratched or damaged, immediately reheal and restore itself in the presence of air or even very small amounts of water. Under many otherwise corrosive conditions passivation with inhibitors is possible.

The corrosion resistance of titanium can be enhanced in reducing environments by alloying with certain noble metals, such as palladium or ruthenium, with molybdenum and nickel, or by impressed anodic potentials (anodic protection).

Titanium alloys are essentially immune to chloride pitting and intergranular attack and are highly resistant to crevice corrosion and stress corrosion cracking. Titanium is immune to microbiologically influenced corrosion (MIC).

Commercially pure titanium and the simple alpha alloys have simple, single phase structures and do not suffer from corrosion caused by secondary phase precipitates in welds, heat-affected areas or cast structures. Many titanium alloys irrespective of phase composition offer similar corrosion-resistant performance other than in more aggressive media.

## 10.7 Erosion resistance

All titanium alloys have excellent resistance to erosion. Impingement attack due to high water velocities, usually caused by partial tube or pipe blockage, has been a persistent problem, typically for copper–nickel alloys in offshore service. Because of the nature of its oxide film, titanium has superior resistance to erosion, cavitation and impingement attack. Titanium is over 20 times more erosion-resistant than the copper–nickel alloys. Seawater flowing at velocities up to 30 m/s can be safely handled by titanium. The presence of abrasive particulates lowers the maximum allowable velocity, but any titanium alloy will outperform most other materials in conditions where its oxide film if damaged will self-repair.

Where adequate pumping capacity exists, system flow rates can safely be increased, permitting design with smaller diameter pipework, and tighter pipework bend radii. The benefits are weight and cost reduction and space saving. Titanium requires no inlet or outlet or bend erosion protection.

Some reported values for erosion of commercially pure titanium are given in Table 10.3.

**Table 10.3** Reported values for erosion of titanium

| *Alloy grade* | *Media* | *Velocity (m/s)* | *Metal loss (mm/year)* |
|---|---|---|---|
| 2 | Seawater | 7 | Nil |
| | Seawater | 36 | .008 |
| | Seawater + 40 g/L 60 mesh sand | 2 | .003 |
| | Seawater + 40 g/L 10 mesh emery | 2 | .013 |
| | Seawater + 40% 80 mesh emery | 7 | 1.5 |
| 5 | Seawater | 36 | .01 |

## 10.8 Fire resistance and fire systems

Titanium has a high melting point, above 1600°C, and the metal in bulk presents no fire hazard. Uninsulated thin wall welded titanium pipes pass the Norwegian Petroleum Directorate (NPD) H-class hydrocarbon fire test. Titanium has been certified as a 'non-combustible material' for roofing and cladding by the Japanese Ministry of Construction. Fines, grinding dust, fine turnings, etc. do represent a hazard and are subject to workplace fire safety procedures which include keeping machines free of accumulated fines, dust and turnings. As with many other metals, exposed surfaces of titanium should be coated in work areas where dynamically induced sparking is a defined hazard.

In the NPD fire test, unsupported 2 m lengths of Grade 2 Schedule 10 pipe stand dry for 10 min. In this time the test furnace and the pipe temperature rises to 1000°C. Cold water at 10 bar pressure is then pumped through the pipes while the furnace temperature further rises to ~1200°C over the remainder of the 2 h test period.

The unique shock resistance and damage tolerance of titanium provide the maximum possibility for survival in the event of explosion, fire or other disaster. Leading fire system manufacturers now offer all titanium sprinkler and deluge systems detectors, nozzles, valves and pipework installed at minimum cost using cold bending. Titanium fire water systems are installed on Frøy/TCP (Elf Petroleum); Sleipner West and Norne (Statoil); and Troll B and Brage (Norsk Hydro).

Increasingly for naval vessels as well as fast ferries and other civilian shipping, the drive for weight reduction, ease of installation, freedom from maintenance and damage tolerance makes titanium a prime candidate for fire systems as well as all water management piping exhaust systems and heavier section structural components. This factor coupled with increasing pressure from insurance underwriters to reduce risk and increase the capabilities of fire protection methods and equipment has brought titanium to the fore in what has been a major re-evaluation of the particular problems posed by shipboard and offshore platform fire protection.

## 10.9 Applications of titanium and titanium alloys

Titanium and its alloys have proved to be technically superior and cost-effective in a wide variety of industrial and commercial applications. These include aero engines, airframes and aircraft landing gear, chemical and petrochemical plant and systems power generation and desalination plant, offshore and marine applications, architectural cladding and roofing, medical applications, implants and instruments, automotive and road transport, sporting goods and personal effects.

## 10.10 Chemical and petrochemical plant and systems

Titanium was first used in chemical plant in the mid-1960s. Its outstanding resistance to corrosion in oxidising chloride environments, seawater and other aggressive media was rapidly established. In several processes titanium was the first and only choice for effective plant performance and acceptable levels of life cycle cost. As shown below, titanium is now more widely used. Titanium equipment, properly designed, can compete with most of the more common but usually less effective corrosion-resistant alloys. A range of titanium alloys is included in the provisions of MRO-175 covering sour service in petroleum refining and extraction (including downhole logging tools and the like).

## 10.11 Electrochemical processes

The use of titanium anodes greatly increases the environmental friendliness of electrochemical processes such as chlorine production. Titanium anodes are more stable than nickel, lead, zinc or mercury. Titanium electrode activating coatings can be replaced several times on the same titanium structure. The process efficiency (energy input per unit of product) and process control (consistency and safety) are significantly higher when titanium electrodes are used. Titanium is also used successfully for fixtures in the electroplating and anodising industry, and as the substrate for precious metal/precious metal oxide electrodes for cathodic protection, electroplating, electrolysis and electro-osmotic damp control (Table 10.4).

## 10.12 Power generation and desalination plant

### 10.12.1 Condensers

Titanium has been proved by the power industry to be the most reliable of all surface condenser tubing materials. Several hundred million metres of thin wall welded titanium tube are now in service worldwide. Titanium is immune to all of the processes of corrosion that occur in condenser operation, which continue to cause damage or threaten operational efficiency for units tubed with less corrosion-resistant metals and alloys. Design to optimise the engineering properties of titanium is possible for new condensing equipment. The retubing of existing units may require modifications to the condenser to compensate for the lower modulus and density of titanium as well as its possible influence as a cathode in a mixed metal system.

Experience gathered from hundreds of condenser units operating in widely varying conditions has progressively exposed practices that are prejudicial to the normal long life expectation of titanium tubes. Condenser performance is a significant factor in the overall operational efficiency of power plant. The widespread use of low-cost titanium welded tube, which is immune to corrosion, has virtually eliminated condenser problems from power station economics.

### 10.12.2 Steam turbine blades

Titanium alloys are candidate materials for advanced steam turbine blades. The advantages are weight reduction (56% the density of 12 Cr steel) and corrosion resistance. Titanium resists attack by reducing acid chlorides in condensate which forms at the Wilson line. Titanium is additionally resistant to corrosion fatigue and stress corrosion cracking. Hydrogen absorption is not a concern at the temperatures in the L-1 and L regions of turbines. As with other blading alloys, titanium requires a leading edge shield to improve the resistance to water droplet erosion.

### 10.12.3 Flue gas desulphurisation

More than 20 years' operational experience with wet FGD scrubbers, ductwork and stacks has shown that titanium can resist the highly corrosive conditions encountered continuously or intermittently in the FGD process. The aggressive conditions experienced include wet/dry interfaces, acidic condensates and flyash laden deposits enriched in chloride and fluoride species. The largest single FGD plant installation in Europe is the Drax plant of National Power PLC in North Yorkshire, England. The

**Table 10.4** Process media where titanium's oxide film is stable (within wide operating boundaries)

1. *Chlorine and chlorine compounds*
   a. Moist chlorine gas
   b. Water solutions of chlorites, hypochlorites, perchlorates and chlorine dioxide
   c. Chlorinated hydrocarbons and oxychloro compounds
2. *Other halogens*
   a. Bromine; moist gases, aqueous solutions and compounds
   b. Iodine; moist gases and compounds
3. *Water*
   a. Fresh water, steam
   b. River water
   c. Seawater
   d. Water containing organisms that cause biofouling
   e. Water containing microorganisms that would typically cause microbiologically influenced corrosion in other materials of construction
4. *Oxidizing mineral acids*
   a. Nitric
   b. Chromic
   c. Perchloric
   d. Hypochlorous (wet chlorine gas)
5. *Inorganic salt solutions*
   a. Chlorides of such minerals as sodium, potassium, magnesium, calcium, copper, iron, ammonia, manganese, nickel
   b. Bromide salts
   c. Sulfides, sulfates, carbonates, nitrates, chlorates, hypochlorites
6. *Organic acids*
   a. Acetic
   b. Terephthalic
   c. Adipic
   d. Citric (aerated)
   e. Formic (aerated)
   f. Lactic
   g. Stearic
   h. Tartaric
   i. Tannic
7. *Organic chemicals (with moisture or oxygen)*
   a. Alcohols
   b. Aldehydes
   c. Esters
   d. Ketones
   e. Hydrocarbons
8. *Gases*
   a. Sulfur dioxide
   b. Ammonium
   c. Carbon dioxide
   d. Carbon monoxide
   e. Hydrogen sulfide
   f. Nitrogen
9. *Alkaline media*
   a. Sodium hydroxide
   b. Potassium hydroxide
   c. Calcium hydroxide
   d. Magnesium hydroxide
   e. Ammonium hydroxide

three 260 m high concrete flues of the stack are lined with a total of over 24,000 $m^2$ of Grade 2 commercially pure titanium sheet. One flue per year was lined in 1992, 1993 and 1994 respectively. An inspection (1999) revealed no significant corrosive attack on the titanium after seven years' exposure to a full range of operating conditions.

### 10.12.4 Desalination applications

Over 14 million metres of thin wall titanium welded tube are in desalination plant service worldwide. One large plant alone uses 5.7 million metres of titanium tube. Most items of desalination plant operate satisfactorily with commercially pure titanium tube, however where design metal temperature exceeds 80°C or where acid injection is used for scale control alloys with enhanced corrosion resistance such as Grade 12, or 16 should be specified. As with steam turbine condensers the benefits, in terms of performance and reliability, can be extended by using titanium tubeplates. Titanium tubed desalination plants operating without corrosion problems have set unsurpassable standards for pure water cost and quality, plant leak tightness and availability in both Multi-Stage–Flash (MSF) and vapour recompression desalination plants.

## 10.13 Offshore and marine applications

Because of high strength, high toughness and exceptional erosion/corrosion resistance in marine environments, titanium is currently being used for:

- offshore fire water piping, pumps and valves
- subsea drilling and production riser pipes and stress joints
- hulls and pressure vessels for deep sea submersibles
- underwater manipulator systems
- underwater bolting and other fasteners
- water jet propulsion systems
- propeller shafts and propellers
- shipboard cooling and piping systems, heat exchangers
- exhaust stack liners
- naval armour
- high-strength fasteners
- yacht fittings.

Offshore platform applications, including titanium stress joints, risers and fire, service and grey water management systems are required to perform in harsh operating environments. The use of titanium for these has clearly demonstrated the cost benefits and energy savings that attach to greater availability and reliability, reduction of unscheduled outages, longer intervals between shutdowns for routine maintenance and longer safe life overall. Such systems, operating for example in the North Sea, with planned lives of up to 70 years and with critical requirements for continuous safe, reliable performance, demonstrate in particular the low life cycle costs and high environmental benefits that result from the near total compatibility of titanium with marine environments.

The first all-titanium fishing boat was launched in Japan in 1998. Weighing 4.6 tonnes, the 12.5 m long vessel can travel at 30 knots with improved fuel efficiency. Operational cost savings include no necessity for hull painting and easier removal of biofouling. The progressive degradation of glass fibre boat hulls by repeated fouling and cleaning is an ongoing penalty for the Japanese inshore fishing fleet.

## 10.14 Architectural applications

Immunity to atmospheric corrosion, high strength and light weight of titanium make it an ideal architectural metal. Applications now include exterior and interior cladding, roofing, fascias, panelling, protective cladding for piers and columns, artwork, sculpture, plaques and monuments.

Innovation and cost-competitive design have been the key to success. Close to 2000 tonnes of titanium has been used in architectural applications worldwide, representing over one million square metres of roofing and cladding. The Guggenheim Museum in Bilbao has focused the attention of European designers and architects on titanium, and there are now several prestigious buildings including the Glasgow Museum of Science, and its adjacent IMAX theatre (Figure 10.1), the Maritime Museum and the Van Gogh Museum in Amsterdam which have made extensive use of titanium for roofing and cladding.

The Guggenheim Museum is clad with 32,000 $m^2$ of Grade 1 titanium sheet, manufactured to the architect's specification with a deliberately introduced ripple and soft textured finish. Roofing and cladding sheet thicknesses are typically 0.3–0.4 mm, weighing 1.35–1.8 $kgm^2$. This dramatically low weight means that even large areas of roofing can be carried on less massive structural members, with corresponding cost savings.

Titanium has a low coefficient of thermal expansion, half that of stainless steel and copper and one third that of aluminium. It is virtually equal to that of glass and granite and close to that of concrete. Consequently the thermal stress on titanium is very small. Longer lengths of roofing sheet can be laid in titanium than in other metals, thus reducing installation cost.

Titanium, with a melting point of 1660°C has been certified as a 'non-combustible material' for roofing and cladding by the Japanese Ministry of Construction. Uninsulated titanium can withstand fire tests in which the metal temperature is sustained at 1100°C.

If scratched or otherwise damaged, the oxide film on titanium is instantaneously self-healing, and no corrosion occurs. Pitting or general corrosion caused by acid rain is no problem for titanium. Service in more unusual and aggressive conditions such as volcanic ash deposition, and high-humidity geothermal spa environments has similarly failed to damage the tough, tenacious, titanium oxide film. The film is colourless and at its normal thickness appears so, but thickening, normally by anodising can produce a range of spectacular refractive colours, whose attractiveness can be enhanced by pre-treating the metal surface. Effects from a soft matte finish to bright, near gleaming reflectivity have been used (Figure 10.2).

The immunity to atmospheric corrosion makes titanium a most environmentally friendly metal: there is no corrosion or harmful run-off to contaminate rainwater. Titanium does not degrade in use and is 100% recyclable.

The combination of low material cost and structural weight reduction when titanium is used can, with appropriate design, give a lower overall building cost. In competition with stainless steel and copper, titanium has on occasions been the lowest first cost option. The long and trouble-free service life of titanium roofs, cladding and other architectural applications points to reduced maintenance cost and low cost of ownership, and means that on a life cycle basis titanium will be among the most competitive of architectural metals.

## 10.15 Other applications based on corrosion resistance

Early exploitation of the resistance of titanium to chlorides guaranteed the metal a leading position in equipment for pulp bleaching for paper production. Much titanium was used in plant for terepthalic acid (TPA) in the manufacture of synthetic fibres. Modern hydrometallurgical plant, geothermal energy extraction equipment and wet air oxidation plant require the use of the most corrosion- and erosion-resistant titanium alloys. Specific applications within the food brewing and pharmaceutical industries

**Figure 10.1** IMAX theatre, Glasgow. © Keith Hunter

**Figure 10.2** Shimane Prefecture Art Museum (courtesy Japan Titanium Society)

have demonstrated the superior performance of titanium in respect of freedom from pitting, easier sterilisation and reduced risks of product contamination.

The immunity of titanium to atmospheric corrosion coupled with the ability to produce a range of textures and durable colours has placed it at the forefront of modern materials for architectural roofing and cladding, with many fine buildings in Europe and Japan now demonstrating the versatility of the metal and the cost savings in building possible due to its lighter weight.

## 10.16 Applications based on damage tolerance

The high strength, low weight and low modulus of elasticity of titanium contribute to its effectiveness in components and structures requiring damage tolerance. Applications include cladding for concrete bridge piers, fighting vehicle and personnel armour and the deck cladding of fighting ships. Titanium also maintains its inherent toughness to very low temperatures. The ASME code allows room temperature properties to be used in design for service temperatures down to – 60°C, but in practice all titanium alloys are mechanically safe at temperatures down to at least –100°C. Alloys with reduced levels of interstitial oxygen (ELI) are specified for liquid gas containers and other cryogenic applications.

## 10.17 Fabrication of titanium

Hot and cold working processes are required in providing the full range of titanium and titanium alloy mill products and finished goods. Casting, classical hot forging in open or closed dies, ring rolling, extrusion, hot and cold rolling and drawing typify the processes for both primary conversion and finishing of mill products. All grades of titanium can be forged. Die forgings currently range in size from precision forged gas turbine blades of less than 0.1 kg to a large engineering components weighing more than a tonne. For small batch sizes, conventional open die forging or machining from solid may be more economical. Isothermal forging can be used to make components with thinner sections and shallower draft angles than can be achieved with conventional forging. Sheet metal work, machining, welding and surface treatment processes may all be used in production of finished components.

## 10.18 Sheet metal work

All traditional forming methods can be used on commercially pure titanium and those alloys with sufficient ductility. Titanium has a low modulus of elasticity; this means it will deform considerably under load and then spring back. This characteristic must be allowed for when fabricating. Springback is not an issue when hot creep forming or superplastic forming is used. Since titanium is more ductile at elevated temperatures, lower forming pressures may be used. Warm pressing, stretch forming and spinning can also be used for forming titanium. Handling annealed commercially pure titanium sheet is similar to working quarter hard stainless steel. Bends can be made without undue difficulty with a radius of 1 to 1.5t, in ductile grades of commercially pure titanium (Grade 1) and cold formable beta alloys such as Ti-15-3 and ASTM Grade 21.

## 10.19 Welding

Welding of titanium by various processes is widely practised and service performance of welds is proven, with an extensive and continuously extending record of achievements. Most titanium alloys can be fusion welded and all alloys can be joined by solid state processes. Unfortunately some engineers still believe titanium is difficult to weld, possibly due to its particular requirements for argon or helium gas shielding, or because in the past it has normally been handled only by specialist fabricators. Titanium is actually easier to weld than many metallurgically more complex metals and alloys (Table 10.5).

Welding and joining methods appropriate to titanium include the following.

- *Arc welding processes:* tungsten inert gas (TIG), metal inert gas (MIG), plasma arc (PAW)
- *Power beam processes:* laser and electron beam (EB) welding
- *Resistance welding:* spot, seam and the Resista-CladTM cladding process
- *Friction welding:* rotary, radial, linear, orbital, stir, stud and other specific jointing techniques
- *Diffusion bonding*
- *Forge welding*
- *Explosive bonding*
- *Brazing, soldering*
- *Adhesive bonding*

## 10.20 Machining

Machining of titanium on conventional equipment is considered by experienced operators as being no more difficult than the machining of austenitic stainless steel. Different grades of titanium varying from commercially pure to complex alloys do have different machining characteristics, as do different grades of steel or different aluminium alloys. Use rigid set-ups, correct speeds, feeds and tooling. Large volumes of metal should not generally be removed without intermediate stress relief.

Disposable carbide tools should be used whenever possible for turning and boring since they increase production rates. Where high-speed steel tools must be used, such as on drilling, super-grades are recommended. Overhang should be kept to a minimum in all cases to avoid deflection and reduce the tendency for titanium to smear on the tool flank.

High-speed drills are satisfactory for the lower hardness commercially pure grades but super-high-speed steels or carbide should be used for the harder alloy grades. Sharp, clean drills just long enough for the hole being drilled and of sufficient length to allow a free flow of chips should be used. A dull drill impedes the flow of chips along the flutes and is indicated by feathered or discoloured chips.

Straight, clean holes must be drilled to ensure good tapping results. Reducing the tendency of titanium to smear to the lands of the tap by nitriding the taps, relieving the land, use of an interrupted tap or providing for a free flow of chips in the flutes will ensure that sound threads are produced. Chip clogging can be reduced by the use of spiral pointed taps, which push the chips ahead of the tool, and more chip clearance can be provided by sharply grinding away the trailing edge of flutes. For correct clearance, two-fluted spiral point taps are recommended for diameters up to 8 mm and three fluted taps for larger sizes.

'Climb milling' should be used to lengthen the life of face milling cutters. 'Climb milling' produces a thin chip as the cutter

**Table 10.5** Weldability of the commonly used ASTM grades

| *ASTM grades* | *Weldability* | *Comments* |
|---|---|---|
| 1, 2, 3, 4, 7, 11, 12, 13, 14, 15, 16, 17, 26, 27 | Excellent | Commercially pure titanium and low alloy grades with minor additions of Pd, Ru, Mo, etc. |
| 9, 18, 28 | Excellent | Ti-3Al-2.5V alloy grades |
| 5, 23, 24, 29 | Fair–good | Ti-6Al-4V alloy grades |

teeth leave the work, reducing the tendency of the chip to weld to the cutting edge. Relief or clearance angles for face milling cutters should be greater than those used for steel. Sharp tools must be used. End milling of titanium is best performed using short length cutters. Cutters should have sufficient flute space to prevent chip clogging. Cutters up to 25 mm diameter should have no more than three flutes.

## 10.21 Chemical milling

Very precise intricate milling of titanium can be effected by using controlled selective acid attack of the surface. The titanium component is placed in a solution of 12–20% nitric acid and 4–5% hydrofluoric acid with a wetting agent. Solution temperature is maintained between 30 and 40°C. At 36°C metal is removed at a rate of 0.02 mm/min. Areas that do not require material removal are masked with either a neoprene elastomer or an isobutylene-isoprene co-polymer.

## 10.22 Titanium and the environment

Titanium's resistance to corrosion in a wide range of aggressive conditions eliminates or substantially reduces metal loss and energy input for repair/replacement. The risk of land, water or air pollution from corrosion failure of process plant is similarly reduced, as is product contamination from metal loss or by cross-stream leakage caused by corrosion. Titanium in architectural applications will not cause pollution of rainwater from run-off of roofs or cladding.

The low weight of titanium reduces energy loss in reciprocating equipment, fuel consumption in aircraft, ships and land vehicles and can compensate for performance shortfalls in payload, range, speed and other critical factors.

The biocompatibility of titanium assures safe use in human bone and tissue replacement, harmlessness to terrestrial and marine flora and fauna, non-interference with microbiological processes and immunity to them.

# 11 Zinc

**Tony Wall** PhD
Consultant to the Zinc Information Centre

## Contents

## 11.1 Introduction

Zinc is the fourth most important metal in terms of tonnages used, after iron, aluminium and copper. Worldwide some 9 million tonnes of zinc is produced and used annually. Basic technical data on unalloyed zinc are given in Table 11.1.

The pattern of zinc use is very different from that of the other industrial metals. About half of the total is used in coatings to protect iron and steel from corrosion. The several methods of applying the coatings and their properties are described in a later section.

About one fifth is used with copper to make brass. Brasses are copper alloys containing between 20% and 40% zinc, sometimes with the addition of other metals to obtain particular properties. The range of applications for brasses – which may be cast, forged, rolled, extruded and drawn – is outlined in Chapter 3.

Some 15% is used in the production of zinc base alloys, which generally contain between 4% and 27% aluminium and small additions of magnesium and sometimes copper. Zinc base alloys are generally pressure die cast to produce accurate components for a wide range of industries – automotives to zip fasteners and many in between.

Zinc sheet containing additions of copper and titanium is widely used for roofing and cladding, mainly in Europe. Zinc oxide is an essential component in the production of the vulcanised rubber needed for tyres and many other products.

There are many very small uses of zinc, particularly in chemical form. These include the addition of zinc compounds to human and animal foods and fertilisers, to ensure that the essential intake is available and to guard against a range of diseases. Taking all the uses together, it has been estimated the almost half the total zinc used goes into construction in some form – protecting structural steel; zinc-coated steel sheet and zinc sheet for roofing and cladding; ducting; plumbing; hardware; and many others.

The performance of zinc as a protective coating for steel depends on two properties acting synergistically. Firstly, zinc in most environments corrodes much more slowly than steel. Secondly, zinc is below iron in the electromotive series, so when the two metals are in contact in the presence of an electrolyte, zinc corrodes preferentially, protecting the iron. This means that zinc-coated steel is protected at any areas where the coating is missing through damage or such operations as drilling or sawing.

Pure zinc melts at 419°C and boils at the unusually low temperature of 910°C. As a result of those properties and its unusual (for a metal) crystal structure, the mechanical properties of unalloyed zinc, particularly at elevated temperatures, are generally unattractive and it is little used in that form. However, when it is alloyed with even relatively small percentages of aluminium the properties are very attractive, especially when pressure die cast (see Section 11.4), and components made that way can compete with those made on other metals as well as with some engineering plastics.

**Table 11.1** Basic technical data on zinc

| | |
|---|---|
| Atomic number | 30 |
| Atomic weight | 65.37 |
| Density (at 25°C) | 7.14 g/cm$^2$ |
| Melting point | 419.5°C |
| Boiling point (at 760 mm Hg) | 907°C |
| Modulus of elasticity | 70,000 MN/m$^2$ |
| Specific heat (at 20°C) | 0.382 kJ/kg K |
| Latent heat of fusion (at 419.5°C) | 100.9 kJ/kg |
| Latent heat of vaporisation (at 906°C) | 1.782 MJ/kg |
| Linear coefficient of thermal expansion (between 20 and 250°C) | 39.7 μm/m K |
| Volume coefficient of thermal expansion (between 20 and 400°C) | 0.89 × 10$^{-6}$ K |
| Thermal conductivity (at 18°C) | 113 W/m K |
| Electrical resistivity (at 20°C) | 5.9 μΩ m |
| Standard electrical potential (against hydrogen electrode) | –0.762 V |

## 11.2 Sources and production

Zinc ores are mined in many countries and in every continent, but more than 40% of the total comes from China, Australia and Canada. The most common ore is sphalerite (also called Zinc Blende), which contains zinc sulphide together with significant amounts of lead and other elements that are separated out during the refining process.

As with most ores, the first stage of processing is crushing followed by flotation, which separates the metal-bearing compounds from the waste or *gangue.* The resultant *concentrate* containing around 50% zinc is then refined to produce pure metal. The refinery may be near the mine, but more generally the concentrate is shipped to refineries nearer the industrial centres where demand for the metal is greatest.

Around 90% of refined zinc is produced by an electrolytic process. The concentrate is roasted to release sulphur as sulphur dioxide, which is used in the production of sulphuric acid. The remainder – essentially a crude zinc oxide – is then dissolved in sulphuric acid and the resulting solution is purified to remove other elements that may be recovered in separate processes. The purified solution is electrolysed to produce zinc metal of 99.99% purity or greater. The depleted electrolyte is recirculated to dissolve more of the concentrate.

Most of the remaining zinc metal is produced by a thermal process – the Imperial Smelting Process. After roasting, the concentrate is mixed with coke and fed into a special blast furnace. The zinc oxide is reduced to zinc metal, but as the process operates at above the boiling point of zinc, it is in the form of vapour. If the vapour were allowed to come into contact with air, it would immediately recombine to form the oxide. To overcome this problem the furnace gas containing zinc vapour is passed through a splash condenser carrying a spray of molten lead. The zinc dissolves in the lead and, as it cools, liquid zinc separates and is drawn off and cast into ingots. The liquid lead then returns to the circuit. The zinc produced by this process contains about 2.5% of lead and may require further refining for some applications.

More detailed descriptions of mining and refining processes are given in Porter (1999).

Recycling is and always has been an important aspect of the zinc industry. Materials for recycling arise from several sources and are returned to the zinc circuit in several different ways (www.zinc.org/info/zinc_recycling). Overall some 30% of zinc used comes from secondary sources.

European standards EN1179 (2003) and EN13283 (2002) govern the purity of both refined and secondary zinc and the forms under which they are traded.

## 11.3 Zinc coatings

The most commonly used zinc coating is hot dip galvanising, although very large components may be flame sprayed while small components may be electro-plated, mechanically plated or sherardised. Galvanising may be split into two categories. Batch hot dip galvanising – also called general galvanising – is used for fabricated parts and is covered by BS EN ISO 1461. Continuous galvanising – hot dip or electrolytic – is applied to steel sheet and is covered by BS EN 10326 for hot dip and BS EN 10152 for electrolytic.

### 11.3.1 Hot dip galvanising fabricated steel

In common with most coating processes, the secret to achieving a good-quality coating lies in the preparation of the surface of the iron or steel. It is essential that this is free of grease, dirt and scale before galvanising. These types of contamination are removed by a variety of processes. Common practice is to degrease using an alkaline or acidic degreasing solution into which the component is dipped. The article is then rinsed in cold water and then dipped in dilute hydrochloric acid at ambient temperature to remove rust and mill scale.

After a further rinsing operation, the components will then commonly undergo a fluxing procedure. This is normally applied by dipping in a flux solution – usually about 30% zinc ammonium chloride at around 65–80°C. Alternatively, some galvanising plants may operate using a flux blanket on top of the galvanising bath. The fluxing operation removes any of the last traces of oxide from the surface of the component and allows the molten zinc to wet the steel.

When the clean iron or steel component is dipped into the molten zinc (which is commonly at around 450°C), a series of zinc–iron alloy layers are formed by a metallurgical reaction between the iron and zinc. The rate of reaction between the steel and the zinc is normally parabolic with time and so the initial rate of reaction is very rapid and considerable agitation can be seen in the zinc bath. The main thickness of coating is formed during this period. Subsequently the reaction slows down and the coating thickness is not increased significantly even if the article is in the bath for a longer period of time. A typical time of immersion is about four or five minutes, but it can be longer for heavy articles that have high thermal inertia or where the zinc is required to penetrate internal spaces. Upon withdrawal from the galvanising bath a layer of molten zinc will be taken out on top of the alloy layer. Often this cools to exhibit the bright shiny appearance associated with galvanised products. The galvanised coating thickness varies with steel section thickness but for steel 6 mm and thicker a minimum average coating thickness of 85 µm is typical.

### 11.3.2 Galvanising steel sheet

The surface must also be cleaned for galvanising steel sheet, although this is usually done by passing the sheet through an oxidising and/or reducing atmosphere. The sheet is then immersed into the molten zinc for a relatively short period of time before being wiped either mechanically or by use of air knives (high-pressure jets of air) to control coating thickness, as it is withdrawn. This being an automated process, careful control of bath composition, bath temperature, immersion time and the wiping procedure allows close control of the coating thickness that is achieved on sheet; the most commonly used grade has an average coating thickness of 20 µm.

While continuously galvanised steel sheet was originally coated with pure zinc, greater use is now made of a zinc–5% aluminium coating, often referred to as Galfan, together with aluminium-rich coatings (aluminium–45% zinc) such as Galvalume. The aluminium addition has been found to increase the level of corrosion protection by 2–3 times when compared to unalloyed zinc coatings of equivalent thickness.

Electro-galvanised steel sheet is produced by passing an electric current through the steel as it passes at a controlled rate through a bath containing a dissolved zinc salt. The coating produced is thin but very even.

Adaptations of the hot dip and electrolytic coating processes for steel sheet are used to produce zinc-coated steel wire.

### 11.3.3 Other zinc coating processes

For zinc flame spraying, which is covered by BS EN 22063, good preparation is essential and requires the steel to be grit blasted to Sa3. The coating process then involves feeding a zinc wire into a gun where it is melted by a flame or arc system before being propelled in an inert gas onto the steel surface. Thick coatings in excess of 100 µm may be achieved by conducting additional passes, but it is essential that the coating, which is porous, be sealed.

Electroplating of small components (BS3382-2 for threaded components) also involves electrochemical deposition of a zinc coating on to steel while immersed in a solution of a zinc salt. The coatings formed on such products are thin typically about 5 µm.

Small components can be mechanically plated by tumbling pre-cleaned articles in a solution containing zinc powder and glass beads. The beads impact the powder onto the steel surface to form a zinc coating.

Small components may be coated by a process known as sherardising, which is covered by BS4921. This involves tumbling articles with zinc dust and sand at a temperature a little below the melting point of zinc (419°C). The coating formed is fairly uniform, with a thickness of 20–30 µm being typical.

### 11.3.4 Properties of zinc coatings

As the galvanised coating is metallurgically bonded to the base steel, it has excellent mechanical properties and is less susceptible to handling damage than other anti-corrosion systems. On batch-galvanised steel the zinc–iron alloy layer is hard and wear-resistant while the soft outer layer of zinc cushions against impact damage. Due to the nature of the zinc–iron alloy layer it is important that all forming is conducted prior to galvanising, otherwise damage to the coating (cracking and flaking) may occur. Continuously galvanised steel sheet has a coating that is almost entirely of pure zinc and can therefore normally be mechanically formed without damage to the coating.

Once applied, the zinc coating has a great advantage over alternative corrosion protection systems as it provides not only barrier protection but also cathodic protection should the coating become damaged. Cathodic protection occurs because zinc is electro-negative to steel and so corrodes in preference to steel, sacrificing itself to protect the steel from corrosion. The corrosion products from the zinc are deposited on the steel, resealing it from the atmosphere and therefore stopping corrosion.

The resistance of galvanised steel to atmospheric corrosion depends on a protective film that forms on the surface of the zinc. When the steel is withdrawn from the galvanising bath, the zinc has a clean, bright, shiny surface. With time this changes to a dull grey patina as the surface reacts with oxygen, water and carbon dioxide in the atmosphere. A complex but tough, stable, protective layer is formed that is tightly adherent to the zinc. Contaminants in the atmosphere affect the nature of this protective film. The most important contaminant for zinc is sulphur dioxide ($SO_2$), and it is the presence of $SO_2$ that largely controls the atmospheric corrosion of zinc, although presence of chlorides and time of wetness also play a role. Once the protective layer is formed, the galvanised coating corrodes very slowly and at a uniform rate.

The life of a zinc coating varies according to service conditions; two good sources of information are the Galvanizers Association's Zinc Millennium Map for the UK (see www.galvanizing.org.uk) and BS EN ISO 14713. Based upon the GA map, a corrosion rate of 1.5 µm or less could be expected in most atmospheric environments in the UK. Table 11.2 summarises the expected life for different coating processes and thickness', assuming an average annual corrosion rate of 1.5µm per year.

Regardless of which zinc coating is applied, zinc corrosion products sometimes known as wet storage stain may be formed if the coated article becomes wet and is unable to dry out before the patina has developed. On the thicker coatings such as batch hot dip galvanising and zinc flame spraying the formation of such

**Table 11.2** Durability of zinc coatings

| *Zinc coating* | *Typical coating thickness (μm)* | *Approximate coating life (years)* |
|---|---|---|
| Electro-plated | 5 | 3 |
| Continuously galvanised sheet | 20 | 13 |
| Sherardised | 30 | 20 |
| Standard batch galvanised | 85 | 56 |
| Grit blast and batch galvanised | 140 | 93 |

products has little or no effect on the level of corrosion protection provided and is only an aesthetic issue. However, on thinner zinc coatings such as electro-plating the presence of wet storage staining is more relevant as it might significantly reduce the level of corrosion protection provided. Some short-term protection against wet storage stain may be provided by passivating the zinc coating.

## 11.4 Die castings

The second major use of zinc is for the production of die castings. This is a true near-net-shape forming process for the automated mass-production of small components such as building hardware, safety equipment, security locks, components for computers, electrical hand tools, motor car components, camera bodies and vending machines. Fine detail, very thin wall sections (less than 1 mm thick when required) and intricate shapes can be produced to close dimensional limits in small castings weighing typically up to about 400 g. Larger castings can be made when required. The great advantage of zinc alloys compared to, say, the aluminium alloys that are produced as die castings is that the zinc alloys have far stronger tensile, stiffness and impact properties at ambient temperatures. Supreme accuracy and dimensional consistency combined with faster production rates, lower costs and recyclability add to the attraction of zinc die-casting. Consequently, the majority of die castings produced for use at temperatures below about 100°C are cast in zinc.

In pressure die casting, a measured quantity of molten metal is injected under pressure into a heated, permanent, split steel mould. The two halves of the mould are clamped together, or closed under hydraulic pressure during injection. The small cross-sectional areas of the castings produced result in very rapid solidification, so the two halves of the die can be opened almost immediately after the metal injection is complete. Zinc die castings are very easily machined but dimensional accuracy is such that no machining to final size is normally required.

Very pure grades of zinc (99.995+% zinc) must be used for the production of die castings. Small quantities of some impurities give rise to embrittlement. In particular, the concentrations of lead and cadmium should be 0.005% maximum and the tin content should be kept below 0.002% in the finished castings as required in BS EN 12844. Otherwise, the castings are susceptible to intercrystalline corrosion, forming bulky corrosion products that cause the castings to swell. Impurity levels in the alloys are even tighter to allow for some very slight contamination during the casting process. The most commonly used alloys are given in Table 11.3.

Optimum mechanical properties in these alloys are only developed by rapid solidification. Hence the die casting process is ideal. Typical minimum values for ultimate tensile strength at room temperature are 280 N mm$^{-2}$ for *ZL3*, and 330 N mm$^{-2}$ for *ZL5*. Freshly cast material in both alloys gives elongation to failure of about 6% and Izod impact values of over 50 J at ambient temperature.

Over a period of time, which may extend to years at room temperature, the alloys age, resulting in a decrease in strength and an increase in ductility. This is accompanied by a small contraction during the first month or so, followed by a very slight expansion. These changes can be accelerated and the properties and dimensions stabilised by elevated-temperature ageing at about 100°C for 3–6 h. Alloys ZL3 and ZL5 have limited creep resistance. Newer alloys such as ZL8 or ZL16 behave much better in this respect.

As with many metallic products, zinc die castings are susceptible to the formation of corrosion films. A tarnish will slowly develop even in a relatively low-humidity indoor atmosphere. Hence it is customary to provide some form of surface treatment to preserve the appearance. For less demanding applications, a simple chromate, phosphate or lacquer dip may suffice. Electroless nickel plating gives a black colouration with enhanced wear resistance. Almost any metal can be electrodeposited on the surface, but copper, brass, nickel and chromium are most commonly applied. Thin chromium platings tend to be inadequate and porous and the zinc corrosion products formed at the pores soon cause surface deterioration. It is common to plate successive thin layers of different metals, such as copper followed by nickel followed by chromium, in preference to a single thicker layer of chromium, particularly for demanding applications.

## 11.5 Wrought zinc

Zinc is produced in several wrought forms: sheet, strip and foil (as thin as 0.02 mm), rod, wire and extruded sections.

Wrought zinc sheet is used for roofing and cladding of buildings (BS EN 6561), and for gutters and downpipes. Unalloyed zinc for these purposes had drawbacks: creep resistance was low and the coefficient of expansion rather high. Today, alloyed zinc sheet is used. This contains at least 0.05% titanium to increase the creep resistance and at least 0.10% copper to improve the mechanical strength. High-purity grades of zinc must be used for the manufacture of these alloys. Corrosion resistance is at least as good as unalloyed zinc. The surface gradually weathers to produce a patina, which has a more greyish colouration than unalloyed zinc. A phosphate pre-treatment can be used to produce a durable darker grey patina.

Typical properties (BS 6561A) are given in Table 11.4 The creep rate at room temperature under a load of 70 N mm$^{-2}$ is not

**Table 11.3** Composition of casting alloys to BS EN 1774

| *Alloy designation* | *Composition (%)* | | | | | |
|---|---|---|---|---|---|---|
| | *Al* | *Cu* | *Mg* | *Pb* | *Cd* | *Sn* |
| ZL3 | 3.8–4.2 | 0.03 max | 0.035–0.06 | 0.003 max | 0.003 max | 0.001 max |
| ZL5 | 3.8–4.2 | 0.7–1.1 | 0.035–0.06 | 0.003 max | 0.003 max | 0.001 max |

**Table 11.4** Typical properties of zinc–copper–titanium sheet

| | *Rolling direction* | *Transverse direction* |
|---|---|---|
| Minimum UTS (N mm$^{-2}$) | 150 | 200 |
| Minimum elongation (%) | 30 | 15 |

greater than $5 \times 10^{-3}$ in the rolling direction, compared with a value of 1% for commercial grade zinc. The coefficient of thermal expansion is also lowered to about $23 \times 10^{-6}$ cm$^{-1}$ °C$^{-1}$. The density (7.18 g cm$^{-3}$) is significantly lower than for galvanised mild steel (7.85 g cm$^{-3}$) and hence requires less support. Zinc alloy sheet was one of the first metallic alloys exploited commercially for superplastic forming. This is a process in which the material is formed at very slow strain rate at temperatures above half the melting point in kelvins, either by blow-forming a clamped sheet on to a female die or advancing a male forming die into the clamped sheet. To obtain the required uniform, controlled metal flow, the material is processed to produce an extremely fine-grain structure, which remains stable throughout the forming operation, which may take several minutes at the elevated temperature. Although the forming process is very slow, it is economically viable because quite complex shapes can be produced in one operation. The production of similar shapes by conventional techniques frequently requires the production and joining of several smaller components.

The aluminium–zinc phase diagram shows a eutectoid point at 22% aluminium. This is the composition chosen for superplastic zinc sheet. A very fine, microduplex eutectoid structure can be formed, which is stabilised by fine precipitates formed by small copper and magnesium additions. The sheet is shaped at about 250°C. During subsequent service, the formed sheet must not be heated above about 100°C to avoid a return to the superplastic state and it must not be subjected to sustained heavy loads even at room temperature unless it is first heat treated to coarsen the grain size.

## 11.6 Other uses

Zinc anodes, connected by a conductor to metallic structures, are widely used for cathodic protection of steelwork in salt water and for pipes, conduits, etc., buried in the soil. In the latter case, care has to be taken to prevent the formation of a protective corrosion coating on the zinc, thereby impeding the electrolytic reaction.

A wide range of zinc compounds have commercial applications. The most important is zinc oxide, which is essential for the production of vulcanised rubber, for some ceramics and as an intermediary in the production of other zinc compounds. Zinc sulphide is used in phosphors for fluorescent lights and cathode ray tubes.

## 11.7 Sustainability

### 11.7.1 Health

Zinc is essential for human health. Adequate daily intake of zinc is vital for the proper functioning of the immune system, digestion, reproduction, taste and smell and many other natural processes (Cunningham-Rundles, 1999). Analysis of diet and nutritional needs has led researchers to estimate that as much as 49% of the world's population is at risk from zinc deficiency (Brown, 2000).

Zinc is also used in a variety of medical and pharmaceutical forms, such as bandages, cold lozenges, skin treatments, sunblock creams and lotions, burn and wound treatments, baby creams, shampoo, and cosmetics.

For many food crops, zinc is an essential micronutrient (Alloway, 2001). Zinc deficiency in agricultural soils is common on all continents and constitutes a major problem in many parts of the world because it causes serious inefficiencies in crop production.

### 11.7.2 Environment

In contrast to man-made chemicals, zinc is a natural element that plays an essential role in the biological processes of all living organisms, including humans, animals and plants.

Industrial emissions of zinc have been decreasing significantly since the 1970s, largely due to better controls by industry. For example, emissions of zinc from zinc production facilities worldwide fell by 43% between 1983 and 1995. The emission factor for zinc production in the 1970s was estimated at 100–180 kg zinc emitted per tonne of zinc produced. By 1995, this had fallen to under 12 kg per tonne (Pacyna and Pacyna, 2002).

Zinc, the 27th most common element in the earth's crust, is fully recyclable. At present, approximately 70% of the zinc produced originates from mined ores and 30% from recycled or secondary zinc.

Zinc recycling is a well-established process. Zinc can be recycled indefinitely without any loss of physical or chemical properties, ensuring its environmental sustainability and cost-effectiveness. Wrought zinc products such as roofing sheet and gutters have very high recycling rates, and the secondary zinc that is produced is commonly used in brass making or general galvanising. Galvanised steel is more difficult to recover but can be recycled with other steel scrap in the steel production process. Zinc volatilises early in the process, and is collected in the electric arc furnace (EAF) dust that is then recycled by introduction back into the primary production circuit after some intermediary steps.

Calculating what is available for recycling is a difficult task because the life of zinc-containing products is variable and can range from 10–15 years for cars or household appliances to over 100 years for zinc sheet used for roofing. Street lighting columns made of zinc-coated steel can remain in service for 40 years or much longer, and transmission towers for over 70 years. All these products tend to be replaced due to obsolescence, not because the zinc has ceased to protect the underlying steel. For example, zinc-coated steel poles placed in the Australian outback 100 years ago are still in excellent condition.

Today it is estimated that over 80% of the zinc available for recycling is recycled.

## 11.8 References


Alloway, B. J. (2001). *Zinc: The essential micronutrient for healthy high value crops*. Brussels, Belgium: IZA.

Brown, K. H. (2000). *Conclusions of the International Conference on Zinc and Human Health*, Stockholm.

Cunningham-Rundles, S. (1999). *Zinc and immune function*. Brussels, Belgium: IZA.

Martin, M. and Wildt, R. (2001). *Closing the loop: An introduction to recycling zinc coated steel*. Brussels, Belgium: IZA.

Pacyna, J. M. and Pacyna, E. G. (2002). *Assessment of global and regional emissions of metals from anthropogenic sources worldwide*. Oslo, Norway: Norwegian Institute for Air Research.

Porter, F. C. (1991). *Zinc handbook: Properties, processing and use in design*. New York, NY: Marcel Dekker.

# 12 Bricks and Brickwork

**Barry Haseltine** MBE FCGI FREng DIC
BSc(Eng) FICE, FIStructE
Consultant

## Contents

## 12.1 Introduction

Bricks are probably the oldest industrialised building material known to man. Surviving examples of brickwork from *c.*1300 BC still retain an attractive appearance, as with the Choga Zanbu Ziggurat in Iran. The earliest bricks were made from clay, taken from close to the surface of the ground, or from river banks, moulded into shape by hand and dried in the sun. 'Adobes', as they are known, have been found in the remains of Jericho dating back to about 8000 BC. The Bronze Age Mycenaeans had learnt how to 'fire' the sun-dried bricks to give them greater permanence. By the time of the Roman occupation of Britain, the technique of burning the moulded bricks to give them durability was well known and practised. Many examples of Roman bricks exist to this day as testimony to the sound techniques that the Romans had in producing their burnt clay products.

Clay bricks over the centuries were traditionally made locally and not transported very far, so that they had widely differing characteristics depending on the material available and the way in which the bricks were treated by the maker; it was quite usual for the bricks to be made on the building site, so that they did not need to be transported at all – assuming that there were suitable supplies of soft mouldable clay available. Much of the attraction of brickwork, both historic and modern, lies in the textures, colours and variations that arise from the use of materials made from various clays and manufacturing processes.

In the 19th century a process was developed for making bricks from sand and lime by subjecting them to high-pressure steam (autoclaving) to form calcium silicate, which binds the materials together. Additionally, bricks are now made from concrete, and there are even plastic imitations.

In Britain, a brick has traditionally been made of such a size that it can be picked up in one hand, to be laid on a bed of mortar. Modern processes permit clay to be formed into much larger units and, while this is done in other countries, it is not practised in the UK. Again in other countries, large calcium silicate units are manufactured and used but they are not made in the UK.

In addition to being used in walls, either for their attractive facing appearance or as part of the structural support system, bricks are used as features in landscaping work and as hard paving, both inside and outside buildings. Features and pavings are not covered in the rest of this chapter.

Modern clay brick-making plants make use of deeply quarried shales and marls as well as soft clays and brick earths; such materials exist naturally over a great deal of the UK. Other materials may be added to the clay in the manufacturing process to improve the properties, for instance sand, chalk or chemical admixtures.

Calcium silicate brick production requires a supply of suitable sand, and this is not as readily available in the UK as it is in some continental countries. Some calcium silicate bricks use a coarse aggregate as well as sand, and these are usually known as flint–lime bricks, as opposed to sand–lime. Apart from the sand, aggregate and lime, other admixtures may be used in the process, particularly colouring agents; the final process is autoclaving.

Concrete bricks are made using the normal materials associated with concrete, i.e. sand, fine aggregate and cement, to which, additionally, colouring agents are added; they gain their strength by the curing of the concrete, in the usual way.

## 12.2 Types of brick

Bricks are classified both by their quality as it would affect their use, and by the manufacturing process as it affects their appearance. BS6100: Section 5.3: 1984: Glossary of Terms gives extensive definitions of the relevant words, but a distinction needs to be drawn between the following usage categories.

### 12.2.1 Common bricks

Common bricks are those that are used where they will not normally be exposed to view and where there is no claim as to their appearance. They are suitable for general use in construction, given that they possess adequate strength and durability for the location. The term relates to appearance only and has no significance in the classification of bricks according to British Standards.

### 12.2.2 Facing bricks

Facing bricks have a suitable appearance for use where they will be exposed to view so that they give an attractive and pleasing effect. Very many types of facing brick are available and the choice can be made either from catalogues and display panels or from examples of work carried out in the particular brick.

Common terms to describe appearance are as follows.

- *Sandfaced.* Sand is incorporated on the surfaces during manufacture so that it affects both colour and texture.
- *Rustic.* A texture is imposed mechanically; a variety of patterns is available.
- *Multi-coloured.* This implies a variation in colour which may occur as a difference from brick to brick or across the faces of individual bricks.
- *Blue, yellow, red, etc.* A direct description of colour which is often added to some other descriptions, e.g. multi-red facings.
- *Smooth.* No surface texture.

The designation of a brick as facing does not imply that it will be durable in all locations (see Section 12.5).

### 12.2.3 Engineering bricks

The term 'engineering bricks' is used almost only in Great Britain. These bricks may be facing bricks, but are not defined with facing qualities in mind. Engineering bricks were required to conform to defined upper limits for water absorption and lower limits for compressive strength in BS3921: 1985, but there is no recognition of the term 'engineering brick' in the European Standard for clay bricks (see below). The National Annex to BS EN 771-1 seeks to keep the term alive in the UK. The bricks are used where high strength and/or resistance to aggressive environments is required.

### 12.2.4 Damp-proof-course bricks

Damp-proof bricks are ones with defined upper limits to water absorption, similar to engineering bricks, but without the strength requirements of engineering bricks; as with engineering bricks, the European Standard does not use the term. Two courses of such bricks are permitted to fulfil the function of a damp-proof course according to BS5628: Part 3: 2005 and are particularly useful at the base of free-standing and retaining walls, where the continuity of the brickwork is maintained without the plane of weakness that a sheet damp-proof course introduces.

### 12.2.5 Terms that describe bricks by the manufacturing process

#### *12.2.5.1 Extruded wire cut*

A large proportion of bricks that are made in the UK involve extruding a column of stiff clay, which is then cut by wires. The header and stretcher surfaces may be left smooth from the steel die or a texture can be applied immediately. Such bricks usually contain formed voids (perforations) to facilitate the extruding, drying and firing processes.

*12.2.5.2 Pressed bricks*

Prepared clay is pressed into moulds; the semi-dry pressing process is used for the type of brick named after the village in which it used to be made, Fletton, and this is a large part of the pressed brick production.

*12.2.5.3 Soft-mud bricks*

Soft-mud bricks are usually machine moulded from soft clays which often have a high water content. Such bricks have long been associated, though no longer exclusively, with those commonly made in the south and east of England.

*12.2.5.4 Stock bricks*

This term is used somewhat loosely and was originally applied to hand-made bricks from the south of England, as they were made on a 'stock' – a piece of wood positioning the mould. Now the term 'stock' is often used to describe bricks made by the soft-mud process.

*12.2.5.5 Hand-made bricks*

These are formed by hand-throwing a clod of soft clay (a lump) into a mould. Sometimes the preparation of the clod may be assisted by mechanical means.

Hand-made, stock and soft-mud bricks are often sought after for their attractive unevenness of colour, texture and shape resulting from the raw materials and the methods of shaping and firing used. They tend to have lower compressive strengths and higher water absorptions than other types of brick but are often frost-resistant.

*12.2.5.6 Special bricks*

Standard bricks are shaped as rectangular prisms, but a variety of other shaped bricks are commonly made. These include bricks with splayed or rounded edges. Forms in regular production are described in BS4729: 2005. Those shapes that are commonly made are known as 'standard specials' to distinguish them from other forms that might be made specially for a particular application.

## 12.3 Manufacture of bricks

### 12.3.1 Clay and shale bricks

There are four main stages in the manufacture of clay and shale bricks:

1 obtaining and preparing the material
2 shaping
3 drying
4 firing.

When a plastic clay with high moisture content is used in the process there is usually an intermediate stage of drying between the shaping and firing processes. For facing bricks the finished appearance may depend on additional work at the shaping stage, usually either by the application of coloured sand to the surface of the wet clay or by mechanical texturing of the surface (see Section 12.2). In some processes other materials are added to the clay, either to improve appearance or to incorporate combustible materials to assist firing.

### 12.3.2 Clay preparation

After the clay has been extracted it is prepared by crushing, grinding and mixing in a variety of ways, depending on the type of raw material and particular requirements of the subsequent shaping process. Water content is controlled so that the material going on to the shaping process may vary from very wet to relatively dry clay dust.

### 12.3.3 Shaping the bricks

There are five systems; one is an extrusion process, while the others all involve material being forced into moulds by pressure.

In the extrusion process holes of a variety of shapes may be formed through the bricks – formed voids. In the moulding process indentations such as frogs and hollows may be formed.

The shaping method used depends on the type of raw material and, in particular, on water content and type of brick required. The five shaping methods are described briefly below.

*12.3.3.1 Hand-made bricks*

Hand-making is the traditional method in which a clod of high-quality, very plastic clay is rolled in moulding sand and hand-thrown into a mould. Although more expensive than machine making, the method gives a particularly attractive surface finish. A special version of this process produces simulated hand-made bricks, known as 'machine hand-made', by using a machine that throws the clay into a mould in a way similar to that of a man throwing a clod of clay into a mould. Many of the machine hand-made bricks now do look very much like traditional hand-made ones.

Because the material is very plastic and has a high water content, considerable shrinkage occurs during drying and firing, with the result that bricks may vary in size and shape. Within limits, these variations are welcomed for the character they provide in facing work, but there are design implications. Thin mortar joints are rarely suitable and, even with fairly wide joints, building to closely controlled dimensions may be difficult on short lengths of brickwork where there are insufficient joints over which to average out the dimensional variation of the bricks. Hand-made bricks usually have one frog.

*12.3.3.2 Soft-mud bricks*

Soft-mud bricks are made from high moisture content clay which is machine formed by mechanically forcing the clay into sanded moulds. A wide variety of bricks is made in this way. Many have very similar properties to hand-made types, although they may not have quite the same character of surface finish.

The well-known yellow London stock brick is a rather special type of this general class, chalk or lime and breeze or other combustible material being added to the clay at the mixing stage. Soft-mud bricks often have one frog.

*12.3.3.3 Semi-dry bricks*

This is the process used in the making of Fletton bricks. Preparation results in a fine clay dust that is delivered to machines where it is pressed to shape in moulds. Usually the pressing process is repeated; for example, most Fletton bricks receive four pressings in one cycle of the process. Because the material is relatively dry the shaped bricks go direct to the kilns for firing, without intermediate drying.

Bricks made by this process are fairly regular in shape and size, but the commons are unattractive in appearance. When intended for facing work they may be either sand faced or machine textured after the pressing process. These surface finishes are often noticeably different in colour from the body of the brick, and care is needed in delivery and site handling to avoid unsightly damage.

Pressed bricks may have one or two frogs or may have a variety of other shaped holes pressed in to produce the type of brick described as 'cellular'.

*12.3.3.4 Stiff-plastic bricks*

This is a similar process to the semi-dry one, but, because the clays and shales used have an inherently low plasticity, some water is added to the clay dust before the material is delivered to an extrusion pug which forces roughly brick-sized clots into moulds. A press die gives the final shape and compaction to each brick. The water content is low enough for the bricks to go direct from pressing to kiln.

*12.3.3.5 Wire-cut bricks*

Most clays, except those that have a high indigenous water content, are used in this method. After preparation the clay is extruded through dies, with or without forming perforations through the full thickness of the bricks. The extruded column of clay is of the correct length and width to give the equivalent dimensions of the finished bricks and is cut with wires to the required brick thickness. The wire-cut process also provides flexibility in producing varying sizes and shapes of brick and is an economical method for mass production.

The wire cutting often leaves drag marks across the bed faces of the bricks and occasionally some distortion to the arris may occur. Facing and engineering bricks are sometimes pressed after being wire cut, to provide smoother faces and sharper arrises.

Wire-cut bricks do not have frogs. Holes formed during the extrusion process vary considerably in number and size. The current British Standard EN for clay bricks does not require any differentiation of description according to the volume of holes that are formed in the bricks, but the Eurocode for masonry design (see below) allows for the categorisation of masonry units by a number of criteria, including percentage of holes.

Until the 1970s, perforated bricks were more widely used abroad than in the UK, but now a large part of the UK production is of perforated bricks. Advantages of perforated bricks include reduction in process times, reduction in weight and some increase in thermal insulation value. The perforations do not appear to affect rain penetration through walls after completion, but, because the holes go right through the bricks, vertically as laid, they are vulnerable to saturation from rain during construction. Saturation will increase drying-out time, may inhibit early decoration and will increase risk of efflorescence and possibly of frost damage. Good protection against excessive wetting during construction, important for all brickwork, is especially necessary with perforated bricks.

*12.3.3.6 Drying*

Drying is necessary in all cases where the brick, after forming, is soft and unable to withstand the weight of other bricks when stacked for firing, and may be used also to reduce the time of firing. This is the case with all hand-made and soft-mud types and also with wire-cut bricks where the moisture content of the extruded column is relatively high. Drying is carried out in a series of chambers or tunnels in which the bricks are arranged so that a flow of heated air can pass over them, the temperature and humidity of the air being regulated to control the shrinkage that takes place during drying.

*12.3.3.7 Firing*

Firing produces a number of complicated chemical and physical changes in clay and the degree of control obtained is important. Ultimate firing temperature and type of atmosphere affect colour. Soluble-salt content in the finished bricks is partially dependent on firing conditions. There are four firing methods, as described below.

- *Clamp.* A centuries-old method, still sometimes used for stocks and hand-made bricks. Green bricks, that is ones that have been dried but not fired, are closely stacked on a layer of fuel while the bricks themselves also contain combustible material. The clamp is set on fire and usually allowed to burn itself out. Control of the firing is achieved by varying the amount of fuel added to the clamp and the clay of the bricks, with the result that bricks may vary considerably in quality. The bricks are usually sorted into grades before sale.
- *Intermittent kilns.* Except for clamps, intermittent kilns were the only method of firing until 1858, when the Hoffman continuous kiln was introduced. Intermittent kilns are now principally used for the manufacture of special types of bricks.
- *Continuous kilns.* A continuous kiln consists of a number of chambers connected in such a way that the fire can be led from one to another, so that the stationary bricks are heated, fired and cooled. The system combines economy and a good degree of control.
- *Tunnel kilns.* Tunnel kilns are a more recent innovation and their use has increased considerably in recent years. In tunnel kilns the fire remains stationary while bricks, carried on kiln cars, pass along a tunnel through preheating, firing and cooling zones. The ability to vary temperatures and track speeds provides optimum conditions for quality control.

*12.3.3.8 Quality control*

Quality is an ever more important aspect of brick manufacture; most works are now 'quality assured' under approved schemes. Testing of individual batches of bricks may be carried out and methods and standards for size, compressive strength, water absorption and soluble-salt content are defined in a number of BS EN standards. While check testing by the user is always an option, the expense of testing may often be avoided by obtaining from the manufacturer data from control charts from its continuous checking.

There is no standard for colour control and, because colour variation is dependent on raw material and processing, variations may occur between batches in addition to those within a single batch. Where a large quantity of bricks is required to be consistent in colour it is important to discuss the requirements with the manufacturer to see if this is feasible; mixing of batches of bricks is often necessary.

Dimensional control for clay bricks, as defined in BS EN 771-1, is sometimes not very satisfactory for users, although the change to the European standard has removed the traditional UK method of the overall measurement of 24 bricks falling within prescribed limits which did not exclude the possibility of some bricks being considerably different from the 'standard' size (see Section 12.4).

### 12.3.4 Calcium silicate bricks

The raw materials used for calcium silicate bricks are lime, silica sand and water. Crushed or uncrushed siliceous gravel or crushed siliceous rock is sometimes used instead of, or in combination with, sand.

The raw materials are mixed in carefully controlled proportions, with colouring materials added as required. The material is shaped by pressing and the bricks are then loaded onto bogies and passed into autoclaves where they are subjected to high-pressure steam. By varying autoclaving time and steam pressure the performance characteristics of the bricks can be adjusted.

Colour and quality are determined by the mix and the autoclaving process. Calcium silicate bricks are a uniform product of fairly regular size and shape, with sharp arrises and little colour variation.

Arrises are susceptible to damage which, because of the inherent regularity of the bricks, may be very noticeable. Care in handling facing bricks is important.

### 12.3.5 Concrete bricks

Concrete bricks are made from a carefully controlled mixture of cement, sand and aggregate together with additives such as colouring agents. The mixture is pressed and/or vibrated into brick-sized moulds in much the same way as calcium silicate bricks are made. The moulded bricks are then cured, either in steam chambers or in the air. The product is regular in size and colour.

## 12.4 Properties of bricks

Bricks are covered by British Standards, most of which are now BS European standards. The so-called coexistence period of British Standards and European ones is now finished, so for all masonry units, the European term for bricks and blocks, only the BS ENs are valid. Clay bricks are covered by BS EN 771-1, wherein two density bands are recognised – HD units and LD units, meaning high and low density, respectively. Calcium silicate bricks are covered by BS EN 771-2 and concrete bricks by BS EN 771-3 (the same standard as covers concrete blocks). The new breed of European Standards makes use of a declaration system, whereby the manufacturer claims, by his declaration, the properties that he expects his bricks to have.

### 12.4.1 Size

The size of a brick is no longer defined in the relevant BS EN. The intended size of a brick is referred to as the work size and the space that it occupies, with its mortar joint, is called the co-ordinating size.

Although not in the BS ENs, the work size of clay, calcium silicate and concrete bricks is still the traditional size, as given in the withdrawn BS3921: 1985 – 215 mm × 102.5 mm × 65 mm with a co-ordinating size of 225 mm × 112.5 mm × 75 mm. Other sizes are made, e.g. thinner bricks and some of the old Northern 3 in. bricks (83 mm with joint) are still used. Attempts have been made to interest the market in larger bricks, or modular co-ordinated bricks, but there has been little take-up. The standard metric size of 112.5 × 215 × 65 mm is fractionally smaller than the old Imperial size of $4^1/_8 \times 8^5/_8 \times 2^5/_8$, so care has to be taken when trying to match new bricks with old (Imperial) ones.

Finished bricks inevitably vary in their actual size depending on the manufacturing process; clay, when dried and fired, shrinks variably, and so clay bricks tend to have greater variation between them than do calcium silicate or concrete bricks. BS EN 771-1 requires that the manufacturer of clay bricks declares the Tolerance Category and the Range Category to which he sells his product; for calcium silicate bricks, the manufacturer also declares the tolerance that he claims to meet. For concrete bricks, the manufacturer is required to declare deviations in relation to a table containing four categories, D1 to D4.

Table 12.1 gives tolerance information in relation to clay bricks. Table 12.2 gives the exceptions to the rule that all tolerances are ±2 mm of the relevant dimension of the product in calcium silicate.

**Table 12.2** Variations in mm from ±2 mm of tolerances for calcium silicate bricks

| | *TLM* | *TLMP* |
|---|---|---|
| Mean height | ±1 | |
| Individual height | ±1 | ±1 |
| Individual length and height | | ±3 |

For concrete bricks, the limits in deviation are as for all concrete masonry products.

### 12.4.2 Water absorption

The water absorption of a clay brick is the increase in weight of a dry brick when it has been saturated, expressed as a percentage of the dry weight. It is one of the parameters for the definition of engineering bricks and damp-proof-course bricks. The water absorption of the bricks used in a wall affects the mode of rain penetration through the outer leaf of a cavity wall and is used to define the flexural strength used in lateral load design. The water absorption of calcium silicate bricks is not as significant in relation to flexural strength as with clay bricks, but their surface texture (very smooth) and capillary attraction to water can affect adversely the flexural bond of mortar to bricks.

### 12.4.3 Compressive strength

A manufacturer will declare the compressive strength of the bricks that it makes, having tested the bricks in accordance with BS EN 772-1, the appropriate test method standard. The compressive strength of a brick is the mean of 10 crushing tests (six for a limited case with calcium silicate bricks), when the failing load is divided by the gross area of the brick for solid and perforated bricks. For frogged bricks that do not have a frog area of more than 65% of the bed face, the crushing strength is divided by the net area of the contact of the crushing press and the brick. The compressive strength is a quantity used in engineering calculations.

As the compressive strength of perforated bricks relates to the gross area of the brick bed, no allowance needs be made for the perforations, which should not be filled in bricklaying. The frogs of frogged bricks should, however, be filled in bricklaying; that is, they should be laid in the wall with the frog uppermost. Different types of clay bricks have compressive strengths ranging from about 7 to well over 100 N mm$^{-2}$, and calcium silicate and concrete from about 21 to nearly 60 N mm$^{-2}$.

In setting certain rules for the use of CEN standards for meeting aspects of the Construction Products Directive, the European Commission required that bricks of all methods of manufacture should be grouped as Category I or Category II, according to the degree of reliability of the strength that the manufacturer claims for its bricks. This differentiation into two categories mirrors the Category of Manufacturing Control used in BS5628: Parts 1 and 2 until the 2005 versions.

**Table 12.1** Tolerance and range category requirements for HD clay masonry units

| *Tolerance category* | *Allowed discrepancy* | *Range category* | *Allowed discrepancy* |
|---|---|---|---|
| $T_1$ | Larger of 3 mm or 0.4√(work size) | $R_1$ | 0.6√(work size) |
| $T_2$ | Larger of 2 mm or 0.25√(work size) | $R_2$ | 0.3√(work size) |
| $T_M$ | Manufacturer's declaration | $R_M$ | Manufacturer's declaration |

### 12.4.4 Soluble-salt content

Most clays used in brick-making contain soluble salts, for example sulphates, that may be retained in the fired bricks. If brickwork becomes saturated for long periods, soluble sulphates may become dissolved in water. Dissolved sulphates may cause mortars that have been incorrectly specified or batched and have a low cement content to deteriorate under sulphate attack. Of course sulphates from the ground or other sources may also be destructive. When BS3921: 1985 was in use, some clay bricks met limits placed on the level of certain soluble salts so that they could be designated 'L', signifying low soluble-salt content. Those that did not meet the limits were designated 'N' for normal soluble-salt content. BS EN 771-1 now requires a manufacturer to declare the active soluble salt content of bricks to be S2, S1, or S0; S2 is approximately equivalent to the old 'L' and S1 to 'N'.

### 12.4.5 Frost resistance

The British Standard for clay bricks (BS3921: 1985) contained no mandatory test method for classifying frost resistance, and manufacturers were required to state the frost resistance of their clay bricks by classifying them from experience in use as follows.

- *F* – which meant frost-resistant, even when used in exposed positions where the bricks would be liable to freezing while saturated. In practice, saturation of brickwork is likely not only in copings, cappings and sills but also in walls immediately below flush cappings and sills as they offer little protection. Protection from saturation is particularly advisable for brickwork immediately below large areas of glass or impervious cladding.
- *M* – which meant moderately frost-resistant and suitable for general use in walling that would be protected from becoming saturated. This would mean the use of features and details such as overhanging verges, eaves, copings and projecting sills that have protective drips.
- *O* – which meant not frost-resistant. Such bricks are seldom made deliberately. They are only suitable for internal use but must be protected when exposed on site so as not to freeze when saturated.

The replacement for BS3921, BS EN 771-1, still has no test method for freeze/thaw resistance, so it remains that the experience of the manufacturer determines its claim for resistance; the classification is now F2, F1 and F0, equivalent to F, M and O.

The frost resistance of calcium silicate bricks generally increases with increasing strength class. However, calcium silicate bricks should not be used where they may be subject to salt spray, e.g. on sea fronts, or to de-icing salts.

### 12.4.6 Efflorescence

Efflorescence is a crystalline deposit left on the surface of clay brickwork after the evaporation of water carrying dissolved soluble salts. Manufacturers used to have to state the category to which the bricks being offered corresponded when subjected to the efflorescence test described in BS3921: 1985, namely the categories nil, slight or moderate. The BS EN contains no requirements for efflorescence. The risk of efflorescence, which is harmless and usually temporary, can best be minimised by protecting the bricks in the stacks, as well as newly built brickwork, from rain (see BS5628: Part 3: Clause 3.5: 1985).

### 12.4.7 Thermal movement

Bricks expand on being warmed, and shrink when cooled. In practice it is the expansion and contraction of the brickwork, rather than of the individual units, that is of interest to the user. Some reversible movement will occur in brickwork exposed to temperature variation. The type of brick will not make any significant difference and the coefficient of thermal expansion in a horizontal direction will be about $5.6 \times 10^{-6}$ for a 1°C temperature change. For a 10 m length of wall, variation in length between summer and winter might be about 2 mm. Vertical expansion may be up to 1.5 times the horizontal value. *BRE Digest* 228 lists thermal movements for many materials.

### 12.4.8 Moisture movement

Clay bricks expand on cooling from the kiln, as some of the molecules of water reattach themselves after being driven off by the heat of the kiln. This expansion is effectively non-reversible, unless the bricks are re-fired. The magnitude of this movement varies according to the type of brick. Fortunately, a considerable portion of the expansion takes place quite quickly, probably at least half of it occurring within a few days. The remainder may take place slowly over a considerable period, but at an ever decreasing rate (see *BRE Digest* 228). In addition, there is a small reversible movement due to wetting and drying of clay bricks.

Calcium silicate bricks tend to shrink as they dry out after manufacture, and then expand again if wetted. The drying shrinkage may range from less than 0.01% to about 0.04%. Limits for drying shrinkage are no longer specified in BS EN 771-2.

Concrete bricks behave like other concrete products, e.g. blocks, in that they shrink on drying and expand on wetting. BS EN 771-3, which covers concrete bricks, gives no requirements for limits of drying shrinkage; it requires the manufacturer to declare the drying shrinkage when the units might be used in structural elements.

As with thermal movement, it is the movement of the brickwork itself that is of interest to the user. Moisture movement in bricks may occur in the form of an irreversible movement that continues throughout the life of the building. Very occasionally, there can be a third type of movement, occurring as a continuing expansion in which bricks expand on drying and expand again if wetted. The amount of information on this type of movement is limited, but it is clear that bricks should not be used immediately after being delivered straight from the kiln.

## 12.5 Brickwork

There would be little point in manufacturing bricks if they were not to be used in brickwork, and it is the varied and attractive appearance of brickwork, together with its long-lasting characteristics, that have made it such a popular material for thousands of years. There are many examples of ancient brickwork that illustrate this point. In addition to an attractive and durable appearance, brickwork can give weather resistance, support loads, in-fill other structural elements, and provide thermal, sound and fire resistance. Historically, brickwork might have been the only element of masonry, meaning brickwork, stonework or blockwork, in a building, but in modern times there has been an increasing amalgamation of brickwork and concrete blockwork to combine the strengths and overcome the weaknesses of each material, so that most brickwork now is a facing material used in combination with concrete blockwork. The most common combination of brickwork and blockwork arises in the cavity wall. This is not to say that brickwork does not still have important structural uses, but it is no longer commonplace for the entire load of a building to be carried on the brickwork. Brickwork is still used structurally for its great strength, in combination with its appearance, when these are required, and new forms of building have been developed to take advantage of this, such as a 'fin'-walled form for a sports hall.

Bricks may be arranged in various patterns called 'bond', and this can affect the strength of a wall, its appearance or its cost.

Walls that are as thick as the thickness of the brick are usually laid in stretcher bond – the familiar half overlap, showing on the 'stretcher' face of the brick. However, one-third overlap is sometimes used. In walls thicker than the thickness of the brick, English and Flemish bond are the most common, with the variants English garden wall and Flemish garden wall. For special purposes, e.g. to incorporate reinforcement in small pockets, bonds such as Quetta (named after the town where it was first used) have been developed. Detailed guidance on bonding patterns is given in textbooks on building construction, and in BS5628: Part 3: 2005.

### 12.5.1 Cavity walls

After initial use in the late 19th century, cavity walls became really popular between the World Wars and are now the universal application of masonry to the external envelope of buildings. It is only since the late 1950s that the brick inner skin of a cavity wall has been replaced with one in concrete blockwork. Initially, when the required U value of an external wall was capable of being provided by a brick inner and outer skin, with a 50 mm cavity, blockwork was used as a cheaper material, perhaps faster to lay. As required U values have fallen, the inner leaf has been built in various forms of insulating concrete block, and there has been a need with some blocks to incorporate further insulation in the cavity itself. As the requirements for minimum U values have been lowered, it has become necessary to use thick insulating block inner skins of cavity walls, or to revert to dense materials, even brick, for the inner leaf skin and to use additional insulating material in the cavity.

The detailing of cavity walls is well covered in text books of building construction and BS5628: Part 3: 2005. A commonly neglected aspect of cavity walls is the provision, and method of fixing, of the ties that link one skin to the other. Wind forces and small movements are resisted by the wall ties, so their type, spacing and embedment are very important. Pushing wall ties into semi-hardened mortar, or bending them up or down in position, leads to understandable problems of strength and durability.

The outer leaf of cavity walls may allow some water to reach the cavity, and this is catered for by using weepholes over cavity trays and at ground level to permit the water to drain out. Cavity insulation has to be used in the knowledge that some water may penetrate the outer leaf. Cleanliness of the cavity – that is, keeping it clear of mortar – is vital in preventing water from bridging to the inner leaf.

### 12.5.2 Mortar for brickwork

Brickwork is the combination of the individual bricks with mortar to form a strong assembly. Variation in the colour or treatment of the mortar joint can have a profound effect on the appearance of the finished work; variation in the strength of the mortar can influence significantly the strength of the brickwork in both compression and flexure. The type of mortar and the joint finish can make a large difference to the rain resistance of the finished wall.

Traditional mortars were made with lime and sand, sometimes with a cement of whatever sort was available. Such lime–sand or weak-cement–lime–sand mortars were very flexible and accommodated the movement that occurred in the brickwork, whether arising from the properties of the bricks or the movement of the building as a whole. It is often said, now as a leftover from times when these mortars were used, that the mortar should be as weak as possible in order to allow differential movement to be accommodated; the mortar must not be stronger than the brick, etc. However, in modern terms, this is really no longer true. The weakest mortar practicable is a 1:2:9, which is very strong in comparison with lime mortar: it sticks to the bricks and holds them together in much the same way as do the stronger 1:1:6 or even 1:¼: 3 mortars. None of the modern cement–lime–sand mortars, or their equivalents, allows any real 'flexing' of the brickwork to accommodate expansion or settlement. Furthermore, it is unlikely that the mortar *will* be stronger than the bricks; a 1:¼:3 mortar is unlikely to be stronger than 15–20 N mm$^{-2}$, whereas almost all bricks, except for some stock or soft mud types, are stronger than this.

**Table 12.3** Average sound insulation value of brick walls

| *Construction* | *Average sound reduction (dB)* |
|---|---|
| Half-brick wall, unplastered | 42 |
| Half-brick wall, plastered both sides | 45 |
| One-brick wall, plastered both sides | 50 |
| Cavity wall of two half-brick skins with butterfly wall ties not more than one per m$^2$; wall plastered both sides | 50 |
| One-and-a-half-brick wall | 52/3 |

Mortar mixes are defined in BS5628: Part 1 and 3: 2005 and in Table 12.3. Many mortars are now factory made, or semi-factory made; these products are covered by BS EN 998-2. The current thinking on specification of mortar is to use a similar system to that used for concrete, that is to differentiate between prescribed mixes or designed mixes. Mortar in BS EN 998-2 is defined by its 28 day strength preceded by the letter M, e.g. M12, being a 12 N/mm$^2$ mortar. The National Annex to BS EN 998-2 gives equivalent traditional designations that give the M number strengths and these are also given in Table 12.4.

The use of thin layer mortars is very common on the Continent, and is becoming more popular in the UK, mainly with aac (autoclaved aerated concrete) blocks, which are made to a close height tolerance. Thin layer mortars are sold to BS EN 998-2.

Mortars are available in many colours to match or contrast with the bricks. Coloured mortar is most easily made from semi-finished factory made mortar, what used to be known as premixed coloured lime:sand, delivered to site and mixed there with the appropriate amount of cement.

### 12.5.3 Properties of brickwork

#### *12.5.3.1 Appearance*

The appearance of brickwork, apart from the obvious effect of choice of brick texture, colour and pattern of arranging the bricks, is affected to a surprising extent by the colour of the mortar. Similarly, appearance is affected by the type of pointing used, so that a far greater texture is seen if the joint is raked out than it if is flush pointed (but see Section 12.5.3.2).

The pattern of laying the bricks, i.e. the bond, affects the appearance, and there is some beautiful brickwork where more than one colour of brick has been used in a regular pattern. Nowadays, so many walls are 'half brick', i.e. the thickness of the width of the brick, that the bond is nearly always 'stretching'.

It is rarely satisfactory to assess the likely appearance of brickwork from a sample of the bricks themselves; at least a sample panel, built in the mortar to be used, and with the chosen type of pointing, needs to be seen. Ideally, examples of other finished work in the brick should be inspected.

Appearance can be adversely affected by batch variations in brick colour; some manufacturers mix their production to even out this effect, but it is usually essential for the contractor to mix bricks from a number of packs as the work is carried out.

**Table 12.4** Proportions and materials for prescribed masonry mortars

| | Mortar designation | Compressive strength class | Prescribed mortars (proportion of materials by volume[a,b]) | | | | Compressive strength at 28 days (N/mm²) |
|---|---|---|---|---|---|---|---|
| | | | Cement[c]: lime: sand with or without air entrainment | Cement[c]: sand with or without air entrainment | Masonry cement[d]: sand | Masonry cement[e]: sand | |
| ↓ Increasing ability to accommodate movement, e.g. due to settlement, temperature and moisture changes | (i) | M12 | 1:0 to ¼:3 | – | – | | 12 |
| | (ii) | M6 | 1:½:4 to 4½ | 1:3 to 4 | 1:2½ to 3½ | 1:3 | 6 |
| | (iii) | M4 | 1:1:5 to 6 | 1:5 to 6 | 1:4 to 5 | 1:3½ to 4 | 4 |
| | (iv) | M2 | 1:2:8 to 9 | 1:7 to 8 | 1:5½ to 6½ | 1:4½ | 2 |

[a]Proportioning by mass will give more accurate batching than proportioning by volume, provided that the bulk densities of the materials are checked on site. [b]When the sand portion is given as, for example, 5 to 6, the lower figure should be used with sands containing a higher proportion of fines while the higher figure should be used with sands containing a lower proportion of fines. [c]Cement or combinations of cement in accordance with Clause 13 of BS5628: Part 1: 2005, except masonry cements. [d]Masonry cement in accordance with Clause 13 of BS5628: Part 1: 2005 (inorganic filler other than lime). [e]Masonry cement in accordance with Clause 13 of BS5628: Part 1: 2005 (lime).

#### *12.5.3.2 Rain resistance*

In detailing, the outer leaf of cavity brickwork should never be considered waterproof but it should be specified and built such that its rain resistance is maximised. All joints must be filled (except weepholes). This means avoiding deep furrowing of bed joints and, in particular, ensuring that all cross-joints are solidly filled and that the mortar is not just applied to front and rear vertical edges. It is generally considered that the use of lime in mortars enhances the bond between bricks and mortars and hence the rain resistance of the brickwork.

Tooled flush joints, for example bucket handle and struck weathered joints, are the most rain resistant. A special tool or the edge of the trowel is used not only to shape the joint but also to press the mortar into intimate contact with the bricks and consolidate the surface of the mortar.

Flush joints are a little less rain resistant than tooled ones. The surplus mortar is cut off using the edge of a trowel in an upward slicing motion. The joint is not subsequently tooled and the surface, therefore, tends to be torn rather than consolidated.

Recessed joints, especially if not 'tooled', considerably increase the amount of rain penetrating the outer leaf and they should never be used with clay bricks having perforations near the surface or with clay bricks whose frost resistance is not adequate.

It should not need stating, but it does, that mortar allowed to bridge the cavity of an external wall can allow water to cross the cavity and negate the rain-resisting function of the wall. Ensuring clean lines of ties and damp-proofing trays is vital.

#### *12.5.3.3 Structural properties*

Brickwork has extremely good load-bearing capacity and many successful high-rise structures have been built relying only on the brickwork to carry the load to the ground. The detailed design of such walls has been covered by BS5628: Parts 1 and 2: 2005. In 2010 the BSs were replaced by Eurocodes, which make a complete suite of design and construction information available on a pan-European level. The structural design of masonry walls is beyond the scope of this book.

However, it is not only the vertical-load-bearing capacity of brickwork that is important; much brickwork is external to a building, and as such, is exposed to the wind. This puts the brickwork into bending, so that it can span as a beam or slab to the supporting framework. The flexural strength of brickwork, as it is known, is affected by the type of brick and, if clay bricks are used, by its water absorption. The mortar used also influences the flexural strength, as does the condition of the bricks being laid, e.g. bone dry or damp. More detail is given in BS5628: Parts 1 and 3.

Brickwork is also used in conjunction with reinforcement or prestressing tendons to act as a strong and economic structural material. More detail can be found in BS5628: Part 2: 2005.

#### *12.5.3.4 Movement of brickwork*

Allowance needs to be made for the movements that occur in brickwork from wetting and drying, long-term irreversible moisture expansion of clay bricks, and thermal expansion or contraction. This is done by breaking the brickwork down into defined lengths, or heights, by a complete break – a movement joint. In clay brickwork, such joints are primarily expansion joints, but in calcium silicate and concrete brickwork they are mostly for shrinkage purposes. Guidance on the spacing and detailing of movement joints is given in BS5628: Part 3. For clay brickwork it is suggested that an allowance of 1 mm per 1 m expansion be made, so that for joints at 12 m centres, 12 mm of expansion must be catered for. Since most fillers (they must be easily compressible and not, for example, cane fibre) can only accept movement of 50% of their width, this can mean rather wide joints that need to be sealed effectively. When clay brickwork is built into a framed structure (particularly, but not exclusively, reinforced concrete), joints are needed to prevent the shortening of the frame and expansion of the brickwork from causing a build-up of compressive stress in the wall. Failure to appreciate this soon enough has led to many failures of 1960s and early 1970s brickwork built on reinforced concrete structures.

#### *12.5.3.5 Thermal resistance*

Brickwork, in itself, is rarely, if ever, called upon to provide a thermal barrier against heat loss in modern buildings. Requirements for energy conservation make mandatory U values far lower than can be achieved with a solid wall alone. Most brickwork in external walls is, therefore, part of a composite system; this may be, and usually is, a cavity wall, but it is also possible to use brickwork alone lined with high-quality polystyrene insulation. To achieve the low U values required in a cavity wall, either lightweight concrete blocks are used as the inner leaf of the cavity wall, or dense aggregate blocks with insulation in the cavity are used. It is sometimes necessary to use brick, cavity insulation and lightweight block to achieve the desired insulation level.

#### *12.5.3.6 Sound insulation*

Brick walls are excellent barriers to sound transmission; the effectiveness of a sound barrier is the amount by which it reduces noise level expressed as decibels (dB). For solid walls, average insulation over the normal frequency range is largely dependent on the weight per unit area of the wall. Each doubling of weight adds about 6 dB to the insulation value. It follows that a change from half-brick to full-brick thickness provides about a 6 dB improvement, but to obtain a further 6 dB improvement would mean doubling the thickness to a two-brick wall. For the same reason it follows that the differences in the weight of differing types of brick are of little practical importance. A change from the lightest to heaviest might just be noticeable in a half-brick wall. Average insulation values for ordinary brick walling are given in Table 12.3.

#### *12.5.3.7 Fire resistance*

As with sound resistance, brick walls give good resistance to the passage of fire. The periods of resistance are given for various thicknesses, plaster types and loading conditions in BS5628: Part 3: 2005, and are available in the National Annex to BS EN 1996-1-2, the fire resistance part of the Eurocode for masonry design and construction. Since spring 2010, only BS EN 1996-1-2 has been valid.

#### *12.5.3.8 Durability*

Brick walls built for internal use only require frost and soluble-salt resistance for the period in which they will be exposed to the weather during construction. Most bricks have sufficient durability for this purpose and good building practice will reduce vulnerability to a minimum.

For walls that may be exposed to the weather during their lifetime, the soluble-salt content of the bricks is important. However, a distinction must be drawn between efflorescence, which is unsightly but rarely harmful, and sulphate attack, which can cause serious damage.

The crystallisation of soluble salts on the exterior of brick buildings causes efflorescence, usually in the form of white patches. For efflorescence to occur there must be soluble salts and enough water drying out to carry the salts in solution to the surface. Given a sufficient supply of salts and water, efflorescence can persist, but it is generally seen in the early life of a building during the initial drying period. To avoid efflorescence, saturation of walls during construction should be avoided and detailing should prevent excessive subsequent wetting. It is noticeable, for example, that efflorescence quite often occurs only in positions such as exposed parapet walls.

Salts may be present in bricks or be introduced from other materials such as mortar or from soil. Calcium silicate and concrete bricks do not contain harmful salts. Sulphate action occurs when water carries sulphates from clay bricks or from contamination into mortar that contains tricalcium aluminate, a constituent of the cement. The resulting expansion of mortar may be serious, causing cracking or spalling of facing bricks. Because fairly persistent wetting of the brickwork is one of the factors, the trouble is most often seen in situations such as chimneys, parapets or free-standing walls exposed on both sides, and in earth-retaining walls, but it may occur anywhere where walls are not protected from excessive dampness, if there are sufficient salts to affect the mortar.

The best precaution against sulphate attack is to keep walls dry, but for many positions this is not feasible. In such situations, therefore, a reasonable combination of brick and mortar must be used. The sulphate resistance of mortars can be increased by specifying fairly cement-rich mixes such as 1:¼:3 or 1:½: 4½ or, more effectively, by using a sulphate-resisting cement instead of ordinary Portland cement.

As explained above, clay bricks are classified as S2 or S1depending on whether they have low or normal quantities of soluble sulphates. For susceptible positions class S2 bricks should be used.

The frost resistance of clay bricks is defined in BS EN 771-1 as F2 and F1 (see above). Calcium silicate and concrete bricks generally have good frost resistance, especially with the higher strength classes. The choice of a brick for durability in a particular building can be made from BS5628: Part 3: 2005: Table 3 Durability of Masonry in Finished Construction; the table covers brickwork and blockwork. Frost resistance of the mortar is also necessary; if 1:1:6 or 1:¼:3 mortars are used, this will normally be achieved. Table 3 of BS5628: Part 3: 2005 also gives recommended mixes.

A comprehensive range of situations is covered as follows:

1. work below or near external ground level
2. damp-proof course
3. unrendered external walls (other than chimneys, cappings, copings, parapets or sills)
4. rendered external walls
5. internal walls and inner leaves of cavity walls
6. unrendered parapets
7. rendered parapets
8. chimneys
9. cappings, copings and sills
10. freestanding boundary and screen walls
11. earth-retaining walls
12. drainage and sewerage.

Care is needed in the choice of brick and mortar for given situations, but when care is given it is repaid in durable brickwork that will last the lifetime of the building of which it forms a part.

## 12.6 Acknowledgements

Grateful thanks are given to the Brick Development Association for its help in providing background material. Helpful advice on the choice of clay and calcium silicate bricks can be obtained from it. Advice on concrete bricks is given by the CBMA.

## 12.7 Bibliography

### 12.7.1 British Standards Institution

BS EN 771 – 1: 2003 including amendment No 1, Specification for masonry units. Clay masonry units.
BS EN 771 – 2: 2003 including amendment No 1, Specification for masonry units. Calcium silicate masonry units.
BS EN 771 – 3: 2003 including amendment No 1, Specification for masonry units. Aggregate concrete masonry units (dense and lightweight aggregates).
BS743: 1970, Specification for materials for damp proof courses.
BS EN 845 – 1: Specification for ancillary components for masonry. Ties, tension straps, hangers and brackets.
BS4729: 2005, Clay and calcium silicate bricks of special shapes and sizes – Recommendations.
BS5628: Code of practice for use of masonry.
BS5628: Part 1: 2005, Structural use of unreinforced masonry.
BS5628: Part 2: 2005, Structural use of reinforced and prestressed masonry.
BS5628: Part 3: 2005, Materials and components, design and workmanship.

BS6100: Glossary of building and civil engineering terms.
BS6100: Part 5: 1985, Masonry.

### 12.7.2 Building Research Establishment Digests

*BRE Digest* 362: 1991, Building Mortars.
*BRE Digest* 441 Parts 1 & 2: 1999, Clay bricks and clay brick masonry.

### 12.7.3 Trade associations concerned with bricks, concrete blocks, brickwork and concrete blockwork

AACPA – Autoclaved Aerated Concrete Products Association Ltd, 60 Charles Street, Leicester LE1 1FB, UK.

BDA – Brick Development Association, The Building Centre, 26 Store Street, London WCIE 7 BT.

CBA – Concrete Block Association, 60 Charles Street, Leicester LE1 1FB, UK.

CBMA – Concrete Brick Manufacturers' Association, 60 Charles Street, Leicester LE1 1FB, UK.

The Concrete Centre Riverside House, 4 Meadows Business Park, Station Approach, Blackwater, Camberley GU17 9AB, UK.

# 13 Ceramics

**Geoff Edgell** BSc PhD CEng MICE FIM
Technical Director, CERAM Building Technology

## Contents

## 13.1 Introduction

In the context of this book this chapter on ceramics covers products manufactured by forming inorganic mineral sources, clay, marl, shale, etc. into the required shapes and then firing the units at very high temperatures to produce hard, durable building elements.

Materials and products considered here are:

- terracotta
- faience
- vitrified clay pipes
- clay roof tiles
- wall and floor tiles.

Chapter 12 deals separately with the much larger subject of bricks and brickwork.

## 13.2 Terracotta

Terracotta is the name given to a dense, hard ceramic formed from once fired clay. It is most commonly seen in the characteristic red colour that derives from the Etruria marl from which it is made. It is however also available in a buff colour which is made from fireclay, and even a dark slate colour.

Historically terracotta has been used either in large slabs bedded in sand and cement and held with metal cramps or as highly decorated pieces for the embellishment of ornate buildings. Consequently a lot of terracotta today is supplied either for the restoration of historic buildings or for mimicking their features. The complex pieces that can be produced are by highly skilled hand moulders, sometimes from several pieces that are subsequently joined together and the joints disguised with wet clay. However, terracotta can be produced by a variety of forming processes, namely stiff or soft extrusion, dry pressing or slip casting. In fact some products are made from several pieces produced by different processes and assembled finally by hand. Some, especially the more complex restoration pieces, are cast in plaster moulds made about models of an original made in clay, plaster or even rubber lubricated with soft soap as a release agent before casting the mould.

The manufacture of ornate chimney pots, finials, etc. is very much a niche activity and there are only three specialist manufacturers in the UK. Surprisingly there are over 400 separately listed designs of chimney pots alone (Valentine, 1968). However, in recent years terracotta has reinvented itself as rainscreen cladding. In this case flat tiles, produced by extrusion and with shaped lugs on the back, are securely fixed by clips to an aluminium rail system. The rails are fixed to the substructure but stand sufficiently forward to allow insulation to be fixed to the outside of the structural wall and for there to be an air gap between the tiles and insulation to drain the water away. Such ventilated rainscreen claddings, as they are known, became popular following their use by the architect Renzo Piano, for example at the Potsdamer Platz in Berlin. They are now commonly used, often in association with glazing, for high-quality buildings where what is a very traditional material provides through the dry fixed, stack bonded pattern a new aesthetic. The rainscreen tiles when fired are machined to very fine tolerances. A similar aesthetic is achieved using stack bonded large-format blocks.

Against the backdrop of making complex architectural features and the modern rainscreen products there is a range of simpler products such as angular or rounded ridge tiles, simple chimney pots and flue liners which are manufactured as stock items in large volumes.

Terracotta is therefore an unusual material in that there is a commodity business, a skilled craft business and a modern quality project business. This is evident in the vastly different quantity of products made in each business – the huge numbers of moulds and patterns retained for in some cases years as opposed to rapid forming of the products with volume sales.

However, once formed all are fired at temperatures in excess of 1045°C to produce a dense, hard and durable material. The fired material has low water absorption, less than 5% by mass of fired clay, and is low in soluble salts, hence there is little likelihood of efflorescence. When tested using the panel freezing test which forms the basis of TSEN772-22 (European Committee for Standardization, 2005), the European test method for clay bricks, the tiles reach the 100 cycle criterion that is used to classify clay bricks as fully frost-resistant. Rainscreen tiles will withstand a lateral load in excess of 2 kN, which means they can easily sustain the wind pressures likely to be achieved in the UK although in practice the ventilated cavity should mean that any differential pressure across them is minimised. Although individual tiles can be replaced if broken accidentally, it would not be common to use them at low levels in areas of significant vehicular or pedestrian traffic. In common with most fired clay products, the performance in fire is good.

Although modern terracotta is supplied with a guarantee regarding its durability and colour fastness, often for 30 years, in practice its use on imposing buildings dating back to the 1860s means that we expect it to last and maintain its appearance much longer. In terms of sustainability, therefore, terracotta has a fine record for its longevity and the usual questions of recycling and reuse are not as relevant as with, for example, bricks and tiles. Numerous buildings give testimony to this argument; for example, the Victoria and Albert Museum, the Grand Arcade, Leeds and the Natural History Museum, built in 1881 and spectacularly revitalised by steam cleaning in 1974.

## 13.3 Faience

Terracotta that has been glazed prior to a second firing is known as faience. Its use is relatively uncommon today. It is usually seen in the form of decorative slabs around the entrances to buildings or some of the grander Victorian shopping arcades. Historically there was a difficulty in controlling the drying shrinkage in the pieces and sizes were commonly limited to 18 × 12 in. (450 × 300 mm) or 12 × 9 in. (300 × 225 mm). The backing of the pieces was often profiled to give a good key to the rendering mortar to which it was fixed. On taller buildings it was common to tie back the pieces with copper wire cramps every few courses.

Although the glaze is itself impervious, this means that any water runoff is focused on the joints and any surface beneath it and is likely to come from a greater area than, for example, in low water absorption clay brickwork. The joints are narrow but any water penetration could lead to freeze–thaw damage, as could penetration at positions that are chipped. Having said that, there are some splendid examples that have stood the test of time.

## 13.4 Vitrified clay pipes

Vitrified clay pipes are made in a variety of sizes ranging from 100 mm bore, which are used for round the house drainage, to 1 m, which are used for mains sewerage. Lengths are typically from 1 m to 3 m. Historically many of the pipes would have been glazed, and although this is relatively uncommon in the UK market it is very common in some mainland European countries, notably Germany. A wide range of junctions, bends, traps, etc. are made so that complete drainage systems in clayware can be provided. The larger sizes are relatively little used in the UK but have been used for example in the Middle East where ground conditions are not suitable for concrete pipes.

Historically pipes were all jointed together by spigot and socket ends, initially using rope and then mortar to seal the connection. This however forms a very rigid joint which is not very tolerant of ground movement, and consequently cracking of the mortar could cause leakage or the ingress of tree roots seeking the moisture. Today there are two types of joint mechanism, both of which introduce some flexibility to the joint. Spigot and socket joints are produced with a polyester lining to the socket and a section on the spigot containing a groove into which a rubber "O" ring is placed and which when the spigot is pushed home seals the gap between the polyester clad pipe ends. A later development was the plastic sleeve joint containing two "O" rings which slides neatly onto the ends of plain ended pipes; these are now the more common and the development of suitable designs of the sleeves that lead to manufacturing efficiency is an ongoing process.

In the UK clay pipes are made from coal measure shales, often blended to give the correct properties for extrusion, drying and firing. The clay is crushed initially, typically down to 100 mm pieces, which are then ground down to around 2 mm in size. Hepworth Building Products is the largest UK manufacturer of vitrified clay pipes and is unique in that at this stage of production the material is fired at a relatively low temperature, typically at around 650°C – a process known as calcining – to produce part-fired material which is mixed with some virgin clay and ball milled to give a very fine material, typically less than 75 μm in size. This, because of the particle size and high surface area, is very reactive. It is also very low in carbon. The material is mixed with water and air is removed under vacuum and the pipe extruded ready to be dried. The pipes enter the drier at a rate of one per minute on individual curved supports and after drying to about 0.5% moisture content enter the roller kiln. Initially the pipes are pushed along the kiln by two vertical arms but when they reach the top temperature zone they roll along the refractory hearth of the kiln and are subsequently cooled, tested for soundness and then packaged. Typically the water absorption of the end product is about 1.5%.

The word "vitrified" is used in relation to clay pipes although the process of vitrification applies to all fired clay products to an extent. Probably the simplest description is that the process involves the melting of a clay mass to become a liquid that coats the remaining solid particles. When cooled the liquid forms a glass that binds the solid particles together

Vitrified clay pipes need to perform a wide variety of functions, and hence the testing regimes applied to them are among the most extensive of those for all clay products (Edgell, 2005). Pipes must perform structurally in the ground when subjected to backfill and surface loadings. The earliest structural designs were based on the Marston-Spangler theory of soil behaviour, which is still used today for design to resist crushing loads (Walton, 1970). Design tables are available for pipes laid on a range of pipe beddings, ranging from a simple trimmed trench bottom to a complete surround of aggregate (Bland, 1991). However, since the 1970s it was recognised that, especially for the smaller diameter pipes, failure was more likely to occur by the pipe breaking like a beam. Pipes with rigid joints had proved to be especially vulnerable, as any leakage at the joints could lead to a leaching away of fine material in the pipe bedding and to voids that the pipe needed to span to continue to perform satisfactorily. Hence today pipes of the various sizes need to achieve a certain 'bending moment resistance' in order to minimise the risk of bending failure. In practice bending failures have been caused by poor site practice, in particular propping pipes on pieces of brick in order to place them to the desired fall.

The pipe walls should not of course leak and neither should the joints. Pipelines prior to being backfilled are often fitted with rubber bungs at either end and either filled with water under pressure from a stand pipe or tested using compressed air. These simple approaches, which do not take up too much time on site, will detect improperly made joints or pipes that have been cracked during carriage or installation.

Pipelines must also perform hydraulically, and although they are designed to operate effectively under gravity there will occasionally be situations after storms where the flow will put the pipes under pressure for a short period; hence the importance of the integrity of joints. In service the inside surface of the pipes becomes coated in a layer of slime, the build-up of which is not dependent on the material of the pipe walls. As a consequence, differences in the hydraulic roughness of the bare material, whether it be clay, concrete or plastic, becomes irrelevant over time.

The flexible joints in pipelines are there to accommodate ground movements which could cause an in-line displacement or draw, a transverse or shear displacement or a rotation. Although the joints are designed to remain watertight if such movements occur, the hydraulic design does not cater for them. It has been demonstrated that the effect on the flow capacity of the introduction of transverse displacement, draw or both is extremely small. This is important as the increased use of closed circuit television (CCTV) surveys to investigate defects does raise awareness of the existence of such features caused by the accommodation of ground movements. CCTV surveys can be most useful in identifying the cause of any blockages and show clearly the effects of defects at the joints, as although leakage can be a problem intrusion of the tree roots seeking moisture can be a greater one. Not only do the roots grow and cause further damage to the joints, but a network of fine roots can lead to serious blockages due to the accumulation of fat, paper, etc. Roots can be quite easily removed using a rotary cutter but the cause of the problem remains and it may well need redoing from time to time. Serious blockages may need removal by water jetting. Jetting at the pressures needed to remove serious blockages can cause minor pitting in the surface of clay pipes, which is not injurious to their performance; however, it can cut through the thin walls of structured wall plastic pipes.

A test method has been developed to measure the resistance of pipe walls to high-pressure low-flow water jetting, although it is as yet not standardised. Discussions about the development of European Standards for pipes also led to the test being capable of investigating the effect of low-pressure high-flow jetting aimed to simulate the effects of cleaning whole sewerage systems by purging them, a practice common in some cities in mainland Europe.

Other properties that have become important through discussions in Europe are abrasion and fatigue. Abrasion resistance is important where flow may regularly contain hard abrasive materials such as chippings from road surfaces. Clay is highly resistant to such action and test procedures do exist. The interest stems more from questions over the performance of glazed surfaces, which are quite common, particularly in Germany. Fatigue has also been a concern raised by Germany, and test procedures have been developed which show that cycled loading has no appreciable effect on clay pipes (Edgell, 2005).

Trenchless construction is a technique sometimes used to minimise surface disruption and pipes are jacked horizontally into the side of a hole in the ground following a pilot hole produced by a mole. In this case the pipes must be able to resist the axial forces needed to push them into the ground. Proof testing of the longer, more expensive pipes is often carried out, which includes some misalignment of the jacking force as might occur in practice.

Modern clay pipes and fittings are extremely durable. The longevity of clay pipes is demonstrated by examples dating back to Roman times and before. Some exist at the Palace of Knossos in Crete, which is some 3500 years old. One of the benefits of the very low water absorption to which they are fired is that they are resistant to frost, which, although rarely a problem in service, has been known when they are on the yard. Modern flexible pipe joints are less susceptible to leakage than the earlier rigid joints. Hence if properly laid and backfilled they should last a very long time, and under normal circumstances require no maintenance (Housing Association Property Manual, 1992).

It is not usual to recycle and reuse clay pipes; however, they are a sustainable product in the sense that little waste is created during manufacture and, like most clay building products, the efficiency of use of raw materials is extremely high. However, they do use a non-renewable resource (clay) and are fired to high temperatures. Trials are in hand to determine whether waste cathode ray tube glass from televisions and monitors can be used as a useful fluxing agent in the firing process. In some cases, and it is very product-specific, good energy savings have been demonstrated using waste container glass in bricks (Smith, 2005). In thinking of sustainability the whole picture needs to be reviewed, and it has been demonstrated on the laboratory scale that crushed concrete and even builder's rubble can be used to provide the bedding for pipes in trenches (Barnes and Beardmore, 1999). As yet this has not been taken up in practice but, like most issues of sustainability, the economics change with time, and pressures from specifiers may cause a take-up.

## 13.5 Clay roof tiles

Clay roof tiles have been, like other clay products, important throughout history. Today they have quite a small share in the overall market for roof tiles in the UK, which is dominated by concrete tiles. However, clay is probably acknowledged to be the quality product for this application.

There are essentially two types of clay roof tile made today: the more common plain clay tile and the interlocking tile. These are also commonly known as double lap and single lap respectively. Double lap tiles are of a standard size which is 265 mm long by 165 mm wide; they are usually 12–15 mm thick with a longitudinal camber and usually two nibs or occasionally one continuous rib at the head and two nail holes near the head. Some tiles also have a cross camber. A common roof construction is for an impervious roofing felt to be placed over the rafters and for timber battens, usually 38 mm × 25 mm, to be fixed on top and perpendicular to the rafters spaced at a gauge of 100 mm centres up the slope. Counterbattens may be used to retain a 50 mm ventilated cavity beneath the tiles. The details vary depending on where the insulation is placed. In view of the length of the tiles and the gauge of the battens, there are parts of the roof where there are three thicknesses of tile or a "double lap". These tiles are laid at a minimum roof pitch of 35°, which is sufficient to ensure that rain is not driven up beneath the tiles so as to penetrate or that water does not remain captured between the tiles where it might freeze and cause frost damage to susceptible tiles.

In practice on modern roofs the underlay provides another line of defence, although historically the minimum pitch was 40° which was arrived at when, in the absence of underlay, lime mortar was used internally to parge the gaps at the head of tiles. Tiles that do not meet the dimensional requirements of the standard are still laid to a minimum pitch of 40°.

Single lap tiles come in a variety of sizes, all of which are larger than plain tiles and typically have a roll profile. There are a variety of descriptions, e.g. Spanish, Italian, and Single and Double Roman, typically with a width of 185 mm to 340 mm and lengths from 340 mm to 420 mm. Importantly, there are interlocking projections between one tile and another, both along the sides and at the top, and this arrangement aids watertightness. Single lap or interlocking tiles may be laid at roof pitches as low as 12.5°. There are a wide variety of special shapes or accessory tiles to finish the roof when differing roof slopes or orientations meet one another (for example, ridge tiles, hips and valleys), which can of course lead to the use of some of the more exotic terracotta for decoration. Modern clay roof systems also include special tiles to, for example, ensure a well-ventilated roof space to minimise the risk of condensation; there are also dry fixed verge tiles to eliminate the use of mortar at the verges, which so often has the tendency to fall out.

The use of plain clay roofing tiles has a long tradition and at roof pitches of 40° or more the double lapping technique is a well proven method of resisting rain, either with or without an underlay. However, the use of interlocking tiles with the safeguard of the interlock led to their use at ever lower roof pitches. Without the benefit of history, the watertightness of this arrangement needed to be proved and hence a pioneering major programme of tests was undertaken by the concrete roof tile industry. In the early days the experimental set-up was fairly crude, an aircraft engine positioned at the end of a rectangular cross-section "tube" with the facility to introduce "rain" from sparge pipes. This work was quite fundamental to establishing confidence in the use of interlocking tiles at reduced pitches.

A consequence of this work was the need to reconsider the likelihood of the tiles being removed by the wind. When the wind blows up the slope of the roof and meets the small step encountered at the tail of a tile, a localised suction pressure is generated at the head of the tile. Traditional guidance for the fixing of double lap tiles has been to fix every fifth course with two nails at the head of each tile in the general areas of the roof and to fix each tile at more vulnerable areas such as the eaves, verges and around chimneys etc. This was not adequate for interlocking tiles, and the roof became something to be engineered in much the same way as masonry or sheet cladding. The Code of Practice (British Standards Institution, 2003) hence became a complex document dealing with wind uplift, dead weight resistance and the design of fixings, which could be simple head nailing but extended to the more effective clip arrangements that could be fitted at the tail of the tile. The enthusiasm for the engineered approach led to its application to double lap tiles and the traditional fixing rules appeared somewhat out of date. Historically reference was always made to calculated fixings to cover unusually exposed situations, but the practice of fifth course nailing was the norm and confidence in it was reinforced by the numerous cases where the roofs survived windy weather despite nails having been omitted or having corroded away.

The majority of clay tiles are made by machine. A typical process for plain clay tiles is that the clay is crushed and rolled with the addition of some sand down to a particle size of about 1 mm. Water is added to bring the moisture content to about 15% when the clay is extruded, cut into tile sized pieces or bats and pressed to give the required curvature and nibs. The tiles are then dried and fired, sometimes in bunds or in cassettes of refractory material. As with other clay products, notably bricks, they may also be hand moulded, which leads to somewhat thicker tiles that usually have sanded surfaces, and those from Kent were typically fixed with wooden pegs. Clay roof tiles need to be impermeable to water and those made in the UK are; however, the British Standard, which is the British published version of a European Standard (British Standards Institution, 1998), does require impermeability testing. This is more of an issue in some European countries (for example Denmark) where permeable tiles can be used on fully boarded and weatherproofed roofs. The other key properties are the flexural strength of the tiles so that they can withstand the weight of a person working on the roof, and their frost resistance.

Tiles in the UK have traditionally been tested using a development of the BCRL Panel Freezing test used for bricks, whereby an array of presaturated tiles was subjected to a freezing, thawing, resaturation and refreezing regime. Tiles that withstood 100 cycles were deemed frost-resistant. Attempts to produce a single test for Europe have not been very successful, and four test methods are in use for an area dominated by the country of the method's origin. A development of the French approach based on the freezing and thawing of individual tiles will be published shortly by the European Committee for Standardization, although it is anticipated that it may be five years before it can be adopted as the single test for Europe. Tiles are classified as being able to survive 30, 90 or 150 cycles in the test, and the 150 cycle criterion will be applied in the UK initially. However it is known that most UK-made tiles can survive at least 400 cycles and this may well become a *de facto* requirement through the market place.

Plain clay tiles may also be used vertically, although this is not common in areas where they may be prone to damage from pedestrian or vehicular traffic. There are nevertheless fine examples and there are numerous decorative tiles for this application, where the tail is shaped to form an arrowhead, beavertail, etc. Guides for the use of clay tiles are available from the Clay Roofing Tile Council (2004).

Clay tiles are a very sustainable product in that the manufacture is very efficient in the use of resources, as any waste in the process is collected and reintroduced to the process; firing is quick and efficient but, most importantly, tiles have a very long lifespan. There is a thriving second-hand market and it is not uncommon to find tiles being reused. A good example is the North Staffordshire Royal Infirmary, where 94% of the 150-year-old tiles were salvaged and reused in a roofing upgrade. There is, however, a word of caution, as some part of the 'frost' life of the tiles will have been used at the point of reuse and, if the new roof is significantly more exposed than the original, failures could occur. It is therefore a very sensible precaution to carry out a freeze–thaw test prior to reuse. Although the test was developed for new tiles and there is no direct correlation with expected lifespan, good or poor performance will be identified. With new tiles, although there is no general relationship between durability (freeze–thaw resistance) and water absorption, for a particular clay and process the manufacturer will be aware of the water absorption (degree of firing) required to ensure the products are durable. Hence day-to-day quality control is often based on water absorption testing.

The long tradition of use of clay tiles in the UK and indeed many other countries derives from their appearance, which tends to improve with age and the effects of the weather. In fact architects have been known to try to encourage algal growth on new roofs by applying mixtures of milk and cow's urine. The point here is that clay is perceived to mellow with age, whereas concrete has the reputation of fading. Some efforts are in hand to quantify these effects by colour measurement on exposed tiles. Tiles in the UK have a history of being of good quality and, because of the relative size of the industries, it has been France that has dominated the development of European Standards. Surprisingly this has led to a great focus on the identification and classification of defects such as cracks and chips that are considered to be acceptable. Until now this has not been a great consideration in the UK, and it is hoped that it will not lead to a reduction of the standards of tiles used in the UK.

## 13.6 Wall and floor tiles

Tiles have been made for some 5000 years or more and their origin was in Ancient Egypt. There are many early examples, as in the highly decorated tiles used on the Mausoleum of Rameses III. Although there are early examples of production in many countries including, for example, China, major developments seem to have taken place around the Mediterranean. In time a strong tradition of wall tiling in Spain around Valencia seems to have merged with the tradition of floor tiling in Italy, possibly around Pisa. The strong tradition of sailing and trade around the Mediterranean led to widespread use of both wall and floor tiles. Probably the most important manufacturing centre now is at Sassuolo in Italy, although Castellon de la Plana in Spain is well known and important manufacturing centres developed in the Netherlands, England, France and Germany (Biffi, 2003).

Tiles are classified in BSEN 14411 (British Standards Institution, 2012) in relation to the way they have been formed. This is usually by granulate (or dust) pressing under high pressure (B) or extrusion (A). Other, less common, methods – for example wet pressing or hard making – are grouped together in a single classification (C). Within each classification is a subclassification based on water absorption. Group I is for tiles with a water absorption of less than or equal to 3%, Group III is for those greater than 10%. Group I is subdivided into Ia and Ib, with 0.5% being the limiting value. Group II is for those between 3% and 10% and is subdivided into Groups IIa and IIb, with 6% as the limiting water absorption. Consequently there is a fairly complicated classification system such as AI through to CIII, added to which is the complication that within each group tiles can be glazed or unglazed and the specifications give different test regimes for each type.

Although the system appears complex, there are within it some common types. For example, the lower water absorption tiles (< 3%), whether extruded or pressed, are normally for flooring applications and unglazed, whereas the pressed high water absorption tiles (> 10%) are usually glazed and for walling applications. A wide range of tests has been standardised at the National, European and international levels covering the main properties of importance in use, e.g. resistance to scratching, slip resistance, chemical resistance and, for external use, frost resistance (British Standards Institution, 1996–2012).

British Standard (British Standards Institution, 2012) provides guidance on the selection of tiles of the various classes for interior and exterior use in floors and walls. This seems to be a considerable step forward where previously the guidance was to consult the manufacturer. However, for some of the requirements clearly agreement could not be reached on numerical values and manufacturer's advice is still to be recommended.

Resistance to wear when trafficked is of course most important for floor tiling, and an abrasion test is standardised in which the tile is offered up to a 10 mm wide rotating steel wheel and an abrasive (white fired aluminium oxide) of precise specification is fed onto the wheel at a prescribed rate. The action leads to an arc being worn into the tile surface, and the greater the volume of material removed the lower the resistance, and hence the lower the suitability for use in heavy traffic. BSEN ISO 10545 Part 7 does give some guidance on the classes of resistance needed for differing degrees of traffic.

In the case of floor tiles, clearly resistance to slipping is the most important property. The test procedures are described elsewhere, but essentially consist either of using a simple pendulum device similar to that for road surfaces or an inclined plane which is raised until an operative wearing a suitable safety harness begins to slip. These methods do lead to usable classifications, but there are differences between the results from the different methods.

One type of tile worth mentioning is vitrified stoneware, or gres porcellaneto in Italian, whose modern development began in the 1960s. It is usually dust-pressed and hence of type BI and made from white clays, kaolins and feldspars, which offer a fast single firing to an excess of 1200°C and forms a white or grey tile body that may be glazed or not. The key characteristics are that the body has very high flexural strength and very low water absorption and is frost-resistant. It is hence very durable and can be used with a wide variety of finishes, for example to resemble pebbles or polished stonework, or to form mosaics or large slabs for shopping plazas. The increase in production in Italy has been quite remarkable, the glazed tiles ranging from an annual production of 24 million square metres in 1997 to 161 million square metres in 2001.

Tile installations are dependent not only on the quality of the tiles but also on the adhesives and grouts used, together with the workmanship. The adhesives are described as cementations, dispersive or reactive, and each type is covered by a series of standards and tests. Guidance on installation is given in BS5385 (British Standards Institution, 2007–2009). The main features are ensuring good adhesion with no voids beneath the tiles, line and level, grouting to ensure an impermeable surface and the location of movement joints. All of these aspects are covered but, as with tile selection, manufacturer's advice should be followed. One situation that requires a specialist contractor is installation of swimming pools; here there may be wall and floor tiling, possibly mosaic, and health and safety are critical – for example, slip resistance in wet situations – and although Code of Practice advice does cover them, there are other specialist volumes that may be helpful (Perkins, 1971).

## 13.7 References


Barnes, J. D. and Beardmore, C. (1999). *Use of recycled aggregates as clay pipe bedding*. Ceram Research Paper 815. Stoke-on-Trent, UK: Ceram Research.

Biffi, G. (2003). *Book for the production of ceramic tiles*. Faenza, Italy: Gruppo Editoriale Faenza Editrice.

Bland, C. E. G (1991). *Design tables for determining the bedding construction of vitrified clay pipes (for BSEN 295: 1991 pipe strengths)*. Chesham, UK: Clay Pipe Development Association.

British Standards Institution (1996–2012). *Ceramic tiles*. BSEN ISO 10545, Parts 1–16. London, UK: BSI.

British Standards Institution (1998). *Clay roofing tiles for discontinuous laying: Product definitions and specifications*. BSEN 1304: 1998. London, UK: BSI.

British Standards Institution (2003). *Code of practice for slating and tiling*. BSEN 1304 1998. London, UK: BSI.

British Standards Institution (2007–2009). *Wall and floor tiling*. BS5385, Parts 1–5. London, UK: BSI.

British Standards Institution (2012). *Ceramic tiles – Definitions, classification, characteristics and marking*. BSEN 14411: 2012. London, UK: BSI.

Clay Roofing Tile Council (2004). *A guide to plain tiling including vertical tiling*. Stoke-on-Trent, UK: CRTC.

Edgell, G. J (2005). *Testing of ceramics in construction*. Caithness, UK: Whittles Publishing.

European Committee for Standardization (2005). *Technical Specification TSEN 772-22: 2005. Methods of test for masonry units. Part 22: Determination of freeze thaw resistance of clay masonry units*. Brussels, Belgium: CEN.

Housing Association Property Manual (1992). *Component life manual*. London, UK: E&FN Spon.

Perkins, P. H. (1971). *Swimming pools*. London, UK: Elsevier.

Smith, A. (2005). Recycled glass. *Clay Technology*, No. 102.

Valentine, F. (1968). *Chimney pots and stacks*. New York, NY: Centaur Press.

Walton, J. (1970). *The structural design of the cross section of buried vitrified clay pipelines*. Chesham, UK: Clay Pipe Development Association.

# 14 Concrete

**C.D. Pomeroy** DSc CPhys FinstP FACI FSS
Formerly of the British Cement Association

Updated for Second Edition by

**Bryan Marsh** BSc PhD MICE FICT FCS
Associate, Arup
(with some additions by the Editors)

## Contents

## 14.1 Introduction

According to the *Oxford English Dictionary*, concrete is 'formed by the cohesion of particles into a mass. It is a composition of stone chippings, sand, pebbles, etc. formed into a mass with cement (or lime).' This definition offers the scope for many variations in the composition, and today these variations are exploited by the producers of concrete to make materials of different strengths and elastic moduli, of different densities or porosities, of different textures or colours, and for resistance to different environments. Because of the large number of options available to the concrete producer and user, the whole subject of concrete technology has moved from one of relative simplicity to one where specialist skills may be required if the best selection or choice of constituents is to be made. There will always be uses for concrete where basic skills are all that is required, but economical production of concrete of the highest quality requires a deep understanding of the subject.

Concrete is not a new material, and a little background history helps to place modern concrete technology into perspective. More concrete has been produced during the past few decades than in the rest of human existence, but it is not a modern invention. The earliest known concrete, which dates from about 5600 BC, formed the floor of a hut in the former Yugoslavia. The 'cement' was red lime, which probably came from 200 miles upstream, and this was mixed with sand, gravel and water before being compacted to form the hut floor. The first written proof of the ancient use of mortars is found in Egyptian hieroglyphics on a wall in Thebes dating from 1950 BC. This shows all the operations of concrete construction from metering the water and materials to the use in the structure.[1] Indeed, it has even been claimed that the pyramids themselves were made from cast concrete blocks and not from cut stone.[2]

There are also existing examples of concrete made by the Romans in many of the countries occupied by them. One sample of concrete unearthed in Lincolnshire had been used to encase a clay pipe. This concrete is interesting, since the aggregate is largely pieces of broken fired clay and carbon waste. The mortar is crushed lime and the specimen had a cube strength of about 40 MPa. It could be argued that the builders were good environmentalists who used up the broken pot, but the Romans learned that the pot could be pozzolanic and add to the strength of the concrete as it reacted with the lime.

Another example of 'modern' thinking is found in the dome of the Pantheon in Rome, built in 27 BC. The dome, made of lightweight concrete, using volcanic pumice as the aggregate, spans almost 50 m. It has stood for over 2000 years and is clearly durable. It was unreinforced, so that steel corrosion or rusting was not a problem to be contended with.

Concrete is thus not a new material, the name itself coming from *concretus*, a Latin word for bringing together to form a composite. Admixtures, such as blood or milk, were even used to improve workability and appearance, and it is very humbling to relate modern sophistication to the practical evidence of the past. Nevertheless, after the fall of the Roman Empire, the skills needed to make effective limes, mortars and concretes seem to have been lost, and it was only some 200 years ago that the modern concrete industry came into existence with the development and evolution of the hydraulic cements used today.

## 14.2 Hydraulic cements

Although Joseph Aspdin is usually credited as the father of the modern Portland cement industry, it is likely that he developed his ideas from the earlier works of John Smeaton and the Reverend James Parker of Northfleet. Smeaton was charged in 1756 with the rebuilding Eddystone Lighthouse off the coast of south-west England and he demonstrated that he could make a durable mortar that could be used to bind interlocking stone segments together. The mortar contained burnt Aberthaw blue lias, a limestone and some natural pozzolanas from Italy. The lighthouse withstood the ravages of the Atlantic storms for over 100 years before it was replaced by the present structure. The original lighthouse was dismantled block by block and re-erected on land in Plymouth, but the original foundations were left and they still exist over 200 years after being built.

The contribution to the development of cement by Parker is attributed to his observation that when a stone, picked up on a beach near his home, was thrown on a fire, it became calcined and, when pulverized and mixed with water, it hardened to form a cement. Parker thought he had rediscovered the ancient 'Roman' cements and hence gave his invention this name.

These cements had little potential for wide commercial development and it was the patented inventions of Aspdin that led to the modern cement industry, although very substantial improvements have taken place since his original work near Wakefield. It was only when in 1845 a bottle kiln was built near Swanscombe in Kent that the firing temperature of the mixture of chalk and clay was high enough to cause the chemical reactions that are fundamental to the production of modern cements.

Since that time, the blends of raw materials that are fired in the kilns have been refined and the manufacturing process has changed from one of batch production to continuous production in the modern rotary kilns, a single one of which can produce 1,000,000 tonnes/year. As the raw materials pass through the kiln, their temperature is gradually increased to a peak in excess of 1400°C when the calcium carbonate from the limestone or chalk is dissociated into calcium oxide and carbon dioxide and the oxides of calcium, silicon, aluminium, and iron are transformed into the active ingredients of Portland cement. The resulting product is a clinker that leaves the kiln at a temperature of about 1000°C and is then cooled to about 60°C, the rate of cooling having a significant effect on the crystallographic form of the clinker constituents.

Portland cement clinkers generally have four principal components:

1 tricalcium silicate, 3CaO.SiO2 (abbreviated to $C_3S$)
2 dicalcium silicate, $2CaO.SiO_2$ (abbreviated to $C_2S$)
3 tricalcium aluminate, $3CaO.Al_2O_3$ (abbreviated to $C_3A$)
4 tetracalcium aluminoferrite, 4CaO.Al2O3.Fe2O3 (abbreviated to $C_4AF$).

These proportions can be varied by the selection of different raw materials and by the kiln temperature profiles. The constituents of Portland cement clinker are not pure compounds but contain minor impurities within the crystal structures, and these can have a significant effect on the performance of the cements. There are also other impurities such as magnesia (MgO) and free lime (CaO), which the cement chemist must ensure are kept within acceptable limits if the cement is to be consistently sound. There will also be small amounts of alkali (KOH and NaOH) present.

The clinkers are ground to a fine powder with the addition of a small proportion of gypsum (calcium sulfate), which is added to control the rate of hardening that will take place when the Portland cement is mixed with water. The fineness of the powder is one of the parameters that can be used to control the strength-generating characteristics of the cement.

Within Europe, a family of cements that covers the main products currently in use has been identified within European Standard EN 197-1[3] (see Table 14.1). Cements of different strength potentials and with different constituent formulations are defined. The standard takes account of differences that have traditionally existed in Europe, such as the combinations of CEM

I Portland cement with natural pozzolanas in Italy, the use of inert fillers such as limestone at different replacement levels, and the inclusion, particularly in France, of more than one addition to CEM I Portland cement to make a composite cement. These differences are not purely academic: because it is important to be able to specify and produce durable concrete and as this relies on cement type, it is essential to know that the cement selected does not have any unsatisfactory or unsuspected attributes. This is discussed further in Section 14.17.

Inclusion of up to 5% of minor additional constituents is permitted in cement by the European Standard. These are usually selected from finely ground limestone, blastfurnace slag or pulverized-fuel ash. A mortar prism test is used to determine the strength potential of the cement and the definitions of strength bands within which cements of specified classes must lie.

In the USA the different types of Portland cement are described and specified in ASTM C150.[4]

**Table 14.1** The types of common or general-purpose cements in EN 197-1

| *Main type* | *Notation* | | *Clinker content (%)* | *Content of other main constituents (%)* |
|---|---|---|---|---|
| CEM I | Portland cement | CEM I | 95–100 | – |
| CEM II | Portland slag cement | CEM II/A-S | 80–94 | 6–20 |
| | | CEM II/B-S | 65–79 | 21–35 |
| | Portland silica fume cement | CEM II/A-D | 90–94 | 6–10 |
| | Portland pozzolana cement[1] | CEM II/A-P | 80–94 | 6–20 |
| | | CEM II/B-P | 65–79 | 21–35 |
| | | CEM II/A-Q | 80–94 | 6–20 |
| | | CEM II/B-Q | 65–79 | 21–35 |
| | Portland fly ash cement[2] | CEM II/A-V | 80–94 | 6–20 |
| | | CEM II/B-V | 65–79 | 21–35 |
| | | CEM II/A-W | 80–94 | 6–20 |
| | | CEM II/B-W | 65–79 | 21–35 |
| | Portland burnt shale cement | CEM II/A-T | 80–94 | 6–20 |
| | | CEM II/B-T | 65–79 | 21–35 |
| | Portland limestone cement[3] | CEM II/A-L | 80–94 | 6–20 |
| | | CEM II/B-L | 65–79 | 21–35 |
| | | CEM II/A-LL | 80–94 | 6–20 |
| | | CEM II/B-LL | 65–79 | 21–35 |
| | Portland composite cement[4] | CEM II/A-M | 80–94 | 6–20 |
| | | CEM II/B-M | 65–79 | 21–35 |
| CEM III | Blastfurnace cement | CEM III/A | 36–64 | 36–65 |
| | | CEM III/B | 20–34 | 66–80 |
| | | CEM III/C | 5–19 | 81–95 |
| CEM IV | Pozzolanic cement[5] | CEM IV/A | 65–89 | 11–35 |
| | | CEM IV/B | 45–64 | 36–55 |
| CEM V | Composite cement[6] | CEM V/A | 40–64 | 36–60 |
| | | CEM V/B | 20–38 | 62–80 |

[1]P = natural pozzolana; Q = calcined natural pozzolana. [2]V= siliceous fly ash; W = calcareous fly ash. [3]L = organic carbon content ≤ 0.5%; LL = organic carbon content ≤ 0.2%. [4]Contains clinker + one of blastfurnace slag, silica fume, natural pozzolana, fly ash, burnt shale or limestone. [5]Contains clinker + one of silica fume, natural pozzolana or fly ash. [6]Contains clinker + blastfurnace slag + either natural pozzolana or siliceous fly ash.

## 14.3 Cement quality

Hydraulic cements react chemically with water to form calcium or aluminium silicate hydrates (CSH or ASH) and these reactions can occur over a long period of time. The hydration mechanism is complex and still not fully understood. Each of the main cement constituents reacts exothermally at a different rate. As the reactions take place, the originally water-filled spaces between the cement grains become progressively filled by calcium silicate hydrate gel to form a strong hardened cement paste. The strength of this paste is controlled largely by its residual porosity: the lower the porosity, the higher is its strength.

The quality of a cement is frequently related to the strength that it will generate at 28 days when a standard mortar or concrete is made, using well-characterized aggregates and a given ratio of water to cement. Nevertheless, strength is only one aspect of quality and cements can be designed to have other special properties as well.

During the second half of the last century, the strengths of UK ordinary (CEM I) Portland cements underwent large increases (see Table 14.2). Nevertheless, the eventual introduction of cement strength classes enabled the user to be aware of cement strength and to optimize mix designs. CEM I Portland cement is, however, by no means the only available type. It is often combined with blastfurnace slag, fly ash or other materials.

## 14.4 Ground granulated blastfurnace slag, fly ash and silica fume

Blastfurnace slag is a waste material or by-product produced during the manufacture of iron in a blastfurnace. These slags are composed predominantly of the oxides similar to those found in Portland cement, calcium, silicon and aluminium, although the proportions are different and depend on the particular raw materials used to make the iron and on the blastfurnace operation. When the slag leaves the blastfurnace, it must be cooled and the rate of cooling has a large effect on the suitability of the slag as a hydraulic cement; the faster the rate of cooling, the better. The slags, often produced as glassy pellets, are ground into a powder to form the basis of a cement.

Although ground granulated blastfurnace slag is cementitious and reacts with water, the rate of reaction is slow and is known as latent hydraulicity. It is necessary to blend the powdered slag with Portland cement to produce a cement that is suitable in practice.

The extent of use of blastfurnace slags has varied from one country to another and depends largely on economic factors. The use of preblended cement in the UK has been modest, even though the first British Standard for Portland blastfurnace slag

**Table 14.2** Change in strength development of Portland cement concrete over three decades: total water/cement ratio = 0.6; cement content 300 kg/m$^3$

| | *Compressive strength (MPa)* | | |
|---|---|---|---|
| | *3 day* | *7 day* | *28 day* |
| Pre-1950 | 13 | 20 | 32 |
| 1950–54 | 14 | 22 | 34 |
| 1960 | 16 | 24 | 35 |
| 1975–79 | 20 | 30 | 41 |
| 1980 | 24 | 33 | 44 |

cement (BS 146) was originally issued in 1923. Nevertheless, the use of powdered slags for inclusion in concrete mixes in combination with Portland cement has proved very popular. Ground granulated blastfurnace slag for use in concrete is covered by the European Standard, EN 15167-1: 2006[5] and preblended cements containing blastfurnace slag are included in the cement standard EN 197.[3]

Ground granulated blastfurnace slag, used either as a component in a preblended cement or combined with Portland cement in a concrete mix, can make excellent concrete which will differ in performance in certain respects from that made with a CEM I Portland cement. As already mentioned, blastfurnace slag tends to develop strength more slowly, although this can be compensated for by grinding the cement more finely and restricting the proportion of slag. The user must, therefore, understand the cement and make the relevant changes to his concreting practice. At higher slag proportions this may involve a longer curing time and possibly an extension of the time before formwork is stripped.

Concretes, properly compacted and cured, made from blastfurnace slag cements will generally have good resistance to chloride penetration and chemical attack. They also produce concrete that is lighter in colour than Portland cement concrete, and this can be attractive architecturally.

Fly ash, previously also known in some countries as pulverized-fuel ash or pfa, is another industrial by-product, formed in electricity-generating stations by burning pulverized coal that is fed into the furnaces. The fly ash is electrostatically precipitated from the furnace flue gases. It consists primarily of alumino-silicate glass spheres, together with other minerals such as quartz, iron oxides and mullite. Fly ash is pozzolanic and not a hydraulic material, i.e. it reacts with lime to form hydrates that are similar to those produced by Portland cement hydration. The combination of Portland cement and fly ash can thus provide the essential ingredients for a composite cement in which the Portland cement reacts initially to form calcium silicate hydrates and lime (Portlandite) and, in turn, the lime will react with the fly ash to supplement the hydration products. The pozzolanicity of fly ashes varies, the reactivity being dependent on the original coal used in the furnace, the efficiency of the burning and the fineness of the powder. Some fly ashes contain high levels of lime naturally, but these are available only in a few places in the world (e.g. the USA and Greece) and derive from the use of lignite or sub-bituminous coals. Within Europe there is a standard for fly ash for use in concrete.[6] Cements made from a combination of Portland cement and fly ash are included within the cement standard EN 197.[3] In the USA, requirements for fly ash are given in the ASTM C618 standard.[7]

As with ground granulated blastfurnace slags, fly ash can be used as a component in excellent concrete. Its performance will depend on the selected Portland cement, the fineness of the combined powders and on the proportion of fly ash relative to the cement. Curing will normally have to be increased to ensure performance comparable with CEM I Portland cement and the effects of change of ambient temperature may be important. It is usually necessary to increase the total cementitious content in a concrete mix slightly to attain the same 28-day strength. Nevertheless, the use of fly ash does offer many benefits. In the fresh state this is particularly with regard to the workability and cohesion (reduced bleeding) of mixes. In the hardened state, the concrete may have increased resistance to chloride ingress and to chemical attack. Fly ash cement can be used effectively to reduce heat build-up during hydration in large sections. As with all concrete, fly ash concretes must be sufficiently well cured to achieve their full potential.

Another industrial by-product that has seen increased usage in recent years is silica fume, known sometimes as microsilica or condensed silica fume. It comprises very fine spherical particles of almost pure non-crystalline silica that are by-products of the ferrosilicon industries. Reduction of quartz to silicon in electric-arc furnaces produces SiO vapours that oxidize and then condense into the microsilica powders. As with fly ash, these powders are highly pozzolanic and react with lime to form silicate hydrates. The powders are one order of magnitude finer than Portland cements and, hence, a proportion can pack between the cement grains and thereby provide the basis for a dense and low-permeability hardened cement paste. Silica fume is not available in vast tonnages, but it has many applications in specialist areas such as abrasion resistance, chemical resistance and in the achievement of high strength. It can be difficult to handle as a powder and is often introduced into concrete in a densified form or as aqueous slurry. Requirements for silica fume for use in concrete are given in EN 13263-1.[8]

As sustainability in construction moves up the global agenda, attention is being turned towards the relative environmental impact of different cements. Portland cement production invariably has high levels of carbon dioxide emissions because of the fundamental chemistry and high temperature of the clinkering process. Nevertheless, much has been done in many countries to reduce energy consumption, particularly in the use of fossil fuels. Significant reductions in environmental impact can be gained by reduction in Portland cement use through the use of cements containing fly ash or ground granulated blastfurnace slag. Much research is under way to find low-impact alternatives to Portland cement.

## 14.5 Cements for special purposes

Sulfate-resisting Portland cement (SRPC)[9] is a special class of Portland cement wherein the tricalcium aluminate content ($C_3A$) is controlled to a low level. This gives a high level of resistance to the ettringite, or 'conventional' form of sulfate attack whereby the sulfate reacts with $C_3A$ to form ettringite, sometimes resulting in disruptive expansion. Recent research has, however, shown that SRPC does not have enhanced resistance to the thaumasite form of sulfate attack (TSA) that can occur in cool ground conditions. Caution should also be exercised in the use of SRPC in sea water, despite the presence of a significant amount of sulfate, as it has a lower level of resistance to chloride ingress compared to normal Portland cements or cements containing blastfurnace slag or fly ash. The level of production of SRPC in the UK has fallen to a low level due to the recognition of the sulfate-resisting properties of cements containing blastfurnace slag or fly ash.

An important type of non-Portland cement is aluminous or calcium aluminate cement,[10] although this is really a family of varying composition cements rather than a single type. They are manufactured from a mixture of bauxite and limestone to give alumina contents in the range 40–80%. Their resistance to high temperatures make them ideally suited to use in refractory concrete, and this is their most common use. Their rapid hardening and high resistance to chemical attack and abrasion are properties put to good use in a wide range of specialist products for applications such as repair and flooring, often blended with other binders.

The hydration products of calcium aluminate cements initially exist in a metastable form but eventually convert to a stable form, generally over many years. The stable hydrates occupy a smaller volume than those that are initially formed, so the conversion process involves an increase in porosity and a corresponding reduction in strength. Any adverse effects of conversion can be avoided by use of a very low initial free water/cement ratio ($<0.4$). Three incidences of collapse during the 1970s involving calcium aluminate cements, known at the time as high alumina cement, have effectively led to their exclusion from use in structural concrete even though the primary cause of collapse was not the strength loss resulting from conversion.

In addition, there are available in some countries a variety of hydraulic cements for special purposes. In particular, there is a family of expansive cements that can be used to offset cracking due to shrinkage.[11] There are oil-well cements that must be usable at the high temperatures prevalent, and there are ultrarapid hardening cements that are very finely ground and in which the constituents are selected to react early with water. Masonry cement is specially formulated for use in mortars and comprises a mixture of Portland cement and other fine minerals, such as calcium carbonate, together with an air-entraining admixture to impart the special properties desirable for rendering, plastering and masonry work.

The specifier of cement thus has a wide number of options and needs to know both what is required to ensure the satisfactory construction of the structure and what is available. Sometimes it is even possible to make special cements to satisfy special manufacturing requirements. Typically, coarse-ground cements are often made to facilitate the production of concrete pipes using a spinning method and other precast products. White Portland cement is often used in precast architectural elements.

## 14.6 Handling and storage of cement

Cement may be delivered in bulk or in bags or sacks. Bulk deliveries are stored in silos, which should be weatherproof to minimize the risk of air-setting in the presence of moist air. Where different types of cement are being used, precautions must be taken to avoid confusion and the inadvertent combination of dissimilar cements when successive deliveries are made. Bags should be stored on pallets or a raised floor, be kept dry and be used in the sequence of delivery, because cements tend to lose 'strength' during prolonged storage. Health and safety considerations are discussed in Section 14.24.

## 14.7 Concrete admixtures

Although it is believed that the Romans added blood and milk to mortars to improve their properties, the more controlled use of chemical admixtures is comparatively recent. The principal developments in this field have occurred since the Second World War, although the first uses were probably made during the 1930s. Hewlett[12] defines admixtures as 'materials that are added to concrete at some stage in its making to give to the concrete new properties either when fluid or plastic and/or in the set or cured condition'. Concrete admixtures are quite distinct from any admixtures made to the cement during its manufacture and from additions such as fly ash, ground granulated blastfurnace slag or silica fume which may be added to a concrete mix.

The purpose of using an admixture is to change the properties of the concrete, either in the fresh state or in the hardened form, thereby making it more suitable for its purpose. The main changes that can be made are:

1. to improve the workability of the concrete without increasing water content, thus making it easier to place without the penalty of reduced strength
2. to reduce the water content while keeping the workability constant, thus increasing the strength or allowing reduced cement content
3. to reduce water content at a given strength, thus allowing a reduction in cement content
4. to accelerate or to retard the setting of the concrete, in order to facilitate the construction schedule
5. to entrain bubbles of air, thereby improving the resistance of concrete to cycles of freezing and thawing
6. to reduce permeability (improve watertightness)
7. to add colour (pigments) to concrete
8. to improve cohesion, thereby reducing segregation in self-compacting concrete and to allow concrete to be placed in flowing water.

The range of available admixtures also includes corrosion inhibitors, shrinkage reducers, viscosity modifiers, fungicides and gas formers, the last of these being used to make low-density and high-air-content materials.

Careful consideration has to be given to the suitability of an admixture for its particular purpose, and also about any deleterious side-effects or the effects of overdosing. For example, air entrainment will greatly improve the resistance of concrete to frost attack, but it will also reduce the strength of the concrete. There is therefore the need for the correct use of admixtures and, in the UK, the Cement Admixtures Association,[13] formed in the 1960s, has given excellent guidance on their use, and their effects on the performance of the concrete. Special care needs to be exercised when using two or more admixtures together, to ensure their compatibility.

Admixtures are an almost essential component of most structural concrete and there is no doubt that that new types of admixtures will continue to be developed to impart special properties.

## 14.8 Aggregates

The essential components of concrete are cement, aggregates and water, the aggregates generally filling the greater proportion of the mix. Aggregates come from many sources and can be selected to make concretes with different properties and performance characteristics. The aggregate for normal concrete will comprise a range of different sized particles, the gradings being chosen to ensure good packing when mixed with cement and suitable rheological properties that allow the concrete to be cast and compacted easily and well.

In the UK, gravels and sands, both from land-based quarries and from marine beds, traditionally provided the main sources of aggregate, although crushed rocks are now widely used. Crushed rocks may be used with natural sands or with rock fines, or a combination of the two, to give a suitable aggregate grading. The choice of aggregates will depend largely on local availability because they are an expensive commodity to transport. Fortunately, there are many types of rock that can be used safely in concrete, although it is necessary to avoid some and to take precautions when using others. In some parts of the world, it is difficult to obtain the good aggregate gradings that are vital to the production of a good workable concrete.

Concrete can be made without a component of sand or fines (no fines concrete), or with very dense aggregates to satisfy special needs (e.g. radiation shielding). There are also lightweight aggregates, often man-made, that can be used to make low-density concretes. These aggregates include sintered fly ash, expanded clay or shale, and naturally occurring volcanic rocks such as pumice. Lightweight aggregates may be used with lightweight fines or with normal sands to produce a range of concretes with different densities or strengths. They can also be used together with normal-weight coarse aggregates to produce concrete of a specific density between that of normal-weight and lightweight concrete. This is sometimes known as modified density concrete or modified density normal-weight concrete.

Care must be taken to use aggregates with low chloride levels when they are to be used in reinforced concrete and, particularly, in prestressed concrete. Stringent levels are set in codes for chloride content in structural concrete.[14] In some parts of the world, particularly the Middle East, the land-based aggregates are heavily contaminated with salts and these can be particularly harmful when used in reinforced concrete. Sea-dredged aggregates are used widely in many countries after thorough washing to reduce levels of salt.

Other aggregate characteristics that require care are their shrinkability and freeze–thaw resistance, the presence of high levels of fine particles of clay and the inclusion of constituents that can react deleteriously with the alkalis in cement. Small proportions of reactive silica, silicates or certain carbonates can react with alkalis. A small amount of reaction can be tolerated and might even be beneficial (i.e. add to the cementing or binding potential), but if there is too much reaction, the concrete might crack. For cracking to occur, not only must critical quantities of the reactive component be present, but there must also be sufficient alkalis and water. In particular, with an alkali–silica reaction, the gel formed will only expand and then possibly cause cracking if plenty of water is available.

There is increasing pressure to meet the ever-increasing demand for aggregates, at least in part, by the use of recycled and secondary aggregates (RSA) and, in consequence, to reduce the ever-increasing waste problem. Recycled aggregates are usually crushed demolition waste whereas secondary aggregates are usually by-products of industrial processes. Crushed concrete can be used, typically as approximately 20–40% by mass of the coarse aggregate, in concrete of characteristic cube strength up to about 50 MPa without detriment. Some secondary aggregates, such as china clay waste (sand and coarse aggregate), can be used to replace natural aggregates completely.

### 14.8.1 Aggregate gradings

The proportions of the different sizes of particle in an aggregate sample to be used to make concrete are found by sieving, using sieves of approved mesh sizes. In Europe, the sieves and testing should conform to the requirements of EN 12620.[15] Although various aggregate gradings may be used to obtain special effects and surface finishes, in the UK it is normal to select an aggregate for use in concrete that satisfies the grading limits described in PD 6682-1.[16] At the concrete mixer, the aggregate may be added as a graded or 'all-in' aggregate or, more usually, in higher quality work it is formed from a combination of fractions of single sizes of aggregate. It is customary to consider the fines separately from the coarser fractions of the aggregate, and PD 6682-1 defines five recommended classes of fines for general uses.

These standard classifications of aggregates, together with information on particle shape, water absorption and density, are used to design concrete mixes that will meet strength and performance goals. It is important to have a consistent supply of aggregates and to store them without risk of contamination if a consistent quality of concrete is to be produced.

## 14.9 Ready-mixed concrete

Although it was possible to buy ready-mixed concrete from a few plants in the UK before the Second World War, it was not until the 1950s that the industry grew rapidly, and now there are plants covering the whole of most developed countries. This change had significant effects on the use of concrete in construction. It resulted in the provision of more consistent and economical mixes, but ensuring that the concrete is right for the job and of the right consistency relies on complete, sound and unambiguous specifications for the required concrete. All too frequently disputes arise because inadequate or incomplete specifications have been provided.

Today, most ready-mixed concrete suppliers have sophisticated computer-controlled plants and it can be possible to provide customers with accurate records of the constituents of the concrete supplied. Nevertheless, if only strength is specified, the supplier has the freedom to use his expertise and select a mix design at the lowest cost. While this may satisfy the specification, it will not necessarily guarantee an adequate performance. Thus, ready-mixed concrete is a valuable construction material that can be supplied very conveniently at a time to suit the customer; however, it can also be inadequate if not properly specified.

In the UK, the Quality Scheme for Ready-mixed Concrete[17] has done much to improve the standards of the industry. This is, in fact, in full agreement with the recommendation of Vitruvius,[18] who in 27 BC said 'When writing the specification careful regard is to be paid both to the employer and the contractor'. The problems of today are not new.

The British Standard for specifying concrete (BS 8500[14]) contains a family of concretes that has been described for specific or 'designated' uses. The UK producers of ready-mixed concrete that satisfy the stringent requirements of approved third-party quality schemes for ready-mixed concrete undertake to supply concretes that conform fully with the given designated mix proportions. These mixes have been selected to satisfy the relevant requirements of the prevailing product and structural British Standards. This system makes the specification of the right concrete for a given use much easier, particularly for unsophisticated specifiers and many routine applications.

## 14.10 Steel reinforcement

Because concrete is weak in tension, it will normally be reinforced with steel. It most reinforced structures, a reinforcement pattern is chosen to provide the greatest contribution to the tensile zones, but not exclusively so. The design of concrete structures is well covered by EN 1992[19] and by the ACI 318[20] building code requirements for reinforced concrete.

During the last century, the strengths of steel for reinforcement rose, the permissible values of stress for design rising from about 100 N/mm$^2$ in the 1920s to the current level[21] of 500 N/mm$^2$. These changes, coupled with the higher strength concretes that are now readily available, enable lighter structures with longer spans to be built. Stainless-steel bars can also be obtained at a cost premium but can offer a cost-effective solution to corrosion risk in severe environments when whole-life costs are taken into account.

Reinforcement is well specified by national standards that cover not only the steel grade and dimensions of the bars, but also the cutting and bending schedules. The most common way to order steel is 'cut and bent'. This is delivered to site in bundles that have been cut and bent ready for assembly into cages or directly into the work. Alternatively, straight bars, usually in 12 m lengths, are obtained directly from stockists or a steel mill. Cutting and bending is then carried out on site; this will normally be undertaken only on large jobs.

Reinforcement for pavements or slabs is usually provided as a welded mesh that will be supported on plastic or steel chairs to ensure that it is laid flat with the correct depth of cover. Unfortunately, steel rusts when exposed to a moist environment and many of the concrete structures that have shown signs of deterioration have suffered in this way. The alkaline level in concrete is normally sufficiently high to passivate the steel, i.e. to stop it rusting, but this protection will be lost if carbon dioxide ($CO_2$) from the atmosphere permeates through the concrete to the steel, carbonating the calcium hydroxide and calcium silicate hydrates as it does so. This process can take many years, particularly in a continually moist environment or if the concrete is of low permeability. Chlorides, either from de-icing salts or from sea spray, can also permeate the concrete and provide an environment in which corrosion will start.

One way to reduce the risk of corrosion is to ensure that there is sufficient concrete cover to the steel and that this is of good quality and is well cured. In addition, special steels (stainless) can be used in critical zones. Epoxy-coated bars became popular for a while, particularly in bridge decks in the USA, but have not proved to be fully corrosion-resistant and are now seldom used.

## 14.11 The use of fibres in concrete

In biblical times straw was used in clay bricks and it was obviously an essential component since the Pharaoh, to punish Moses and Aaron, cut off the supplies of straw and made the people collect their own. Why should this be? Clay bricks, like concrete, are brittle and weak in tension and it is necessary to provide something that will limit the effects of cracking. Conventional steel reinforcement carries the tensile loads in a concrete element, but the use of reinforcement imposes limitations to the geometrical size of the unit. The steel must normally be embedded with sufficient cover for protection from the environment.

The only way to allow the fabrication of thin, yet tough, mortar or concrete products is to use thin fibres to control the cracking. In the past, asbestos fibres were widely used to provide reinforcement of cement bound boards, corrugated sheeting and for cladding panels. The very fine asbestos fibres were found to cause serious health problems in those exposed to them; in particular, crippling lung disease, asbestosis, and mesothelioma, a cancer of the lining of the chest. Ingestion of food containing asbestos dust can also cause cancer of the stomach. The use of asbestos in new products is prohibited and the handling or working of existing installation subject to considerable regulation.[22]

Other fibres, which do not have the health hazards, used in cement-based products include glass, steel, polymer, cellulose, carbon and a variety of natural materials such as coconut. Each type of fibre imparts different characteristics to the concrete but also some limitations.

Briefly, although not exclusively, the principal virtues of using fibres in concrete are given below. Whether each of these enhancements is achieved depends on the type of fibre and the quantity in which it added. Indeed, no single type of fibre concrete will be capable of delivering all of these benefits.

- To increase the cohesion of a fresh mortar or concrete mix, enabling special manufacturing methods to be used and, in particular, the manufacture of thin sheets.
- To limit the growth of cracks in the plastic state.
- To limit crack growth in the hardened state and help to carry the tensile forces on the element.
- To prevent brittle failure and to provide ductility.
- To absorb energy and thereby provide impact resistance.
- To mitigate spalling of high-strength concrete subjected to fire.

### 14.11.1 Glass fibres

Normal glass, used in fibre reinforcement applications, is chemically attacked by the alkalis in Portland cement. A new type of glass fibre based on a glass composition containing greater zirconium oxide content, designed specifically to be resistant to alkali attack, was developed in the early 1970s and has achieved a considerable market for non-primary structural applications such as building façade panels and street furniture. Even with the special glass there is still a measurable but predictable deterioration in mechanical performance of a fibre cement composite with time. It is important to allow for this in design, and the available authoritative guidance does this.

The deterioration takes the form of a reduction in the post-crack ductility and results from the continued growth of hydration between the individual glass strands, restricting their ability to absorb crack energy and creating stress raisers on the glass surface. The properties of glass fibre concrete (grc) products prior to the point of matrix cracking are unaffected by ageing.

Since its original adoption, improvements have been made to both the glass formulation and the recommended concrete matrix design which have improved property retention.

For a fuller treatment of the use of glass fibre reinforcement the reader is referred to the GCRA (Glassfibre Reinforced Concrete Association, Camberley, UK) publication *Practical Design Guide for Glassfibre Reinforced Concrete 2005*.

### 14.11.2 Steel fibres

The use of steel fibres has a long history; the first patent for their use was in 1874. One interesting use of the technique was for the repair of concrete runways damaged during the Second World War.

Steel fibres (mild, high tensile or stainless) are usually in the order of 25 to 60 mm in length with diameters between 0.4 and 1.3 mm. They are produced in a range of shapes including smooth round or flat, indented, with end paddles, buttons or hooks, crimped or polygonal twisted; the last of these types are designed to enhance bond and/or anchorage.

The dosage would normally be around 40 kg/m$^3$, but dosages up to 100 kg/m$^3$ are not uncommon in suspended slabs. The inclusion of fibres should not, however, be regarded as a substitute for conventional reinforcement to resist tensile forces without considerable evidence of real performance.

In more specialist applications, manufacturers claim the following improvements in performance related to the use of steel fibres in concrete:

- fatigue resistance
- impact resistance
- restrained shrinkage
- thermal shock resistance
- toughness.

For a fuller treatment of the use of steel fibre reinforcement in concrete the reader is referred to *Concrete Society Technical Report No. 63*.

### 14.11.3 Polymeric fibres

The period since about 1980 has seen the growth of the use of polymeric material fibres (usually polypropylene, but nylon and other materials have been used) as an addition to concrete to modify properties. The fibres are either 'straight' profile or 'fibrillated', the latter having considerable side branching between fibre bundles. These fibres are normally 6–20 mm long and a few tens of micrometres in diameter.

The majority of the polymer fibres have the disadvantage of a low elastic modulus and so are of low effectiveness in conventional mechanical 'reinforcement' performance, crack control, etc. Therefore polymer fibre–cement composites may crack at low loads.

They are used in relatively small dosages (0.9 kg/m$^3$ or 0.1% by volume) primarily as a means of controlling plastic settlement and plastic shrinkage cracking. Micro fibres may also affect the bleeding rate and lead to a consequential improvement in some of the near-surface properties of the hardened concrete, although they do not provide any significant load-bearing capacity after the concrete matrix has cracked.

The low melting point of polymer fibres can be advantageous in fire, where the channels left by the melting fibre can provide an exit for gases and hence prevent explosive spalling, particularly in higher strength concretes. Typically between 1 and 2 kg/m$^3$ of the fibres are incorporated in the concrete.

There are claims that reduced deterioration due to freeze–thaw cycling can be achieved by the use of polypropylene fibre-reinforced concrete, but the claims are disputed by others.

Some polymer fibres have been produced with higher moduli (such as Kevlar, a polyamide), but these tend to be more expensive and so have limited, although useful, applications.

Since about 2000 and on the basis of further research and experience, a new range of fibres known as macro-synthetic fibres has been made available. The fibres, with higher modulus than other polymer fibres but lower modulus than steel, are intended to generate some of the benefits. These have dimensions of a similar range to that of steel fibres and are used in dosages up to 12 kg/m$^3$ (1.35% by volume) but are less strong/stiff and therefore demand higher dosage to achieve required performance. The inclusion of fibres should not be regarded as a substitute for conventional reinforcement to resist tensile forces without considerable evidence of real performance.

For a fuller treatment of the use of macro-synthetic fibres the reader is referred to Concrete Society Technical Report No. 65, *Guidance on the use of Macro-synthetic-fibre-reinforced Concrete* (Camberley, UK)

### 14.11.4 Carbon fibres

Conventional carbon fibres, widely used in high-technology resin-bound composite elements, have been incorporated into cement-bound systems with some technical success. The high cost of the fibres has however limited the opportunities for commercial scale application. There has been some experience in the use of carbon fibres derived from pitch. These fibres are much cheaper than the more conventional but, being much shorter, are less effective in reinforcing.

## 14.12 The use of polymers in concrete

Organic polymers are used in concrete for a variety of reasons. One reason is to change the rheological behaviour of fresh concrete in order to make it flow more easily or to allow lower water contents to be used. Such applications are considered in Section 14.7. Polymers are also used to alter the behaviour of hardened concrete and, in particular, to make concretes that are more resistant to chemical attack, which have a higher tensile and compressive strength than conventional Portland cement concretes, or which are more able to accommodate strains without cracking.

Polymer concretes can be made in which the aggregates are embedded totally within the polymer matrix. Such concretes have been used to make architectural panels, floors, work-bench surfaces and even machine-tool beds. This kind of concrete does not fall within the scope of this chapter.

Polymers can be added to conventional concrete in a number of ways. The most common is to form an aqueous polymer dispersion which is added to the Portland cement and aggregate during mixing. The polymers include natural rubbers, styrene butadiene rubber (SBR), acrylic and epoxy dispersions. Care must be taken during mixing to avoid the entrapment of excess amounts of air. Often an antifoaming agent will be required. These concretes harden normally and will generally have a lower permeability than conventional concretes, a better durability against aggressive environments and an increased failure strain. Such concretes have been used in repairs and here, the bond between the old and new concretes is generally high. These concretes can be used on overlays on factory floors, in garages and as patching repairs. They are usually known as polymer-modified-cement concretes.

Polymers can add useful properties to concrete, but they are expensive as compared with the price of cement. Thus they will only be used in selected applications where their special attributes can be beneficial.

As with all special concretes, their economic use requires a proper appraisal of the requirements. As an example, if a higher strength but higher cost product allows weight and size reductions, there can be savings that offset the material costs. With tunnel linings, for example, thinner sections mean smaller excavations and easier segment handling.

## 14.13 Mix design

The first serious attempt to understand the factors that govern the strength potential of concrete was made in the early years of the 20th century by Duff Abrams,[23] who showed experimentally that the strength of fully compacted concrete was inversely related to the water/cement ratio. This principle is still relevant today, but must be used with circumspection because of the wide variety of hydraulic cements available and because combinations of hydraulic and pozzolanic cements are used. Nevertheless, the basic tenet still dominates the rules that must be applied if good-quality, dense and strong concrete is to be produced, namely that for any given cement or cementitious combination, the lower the water/cement ratio, the higher will be the strength and the lower the permeability of the concrete. There are practical limits to the reduction in water content in a concrete because the mix must be sufficiently workable to be placed and compacted fully to displace virtually all the entrapped air in the mix. These limits have, however, fallen significantly with the introduction of high-performance water reducing admixtures.

In practice, the highest possible strength concretes will seldom be required so that the task of the concrete producer is to select a concrete mix that economically fulfils all specified requirements. These requirements will include:

1. the workability of the fresh concrete
2. the strength (usually, but not exclusively, at 28 days)
3. the durability
4. possibly the density, the air-entrainment and the choice of aggregates.

The variables that the mix designer must consider are the choice of aggregate and the grading of the particles, the sand, the cement, the water/cement ratio and the inclusion of admixtures. The proportions of each component can be based on volumes or on weight, and while the former has some attractions, most mixing plants rely on weight batching. It is now general practice to refer to the weights of the different concrete constituents that, in combination, will produce a unit volume of concrete. For example, the cement content might be given as 300 kg/m$^3$, the aggregate content as 1900 kg/m$^3$ and the water content as 180 kg/m$^3$.

In the UK, one of the earliest mix-design publications was *Road Note No. 4, Design of Concrete Mixes*, prepared at the Road Research Laboratory.[24] This was superseded in 1975 by the Department of the Environment (DOE) publication *Design of Normal Concrete Mixes.* These publications were concerned solely with the use of Portland cements in combination with different aggregates and gradings. The latter publication now includes advice on the use of fly ash as partial cement replacement and, to a lesser extent, the use of ground granulated blastfurnace slag.[25] In the USA, both the ACI and the Portland Cement Association have published recommended procedures for mix design and these broadly parallel those used in the UK. These methods provide a sound basis for basic mix design but, in practice, much use is made of more sophisticated computer-based systems.

Variations in the batching of the components in a concrete mix, in the way it is mixed and in its placing, compaction and curing, result in a variable product. It is essential to allow for such practical effects when selecting a concrete for a defined purpose. The better the control of the total system of manufacture and use, the smaller need be the margins between the specified concrete mix and the target levels chosen by the supplier. Sophisticated statistical procedures must be adopted, with careful monitoring of the concrete produced, to ensure that the concrete supplied is both economical and fit for its purpose. The mix design process should be carried out in a logical way. There are variations in the sequence of the steps followed, but basically they are as follows:

1 The selection of a water/cement ratio that is likely to give the required strength. This will be the free water/cement ratio after allowing for the proportion of water that is absorbed by the aggregates.
2 The selection of a water content that will give the desired workability of the mix, taking account of the use of any plasticizing admixtures.
3 Calculation of a cement content from the water/cement ratio and the water content.
4 Calculation of the total aggregate content, from knowledge of the cement and water contents.
5 Selection of the aggregate grading, taking account of the specified top sizes and the aggregate shape and type. This will include both the coarse and fine fractions.

The selected values are obtained from graphical and tabular presentations in the mix design publications. It will normally be necessary to carry out trial mixes and, where necessary, to make adjustments to ensure that all specified requirements are met. Frequently, a maximum water/cement ratio and a minimum cement content will be specified. It may be found that, because of durability requirements, when these requirements are met the strength will exceed that specified. This should be seen as a bonus and not a justification to reduce the cement contents or to add more water to the mix.

When fly ash is used, it has been suggested[25] that only a proportion can be equated directly to the Portland cement that it replaces and a cementing efficiency factor, $k$, of 0.3 is adopted using

$$\frac{W}{C+kF}=\frac{W_1}{C_1}$$

where $W$, $C$ and $F$ are the free water, the cement and the fly ash contents (by weight), respectively, and $W_1$ and $C_1$ are the free water and CEM I Portland cement contents of a concrete of similar workability and 28-day strength.

The performance of ground granulated blastfurnace slags is variable between sources and depends on the particular CEM I Portland cement used in combination; the use of such a consistent method as that used for fly ash has not yet been found possible and values of $k$ between 0 and 1 have been found to apply. It is necessary to determine the relevant factor for each selected CEM I Portland cement/slag combination. This factor can then be used to design other concrete mixes using the same CEM I Portland cement/slag combination.

It is important to note that, in this basic approach, no account has been taken of plasticizing admixtures, although the DOE *Design of Normal Concrete Mixes*[25] does include consideration of air-entrained concrete mixes for CEM I Portland cement concretes.

Some specifications for concrete have been known to be ambiguous. They should be as simple as possible to avoid difficulties. If a minimum cement content and maximum water/cement ratio are essential for the assurance of durability, this should be made abundantly clear. If strength is all that matters, the cement content can often be reduced, thereby cutting the cost of the concrete, but this should only be done when it is safe to do so. If a particular type of cement is required, such as one containing ground granulated blastfurnace slag or fly ash, this too should be stated clearly.

Mix design, for all but the most simple jobs, requires sensible care by someone who is aware of the options available and knowledgeable about the properties of the materials available for use.

## 14.14 Mortars, screeds and renders

Mortars are used principally in the construction of masonry walls. The type of mortar mix used will depend on the type of brick or block used and the degree of exposure to which it is subjected. Blocks are ideally laid on weaker mortars in order to confine cracks to the joins and not within blocks. A mortar must remain easily workable for a reasonable length of time, but also develop strength quickly enough to allow construction to continue uninterrupted. It should bond well with the blocks or bricks, be durable and, where necessary, be resistant to attack from the sulfates that can occur in some clay bricks.

Mortars often comprise:

1 cement, lime and sand (1:1:5 to 6)
2 cement, mortar plasticizer and sand (1:5 to 6)
3 masonry cement, sand (1:4 to 5).

Stronger mixes will be required when dense masonry is used. EN 934-3[26] describes some of the admixtures that can be used in mortars. These include air-entraining agents to entrain air, and thereby reduce the risk of danger from freezing and thawing, and to retard set. Also included are mortar plasticizers which are used to reduce the amount of water required in the mix, and this will significantly reduce the shrinkage and improve the bond with masonry. For small jobs it is convenient to use premixed bags of mortar.

Screeds are used to provide a level surface on a concrete raft or slab. They are not typically used as the wearing surface but provide a surface on which tiles, flexible coverings (plastic or rubber), or carpeting can be laid. Screeds must be laid onto a clean and wetted surface to obtain a good bond. A typical mix is one part cement to three to four parts of sand by weight and the water/cement ratio should be kept below 0.45. As with all good concreting, the screed must be cured by keeping it damp for several days. Screeds are usually less than 75 mm thick and may be laid on waterproof membranes (plastic sheets) that separate them from the concrete base. Thinner screeds can be bonded to a suitably prepared base. Unbonded screeds are liable to curl if allowed to dry out too quickly.

Renders are mortars that are normally applied to vertical walls to produce various visual finishes or protection against rainwater ingress. EN 13914-1[27] describes a family of mortar mixes that can be used for a diverse range of applications. Renders will usually be built up in two or three layers, depending on the nature of the substrate. Monks[28] described the different sequences in detail. Frequently, a stipple coat is applied that is cement rich (water/cement = 1:1:5 or 2) to improve bond or reduce suction. This coat will be covered with a thicker undercoat that might be a 1:1:6 cement–lime–sand mix on top of which a weaker finishing coat will be applied (1:2:9, say). These proportions are only typical examples because there are many factors that must be considered.

## 14.15 Grouts

Grouts are cement-based fluid mixtures used to fill voids. In particular, they are pumped into ducts in structural elements through which prestressing or post-tensioning reinforcement passes. The grout has two functions: it prevents corrosion of the steel tendons and provides an efficient bond between the tendon and the structural member.

Grouts are also used to fill voids between structural members, such as tunnel-lining segments and the surrounding rock. Another example is the filling of the gap between steel sleeves used to repair the legs of steel drilling platforms in the North Sea. Grouts may be made simply from cement and water but often also contain sand and/or admixtures. The choice of cement type is likely to depend on the properties required for the particular application. Admixtures are frequently used to aid workability and to minimize segregation of the constituents as the grout is forced into place. Grouts are normally pumped into place from a low position in the void or duct to be filled. Bleed holes are employed to help the upward flow and to ensure that void filling is effective. High-speed mixers are frequently used. These may 'shear' the cement grains, which helps to create a stable and effective grout.

The greatest problem is to ensure that the voids or ducts are completely filled and that filtration of the solids from the water does not occur as the grout is forced into narrow ducts or through areas of reinforcement congestion. It is also essential to fill a series of ducts systematically. On some occasions unfilled ducts have been found, with the consequence that the steel reinforcement is more prone to early corrosion.

## 14.16 Concrete properties

### 14.16.1 Workability

The need to be able to place concrete within forms or moulds is obvious, but there is more to this statement than may be inferred. The concrete must be sufficiently fluid to flow within the moulds and within and around the steel reinforcement cages without significant segregation of the aggregate from the cement-paste matrix. Once in place, the concrete needs to be fully compacted in order to remove unnecessary air, and this is particularly important below the steel and between the steel and the formwork. If the mix is too stiff it will hang-up and leave air pockets which are aesthetically unacceptable on the surface and which do not provide adequate protection to the reinforcement. If the concrete has too high a water content segregation can occur, and where the concrete flows over horizontal bars plastic settlement cracks may form. The workability required will depend on the chosen compaction methods. Where highly fluid concrete is needed, for pumping to a high level, say, it is important that the mix design be suitable and selected to minimize bleeding effects. In such cases, it is helpful to rework (revibrate) the concrete surface before the concrete finally hardens.

The methods used to check workability are mentioned below. The one chosen must be relevant to the need. Where high workability is needed, the slump test is not suitable and flow tests should be used.

Another important factor is the duration for which the concrete stays workable, particularly in hot weather. As hydration occurs, or as moisture is lost, the concrete will stiffen and become more difficult to place in the forms around the steel. Under some circumstances, the concrete may be brought back to a workable state by the addition of water, although this practice is strongly deprecated if it effectively increases the water/cement ratio. Nevertheless, concrete badly compacted because of its stiffness can be more unsatisfactory than concrete to which small water additions have been made. A preferred alternative is the addition of a plasticizer or workability aid that restores mobility without adjusting the water content. Perhaps better still is the avoidance of unnecessarily low workabilities. Traditionally, low workability was used to minimize water content and ensure high quality. With the almost universal use of plasticizing admixtures this is no longer appropriate, and higher workability is likely to result in easier placing and more complete compaction.

### 14.16.2 Setting time

After hydration of Portland cement has started, there is a dormant period when nothing appears to be happening. This is fortunate because it provides a period during which concrete can be transported to the site and be placed. After a period of about two to four hours, the concrete will begin to set, reaching final set about two hours later. These times depend significantly on the prevailing climate and on the selected cement type. Special cements such as high alumina cement, ultrarapid hardening Portland cement and specially formulated products for particular applications react much more quickly and can harden very quickly after the addition of water. Setting times for cements containing fly ash or blastfurnace slag may be greater than for Portland cement, especially in colder weather.

### 14.16.3 Early-age behaviour

As soon as water is added to concrete, the cement starts to react exothermally and the temperature will soon begin to rise once the initial dormant phase has been passed. The rate of temperature rise will depend primarily on the temperature when the concrete is cast, the type and amount of cement and the size of the pour. Once the concrete hardens, thermal-stress gradients will be set up and at early stages where the concrete is restrained these will be predominantly compressive. Later the concrete will cool and tensile strains will be induced that can be large enough to cause cracking. Where this risk is high, it is possible to specify the use of a low-heat cement, but coupled with the low heat will be a low rate of strength gain which may not be acceptable. An alternative method for reducing the risk of cracking is to lay thermally insulating mats on the surface of the concrete and, while this may mean an even higher temperature increase in the concrete as hydration proceeds, it can also ensure that thermal-strain gradients are kept sufficiently low and thereby minimize the risk of cracks occurring.

The use of fly ash and slag cements can be a very effective way of reducing the temperature rises in concrete. The heat evolution of both these additions is, however, affected to some degree by the temperatures at which they are used, and temperature rises may be greater than expected. In particular, the slag reaction is accelerated by temperature to a greater degree than Portland cement and its relative effectiveness in reducing heat may be lessened to some degree in very large sections. Careful mix design can help reduce heat by minimizing the cement content.

### 14.16.4 Strength

In most structures the concrete is required to carry structural loads. For this reason, concrete is frequently specified by strength, usually compressive strength. The compressive strength of concrete is usually measured using either cast concrete cubes of 100 or 150 mm side or cylinders of 300 × 150 mm diameter or 200 × 100 mm diameter. Strengths measured on cylinders of normal-strength concrete are about 80% of those of cubes made from the same concrete; a slightly different relationship occurs for lightweight concrete. For design purposes, the characteristic strength is used. This is the value below which 5% of the measured strengths may lie, due to the inherent variability of the concretes and of the testing itself. Within Europe the required compressive strength is specified as a strength class which is expressed as a composite reference first to the characteristic cylinder strength and then to the characteristic cube strength,[29] e.g. C30/37 = 30 N/$mm^2$ cylinder, 37 N/$mm^2$ cube.

The compressive strength can also be determined from cores cut from hardened concrete in a pavement or structure. The results can, however, vary with the height of the casting (e.g. the top of a wall or column will often be weaker than the area close to the foot). The strength of concrete in a core will differ from that

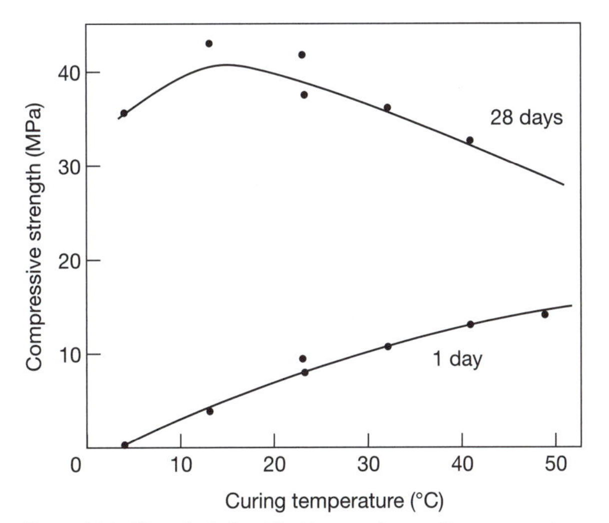


**Figure 14.1** The effect of ambient temperature on the compressive strength at 1 and 28 days of an ordinary Portland cement concrete kept wet

in a standard cube or cylinder test due to differences in age, curing, compaction and moisture state at test. Cores of 1:1 length–diameter ratio should be used in the estimation of *in situ* cube strength; 2:1 cores should be used if estimating cylinder strength.

The tensile strength of most concrete is about one-tenth of the compressive strength. This strength is seldom specified but can be measured in a flexural test or by splitting a cylinder by a diametrical compressive load. All these tests are described in detail in BS EN 12390.[30]

The rate of gain in strength depends not only on the selected concrete mix and the chosen cement, but also on the ambient temperature and on the curing regime. Figure 14.1 shows one of these effects for CEM I Portland cement. At low temperatures, the early strengths will be low, but if a sufficiently long curing time is allowed, the concrete may finally exceed the strength of concrete cured at higher temperatures and thereby develop strength early.

### 14.16.5 Penetrability and porosity

Another property of importance is the penetrability of concrete. Penetrability is an undefined composite parameter which covers various transport processes including permeability, capillary absorption, wick action and diffusion. Powers[31] suggested that the pore structure of concretes largely controls both its strength and durability and that it is essential to keep the water/cement ratio low if the pore structure is to become discontinuous with no channels through which fluids or gases could permeate. For water/cement ratios of about 0.7, Powers showed that there was insufficient potential hydration to provide a discontinuous or segmented matrix. At the other extreme, for water/cement ratios of 0.4, the matrix could become a closed-pore system after about three days' hydration. Modern cements, both CEM I Portland and the composite varieties, do not satisfy the exact pattern described by Powers, but the general principle still applies, namely that at high water/cement ratios it is not possible to make very impermeable concrete. With composite cements, because hydration may be slower than that of a CEM I Portland cement, it may be necessary to extend the hydration period (the duration of the wet curing) to establish the same result. Nevertheless, if prolonged curing occurs, in a marine environment or in buried concrete for example, concretes made from composite cements will often have lower penetrability and hence improved durability.

**Table 14.3** Indicative values of elastic moduli of concretes related to concrete strengths and aggregate type[a]

| *Concrete strength class (cylinder/cube)* | *Indicative value of elastic modulus* | | | |
|---|---|---|---|---|
| | *Quartzite* | *Limestone* | *Sandstone* | *Basalt* |
| C12/15 | 27 | 24 | 19 | 32 |
| C16/20 | 29 | 26 | 20 | 35 |
| C20/25 | 30 | 27 | 21 | 36 |
| C25/30 | 31 | 28 | 22 | 37 |
| C30/37 | 33 | 30 | 23 | 40 |
| C35/45 | 34 | 31 | 24 | 41 |
| C40/50 | 35 | 31.5 | 24.5 | 42 |
| C45/55 | 36 | 32 | 25 | 43 |
| C50/60 | 37 | 33 | 26 | 44 |
| C55/67 | 38 | 34 | 26.5 | 45.5 |
| C60/75 | 39 | 35 | 27 | 47 |
| C70/85 | 41 | 37 | 29 | 49 |
| C80/95 | 42 | 38 | 29.5 | 50.5 |
| C90/105 | 44 | 39.5 | 31 | 53 |

[a]From information in ref. 19.

Capillary absorption and wick action can be reduced by the use of surface treatments and admixtures such as silanes and integral waterproofers respectively.

### 14.16.6 Elastic modulus

The elastic modulus of concrete is closely related to the strength of normal concrete and to the modulus of the aggregate used and its volume concentration. In general, the elastic modulus $E_c = f(E_a, E_p, V)$, where $E_a$ is the modulus of the hardened cement paste and $V$ is the volume proportion of aggregate. In practice, it is seldom necessary to be too precise and Eurocode 2,[19] for example, gives indicative values for concrete of different aggregate types and strength classes as indicated in Table 14.3. Additional information is provided in the Eurocode for calculation of modulus at different ages or, for more demanding situations, it can measured[32] on the proposed concrete mix where a precise value is required. Lightweight aggregate concretes have lower moduli and the relevant manufacturer's data sheets should be consulted.

### 14.16.7 Shrinkage

As concrete dries it shrinks, the rate and extent of shrinkage depending on the size of the structural member and the ambient environment.[33, 43] Figure 14.2 provides a representation of these interactive effects and enables shrinkage strains to be estimated at six months and virtually at completion of movement (30 years) for normal dense aggregate concretes. Aggregates that have a high moisture movement, such as Scottish dolerites, have a higher shrinkage at early ages and can cause problems in concrete.

In practice, the temperature and humidity around the concrete will change periodically and if rain falls onto the concrete the rate of shrinkage will be greatly reduced. In some instances dry concrete will expand if it is suddenly soaked. Shrinkage can also be caused by the carbonation of the concrete by the carbon dioxide in the atmosphere, and by hydration itself in very low water/cement ratio concrete.

Shrinkage is a particularly important consideration for large area floors but can be accommodated by the use of joints to allow movement. Shrinkage can be reduced by the use of concrete with low initial free water content and through the use of specially

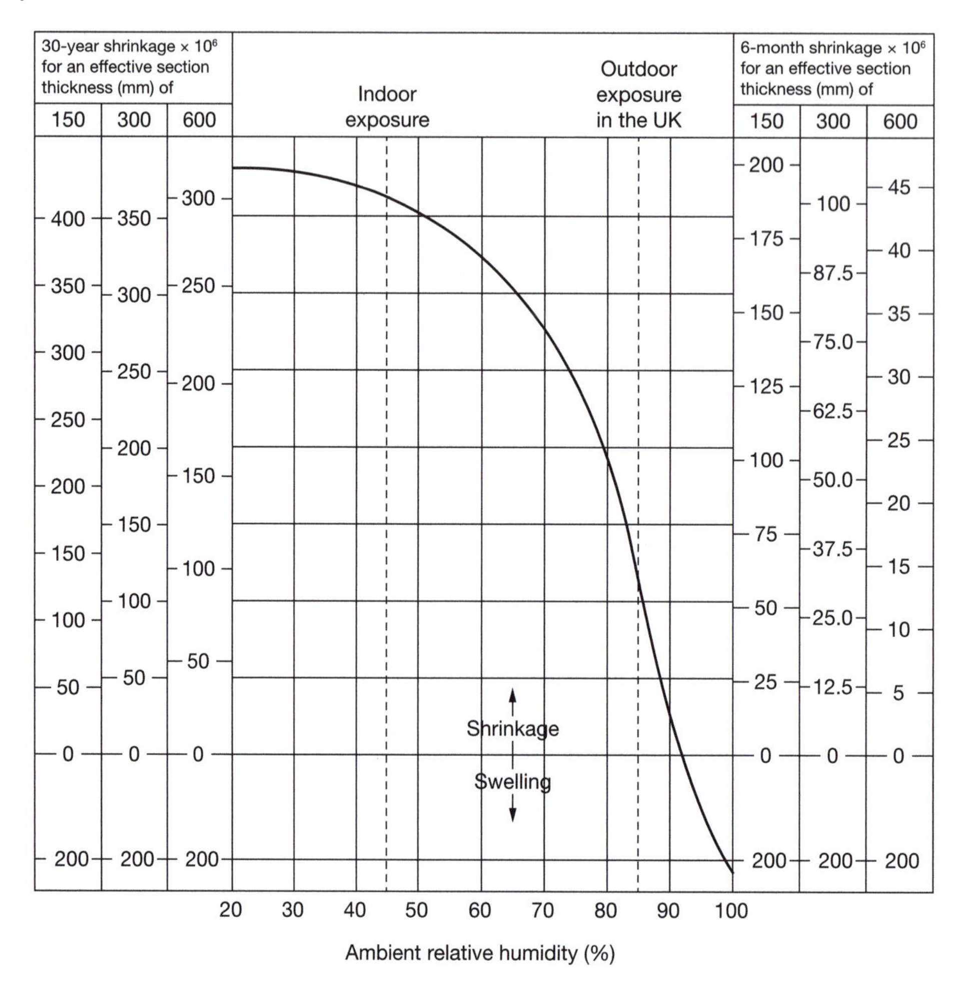


**Figure 14.2** Shrinkage strains likely to be reached by concretes of different thicknesses exposed to different environments: values are given for 6-month and 30-year exposures

formulated admixtures. Shrinkage-reducing admixtures reduce the surface tension of water in the pore structure and hence reduce the tensile forces developed by the formation of menisci in the pores as the water is lost.

### 14.16.8 Creep

Concrete subjected to stress will deform with time, the deformation being known as creep. The creep strains are very dependent on the applied stress, the concrete mix, aggregate and aggregate concentrations. If the concrete is always saturated with water, the creep, known as basic creep, will be much smaller than that when drying occurs concurrently, the total movement then being greater than the basic creep and shrinkage strains combined. Because environmental effects are very dependent on the size of the structural elements, the estimation of the creep of concrete is not always very precise and where annual cycles of weather occur, there can be large variations in structural movement.[34]

Various methods have been proposed to help the concrete structural designer. Eurocode 2[19] uses a graphical method for estimation of the creep coefficient for normal indoor (RH = 50%) and outdoor (RH = 80%) environmental conditions and strength classes between C20/25 and C90/105. As an alternative, mathematical models can be used in computer programs. None of the methods are likely to give results that are more accurate than about 15–20%, but even this is adequate for most design needs, because design allowances must be made for structural movements.

### 14.16.9 Thermal expansion

Allowance may also have to be made for thermal expansion. A value of 8–12 × $10^{-6}$ per °C, depending on the coarse aggregate type, is usually adequate. Where concrete is painted black or other dark colours, there can be a significant fluctuation in temperature when it is exposed to intermittent sunlight. In such circumstances, particularly with thin section sizes, movements may be larger than expected with a consequent risk of cracking. Choice of a coarse aggregate with low thermal expansion, such as many limestones, may be advantageous when trying to minimize the effects of early age thermal contraction, e.g. liquid-retaining structures.

## 14.17 Durability

Like all materials, concrete is affected by the environment in which it is used and, in particular, corrosion of the reinforcing steel can reduce the serviceability of a structure. Many volumes have been written on this subject alone, so that only brief mention of some important considerations will be made here.

The causes of deterioration in concrete include the ingress into the concrete of chlorides from marine spray or from de-icing salts or carbon dioxide from the atmosphere followed by the onset and development of corrosion of the reinforcement. These are probably the two more serious causes. The effects are minimized by choosing a good-quality, low-permeability concrete, using an appropriate cement and aggregate, and ensuring that there is an adequate depth of cover to the steel, including any secondary stirrups or ties, and that this cover concrete is properly compacted and, in particular, is adequately cured. Failure of any one of these requirements will reduce the quality of the concrete. Frequently, the first signs of spalling caused by steel corrosion occur at locations where the specified cover to the concrete has not been satisfied. Such a failure is not a criticism of the materials but of the workmanship on site or in the factory and of the supervision. Due allowance must be made in design for the accuracy with which reinforcement can be placed under practical conditions, but a high level of workmanship remains essential.

Chlorides can also come from the concrete constituents, from the cement, from the aggregates or from any admixtures used, although these should be controlled to the acceptably low levels given in standards.[15]

### 14.17.1 Acid and sulfate attack

Concrete can be damaged by acid waters and by the sulfates present in groundwaters. If acids, or even neutral flowing water, are to be handled, it is essential to take suitable precautions. Sulfate resistance was traditionally provided by the use of sulphate-resisting Portland cements (SRPCs) with a $C_3A$ level below about 4–5%, but such cement is no longer available from many ready-mixed concrete suppliers in the UK. Portland cements containing a sufficiently high level of fly ash (>25% by mass) or ground granulated blastfurnace slag (>40% by mass, or >65% by mass for very high sulfate resistance) can provide an equally effective and more practical means of protection. Moreover, recent research has shown that although SRPC is highly effective in resisting the ettringite form of sulfate, it can have poor resistance to the potentially highly aggressive thaumasite form of attack (TSA) that can occur under certain circumstances in the cool ground conditions in climates such as the UK. In addition to the choice of an appropriate cement type it is essential to use a sufficiently low water/cement ratio to prevent the aggressive waters from penetrating beyond the concrete surface. The use of fly ash or, in particular, blastfurnace slag, cement concrete can give much greater resistance. More extensive information is provided in BRE Special Digest 1[35] and summarised in BS 8500-1.[14]

Resistance to acidic waters does not vary widely between the common cement types so it is the provision of a low water/cement ratio that is crucial to reduce penetrability. None of the common cements are inherently acid-resistant so it may be necessary to use a protective barrier in highly aggressive conditions as may be met in some industrial processes. High alumina cement (HAC) may give enhanced resistance to acid attack but the use of concrete made with HAC is not covered by current standards.

### 14.17.2 Alkali–silica reaction

The reaction between the alkalis present in the pore fluid of damp concrete and certain silicaceous aggregate constituents may generate stresses within concrete that are large enough to cause cracking. There are also reports from some parts of the world of alkali–carbonate reaction. The Building Research Establishment (BRE) Digest 330[36] provides guidance on the precautions that can be taken to minimize the risk of adverse effects of alkali–silica reaction (ASR). The Institution of Structural Engineers has also provided guidance on the structural effects of alkali–silica reaction.[37]

The risk of cracking due to ASR is minimal if any of the following apply:

1 The source of aggregate has been used satisfactorily over a long period of time in comparable situations.
2 The quantity of alkali ($Na_2O_{eq}$) in the concrete mix is below about 3.5 kg $m^{-3}$.
3 A low-alkali cement, below 0.6% alkali, is used.
4 The concrete is consistently exposed to a dry environment.
5 The concrete is made with cement containing a suitably high proportion of fly ash or blastfurnace slag.

Even when there is evidence of ASR in concrete, usually obtained by the petrographic examination of thin sections cut from concrete cores, the likelihood of serious cracking will remain low because small quantities of the reaction product, a gel, will fill pores or may even exude from the concrete without causing damage. Nevertheless, even when significant cracking has been observed, structural load tests generally show the effects to be less serious than they may appear.[38]

ASR will continue until either the alkalis in the concrete or the reactive constituent in the aggregate have been depleted by the reaction. The implication is that, in most cases, there is a finite limit to the amount of damage or cracking that can occur. Once this stage is reached, normal remedial work can be carried out, provided that no extra alkalis are made available to react with any residual reactive aggregate. Again, it must be emphasized that water is needed to cause the reaction product, the gel, to expand and it is possible that an unexpected soaking of dry concrete, in which ASR has occurred, could result in cracking, but this will be very rare.

### 14.17.3 Freeze–thaw damage

Damage can occur when wet concrete is subjected to freezing and thawing cycles. The normal way of avoiding this form of damage is to entrain air into the concrete. A chemical admixture is used that introduces a stable family of small, closely spaced air bubbles into the concrete. This entrainment of air will, however, lower the strength of the concrete so that some adjustment to the mix may be required to meet strength requirements. The inclusion of about 5% air will provide good freeze–thaw resistance, but will also reduce the strength of the concrete by about one-quarter. It is normally possible to reduce the water content of air-entrained concrete and thereby regain some of the strength loss through the lower water/cement ratio.

The air bubbles should be small, about 0.5 mm, and evenly dispersed in the paste. The bubbles tend to add cohesivity to the mix and increase the workability. Air-entrained concrete is recommended for all forms of exposed paving where freezing cycles are likely to occur. Air entrainment also reduces the likelihood of plastic settlement and the formation of plastic settlement cracks. The amount of air-entraining agent required will depend on the cement used; in particular the dosage may be higher when fly ash is included in the mix. Over-vibration should be avoided as this can reduce the air content of the concrete.

Resistance to freeze–thaw conditions in moderate climates such as that in the UK can also usually be achieved by the use of concrete with sufficient strength to resist the internal stresses developed. Guidance on the required strength class is given in standards[15] but is typically C40/50.

## 14.18 Concrete practice

Mention has already been made of the use of highly workable concretes that can be made using organic plasticizers. This is but one change in the concrete industry that has altered site practice. Most concrete used on medium and large sites is delivered from a ready-mixed concrete plant and it will be transported to the place of work by skips, trucks, or, frequently, by pump. Because several different mixes may be used on a site, it is vital to install a good acceptance and control procedure to ensure that the correct concrete is used on all occasions. The concrete will normally be vibrated into place and this must be done thoroughly and consistently if all entrapped air is to be expelled and if voids and honeycombed areas to be avoided.

Compaction reduces the permeability of the concrete and improves the bond between the concrete and the steel reinforcement. Concrete should be placed in layers less than 0.5 m thick and each layer vibrated in turn before further concrete is added. Before any concrete is placed, the forms should be closely examined to ensure that they are clean, that no water is lying at the bottom and that the reinforcement is properly fixed. Spacers should be used to provide the required cover to the steel.

For most concrete placed in forms, poker vibrators are used. These must be moved systematically to ensure that all parts of the casting are uniformly compacted. Care must be taken to remove the poker slowly in order to avoid weak zones being left near the surface. For slabs, it is preferable to use beam vibrators to obtain a satisfactory flat surface. In addition, poker vibrations should be used close to the edges of slabs to ensure proper compaction.

If plastic settlement occurs around the reinforcement, the concrete should be revibrated while it is still workable. Reworking the surface of concrete in this way is usually beneficial.

Self-compacting concrete can be advantageous where placing or compaction of conventional-workability concrete is difficult, such as elements with very congested reinforcement. It can also make concrete placement less labour-intensive and can reduce site noise levels by the elimination of poker vibrators. Self-compacting concrete should not be deliberately vibrated during placing, as this may cause segregation.

## 14.19 Striking of formwork

The period that should elapse before formwork is struck will vary from job to job and according to the prevailing temperature. As already mentioned, the types of cement used have a large influence on the early-strength development and it is prudent to find out these effects if the cast concrete is to be strong enough to be self-supporting and able to carry imposed loads. In winter it can be helpful to insulate the moulds both to avoid freezing and to reduce the stripping times. In winter, the removal of an insulated or timber form can subject the concrete to a thermal shock (a sudden drop in surface temperature) that could result in cracking. This can be avoided by releasing but not removing forms or by the use of insulating mats.

## 14.20 Curing

It is fairly common to hear people talking about concrete drying out. If this is allowed to occur at early ages, the quality of the concrete will be poor. Concrete gains strength by hydration, which is a chemical reaction between cement and water and as soon as the forms have been removed or, with slabs, immediately after finishing the concrete, it is essential to start a curing routine. That is, precautions must be taken to avoid the premature drying of the concrete and to replenish any moisture losses that do occur. The quality of the surface layer is particularly important since this protects the steel reinforcement from the effects of carbon dioxide from the air and from the chloride ions that may come from de-icing salts or marine spray. The rate of moisture loss is highly dependent on the local weather, being particularly severe on a hot and windy day.

There are two main methods of curing, as follows:

1 Water or moisture is kept in close contact with the surface by ponding, applying water with a sprinkler, covering with damp sand or with damp hessian.
2 Evaporation is prevented by covering with plastic sheeting or by the application of a chemical curing membrane.

The choice of method will depend on the structural element in question. With a slab or pavement, spraying can be used, but this is not possible for the soffit of an *in situ* cast beam. Again, it is emphasized that curing must start as early as possible, because once concrete has dried out some permanent changes occur that are likely to leave the concrete more porous and, therefore, prone to the ingress of carbon dioxide or chlorides. Guidance on minimum curing times, which takes account of the prevailing ambient conditions and the differences between cements, is given in the European Standard for execution of concrete works.[39]

The application of curing is relatively complex, particularly if the appearance of the concrete is to be uniform and free of blemishes. Furthermore, if screeds are to be applied to a horizontal slab, the use of curing membranes on the substrate can cause poor bond. Thus, it is important to choose the correct method of curing, but whatever method is selected it must be applied consistently because failure to do so is the cause of many of the problems that occur in concrete practice including shrinkage cracking, poor-quality 'covercrete' and variations in colour that are unsightly.

The process of curing also involves the protection of concrete at early ages from the adverse effects of shock, vibration and freezing.

## 14.21 Exposed aggregate finishes to concrete surfaces

There are many ways in which the surface of concrete can be finished to provide an attractive visual appearance. Plastic moulds have been used to form surfaces that resemble the contours of rocks, heavily grained timber forms are used to produce 'board-marked' finishes and ribs have been cast in. The possibilities are endless. and include exposed aggregate finishes obtained by the removal of the surface mortar without significant damage to the body of the concrete. There are two common methods:

1 exposure of the aggregate by washing and brushing – normally a retarder will be painted onto the formwork to keep the surface soft while the bulk of the concrete hardens
2 abrasive blasting of the surface.

The choice of size and type of aggregate is obviously important in order to provide the required effects. In some instances, large pieces of aggregate will be laid on the base of horizontal formwork and concrete will be cast onto this. The mortar surface will be removed soon after the forms are struck by wet brushing. When any of these special surface effects are used, the depth of cover to the steel reinforcement must be measured from the deepest indentation or deformation into the surface.

## 14.22 Precast concrete

Many concrete products are factory-made. These range from concrete bricks and blocks to large precast concrete beams, façade panels or segments of tunnel linings. An advantage of factory production is the greater degree of uniformity and control that can be achieved. This is particularly true when special cladding panels are designed and manufactured to add a quality visual appearance to a structure. As already discussed, special finishes can be given to exposed concrete panels and these are best made in a controlled environment.

The concrete mixes used for precasting may differ from those that are suitable on site. Drier mixes can be hydraulically rammed into moulds, vacuum dewatering is possible and spinning techniques can be used in the manufacture of pipes. The use of self-compacting concrete has become popular in the precast concrete industry because of the elimination of expensive compaction equipment and the improvement in the working environment through dramatic reduction in factory noise levels. Because fabrication methods are so different from conventional concreting practice, it is often more convenient to undertake product testing instead of the material tests that are used to measure the quality and consistency of concrete used on site or delivered from a ready-mixed plant.

A major difference in the factory can be the use of warmth to speed up the hydration of the cement and, therefore, speed up the throughput of the products. There are too many variations in procedure for discussion here, but they include steam curing and autoclaving, as well as the use of milder heating regimes. A large market has been developed for aerated autoclaved concrete blocks that are of very low density and are used to provide good thermal insulation in buildings.

## 14.23 Testing methods, quality and compliance

All products should be tested and checked for compliance. Unfortunately, a concrete structure cannot be tested in its entirety, so it is necessary to make checks on the various elements that go into the construction. These checks start with the materials. It is normally accepted that the reinforcing steel delivered to a site will comply with specification, and it is unusual to sample and test it. In the UK, a quality-assurance scheme known as CARES[40] provides protection to the consumer and conformity can be checked through the delivery documentation.

With concrete, the situation is more confusing because not only must the concrete be able to generate the desired long-term properties such as strength, stiffness and durability, but it must also be in a state in which it can be delivered to the job, by truck, skip or pump, be placed in the forms and compacted into place. It may also be necessary to check the air content of the mix. Site tests to check these early-age properties are described in EN 12350-7.[41] A similar range of tests is covered in other parts of the world, many of the tests now being International Standards. The most common site tests are the slump test for workability and the test for air content.

Concrete is commonly pumped to the place of use. A special mix must be chosen that does not segregate or bleed under the action of the pumping pressure. The mix must also be adequately fluid. These characteristics are controlled primarily by the grading and proportions of the sand in the mix. Normal workability tests do not characterize a pumpable concrete nor are there, at present, any widely accepted tests. It is necessary to observe the pumping operation and to adjust the sand component and to use plasticizers to achieve a satisfactory mix. Visual monitoring of the operation is currently the most effective way to ensure consistency and absence of segregation.

Tests on site to check the composition and all but the most basic properties of fresh concrete require skilled operatives, are often too late to provide the best control of production, and are rarely use in practice for these reasons. Third-party quality-assurance schemes are used to monitor the quality of concrete produced from ready-mixed plants. In the UK, the Quality Scheme for Ready-mixed Concrete (QSRMC) does much to maintain the consistency of supply of concrete to jobs, both large and small, so that gradually the need to test concrete on site has been reduced.

Site conditions and the prevailing weather will have a large effect on strength development. Another family of tests is available to help monitor this stage of concrete development. The tests include measuring the maturity of concrete (by integrating time and temperature in the concrete), by matching the strength gain of a companion concrete specimen subjected to the temperature profile within the structural member to the *in situ* strength and also by measuring the actual strength of a structure by pull-out tests.

Another important check is that of cover to the steel reinforcement. Once the concrete has been cast, it is really too late to realign the steel. Nevertheless, the steel can be displaced during casting and some checks are sensible so that errors can be detected early, remedial measures applied and repetition avoided. Electromagnetic devices are available to measure the cover to the steel and these are described, for example, in BS 1881.[42] The use of such a device on site will often have the effect of raising site standards, thereby ensuring satisfactory construction.

Most structural design is based on the 28-day compressive strength of concrete so that the most common test of concrete is that of the compressive strength of a cube or cylinder of concrete. A badly made and unrepresentative sample of concrete can be misleading and in extreme cases can result in concrete being wrongly rejected, with the consequent expensive removal and replacement of completed work or of time-consuming further testing or re-evaluation of the design. Good sampling and testing practice is described fully in Standards and this should be followed closely.

When disputes do occur, it is normal to carry out further tests on the structural concrete, i.e. non-destructive tests or surface-strength tests such as indentation or pull-out tests. Ultrasonic pulses can be transmitted through concrete to show up large flaws or gross variations in concrete quality throughout the element or structure. Frequently, cores are cut and these can be used to obtain a direct measure of the *in situ* strength in compression and degree of compaction. They can also be sliced to study the gradation in quality, such as permeability or strength.

Any faults found late in the construction process will be expensive to remedy; it is much better and cheaper to 'get it right' first time. This requires adequate training of the staff involved and sufficient checks to ensure that the main requirements of good practice are being supplied. These can briefly be summarized by the 'Cs': i.e. the *constituents* in the mix, the *cover* to the steel, the *compaction* of the concrete and, finally, the *curing*. The quality-assurance schemes should provide a guarantee that the constituents *specified* are delivered. It is essential, therefore, to specify what is required for the particular job and it is sensible to discuss this with the concrete supplier to ensure that he knows all the facts. The use of 'designated mixes' for defined applications can make life less complicated in appropriate applications.[15]

The steel must be correctly fixed and held in the forms so that it does not displace during casting of the concrete. The concrete must be able to flow around the steel without segregation and must be uniformly compacted through the mix, particularly in areas close to the corners of the moulds. Where air entrainment is required, special care is needed to ensure good compaction without excess loss of air. Finally, the benefits of all these stages will be nullified if the concrete is not properly cured.

All of these stages add up to good site practice. The test methods and quality-assurance systems are there to help the contractor to check that the job is well done. This will ensure a good and durable concrete structure.

## 14.24 Health and safety

Portland and other hydraulic cements are harmless in normal use. Nevertheless, when cement is mixed with water such as when making concrete or mortar, or when the cement becomes damp, a strong alkaline solution is produced. If this comes into contact with the eyes or skin it may cause serious burns and ulceration.

The eyes are particularly vulnerable and damage will increase with contact time. Strong alkaline solutions in contact with the skin tend to damage the nerve endings first before damaging the skin, therefore chemical burns can develop without pain being felt at the time.

Cement, mortar and concrete mixes may, until set, cause irritant dermatitis due to a combination of the wetness, alkalinity and abrasiveness of the constituent materials. The level of soluble chromium VI in cement is required to be less than 2 parts per million to control the risk of allergic dermatitis caused mainly by the sensitivity of an individual's skin to the chromium salt. To achieve this it may be necessary for the cement manufacturer to add small quantities of a reducing agent, but these materials are active for a limited period. If cement is used outside of the declared shelf life there may therefore be a risk of allergic dermatitis.

Any concrete or mortar on the skin should be removed with soap and water and medical advice sought if irritation, pain or other skin trouble occurs. If cement enters the eye, a speedy response is essential to avoid permanent damage. It should be washed out with plenty of clean water for at least 15 minutes and medical treatment sought without delay.

Protective clothing should be worn which ensures that cement, or any cement/water mixture, e.g. concrete or mortar, does not come into contact with the skin. In some circumstances such as when laying concrete, waterproof trousers and wellingtons may be necessary. Particular care should be taken to ensure that wet concrete does not enter the boots and persons do not kneel on the wet concrete so as to bring it into contact with unprotected skin. Should wet mortar or wet concrete get inside boots, gloves or other protective clothing, this clothing should be immediately removed and the skin thoroughly washed as well as the protective clothing/footwear.

Dust-proof goggles should be worn wherever there is a risk of cement powder or any cement/water mixture entering the eye.

Suitable respiratory protection should be worn to ensure that personal exposure is less than permissible exposure levels.

## 14.25 Future opportunities

The expectations of our society are always changing or being affected by events of the world. For example, the recent increase in concern about the environment has brought consideration of sustainability to the fore. This is increasing the demand for alternative aggregates, from recycled or secondary sources, and for alternatives to Portland cement.

Some of the ways in which concretes with widely different properties can be made have already been described. A number of very strong cements have been made with strength that can be several times that of conventional cements and with a hydrate matrix that is extremely dense and virtually impermeable. Steel fibres can be incorporated to provide improved ductility. In this way, concrete can be tailored for special purposes.

One family of materials using a high strength (>150 MPa) cement-based matrix and high-volume steel fibre reinforcement (2–6% by volume) have been termed ultra-high performance fibre-reinforced concretes (UHPFRCs). Rather than being specified in conventional concrete terms, these 'new materials' have largely been developed and promoted as proprietary products or systems. Unlike other steel and other fibre 'reinforced' concretes, UHPFRC products under load can leave a cracked section that is stronger than the uncracked section, leading to increased overall capacity and ductility.

These new concretes open up a wide range of challenging and interesting design solutions. Projects so far have included:

- prestressed beams
- columns
- arched slender footbridges
- façade and other architecturally driven panels
- street furniture
- canopy roofs
- feature staircases
- grouting for demanding jointing details on-shore and off-shore.

Further reading on UHPFRC materials and experience can be found in the bibliography to this chapter.

Designers will have to think afresh about the best ways to exploit the possibilities concrete technology has made available. Solutions completely different from those traditionally used will be necessary, but such opportunities will only result if there is a wider perception and knowledge of what can be done. Architects, structural designers and materials specialists will have to work in close harmony if society is to gain from the possibilities on offer.

Concrete, as we currently know it, can be used better and in more aesthetically acceptable ways. More emphasis must be placed on the provision of long service life, with proper allowance for maintenance and repair. Some of the basic principles have been discussed and these should be followed meticulously.

## 14.26 References

1. Stanley, C. C., *Highlights in the History of Concrete*, Cement & Concrete Association, Slough, UK, 44 (1979).
2. Morris, M., 'Archaeology and technology', *Journal of the American Concrete Institute*, 9, 28–35 (1987).
3. European Committee for Standardization, EN 197-1, *Cement. Composition, specifications and conformity criteria for common cements*. Brussels, Belgium (2000).
4. ASTM International, *Standard specification for Portland cement*, ASTM C150-07, Philadelphia, PA (2007).
5. European Committee for Standardization, EN 15167-1, *Ground granulated blastfurnace slag for use in concrete, mortar and grout. Definitions, specifications and conformity criteria*, Brussels, Belgium (2006).
6. European Committee for Standardization, EN 450-1, *Fly ash for concrete. Definitions, specifications and conformity criteria*, Brussels, Belgium (2005).
7. ASTM International, *Standard specification for coal fly ash and raw or calcined natural pozzolana for use in concrete*, ASTM C618-05, Philadelphia, PA (2005).
8. European Committee for Standardization, EN 13263-1, *Silica fume for concrete. Definitions, requirements and conformity criteria*, Brussels, Belgium, (2005)
9. British Standards Institution, BS 4027, *Specification for sulfate-resisting Portland cement*, London, UK (1996).
10. Scrivener, K., Calcium aluminate cements, in *Advanced Concrete Technology – Constituent Materials*, Newman, J. & Choo, B. S. (Eds), Butterworth-Heinemann, Oxford, UK (2003).
11. Mehta, P. K., *Concrete: Structure, Properties and Materials*, Prentice Hall, Englewood Cliffs, NJ (1992).
12. Hewlett, P. J. (Ed.), *Cement Admixtures: Use and Applications*, Longman Scientific and Technical, Harlow, UK (1988).
13. Cement Admixtures Association, *Guidance on the use of BS EN 934 – admixtures for concrete, mortar and grout*, Admixture Information Sheet AIS 1, Cement Admixtures Association, Knowle, UK (2006).
14. British Standards Institution, BS 8500, *Concrete*. Complementary British Standard to BS EN 206-1 (2006).
15. European Committee for Standardization, EN 12620, *Aggregates for concrete*, Brussels, Belgium (2002).
16. British Standards Institution, PD 6682-1, *Aggregates – Part 1: Aggregates for concrete – guidance on the use of BS EN 12620*, London, UK (2003).
17. Quality Scheme for Ready-Mixed Concretes, *The QSRMC quality and product conformity regulations*, QSRMC, London, UK (2003).
18. Vitruvius, *De Architectura* (27 BC). Grange, F. (Translator), Harvard University Press, Cambridge, MA (1931).
19. European Committee for Standardization, EN 1992, Eurocode 2, Design of concrete structures, Brussels, Belgium (2004).

20. American Concrete Institute, ACI 318-05, *Building code requirements for structural concrete*, Farmington Hills, MI (2005).
21. European Committee for Standardization, EN 10080, *Steel for the reinforcement of concrete. Weldable reinforcing steel. General*, Brussels, Belgium (2005)
22. Health and Safety Executive, *A Comprehensive Guide to Managing Asbestos in Premises, Hsg 213*, HSE Books, Sudbury, UK (2002).
23. Abrams, D. A., *Design of concrete mixtures*, Bulletin 1, Structural Materials Research Laboratory, Lewis Institute, Chicago, IL (1918).
24. Road Research Laboratory, *Design of Concrete Mixes*, Road Note No. 4, 2nd edn, HMSO, London, UK (1950).
25. Department of the Environment, *Design of Normal Concrete Mixes*, 2nd edn, BR 331, Building Research Establishment, Garston, UK (1997).
26. European Committee for Standardization, EN 934-3, *Admixtures for concrete, mortar and grout. Admixtures for masonry mortar. Definitions, requirements, conformity, marking and labelling*, Brussels, Belgium (2003).
27. European Committee for Standardization, EN 13914-1, *Design, preparation and application of external rendering and internal plastering. External rendering*, Brussels, Belgium (2005).
28. Monks, W. *External rendering*, Publication No. 47.102. British Cement Association, Slough, UK (1988).
29. European Committee for Standardization, EN 206-1, *Concrete. Specification, performance, production and conformity*, Brussels, Belgium (2000).
30. European Committee for Standardization, EN 12390, *Testing hardened concrete*, Brussels, Belgium (2000).
31. Powers, T. C., 'The non-evaporable water content of hardened Portland cement pastes', *ASTM Bulletin*, 158, 68 (1949).
32. British Standards Institution, BS 1881-123, *Testing concrete. Method for determination of the static modulus of elasticity in compression*, London, UK (1983).
33. Parrott, L. J., *Simplified methods of predicting the deformation of structural concrete,* Development Report 3, Cement and Concrete Association, Slough, UK (1979).
34. Parrott, L. J., 'A study of some long-term strains measured in two concrete structures', *RILEM International Symposium on Testing of in-situ Structures, Vol. 2*, Budapest, p. 140 (1977).
35. Building Research Establishment, 'Concrete in aggressive ground', *Special Digest 1*, Building Research Establishment, Garston, UK (2005).
36. Building Research Establishment, 'Alkali–silica reaction in concrete', *Digest 330*, Building Research Establishment, Garston, UK (2004).
37. Institution Of Structural Engineers (ISE), *Structural Effects of Alkali–Silica Reaction*, ISE, London (1988).
38. Hobbs, D. W., Alkali–Silica Reaction in Concrete, Thomas Telford, London (1988).
39. European Committee for Standardization, EN 13670, *Execution of Concrete Structures*, Brussels (2009).
40. UK Certification Authority for Reinforcing Steels (CARES), The product certification scheme for steel for the reinforcement of concrete, *The CARES Guide to Reinforcing Steel*, Part 1, CARES, Sevenoaks, UK (2004).
41. European Committee for Standardization, EN 12350-7, Testing fresh concrete. Air content. Pressure methods, Brussels, Belgium (2000).
42. British Standards Institution, BS 1881-204, *Testing Concrete. Recommendations on the use of electromagnetic covermeters*, London, UK (1988).
43. British Standards Institution, BS 8110, *Structural use of concrete Part 2. Code of Practice for special circumstances,* Milton Keynes, UK, 1985.

## 14.27 Bibliography on UHPFRC materials

- Concrete Society Technical Report No. 63. *Guidance for the Design of Steel-Fibre-Reinforced Concrete.* March 2007. Concrete Society, Camberley, UK.
- Concrete Society Technical Report No. 65. *Guidance on the Use of Macro-Synthetic-Fibre-Reinforced Concrete.* April 2007. Concrete Society, Camberley, UK.
- Cather, R. and Jones, A. Spring 2005. Ultra-high performance fibre-reinforced concrete. *Concrete Engineering International*. Concrete Society, Camberley, UK.
- Rossi P. (2008). Ultra high-performance concretes. *Concrete International*. American Concrete Institute, Farmington Hills, MI.

# 15 Glass

**Chris Brown**
Formerly of Arup

**Edwin Stokes** BEng CEng MIMMM
Stokes Consulting

**Contents**

## 15.1 Introduction

The most common type of glass used in construction is soda-lime silicate glass, made up of a combination of oxides linked by covalent bonds. The principal constituent is silicon dioxide, obtained from sand, mixed with other oxides in the following proportions:

- silicon dioxide (SiO2) (69%–74%)
- calcium oxide (CaO) (5%–12%)
- sodium oxide (Na2O) (12%–16%)
- magnesium oxide (MgO) (0%–6%)
- aluminium oxide (Al2O3) (0%–3%)

## 15.2 Structure

The basic structure of pure silicon dioxide glass and soda-lime silicate glass is illustrated in Figure 15.1. The amorphous orientation of the molecules resembles the random atom structure seen in a liquid. For this reason, glass may be referred to as a 'supercooled liquid'. It should be noted that this designation is based on the molecular structure of the material and does not imply it behaves as a liquid. Reports of glass gradually flowing at ambient temperature, supposedly evident in ancient church windows, are a myth.

## 15.3 Properties

Probably the most notable characteristic of glass is its transparency to visible light radiation. This arises as a result of its amorphous structure and predominant covalent bonding, as this binds the electrons in the material so that there are none available to absorb light as it passes through.

The properties of glass can be engineered by manipulating the concentration in which various oxide compounds are present, for example the addition of boron oxide creates borosilicate glass, which is thermally more stable than basic soda-lime silicate glass. However, as a guide the following properties are typical of clear float glass:

- Density: 2500 kg $m^{-3}$
- Young's modulus ($E$): 70 GPa
- Poisson's ratio: 0.2
- Specific heat capacity: 720 J $kg^{-1}$ $K^{-1}$
- Thermal expansion coefficient: $9 \times 10^{-6}$ $K^{-1}$
- Thermal conductivity: 1 W $m^{-1}$ $K^{-1}$
- Refractive index: 1.5 (average for visible wavelengths).

Glass is a brittle material – it behaves elastically at ambient temperatures and does not exhibit any plastic deformation before failure. It has a relatively low strain capacity prior to failure compared to many other materials such as metals or plastics, and it will not accept plastic deformation at ambient temperatures. The strength of glass it not an intrinsic property; it depends on the surface condition and the residual compressive strength.

## 15.4 Manufacturing of glass

During manufacture the raw materials are heated to a molten state (above 1000°C), causing them to decompose and produce the required oxides. The mix is then formed to the required shape, cooled and solidified. Solidification occurs too quickly for the molecules to crystallize, giving rise to the characteristic amorphous structure.

Historically glass has been produced by processes such as blowing, spinning, casting and drawing (where molten glass is drawn from a tank using a metal 'bait' and rollers). However, these processes have limitations, not least in terms of the visual quality that can be achieved. To remove visual surface blemishes and distortions, expensive grinding and polishing after manufacture is required. Today most glass used in construction is manufactured by the float process, introduced in the UK in the late 1950s, which produces flat glass of much higher visual quality and uniformity. Some glass is also produced by rolling, though this tends to be for specialist glass products and patterned glass.

### 15.4.1 The float process

The float glass process accounts for more than 90% of flat glass produced. The molten glass is poured continuously over a bed of molten tin, and as the two are insoluble, the glass forms a layer of constant thickness over the surface. The glass then solidifies as it is drawn off the tin bed, with high consistency of thickness across the ribbon.

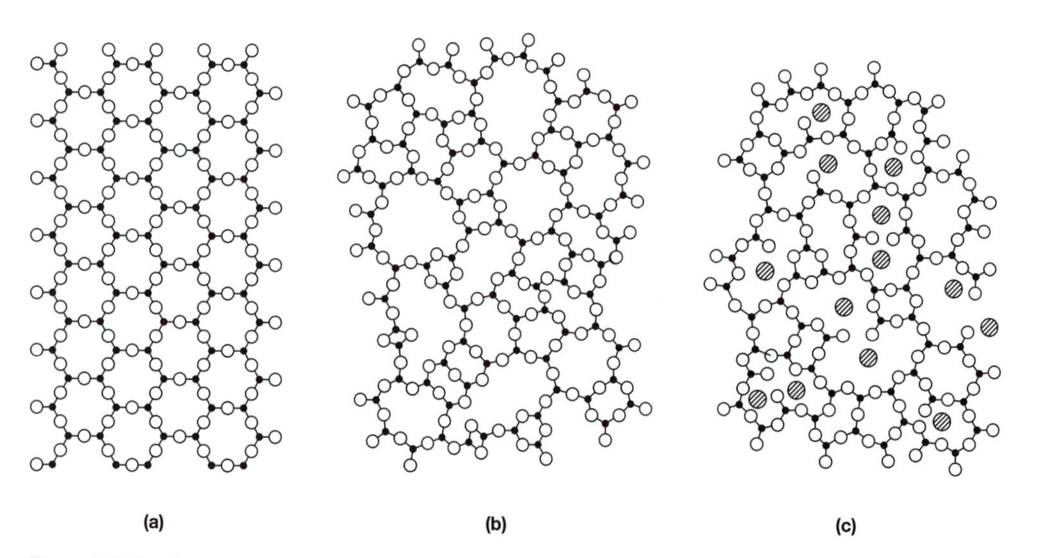


**Figure 15.1** Atomic arrangement of: (a) crystalline silicon dioxide; (b) pure silica glass; (c) soda-lime silicate glass

Float lines produce glass continually which is either immediately supplied to the market or stockpiled. The cost of stopping and reinitiating the float process means that it is prohibitive to do so except where unavoidable to carry our maintenance on the machinery. Periodically the thickness of the glass produced may be changed, which requires reconfiguration of the processing plant (generally the thickness will be progressively stepped up and then down in what is known as a 'campaign').

### 15.4.2 Rolled glass

Rolling is mainly used where some level of opacity or texture is required, without full transparency. It involves passing a ribbon of semi-molten glass through a pair of rollers, which impart a surface pattern on the glass. This is typically used where light transmission is desired while maintaining a level of privacy, such as in bathroom windows. Wired glass is also produced by rolling.

## 15.5 Glass types

### 15.5.1 Annealed glass

The basic glass product, which has not been heat treated following the float process, is commonly referred to as annealed glass.

### 15.5.2 Thermally toughened glass

Thermally toughened glass (frequently referred to as 'toughened' or 'tempered' glass) is heat treated after the float process to modify the residual stress fields present in the material. Typically, the toughening process involves heating the glass to a temperature of around 700°C, and then quenching it rapidly by forced convection. As heat is extracted from the surfaces of the glass, the core is the slowest region of the glass to cool. As a result, contraction of the core is restrained by the surface, which cools and contracts first. This restrained contraction develops tensile stress in the central zone of the glass and compressive stress of equal magnitude at the surface. This residual compression stress in the surface opposes the propagation of surface defects, and consequently the strength of the glass is increased to several times the typical strength of annealed glass.

The toughening process has several effects other than increasing the glass strength. The behaviour of the material on breakage is very different to that of annealed glass, as the whole panel shatters into numerous small fragments as the substantial strain energy present in the material is released. Annealed glass on the other hand typically forms several large shards. Also, toughening can affect the surface profile of the glass due to sagging of the material during heating ('roller wave'). This affects the visual appearance of flat glass products and therefore needs to be carefully controlled. Additionally the residual stress fields often introduce localized variations in the refractive index of the material, with the consequent formation of visible diffraction patterns in some light conditions. In order to minimize such visual effects, toughening is normally carried out to well-defined 'recipes', which define the parameters of the process – dwell times, cooling rates and so on – so that the visual quality of the end product is optimized.

One other feature that has arisen with the increased prevalence of toughened glass in construction is the risk of spontaneous failure due to nickel sulphide (NiS) inclusions. Residual levels of nickel invariably lead to NiS inclusions forming in the material as it reacts with sulphur during melting. In annealed glass they are of no consequence. However, during the toughening process the inclusions transform into a high-temperature form of nickel sulphide, only to revert to their low-temperature form at some subsequent time with an associated volume expansion. This can initiate a crack in the tensile zone of the glass and lead to failure. Unfortunately, reversion to the low-temperature form may be retarded due to the rapid quenching of the glass during toughening, and this can be problematic when toughened glass is used in construction applications as failures can occur spontaneously at unpredictable times during the service life of the glazing. The incidence of the problem can be reduced by controlling the manufacturing process to eliminate sources of contamination, and screening the toughened glass at the factory by heat soak testing, which identifies and rejects affected units.

### 15.5.3 Heat-strengthened glass

Heat-strengthening is a similar process and effect to toughening. However, the heating and cooling regime is controlled so that the residual stresses created in the material are less than for toughened glass. Consequently, heat-strengthened glass is designed to break into shards similar to annealed glass, but with mechanical strength between that of annealed and toughened glass. Heat-strengthened glass is also unaffected by NiS inclusions and is less prone to optical distortion effects than toughened glass.

### 15.5.4 Laminated glass

Laminated glass is formed of two or more layers of glass bonded in a stack arrangement with an adhesive. Typically this adhesive is either a polyvinyl butryal (PVB) film, which is bonded to the glass when heated and pressurized in an autoclave, or a laminating resin, which is injected into the cavity between two panes of glass and sets to form a bond.

### 15.5.5 Other glass products

As well as flat glass panels formed by the float process, other glass products are also used in construction. Prominent among these are glass blocks, formed of two pieces of cast glass assembled into a hollow block, which is typically used as a walling module. Glass fibres are used as a constituent of several composite materials, including glassfibre reinforced concrete (GRC) and glassfibre reinforced plastics (GRP). GRC and GRP are described in depth in other chapters in this book.

## 15.6 Performance

Of the advantages that glass offers as a construction material, it is generally its visual transparency that makes its use so desirable in many buildings and structures. However, in some cases the use of a transparent material may be less than ideal in terms of performance factors other than visual appearance. In order to overcome these issues, processes have been developed to modify the performance of the basic glass product in order to improve the overall performance of the material for specific uses.

### 15.6.1 Thermal performance

As well as transmitting visible light, glass is also transparent to other wavelengths of solar radiation. Short wavelength infra-red (IR) radiation from the sun transmitted through glass is absorbed by IR absorbent surfaces in a building, causing them to heat up. This absorbed solar energy is re-radiated from the heated surfaces as longer wavelength IR radiation, which is not transmitted by the glass. This can lead to excessive heat gain in a building during the summer – the 'greenhouse effect'.

During the winter there may be less heat gain due to lack of solar radiation, and heat loss can then be a problem because heat can be lost by thermal conduction through the glass.

In order to address these issues, various glass processes have been developed to improve the thermal performance of glass.

### 15.6.2 Double and triple glazed units

In order to reduce conduction of heat across glazed areas, glass units can be used that incorporate a sealed air cavity between two pieces of glass. This can greatly reduce heat loss when internal conditions are warmer than external. The air cavity is formed by spacing the two panes of glass with a spacer at their edges, and gluing the units together with a wet-applied edge seal on all sides. The width of the cavity is kept fairly low (typically below 20mm), and as a result convection currents in the air gap cannot form and heat transfer through the unit is largely limited to radiation between the panes. Further performance gains can be achieved by creating more than one air gap, for example in triple glazed panels, and by filling the cavity with an inert gas such as argon.

### 15.6.3 Solar control coatings

The problem of controlling heat gain through glazing elements has been addressed by the development of solar control coatings. These coating are more sophisticated than low-E coatings, and are formed of layers of metal and metal oxide, deposited sequentially on the glass surface. They are designed not only to provide low emissivity, but also to filter out unwanted wavelengths of electromagnetic radiation and reduce solar gain in hot weather. Specifically, UV emissions are filtered whereas the visible wavelengths are transmitted.

The coating technology does not allow a perfect delineation between wavelengths such that full visible transparency can be achieved with the elimination of other wavelengths, and as a result there is a trade-off between transparency and the amount of heat absorbed or reflected by the coating. The precise characteristics of a coating can be engineered by varying the layers that are deposited, and the layer thicknesses. A wide range of products are available to give different combinations of performance, and therefore the product most suited to any particular application can be selected from the options available.

As an alternative approach to managing thermal performance, special interlayer products are also available which provide solar control properties in laminated glass by absorbing radiant heat energy.

### 15.6.4 Security

Due to the transparency of glass and the brittle manner in which it fails, it is often seen as a weakness in a building or structure which may be maliciously targeted by vandals or persons seeking unauthorized access. Specialist glass products have been developed to address this.

#### *15.6.4.1 Anti-bandit glazing*

Anti-bandit glazing is specifically designed to resist manual attack, and to delay access into a building or structure. Various approaches to achieving this performance may be adopted, and classification as anti-bandit glazing depends simply on compliance with standard performance requirements rather than any specific characteristics of the glass itself. However, typically robustness against manual attack is achieved by the use of a laminated product, which retains some coherency on breakage of the glass, and by selecting glass of a sufficient strength and thickness. Annealed glass may be a better deterrent than toughened glass, due to the large jagged shards into which it breaks.

#### *15.6.4.2 Bullet-proof glazing*

Bullet-proof glazing is similarly classified based on performance, and a range of standard levels of bullet resistance relating to different types of weapon are referred to as standard. As well as preventing entry of a bullet, the glazing must achieve set requirements for the extent and size of splintering of glass from the rear face. Again laminated glass is typically used; however, the more onerous performance requirements are likely to require increased numbers of layers of glass and/or thicker sheets. It is also important to note that the glass framing also needs to provide the required level of performance.

#### *15.6.4.3 Blast-resistant glazing*

In some instances glass installations are designed to resist blast loading, where there is an increased risk of bomb blast or other threat to the building or structure. Such glazing is designed to be retained by the window framing in the event of a blast and prevent broken fragments from escaping as dangerous projectiles.

### 15.6.5 Fire performance

Glass that is typically used in buildings and structures is prone to fail when exposed to a fire as a result of imposed thermal stresses due to rapid heating, and it is ill equipped to provide containment due to its transparency to radiant heat energy. However, a variety of specialist fire-rated glass products can provide both enhanced integrity and insulation in a fire. Borosilicate glass exhibits much lower thermal expansion than standard soda-lime silicate glass, and is therefore more stable against rapid heating. Alternatively, fire-resistant glass may incorporate intumescent material in the cavity of a double glazed unit or in the interlayer of laminated glass, which expands and opacifies on heating to provide insulation and contain the fire on one side of the glass. (Traditionally wired glass has been used for its fire resistance; however, the performance offered is inferior to what can now be achieved by alternative products.)

### 15.6.6 Cleaning

Glass is often used in applications where its transparency is key to its role, and therefore the loss of transparency caused by soiling necessitates cleaning much more frequently than may be required for opaque elements. To compound this problem, glass is commonly installed in external applications such as building façades, where atmospheric pollution is likely to increase the rate at which soiling occurs, and access to the glass for cleaning may not be straightforward. This can add significantly to the cost of running a building or structure due to the need to clean the glass regularly throughout its life.

Various methods have been developed to reduce the need to clean glazing. Possibly the most practical solution that has reached the market is self-cleaning glasses, which incorporate titanium dioxide based photocatalytic coatings. The coating is applied to the outside surface of window glazing, and if exposed to ultraviolet radiation will accelerate the breakdown of organic deposits on the surface. Additionally it is hydrophilic, so rainwater on the surface naturally spreads and runs off, carrying away the dirt as it breaks down. The required frequency of cleaning windows can be much reduced by self-cleaning coatings.

## 15.7 Processing and fabrication

Various processes exist for processing glass after it has been manufactured by the float process.

### 15.7.1 Cutting and edge working

Glass comes off the float line in sheets of standard size, which are generally cut down to project-specific dimensions. Typically cutting is carried out horizontally on a roller bed, most commonly by diamond cutting although processes such as water jet cutting and laser cutting can also be used for more complex shapes.

The cut edge of the glass is rough and inconsistent, therefore needs to be polished. A variety of edge profiles are possible.

### 15.7.2 Drilling

Drilling holes for the introduction of fixings into the glass is also best carried out on a flat bed drilling machine. Holes lead to stress concentrations in the glass and therefore weaken the panel.

### 15.7.3 Heat treating

Toughening and heat strengthening are carried out in toughening ovens, where the glass sheets are passed through the oven on roller beds. After introduction to the furnace, the glass is heated while constantly being rolled back and forth to avoid sagging as the glass softens. It then passes to the quench stage, where it is rapidly cooled by air jets above and below the glass.

### 15.7.4 Shaping

Although the float process is limited to the production of flat glass sheets, it can subsequently be shaped into curved forms. Various processes exist for this. All glass shaping processes require the glass to be heated to a point at which it softens, without completely melting (around 1000°C).

Vertical shaping involves suspending the glass in tongues that grip one edge, and pressing the glass into shape.

Horizontal shaping is generally a more versatile method, in which the glass is either slumped or pressed into a mould on heating, or is shaped into the required form by the movement of a roller bed on which the glass is supported. In this process, shaping can be combined with toughening if the glass is quenched after shaping.

### 15.7.5 Patterning

Various processes exist for imparting a pattern or texture to the surface of glass. These may be required, for example, to create spandrel areas in an all-glass wall, or simply for decorative effect.

Acid etching, as the name implies, is the controlled reaction of the glass surface with acid to create a uniform frosted appearance, while maintaining reasonable light transmission.

Sand blasting involves creating a pattern on the glass by impacting the surface with a constant flow of sand. Images can be created on the glass by hand, or by an automated process.

Ceramic fritting is also commonly used. This is a ceramic layer applied to the glass surface, again with the potential to create images or patterns. The ceramic is fused to the glass by heat treating, therefore is it normally applied to toughened glass when the toughening process serves to fuse the glass and ceramic frit, as well as strengthening the glass itself.

## 15.8 Glazing in buildings and the built environment

### 15.8.1 Glass houses

As a construction material, glass offers the designer many benefits, perhaps most notably its transparency and perceived architectural qualities. Traditionally this has often led to glass being used simply as infill for windows. During the 19th century the use of glass in buildings became more ambitious. Large, sophisticated glazed buildings such as Crystal Palace (1850) and the Palm House at Bicton (1890) were built. The glass in these buildings had structural function, bracing the lightweight iron frame into which it was fixed against self-weight and imposed loads.

### 15.8.2 Curtain wall façades

With the advent of the float process in the 1960s, high-quality glass became more readily available at a lower cost. Additionally, technical advances such as toughening and laminating, glass coatings such as low-E and solar control pushed the bounds of what could be achieved. Since the 1970s there has been a strong drive from architects to maximize the transparency of façades of buildings to transmit more light into them. Large glazing panels in façade curtain walls are now commonplace. The use of ventilated double-skinned façades in the design of office buildings has also been common. Such double-skinned façades typically utilize large quantities of glass with minimal structural framing.

### 15.8.3 The structural use of glass

The traditional method of incorporating glass into a building or structure is a framed window, where the glass sits in a rebate around all sides. An alternative to reduce the amount of visible framing is to use structural silicone to connect aluminium structure to the back of the glass. This is effectively an adhesively bonded fixing system, for which specialist silicone sealants have been developed and controlled application processes defined that will ensure many years of reliable service.

Point fixing systems offer the possibility of creating glass walls with minimum visible support structure. A number of proprietary fixing systems have been designed by glass suppliers such as Pilkington and Saint-Gobain. The glass is supported in mechanical fixings bolted through holes drilled in the material, and supported on cast metal brackets known as 'spiders'. Alternatively, patch fixings may be used which form nodes at the corners of the glass panels, effectively creating a net structure.

As the understanding of the properties of glass and ways of using it has developed, glass has been used for more ambitious structures such as floors, bridges, staircases and sculptures.

## 15.9 Designing with glass

Due to its amorphous structure, glass is intolerant of plastic deformation and it fails in a brittle manner. As a result particular considerations should be applied when designing with glass, and these principles are introduced here.

### 15.9.1 Safety

Due to the behaviour of glass upon breakage, the risks associated with glazing installations warrant specific consideration during the design period. Inappropriate use of glass may lead to several types of hazard to safety, such as unprotected drops, broken glass falling from height and jagged shards of glass becoming exposed.

Safety glass is defined in BS 6206 as glass that meets the requirements for the given safety class when tested according to the methods described, and either does not break or 'breaks safely' (for example by disintegrating into small particles, which are usually considered to be less hazardous than larger shards of glass). Monolithic toughened glass meets the requirements for safety glass as it disintegrates into small fragments on breakage. Laminated glass can also achieve a safety glass classification if the glass remains adhered to the interlayer on breakage and the glass panel does not fall out of its fixings or framing.

Where glass performs the role of a barrier, it is required by Part N of The Building Regulations to be safety glass. Where it protects a drop it should also provide containment after breakage.

In addition to the issues addressed in the Building Regulations and British Standards, the possibility of broken glass fragments falling out and dropping large distances into occupied spaces should be addressed by designers as this may be hazardous to

**Table 15.1** Typical allowable stresses in glass

| *Duration of load* | *Typical allowable stresses (N/mm²)* | | |
|---|---|---|---|
| | *Short (seconds)* | *Medium (hours)* | *Long (years)* |
| Toughened glass | 50 | 50 | 50 |
| Heat-strengthened glass | 25 | 25 | 25 |
| Annealed glass | 17 | 8.4 | 7 |

safety. Where building-specific safety requirements apply, such as bullet resistance or blast resistance, specialist glazing products and systems are required.

### 15.9.2 Strength

The strength of glass it not an intrinsic property; it depends on the surface condition, the residual compressive strength and the load duration. As the surface of the material always contains random flaws, the strength values across a number of samples of similar dimensions will exhibit a statistical variation, and this must be taken into account when designing with glass. However, as a guide, typical allowable stresses may be taken as in Table 15.1.

### 15.9.3 Durability

Glass is an inert and stable material and does not degrade or change when exposed to sunlight and rainwater. Therefore monolithic single glazing is generally maintenance-free for the lifetime of a building, aside from routine cleaning and replacement in the event of breakage.

More unusual environments can have a detrimental effect on glass. Most notably, alkaline environments created by contact with cementitious materials can lead to a chemical reaction with the silica in glass, etching the surface.

Where glass is assembled into a double glazed unit, the edge seals are expected to offer a limited service life up to 25 years or so. Similarly, interlayers in laminated glazing may not be as durable as the glass itself due to mechanisms such as yellowing when exposed to UV radiation.

### 15.9.4 Sustainability

The raw materials required to produce glass are in abundant supply. However, the energy cost of actually producing glass from the raw materials is high due to the temperatures involved. Further high-energy processes may occur once the float process is complete, such as toughening and heat soak testing. This should be set against the fact that glass is a durable material, and therefore offers the benefit of prolonged, low-maintenance service superior to many of the possible alternatives. This is seen in many old buildings, such as historic stained glass windows, where material hundreds of years old has survived since construction and continues to function as intended.

The stability of glass makes disposal difficult as it will not readily break down. Recycling is a viable option, by crushing, remelting and reforming waste glass into a new product. This process is not difficult to carry out, but it tends to lead to contamination and it is difficult to produce recycled glass of the highest optical quality. Therefore recycled material is mostly used in non-architectural applications, such as coloured drinks bottles, where visual quality is less critical. Reusing crushed glass directly, for example as an aggregate in concrete or screed, is also possible, though the level of demand is limited.

## 15.10 Further reading

*Glazing for buildings*, BS 6262, parts 1–7. 2004. London, UK: BSI.

*Glass in building. Basic soda lime silicate glass products*. BS EN 572, parts 1–9. 2004. London, UK: BSI.

*Structural use of glass in buildings*. Institution of Structural Engineers. 1999. London, UK: ISE.

*Guidance on glazing at height*. CIRIA (Report C632). 2005. London, UK: CIRIA.

*Glass structures: Design and construction of self-supporting skins*. J. Wurm. 2007. Basle, Switzerland: Birkhauser.

# 16 Grouts and Slurries

**Stephan Jefferis** MA MEng MSc PhD CEng CEnv FICE CGeol FGS FRSA
Director Environmental Geotechnics Ltd; Visiting Professor, Department of Engineering Science, University of Oxford

## Contents

## 16.1 Introduction

Grouts and slurries are two special material systems much used in civil engineering but often poorly understood. The diversity of uses and properties of these systems is so great that it is difficult to give a concise overview. By way of introduction it is appropriate to consider some of the fundamental similarities and differences between these two systems. Similarities are:

- with the exception of high-strength structural grouts, water is volumetrically the major component of both grouts and slurries
- many grouts and slurries are based on active swelling clays such as bentonite
- both grouts and slurries tend to be complex non-Newtonian fluids, that is they do not show simple linear relationship between shear stress and rate of shear strain.

A distinction between grouts and slurries is that slurries are specifically used in slurry-supported excavations such as diaphragm walling, piling, slurry tunnelling and borehole drilling. Grouts are generally placed by injection and find a wider range of application from geotechnical engineering to structural engineering. A logical distinction between grouts and slurries would be to refer to all non-setting systems as slurries and to reserve the term 'grout' for setting materials. However, the term 'self-hardening slurry' has been coined for cut-off wall slurries which are setting materials. Thus grouts and slurries could be regarded as a spectrum of materials with structural grouts which develop substantial strength at one extreme and non-setting slurries at the other. Self-hardening cut-off wall slurries are thus intermediate systems; though in practice the formulations for some self-hardening slurries may be similar to those of geotechnical grouts.

An apparent similarity between grouts and slurries is that clays, Portland cements and the silicate solutions much used in chemical grouting all involve silicon–oxygen chemistry (for example, the silica layer of clays and the calcium silicate hydrates of cements) and to a lesser extent aluminium–oxygen chemistry. Despite this common chemistry the engineering behaviour of the materials is quite distinct. However, the common fundamental chemistry does suggest some potential for chemical compatibility of the materials and the potential for mixed systems. Indeed, some of the most ancient and enduring construction materials are based on mixtures of clay and pozzolanic cements.

For both grouts and slurries there is a wide range of special chemical systems not involving clays or hydraulic cements. In grouting these are now quite widely used, though their relatively high cost tends to restrict their application to situations requiring their special properties as a secondary materials following bulk grouting with clay-cement-based systems. There are also chemical-based slurries originally developed from experience in the oil well drilling industry. There are now specific civil engineering slurries based on synthetic water-soluble polymers which are gaining considerable acceptance for slurry piling where tailored systems are available for a range of ground conditions. These systems offer a smaller on-site footprint than bentonite-based systems and easier and more flexible re-use of the slurry. Specialist polymers can 'encapsulate' the excavated soil and prevent the slurry wetting it as happens with bentonite. If wetting occurs, it is difficult to re-use the excavated material, which, from considerations of economics and more importantly sustainability, should be re-used and not disposed to landfill. Under some regulatory regimes (e.g. current UK waste regulations) it may be inadvisable to refer to excavated soil as spoil. Spoil suggests waste and as a result the excavated material may be classed as waste and therefore subject to waste regulation and fiscal regimes [1]. In this text the soil from slurry-supported excavations is referred to as 'excavated material' or 'process arisings' to encourage a mindset of sustainable re-use. The terms 'spoil', 'contaminated slurry' and 'cleaning' are avoided to further encourage a mindset of re-use.

Another link between slurries and grouts is that much of the equipment for testing fluid properties is common to the two and is based on the equipment used for oil well drilling fluids. Oil well drilling fluids have developed from simple native clay systems to complex chemical systems, and there are similar trends with civil engineering grouts and slurries. Grouts are also used in oil well operations for cementing conductor pipes to the formation. The technology of well cementing has not significantly influenced geotechnical grouting or vice versa, but there are close parallels between oil well grouting and the filling of ducts in prestressed, post-tensioned concrete and the specialist area of offshore grouting for structural connections and repairs etc.

## 16.2 Overview of applications

The first application of a slurry for trench excavation was to form a soil–slurry cut-off wall on the levees of the Mississippi River [2]. However, the development of concrete diaphragm walling in the 1950s overshadowed the cut-off wall applications and for many years slurries were synonymous with diaphragm wall construction. In recent years much more cut-off wall work has been undertaken, particularly for the control of leachate from landfill sites and contaminated land. It is interesting to note that in the UK cement–bentonite cut-off walls are most widely used whereas in the USA soil–bentonite and soil–cement–bentonite walls are more popular, though cement–bentonite walls are now becoming more widely used. These national preferences appear to relate to the relative cost of materials, the availability of plant and space on site, and particularly the availability of suitable backfill material [3, 4]. Soil backfill materials are often prepared by blending the excavated material with bentonite and if required extra fines, at the side of the trench – an operation that requires more space than is necessary for a cement–bentonite wall, where the slurry may be pumped from a central mixing plant.

Applications for slurries thus include slurry-supported excavations for diaphragm walling, piling, slurry tunnelling, borehole drilling and cut-off walls. The applications of slurries in piling and diaphragm walling are very similar and need not be considered separately. Slurry tunnelling involves rather different constraints; excavated soil conveyance from tunnel face to the ground surface and then soil/slurry separation (there is, of course, face sealing in permeable grounds but this is a requirement common to all excavation support slurries). Cut-off wall slurries are again a special class of materials for which the hardened properties rather than the fluid properties are of fundamental importance.

The range of applications of grouts is much wider and includes geotechnical grouting for soil improvement and soil modification. There are also applications in piezometer and borehole sealing – operations that are often left to the discretion of site staff and for which it is difficult to obtain even typical data on mix proportions, performance, etc. Indeed, specifications are often very loosely worded so that neither mix proportions nor performance are adequately quantified – an unsatisfactory situation which must lead to unsatisfactory grouts. Structural grouting is much used for corrosion protection and bonding of tendons in post-tensioned prestressed concrete. Grouts are also used in the repair of damaged concrete to replace cover etc. In the offshore industry, structural grouting found considerable application for pile–sleeve and pile–socket connections and repair clamps [33]. A continuing application of cement grouts is the stabilisation of toxic and radioactive waste to reduce risks of hazardous matter entering the biosphere. For these applications it is important to minimise the volume increase on stabilisation as space in the landfill/repository will be at a premium. Furthermore, as both applications require much longer term durability than most civil engineering applications, thermodynamic and other modelling of the resulting mineral systems is fundamental to their design.

Self-hardening slurries were regularly used in a secant piling system to produce an economic combination of a cut-off wall and a retaining wall consisting of primary cement–bentonite piles and secondary structural concrete piles – the hard–soft piling system [5]. However, the procedure is now used more rarely in the UK.

There is of course a very wide range of other applications of slurries and grouts. It is not possible or practicable to list all uses, and thus only a selection of applications has been included to show the diversity of uses for such materials. In the following sections a few of these applications will be discussed in more detail to bring out particular features of the systems, materials, test procedures, etc.

## 16.3 Materials

Many different chemicals may be used to formulate grouts and slurries, but the materials that are used in the greatest quantity are swelling clays and hydraulic cements. For economy and to obtain special properties, materials such as pulverised fuel ash or ground granulated blastfurnace slag also may be used (these materials were formerly known as cement replacement materials but are now properly known as additions or supplementary cementing materials, SCM). There are also applications for chemical admixtures to modify the properties of clay and cement systems as well as purely chemical systems. As many aspects of the materials used in slurries and grouts are common to the two systems it is convenient to discuss them together. It is also necessary to consider mixing, as the properties of both grouts and slurries can be strongly influenced by the mixing equipment and mixing sequence. It follows that an understanding of grouts and slurries requires an understanding of the constituent materials and also of the effects of mixing.

### 16.3.1 Clays

The traditional base of slurries and many grouts is the smectite clay, bentonite. Occasionally attapulgite has been used [6] as it will disperse in salt water, though the suspension properties are often not as good as can be achieved with bentonites in fresh water. For cut-offs a slag–attapulgite clay system has been developed which gives very low permeabilities ($<<10^{-9}$ m $S^{-1}$ [7]). Attapulgite is less widely available than bentonite and can require a longer mixing time to achieve useful slurry viscosity – unlike bentonite, the key issue is not hydration of the clay to form more and smaller particles but mechanical shearing to split the lath-shaped attapulgite particles into more and narrower laths that tangle to develop viscosity. Non-swelling clays may be used to form slurries but relatively high clay concentrations can be necessary to produce stable slurries that do not bleed excessively (see Section 16.12.6). This in turn makes the slurries more viscous and difficult to use. An alternative is to use a blend of swelling and non-swelling clays but the extra materials handling, batching, and quality control can make this uneconomic or impracticable.

All clays' particles carry a negative charge and this is balanced by positively charged cations that are held by electrostatic forces. As these ions are held only by electrostatic forces they can be exchanged for other cations provided the overall charge balance is maintained. Factors that will influence the exchange are the relative concentrations of the ions and their valency. Generally higher valency ions will be more strongly held than lower valency ions. The size and chemical coordination number of the ions is also important: large ions tend to be held more strongly than smaller ions of equal charge. In addition to these general effects there are specific effects. For example, potassium and caesium ions tend to be strongly held as they can fit into holes in the surface of many clay mineral plates and act as bridges to link particles together (it is the bridging action of potassium ions that makes illite clays non-swelling despite the fact that otherwise they are quite similar to the swelling smectite clays). The valency of the exchangeable cations modifies the clay particle–particle interaction. In particular the presence of divalent and trivalent ions (e.g. $Ca^{2+}$ and $Al^{3+}$) allows closer approach of individual clay particles and so promotes edge-to-face flocculation and face-to-face aggregation. For use in slurries and grouts it is important that the clay is dispersed or loosely flocculated. Face-to-face aggregated clay particles (which may be likened to 'packs' of cards rather than the loose 'house of cards structure' of flocculated systems) will be coarse and of little colloidal activity. Thus they will be rapid settling and will not control bleed of grouts etc. or develop useful suspension viscosity.

For almost all bentonite-based slurries and grouts the clay is used in the sodium form. Fundamentally there are two classes of sodium bentonite: natural sodium bentonites (for example, many of the bentonites from Wyoming in the USA) and sodium ion exchanged calcium bentonites (which are available in many parts of the world). The bentonites produced in the UK and many parts of Europe are ion exchanged bentonites in which the calcium of the natural clay has been exchanged for sodium. Sodium carbonate is used for the exchange and typically a slight excess is used so that ion exchanged bentonites have an alkaline pH in the range 9.5 to 10.5. The excess sodium carbonate tends to flocculate the clay so that ion exchanged bentonites can show relatively high gel strengths and viscosities. Natural bentonites tend to have a nearly neutral pH and to show lower gel strengths. However, any contamination with cations (for example, from salt waters or cement) may lead to a very substantial thickening of natural sodium bentonites. It should be noted that as a result of the ion exchange with sodium carbonate, the calcium initially present on the clay will be converted to calcium carbonate (chalk). Consequently calcium exchanged bentonites will contain a few per cent of calcium carbonate. This is of very low solubility and in normal circumstances it has no effect on the performance of the clay. However, its presence should be considered if chemical interactions are being analysed – the calcium carbonate may be found to be beneficial or slightly harmful depending on the specific circumstances. Many bentonites are now treated with polymer materials to develop or enhance their properties – a practice that can be expected to develop as resources of good-quality bentonite become scarcer.

An important role of the bentonite is to stop a slurry or grout bleeding (see Section 16.12.6). As bentonite is a natural material it should be expected that bentonites from different sources will behave differently. For example, cement–bentonite slurries may show substantial bleeding when prepared with bentonite from one source but minimal bleeding with another bentonite. It can be difficult to identify the underlying cause of such behaviour. Excessive bleed suggests an unstable system. Bleed water is particularly significant for grouts, where it may lead to voids, and for slurry trench cut-offs, where it represents wasted cut-off volume. Furthermore as bleed water can escape rapidly from a slurry trench (it does not deposit a filter cake), the resulting drop in the slurry level in the trench can lead to instability of the trench. Slurries that show high bleed are likely to show high filtration losses (the processes have similarities: see Section 16.12.5). It should be noted that the filter loss for cement–bentonites is very substantially greater than that of pure bentonite slurries and thus some drop in slurry level must be expected for cut-off walls after the end of excavation and prior to set of the slurry (the effect is more pronounced for short trench lengths as the filter cake will have been less developed before the end of excavation and the final filling of the trench with slurry).

As noted, for almost all applications sodium bentonites are used but more specialised mixes have been developed for cut-off walls [8]. These involve calcium bentonite and can give a lower permeability than the equivalent cement–sodium bentonite slurries. However, if calcium bentonites are used the clay content may be much higher at up to 50% by weight of water and a different design philosophy is necessary for these dense mixes. Furthermore special excavation plant may be needed because of the high density of the slurry and chemical dispersing agents may be required to produce a mix that is suitably fluid. The resulting cut-off walls are potentially more resistant to aqueous leachates than conventional mixes – though costs are considerably higher and it may be more economic to employ a conventional cut-off wall with an hdpe geomembrane.

### 16.3.2 Cements

#### *16.3.2.1 Structural grouts*

Structural grouts are almost exclusively based on hydraulic cements. The cements that may be used will depend on the application but include Portland, sulphate resisting Portland, calcium aluminate (high alumina) and oil well cements. Oil well cements were developed for grouting the oil conductor pipes in wells to the formation to prevent fluid migration between strata. A range of such cements has been developed for the special temperature and pressure conditions that can occur in oil wells [9]. Class B oil well cement has been much used for offshore structural grouting. This is a sulphate-resisting cement which is slightly slower setting than many CEM 1 cements. In structural and oil well grouting, the grout may have to be pumped substantial distances and thus fluid behaviour (rheology) is important. However, specifications hydraulic cements (e.g. CEM I) [14] typically do not consider rheology and differences in grout rheology must be expected between different cement batches/works. If repeatable rheology is important the use of oil well cements should be considered as a rheology-related parameter (thickening time – an oil industry parameter) will be included in specifications. Class G, a tightly specified cement [9], may be appropriate provided that the set time is acceptable.

Aeration of the cement (exposure to the atmosphere) prior to use can markedly change the rheology of both geotechnical and structural grouts [9]. For critical applications there is no substitute for trial mixes and if possible all cement should come from a single batch. The samples for the trial mix should of course come from this batch, though this may cause practical difficulties if 28 day strengths are to be confirmed and aeration is to be avoided.

#### *16.3.2.2 Geotechnical grouts*

For geotechnical grouting most suspension grouts will contain a hydraulic cement to provide set. As large volumes may be used, cost will be an important parameter and this tends to limit the application of special cements (though for many such cements the extra cost may be quite modest). In general, if modified cements are required these are achieved by the use of additions (supplementary cementing materials). However, a special class of cements that is used is the microfine cements (see [10, 11] for early developments). These are cements of finer particle size than Portland cements (for example, a $D_{95}$ < 16 μm). Grouts made from these cements may penetrate fine soils and fine rock fissures in a manner comparable to chemical grouts. Particle size distribution is very important for such cements; a small amount of oversize may lead to clogging and failure to penetrate. The increased fineness will give a more rapid set than ordinary Portland cement so that it is often necessary to add a retarding agent, and indeed this may be included in the material as sold.

### 16.3.3 Cement additions – supplementary cementing materials

'Additions' is now the preferred term for those mineral materials formerly known as cement replacement materials. They may also be known as supplementary cementing materials (SCM). Original reasons for the use of additions such as such as ground granulated blastfurnace slag or pulverised fuel ash (pfa) in grouts may have been focused on cost savings. However, it is the author's experience that they have quite distinct influences on the behaviour of grouts and slurries (and in general all high water–cement ratio systems). Additions cannot be used interchangeably in grouts and slurries.

Pfa/cement grouts are widely used for void filling; for example, old mine workings, collapsing sewers. They also may be used in specialist formulations for concrete repairs etc. In such applications, the largely spherical nature of the pfa particles may be significant in improving workability, though curiously, pumping pressures for pfa/cement grouts may be greater than those of purely Portland cement grouts [12]. Pfa is a product of coal burning and so may be classed as a waste and subject to waste management regulations until it is 'recovered' so that it is no longer a waste (e.g. until it has set in a cementitious system). Therefore before using pfa it will be important to consider the relevant waste management regulations.

For structural grouting, the cost of grout materials usually will be small compared with other costs, so that additions will offer no useful economy and indeed the extra batching and quality control measures may lead to an overall increase in cost. Furthermore, additions may not offer the same chemical environment for the prestressing strand etc. as is achieved with a purely Portland cement grout. Thus additions should be used only if required for some special technical feature of the work.

Additions may have a special role in the cementation of toxic and radioactive waste where the reduced setting exotherm can be very important in the control of thermal cracking. Additions, particularly slag, may improve the leaching behaviour of the cemented material – though other effects on the chemical environment (e.g. reducing conditions) are sometimes less desirable and will require analysis.

For cut-off wall slurries, the replacement of up to about 30% of the Portland cement with pfa has rather little influence on the strength, permeability or stress–strain behaviour. Above 30% the permeability can increase and the strength might reduce. If pfa is used as an additional material or with less than a pro-rata reduction in the cement content it serves to increase the slurry density (which may help displacement with replacement mixes, see Section 16.7) and to reduce sensitivity to drying because of the extra solids in the mix. Cut-off wall mixes with high pfa contents but low cement contents can give much-improved chemical resistance, though to achieve permeabilities below $10^{-9}$ m $s^{-1}$ and chemical resistance it may be necessary to use much higher than normal pfa contents and to select the cement content very carefully [13].

In contrast to the sometimes rather modest influence of pfa, ground granulated blastfurnace slag (ggbs) when used in cement–bentonite slurries can produce striking changes in the properties if a replacement level of over about 60% is used. Below 60% slag has rather little effect. Above 60% slag replacement progressively increases strength and reduces permeability. However, at very high replacement levels the set slag–cement–bentonite materials tend to become rather brittle though this can be avoided in a number of ways [7]. The reduction in permeability may be by an order of magnitude or more and thus substantially improves the cut-off performance of the material. The increase in strength may be less desirable, particularly in weak grounds (typically cut-off materials are specified to have a strength comparable to that of the adjacent ground, though the reason for this is seldom stated: see Sections 16.7.1 and 16.13.2). To control the strength and maintain the low permeability, the best procedure is to reduce the total slag and cement content while maintaining the slag–cement ratio (this also gives a useful economy of materials). In this way, mixes of permeability $10^{-9}$ m $s^{-1}$ and lower can be achieved with total slag plus cement contents of 100–150 kg $m^{-3}$. With purely Portland cement mixes such permeabilities may not be achieved even at 250 kg $m^{-3}$ cement and thus for equal permeability mixes, slag can give substantial savings.

Slag–cement–bentonite mixes at high slag replacement levels (>60%) also show markedly lower bleed than equivalent 100% Portland cement–bentonite mixes. The reason for this is that slag, unlike cement, does not cause rapid and damaging flocculation and aggregation of bentonite slurries. The bentonite particles thus remain in a finely dispersed state. The role of the cement becomes merely to set the slag, and the smaller the amount that is used, the less flocculation/aggregation occurs and the finer the microstructure that develops in the set system (readily demonstrated by mercury intrusion porosimetry). It should be noted that the fine and uniform microstructure of the system is the key to the low permeability discussed above. The low permeability is not due to a reduced porosity of the system. For equal total cementitious content, slag replacement scarcely changes the overall porosity. Extremely low permeabilities (for such high water–cement ratio systems, $<10^{-12}$ m $s^{-1}$) can be achieved if more subtle setting agents are used for the slag. However, it should be noted that bentonite is not stable in the strongly alkaline environment of a cement–bentonite or slag–cement–bentonite mix. It will react to form calcium silicate and calcium aluminate systems broadly similar to those found in hydrated cements or other clay minerals. At the mix proportions typical of cement–bentonite cut-off walls slurries, it is unlikely that any bentonite will remain in the final, hardened material – though its initial presence does have an impact on the microstructure. For example, to reduce the permeability of a cement–bentonite material it is more effective to add a few additional kilograms of bentonite to the system than an equal weight of slag plus Portland cement.

Experience suggests that the increased strength and reduced permeability that can be achieved with slag mixes is not restricted to cut-off-wall mixes. These facets of slag replacement can be exploited in many high water–cement ratio systems. For example, in the stabilisation of high water content peats by soil mixing, slag mixes may give markedly improved mechanical properties (strength and permeability). For lower water–cement ratio systems, perhaps below about 4:1, the effects of slag replacement become progressively less marked (a cut-off wall slurry may have a water–cement ratio >6).

As both slag and pfa are by-product materials, some variability between sources is to be expected – particularly for pfa. When developing new systems, trial mixes can be important to optimise a slag–cement–bentonite or cement–pfa–bentonite system (note when referring to mixed systems it can be convenient to list the major component first – generally this will be slag for slag–cement–bentonites and cement for cement–pfa–bentonite cut-off materials).

### 16.3.4 One-bag grout and cut-off mixes

As discussed above, for cut-of wall slurries and for high water–cement ratio geotechnical grouts, it is necessary to hydrate the bentonite in fresh water prior to the addition of cement. This requires that hydration tanks are available and that the bentonite slurry can be left to hydrate for at least a few hours after mixing and before use. This represents a major complication for small projects. However, pure slag has minimal effect on the hydration of bentonite and thus if a pre-blend of slag and bentonite powders is added to water, the bentonite will hydrate almost as in fresh water. This untrammelled development of bentonite properties can be exploited to achieve one-bag mixes for use on small cut-off wall jobs, soft piles in hard–soft piles and geotechnical grouts. A pure slag–bentonite system will set rather slowly and in proprietary one-bag mixes, stiffening and setting agents are included (the setting agent may include a small amount of Portland cement).

### 16.3.5 Admixtures

Admixtures such as superplasticisers, dispersing agents, thinners and retarders may be used to manage the properties of grouts. However, if the grout system is based on hydrated bentonite then the timing of the admixture addition can have a profound effect. Clays are of very high surface area and can sorb admixtures. It follows that an admixture added to a clay slurry before the cement is added may produce very different effects to the same admixture added after cement addition. For example, if a superplasticiser is added to a bentonite it will tend to disperse the bentonite so that when cement is added the clay–cement interaction will be much stronger than normal and the mix may become very thick and viscous. However, if the superplasticiser is added after the cement it may have its desired effect – a substantial thinning of the mix.

### 16.3.6 Mix water and groundwater

Water will be a major component of almost all grouts and slurries. As a general rule water of drinking quality is satisfactory for the preparation of grouts and slurries. If there is concern about the suitability of a water, the best procedure is to prepare trial mixes. Chemical analysis of a water can provide some steer but it is not always possible to predict how a grout or slurry system will behave with a particular water. Particulate grouts and slurries will contain material of colloidal size (0.001–1 μm). These materials are notoriously sensitive to salts, and without detailed investigation it is often impossible to predict responses. Thus for commercial practice with grouts and slurries it is generally expedient to look for effects rather than to try to establish causes.

For polymer slurries, the polymer concentration is often much lower than the bentonite concentration in a bentonite-based slurry. Polymer concentrations might be in the range 0.2% to 1% by weight of water. Polymers can be broken down by the residual chlorine in mains water. The effect will depend on the sensitivity of the polymer and the relative concentrations of polymer and chlorine. For some sensitive polymer slurries, it can be economic to pre-treat mains water to remove chlorine prior to slurry mixing polymer addition.

#### *16.3.6.1 Dispersion of clays*

Typically bentonite will be the principal colloidal component of grouts and slurries and the most sensitive to water chemistry. Mains waters usually are suitable for the dispersion of bentonite into water. However, groundwater and river, lake and pond waters should be not be used without prior investigation. For example, problems have been encountered with magnesium ion concentrations of 50 mg $l^{-1}$, with calcium at above 250 mg $l^{-1}$ and with sodium or potassium concentrations above 500 mg $l^{-1}$. These figures are examples and should not be taken as safe limits. The sensitivity of clays from different sources may be different and the mixing sequence may be of considerable significance. For example, if the clay is added in increments over a period of time it has been suggested that the first clay added may absorb and attenuate some of the damaging ions so as to allow better dispersion of the later clay. This is a labour-intensive procedure and expensive, as some clay will be wasted. Clays are relatively insensitive to anions such as chloride or sulphate.

To obtain the most benefit from the use of bentonite it is important that it is well dispersed. However, for some clay–cement grouts the role of the clay may be more as a fine filler to reduce the potential for segregation of solids and water. For these grouts the clay is not allowed any time to swell and hydrate prior to the addition of cement. Once cement is added to bentonite, further dispersion of the clay will be inhibited by the cations from the cement.

Clay slurries may also be affected by salts in the soil or groundwater during excavation. Contamination usually will lead to an initial thickening (caused by edge-to-face flocculation) and ultimately, if very severe, to aggregation of the clay particles. Aggregation (face-to-face bonding) produces fewer thicker clay particles with a consequent thinning of the slurry (reduction in viscosity and gel) and much increased bleeding and filter loss. Contamination at this level may effectively destroy the slurry properties. In practice such contamination is rather rare in slurry-supported excavations as the hydrostatic pressure of the slurry should prevent groundwater entering the excavation though some pore water from the excavated material may mix with the slurry (see Section 16.6.1). In oil well drilling it is sometimes necessary to drill through solid salt layers, and this does cause substantial contamination. However, such a situation would be unusual in construction work. In general, the effect of salt contamination in civil engineering work can be limited by careful selection of the initial slurry concentration. Occasionally it may be necessary to use a dispersing agent. Formerly ferrochrome lignosulphonate was much used as a dispersant but now a wide range of special chemicals is available including lignosulphonates without the environmentally unacceptable heavy metal chromium. However, if severe salt contamination is expected, polymer-based slurries rather than clay-based slurries should be considered, though many polymer systems are not immune to salt water and the supplier's advice must be followed.

*16.3.6.2 Influence of water chemistry on cement-based grouts*

For cement grouts the presence of moderate quantities of dissolved salts in the mix water usually can be tolerated without detriment to the material, though in extreme situations [15] the results may be quite unusual (of course if pre-hydrated bentonite is to be used the guidance on water quality set out at Section 16.3.4.1 should be considered). In many offshore structural grouting operations sea water is used to prepare the grout without detriment to the set or hardened properties, and indeed the use of fresh water for such grouts could be seriously questioned as, in use, the grout will be immersed in seawater.

The effects of salts present in the mix water may have limited effect if any induced strains etc. can be accommodated by the material while in the plastic state. However, the effect of salts present in the groundwater on hardened materials always must be considered. For cement grouts any material that damages concrete (for example, sulphate) must be considered as potentially damaging. For dense low water–cement ratio structural grouts the low permeability may limit the rate of attack. Despite this the effects may be more severe than for concrete as the grout may be more brittle; that is, have a lower strain at failure. It is very difficult to predict how structural grouts will react to chemical attack and every situation must be examined on its own facts. Clearly as structural grouts often will be required to provide corrosion protection for steel, their application in chemically contaminated environments must be very carefully considered.

In geotechnical grouting, salts in the ground that are known to affect concrete must also be expected to affect cement-based grouts. However, many grouts, particularly those containing clay, will have a higher strain at failure than concrete and thus may not disintegrate when subjected to expansive attack, for example, by sulphate provided that there is a confining pressure comparable to the strength of the material. Thus clay–cement grouts and slurries may be softened by sulphate attack but remain effectively intact provided they are subject to sufficient confining stress as will be usual in the ground. If they are not restrained severe cracking may occur. While clay–cement grouts and slurries may be much more permeable than concrete (because of their higher water–cement ratio), the actual rate of attack still may be very small [16, 17].

*16.3.6.3 Chemical grouts*

Sodium silicate may be set by the addition of electrolyte (salt) solutions; for example, calcium chloride. Thus some caution must be exercised when preparing silicate grouts (and indeed any chemical grout) with waters containing significant levels of salts. There may be early precipitation of some gel product prior to the normal setting of the grout. This precipitate may adhere to the walls of the grout lines and tend to block them. Also it will clog the soil surface at the point of injection and substantially reduce or prevent injection.

### 16.3.7 Polymer slurries

Slurries prepared with polymers derived from natural materials such as xanthan and guar gums, and cellulosics (carboxymethyl and hydroxyethyl celluloses) and synthetic partially hydrolysed polyacrylamides have been used quite extensively for tunnelling and to a more limited extent in diaphragm walling and piling. However, the use of these 'simple' polymers is now being superseded by the introduction of bespoke systems containing a mixture of polymers with 'designer' properties. Polymer slurries can be designed to:

- inhibit clay dispersion and thus prevent the build-up of fines in an excavation slurry
- be used at slurry density close to that of water, with only very limited soil solids in suspension
- be more tolerant of salt and cement contamination than bentonite
- 'encapsulate' excavated materials so that, for example, sands excavated with a piling auger come out of the hole with minimal slurry contamination and are effectively at/near their natural in-situ moisture content. As a result the excavated material need not be disposed as wet spoil (see discussion at Section 16.1 regarding the re-use of excavated soil)
- be broken down to water viscosity by simple chemical treatment (for example, with an oxidising agent such as bleach). Soil solids then may settle rapidly so that the supernatant water can be disposed to foul sewer – with the permission of the sewerage undertaker. This can much reduce slurry disposal costs. For bentonite slurries settlement is not a practicable method for soil/slurry separation as the gel will prevent the settlement of fines
- for vertical excavations such as piles, be treated, in hole, at intervals to promote rapid settlement of any soil that has dispersed in the slurry. In this way the excavation hole itself can act as a 'settling' tank for soil slurry separation.

Salt tolerance can be useful in saline soils particularly if slurries are to be repeatedly re-used, as will be the normal practice with polymer slurries. Cement tolerance can also be very useful as cement contamination can occur during concreting of diaphragm walls or grouting behind the lining in slurry tunnelling work. For bentonite slurries a small amount of cement may lead to some thickening/damage to the whole volume of the slurry. Some polymers are effectively inert to cement, but for others such as carboxymethyl cellulose the slurry properties are effectively destroyed by it. However, the destruction is stoichiometric so that, for example, 1 kg of cement might destroy a few hundred grams of carboxymethyl cellulose and in a slurry circuit containing perhaps hundreds of kilograms of polymer this may have no measurable effect – though the polymer lost from the circuit to cement contamination and sorbed onto the soils must be made-up at regular intervals, otherwise the system will rapidly become a soil water system with no active polymer – a problem that appears to have happened all too frequently.

Many polymers derived from natural materials will degrade rather rapidly due to bacterial attack unless used in conjunction with a biocide. Synthetic, man-made polymers tend to be more resistant to biodegradation.

In a slurry-supported excavation, a filter cake of slurry and soil fines will build on the formation during drilling and if not removed this cake will reduce well performance in service. With a polymer slurry, the cake can be destroyed with an oxidising agent during back flushing so that the soil fines are removed from the hole and do not migrate into the formation. A similar procedure can be used to form drainage trenches: a trench is formed under polymer slurry and then backfilled with a free draining aggregate.

Some clays and polymers can act in synergy so that the viscosity/gel of the clay–polymer mix can be higher than that of the clay or polymer slurry alone. Occasionally this can be useful, but advice should be sought from suppliers before clay/polymer blends are used. In use polymer slurries may pick up clay and fines from the excavation and thus when developing polymer systems it is appropriate to test all polymer slurries with a small amount of added clay, particularly if synergy is expected. The added clay need not be a swelling clay such as bentonite. Clay polymer synergy may be usefully exploited when developing systems to build a plug in the screw conveyor of earth pressure balance (EPB) tunnelling machines. In contrast, for piling and diaphragm walling in clay or silty-clay soils it may be desirable to use a polymer that specifically inhibits soil dispersion, as this will make soil/slurry separation much simpler and prevent the slurry becoming too thick on repeated re-use.

In summary, as is so often the case in civil engineering, there is no simple rule for selecting bentonite or polymer systems. Each system must be judged on its merits. Cost of purchase of the slurry materials may be an apparently important factor but it should be remembered that overall slurry-related costs also will include: hire/purchase of slurry preparation plant and soil separation plant (if required), disposal of used slurry and disposal/potential for re-use of excavated material. The plant footprint, the area needed for the slurry plant, also will be important on congested sites. Technical considerations therefore ought to be fundamental. However, ultimately the choice is often based on raw material cost, though conservatism can be an important influence.

### 16.3.8 Chemical grout materials

A very wide range of chemical grouts is available [18], and it is not possible to provide any general guidelines. However, many chemical grout systems are based on sodium silicate, and discussion of the silicate system can be used to highlight some of the parameters that must be considered when using chemical grouts.

Sodium silicate is a highly alkaline viscous liquid. Conceptually, it may be regarded as a mixture of sodium and silicon oxides in water. Important parameters are the solids concentration and the silicon oxide to sodium oxide ratio (the silica–soda ratio), often referred to as the n value. The n value may be expressed as a weight ratio or as a molecular ratio; that is, molecules of silicon oxide ($SiO_2$) per molecule of sodium oxide ($Na_2O$). It is the silica component of the mix that polymerises to form a solid structure, the role of the soda being to bring the silica into solution. Thus sodium silicates of high n value are to be preferred for grouting as the quantity of the non-hardening sodium salt is kept to a minimum. However, for values above about 2, increasing the n value increases the viscosity of the solution and as a compromise, typically for grouting work n values of 3 to 3.8 (molecular ratio) are used, though slightly lower values may be acceptable. As the sodium hydroxide maintains the silica in solution if it is neutralised, for example by an acid, the silica will precipitate as a solid mass. This is the basis of many silicate grout systems, though silicates may also be hardened by reaction with strong electrolyte solutions such as calcium chloride or, as will be shown, by polymerisation with sodium aluminate.

## 16.4 Mixing

Good mixing is fundamental to the preparation of all particulate grout and slurry systems. At the most basic, poor mixing may lead to undispersed materials, such as lumps of cement. Such lumps will not contribute to the bulk performance of the system and thus will represent wasted material with the potential for blockages during grout injection etc. Thus at the macro level poor mixing may be manifested by large lumps of undispersed material of dimensions millimetres or greater. At the micro level the result will be poor dispersion of cement grains or clay particles etc. At this level, dispersion may be strongly influenced by the chemistry of the suspension as well as the mixing. However, mixing may still be of sufficient influence to affect the colloidal stability of the system and thus, for example, the filtration and bleeding behaviour and the permeability of the set material. Indeed, effective high shear mixing of cement–bentonite slurries may reduce permeabilities by at least an order of magnitude compared with low shear mixing. This needs to be considered when undertaking laboratory trials, as laboratory mixers tend to have much higher energies per unit volume of mix than site mixers and there are other more subtle issues relating to mixer design. As a result it can be difficult to simulate field mixing fully in the laboratory – and in mixer-sensitive systems laboratory mixers may not provide a conservative design – see, for example, [15], a powerful example but for a very special grout prepared with water from the Dead Sea.

With polymer slurries and chemical grouts it is important that the manufacturer's instructions regarding mixing are followed. Excessive shear during mixing may damage a polymer or modify the behaviour of a grout.

When dispersing solids such as clay or cement, it is important that the mixer offers a reasonably high level of shear and that the slurry is circulated through the mixing zone, otherwise the solids may become a sticky and intractable mass adhering to the walls etc. of the mixer. The situation with polymers can be much worse. If they are simply poured into water many polymers tend to lump, and subsequent swelling and hydration may be very slow so that lumps (fish-eyes) persist effectively indefinitely and as a consequence much of the polymer is wasted. It is therefore essential that lump-sensitive polymers are added to stirred water or that a venturi eductor is used. If properly dispersed into the water, many polymers will reach effectively full dispersion within about 1 hour. Almost instantaneous hydration can be achieved if the polymer is available in a water in oil emulsion. The polymer will not hydrate (swell) in the oil and so the emulsion remains of moderate viscosity. On contact with water, the polymer in the fine emulsion droplets hydrates rapidly and without lumping. However, the oil (and the surfactants necessary to stabilise the emulsion and enable hydration in water) may have detrimental properties in-hole. Furthermore it should be noted that the active polymer content of an emulsion may be only about one third of the weight of the emulsion.

Bentonites may achieve sufficient dispersion and hydration to be usable within 1 to 2 hours, though considerable further development of properties will occur after this. Testing of aged slurries suggests that the properties continue to develop on a logarithmic time scale and thus it is not possible to identify

## Box 16.1 Density calculations using simple spreadsheets

### Example 1

Calculate the particle density of an 'as received' bentonite powder containing 11% moisture by dry weight. Assume that the particle specific gravity of the bentonite when oven dry is 2.78.

| | *A* | *B* | *C* | *D* |
|---|---|---|---|---|
| 1 | Material | Mass, kg | Volume, $m^3$ | Density, kg $m^{-3}$ |
| 2 | Bentonite, dry powder | *100* | 0.0360 | *2780* |
| 3 | Water | 11 | 0.0110 | *1000* |
| 4 | Bentonite at 11% moisture content | 111 | 0.0470 | 2363 |

***Procedure***

Prepare a simple spreadsheet as above and enter known data: the densities of the dry bentonite and water (the density of water is 1000 kg $m^{-3}$ and thus the density of the dry bentonite particles is 2780 kg $m^{-3}$). These figures are shown in italics (note row and column headings are included for ease of explanation and otherwise can be ignored). Enter 100 for the mass of dry bentonite (Cell B2, figure in italics, any convenient number will do). As the moisture content of the bentonite powder is 11%, set the weight of water, Cell B3 = 11% × B2 = 11 kg. Enter the formula for the volume of dry bentonite in Cell C2 (volume = B2/D2). Similarly in Cell C3 enter formula for volume of water (B3/D3). In Row 4, sum masses and volumes in Columns B and C. Enter the formula for the density of the mixture (B4/ C4) in Cell D4.

### Example 2

Calculate the density of a slurry containing 50 kg of the above 'as received' bentonite of 11% moisture content per 1000 kg of water. Also calculate the volume of the resulting slurry.

| | *A* | *B* | *C* | *D* |
|---|---|---|---|---|
| 1 | Material | Mass, kg | Volume, $m^3$ | Density, kg $m^{-3}$ |
| 2 | Bentonite powder at 11% moisture | *50* | 0.021 | *2363* |
| 3 | Water | 1000 | 1.000 | *1000* |
| 4 | Bentonite at 11% moisture content | 1050 | 1.021 | 1028 |

***Procedure***

Prepare spreadsheet as above and enter known data (shown in italics). The density of bentonite powder is 2363 kg $m^{-3}$ as obtained from the first spreadsheet. Enter 50 kg for the weight of as-received bentonite and 1000 kg for the weight of water. Then follow the same steps as in Example 1.

From the spreadsheet it can be seen that density of the slurry is 1028 kg $m^{-3}$. The volume of the slurry produced is 1.021 $m^3$.

### Example 3

Calculate the mass and volume of soil (excavation material dispersed in the slurry) of density 2000 kg $m^{-3}$ necessary to raise the density of the above 5% bentonite slurry to 1250 kg $m^{-3}$. Results should be per cubic metre of soil/slurry mix.

| | *A* | *B* | *C* | *D* |
|---|---|---|---|---|
| 1 | Material | Mass, kg | Volume, $m^3$ | Density, kg $m^{-3}$ |
| 2 | 5% bentonite slurry | 794 | 0.772 | *1028* |
| 3 | Dispersed soil | 456 | 0.228 | *2000* |
| 4 | Slurry plus dispersed soil | 1250 | 1.000 | *1250* |

***Procedure***

Insert slurry and soil densities into the spreadsheet. Enter 0.5 (a dummy number) as the volume of slurry in Cell C2. Set Volume of soil, Cell C3 = 1 – C2 (total volume of slurry + soil is to be 1 $m^3$). Set soil and bentonite slurry masses in Column B equal to volumes (column C) × densities (Column D). Set Cell B4 = B2 + B3 and C4 = C2 + C3. Set density Cell D4 = B4/C4. Adjust the volume in Cell C2 until the density shown in Cell D4 = 1250 kg $m^{-3}$ as required. There is a function to do this adjustment automatically in some spreadsheet software.

a time for full hydration. High shear mixing will break up clusters of clay particles and thus lead to more rapid development of properties (and higher gel strengths and viscosities) than low shear mixing, and this effect appears to persist indefinitely after mixing [19]. Thus the general rule is to use high shear mixing and to hydrate for as long as practicable, though after 24 hours the further benefit will be rather limited. Generally after mixing, the clay slurry will be pumped to a storage tank to hydrate. The freshly mixed slurry will have little gel strength and thus there will be a tendency for the clay solids to settle. It is therefore good practice to homogenise the stored slurry by recirculation before it is drawn off for use.

If a bentonite slurry is likely to come into contact with salts or cement (as it will be when preparing a cement–bentonite slurry or grout) then at least 4 hours' hydration generally should be allowed as the contamination may effectively stop further hydration and development of the clay slurry properties.

Sodium bentonites as natural materials vary to some extent in their compatibility with cement, but if pre-hydrated clay is used there will be initial and substantial thickening when cement is added as the clay is flocculated by the cement. After a few minutes of mixing considerable thinning will occur (due to aggregation of the clay), but this thinning will not occur unless mixing is continued. It is therefore important that the mixers should be capable of handling the stiffening without loss of circulation, stalling, blocking, etc.

For structural grouts good mixing is fundamental. Low shear paddle-type mixers will produce very different grouts from high speed, high shear mixers unless extended mixing times are used. High shear mixes will show good cohesion and reduced tendency to segregate/bleed. Low shear mixes may show substantially greater bleeding and if they come into contact with water may show insufficient cohesion to prevent the grout and water intermingling.

An apparent disadvantage of high shear mixing is that the extra energy input heats up the grout and this and the extra dispersion tend to reduce the setting time and hence the pumpable time. However, with appropriate design pumpable time should not be a limiting factor and the faster setting can be an advantage as it allows less time for bleeding. High shear mixes do not necessarily show higher pumping pressures. Indeed, the initial pumping pressure of low shear mixes may be higher than for high shear mixes. Actually pumping pressures seldom appear to be a problem unless high pipe flow velocities are required [12].

High sheared mixes are also more homogeneous and thus there is less risk of blockages caused by lumps of undispersed cement. Such blockages can cause serious problems during duct filling etc. Blockages may also be caused by old set material dislodged from the mixer and pipework. An easily cleaned in-line screen may be necessary in grout lines.

## 16.5 Grout and slurry composition

For polymer slurries and chemical grouts it will be necessary to follow manufacturers'/suppliers' guidelines when preparing the materials. For clay–cement–water systems it is more difficult to find guidelines as a very wide range of compositions may be used depending on the application, the type of mixer and, if bentonite is used, whether it is to be hydrated prior to the addition of the cement. Figure 16.1 shows a triangular diagram indicating the main regions in the cement–water–hydrated bentonite system.

For pre-hydrated bentonite systems it is useful to consider the bentonite concentration in the mix water. Below about 2% by weight, bentonite will have rather little influence. Above about 6% bentonite the slurry is likely to become too thick to use (this range of 2–6% is typical for UK and European bentonites, but the limits will depend on the source and the quality of the bentonite and may

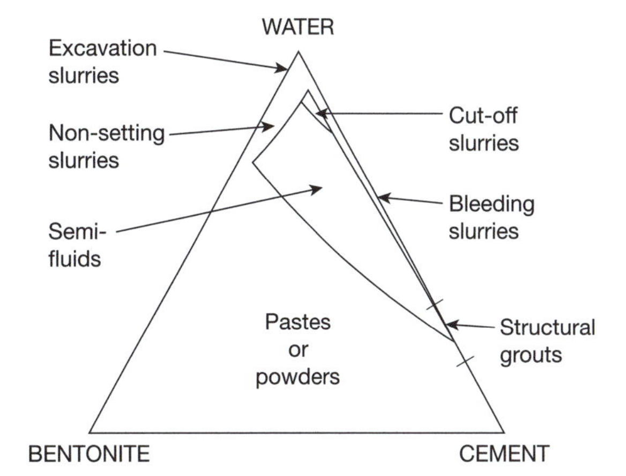


**Figure 16.1** Principal regions in the bentonite–cement–water system for hydrated bentonite

be influenced by polymer pre-treatment of the bentonite by the supplier).

For many grout systems bentonite is used without pre-hydration as this means that hydration tanks are not required (in a system using hydrated bentonite, tanks equal in volume to rather more than one day's working will be required if 24 hours' hydration is to be allowed). Also hydration requires extra time. If the bentonite is not hydrated, much higher concentrations (by weight of mix water) will be necessary to prevent bleeding, though the use of bentonite will allow a higher water–solids ratio than would be possible with cement alone. The effect of the bentonite on the hardened grout will be to reduce strength and possibly add some plasticity. For cement–bentonite grouts using non-hydrated bentonite, the design procedure is to select a cement–bentonite ratio and then add sufficient water to give a pumpable mix. If this leads to a mix that shows excessive bleeding then the bentonite–cement ratio must be altered or the type of mixer or mix procedure changed. Selection of an appropriate cement–bentonite ratio may have to be by trial and error. A very wide range of ratios can produce effective grouts, and either bentonite or cement may be the major component.

For plastic concretes (see Section 16.7), which are effectively a cement–bentonite slurry plus aggregate, the cement and bentonite contents, per cubic metre of mix, will be lower than in a slurry without aggregate (because of the volume occupied by the aggregate). Mix proportions expressed per cubic metre of plastic concrete give little indication of the mix proportions of the underlying cement–bentonite slurry. When designing or analysing plastic concretes it is essential to consider the bentonite–water and cement–water ratios in the light of typical values for cement–bentonite slurries and then to consider the volume ratio of cement–bentonite slurry to aggregate.

For structural grouts water–cement ratio will be the most important parameter. Above a water–cement ratio of about 0.45 bleeding will be severe and will limit the applications of the grout [21]. Below a water–cement ratio of 0.3 the grout is unlikely to be pumpable; indeed it may be necessary to add a superplasticiser for water–cement ratios less than about 0.35. The effects of water–cement ratio on structural grouts are discussed more fully in Section 16.8.

### 16.5.1 Drying

A necessary feature of all grouts and slurries is that they are initially fluids. With time grouts and self-hardening slurries must set. However, the requirement for fluidity means that the materials must contain a substantial amount of water. If this water is lost, shrinkage and cracking can occur and on prolonged drying many geotechnical grouts and self-hardening slurries may be reduced to coarse crumbs or even powder. Typically only a small amount of water is required to maintain integrity of the solid material, and thus when assessing the worst-case effects of drying it is important that tests are continued until all evaporable water is lost. Underground in the UK and in similar climates, below a depth of about 1 metre, there is likely to be sufficient moisture to prevent cracking. However, grouts and self-hardening slurries if exposed to the effects of sun and wind may crack severely. Thus cut-off walls must be capped as soon as the slurry is sufficiently set if cracking is to be avoided. Sensitivity to drying tends to reduce with repeated wetting and drying cycles, perhaps as result of carbonation.

Structural grouts, although of low water–cement ratio, are not immune to cracking. Indeed they tend to show worse drying shrinkage than concretes. Samples of structural grouts left in free air ultimately tend to crack and break up. Such effects are not seen in concrete, as the aggregate limits the propagation of cracks. Thus structural grouts must be maintained in a sealed environment or submerged under water if cracking is to be avoided.

### 16.5.2 Freezing

Bentonite slurries will freeze at effectively the freezing point of water (0°C) – the clay solids having minimal effect on the freezing point. Thus in cold weather slurry plant and slurry lines may need to be protected from freezing. Specifications may limit working during cold weather to ensure that frozen material is not incorporated in the works; for example, in soil–bentonite or soil–cement–bentonite cut-off walls.

Cement–bentonite materials will also be sensitive to freezing and are unlikely to develop sufficient strength to prevent ice formation leading to spalling of the slurry surface. For in-ground material, for example, in a cement–bentonite cut-off wall during construction the affected depth may be quite slight and easily removed during capping operations. However, for exposed materials – for example, the soft piles in hard–soft secant walls – the spalling may be substantial and protection of exposed piles may be necessary – weakened surface material should not be removed during freezing weather as it may provide some slight protection from further frost action.

## 16.6 Excavation slurries

For excavation slurries the role of the slurry is to support the excavation until a permanent structural material is installed. Thus in a diaphragm wall or pile formed under slurry, the fluid must support the excavation until the concrete is poured. In a slurry tunnelling operation the slurry must support the face of the excavation and any unlined tunnel within the slurry chamber of the machine. The fundamental parameters for excavation support and construction expediency are:

(a) slurry pressure
(b) slurry loss control: penetration, rheological blocking, bleeding and filtration
(c) slurry rheology and slurry displacement
(d) excavated soil/slurry separation.

These parameters are closely interlinked but can run counter to each other so that slurry design has to be a compromise.

### 16.6.1 Slurry pressure

In a slurry trench excavation the stabilising pressure on the formation will be the hydrostatic pressure of the slurry less that of the groundwater. Thus the head of slurry over the groundwater and density of the slurry will be important parameters. For slurry tunnelling the pressure in the face generally will be controlled by regulating the slurry supply and discharge pumps, and thus slurry density is not of direct significance for excavation support though it must be carefully controlled to avoid clogging the machine.

For bentonite clay-based slurries the bentonite concentration typically will be in the range 20 to 60 kg of bentonite per cubic metre of water, depending on the quality of the bentonite and the application. The particle specific gravity of oven dry bentonites measured by water displacement generally will be of order 2.78. However, as supplied to site the clay will contain some moisture, typically of order 10% by wet weight of clay which corresponds to 11.1% by weight of dry solids (in soil mechanics the moisture content of soils is normally expressed by weight of dry solids; manufacturers' specification for bentonite (see for example [23]) typically use wet weight). It is difficult to predict precisely the combined density of the clay and moisture as the water may become partially bound up in the clay mineral structure. However, for slurry work, the density can be predicted with sufficient precision from the density of water and the oven dry clay particle density using a simple rule of mixtures (see Example 1 of Box 1).

In Example 1 as shown in Box 16.1, the particle specific gravity of the dry bentonite has been assumed to be 2.78, which appears to be a reasonable average value. It can be seen that the moisture in the bentonite (taken as 11% in the example) significantly reduces the particle density of the bentonite as supplied to site (2363 kg $m^{-3}$).

The calculation procedure to determine the density of a bentonite slurry containing 50 kg of bentonite dispersed in 1000 kg of water is shown in Example 2 of Box 16.1. The density is 1028 kg $m^{-3}$. Had the mix been 50 kg of bentonite per cubic metre of final slurry then the density would have been 1029 kg $m^{-3}$ and the required volume of mix water would have been 979 litres. This can be seen by reducing the value of Cell B3 until Cell C4 is equal to 1.0 – with some spreadsheets this can be done automatically.

Using the same procedure it can be shown that the density of slurries containing 2%–6% bentonite (20–60 kg of bentonite as received per 1000 kg of water) will be in the range 1011–1034 kg $m^{-3}$. It follows that the typical density of 'as mixed' (unused) bentonite slurries is only very slightly greater than that of water.

For polymer-based slurries the typical range of concentrations might be 0.3% to 2% and as the density of the polymer is likely to be similar to that of water, polymer slurries are effectively of water density.

During excavation slurries will pick up soil and it is this material that makes the major contribution to slurry density. It is instructive to investigate the relationship between soil concentration and slurry density. For a slurry of initial density $\rho_s$ mixed with a soil of density $\rho_g$ the density of the mixture $\rho_m$ is given by:

$$\rho_m = \rho_s + C\left(1 - \frac{\rho_s}{\rho_g}\right) \quad (1)$$

where $C$ is the quantity of soil dispersed in the slurry expressed as kilograms of soil per cubic metre of soil-laden slurry. The soil dispersed in the slurry will be that from the excavation and it should be noted that both soil grains and some or all of the water originally present in the pores of the soil will be dispersed into the slurry (in coarse grounds the slurry may have displaced some of the pore water prior to excavation but in clays, silts and fine/medium sands it is likely that all water associated with the dispersed soil grains will enter the slurry). Table 16.1 shows the soil content to produce a range of slurry densities assuming that the soil is saturated and has a density of 2000 kg $m^{-3}$ and that all

**Table 16.1** Effect of dispersed soil on slurry density

| | | | | | | | |
|---|---|---|---|---|---|---|---|
| *Bentonite slurry* | | | | | | | |
| Total weight (kg) | 1028 | 1005 | 952 | 899 | 847 | 794 | 741 |
| Bentonite (kg) | 50 | 47.87 | 45.35 | 42.83 | 40.31 | 37.79 | 35.27 |
| Water (kg) | 1000 | 957.36 | 906.97 | 856.58 | 806.19 | 755.81 | 705.42 |
| Volume ($m^3$) | 1.00 | 0.98 | 0.93 | 0.87 | 0.82 | 0.77 | 0.72 |
| *Dispersed soil* | | | | | | | |
| Total weight (kg) | 0 | 45 | 148 | 251 | 353 | 456 | 559 |
| Solids (kg) | 0 | 36 | 119 | 201 | 284 | 367 | 449 |
| Water (kg) | 0 | 9 | 29 | 49 | 70 | 90 | 110 |
| Volume ($m^3$) | 0.00 | 0.02 | 0.07 | 0.13 | 0.18 | 0.23 | 0.28 |
| *Bentonite slurry plus dispersed soil* | | | | | | | |
| Slurry density (kg $m^{-3}$) | 1028 | 1050 | 1100 | 1150 | 1200 | 1250 | 1300 |
| Bentonite concentration by total weight of water (%) | 5.00% | 4.95% | 4.84% | 4.73% | 4.60% | 4.47% | 4.32% |
| Soil solids concentration by weight of water | 0.00% | 3.72% | 12.67% | 22.21% | 32.41% | 43.34% | 55.07% |

This table considers the effects of the dispersion of a wet soil (soil grains and associated pore water) into a bentonite slurry. The first set of data in the table shows masses and volumes for the bentonite slurry. The second set of data shows masses and volumes for the dispersed soil. The total volume of bentonite slurry plus soil is always 1.00 $m^3$. The final set shows data for the combined slurry and soil. The initial bentonite slurry concentration is 50 kg of bentonite per cubic metre of water. The soil density has been assumed to be 2000 kg $m^{-3}$ – a soil of particle specific gravity 2.65 at a moisture content of 24.5%. The soil additions are such as to increase the slurry density to the figures shown in the slurry density row of the table: 1050, 1100 . . . 1300 kg $m^{-3}$.

the water in the soil is dispersed in the slurry. The reader may wish to confirm the figures given in Table 16.1 using the procedures shown in Box 16.1.

As a slurry picks up soil, its volume will increase in line with the volume of soil dispersed into it. Thus the quantity of bentonite per cubic metre of slurry will decrease as the level of soil contamination increases. Because bentonite is an active swelling clay, its properties will dominate the behaviour of slurries containing small amounts of inactive native clays/fines. However, as can be seen in Table 16.1, as the density rises so the fraction of soil in the slurry increases substantially. For example, from Table 16.1 it can be seen that at a density of 1200 kg $m^{-3}$ the slurry contains 291 kg of soil solids per cubic metre of slurry or 32.4% soil solids by weight of water, which may be compared with 5.0% bentonite by weight of water in the original slurry. Ultimately the slurry will become unacceptably thick: the dispersed soil level at which this occurs will depend on the nature of the soil; clays, if dispersed into the slurry, will produce much greater thickening than sands. However, if shear levels during excavation are low, much of the clay may remain as lumps that can be readily removed from the slurry. Fine, non-cohesive silts may be particularly troublesome as there is little to inhibit their dispersion. The combination of high density and high viscosity can lead to poor displacement of slurry by the tremied concrete in diaphragm wall and pile excavations. The Federation of Piling Specialists, 1977 Specification [22] recommends that the slurry density should not exceed 1300 kg $m^{-3}$ prior to concreting and the ICE Specification for Piling and Embedded Retaining Walls [20] requires a density $<1100$ kg $m^{-3}$ (coupled with controls on the sand content, the potentially settleable material) for diaphragm walls but does not specify a maximum density for piling works. In addition to controls on the slurry prior to concreting, careful attention must be paid to the reinforcement details near the base of a pile or diaphragm wall and also to the tremie technique to ensure full slurry displacement at the base of the excavation.

If densities at or over 1200 kg $m^{-3}$ are necessary for wall stability then high-density additives such as barites (barium sulphate, density 4500 kg $m^{-3}$, though commercial grades may be of slightly lower density) should be used to limit the viscosity rise that would occur if such a density were achieved by dispersion of soil. Even if barites are used it may be extremely difficult to achieve full displacement of a high-density slurry by concrete. Weighted slurries should be used with great care and the necessary quantities of weighting materials can add very substantially to the cost of the slurry.

Table 16.1 also shows the effect of soil contamination on the bentonite concentration. For the calculations, an initial slurry concentration of 5% bentonite by weight of water was used and for the soil-containing slurries the bentonite concentration has been calculated from the total weight of water in the system (ignoring the moisture originally present in the bentonite – i.e. the concentrations are based on bentonite at the 'as received' moisture content). For small increases in slurry density the dilution of the bentonite can be seen to be rather limited. However, the effect becomes more significant at higher levels of contamination. Dilution can become very serious if the slurry is repeatedly used (soil dispersed into the slurry during the excavation and then removed in a soil/slurry separation plant). Whenever the slurry is used some will be lost and the volume must be made up. If fresh slurry is used to make up the volume losses, the effects of dilution may be largely mitigated though it is important to realise that the in use bentonite–water ratio will be slightly lower than the as-mixed ratio. If the lost volume is made up with water in a misguided attempt to offset the thickening caused by contamination, the effects may be very serious. The extra water will add to the dilution and after some uses the slurry may contain effectively no bentonite.

The slurry at the base of an excavation normally will show the highest density due to settlement of solids from higher regions. These soil solids will settle without any associated water and thus the dilution of the bentonite in the slurry will not be as severe as suggested by the density at the base.

In slurry tunnelling operations it may be appropriate to allow a high density for the slurry output from the machine so as to maximise the excavated soil carrying capacity of the slurry pipes from the tunnelling machine to the soil/slurry separation plant.

Densities up to 1250 kg $m^{-3}$ often are quoted as typical for the slurry discharge from a tunnelling machine. However, a slurry of this density must contain over 400 kg of solids per cubic metre of slurry (assuming an average particle specific gravity of 2.65). At such a solids loading, tunnelling machines may struggle and blockages may occur within the machine. The density of the slurry return from the soil/slurry separation plant to the tunnelling machine should be as low as possible. A density of 1050 kg $m^{-3}$ (80 kg $m^{-3}$ solids) often is quoted and should be regarded as a design maximum. It can be difficult to achieve 1050 kg $m^{-3}$ in clay or silt-rich soils. In these soils very careful attention must be paid to the design of the soil/slurry separation plant and especially the selection and sizing of the fines treatment plant.

## 16.6.2 Slurry loss

In any slurry excavation process, some slurry will be lost with the excavated material. Such losses may be expensive, so it is important that trench excavation slurries should drain freely from the grab or auger and that for slurry tunnelling the excavated soil is readily separated from the slurry. Minimisation of slurry loss with the excavated soil has the added advantage of ensuring that the arisings are as dry as possible, so promoting the potential for re-use and minimising costs if the material must be disposed to landfill.

However, slurry loss with the excavated soil does not compromise the stability of the excavation as may occur with slurry losses to the ground, which may raise local pore pressures and soften adjacent soils. In slurry trench or pile excavations losses to the ground can be very serious. In slurry tunnelling it may be possible to control the face pressure to reduce the influence of losses. In cut-off walls modest loss of slurry to the ground may be beneficial as it will effectively thicken the wall in open grounds.

Loss of slurry to the ground may occur by bulk penetration of slurry into large voids and/or filtration of slurry solids against finer soils so that only liquid is lost and the solids are deposited at the interface.

In slurry tunnelling it is very important to prevent slurry loss to the ground to ensure:

(i) that the face is sealed so that the slurry pressure can stabilise it
(ii) that there is sufficient slurry leaving the machine to carry the excavated soil.

In some soils it is not necessary to use bentonite in tunelling slurries. However, in permeable, fissured or unstable grounds it can be necessary to use bentonite, a polymer or a polymer/bentonite combination at concentrations sufficient to develop the gel necessary to seal the face and develop a tight filter cake. Simple calculations based on fissure sealing show that it may not be practicable to seal large open fissures unless the slurry contains some coarse material to promote bridging and blocking in the fissures (openings over perhaps 10 mm will require detailed analysis and consideration should be given to prior grouting). In other ground types the mix of soil types in the face or within a reasonable length of tunnel (this will depend on the volume of slurry in the circuit) may be such that satisfactory slurry performance can be maintained, during machine driving, without adding any bentonite or polymer and the circuit can be operated with just water. However, at any significant stoppages – overnight, weekend, maintenance, etc. – it may be necessary to fill the face with bentonite slurry. The solids in a simple soil–water slurry will not form a low-permeability cake, and such a slurry may permit considerable water losses into the ground leading to softening, ravelling and collapse of the face. Furthermore, the solids may settle from a soil–water slurry, leading to blockages in the machine. In coarse grounds, a polymer such as xanthan gum may be used to help carry the solids in the discharge pipe from the machine.

### *16.6.2.1 Penetration of slurry into soils*

In coarse grounds the slurry will penetrate into the ground until either the voids are blocked by deposited solids or the gel strength of the slurry acting over the surface area within the penetrated void system is sufficient to resist the effective pressure of the slurry (the differential pressure between slurry and pore pressure in the soil). This latter process is known as rheological blocking (see Section 16.6.2.2). For fresh, unused slurries (i.e. containing no suspended soil particles from the excavation) in coarse grounds, blocking by solids deposition may be rather limited. However, once excavation has progressed for some time the slurry will contain soil of a range of sizes especially the finer sizes present in the ground (the coarse fraction may settle to the base of the excavation in slurry trenching or piling or be completely removed in the soil/slurry separation plant of a slurry tunnelling circuit). The presence of this dispersed soil will normally ensure that deposition of solids severely limits slurry penetration [24]. It follows that in normal excavation substantial penetration of the slurry into adjacent ground should occur only in coarse strata within otherwise rather fine grounds or in fissured soils as discussed in Section 16.6.1. In such grounds the slurry may not be able to pick up fines of a size that will block the fissures and it may be necessary to ensure that the slurry has sufficient gel strength to limit penetration or to add polymers that promote clogging and filter cake formation (xanthan gum and carboxymethyl cellulose have been used but today synthetic polymers may be preferred). In certain parts of the world there is a preference for thin slurries, perhaps because of fears concerning displacement of slurry from reinforcing steel (though this should not be a problem if there is satisfactory control of the slurry; the choice of bar type may be more important). Great care must be taken when designing thin slurries for use in soils sensitive to softening etc., as slurry penetration and filtration losses can be severe.

### *16.6.2.2 Rheological blocking*

In coarse grounds or if high pressures are used, for example in slurry tunnelling with an entirely slurry-supported face (that is, if there is no mechanical support from the machine) or in deep slurry-supported excavations with a low external groundwater level, significant slurry penetration can occur. For a slurry of yield value $\tau$ (see Section 16.6.3) penetrating a soil of effective particle diameter $D$ and porosity $n$ under a pressure $P$, the distance penetrated $L$ is given by [25, 26]:

$$L = f\,D\left(\frac{P}{\tau}\right)\left(\frac{n}{1-n}\right) \qquad (2)$$

Where $f$ is a factor to take account of the geometry and tortuosity of the flow path ($f$ may be of order 0.3). Thus for a slurry of yield value 10 N $m^{-2}$ penetrating a soil of porosity 40% with an effective particle size of 10 mm, the slurry may penetrate of order 2 m per metre of (water) head difference between slurry and groundwater.

A problem with application of the formula is the selection of a value for $D$. A reasonable value could be the $D_{10}$ of the soil (the particle diameter for which 10% of the soil is finer), though for grout penetration $D_5$ may be preferred as more conservative. Penetration into the soil is not the situation addressed in [25, 26], where the blow-out pressure to force a gelled slurry from the pores of a soil–bentonite mix was examined. For this, the gel strength (see Section 16.6.3) is a more appropriate parameter than the yield value and a different value must be used for $f$. For a conservative assessment of the blow-out pressure the $D_{15}$ size also might be considered. Overall when assessing penetration distance or blow-out pressure, the influence of sizes in the range $D_5$ to $D_{15}$ should be considered, or other sizes if the soil is gap graded.

Equation 2 represents a substantial simplification of a complex flow problem, but for clean coarse soils it can give reasonable estimates. In particular, it shows that for coarse soils penetration distances can be substantial.

In fissured rock or clay (a situation that may be significant in slurry tunnelling), the penetration distance is given by:

$$L = f\,d\left(\frac{P}{2\tau}\right) \qquad (3)$$

where $d$ is the fissure opening and $\tau$ is the yield value (for penetration) or gel strength (for blow-out). Equation 3 is perhaps most useful when assessing whether a fissure of a particular size can be sealed by slurry alone (without additional blocking aids). For example, for a fissure of opening 5 mm subject to a slurry pressure of 50 kN m$^{-2}$ (a 5 m water head) and assuming $f = 1$, the gel strength necessary to limit the slurry penetration to 10 m is 12.5 N m$^{-2}$, which should be practicable with a thick slurry but sealing an open 50 mm fissure with bentonite alone could be impracticable.

Excavation slurries may carry cut material with a range of sizes in suspension and this can help to reduce slurry penetration or increase blow-out pressure.

*16.6.2.3 Filtration*

In granular or fissured soils, once the slurry has ceased to penetrate due to rheological blocking or a combination of mechanical and rheological blocking, filtration will occur. If rheological blocking has occurred the gelled slurry will behave as a region of low-permeability soil (perhaps of order $10^{-9}$ to $10^{-8}$ m s$^{-1}$) and thus filtration may be very slow, particularly if the penetration has been extensive. Although compressed bentonite is of very low permeability, the relatively loose filter cakes deposited under slurry excavation conditions may have a permeability of order $10^{-10}$ m s$^{-1}$. While this a relatively low permeability, the cakes may be only a few millimetres thick and thus may not represent a particularly high resistance to flow (1 mm cake of permeability $10^{-10}$ m s$^{-1}$ will have the same resistance to flow as 1 m of soil of permeability $10^{-7}$ m s$^{-1}$; both have a permittivity (permeability/cake thickness) of $10^{-7}$ s$^{-1}$). The cake resistance will increase with the volume of water that has been lost through it as the cake gets thicker and thus filter losses will decrease with time provided the filter cake permittivity is much less than that of the ground – and this may not always be the case. If there is no flow there is no cake. Thus in low-permeability soils such as clays a minimal thickness of cake will develop as the rate of loss of water from the excavation will be very small. The cake will have little effect in controlling filter loss. In open gravels slurry loss may be controlled by rheological blocking (though often only after significant penetration) and because of the relatively low permeability of gelled slurry in the soil voids again rather little cake may develop at the trench face. In sands a significant cake may develop so that almost the full hydrostatic pressure of the slurry is dropped across the cake and there is little rise in the pore pressure in the sand. In silts, their rather low permeability will mean that any cake will develop slowly and furthermore even small rates of water loss may give rise to substantial increases in pore pressure. It follows that in silty soils filter loss control is particularly important and poor filter loss behaviour can lead to instabilities. Indeed, filter loss control alone may not be sufficient, particularly if the silt is cohesionless. The swabbing action of the grab during excavation may lead to liquefaction, particularly if the external groundwater level is high. Dewatering to lower the groundwater level to perhaps 3–5 m below the slurry level may be necessary to ensure stability in cohesionless silty soils [27]. In other soils 1.5–2 m may be found sufficient.

### 16.6.3 Slurry rheology

For simple (Newtonian) liquids such as water there is a linear relationship between shear stress and shear rate and the fluid behaviour can be described by a single parameter, the viscosity. Slurries show more complex shear stress–shear rate behaviour. Many particulate slurries (e.g. bentonite but not polymer slurries) behave as solids at low shear stresses and flow only when the shear stress exceeds a threshold value which will be dependent on the shear history of the sample (as will the shear stress–shear rate relationship itself).

The simplest rheological model that may be used to describe clay slurries is the Bingham model:

$$\tau = \tau_0 + \eta_p \dot{\gamma} \qquad (4)$$

where $\tau$ is the shear stress, $\gamma$ the shear rate, $\eta_p$ the plastic viscosity and $\tau_0$ the yield value. In practice the model is a substantial over-simplification. The shear stress–shear rate plot is not a straight line as implied by Equation 4 but a curve as shown in Figure 16.2. At any shear rate the tangent to the curve gives the plastic viscosity and the extrapolation of this tangent to the shear stress axis gives the yield value for the shear rate. The slope of the line joining any point on the curve to the origin gives the apparent viscosity – the equivalent Newtonian viscosity. It should be noted that all these parameters are shear rate dependent.

When at rest a slurry may behave as a solid and the shear stress to break this solid is known as the gel strength. However, there is a further complication: the rheology of clay slurries is time-dependent and shear history dependent. Thus a slurry investigated after a period of high shear agitation will be found to be thinner (lower values of plastic viscosity, yield point and gel strength) than a slurry that has been left quiescent. This effect is known as

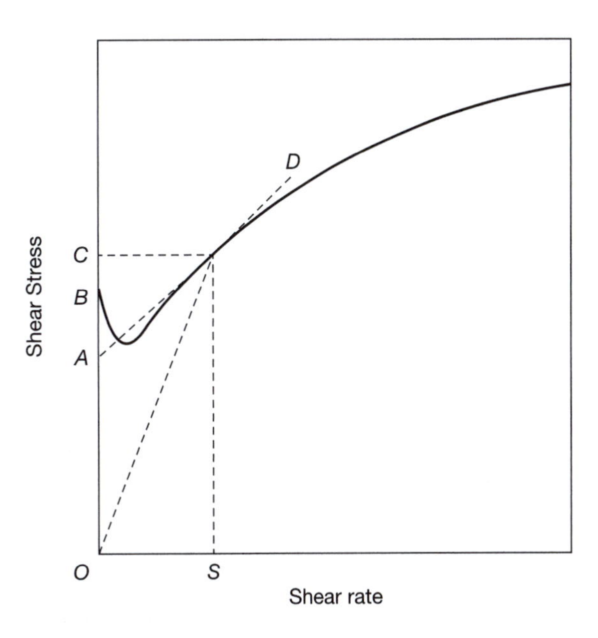


**Figure 16.2** Shear stress against shear rate of slurry: *OB* is the gel strength of the slurry at rest; *AD* is the tangent of the curve at shear rate *S*. At shear rate *S*: plastic viscosity = *CA/OS*; apparent viscosity = *OC/OS;* yield value = *OA*

thixotropy and will affect the way in which solids are held in suspension, and for rheological blocking the pressure to displace a gelled slurry will be greater than that which originally caused the penetration.

As a slurry can behave like a solid, small soil particles can be held in suspension. If the particles are stationary then the shear rate will be zero and thus the gel strength is the important parameter. In theory, a slurry of gel strength τ (see Section 16.3.1 for the determination of gel strength) will hold a particle of diameter *D* in suspension where:

$$D = \frac{6\tau}{g\,(\rho_g - \rho_s)} \qquad (5)$$

where $\rho_g$ is the density of the particles, $\rho_s$ is the density of the slurry and g is the acceleration of gravity. However, in a thixotropic suspension such as a bentonite slurry, τ is not a constant. During excavation the slurry will be continually disturbed and thus τ will be low so that only relatively small particles are held in suspension. Once excavation ceases τ will tend to increase so that rather larger particles can be held in suspension. It follows that some of the particles that were settling during excavation but that had not reached the base of the excavation will be held in the slurry once excavation ceases. Relatively large particles may move sufficiently fast to maintain a locally low value for τ and so will continue to settle. Overall the effect of the gel strength and thixotropy will be that after excavation is complete (and prior to concreting) relatively little entrained soil may settle from the slurry. This is advantageous as it tends to keep the base of the excavation clean. Despite this it is important to clean the base of an excavation prior to concreting, especially if there has been a significant delay after the completion of excavation.

Polymer slurries show rather different behaviour in that few show a true yield value; most are pseudoplastic, for which the simplest rheological model is the power law model:

$$\tau = k\dot{\gamma}^n \qquad (6)$$

where *k* is a parameter describing the consistency of the fluid (a parameter used in place of viscosity when $n \approx 1$) and the exponent *n* is a function of the type of polymer and the shear rate though it may be sensibly constant over a range of shear rates. For pseudoplastic materials *n* will be less than 1. Materials with *n* greater than 1 show dilatant or shear thickening behaviour. Excavation polymers will be pseudoplastic.

The general shape of the pseudoplastic shear stress – shear rate curve is shown in Figure 16.3. Extrapolation to zero shear rate from any point on the curve will give an apparent yield stress. However, the curve actually passes through the origin. Such materials therefore do not have a true yield value but show increasing viscosity with decreasing shear rate. Thus entrained soil will eventually settle from such slurries but the settlement time may be very extended for fine particles and yet rapid for coarse particles (pseudoplastic polymers are much used in domestic products to limit settling of solids or the breaking of emulsions such as vinaigrettes without the need to use very viscous solutions).

Viscosity and gel strength (rather than yield value) are perhaps the most obvious and easily measured properties of slurries and therefore tend to be used for control purposes. However, there is little consensus as to what are acceptable values or even required values. The Federation of Piling Specialists offers some limited guidelines [22] but these are very broad, reflecting the wide range of possible slurries. Furthermore bentonites from different parts of the world may show quite different gel strength and viscosity values (see Section 16.3.1).

Thus when specifying rheological parameters for slurries it is important to consider the source of the bentonite and local practice. In some parts of the world there is a preference for relatively thick slurries whereas in others the local practice is to use thin slurries. As a result severe problems may occur if specifications are lifted from one country to another without regard to the properties of the local bentonite, or if the source of bentonite is changed without considering the necessary changes to the specification. Particular care must be taken with specifications that call for either very thick or very thin slurries.

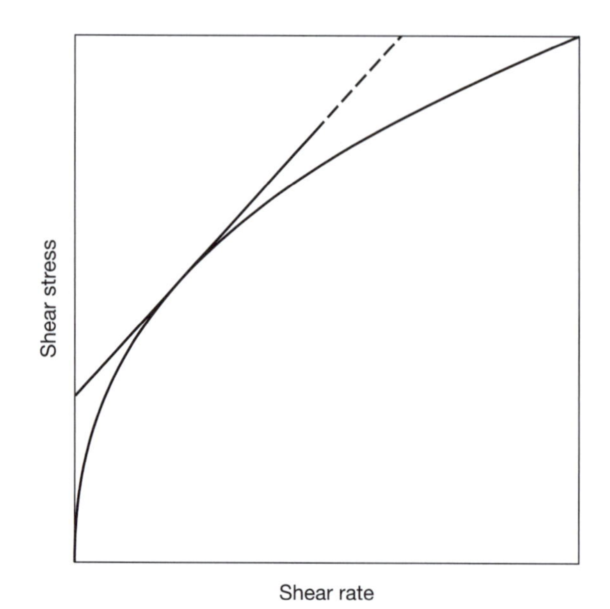


**Figure 16.3** Shear stress–shear rate plot for a pseudoplastic fluid

### 16.6.4 Soil/slurry separation plant

Bentonite powder for slurry preparation was a relatively low-cost material (costs are rising) and for many years slurries were removed from site after a single use or very limited re-use. However, it has become very difficult and expensive to dispose of liquid wastes. Indeed in the UK, where the landfilling of liquid waste is banned, it is very often more expensive to dispose of the bentonite in a liquid slurry than to buy it in the first place. This, and a better understanding of slurry management, has encouraged much greater re-use of slurries. As a result some form of soil/slurry separation plant (formerly known as a slurry cleaning plant – a term now to be avoided as it suggests that the material removed may be 'dirty' and a waste) is now to be found on most sites using bentonite slurries. Recent developments in polymer slurries that inhibit soil dispersion and allow in-hole soil/slurry separation have reduced the necessary plant to very simple settlement tanks (see Section 16.3.7). Soil/slurry separation is of course fundamental for slurry tunnelling, otherwise the volumes of liquid waste would be enormous. Separation plant that may be used includes:

- coarse screens
- fine screens
- hydrocyclones
- dewatering screens
- centrifuges
- filter belt presses
- plate and frame filter presses.

Coarse and fine vibrating screens are widely used and are essential to remove material that might overload, damage or block finer treatment equipment. Screens down to about 2 mm opening are straightforward, but for finer mesh sizes the screen area may have to be rather large if a reasonable throughput is to be achieved.

Hydrocyclones are effectively centrifuges in which the centrifugal effect is generated by the flow of the fluid itself. The hydrocyclone consists of an upper cylindrical section connected to a lower tapered section which terminates in a discharge spigot. Often the spigot is designed to be removable from the tapered section so that it can be replaced when worn. Also for a particular cyclone size, spigots of a range of diameters may be available so that solids removal performance can be optimised. The feed slurry is fed tangentially into the upper cylindrical section so that the fluid spins within the section, causing the coarse solids to migrate towards the wall. The solids then flow into the tapered section and thence to the discharge (underflow) orifice. The separated slurry is forced towards the centre of the cylindrical section and is drawn off from a central (overflow) orifice at the top of the hydrocyclone.

Hydrocyclones are standard treatment plant for slurries. The classification or cut of a hydrocyclone improves as the cyclone diameter is reduced. Thus if fine particles are to be removed, a small-diameter cyclone must be used. This in turn means that the throughput per cyclone will be small and thus a number of cyclones may be required. Also small size means an increased risk of blockage by oversize material. On large sites two sizes of hydrocyclone will be used. Typically there may be a single large cyclone (perhaps 600 mm diameter) and a bank of 100 mm small cyclones per soil/slurry separation unit. Typically particles over about 50 μm will be preferentially removed in a hydrocyclone system, though it is possible to achieve a cut below 20 μm and, if many small cyclones are used, perhaps better than 10 μm.

A problem with hydrocyclones is that the underflow, the stream concentrated in soil from the excavation, is still a fluid and thus re-use is seldom possible and disposal is expensive (in the UK liquid wastes may not be landfilled and must be treated before disposal). Some further treatment of the underflow generally is necessary. A typical method of treatment is with a dewatering screen.

The dewatering screen is similar to a standard fine screen save that the screen may have wedge slots rather than square meshes and the opening may be smaller (perhaps of order 0.25 to 1.0 mm). The wedge shaped slots widen downwards to minimise potential for blocking. The underflow from the hydrocyclone(s) is arranged to fall onto the screen which is vibrated with an eccentric motion so as to convey the solids slowly along the screen to the discharge. The passage of the solids may be slowed by a slight upward slope on the screen or dams across the screen. In operation there should be a bed of soil solids along the length of the screen so that fresh slurry falls onto the bed rather than onto the screen. The deposited bed acts as the filter and so removes material significantly finer than the screen opening. The underflow from the screen is the separated slurry and it is usually arranged that it falls directly into a holding tank. In a hydrocyclone–dewatering screen combination solids are removed only at the dewatering screen and thus it is performance of the bed on this screen that controls the particle size that can be removed. The role of the hydrocyclone becomes that of a concentrator so as to limit the required screen area and to produce a slurry of sufficient solids content that a bed will form (without concentration the high liquid flow rate could flood the screen and/or wash all fines through it). A dewatering screen will not work unless there are sufficient solids, in the combined underflows of all the hydrocyclones discharging on to it, to build a bed. It is therefore important that all the hydrocyclones are working efficiently to produce high solids underflows. Just one hydrocyclone, if worn or poorly adjusted and producing excessively wet underflow, may wash out the bed and destroy the dewatering action.

An indication of the performance of a hydrocyclone can be obtained by looking at the shape of the underflow. If it is in the shape of a wide umbrella it will be too wet. If it is as a thin rope the hydrocyclone may be operating inefficiently as there will be no air core in it. The optimum may be a slight umbrella, and this can be checked by measuring the density of the underflow. A simple check with a cup and scales will suffice, or simply observation of the bed developing on the dewatering. As bed formation is fundamental to the effective function of a dewatering screen, repeated recycling of a slurry through the soil/slurry separation circuit will remove little further solids once a bed has ceased to build. The overflow solids discharging from the dewatering screen are often dry enough to be removed by open truck. The advantage of the dewatering screen is that a fine screening effect can be achieved without the need for a screen of fine opening, which would be rather fragile, though fine screen systems (with openings down to 35 μm) are sometimes used.

On some sites the slurry may become so loaded with clay or silt that it becomes excessively thick and ultimately unsuitable for use in the excavation process. In general, it will not be possible to remove sufficient clay or fine or medium silt with a hydrocyclone and dewatering screen alone (though as noted above the screen may remove a little of this material so long as there is sufficient coarse material to build a bed). A more effective method for the removal of such fines is to use a centrifuge, filter belt press or plate and frame press. These are high-cost items and although often used on slurry tunnelling sites they are seldom found on slurry trench sites unless very large volumes of slurry are to be handled. Centrifuges and filter presses effectively separate the slurry into nearly clear water and a cake of solids. The solids from a filter press (particularly a plate and frame press) may be sufficiently dry for direct removal from site by open truck. However, the cake from a centrifuge may be wetter and more of a sludge than a soil. The grading of the soil in the ground, and its variability during the works, will be a key parameter for slurry plant design and the re-use or disposal of the process arisings. All the plant items listed above will have limiting throughputs and must be appropriately sized. However, a point that can be overlooked is that the plant items will have limited capacities in terms of both liquid flow rate and solids handling. Thus a centrifuge might have a liquid handling capacity of 30 $m^3 h^{-1}$ and a dry solids handling capacity of 12 $t h^{-1}$. From these figures, for this particular centrifuge the liquid and solids handling capacities would be equal only for a slurry of density about 1250 $kg m^{-3}$. For lower densities, as are likely to be the norm downstream of a dewatering screen, the liquid handling capacity would be the controlling parameter for this centrifuge. For example, for a slurry of density 1125 $kg m^{-3}$ the solids handling capacity would be just 6 $t h^{-1}$ at the maximum liquid flow rate of 30 $m^3 h^{-1}$. It follows that it is very important that both the liquid and the solid handling capacities of all plant items are considered when designing cleaning systems.

In principle the water recovered from a filter system or centrifuge may be returned to the slurry circuit. However, it should be used with caution as it may contain a variety of dissolved ions plus coagulants and flocculants used to improve the filtration process. It is therefore unlikely to be suitable for the preparation of fresh bentonite slurry as the chemicals may inhibit the bentonite hydration. However, it is likely to be suitable for return to the slurry circuit as a diluent or for mixing coagulants or flocculants.

If, on site, there is no effective plant for the removal of clay the only option is to dispose of part of the soil-laden slurry and replace it with fresh slurry or water. Disposal usually will have to be as liquid waste, though occasionally and not unreasonably it may be acceptable as an agricultural soil amendment. If the slurry volume is repeatedly made up with water, any bentonite or polymer in the slurry may be substantially diluted (see Section 16.6.1). This can be dangerous for slurry trench excavations, but for slurry tunnelling in grounds that give clay it may be possible to operate with a slurry of the native clay with no added bentonite or polymers during drives, although the face may have to be filled with a bentonite or polymer slurry at any significant stoppages.

## 16.7 Cement–bentonite self-hardening slurries

Cement–bentonite slurries may be used to form cut-off walls by the slurry trench technique [28, 29]. For walls up to about 20 metres deep, excavation can be by backhoe. For deeper walls it will be necessary to excavate in panels using a cable supported or kelly mounted grab as used for structural diaphragm walls or a hydromill. For most cut-off walls, the trench will be excavated under a cement–bentonite slurry which is simply left in the trench to set and harden. However, in some circumstances, for example for very deep walls where extended excavation times will be required, the trench may be excavated under a bentonite slurry which is subsequently displaced by a cement–bentonite slurry or, if a more rigid wall is required, a cement–bentonite–aggregate mix, that is a plastic concrete. For cement–bentonite slurries the typical range of mix proportions will be:

- *bentonite* – 20–60 kg
- *cementitious material* – 100–350 kg
- *water* – 1000 kg.

For the above mix proportions, the bentonite must be hydrated before cement is added. Typically the cementitious material will be a blend of ground granulated blastfurnace slag (ggbs) and cement. If the ggbs proportion in the cementitious material is greater than about 60%, the permeability reduces by about an order of magnitude compared with that of a purely Portland cement mix [29]. Above 60% slag, the permeability continues to reduce as the proportion of ggbs increases, but only rather slightly. Mixes up to about 80% slag may be used but above this the mix can become rather brittle. A slag–cement–bentonite mix with 75% ggbs might be:

- *bentonite* – 40–55 kg
- *ggbs* – 135 kg
- *cement* – 45 kg
- *water* – 1000 kg.

Note that this mix has a total cementitious content of 180 kg per cubic metre of water. To achieve a material of the same permeability with a purely Portland cement, the cement content might have to be 350 kg $m^{-3}$. Both mixes might have a permeability of $<1 \times 10^{-9}$ m $s^{-1}$, though this will depend very much on the mixer used. Higher shear mixers give lower permeability.

If aggregate is to be added to the mix then the bentonite and cement concentrations generally will be lower than the figures given above. As an aggregate seldom will be dry when used, it is most important that its moisture content is measured and allowed for in the mix design process. 10% moisture in an aggregate may reduce the effective bentonite and cement concentrations by of the order of 30%.

Aggregates for cement–bentonite slurries (plastic concretes) should be designed as for normal concrete, that is to give good workability and avoid segregation. However, there must be further consideration of the behaviour of the aggregate in the slurry, and two design philosophies are possible:

(a) For a dense mix, the ratio of slurry to aggregate is arranged so that the aggregate approaches grain-to-grain contact with the slurry just filling the voids. This will give a dense, strong mix but one that has a rather low strain at failure. Typically the slurry for such a mix will occupy perhaps 40% of the total mix volume, and thus by weight the aggregate is very much the dominant component of the mix.

(b) To develop a more plastic material, the cement–bentonite slurry may be designed to have a sufficiently high gel strength to hold the aggregate in suspension and limit grain-to-grain contact. Such a mix will require careful design of the cement–bentonite slurry and selection of the aggregate grading to avoid the aggregate settling.

Usually it is necessary to place plastic concretes by tremie to avoid segregation. Soil–bentonite and soil–cement–bentonite cut-off materials, which are little used in the UK but have been widely used in the USA, are often placed by filling from one end of the trench by grab or tremie to create a sloping surface and then dozing material in at the top so that it slides down the slope [26, 29].

When placing a plastic concrete it is most important to ensure that the aggregate is filled up to the required top level in the trench. After placement it is likely that some settlement of solids will occur so as to leave a surface layer of thin cement–bentonite and fines. The aggregate level should be checked after placement and prior to set and, if necessary, further plastic concrete added. A simpler procedure may be to accept the settlement and scrape off the thin cement–bentonite and replace it with clay as part of the trench capping process. It must of course be ensured that the settlement is not greater than the proposed depth of capping (capping is essential to prevent the surface region of the cut-off wall from drying out – see Section 16.5.1 – and to ensure that it is subject to some effective stress – see Section 16.13.2).

### 16.7.1 Properties

For a cut-off wall a permeability requirement of less than $10^{-8}$ m $s^{-1}$ used to be typical for water-retaining structures (e.g. as a cut-off beneath an earth dam). However, for the management of contamination – for example, the containment of contaminated land – a permeability of order $10^{-9}$ m $s^{-1}$ will be required [30]. Lower permeability also will give better durability [17]. It is sometimes required that cement–bentonites should be designed to have a strength similar to that of the ground in which they are placed or, as a variant on this, to have similar elastic modulus – though the reasons for this are obscure as many cut-offs will not be subject to any significant deformation. Furthermore, in practice the in-situ soil strength is seldom determined, especially for non-cohesive soils where cut-offs find considerable application; as a result a minimum unconfined compressive strength of a few hundred kN $m^{-2}$ tends to be specified regardless of the soil strength.

In addition to a strength criterion, specifications used to call for a strain of 5% without failure. However, this cannot be achieved except under consolidated drained triaxial conditions at a confining stress of the order of the strength of the material – a confining stress that will be available only at considerable depth in a slurry trench. A fuller discussion of the behaviour of cement–bentonite materials is given in the UK National Specification for cement–bentonite cut-off walls for pollution migration control [30].

## 16.8 Structural grouts

The fundamental requirements for a structural grout are likely to be that it completely fills the void into which it is placed, is of appropriate strength, is durable and provides a suitable chemical environment to prevent the corrosion of any stressing tendons etc. Placement generally will be by pumping, and thus pumpability and the time for which the grout is pumpable become additional parameters. A substantial amount of research work on structural grouts was carried out in the 1980s under the auspices of the Marine Technology Directorate [12, 31, 32].

As noted in Section 16.5, typically structural grouts will have water–cement ratios in the range 0.3 to 0.45, which corresponds to cement contents of 1620 to 1303 kg $m^{-3}$ (assuming a particle density of 3150 kg m$^{-3}$ for the cement). Above about 0.45 severe bleeding may occur (and indeed substantial bleeding can occur even at 0.3 in adverse circumstances). Below 0.3 the grouts probably will be unpumpable and it may be necessary to use a superplasticiser even to achieve a water–cement ratio of 0.35.

When pumping grouts, blockages may be a greater problem than grout rheology. Blockages can occur if there are leaks/weeps in the pipework or ducts to be grouted. Leaks/weeps will lead to loss of water by filtration and hence the build-up of a cake of cement solids. This may lead to local blockage of the line or the cake may be dislodged and cause blockage elsewhere. Sharp bends and changes in pipe section should be avoided, as blockages very often occur at such points. Blockages even may occur as a result of sudden expansions in pipe section, as there is a tendency for a transition region of stiffened grout to form at such points. If any of this stiffened grout is dislodged it may cause mechanical blockage elsewhere in the pipework. The hardened debris of earlier mixes, if dislodged from the mixer, also can cause blockages. An in-line strainer can be useful.

Strength is seldom a problem for duct grouts. The necessary strength generally can be achieved at a water–cement ratio of 0.4 to 0.45 and thus there should be no problems with pumping. Bleeding then becomes the controlling parameter. However, for offshore grouted connections higher strengths can be required (particularly early strengths) and thus strength may become the controlling factor for water–cement ratio. Pumping pressure and pumping time thus become more significant, particularly as pumping distances may be quite considerable. Basic data on pumping pressure are outlined in [12, 33].

The complete filling of ducts, clamps or other voids with grout can be a delicate operation requiring careful attention to procedure and to grout bleed and rheology. In a vertical duct or inclined duct where the grout flow is upwards, the hydrostatic pressure of the grout column may ensure that the full cross-section of the duct is filled. However, in horizontal ducts or ducts where the grout flow must be downwards in parts of the duct (for example, ducts in bridge decks may have a sinusoidal profile), it may be difficult to avoid trapping air and so creating ungrouted voids. Very careful attention must be paid to grout vents and venting. Furthermore, for duct grouting it should be remembered that the tendon bundle is unlikely to be central and its position will vary along the length of the duct. The problem therefore is to grout an annulus of varying eccentricity, which in general will not be horizontal but follows the line of tension in the structure. Ideally it would seem appropriate to grout such ducts from both ends so that air trapped in downward sloping sections can be displaced by flow reversal to vents placed at high points. In practice the extra plant, personnel and time involved will normally preclude this. Grouting thus becomes the province of the expert, and trials may be essential if the layout is complex or unusual. Such trials will be expensive. However, it should be remembered that the total cost of the grouting work is likely to be a fraction of the cost of the structure, but a very limited region of corrosion could put the structure at risk [34].

In some grouting specifications there are requirements for grout reservoirs at the high point(s) of the duct or void to be grouted on the assumption that the grout will flow in from the reservoir and displace any bleed water that collects. In practice this may work for a few moments but thereafter the grout stiffens sufficiently that flow is impossible (unless the pipework is of very large diameter). Almost invariably a water-filled void will be found beneath reservoirs and they will have served no useful purpose (see Section 16.12.6, bleed measurement with measuring flasks). It follows that grouts must be designed to be non-bleeding (in the duct/void geometry) and must initially fill the entire duct or void etc.

It is common practice to pressurise ducts to reduce the volume of trapped air and to improve grout penetration. However, there are conflicts. The pressure will reduce the volume of any air voids and, for example, an air void initially at atmospheric pressure may be reduced to about 1/7 of its initial volume by pressurising to 100 psi (689 kN m$^{-2}$), which might be a typical lock-off pressure for a duct. However, any leakage from the duct after pressurisation and prior to set will allow the trapped air to re-expand. Generally the surrounding concrete will ensure that a duct is reasonably watertight, but there may be some filtration of water into the stressing strands, particularly if they are not die-formed. Water then may drain from the ends of the strand. It is important to keep in mind that all water loss from a duct after the completion of grouting is producing a void or voids somewhere within the duct. Voids without any external loss of water can also occur within a duct as a result of bleed; that is, the settlement of grout solids with the expulsion of water. In horizontal sections of duct a bleed track may form along the top surface of the duct. In inclined ducts, such bleed water will migrate towards high points or in vertical ducts may collect at the top or as intermediate lenses. Particularly severe damage can occur on inclined surfaces. The bleed forms an inclined void and eventually the grout collapses to fill the void. The escaping water may also carry away some of the cement, which deposits as little 'volcanoes' at the free surface. Similar severe damage can occur if bleed lenses develop within a vertical duct and then begin to migrate upwards through the stiffening grout. In all these situations, slumping flow of a partially stiffened grout may lead to tearing damage which is not healed by the continuing hydration of the cement.

It should be recognised that all grouts have a substantial potential for bleed. In a 0.32 water–cement ratio grout, the water occupies rather more than 50% of the total volume. In a typical powder the volume of voids might be 40% of the total volume (and very often substantially less than this). If a 0.32% w/c grout were to bleed so that the water occupied just 40% of the total volume rather than the initial 50% then the bleed would be 17% of the initial volume. For a 0.5 w/c grout the corresponding bleed would be 35%. It is therefore most important that duct and tendon design should be such as to minimise the rate of bleed and hence to ensure that set occurs before the full bleeding capacity of a grout has been exerted [32].

It is shown [12, 20] that bleed can be analysed as a consolidation phenomenon similar to the self-weight consolidation of soils but limited by setting. Drainage path length is therefore fundamental. In a vertical duct the drainage path is the height of the duct (if there are no intermediate lenses) and thus perhaps many metres. Whereas if there is leakage into a tendon bundle the drainage may be radial, with a path length of perhaps 10 to 30 mm. Bleeding in a duct with such a tendon bundle will occur many orders of magnitude more rapidly and thus the amount of bleed may be very much more severe. Hence where sealed tendons may give millimetres of bleed, a duct that contains tendons into which water can leak may give bleed of order 10% or more of the duct height (see Section 16.12.6).

### 16.8.1 Thermal effects

The hydration of cement is highly exothermic and in thick grout sections there may be a very substantial temperature rise – even more severe than large concrete pours as the cement content of a structural grout will be much higher than a concrete. In large grout masses – for example, support bags for offshore pipelines [12] – grout temperatures may exceed 100°C unless cementitious additions are used. Temperatures up to 140°C are not unusual. Once the temperature exceeds 100°C it is possible for the grout to be cracked by steam pressure (this will depend on the strength developed by the grout and the steam pressure and hence the grout temperature). Once the grout has cracked, loss of steam, because of its high latent heat, may become an important mechanism in limiting further temperature rise. Thus if a grout mass shows a rapid increase in temperature to over 100°C and then a plateau, the presence of steam-induced cracking must be suspected. The temperature rise may also cause substantial temperature differentials and thus cracking on cooling. Partial replacement of cement by pfa may reduce exotherms. Slag may also be used, but caution must be exercised. At high ambient temperatures or in large masses the hydration exotherm of slag mixes can be more severe than for purely Portland cement. With calcium aluminate (high alumina) cement grouts, severe exotherms will occur unless the section thickness is very limited or the grout is cooled [35]. Calcium aluminate cement grouts can be very useful in the relatively low temperatures of the sub-sea offshore environment but may be difficult to control at higher ambient temperatures. High temperature rises may also need to be controlled when cement-based systems are used for toxic waste encapsulation. In the relatively small sections of grout in prestressing ducts, temperature effects are likely to be limited.

## 16.9 Geotechnical grouts

There is an enormous range of geotechnical grouts [18, 36, 37] and it is not possible to provide more than a brief introduction in a text of this nature. It is therefore proposed to give a very brief overview of some grout types and to indicate how they differ from slurries and structural grouts. Broadly grouts may be divided into three categories: suspensions, emulsions and true solutions.

### 16.9.1 Suspension grouts

Typically suspensions will be mixtures of cement and water or of cement, clay and water. Thus they can be similar to cement–bentonite slurries. However, mix proportions, mix procedure and clay type all can be different. Suspension grouts may be used to reduce permeability and strengthen coarse soils. Typically suspensions grouts (except microfine cements, see Section 16.3.2) will not penetrate soils much finer than a coarse sand. Information on cement grouts for soil and rock grouting is given in [38] and also generally in the American Society of Civil Engineers grouting conferences [39, 40, 41].

If clay is added, its role is generally to reduce bleed and any tendency to segregation and to increase cohesion of the grout rather than act as a nucleation site for cement hydrates, as appears to be its key role in cement–bentonite cut-off slurries (a role that requires pre-hydration of the bentonite before the cement is added). Increased cohesion can be regarded as either a benefit or a problem depending on the application. Increased cohesion can reduce sensitivity to groundwater, intermingling with groundwater during injection and wash-out by flowing groundwater after injection. However, increased cohesion will also bring increased pumping pressures and potentially reduced grout penetration. For example, Warner [42] states of bentonite: 'While cheap and convenient to use the ratio of reduced bleed to increased grout cohesion is poor compared to other readily available viscosity modifying compounds. Further, the adverse effect on strength and especially durability, should preclude its use in most applications.' The comment on durability is included here for completeness but the author of this text disagrees that bentonite reduces durability. It depends on the system in use. If high water cement ratio is essential then a bentonite system can improve permeability and durability. If low water–cement systems can be employed then durability will be better than high water–cement systems but there will be cost penalties and the grouted ground may behave more as rock than soil.

Clay–cement grouts show much higher filter loss than bentonite slurries and water loss during injection into ground/fissures can hinder penetration. As a result filter loss control agents are regularly included in grouts formulations for challenging projects. The effects of these agents can be assessed with a filter loss test (see Section 16.12.5 and [37]).

Local native clays are sometimes used instead of/in addition to bentonites, though the practice is now rare except on very large or remote projects, perhaps because of solids handling and quality assurance issues. When bentonite is used in geotechnical grouts, very often it is not allowed to hydrate for any significant time before the cement is added (or rather no hydration tanks are provided and the mixing cycle is so short as to allow no significant hydration). Thus the bentonite does not swell to produce a thixotropic gel but behaves rather as a non-swelling clay (swelling sodium bentonite will be converted to the non-swelling calcium form by the calcium ions present in cement). As a result the solids contents of these mixes and hence their strength tends to be higher than for mixes using hydrated bentonite. Thus if bentonite is not hydrated a stronger material can be obtained (though of course at the cost of additional materials – if the bentonite were allowed to hydrate the mix could become too thick if the same solids content were used). A penalty for not hydrating the bentonite is that the bleeding of the resulting grouts is often severe unless low water–solids ratios are used and the grout texture can be rather coarse compared with that of hydrated cement–bentonite slurries.

Mixing (see Section 16.12.4) will be important and high shear is fundamental if a cohesive mix is to be obtained. It may be possible to obtain satisfactory grouts with low shear mixing if long mixing times are used, but on site it is generally impossible to police mixing time and thus high shear mixing is much to be preferred – though with the stipulation that a minimum mix time is observed, perhaps 2 minutes.

### 16.9.2 Emulsion grouts

Materials such as bitumen, which would be too viscous for use as a grout, can be rendered less viscous by dispersion as an emulsion in water (fine droplets of bitumen in a continuous phase of water). The emulsion may be designed to break naturally in the soil as the water is drawn out by capillarity so that the bitumen droplets coalesce to form a continuous system. For other grouts an emulsion breaking agent may be added to promote coalescence. Emulsions are used rather rarely, some further information can be obtained from [43, 44, 45].

## 16.10 Chemical grouts

Chemical grouts are widely used and a substantial range of materials is available [18]. The behaviour of chemical grouts tends to be quite specific to the chemical system involved, and it may take some time to develop an understanding of the special features of a new grout system. Traditionally chemical grouts have been based on sodium silicate (effectively silica dissolved in sodium hydroxide), and two silicate systems will be discussed to highlight the wide range of behaviour that can be obtained even with one basic grout type.

If an acid or strong electrolyte solution is added to sodium silicate, there is a near instantaneous gelation. Such a system is therefore useless for grouting work unless a two-shot process of the Joosten type [46] is used (separate injection of silicate and electrolyte or acid so that the solutions interact and gel in the area to be grouted – an expensive technique which may leave some of the silicate unreacted). The Joosten process would not be used today except in very particular circumstances but the process was an important step in the development of chemical grouting.

### 16.10.1 Silicate ester grouts

To produce a delayed gelation, organic esters that slowly hydrolyse to produce acids to neutralise the sodium hydroxide are used (see Section 16.3.6). Typically esters of polyhydric alcohols are used. The hydrolysis reaction is of the form:

ester + water → alcohol + acid

In the alkaline environment of a sodium silicate solution the equilibrium moves in favour of alcohol + acid as the acid is removed by reaction with the sodium hydroxide (from the sodium silicate) and the alcohol remains (this could leave some toxicity, depending on the particular ester – food grade esters can be produced). The rate of hydrolysis of the ester controls the rate of neutralisation and hence the set time of the grout. The set time may be varied to a limited extent by varying the ester and silicate concentrations. For major variations in set time the type of ester must be varied and manufacturers may supply a range of esters. For optimum gelation a substantial proportion of the alkali in the silicate should be neutralised. However, as the esters tend to be more expensive than sodium silicate there can be a temptation to use as little ester as will give some set. While this may produce a gelled product it is important to realise that this will not make optimum use of the sodium silicate and produce the most durable product. Therefore when designing silicate/ester systems it is important to consider the underlying chemistry. To make a full study of the reaction it is necessary to obtain data on the quantity of acid that is produced per unit weight of the ester (the saponification value) and also the alkali content of the silicate solution.

Sodium silicate will be shipped as a concentrated solution to avoid shipping unnecessary quantities of water. Typically the specific gravity of the concentrated solution might be of order 1300–1400 kg $m^{-3}$. Silicate solution densities may be quoted as degrees on the Twaddell or Baume hydrometers. To convert these degree values to specific gravity, use the following formulae:

Specific gravity = 1 – (degrees Twaddell/200)
Specific gravity = 145/(145 – degrees Baume)

For use on site, the sodium silicate solution may be diluted to perhaps 20–70% of the as-supplied concentration. However, the strength of the gelled product is very sensitive to the silicate concentration and so the appropriate dilution must be carefully selected.

The ester and sodium silicate are immiscible until the ester has hydrolysed. There is therefore no problem of flash reaction, though it is important that the two phases are well homogenised and injected rapidly after mixing to avoid their re-separation (the ester will float on the silicate). When formulating the ester component, a surface active agent (a surfactant) usually is added to promote intermixing of the ester and silicate. On occasion, in poorly formulated grouts, this surface active agent has been omitted and there have been problems of segregation of ester and silicate in the ground during injection and perhaps after injection. If the ester and silicate segregate there is potential for the ester to biodegrade to methane and carbon dioxide. Also as part of this process, any sulphates naturally present in the groundwater may be reduced to sulphide species which are intensely smelly and can be toxic by inhalation. The surfactant itself also can cause problems. On a slurry tunnelling project the circulating slurry suffered severe problems of foaming when the machine passed through regions of silicate–ester grouted ground.

A further and potentially very significant problem that may be seen in chemical grouts is syneresis; that is, a shrinkage of the grout skeleton with the exudation of fluid. Generally syneresis is quite small and may be likened to bleeding, but it is not driven by gravitational settlement of solids but chemically induced shrinkage of the grout structure. With some dilute grout formulations it can be very severe. The shrinkage will be in three dimensions and not just vertically as in bleeding. The syneresis that occurs in soil pores may be less than that seen in large grout samples because of the high surface area to volume ratio of soil pores. In this situation it can be hypothesised that the bonding energy of grout to soil is greater than the energy derived from shrinkage forces within the grout mass, which will be proportional to grout volume. This is considered in more detail in [47].

### 16.10.2 Silicate aluminate grouts

Sodium silicate may also be gelled by reaction with a small amount of sodium aluminate. The reaction is more complex than the ester neutralisation reaction and it is more difficult to identify any stoichiometry (the amount of aluminate required to set a given amount of silicate).

If concentrated sodium silicate and aluminate are mixed there is a very rapid gelation. To avoid this gelation it may be necessary to dilute both the silicate and aluminate solutions from their as-supplied concentrations. The system must be kept stirred as the two components are mixed, otherwise localised excess concentrations may occur so that a gel product rapidly develops as small flakes. If such flakes develop before injection they may clog the soil surface and so reduce or prevent grout injection. As the solutions have to be relatively dilute to avoid gelation on mixing, silicate–aluminate grouts may be weaker than silicate–ester hardened grouts. However, the reaction product can be non-toxic and thus particularly suitable for use near water wells etc. The set time may be varied by altering the aluminate–silicate ratio, but it is also sensitive to the silicate concentration.

Because of the differing nature of the setting reaction for aluminate and ester systems, measurement of set time etc. needs to be slightly different. This is considered in Section 16.12.7. Aluminate gelled grouts do not appear to show the rapid and substantial syneresis shown by some dilute silicate–ester grouts. However, large samples can show some slow shrinkage that continues for a long time [47].

Chemical grouts are relatively expensive and thus injection procedures must be carefully designed. Furthermore their low viscosity means that they may penetrate substantial distances, perhaps much greater than actually required. It is therefore common practice to use chemical grouts only after a first stage of cement or clay–cement grouting. Also chemical grouts will not be injected to rejection (limiting pressure) but rather a predefined, calculated quantity will be injected at each location. In principle the set time of the grout should be as short as possible permitted by the surface plant, injection rate and required penetration, as ungelled grout may tend to sink (it is likely to be denser than the groundwater) or disperse and be diluted by the groundwater, or be displaced if there is a groundwater flow.

## 16.11 Sampling slurries and grouts

Samples for testing the properties of excavation slurries must be taken from the trench during the works. A number of sampling systems are available and the only requirements for site operation are simplicity, reliability and the ability to perform at depth in a trench (the hydrostatic pressure can jam some valve mechanisms). The sampler should be of reasonable volume as significant quantities of slurry are needed for tests (for example, the Marsh cone test requires a 1.5 litre sample). The properties of the slurry will vary with depth in the trench due to the settlement of solids. It is therefore common practice to test samples from different depths in the trench. For diaphragm wall excavations a sample should be taken from near the base of the excavation prior to concreting as this slurry will be the densest and thus the most difficult to displace with concrete. For cut-off walls it is sometimes suggested that samples for testing the hardened properties should be cored from the wall after the slurry has set. However, experience shows that such samples are almost invariably so cracked/remoulded as to be unrepresentative. Samples of the fluid slurry should be taken after the end of excavation and prior to set and poured into cylindrical moulds. Typically samples may be taken from the top 1 m, the middle and the bottom 1 m of the trench. Further advice on sampling slurries from cut-off wall trenches is given in the notes for guidance of [30].

For geotechnical grouts normally it will be possible only to take samples of the pure grout, as injected. Such samples may be used for checking set time and performance of the bulk grout. However, they will not give an indication of the behaviour of the grouted ground. For this samples must be cored from the grouted region, a procedure that may also yield shattered or substantially damaged specimens unless the grout is rather strong. Thus in general samples for testing the properties of injected soils must be prepared in the laboratory by injecting soil samples. This can be a tedious and time-consuming process, especially if a number of grout formulations and soil types need to be tested. It should be noted that for some chemical grouts the strength of grouted soils can be very sensitive to the grain size. Fine sands give much higher strengths than coarser soils.

Samples for testing structural grouts may be taken from the fluid material during grouting or more rarely they may be cored after the grout has set, provided that this does not damage the structure etc. – typically this would be undertaken only if there were concerns about failure of the grouting operation. Samples may be taken for strength, bleed and rheological measurements. Generally, in the UK, the samples for strength testing will be taken in cube moulds, though cylindrical moulds may also be used [57]. With low water–cement ratio grouts the setting exotherm can lead to substantial temperature rises (see Section 16.8.1 and [32]) in the mould and indeed in the field if the grout cross-section is more than perhaps 15 cm.

As previously discussed, if cement–bentonite slurries are prepared with bentonite that has not been hydrated prior to cement addition then high solids contents will be necessary to avoid excessive bleeding. As a result they can be relatively strong, so that cube moulds may be appropriate for strength tests. The permeability of such grouts is seldom measured unless unusual mix designs have been adopted or there is some special reason for concern. If permeability tests are required then samples must be cast into cylindrical moulds. The fluid properties of such grouts normally will be tested with a flow cone immediately at the time of injection.

### 16.11.1 Sample containers, storage and handling

When preparing any grout samples for testing of set properties, care should be taken to avoid trapping air in the samples. During filling, moulds, sample tubes, etc. should be held slightly off vertical and the slurry slowly poured down the side so that air bubbles can escape. At regular intervals pouring should be stopped and the mould tapped to release bubbles. After filling the moulds should be capped and stored upright until set. Thereafter they should be stored under water in a curing tank until required for test (unless immersed immediately after moulding to control thermal cracking – which will be necessary for rich cement grouts and calcium aluminate cement grouts).

It is most important that moulds are watertight. If there is any leakage prior to set it is very possible that only water will be lost so that the solids become more concentrated. Also, leakage may cause cracks to develop in the sample.

For all grouts and slurries care must be taken to ensure that the sample containers are compatible with the grout. For example, aluminium components must not be used with alkaline systems such as sodium converted calcium bentonites, cement, ggbs, sodium silicate and aluminate as they will react to liberate hydrogen gas. Plastic tubes are often satisfactory but may be unsatisfactory for some chemical grouts. If samples are cast into tubes then the tube should be wiped with mould release oil prior to casting to ease de-moulding. When preparing samples for testing in the laboratory it may be necessary to trim a significant amount of material from the ends of the samples and so sample tubes should be about 50% longer than the required length of test specimens. However, they should not be over-long. Longer samples require greater extrusion pressures, with the result that weak grouts in long sample tubes may be crushed during extrusion. Note: some contractors prefer long sample tubes so that two samples can be obtained per tube. Such long tubes should be cut into two equal lengths before extrusion.

Great care should be taken when transporting samples from site to the test laboratory. They must not be dropped or subjected to other impact. Ideally the sample tubes should be packed in wet sand during transit for protection against drying and mechanical damage. Once in the laboratory, samples of weak grouts should not be de-moulded until required for test. Samples of these grouts will be fragile and should be treated as sensitive clays. Before samples are extruded the ends of the sample tubes should be checked for burrs etc. to ensure that the samples can be extruded cleanly. Specimens must not be sub-sampled, for example to produce three 38 mm diameter samples from a single 100 mm sample, as this will lead to unacceptable damage.

Cement–bentonite slurries and most grouts are sensitive to drying and even in a nominally sealed tube some drying may occur. Storage under water is the only sure procedure.

## 16.12 Testing grouts and slurries

As previously explained, much of the early technology for civil engineering slurries was developed from oil well drilling practice. Not surprisingly, test procedures were also taken from the oil industry; details of most of the tests applied to civil engineering slurries can be found in [48].

Density and viscosity are often regarded as the fundamental quality control parameters for both grouts and slurries. However, as explained below, field measurement of both these parameters can present difficulties.

### 16.12.1 Density

For grouts and slurries the most usual instrument for density measurement is the mud balance. This is an instrument rather like a steelyard save that the scale pan is replaced by a cup. The instrument thus consists of a cup rigidly fixed to a scale arm fitted with a sliding counterweight or rider (see Figure 16.4). In use the whole unit is mounted on a fulcrum and the rider adjusted until the instrument is balanced. Specific gravity then can be read from an engraved scale. As the balance was developed for the oil industry, the range of the instrument is wider than is necessary for civil engineering work (typically 0.72 to 2.88 whereas construction grouts and slurries will usually have specific gravities in the range 1.0 to 2.0). The smallest scale division is 0.01 and with care the instrument can be read to 0.005, though the repeatability between readings is seldom better than 0.01. A resolution of ±0.005 may allow the estimation of the solids content of geotechnical grouts to an accuracy of only 10% or worse. For high solids structural grouts the instrument will not allow a resolution of better than 0.01 for water–cement ratio (that is, it would be difficult to distinguish water–cement ratios of 0.32 and 0.33). Thus although the mud balance can be used to identify gross errors in batching, it is not suitable for accurate quality control work.

It should be noted that the instrument may have three scales in addition to the specific gravity scale. These are: pounds per cubic foot, pounds per US gallon (0.833 of an Imperial gallon) and

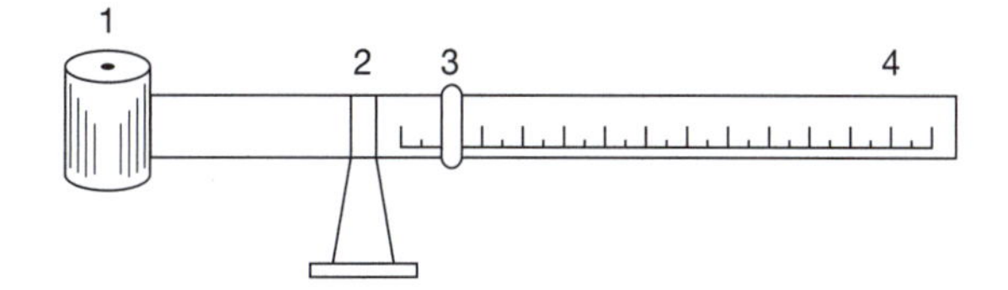


**Figure 16.4** Schematic diagram of the mud balance: (1) mud cup, (2) fulcrum, (3) rider, (4) scale.

pounds per square inch per 1000 foot depth (a one thousand foot column of water exerts a pressure of 433 psi, 2989 kN $m^{-2}$). None of these scales is required for civil engineering work and the specific gravity scale should be used.

To familiarise oneself with the instrument it is good practice to do several tests on a single batch of unused slurry. The results should agree to 0.01. With soil-laden slurries it may be difficult to get good repeatability as the coarser soil solids may tend to settle etc. Such slurries should be well stirred before a sample is taken for testing. The calibration of the instrument should be regularly checked. The procedure is simply to test a sample of clean water. This should show a density in the range 0.995 to 1.005 g $ml^{-1}$. If the reading is outside this range, set the rider to a density of 1.000 and adjust the counterweight until the beam is in balance. (The counterweight, formed of small metal beads, is at the far end of the beam from the cup and is usually a small recess closed by a screw plug.) There is a pressurised version of the mud balance that is useful for thick grouts which may tend to hold air.

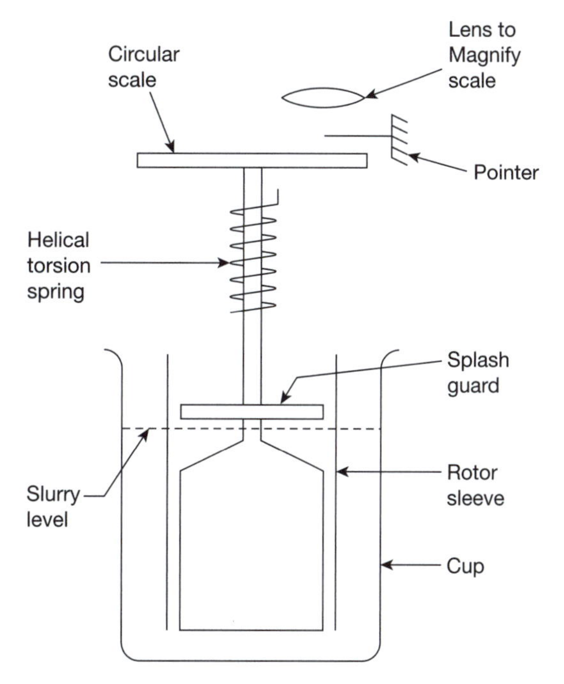


**Figure 16.5** Schematic diagram of Fann viscometer

## 16.12.2 Sand content

During excavation with a slurry the density will increase due to suspension of soil. The density of an excavation slurry provides a measure of the total amount of soil in the slurry but no information as to whether this is sand, silt or clay.

The sand content rig is designed to measure the bulk volume of sand (strictly material coarser than 200 mesh US, 0.075 mm, 75 μm) in a given volume of slurry. The apparatus consists of a tapered graduated tube, a small 200 mesh US sieve and a funnel. To carry out the test, a fixed volume of slurry is washed on the screen and the volume of retained soil is measured as a fraction of the original slurry volume. The result of the test is quoted as the sand content (per cent bulk volume of sand by volume of slurry; to convert this to kg of sand per cubic metre of slurry it is necessary to assume or measure a bulk density for the settled sand).

## 16.12.3 Rheological measurements

The most common instrument used for measuring actual rheological parameters (rather than ranking slurries, for example, by flow time from a funnel) is the Fann viscometer.

*16.12.3.1 The Fann viscometer*

The Fann viscometer (sometimes referred to as a rheometer) is a co-axial cylinders viscometer specially designed for testing oil well drilling fluids. Two general versions of the instrument are available: an electrically driven instrument and a hand cranked instrument. All versions of the instrument can be operated at 600 and 300 rpm and have a handwheel so that the outer cylinder (the sleeve) can be rotated slowly for gel strength measurements. Some have additional speeds of 3, 6, 100 and 200 rpm. For all versions of the instrument there is a central bob connected to a torque measuring system and outer rotating sleeve (see Figure 16.5). The gap between bob and sleeve is only 0.59 mm and thus it will be necessary to screen any soil containing slurries before testing. The full test procedure is given in [48]; see also the notes for guidance of [30]. Fundamentally for both types of instrument four measurements can be made:

- dial reading at a rotational speed of 600 rpm (the 600 rpm reading)
- dial reading at a rotational speed of 300 rpm (the 300 rpm reading)
- 10 second gel strength obtained by slowly rotating the gel strength knob until the gel breaks, after the slurry has been agitated at 600 rpm and then left to rest for 10 seconds
- 10 minute gel strength determined after a rest time of 10 minutes.

*16.12.3.2 Calculation of results*

The Fann viscometer is designed so that:

- apparent viscosity in centipoise (cP) = 600 reading/2
- plastic viscosity in cP = 600 reading – 300 reading
- gel strength, 10 s or 10 min in lb 100 $ft^{-2}$ = dial reading
- yield value in lb 100 $ft^{-2}$ = 2 × 300 reading – 600 reading.

Note: lb 100 $ft^{-2}$ is a unit used in the oil industry; to convert to N $m^{-2}$ multiply the result in lb 100 $ft^{-2}$ by 0.48.

The yield value is the extrapolation, on a shear stress–shear rate plot, of the line passing through the 600 and 300 points to the shear stress axis. Thus for the range of shear rates between 300 and 600 rpm the yield value is the yield stress of equation 4. In contrast gel strength is the stress necessary to cause flow in a slurry that is at rest. For an ideal Bingham fluid the yield value and gel strength would be equal. Clay slurries are not ideal Bingham fluids and they show thixotropy. As a result the yield value will not be equal to the gel strength (either 10 s or 10 min). The yield value is rather seldom used in civil engineering.

*16.12.3.3 Testing grouts with the Fann viscometer*

The Fann viscometer also can be used for measurement of viscosity of chemical grouts. However, the resolution is at best ±1 cP and, as many chemical grouts may have an initial viscosity of just a few cP, it is not well suited to measuring the initial viscosity of these grouts.

The instrument is also somewhat unsatisfactory for cement-based systems. Curiously, many cement–bentonites have lower gel strengths than the corresponding bentonite slurry. However, the slurries may be very viscous and thus gel strength measurements may be strongly influenced by the speed at which the gel strength handwheel is turned (clay slurries tend to show a lower viscosity and higher gel which breaks sharply so that the gel strength is easily identified). For cement-based grouts some semblance of gel strength repeatability can be obtained with the electrically driven instrument by regarding the 3 rpm reading, measured after the appropriate rest time, as the gel strength. Strictly this is merely an apparent viscosity at a low shear rate. A further curious feature of cement systems is that the 10 min gel strength is often slightly lower than the 10 s value despite the fact that continuing hydration of the cement should be increasing the gel structure. This suggests the formation of a water-rich layer at the bob. Overall gel strength measurements are of such poor repeatability for cement–bentonite slurries and cement grouts that there is little point in their measurement, especially the 10 min value, measurement of which much increases the test duration.

For cement-based systems viscosity is rather more repeatable than gel strength though now the effect of hydration is seen and the age of the sample can strongly influence the result. Thus the Fann viscometer is not well suited for quality control of cement grouts. However, it can be useful in identifying gross variations in mix proportions etc., but for this a much cheaper test such as a flow cone time could be equally effective.

For low water–cement ratio structural grouts it is always doubtful whether the grout can fully penetrate the narrow annulus between rotor and bob of the instrument (the gap is only 0.59 mm). Certainly any results obtained should not be regarded as a true indication of the rheology of the grout. Slip and segregation may be dominant.

*16.12.3.4 Flow cones*

The Fann viscometer is an expensive instrument and must be used with care by a trained operator if reliable results are to be obtained. In the laboratory the detailed information that can be obtained from it may be invaluable in the investigation of different mix designs, admixtures, etc. However, there is often a need for a simple test that can be used for compliance testing at mixers, the trench side, etc. and typically a flow cone is used. For slurries the most common cone is the Marsh funnel (see Figure 16.6). However, there is a wide variety of different cones and it is important that the type of cone is specified when reporting results, together with the following information:

- the outlet diameter of the cone
- the volume of grout/slurry to fill the cone to the test level
- the test volume to be discharged (this may not be the full volume of the cone)
- the flow time for water.

Test procedures are similar for all cones and thus the procedure for just the Marsh funnel is detailed.

*16.12.3.5 The Marsh funnel*

The Marsh funnel is the simplest instrument for routine assessment of slurry flow behaviour. The test procedure is simply to pour a freshly stirred sample of the slurry through the screen to fill the funnel to the underside of the screen (a volume of 1.5 l).

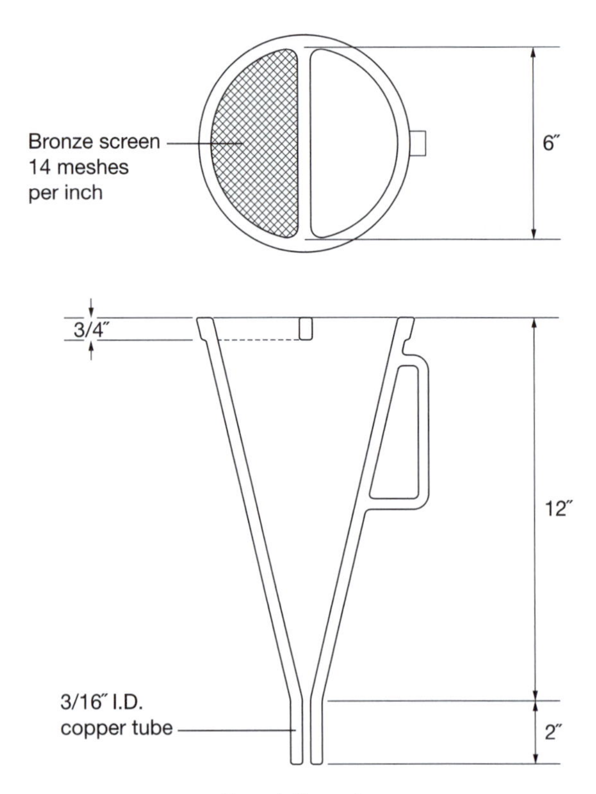


**Figure 16.6** The Marsh funnel

Then measure the time for the discharge of either 946 ml (1 US quart) or 1000 ml of slurry from the funnel. The result is quoted as Marsh funnel seconds; the volume discharged also should be quoted. The Marsh funnel time in seconds cannot be directly converted to a viscosity.

The funnel may be checked by measuring the flow time for water. For clean water at 21°C (70°F) the times should be as follows:

- for 946 ml – 25.5 to 26.5 s
- for 1000 ml – 27.5 to 28.5 s.

No adjustment of the funnel is possible, and if readings outside the above ranges are obtained it must be assumed that the funnel (or the stopwatch) is damaged.

The Marsh funnel is suitable for testing most bentonite and cement–bentonite slurries. However, for cement grouts the flow time may be very extended or the flow may stop before the required volume has been discharged. It is therefore often necessary to use a cone of larger outlet diameter (10 mm or 12.5 mm).

*16.12.3.6 Flow troughs*

The flow trough is an empirical device used to assess the fluidity of cement-based grouts. In the test, a measured quantity of grout is allowed to flow from a wide-necked funnel into a trough and the distance the grout spreads along the trough is recorded. The trough should be wetted and then allowed to stand for some moments prior to use as the reading is somewhat sensitive to the surface condition of the trough. Stiff grouts may show no flow and fluid grouts may flow over the end of the trough. However, for many grouts the instrument can give a useful guide to fluidity. The instrument is robust and very simple to use. Reference [60] gives a specification for a flow trough (the Colcrete flow trough). There are other flow troughs.

*16.12.3.7 Other rheological tests*

A very wide range of rheological test equipment is available. Instruments that have been used for grouts include the Rion hand-held viscometer and the atmospheric consistometer [49]. The Otto Graf viscometer consists of a torpedo-shaped plunger that is allowed to fall under gravity through a tube of grout. Use of the instrument was specified in the 1997 edition of [57] but it is not included in the 2007 edition and it is seldom used in the UK.

## 16.12.4 pH

pH is a measure of the acid or alkaline nature of a material. pH 7 is neutral; below 7 is acid, above 7 is alkaline. pH may be measured with a glass electrode and a matched millivolt meter or with pH papers. With an electrode it should be possible to measure the pH of pure solutions to a repeatability of better than 0.05 pH unit, though it will be necessary to calibrate the electrode with a buffer solution prior to test. Ideally two buffer solutions should be used with a pH range that brackets the expected pH of the slurry (pH electrodes can have a rather short useful life and as they age the response becomes limited so that a single-point calibration can lead to significant errors if the pH of the slurry is much different from that of the calibration solution). As pH is often of rather minor significance for slurries in the field, the use of an electrical pH meter may be regarded as excessive and to be reserved for laboratory situations. If pH papers are used they must be stored in a sealed container to protect from ambient humidity.

With pH papers, by selecting narrow range papers it is, in theory, possible, to measure pH to 0.1 unit. However, there often can be doubts about the indicated colour. When testing suspensions, to avoid masking the colour with deposited solids apply the suspension to one side of the paper and read the colour from the other.

*16.12.4.1 Typical pH values*

Most fresh slurries made with sodium bentonites converted from the calcium form will have pHs in the range 9.5 to 10.0 because of the sodium carbonate added for the conversion. Used bentonite slurries (except those that have come into contact with cement or alkaline soils) usually have a slightly lower pH than their fresh counterparts and the pH tends to continue to reduce towards that of the soil if the slurry is repeatedly re-used. This process can occur without any obvious degradation of slurry properties provided the cautions, with respect to dilution, are respected (see Section 12.6.1). A specification for bentonite excavation slurries is a minimum pH for re-used slurries, and pH can become the controlling parameter for slurry re-use – an inappropriate restriction. Slurry re-use should be controlled by its required physical function, filtration control, viscosity, gel, etc. If pH is to be used as a control on re-use the limiting value should be evaluated in the field. Also it should be noted that a limiting, lower bound, pH may be entirely inappropriate for natural sodium bentonites such as Wyoming products, which even as fresh slurries may have a near neutral pH.

Excavation slurries that have been in contact with cement may have a pH of order 11.5 to 12.5 and thus pH can be a useful test for cement impact on these slurries.

High pH can be a useful indicator that a bentonite excavation slurry has come into contact with cement and thus it may be appropriate to test slurries for high pH.

For grouts and cement–bentonite slurries containing substantial amounts of cement, the pH usually will be over 12.5 and the actual value will give little information about the grout and is not of use except in rather special circumstances. For silicate and aluminate grouts the situation is similar; both are strongly alkaline. Furthermore, such solutions can damage the glass electrode of an electrical pH meter and thus tests should be carried out with caution. Concentrated silicate and aluminate solutions are potentially more damaging than the diluted solutions generally used for actual grouting.

For other grout systems pH measurement is unlikely to be of use unless specifically recommended by the suppliers as a test parameter.

## 16.12.5 Filter loss

Filter loss, bleed and syneresis all will be significant for cement–bentonite slurries and grouts as they are all segregation processes that lead to loss of product volume. At its simplest filtration is a pressure-driven process whereby a particulate system, when pressurised against a permeable formation, deposits a filter cake which accretes with time under pressure and which controls the rate of loss of water. Bleeding is a gravitationally driven process whereby solids settle to create a surface layer of water – bleed water. Syneresis (see Section 16.10.1) is a chemically driven shrinkage of a solid system that may occur during or after setting. A fundamental difference between bleed and filtration is that bleed is controlled by the behaviour of a fluid system and filtration is controlled by the loss of liquid (water) through a solid system – a filter cake. As systems become more concentrated the distinction becomes less clear, and for low pressures/stiff grout fluids a large strain consolidation model can be more appropriate to both. As segregation of the solid and liquid phases is common to bleed and filtration, there may be common causes at the microstructural level. Thus grouts and slurries that show high values for one of these parameters may show high values for the other.

The standard apparatus used for filter loss measurement is the American Petroleum Institute fluid loss apparatus developed for testing drilling fluids. The instrument consists of a 3 inch diameter cell with a detachable base in which a filter paper supported on a wire mesh can be fitted as shown in Figure 16.7. Details are given in [48]. In the standard test the volume of filtrate collected from a slurry sample subjected to a pressure of 100 psi (689 kN m$^{-2}$) for 30 min is measured.

For most slurries and for particulate grouts that deposit a filter cake, the volume of filtrate, $V$, is proportional to the square root of the time, $t$, under pressure as follows:

$$V = Abt^{0.5} \tag{8}$$

It may be noted that Equation 8 implies that the cake is of constant permeability – that is, that it is incompressible in chemical engineering terms.

In Equation 8, $A$ is the cross-sectional area of the filter cake and $b$ is independent of the time, $t$ but is a function of the applied pressure, $P$ and the permeability and solids content of the deposited cake (which in turn will be a function of the applied pressure, the grout/slurry materials and the water chemistry, etc.) and the solid content of the initial grout/slurry. By comparison of Equation 8 with Darcy's equation for flow in porous media, it can be shown that:

$$\frac{k}{d_t} = \frac{b\rho g}{2P} t^{-0.5} \tag{9}$$

where k is the permeability of the cake formed under pressure $P$ and $d_t$ is the thickness of the cake at time $t$, ρ is the density of the filtrate (usually water) and $g$ is the acceleration of gravity. Hence if the thickness of the cake is measured at the end of the test the permeability of the cake can be estimated. Equation 9

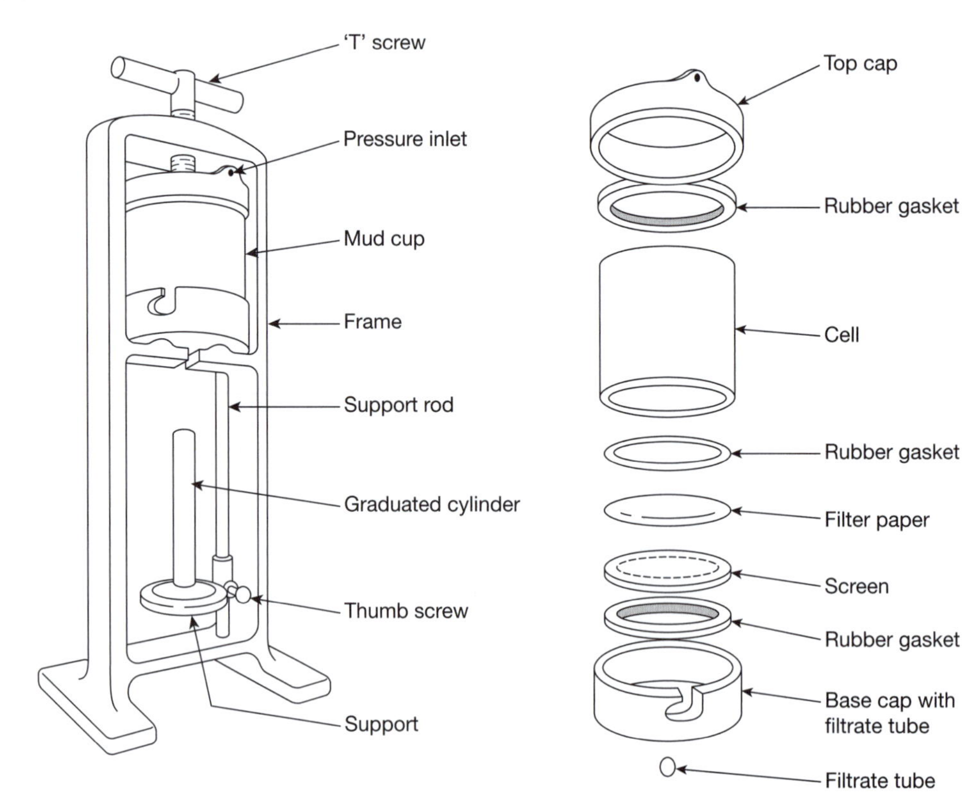


**Figure 16.7** Standard filter press and mud cell assembly

also shows that the permittivity of the cake ($k/d_t$), the inverse of its resistance to flow, decreases as time increase – the cake becomes a better barrier to flow as time increases. It should be noted that the permeability of the cake, $k$ will be a function of the applied pressure, $P$, though not a linear function – a logarithmic function may be a reasonable approximation for some situations.

## 16.12.6 Bleeding

Bleeding may be defined as the separation of water from the solids in a grout or slurry due to gravitational settlement of the solids. In 'thick' grouts and slurries the particles may not settle as independent particles but as a coherent, structured mass. For such systems, bleeding can be modelled as a self-weight consolidation process that continues until the material has sufficient strength to support its own weight. There may be secondary effects due to volume change on hydration of the cement etc.

For cement grouts, self-weight consolidation theory can give useful insights on bleeding behaviour. From consolidation theory it can be shown that the time for full consolidation for a depth $h$ of grout will be a function of $h^2$ and it has been found that except for very shallow cement grout masses full bleed (consolidation) may be prevented by set. Thus if bleed is calculated as a function of grout depth it will be found that the percentage bleed will decrease with the grout depth. Indeed, bleed is approximately independent of depth for deep grout masses [12] and thus it is misleading to quote bleed as a percentage of depth. What is a deep grout mass will depend on the rate of setting of the grout (and hence on the cement type, water cement ratio, admixtures, etc.). For many low water–cement ratio grouts, bleed will be independent of depth for all grout depths greater than about 0.5 m. Thus for studies of bleed it is best to measure bleed for a series of depths from perhaps 10 cm to 1 m to see at what height it becomes independent of depth. For quality control purposes a single fixed container height and diameter may be used. It also useful to investigate the effects of drainage and it is encouraging to see that the 2007 version of [57] includes a wick drain test in a vertical column and in an inclined column test – though for the latter there is little discussion of the tendency of grout to bleed vertically and accumulate on the upper surface of the column with the potential for collapse if a sufficient void forms. Reference [32] gives a fuller discussion on this.

While it is convenient to use the term 'bleeding', it is actually the loss of solid volume that will be important in many grouting operations. Thus surface settlement is the important parameter as it will include not only the effects of exudation of bleed water but also any movements due volume changes on hydration of the cement. Thus a convenient form of bleed test is to fill a container with grout and strike it off to a level surface, then after 24 hours measure the drop in surface level (the surface should be covered to prevent evaporation but the cover should not touch the grout). This is much more satisfactory than measuring the volume of bleed water, as some or all of it may have been re-absorbed (many specifications require that it is fully re-absorbed, though the reasoning behind this is not obvious save that it demonstrates that the bleed has been relatively limited and that there is no risk of cracking from ice formation in freezing weather). Measurement of the volume of bleed water is unsatisfactory as the peak value will occur at an unknown time after the start of the test, depending on the net effect of bleeding and removal of water by hydration. Many cement grout specifications allow 2% bleed but, as already noted, all practical grouts have a much greater potential for bleed and a specification for, say, 2% bleed measured in a small sample will not ensure satisfactory performance at full scale unless the drainage conditions are appropriate. Test procedures for structural grouts are given in [57]. Reference [58] gives recommended tests results and [59] associated grouting procedures.

Some specifications require that the measurement of bleed is carried out in a volumetric flask with a narrow neck, the intention being to concentrate the bleed from a large volume of grout into a narrow column and so improve the accuracy of reading. Unfortunately this procedure does not work as the bleed water will not migrate up the neck of the flask but remains on the upper surface of the bulb. Thus the procedure gives a false low reading. For a fuller discussion of the significance of bleeding for structural grouts see [31, 32].

For clay–cement grouts it is normally satisfactory to limit bleed to 2% as measured with a 1000 ml measuring cylinder. This does not mean that problems will not occur, and in particular problems of bleed and filtration may be seen when cut-off wall trenches are excavated in short lengths unless the trench is topped up after the end of excavation.

### 16.12.7 Measurement of set time

For cement-based grouts and slurries it is difficult to identify a precise set time as these materials have a gel strength from first mixing that increases with time until a hardened material is obtained. The setting/hardening process may be continuous with no obvious region of accelerated strength development. Set therefore tends to be defined arbitrarily as the attainment of some specified strength or penetration resistance (for example, a cone or needle penetration). Actually there may be rather little interest in the set time of many cementitious systems as they normally offer ample pumping/working time unless special cements are used or special conditions prevail.

Chemical grouts may start as effectively ideal (Newtonian) liquids with no gel strength and thus it may be possible to identify a time at which they begin to take on solid type properties such as gel strength. For single phase systems (that is, grouts for which all components are miscible) the onset of gelation may be identified by drawing a wire across the surface of the grout. For pure fluids the wake closes immediately so that the wire cuts no furrow. Once the grout has begun to set a furrow will persist, though initially only momentarily. This is a very simple and highly repeatable test and perhaps the most satisfactory of all procedures, as it actually identifies a change of rheological state. Sadly it will not work for a number of grouts, particularly silicate/ester systems in which the ester will float on the surface of the silicate and obscure the behaviour of the underlying material. Set time for these materials may be shown by a change of colour and quite simply that they become solid and will no longer flow, for example, from a test tube – a rather less repeatable procedure but amply accurate for practical purposes.

A much less satisfactory procedure for assessing set is to specify it as the time to develop a particular viscosity; for example, a specification might define set as the time to reach 100 cP on the assumption that injection will be exceedingly slow/impossible at such a viscosity. This type of definition of set is most unsatisfactory. As set occurs, bonding will develop within the grout mass – bonding that can be destroyed by small strains, in common with all solid bonding. Measurement of viscosity requires flow and therefore will destroy the developing bond structure. Depending on the geometry of the viscometer head and the shear rate, more or less structure may persist in the grout but for many systems the shear will simply break up the setting grout into independent lumps which will produce a meaningless viscosity reading. Unless very low shear rates are used, the maximum recorded viscosity may be quite low. Unfortunately low shear rates are seldom suitable for the measurement of initial viscosity of chemical grouts, which may be only a few cP, and thus the corresponding shear stresses will be too low to measure.

Clearly viscosity is important for injection and therefore when assessing a grout formulation the viscosity should be measured as a function of time. Readings should start as soon as the grout is mixed and may continue until after set. Set should be determined by a separate procedure; for example, by wire or simply when the grout no longer will pour. The set point then should be marked on the viscosity curve. Different grout systems must be expected to show different setting behaviour and thus viscosity and set time measurements must be tailored to individual systems (not only will there be gross effects due to the nature of the setting reactions but also there may be minor influences such as the effect of the shearing action of the viscometer on set time).

## 16.13 Testing hardened properties

For structural cement grouts generally the only test of hardened properties will be cube strength at a specified age or ages. Other tests on the hardened grout will not be necessary unless the application is rather special; for example, the encapsulation of wastes. For geotechnical grouts and cut-off wall slurries, tests that are often specified include: unconfined compressive strength, confined drained triaxial behaviour and permeability.

### 16.13.1 Unconfined compression tests

For cement–bentonite and chemical grouts the samples often will be in the form of cylinders. The sample should be trimmed so that the ends are smooth, flat and parallel and then mounted in a test frame such as a triaxial load frame. The sample should not be enclosed in a membrane but should be tested as soon as possible after de-moulding so as to avoid drying. The loading rate is often not especially important; rates of about 1 mm per minute are frequently used as these give a rapid test. However, this gives failure within a few minutes and only peak load can be recorded unless automatic logging of stress and strain is available. Stress–strain plots from the test may be of interest but are seldom formally required, as under unconfined conditions the strain at failure generally will be rather small. For example, for cement–bentonite slurries the strain at failure typically will be in the range 0.2–2%. It is only under confined drained conditions that these slurries show failure strains of order 5% or greater. The unconfined compressive strength test should be regarded as a quick and cheap test of the repeatability of the material, rather like the cube strength test for concrete. It does not show the stress–strain behaviour that the grout will exhibit under in-situ confined conditions. Curiously, test results from field samples of cement–bentonite cut-off wall slurries can show a surprisingly wide variability and may be of little use for quality control purposes. This may be in part due inherent variability because of soil from the ground trapped in the slurry. However, sample preparation also must be questioned. The ends of cement–bentonite slurries must be trimmed with great care, working gently with a palette knife and removing only a small amount at a time. During trimming the sample must be restrained in a close-fitting sample holder. If edges are broken off by the trimming process then it must be continued until a full flat face is achieved. Damaged areas must not be filled with remoulded material as is sometimes done with clay samples, as the remoulded material will be very weak and lead to false low strength results.

### 16.13.2 Confined drained triaxial testing

To investigate the behaviour of the material under the confined conditions of a trench etc. it is necessary to carry out tests under confined drained triaxial conditions. The sample is set up as for permeability testing with top and bottom drainage, and allowed to equilibrate under the applied cell pressure (which should be matched to the in-situ confining stress) for an appropriate time which will be at least 12 hours. It is then subjected to a steadily increasing strain as in a standard drained triaxial test. To ensure satisfactory drainage a slow strain rate must be used. 1% to 2% strain per hour is often satisfactory for cement–bentonite slurries, though this is quite fast compared with the rates used for clays. The stress–strain behaviour will be sensitive to the effective confining stress. The following behaviour is typical.

(a) *Effective confining stress < 0.5 × unconfined compressive strength.* Brittle type failure at strain of 0.2–2% with a stress–strain curve rather little different from that for unconfined compression save that the post-peak behaviour may be less brittle.
(b) *Effective confining stress ≈ unconfined compressive strength.* The initial part of the loading curve is comparable to (a) above. This is followed by plastic type behaviour with the strain at peak stress increased significantly, typically to greater than 5%.
(c) *Effective confining stress > unconfined compressive strength.* Strain hardening observed in post-peak behaviour and possibly some increase in peak strength, initial stages of loading curve similar to (a).

Thus the effective confining stress is very important. If plastic behaviour is required at low confining stresses – for example, near the surface of a cut-off wall – then a low-strength material is required unless it can be shown that any ground movements that require plastic behaviour will lead to increased stresses. The lack of plasticity at low confining pressures is an important reason for using a capping layer of clay on top of a cut-off wall.

### 16.13.3 Permeability tests

For geotechnical grouts and cut-off wall materials, permeability tests must be carried out in a triaxial cell (for low water/cement grouts permeabilities will be very low and quite special techniques are necessary [52]). Samples for permeability testing should not be sealed into test cells with wax etc. Invariably this leads to false, high results. The procedure is therefore to use a standard triaxial cell fitted with top and bottom drainage. An effective confining pressure of at least 100 kN $m^{-2}$ should be used to ensure satisfactory sealing of the membrane to the sample. The sample may be back-pressured to promote saturation if there has been any drying. Generally a test gradient of order 10 to 20 (a head of 10 to 20 times the length of the sample) is satisfactory but the sample should be allowed to equilibrate for at least 12 hours (or longer as appropriate) before flow measurements are taken (there should be rather little actual consolidation). Measurements then should be made over a period of at least 24 hours. It should be noted that there is often a significant reduction in permeability with time if the test is continued for many days. This occurs irrespective of the age of the sample at the start of the test. For critical tests it may be appropriate to maintain permeation for days or weeks, though this does make for an expensive test. While a saturation stage is normally specified in most permeability testing procedures, the author's experience is that with cement–bentonite grouts this has little influence on the measured permeability provided the samples have been kept in sealed tubes under water.

The permeant from many grouts may be markedly alkaline and thus aluminium or other alkali-sensitive materials should not be used in the test equipment exposed to the permeant (see Section 16.11.1).

## 16.14 Specifications

Specifications for grouts and slurries should include the following components:

- specifications for materials as supplied
- specification of fluid properties
- specification of hardened properties
- and, if required, specification for the process by which the grout or slurry is to be used.

As the properties of many grouts and slurries can develop quite slowly, it is often necessary to have two further components to the specification of properties, a specification for quality control on site and a specification for confirmation of hardened properties by laboratory testing at an appropriate age. For example, a 28 or 90 day strength or permeability is useless as a quality control measure for a job that may be finished before the first results become available. Very often site control specifications will be concerned with fluid properties and those for confirmation of properties with set performance.

References [57–66] are examples of standards relevant to grouts and slurries.

### 16.14.1 Materials specifications

#### *16.14.1.1 Clays*

The Engineering Equipment and Material Users Association specification for bentonite [23] (formerly the Oil Companies Materials Association Specification) is often included in specifications as a requirement for the bentonite. This is a basic specification setting limits on moisture content, dry and wet sieve analysis (on 100 and 200 US mesh sieves respectively), apparent viscosity, and filter loss. For some bentonites the moisture content may be above the specified limit of 15% by weight of moist clay since further drying may damage or impair the properties. The dry screen analysis is not critical provided the wet screen analysis is reasonable. The viscosity specification requires that 1 tonne of clay should produce at least 16 $m^3$ of slurry of apparent viscosity 15 cP. The quantity of 15 cP slurry that can be produced is referred to as the Yield and should not be confused with the yield value (see Section 16.6.3). The American Petroleum Institute Specification API 13A [50] may be used in some specifications and reference should be made to the associated test procedures given in RP13B [51].

#### *16.14.1.2 Cements*

There are standards for cements, for example, BS EN 197-1 [66]. As noted in Section 16.3.2, cement standards may not address problems specific to grouting and thus for critical structural grouting operations it would be preferable to use a cement that has been specifically formulated and quality assured for duct grouting [53], but such a cement may not be available. In general it will be necessary to carry out trial mixes on samples of cement from the proposed supplier and optimise the design for the particular cement [54]. It must be allowed that the cement from some works may be intrinsically better suited to particular grouting operations than cement from other works.

#### *16.14.1.3 Other materials*

Specifications or standards may exist for many of the materials used in grouts and slurries. However, when using such specifications it is important to keep in mind that they may not have been written with grouting as their focus. Furthermore standards prepared in one country should not be used in another unless the materials from that country also are to be used – standards can be country-specific.

### 16.14.2 Specification of fluid properties

The Federation of Piling Specialists (FPS) produced a series of specifications for excavation slurries for diaphragm walls and piles [22]. A proposal for a more general specification in relation to slurry properties was presented by Hodgson [55]. The FPS specification has now been withdrawn and replaced by the ICE Specification for Piling and Embedded Retaining Walls, referred to as SPERW [56]. There are of course other specifications; see for example the series of specifications on the execution of special geotechnical works [61–65]. However, as only a brief discussion of specifications can be given, it will be limited to the SPERW. The SPERW does not specify test procedures or compliance values. However, it does give a table of typical tests and compliance values. It should be noted that the typical values presented in the SPERW [56], [62] and [63] are all slightly, though scarcely significantly, different.

As a general note it should be recognised that the properties of 'local' bentonites (they are natural materials) may vary around the world and national specification requirements for slurry properties may reflect this. As a result there can be a synergy between national specifications and the locally available bentonites – a relationship that should be respected.

There is no generally agreed form of specification for geotechnical grouts. Because of the widely differing applications for which these materials may be used, specifications tend to be produced on an ad hoc basis. This can lead to confusion, especially when specifications are prepared from a blend of typical clauses. For structural grouts there is more consensus. Specifications generally will include density, some measure of rheology, bleed and strength [57, 34]. For cement–bentonite cut-off walls used for contaminant containment there is a UK national specification [30].

#### *16.14.2.1 Density*

The SPERW suggests that as a typical compliance value the slurry supplied to a pile or diaphragm wall excavation should have a density not exceeding 1100 kg $m^{-3}$ when freshly mixed and not greater than 1250 kg $m^{-3}$ when ready for re-use. Furthermore, prior to concreting, the density of the slurry in the excavation should not exceed 1150 kg $m^{-3}$ to ensure satisfactory displacement of slurry by concrete. The slurry re-use density of 1250 kg $m^{-3}$ allows significantly soil-laden slurries to be re-used (a fresh bentonite slurry might have a density of order 1025 kg $m^{-3}$ and the total solids content at 1250 kg $m^{-3}$ might be 400 kg $m^{-3}$ (see Section 16.6.1)

For structural and geotechnical grouting applications, density almost always will be an important part of the specification – as a check on solids content even though the common test instrument, the mud balance, may not provide sufficient resolution for good quality control (see Section 16.12.1). As most grouts and slurries will be mixtures of at least two components it is normally necessary to calculate densities by dividing the total weight of materials in the mix by the total volume assessed from the weights and densities of the individual components. If materials are to be batched on site to a particular density then great care must be taken when calculating the grout density, and it is good practice to check the density of a batch of known proportions prepared under laboratory conditions (the actual and theoretical densities may differ very slightly because of the interaction of the grout components, or somewhat more significantly because of air trapped in the grout).

#### *16.14.2.2 Sand content*

The SPERW suggests that a typical compliance value for the sand content prior to concreting might be <4%, with a requirement of <2% 'if working loads are to be partly resisted by end bearing'.

#### *16.14.2.3 Rheology*

The Fann viscometer can be used to test excavation slurries as it is well suited for testing clay slurries and it should be possible to obtain repeatable readings. Despite this, very few specifications actually detail what readings are required. This is because of the very wide range of results that can be obtained with different bentonites and different slurry concentrations. The normal role of the instrument is therefore to confirm repeatability of behaviour from mix to mix, day to day, etc. For this it can be very valuable. If changes are observed in slurry behaviour the cause often can be identified by consideration of the various viscosity and gel readings (a single point test such as a flow cone cannot give so much information).

The SPERW [56] suggests typical compliance values of viscosity as 30 to 50, 30 to 60 and 30 to 50 for the fresh slurry, the slurry ready for re-use and the slurry in the excavation prior to concreting respectively. It is not stated whether these are to be the apparent or plastic viscosity (see Section 16.6.3), but from the proposed values they must be apparent viscosities.

The specification also requires that the 10 minute gel strength (stated as shear strength in SPERW) of the slurry is in the range 4 to 40 N $m^{-2}$ for all three slurry types.

For geotechnical grouts the rheology may be specified in terms of a flow cone time, or Colcrete flow trough reading or a Fann viscometer reading, etc. For structural grouts for prestressing tendons the standard [57] permits the use of either a flow cone or a spread test in which a 39 mm diameter by 60 mm high cylinder is filled with grout and then lifted and the spread of the grout measured in two directions at right angles.

#### *16.14.2.4 pH*

The FPS specification used to recommend that the pH for slurries supplied to the trench should be in the range 9.5 to 12 for bentonites of UK origin (natural sodium bentonites such as those from Wyoming may have a near neutral pH and polymer slurries different, often higher, values again). UK bentonites typically had a pH of about 10.5, but most sodium ion exchanged calcium bentonites in use in the UK are of European origin and have a pH in the range 9.5 to 10.0. In use the pH of a slurry tends to drop and in some soils the lower limit of 9.5 may be found to be unnecessarily restrictive. If pH is low but rheology and filter loss are acceptable, the slurry should be usable. The upper limit of pH 12 allows quite significant cement content. A slurry of pH 12 may show unsatisfactory rheology or fluid loss. Because of the effects of cement pick-up on bentonite slurries (see Section 16.3.1), it may be preferable to dispose of a slurry of pH 12 rather than to continue to use it and possibly damage much other bentonite. The SPERW has recognised that pH can drop in use and suggests typical compliance values of pH 7 to 10.5 for freshly mixed bentonite and 7 to 11 for slurry ready for re-use. There is no suggested value for pH prior to concreting, which is reasonable.

For structural and geotechnical grouts, pH is unlikely to be specified or useful as a control parameter.

#### *16.14.2.5 Filter loss*

Filter loss from particulate grout systems can inhibit grout penetration – a feature of grouts that is becoming more generally recognised – and as result some specifications for cement grouts now include a fluid loss test [37] based on the American Petroleum Institute test rig [48]. For cement grouts and cement–bentonite slurries the filter loss will be much higher than for bentonite slurries, and if a low value is to be achieved filter loss control material(s) must be included in the mix.

For diaphragm walling slurries, filter loss can influence stability in some grounds and thus there is a case for including it in specifications. The SPERW suggests typical compliance values of filter losses of < 30 ml and < 50 ml respectively for freshly mixed bentonite and slurry ready for re-use. The associated filter cake thicknesses are suggested as < 3 mm and < 6 mm respectively. There is no suggested value for filter loss value prior to concreting. These values seem reasonable, though more stringent requirements may be appropriate for excavations in cohesionless silts; see Section 16.6.2.3. In practice these values should not be difficult to maintain unless very severe dilution (see Section 16.6.1) of the slurry has occurred or there has been substantial contamination with cement. (These situations would almost certainly lead to unsatisfactory rheological properties anyway.)

#### *16.14.2.6 Bleed*

For bentonite slurries the bleed should be minimal when hydrated slurries are tested (though there may be a slight long-term exudation of water (syneresis) in static slurry sample – this is irrelevant for slurry excavation work). Some bleed may occur in storage tanks after mixing and before the bentonite has hydrated. If this is observed then the slurry should be homogenised by recirculation before any slurry is used (it is good practice always to do this prior to use). If severe bleed is observed in slurries that have been allowed to hydrate and recirculated then it is probable that the mix water contains dissolved salts at an unacceptable level (see Section 16.3.4.1) or the bentonite is of poor quality.

For structural and geotechnical grouts, bleed tests should be carried out in a cylindrical container of defined dimensions. Typically for cement-based materials bleed may be specified to be less than 2%, though this may not always ensure satisfactory performance. It can be difficult to measure bleed below 1% on site.

### 16.14.3 Specification of hardened properties

#### *16.14.3.1 Compressive strength*

For structural grouts a principal parameter will be the unconfined compressive strength. This may be measured as for concrete on a cube (generally 100 mm, not 150 mm) or on a cylindrical sample [57]. It should be noted that grout specimens may be much more brittle than concrete cubes and may fail near explosively.

For cement–bentonite cut-off wall materials, it was the custom to specify a maximum strength to ensure plastic behaviour (see Section 16.13.2). Now it is more common to specify a minimum strength (strength has become synonymous with quality, and the more subtle requirements for plasticity have been ignored). The minimum strength will depend on the application: typically it might be 100 kN m$^{-2}$. Lower values at perhaps 20 or 50 kN m$^{-2}$ may be specified. However, it should be remembered that these are very low values for a Portland cement containing materials and it may be difficult to maintain such strengths repeatably. Small variations in mixing or mix proportions may cause substantial variation in strength. For some cut-off wall slurries the properties develop rather slowly, so the specified age for testing may be 28 days or 90 days. It is seldom useful to measure unconfined compressive strength at ages less than 7 days unless a rather strong mix is being used. If an early indication of strength is required, it is best to use a hand vane to measure shear strength.

For cement–bentonite grouts with non-hydrated bentonite, the water–solids ratio will be much lower than for cut-off wall slurries. Thus they will tend to be stronger. For strength measurements it may be practicable to test cubes (that is, the strength may be sufficient that the cubes will give a reasonable reading on a cube crushing machine). For softer grouts it is probably best to use the procedures set out for cut-off wall slurries. It is not practicable to give typical results as the range of possible mix designs is so great.

For chemical grouts the important strength value is that of the in-situ soil when injected with grout. The strength of the grout alone will be a poor indicator of the grouted soil strength, though it may be possible to establish some guidelines [47]. Normally it will be necessary to prepare injected samples in the laboratory using either standard sands or soils from site or simulated soils. Strength values may vary enormously, and again no useful data on typical results can be given.

#### *16.14.3.2 Permeability testing*

The water permeability of low water/cement structural grouts may be of order 10$^{-13}$ m s$^{-1}$ or lower. Measurement of permeabilities of this order is extremely difficult and thus permeability is unlikely to be directly specified for a structural grout; rather it will be implied from the water–cement ratio.

For many years cut-off walls for hydraulic control, for example, beneath earth dams, typically were required to have a permeability of less than 10$^{-8}$ m s$^{-1}$ at 28 or 90 days. However, more stringent requirements were introduced for pollution cut-offs and it is now normal practice to require a permeability of no more than 10$^{-9}$ m s$^{-1}$ at 90 days or longer (see, for example, [30]). The reason for a test at 90 days is that the permeability of a cement–bentonite cut-off material reduces with time and a 28 day permeability may be above 10$^{-9}$ m s$^{-1}$. The relation between permeability and time is typically of the form:

$$k_t = k_o\left(\frac{t_o}{t}\right)^n \tag{10}$$

Where $k_t$ is the permeability at time $t$ and $k_o$ that at time $t_o$. The exponent $n$ is typically in the range 1 to 2 and thus, for example, there may be a reduction of the order of 3 to 10 times in permeability between 28 and 90 days of permeation. However, the equation must be used with caution as the permeability will tend to a limiting value which will be achieved once the full length of the sample is in equilibrium with the permeant. The time to reach this value may be measurable in years.

For low water–solids ratio grouts prepared with unhydrated bentonite the permeabilities of order 10$^{-8}$ m s$^{-1}$ may be realistic depending on the mix design, application and required life.

For chemical grouts, fully injected soils should be effectively impermeable. In-situ it may not be possible to achieve full injection and thus grouted soils may show some remanent permeability. Laboratory tests may be of little help in assessing this permeability.

## 16.15 Conclusions

At the start of this chapter grouts and slurries were said to be generally complex non-Newtonian fluids. The reader who has reached this final section may conclude that grouts and slurries are complex in many other respects. Unfortunately this is true, as the study of grouts and slurries involves many disciplines including soil mechanics, cement technology, chemistry, materials science, rheology, etc. It is hoped that this discussion, by examining the behaviour of grouts and slurries from the standpoint of a few of the disciplines involved, will enable the reader to avoid some of the pitfalls of grout and slurry design and also to distinguish between the many different test parameters and test procedures.

Grouts and slurries are some of the most versatile materials that the civil engineer can employ, and a full understanding of their characteristics will open the door to many new applications and processes.

Finally a note of caution: in the coming years it may become progressively more difficult to obtain good-quality bentonite as reserves are exhausted. Also, obtaining the specialist cementitious materials necessary for some grout formulations may become very difficult/expensive as manufacturers focus on bulk markets such as concrete production to the detriment of specialist markets such as grouting.

## 16.16 References


1 Potter, A. and Jefferis, S.A. *The management of process arisings from tunnels and other earthworks: a guide to regulatory compliance*, The Pipe Jacking and Tunnelling Research Group, Oxford, UK, 2005.

2 Kramer, H. Deep cut-off trench of puddled clay for earth dam and levee protection, *Engineering News Record*, June 27, pp. 986–990, 1946.

3 D'Appolonia, D.J., Soil–bentonite slurry trench cut-offs, *J. Geotech. Eng. Div., ASCE*, 106(4), pp. 399–417, 1980 also 107, No. GT11, Nov. 1981, pp. 1581–1583.

4 La Grega M.D., Buckingham, P.L. and Evans, J.C. *Hazardous Waste Management* 2nd edn, McGraw Hill Higher Education, London, UK, 2001.

5 Anon, First 'stent wall' installed at Kingston upon Thames, *Ground Engineering*, Vol. 18, No. 7, October 1985.

6 Brice, G.J. and Woodward, J.C., Arab potash solar evaporation system; design and development of a novel membrane cut-off wall, *Proceedings of the Institution of Civil Engineers* Part 1, Vol. 76, pp. 185–205, 1984.

7 Tallard, G. *Very low conductivity self-hardening slurry for permanent enclosures*, First International Containment Technology Conference and Exhibition, St Petersburg, FL, 9–12 February 1997.

8 Hass, H., and Hitze, R., *All round encapsulation of hazardous wastes by means of injection gels and cut-off wall materials resistant to aggressive agents*, Seminar on Hazardous Waste, Bergamo, Italy, October 1986.

9 Bensted, J., Special cements, Chapter in Lea's *Chemistry of Cement*, 4th edn, Hewlett, P.C., Edward Arnold, London, UK, 1998.

10 Shimoda, M. and Ohmori, H., *Ultrafine grouting material*, ASCE Specialty conference on Grouting in Geotechnical Engineering, New Orleans, LA, 1982, pp. 77–91.

11 Clarke, W.J., *Microfine cement technology*, 23rd International Cement Seminar, Atlanta, GA, December 1987.

12 *Grouts and grouting for construction and repair of offshore structures*, HMSO Books, London, UK, 1988, pp. 91–110.

13 Jefferis, S.A. and Fernandez, A. *Spanish dyke failure leads to developments in cut-off wall design*, International Conference on Geotechnical and Geological Engineering, Melbourne, Australia, November 2000.

14 British Standards Institution, *BS EN 197-1:200 Cement – Part 1: Composition, specifications and conformity criteria for common cements*, BSI, London, UK.

15 Jefferis, S.A., The Jordan Arab Potash Project. *Proceedings of the Institution of Civil Engineers* Vol. 78, June 1985, pp. 641–646.

16 Jefferis, S.A., The design of cut-off walls for waste containment, In *Land Disposal of Hazardous Wastes, Engineering and Environmental Issues*, Editors J. Gronow, A. Schofield, R. Jain, Ellis Horwood, London, UK, 1988, pp. 225–234.

17 Jefferis, S.A., *Long term performance of grouts and the effects of grout by-products*, ASCE Specialty Conference, Grouting and Ground Improvement, New Orleans, LA, February 2003.

18 Karol, R.H. *Chemical Grouting and Soil Stabilization*, 3rd edn, Marcel Dekker, New York, NY, 2003.

19 Jefferis, S.A., *Effects of mixing on bentonite slurries and grouts*, ASCE Specialty conference on Grouting in Geotechnical Engineering, New Orleans, LA, 1982, pp. 62–76.

20 Institution of Civil Engineers, *Specification for Piling and Embedded Retaining Walls*, Thomas Telford, London, UK, 2007.

21 Doran, S.M., *Surface settlement and bleeding in fresh cement grouts*, University of London Ph.D. thesis, 1985.

22 The Federation of Piling Specialists, *Modifications (1977) to specification for cast in place concrete diaphragm walling (published 1973) also specification for cast in place piles formed under bentonite suspension* (published 1975, revised 1985), FPS, Beckenham, UK.

23 The Engineering Equipment and Material Users Association, Publication no. 163, *Drilling fluid materials*, EEMUA, London, UK, 1988.

24 Hutchinson, M.T., Daw, G.P., Shotton, P.G., and James, A.N., *The properties of bentonite slurries used in diaphragm walling and their control*, Conference on Diaphragm walls and Anchorages, Institution of Civil Engineers, London, UK, 1975, pp. 33–40.

25 Jefferis, S.A. *The composition and uses of slurries in civil engineering practice*, University of London PhD thesis, 1972.

26 Xanthakos, P.P., *Slurry Walls*, McGraw-Hill, London, UK, 1979

27 Institution of Civil Engineers, *Conference on Diaphragm Walls and Anchorages, Discussion on papers* 10–13, p. 105, 1975.

28 Jefferis, S.A., Bentonite–cement slurries for hydraulic cut-offs. *Proc. 10th Int. Conf. Soil Mech. and Foundation Engineers*, Stockholm, June 1981, Vol. 1, pp. 435–440.

29 Jefferis, S.A., *The origins of the slurry trench cut-off and a review of cement–bentonite cut-off walls in the UK*, First International Containment Technology Conference and Exhibition, St Petersburg, FL, 9–12 February 1997.

30 *UK National Specification for the construction of slurry trench cut-off walls as barriers to pollution migration*, Institution of Civil Engineers, London, October 1999 (with Doe, G. and Tedd, P.).

31 Domone, P. and Jefferis, S.A., *Structural Grouts*, Blackie Academic and Professional, Edinburgh, UK, 1994.

32 Jefferis, S.A., Grouts and grouting, Chapter in *Advanced Concrete Technology, Processes*, Newman J., and Choo, B.S. Eds, Elsevier, London, UK, 2003.

33 Jefferis, S.A. and Moxon, S. *Selection of grouts for offshore applications*, International Congress, Concrete in the Service of Mankind, University of Dundee, June 1996.

34 Woodward, R.J. and Williams, W.F., Collapse of Ynys-y-Gwas Bridge, West Glamorgan, *Proceedings of the Institution of Civil Engineers*, Part 1, Vol. 88, August 1988, pp. 635–699 (see also the written discussion on the meeting).

35 Jefferis, S.A. and Mangabhai, R.J., *Effect of temperature rise on properties of high alumina cement grout*, The Midgley Symposium on Calcium Aluminate Cements, London, UK,1990.

36 Warner, J. *Practical Handbook of Grouting: Soil, Rock and Structures*, Chichester, UK, Wiley, 2004.

37 de Paoli, B., Bosco, B., Granata, R. and Bruce D.A. *Fundamental observations on cement based grouts (1): Traditional materials*, Proceedings of the American Society of Civil Engineers Conference, Grouting, soil improvement and geosynthetics, New Orleans, LA, 1992.

38 Littlejohn, G.S., *Design of cement based grouts*, ASCE Specialty Conference on Grouting in Geotechnical Engineering, New Orleans, LA, 1982, pp. 35–48.

39 Hayward Baker, W. Ed., *Proceedings of the American Society of Civil Engineers Specialty Conference, Grouting in geotechnical engineering*, New Orleans, LA, 1982.

40 Borden, R.H., Holtz, R.D. and Juran, I, Eds, *Proceedings of the American Society of Civil Engineers Conference, Grouting, soil improvement and geosynthetics*, New Orleans, LA, 1992.

41 Johnsen, L.F., Bruce, D.A and Byle, M.J., American Society of Civil Engineers Geotechnical Special Publication No. 120, *Grouting and Ground Improvement*, New Orleans, LA, 2003.

42 Warner, J. Discussion of 'How Many Components in a Grout Mix?', *The Grout Line*, September 2006.

43 Hilton, I.C., Classification of grout, in *Methods of Treatment of Unstable Ground*, F.G. Bell, Ed. Newnes-Butterworths, London, UK, 1975, pp. 84–111.

44 Bowen, R., *Grouting in Engineering Practice*, Applied Science Publishers, London, UK, 1975.

45 Verfel, J., *Rock Grouting and Diaphragm Wall Construction*, Elsevier, Oxford, UK, 1989.

46 Scott, R.A., Fundamental considerations governing the penetrability of grouts and their ultimate resistance to displacement, in *Grouts and Drilling Muds in Engineering Practice*, Butterworths, London, UK, 1963.

47 Jefferis, S.A. and Sheikhbahai, A. Investigation of syneresis in silicate–aluminate grouts, *Géotechnique*, March 1995.

48 Rogers, W.F., Composition and properties of oil well drilling fluids, 3rd edn, Gulf Publishing, Houston, TX, 1963.

49 American Petroleum Institute, RP 10B-2/ISO 10426-2, *Recommended Practice for Testing Well Cements*, Washington, DC, 2005.

50 American Petroleum Institute, Specification 13A/ISO 13500, *Specification for Drilling Fluid Materials*, Washington, DC, 2004.

51 American Petroleum Institute, Specification RP 13B-1/ISO 10414-1, *Recommended Practice for Field Testing Water-Based Drilling Fluids*, Washington, DC, 2003.

52 Jefferis, S.A. and Mangabhai, R.J., The divided flow permeameter, *Materials Research Society, Symposium proceedings*, Vol. 137, pp. 209–214, December 1988.

53 Concrete Society, Technical report 47, *Durable bonded post-tensioned concrete bridges*, Second edn, Camberley, UK, 2002.

54 Muszynski, W. and Mierzwa, J., *Possibilities of optimising the properties of prestressing grouts*, International Conference on the Rheology of Fresh Cement and Concrete, University of Liverpool, 1990.

55 Hodgson, F.T., Design and control of bentonite/clay suspensions and concrete in diaphragm wall construction, in *A Review of Diaphragm Walls*, Institution of Civil Engineers, London, UK, 1977, pp. 79–85.
56 ICE Specification for Piling and Embedded Retaining Walls, 2nd edn, Thomas Telford, London, UK, 2007.
57 British Standards Institution, BS EN 445 : 2007, *Grout for prestressing tendons – Test methods*, BSI, London, UK.
58 British Standards Institution, BS EN 446 : 2007, *Grout for prestressing tendons – Grouting procedures*, BSI, London, UK.
59 British Standards Institution, BS EN 447 : 2007, *Grout for prestressing tendons – Basic requirements*, BSI, London, UK.
60 British Standards Institution, BS 3892-3:1997, *Pulverised-fuel ash specification for pulverized-fuel ash for use in cementitious grouts*, BSI, London, UK, 1997.
61 British Standards Institution, BS EN 12715: *Execution of special geotechnical work – Grouting*, BSI, London, UK, 2000.
62 British Standards Institution, BS EN 1536: *Execution of special geotechnical work – Bored piles*, BSI, London, UK, 2000.
63 British Standards Institution, BS EN 1538: *Execution of special geotechnical work – Diaphragm walls*, BSI, London, UK, 2000.
64 British Standards Institution, BS EN 12716: *Execution of special geotechnical work – Jet grouting*. BSI, London , 2000.
65 British Standards Institution. BS EN 14679: *Execution of special geotechnical work – Deep mixing*, BSI, London, UK, 2000.
66 British Standards Institution. BS EN 197-1: *Composition, specifications and conformity criteria for common cements*, BSI, London, UK, 2000.

## 16.17 Bibliography

Bell, A., Editor, *Grouting in the Ground*, Proceedings of the Institution of Civil Engineers Conference, London, UK, 1994.
Bowen, R., *Grouting in Engineering Practice*, 2nd edn, Applied Science Publishers, London, UK, 1981.
Boyes, R.G.H., *Structural and Cut-off Diaphragm Walls*, Applied Science Publishers, London, UK, 1975.
CIRIA Project Report 60, *Geotechnical grouting: a bibliography*, London, UK, September 1997.
CIRIA Project Report 61, *Glossary of terms and definitions used in grouting: proposed definitions and preferred usage*, London, UK, September 1997.
CIRIA Project Report 62, *Fundamental basis of grout injection for ground treatment*, London, UK, September 1997.
Concrete Society, Grouting specifications, *Concrete*, July/August 1993, pp. 27–32.
Ewert, F.-K., *Rock Grouting with Emphasis on Dam Sites*, Springer-Verlag, London, UK, 1985.
FIP Guide to good practice, *Grouting of tendons in prestressed concrete*, Thomas Telford, London, UK, 1990.
Hajnal, I., Marton, J. and Regele, Z., *Construction of Diaphragm Walls*, Wiley, Chichester, UK, 1984.
Highways Agency, *Interim advice note 47/02 DMRB Volume 2: Section 3: Special structures post tensioned grouted duct concrete bridges*, London, UK, 2002.
Houlsby, A.C., *Construction and Design of Cement Grouting*, Wiley, Chichester, UK, 1990.
*Hydraulic Barriers in Soil and Rock*, ASTM Special Technical Publication 874, Philadelphia, PA, 1985.
Moseley, M.P. and Kirsch, K., *Ground Improvement*, 2nd edn, Spon Press, London, UK, 2004.
Nonveiller, E., *Grouting Theory and Practice*, Elsevier, London, UK, 1989.
Shroff, V.A. and Shah, D.L., *Grouting Technology in Tunnelling and Dam Construction*, A.A. Balkema, Rotterdam, The Netherlands, 1993.
Weaver, K. and Bruce, D., *Dam Foundation Grouting*, ASCE Press, Reston, VA, 2007.
Widman, R., Editor, Grouting in rock and concrete, A.A. Balkema, Rotterdam, The Netherlands, 1993.
Woodward, R.J. *The grouting of five prestressed concrete bridges during construction*, TRL Supplementary Report 668, Wokingham, UK, 1981.
Xanthakos, P.P. Abramson, L.W. and Bruce, D.A., *Ground Control and Improvement*, Wiley, Chichester, UK, 1994.

# 17 Gypsum and Related Materials

**Charles Fentiman** MSc PhD
Director, Shire Green Roof Substrates

## Contents

## 17.1 Introduction

Gypsum is one of a number of materials that are mineralogical forms of the chemical compound calcium sulfate. These forms of calcium sulfate have a wide variety of uses in construction materials, some of which are very well known, such as plasterboard and covings. Yet others are virtually unknown, for instance that it is a component in Portland cement and an essential ingredient in many formulated dry mix mortars.

The chemical nomenclature can be rather complex, hence the mineral name 'gypsum' sometimes being used loosely for all the forms, which have different amounts of water in the mineral structures and are distinct minerals with different characteristics.

Anhydrite is the anhydrous (or near anhydrous) form ($CaSO_4$) and the hemihydrate form has half a molecule of water in its formula ($CaSO_4.0.5H_2O$), whereas gypsum has two molecules of water in its molecular formula ($CaSO_4.2H_2O$). Of these the latter is the long-term stable form in the presence of water as the other phases react to form gypsum, generally very slowly in the case of anhydrite but quickly in the case of hemihydrate:

$$CaSO4 + 2H_2O \rightarrow CaSO_4.2H_2O \text{ (very slow)}$$

$$2CaSO_4.0.5H_2O + 2H_2O \rightarrow 2CaSO_4.2H_2O \text{ (rapid)}$$

Hemihydrate (the mineral bassanite) is much better known as plaster of Paris. This occurs in two forms known as α- and β-hemihydrate which seem to differ only in terms of crystal shape but with no significant crystallographic difference between them, which is why this differentiation has sometimes been said to be an unnecessary confusion.

## 17.2 Natural occurrence

The various forms of calcium sulfate occur in nature, with widespread deposits mainly of gypsum and somewhat less of anhydrite; however, due to its inherently unstable nature, hemihydrate is far less frequently found in nature.

Gypsum occurs in columnar to tabloid crystals, generally white or yellowish, and these are sometimes particularly large (some of the largest crystals recorded in nature). In crystalline form gypsum is often referred to as selenite, and small, perfect single crystals can sometimes be found in clay deposits. However, commercial deposits are in massive form derived from deposition from sea water and often intermingled with anhydrite deposits. When the individual granules of gypsum are sufficiently fine the rock can be carved, and is also widely known in that form as alabaster. Although commercial deposits of gypsum are found throughout the world, some major deposits are in Italy, Spain and Egypt. Globally it is a very important commodity, with worldwide production exceeding 100 Mt/year.

Anhydrite is usually compact and whitish to bluish. As with gypsum, it may be formed by deposition from seawater in shallow lakes dried by intense sunshine or by dehydration of gypsum itself. Large anhydrite deposits may alter to gypsum if exposed on the surface of the earth and not collected right away. It also occurs as a cap rock to salt domes. Deposits of anhydrite are fairly widespread and commercial deposits are found in China, Poland, Mexico and Slovakia.

Bassanite occurs as pseudomorphs of gypsum, but usually in small amounts, and generally forms by the dehydration of naturally occurring gypsum. It is found in the USA, Austria, Bolivia, Chile, Czech Republic, Germany, Hungary, Italy, Poland, Iceland, Russia, Slovakia, the UK, Switzerland and Ukraine.

## 17.3 Industrial production

For many applications, such as plaster of Paris and the manufacture of plasterboard, the raw material required is hemihydrate, but with the relative scarcity of suitable natural deposits much of it is manufactured by the dehydration of gypsum.

In this process gypsum is heated to between 100°C and 170°C to partially dehydrate the mineral by driving off most of the water contained in its chemical structure.

$$CaSO_4.2H_2O + \text{heat} \rightarrow CaSO_4.\tfrac{1}{2}H_2O + 1\tfrac{1}{2}H_2O \text{ (steam)}$$

Although there are various sources of anhydrite found in mineral deposits, there is also industrial manufacture which takes place by heating finely ground gypsum to 180–200°C. This results in a nearly water-free form, known as γ-anhydrite ($CaSO_4.nH_2O$ where $n = 0$ to 0.05). This form reacts slowly with water to form gypsum – a property that leads to its use as a desiccant. On heating above 250°C, the completely anhydrous form called β-anhydrite is formed. This is the form most commonly found in natural deposits and therefore is sometimes referred to as 'natural anhydrite' when sourced from natural deposits. This form does not react with water, even over geological timescales, unless ground to extreme fineness or mixed with reactive components. Some further 'dead burned' forms are made by heating to greater temperatures and at 1100–1200°C some lime is also formed, which can also be useful for some applications.

### 17.3.1 Secondary sources

There are many sources of secondary calcium sulphate – after all, it is a relatively simple compound that can be formed in many ways – but the following are some of the more important industrial sources.

- Calcium sulfates are often formed as industrial by-products of various industries, notably the product of desulfurisation of exhaust flue gases from fossil fuel burning power stations. The absorption of waste sulfur oxides in these gases is treated by reaction with lime (calcium hydroxide) to form calcium sulfate in its various hydrated states. Industrial countries produce large quantities of calcium sulfate by desulfurisation and in some countries these can be the primary source of these products; in fact demand can in some cases outstrip supply.
- Similar processes are used to obtain calcium sulfate from waste sulfuric acid wherein calcium sulphate is the precipitated form. It is also produced as a by-product of reacting sulfuric acid with fluorspar in the industrial manufacture of hydrofluoric acid.
- Calcium sulfate is formed as a by-product of processing phosphate ores into fertilizer with sulfuric acid. However, these are associated with radioactivity because of the small amounts of naturally occurring radioactive elements (including uranium and radium) normally present in the phosphate ores. Such products are of limited usefulness despite a worldwide production of some 200 Mt per annum.
- Recycling of plasterboard also generates useful secondary gypsum that can be reprocessed or used as a source of calcium sulfate, e.g. in cement manufacture.

The above sources tend to produce mixed material but these can be purified into the desired grades by relatively straightforward industrial processes, as already indicated.

## 17.4 Hydration of the forms

- *β-anhydrite (natural anhydrite):* Does not react with water, even in geological timescale, and will only react with water when reduced to a fine powder and activated.

- *γ-anhydrite (soluble anhydrite):* Reacts slowly with water to form gypsum.
- *α- and β-hemihydrate (i.e. plaster of Paris):* React quickly with water to form gypsum, typically setting within 15 to 30 minutes. Overcooking during manufacture can give the surfaces of particles a temporarily impervious glassy coating that retards setting by several hours. Once setting starts, strength development is rapid and relatively high strength can be achieved.
- *Gypsum:* Does not react with water, as it is already fully hydrated.

## 17.5 Industrial uses

Calcium sulfate minerals are used for a wide range of applications and most people are familiar with plaster of Paris (which is hemihydrate), used in various plastering grades and plasterboard. However, many people would not realise that calcium sulfate is a key set regulating component of Portand cement. Since the worldwide production of Portland cement is around 2.8 billion t, containing 4–5% of calcium sulphate, this amounts to a worldwide need for cement manufacture of around 112–140 Mt of calcium sulfate (both gypsum and anhydrite are used). This compares with a worldwide requirement of about 39 Mt of calcium sulfate for plasterboard production. With such large applications, it can be seen that these are very important industrial minerals, requiring large-scale mining and quarrying using some impressive extraction operations.

### 17.5.1 Dry lining industry

The use of plasterboard dry lining has become the predominant method of finishing walls, ceilings and even floors in industrialised countries. It is quick, economical to install and, compared to plastering, it greatly reduces the amount of water used in wall or ceiling construction. Generally it provides better thermal properties than conventional wet constructions and is hence less prone to surface condensation. It also allows relatively easy installation, and partition walls can be constructed with a framework onto which the plasterboards can be fixed, giving a smooth surface suitable for direct decoration.

In addition to plasterboard, dry lining requires a wide range of ancillary products which frequently use hemihydrate-based formulations. These include finishing plasters, jointing compounds, covings, and decorative items and plaster-based adhesive for fixings.

Estimates suggest that the world production of plasterboard is at least 5.6 billion square metres produced by more than 250 plasterboard manufacturing plants. Approximately half of these are in the USA, and it is said that an average new American home contains more than 7.3 t of calcium sulfate. Plasterboard is a relatively low-value product and as a result most of the hemihydrate used comes from the country where it is produced.

The manufacture of plasterboard is a continuous process where a slurry of hemihydrate mixed with water and various solid and liquid admixtures are introduced between two layers of paper. The admixtures used serve different purposes; for instance, starch is used to improve adhesion with the cardboard and retarding agents to give the mix a sufficient setting time for the manufacturing process. Indeed, lightweight aggregates may also be introduced to lower the unit weight of the boards and improve insulation.

The two layers of paper are continuously unrolled, one below and one above the mixer and the slurry spreads from multiple outlet hoses onto a moving lower sheet of paper. As the board is actually produced upside down, this paper will form the front face of the plasterboard. This paper is disc-scored, allowing it to be easily folded at the edges. The "back face" paper is fed from above and is applied to the slurry to the desired board thickness. At this point the front face paper is folded at the edges, producing an enclosed envelope of paper for the slurry.

The board is conveyed along the production line on a series of belts and rollers and during this journey the plaster core has time to set, or harden, and the required product information (including product name and type, relevant Standards, date and time of manufacture) is printed on the back of the boards via an ink jet printer. After drying, the sheets are cut into standard lengths and turned and passed into a multi-level dryer.

During the drying period, the excess water, which was required to form the initial slurry, is gently evaporated and, once dried, the boards are trimmed and stacked to form pallets and stored, ready to be shipped to building sites.

### 17.5.2 Plastering

The use of hemihydrate 'gypsum' plasters goes back to antiquity, when, depending on locally available raw materials, gypsum plasters and lime plasters were widely used and it was also common practice to add some gypsum to them (a practice known as 'gauging') in order to speed up the set of a lime plaster in order to keep to a tight programme. Plastering with gypsum plaster is still the common practice in many countries where good plastering skills remain, but where mass building is the norm plastering has become restricted to finish coats on dry lining and where ornate wall and ceiling designs apply. They are also used for historic buildings where original plasters need to be repaired or replaced.

As with dry lining, there are a wide range of ancillary calcium sulfate products for use with the plasters, including decorative mouldings, and the latter are becoming increasingly sophisticated with large architectural items becoming available. These ancilliary products also include special barium-based plaster for the walls of X-ray laboratories, because it prevents the transmittance of X-rays.

Importantly, undercoat plasters tend to be made with some coarser granular material whereas finish coat plasters need to have a smoother finish and therefore have finer particle sizes. Various bonding plasters are also available, including some based on retarded hemihydrate used for bonding, where there is plenty of working time to allow for adjustments to be made to the position of the bonded item, and one mix can last long enough for the area in question to be completed.

Among decorative products are refined formulations designed for particular appearance, e.g. texturing coatings for ceilings. Many mouldings are also used, including covings, and these are largely made from cast hemihydrate plaster compositions, often made lightweight by the introduction of air bubbles or lightweight aggregates, e.g. polystyrene.

### 17.5.3 Anhydrite screeds

During the 1970s and 1980s new more rapid construction systems were developed and among these was a floor screeding system based on mixtures of finely milled anhydrite blended with a lesser amount of alkali sulfate that activate the system, in the case of potassium sulfate by reacting to form syngenite, the double salt $K_2SO_4.CaSO_4.H_2O$ as well as the rapid hydration of the mix to gypsum.

Anhydrite screeds are poured self-levelling mixtures based on anhydrite and sand (typically 0–4 mm) with admixtures (e.g. plasticisers) and water. During the hardening process the anhydrite is reacted with water in the presence of alkali hydroxides which make the anhydrite hydraulic and so hardening by hydration to gypsum. Anhydrite screeds reach relatively high strength, typically compressive strength of at least 20 MPa and flexural strength of at least 5 MPa. They have self-levelling characteristics and therefore seem to offer some definite benefits, as claimed by suppliers:

- shorter period of total laying time – self-levelling mixture; it is possible to use it even in very indented spaces, up to 1000 $m^2$ daily
- exceptionally low shrinkage with minimal risk of cracking and curling
- smaller thickness of floor = less loading on construction
- high strength in compression and bending strength
- high fire resistance
- very level surfaces can be achieved
- can be used with underfloor heating
- claimed to be sustainable owing to secondary origin of anhydrite.

However, it has been reported that drying time is too slow for these screeds to accept some floor finishes, and concerns have been raised by some concerning the possibility of degradation if the system remains wet.

### 17.5.4 Cement applications

Calcium sulfate is also an important component of Portland cement, where it is used to provide an adequate setting time for the cement – without this, most cement compositions would flash set, making them very difficult to use for most applications. The set regulation comes from the reaction between calcium aluminates and calcium sulfate to form the mineral ettringite (sometimes denoted as Aft phase, which is a term for the mineral family), which slows the early hydration of the other phases and by so doing provides ample working time for the mixing and placing of concretes made with it. As already said, this use as set regulator for Portland cement constitutes the largest volume of calcium sulfate in construction products. Late stage formation of ettringite and related phases is deleterious to concrete and for this reason this mineral is often thought of as harmful to concrete, but it is also a key phase formed in many cement-based formulated products based on blends of cements such as calcium aluminate cements with calcium sulphate, where the formation of ettringite and other hydrated sulfates of calcium aluminate provide specific beneficial attributes to such compositions.

Ettringite nucleates very readily from solutions containing calcium, aluminium and sulfate ions, and precipitation starts immediately and continues until one of the reactants is no longer available in solution. Various equations apply to ettringite-forming reactions, of which the following is the one that applies in Portland cement. It is shown long hand and also in cement chemist notation, which is a shorthand method commonly used in literature about cement.[1]

$$Ca_3Al_2O_6 + 3CaSO_4.2H_2O + 30H_2O \rightarrow 3CaO.Al_2O_3.3CaSO_4.32H_2O$$

tricalcium aluminate — gypsum — water — ettringite

Cement chemist notation: $C_3A + 3C\check{S} + 30H \rightarrow C_6A\check{S}_3H3_2$

Compared to most other cement hydrates, the ettringite structure holds a great deal of water and forms prismatic fibrous crystals of relatively low specific gravity, so this mineral can occupy a relatively large amount of space. In formulations containing sufficient calcium aluminates the ettringite reacts with this to form the denser mineral commonly known as monosulfate, often with aluminium hydroxide (amount depending on the calcium aluminate). For example:

$$C_6A\check{S}_3H_{32} + 6CA + 19H \rightarrow 3C_4A\check{S}H_{12} + 5AH3$$

monosulfate

Ettringite and monosulfate represent the pure calcium sulfoaluminate hydrates, but each represents a family of solid solutions and related phases that are broadly known as Aft and Afm phases respectively that depend on the inclusion of calcium sulfates in the formulation. The benefits of forming ettringite can be an acceleration of setting time as well as shrinkage compensation, which can be an important attribute for a formulation, whereas the formation of monosulfate and related Afm phases provides for relatively high density and good durability.

Companies that sell such formulated blends explore these complex interactions whereby different ratios of the raw components are investigated for key attributes, such as setting time, strength, expansion and durability. Typically this is done using ternary diagrams that have calcium sulfate as one apex with Portland cement and calcium aluminate (or CSA) as the others. These constructs can even include a fourth axis, if another component forms part of the cement content (Figure 17.1).

Essentially, these blends are bespoke cements for the applications for which they are intended and are further formulated with sands and admixtures to form the desired dry mix product. These are sold via distribution networks to professional contractors or to the DIY market via the well-known stores supplying that market. This is a high-value, growing market for which global quantities are difficult to estimate because of their diverse nature but nevertheless it amounts to millions of tonnes, of which calcium sulfate represents a relatively small part. Such compositions are also used as an essential component in prefabricated building components, where the composition adds to the performance of the product in question.

### 17.5.5 Some examples of familiar formulated systems

- *Cement-based adhesives.* These are adhesives that require mixing with water. They are compositions containing fine sand and admixtures aiding bonding and trowelabilty and non-slumping properties. Similar adhesives are also used with insulating aircrete blocks, which are manufactured to high tolerances; rather than being laid into sand and cement mortars, which is bad for insulation, the use of cement-based adhesives helps maximise the insulation of the overall blockwork.
- *Self-levelling floor screeds.* Most floor finishes require hard-wearing flat and smooth floors. This can be achieved by various means including these special screeds that are so flowable that they find their own level. Because the surface needs to be hard and dust-free the compositions must remain fluid without water separating on the surface. Hardening needs to be sufficiently fast to permit further work or being put into service within hours of laying (Figure 17.2).
- *Repair mortars and concrete.* Various repairs of concretes are necessary, often where reinforcement has caused corrosion, physical damage or for fixing holes made as part of other works, such as plumbing or electrical work. Typically such mixtures need to be rapid setting and non-slumping to allow for vertical and overhead work. Repair of concrete carriageways is also an important application where the roads need to be quickly returned to service and for these more fluid characteristics are important.
- *Grouts.* Installation of plant and equipment leaves unfilled spaces underneath that provide insufficient support for the high loads imposed by heavy machinery. These compositions generally need to flow well to occupy the space and develop relatively high strength to provide the equipment with sufficient strength.
- *Bedding mortars for street furniture.* Some street furniture sometimes needs to be bedded in low-slump, high-early-strength systems. The bedding of manhole covers is an important application where a failure of these can lead to serious accidents.

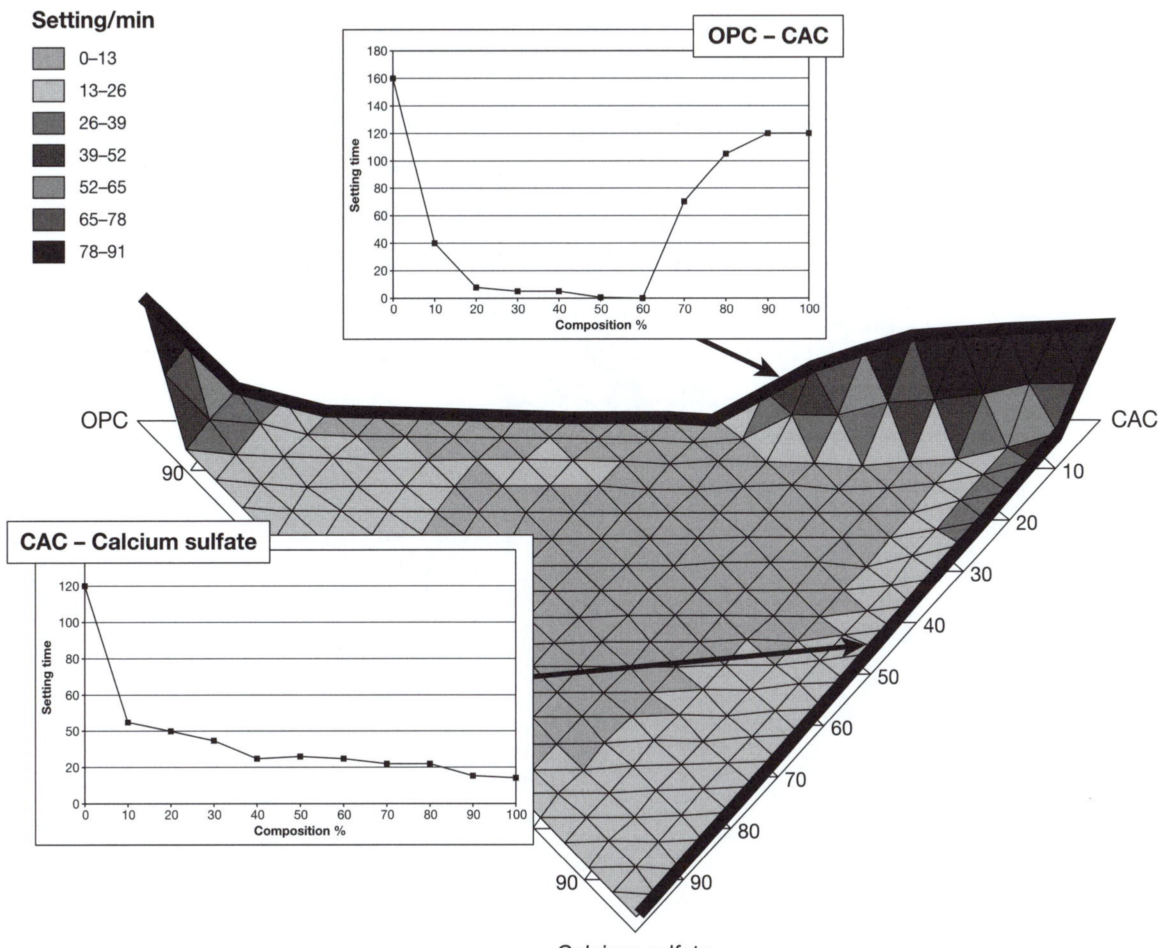


**Figure 17.1** Setting times for blends of plaster with OPC and CAC plotted on a tertiary diagram to enable the formulator to follow the effect of blend composition on setting times. Such charts are also used for other attributes of the blends such as dimensional stability and strength. This chart is reproduced with kind permission of IHS BRE Press and was taken from a paper by S. Maier of Calucem GmbH: 'Ternary system: Calcium aluminate cement–Portland cement–gypsum', *Calcium Aluminate Cements*, Proceedings of the Centenary Conference 2008. Edited by C. H. Fentiman, R. J. Mangabhai and K. L. Scrivener.

The above are just a few examples, of which there are many where formulations often contain a small amount of calcium sulfate. Depending on the formulator gypsum, anhydrite and hemihydrate are used: the choice is somewhat interchangeable, and local availability and formulator preference can be important factors in the choice.

## 17.6 Acknowledgements

The author would like to acknowledge gratefully the help provided by Ron Montgomery (Kerneos Ltd, retired), Steve Brooks, Technical Director of Ardex UK Ltd, and Dr Emmanuel Geeraert of Saint-Gobain Gyproc in preparing this chapter.

## 17.7 Note

1 In cement chemist shorthand oxides are represented by the first letter of the oxide, and where the first letter is the same, one of the oxides (normally the anion) is modified for differentiation. Examples are CaO = C, $Al_2O_3$ = A, $SiO_2$ = S, $SO_4$ = Š and $H_2O$ = H.

**Figure 17.2** A pump-applied, self-levelling mortar, specifically developed for the rapid resurfacing and levelling of existing concrete floors, providing a hard, smooth, flat surface, resilient enough to accept the heavy wheeled traffic typical of industrial workshops and warehouses. This formulation chemically binds the mix water without evaporation, allowing further work, including the application of water-based systems such as epoxy paint coatings, in as little as 6 hours. Reproduced with kind permission of Ardex (UK) Ltd

# 18 Natural Building Stone

**Tim Yates** BTech BA PhD
Technical Director, BRE Ltd

## Contents

## 18.1 Introduction

Stone has been used in buildings as both a structural and a decorative material for a very long time. In medieval times, the stone used for a particular building tended to be that most convenient from a nearby quarry (e.g. Wells and Lincoln Cathedrals). This means that mainly indigenous materials were used, although there were exceptions (e.g. Canterbury and Chichester Cathedrals are built from French limestone). Today, a wide variety of stones are used in construction. This is due to a number of factors such as supply from certain quarries being exhausted leading to less local stone being available, better methods of transport allowing stone from a wider area to be used, and the importation of stone such as Italian marbles, German limestone, and Chinese granite.

This chapter is concerned only with the origin and specification of stone for the structure and cladding of buildings. It does not address the use of stone or rock as aggregates or road stone. Further, it deals mainly with stone indigenous to the British Isles, and stone commonly imported into the UK.

## 18.2 Types of building stone

Natural building stones are obtained from three basic types of rock: igneous, metamorphic and sedimentary.

### 18.2.1 Origins of stone

#### *18.2.1.1 Igneous rocks*

Igneous rocks are those that have crystallized from molten rock (magma). The grain size of igneous rocks is determined by the rate at which the molten rock cools – a slow cooling rate producing finer grained rocks. The most common igneous rock used for building in the UK is granite. Granites are coarse-grained acidic rocks (i.e. they contain at least 66% silica) and consist mainly of quartz, mica and feldspars. Their porosity is usually very low and this, together with their mineral composition, makes them very resistant to weathering. In the building industry the term "granite" is sometimes used to describe a number of igneous rocks that are not technically true granites. The wide range of colours and appearances, combined with strength and hardness, means that granite is commonly used in buildings and elsewhere including for façade cladding panels, paving (both paving slabs and setts), internal floors, and accessories such as kitchen worktops.

#### *18.2.1.2 Metamorphic rocks*

Metamorphic rocks were produced from sedimentary deposits that recrystallized due to the action of heat and pressure on the earth's crust. The most common building stones that have undergone metamorphosis are slate and marble. Slates originate from mudstones, and marble from limestone. Slate, which occurs in many areas of the UK, has traditionally been used for roofing, as well as paving, taking advantage of the way that it splits on its natural cleavage. Marble has been considered a prestigious material and its qualities have made it very popular for carving. It is rare in the UK, but has been imported for more than 2000 years. In more recent years marble has been used for external cladding, where it generally performs well but occasionally undergoes an irreversible expansion creating distortion of the original geometry.

#### *18.2.1.3 Sedimentary rocks*

Sedimentary rocks (e.g. sandstones and limestones) are the main source of indigenous British building stones. The formation of these rocks is a two-stage process. First a sediment is deposited; this can originate from a number of sources such as fragmentation of an earlier rock, the accumulation of animal skeletons, or by chemical deposition in lakes or seas. The second stage is cementing the sediment to form a hard rock; this is assisted by compaction and pressures generated from movement of the earth's surface.

Sandstones were formed by fragmentation of earlier rocks such as gneisses and igneous rocks. The particles of parent rock were transported mainly by the action of water and deposited in layers on the floor of seas and lakes, although transportation could also occur by other mechanisms such as wind action and glaciers. During transportation the sediments were sorted in size by the action of the water; in general, the finer particles would be carried further than coarser ones, and the longer the sediment took to be deposited the more well sorted it would be. The degree and method of sorting determined the texture of the sandstone.

The parent rock determines the mineralogy of a sandstone, but, because the sediments can be transported over large distances, one sandstone can contain sediments from many parent rocks. All the sandstones contain silica along with a variety of other minerals such as mica and feldspar. Once the sediment has been laid down it can be bound together by a number of different cements:

1 *siliceous* – containing silica
2 *calcareous* – containing calcium carbonate
3 *dolomitic* – containing dolomite (a double carbonate of calcium and magnesium)
4 *ferruginous* – containing iron oxide
5 *argillaceous* – containing clay.

The nature of the cementing material has a profound influence on the durability of sandstones. The nature of the cementing can be used to further divide sandstones into "clast" supported stones, that is where the sediment grains are in contact with each other, and "matrix" supported, where the sediment is contained within the matrix.

Limestone sediments are usually formed from the skeletons and shells of aquatic animals or from chemically formed grains such as ooliths ("egg-like"). Ooliths are formed by the crystallization of calcium carbonate from solution around a nucleus which could be a fragment of a shell or grain of sand. Oolitic deposits are formed in seas where there is tidal action; the action of tides is essential for the formation of ooliths because unless they are constantly agitated they cannot grow as individual grains. Limestones whose sediments originate from aquatic organisms are usually forced under marine conditions but can be produced in freshwater (e.g. Doulting limestone or Purbeck marble)

Limestones are all cemented with calcium carbonate (calcite) and their durability is determined more by the structure of the rock than by its chemical nature.

In some cases limestones are converted to magnesium limestones by a process called "dolomitization". In this process the calcite is gradually replaced by dolomite (e.g. Cadeby stone).

The extensive deposits of sedimentary rocks in the UK have naturally led to their use in many buildings, and their properties and uses have shaped much of the vernacular architecture in many areas. Sandstones and limestones can be used for all parts of a building from the plinth to the roof, but are most often seen as wall and paving. The properties of the different stones vary greatly and so it is essential to establish that the selected stone has the correct properties for the intended use, for example whether it is sufficiently durable, and also to use the stones so that their bedding is used to best effect.

### 18.2.2 Sources of stone

The production of granite for building in the British Isles is concentrated in Cornwall and Scotland, with other quarries in Ireland, Cumbria and Devon. A vast amount of granite and other types of igneous rocks is imported into Britain, with many thousands of types available from countries all over the world.

Only two true marbles are quarried on a commercial scale in the British Isles, one from Ireland (Connemara) one from Scotland (Ledmore). Some limestones that will take a polish are often referred to as marbles, the most well known being "Purbeck marble". Because of the lack of indigenous marble, most of that used in Britain is imported mainly via Italy but with the stone originating from many different countries, including Portugal, Spain, Turkey and Greece.

Slates are quarried in Wales, Cumbria and Cornwall. A considerable amount of slate is also imported into the British Isles, mainly from Spain and, increasingly, from China and Brazil.

Sandstones are quarried in two main regions in the UK. The most concentrated area of quarries is that encompassed by a circle of diameter 80 km centred on Manchester, with another less concentrated area of quarries between the England–Scotland border and a line between Barrow in Furness and Middlesbrough. There are also two small groups of quarries, one near Elgin in Scotland and the other near Monmouth. Sandstones from Germany, India and many other countries are also imported.

Most of the limestone quarries lie on a 40–50 km wide belt running from Tiverton in Devon to the Wash, with other quarries as far north as Cumbria, and the Portland and Purbeck quarries to the south. There are also quarries in Ireland. The Magnesian limestone quarries lie in a 10–20 km wide strip to the east of the Pennines running from Nottingham in the south to Ripon in the north. A substantial amount of limestone is imported into Britain, mainly from France, Germany, Bulgaria and India.

### 18.2.3 Extraction of stone

The traditional technique used in the British Isles for quarrying large blocks is "plug and feathers". In this technique a row of holes is drilled along the line where the fracture is required. For "plug and feathers" two metal plates (the "feathers") are put down each hole with a wedge (the "plug") between them. All the wedges are driven in at the same time causing the block to split off along the line of the holes. Some quarries produce blocks that can be extracted with nothing more than a powerful mechanical excavator. There are an increasing number of locations in the England where the stone is mined using large mechanical cutting saws. In mainland Europe, and increasingly in the British Isles, quarries use wire saws to cut blocks and to trim them prior to further processing.

Quarried blocks are further worked using a frame saw to produce slabs or using diamond-tipped circular saws. The slabs are then further processed and ground to the required thickness using multi-headed grinding and polishing machines.

## 18.3 The use of stone

One of the major requirements of architects and specifers of building stones is some measure of the long-term performance of a given stone related to its environmental exposure, and to its position and function on the building in question. This section deals with the selection of stone and with tests for its durability. First, however, it is useful to summarize the decay mechanisms in stone so that the reasons why particular tests are used can be better understood.

### 18.3.1 Decay of stone

The mechanisms by which a stone decays is dependent on three factors: the weathering agent, and the chemical and physical structure of the stone. There are four main weathering agents that affect stone in the UK:

1 frost
2 soluble salts
3 acid deposition
4 moisture and temperature cycles.

#### *18.3.1.1 Frost*

It is well known that when water freezes to form ice the process is accompanied by an increase in volume. The traditional view of frost damage is that when this occurs within the pores of a stone the structure will suffer damage if the stone is not strong enough to withstand the pressure. While the fact that water expands when it freezes will contribute to any damage, the actual mechanism of attack appears to be more complicated. It has been demonstrated that liquids such as nitrobenzene that contract on freezing can also cause damage when frozen in the pore structure of stone. Whether or not a stone is resistant to frost appears also to be related to the distribution of the size of the pores in the stone.

The pore structure affects freezing in two ways: firstly, it determines how much water is absorbed and retained under a given set of conditions; stones with a predominance of micro-pores tend to become more saturated than those that are more macro-porous. If insufficient water is retained, damage will not occur. Thus, as a general rule, macro-porous stone tend to be more durable than micro-porous stone. Secondly the thermodynamics of ice nucleation (i.e. the formation of an embryonic ice crystal and its subsequent growth) is governed by the pore structure. Put simply, once a crystal has been formed it is more favourable for that crystal to continue growing than for another crystal to form. The more water that is available to allow the single crystal to continue growing, the greater is the chance of frost damage. It is this type of "crystallization" mechanism that is thought to be mainly responsible for frost damage.

#### *18.3.1.2 Soluble salts*

Soluble salts can contribute to stone decay in two ways. Firstly, they can cause crystallization damage by a mechanism similar to frost damage. In this case it is the evaporation of water leading to a saturated solution of salt that causes the first crystal to form, but the growth of that crystal is controlled by the same factors as in frost damage.

The second mechanism can take place in "dry" conditions. Many salts have a number of molecules of water attached to each molecule of salt, called the "water of crystallization". The number of molecules can vary according to the ambient temperature and humidity. At higher temperatures and lower humidities salts tend to lose their water crystallization; at lower temperatures and higher humidities they pick up moisture. As a salt picks up moisture its volume increases; if this expansion is restricted by the pore walls within a stone then pressures can be generated that are sufficient to cause disruption of the stone surface.

#### *18.3.1.3 Acid deposition*

Acid gases (such as sulphur dioxide and the oxides of nitrogen) in the atmosphere can attack stones containing carbonate materials, e.g. calcite and dolomite. In the past this was the most common decay mechanism in urban areas, but over the period since 1970 the reductions in atmospheric sulphur dioxide have resulted in it being far less important, with rates of deterioration of limestone now 95% lower than in the 1950s. Where it does occur, this decay mechanism causes damage for two reasons. Firstly, the stone is gradually eaten away with a loss of the surface, and secondly the reaction products are soluble salts that can cause crystallization damage, particularly from stones containing dolomite.

The statues on many cathedrals are made of limestone and are often gradually eaten away (dissolved) partly by acidic species in rainfall at the surface. The stone that remains is still quite sound, however, and will continue to give good service for many years to come.

The worst case of decay from acid attack can occur with stone that contains small amounts of carbonate minerals such as the cementing materials that hold the grains together (e.g. calcareous sandstones and some slates). These materials can decay very rapidly because the dissolution of a small amount of carbonate binder can dislodge a large number of grains.

*18.3.1.4 Moisture and temperature cycles*

Moisture plays an essential role in the deterioration of stone; the mechanisms outlined above all rely on the presence of moisture. It is also possible for decay to occur when stone undergoes simple wetting and dry and thermal cycles. Attack by moisture can be chemical or physical in nature.

Physical damage can result if the cementing material of a rock contains large amounts of clay, due to the expansion of some clays when they absorb water. A combination of wetting and drying cycles and temperature cycles is also thought to contribute to contour scaling. In this type of mechanism it is believed that the wetting and drying cycles result in the migration of iron minerals towards the surface of the stone. This results in a reduced permeability and the build-up of salts and moisture immediately beneath the surface. Periodically the build-up of stresses leads to spalling of the surface.

Wetting and drying cycles can cause chemical damage to occur in some pyrite-containing rocks due to oxidation of the pyrite. Some slates are prone to this type of decay, as are some imported igneous rocks.

Thermal and moisture cycles are also known to affect some types of marble and result in the distortion of cladding panels (usually known as bowing or dishing) and an associated loss of strength as the boundaries between the calcite crystals open. There are now a considerable number of documented examples but the most famous remain the Finlandia Halle in Helsinki and the Amoco Tower in Chicago.

### 18.3.2 Selection of stone

Traditionally, building stone has been extracted and used within the immediate local region – which resulted in the evolution of building design to take account of the materials available and their qualities. Vernacular architecture reflected the properties of the stone, and selection of stone for a project was based on experience and a degree of trial and error. The exceptions were major buildings where stone was imported for architectural or aesthetic reasons or because of the absence of suitable local materials. Many buildings were very conservative in their use of materials, with many structures "over-engineered"; some were more innovative and used new stone in new ways – sometimes successfully but there must also have been many failures that are not recorded but that influence the choice and use of stone.

Over the past 30 years however, stone has become a global industry and raw blocks and finished products are transported great distances and imported from almost any country in the world – leading to their use in regions or countries that are significantly different from their traditional areas. Consequently stones are used in locations where there is no experience of their performance and in new or innovative ways – and sometimes both, with new stones being used in innovative ways. As a result some balance is needed in which the design takes account of the stone's properties or that only stones of suitable quality are used if the design is fixed. But this leaves open the question of how quality can be assessed and which qualities are important for a particular stone or design. It is this that has led to an increased demand for the sampling and testing of stone at various phases in a building project. In response to this need for testing and assessment, many new test methods and codes of practice have been developed at both national and international levels.

*18.3.2.1 Codes and Standards*

A number of British and European Standards and codes of practice deal with the use of stone in buildings, providing a range of guidance on the design and specification of natural stone, as well as tests that can be used to determine the likely performance of the stone in terms of its strength, safety and durability. In general, the specification and testing of natural stone is covered by a series of European Standards while the design and installation are covered by British Standard Codes of Practice. Workmanship is covered by both the codes of practice and a separate series of British Standards. The standards that deal with stone are listed below by building element, along with a brief description of the information they contain.

*18.3.2.1.1 Cladding, lining and tiling* The latest edition of BS 8298, *Design and installation of natural stone cladding and lining*, published in 2010, has a wider scope than its predecessors and addresses the requirements of the currently available range of cladding systems. The new version is made up of the following parts:

- *Part 1:* General – covering items common to the other parts
- *Part 2:* Traditional hand-fixed stone (held to a structural background by metal fixings and used externally)
- *Part 3:* Hand-fixed to pre-cast concrete units (i.e. stone-faced concrete cladding units and used externally)
- *Part 4:* Rain screen and prefabricated systems.

The parts are written to allow the designer to be more innovative while still having provision for more traditional solutions that have been provided by a long history of use. Part 1 contains reference to the European Specification for internal and external cladding in natural stone (BS EN 1469) and to the test results that may be required to support an innovative cladding design. The main tests referred to are strength in bending, strength around the fixings, and durability. In addition to BS EN 1469 there are a number of other European Specifications that are relevant to natural stone cladding and lining when it is being considered as part of a system – for example the air permeability, watertightness, wind load resistance, and impact resistance of a façade.

Tiling, which is units less than 600 × 600 × 12 mm, is covered by two parts of BS5385 *Wall and floor tiling.* Part 1, issued in 2009, covers the design and installation of ceramic, natural stone and mosaic wall tiling in normal internal conditions while Part 3 covers the same products in "special conditions", for example locations where the stone would be wetted on a regular basis.

*18.3.2.1.2 Paving, setts and kerbs* This area is currently very well covered by both codes and standards. The specification and testing of this product group are covered by BS EN 1341 (slabs), BS EN 1342 (setts) and BS EN 1343 (kerbs), which include reference to the main test requirements including slip resistance, abrasion resistance, strength (either in bending or compression) and resistance to freeze/thaw action.

In addition, the extensive series of British Standard Codes of Practice issued as BS7533, *Pavements constructed with clay, natural stone or concrete pavers*, Parts 1–13 cover almost every aspect of the design, installation and reinstatement of lightly and heavily trafficked pavements. The main parts that are relevant to natural stone are:

- *Part 4: Code of practice for the construction of pavements of precast concrete flags or natural stone slabs*
- *Part 5: Guide for the design of pavements (other than structural aspects) (in preparation)*
- *Part 6: Code of practice for construction of pavements – kerbs, channels and edgings*
- *Part 7: Code of practice for the construction of pavements of natural stone setts and cobbles*

- *Part 8: Guide for the structural design of lightly trafficked pavements of precast concrete flags and natural stone slabs*
- *Part 10: Guide for the structural design of trafficked pavements constructed of natural stone setts*
- *Part 11: Code of practice for the opening, maintenance and reinstatement of pavements of concrete, clay and natural stone*
- *Part 12: Guide for the structural design of trafficked pavements constructed of concrete paving flags and natural stone slabs*
- *Part 13: Guide for the design of permeable pavements constructed with concrete paving blocks and flags, natural stone slabs and setts and clay pavers.*

*18.3.2.1.3 Internal flooring and stairs* The specification of stone for internal flooring and stairs is covered by EN 12057 Specification for Modular Tiles and EN 12058 *Specification for Slabs for Floors and Stairs*. The first of these covers modular tiles for both floors and walls. There is also guidance on the design and installation of these products in BS5385 Part 5: 2009 *Wall and Floor Tiling: Design and installation of terrazzo tile and slab, natural stone and composition block flooring.* Design guidance can also be found in the Code of Practice produced by the Stone Federation Great Britain (2005), which formed the basis for many of the amendments found in the 2009 version of BS5385 Part 5.

*18.3.2.1.4 Discontinuous roofing* The strangely named discontinuous roofing deals with traditional roofing slate only, as at present there is no equivalent for traditional sandstone or limestone roofing, though some guidance on the last two groups can be found in English Heritage (2003).

For many years metamorphic slates were covered by BS680, which contained three durability tests for acidic gases, frost, and wetting and drying cycles. However, this has now been replaced by the European standard BS EN 12363, which covers product specification for the materials in Part 1 and the relevant test methods in Part 2. The tests that are referred to include geometry, failure load in bending, water absorption, freeze–thaw, non-carbonate carbon content, sulphur dioxide exposure, and petrographic examination.

*18.3.2.1.5 Masonry* This traditional use of stone is covered by BS EN 771-6, a number of test methods in the BS EN 772 series and BS 5628 Parts 1, 2, and 3. The British Standards and the Eurocodes rely on the application of test methods for individual components and units in order to demonstrate their suitability. The British Standard consists of three parts:

- BS 5628-1:1992, Code of practice for use of masonry. Structural use of un-reinforced masonry
- BS 5628-2:2000, Code of practice for use of masonry. Structural use of reinforced and pre-stressed masonry
- BS 5628-3:2001, Code of practice for use of masonry. Materials and components, design and workmanship.

These British Standards use limit state approach and partial safety factors, and a similar approach is found in the Eurocodes. However, the parts of EN 1996 go further in their specification of test methods for individual products groups and also by having a greater emphasis on structural fire design. In general terms, BS5628-1 and BS5628-2 are covered by EN 1996 Part 1-1 and BS5628 Part 3 by EN 1996-2; part of EN 1996-1-2 will be covered by BS5628-3, with some requirements for fire resistance being given in Building Regulations Approved Document B.

The European specifications and test methods fall into five areas:

- standards and testing methods for masonry units (EN771 and EN772 series)
- standards and testing methods for masonry mortars (EN 998 series)
- standards and testing methods for ancillary components for masonry (EN 845 series)
- methods of test for masonry (EN 1052 series)
- other guidance such as determination of thermal values and application of brickwork finishes including external rendering, internal plastering, etc.

*18.3.2.1.6 Other uses of stone* Stone has many other uses but they are not necessarily covered by codes of practice or standards. Many traditional uses of stone, for example as quays, boundary walls or bridges, are not covered by standards specific to the use of stone but rather by standards for the construction of the particular structure. There are standards for the assessment of armour stone used in coastal sea defences (BS EN 13383: 2002 Armourstone Part 1 Specification and Part 2 Test Methods). Currently there is no standard for stone used as worktops or as vanity units, but there is guidance published by the Stone Federation and there are proposals for a possible European Standard.

*18.3.2.2 Selecting stone: Sustainability and ethical considerations*

Stone extraction and quarrying have a high environmental impact, and stone is an expensive natural material. Therefore it is natural that clients will wish to assess whether stone is an appropriate material to specify. However, stone is a highly durable and practical material, and therefore the negative initial impacts (cost and environment) can often be demonstrated to be offset by long-term benefits. To do this, it is necessary to consider the long-term performance of stone as a material, as well as the specific impacts at various stages of the life cycle.

When it comes to costs, stone is characterized by having high capital costs and low running costs, with a long life. This is a typical profile for a material where it will be highly relevant to demonstrate the relative cost of ownership (whole-life cost) in comparison with competing alternatives.

It should be noted, however, that whole-life costs are considered over different periods of time, and a short period of assessment (e.g. 25 years) or a low discount rate is likely to make stone appear relatively expensive in comparison to low capital cost, high-maintenance alternatives.

Once in use, stone has little negative social impact and in most cases it has a very positive impact, but in the extraction and transportation phase many potential problems are perceived by nearby communities. The potential impacts that are highlighted during the planning stage are usually related to environmental impacts associated with extraction and transportation – destruction of the countryside, dust and waste, and the need for large lorries to transport the blocks of stone. The situation is particularly acute in National Parks or Areas of Outstanding Natural Beauty (AONBs), where the planning controls and lobbying appear particularly severe. However, there are examples where impacts have been minimized, with very limited long-term impacts on the environment. It must also be remembered that for many people stone extraction and processing (with the associated industries) is one of the few profitable primary industries remaining in rural areas, and as such there are positive social impacts as well as the more publicized negative impacts.

One potential reaction to potential social and environmental impacts in the UK is to seek sources of stone outside the UK – usually in less developed countries where labour costs are low, for example India and China. However, there is now a move among some local authorities to question the impacts in the less developed countries – in terms of the destruction of the environment, the social impacts of using child workers, and the exploitation of workers where they are not paid a "fair" wage. In addition, the energy used in transportation and the impacts of this transport must be considered. As a consequence it is important to take a holistic approach to any situation and to attempt to weigh up the positive and negative impacts both in the UK and in the production country if this is outside the UK.

## 18.4 Testing stone

In the past 20 years the testing of stone has become an increasingly important part of the design process for elements in natural stone (Table 18.1). In the past in the UK the only testing carried out related to the durability of the stone, particularly its likely behaviour in a polluted urban environment. But the trend towards more innovative use of stone and the greater range of unfamiliar stones has led to a more rigorous approach to the selection and testing of stone. Most of the tests required are included in the British and European specifications and codes of practice, described above, which have been drafted in response to the "essential requirements" of the Construction Products Directive. The basis for most of the requirements can be seen by undertaking a failure mode effect analysis, which should consider:

- What can go wrong?
- What can we test for this?
- How can we manage the risk?

The tests that can be seen as supporting this analysis fall under three general headings, as follows.

- Is it strong enough (structural)?
  - *Flexural strength/breaking load*
  - *Compressive strength*
  - *Strength around fixings*
  - *Additional sources of information:*
    - BRE Information Paper IP 7/98 "External cladding: how to determine the thickness of natural stone panels"
    - 'Guide to the selection and testing of stone panels for external use', CWCT
    - BS 8200 External envelopes.
- Is it safe (safety in use)?
  - *Breaking load/flexural strength*
  - *Abrasion resistance*
  - *Slip resistance*
  - *Polished paver value*
  - *Additional sources of information:*
    - BRE Information Paper IP 10/00 "Flooring, paving and setts –Requirements for safety in use". 2000
    - SFGB "Natural stone flooring – Code of Practice for the design and installation of internal flooring". 2001
    - CIRIA Report C652 "Safer surfaces to walk on, reducing the risk of slipping". 2006
    - United Kingdom Slip Resistance Group, "The assessment of floor slip resistance" Issue 3. 2006.
- Will it last (durability)?
  - *Changes in properties with time – e.g. flexural strength*
  - *Resistance to frost action*
  - *Resistance to salt*
  - *Resistance to thermal cycles*
  - *Resistance to abrasion*
  - *Additional sources of information:*
    - BRE Information Paper IP 18/98 "Stone cladding panels: in situ weathering"
    - BRE Digest 420 "Selection of natural stone"
    - "Natural stone flooring – Code of Practice for the design and installation of internal flooring" SFGB 2001.

### 18.4.1 Tests appropriate to all stones and all uses

- Examination of petrographic thin sections (BS EN 12407).
- Determination of porosity, saturation coefficient, density and water absorption (BS EN 1936, BS EN 13755 and BRE BR 141).

#### *18.4.1.1 Additional tests for cladding*

- Flexural strength (BS EN 12372 or BS EN 13161) *The specimens should be cut so that the load is as in the proposed use.*
- Strength around the fixing (BS EN 13364) – required if the proposed fixing/panel thickness is outside the requirements of BS8298. *The specimens should be cut so that the load is as in the proposed use.*
- Large-scale impact resistance test to demonstrate compliance with BS8200. *One or two panels with the dimensions of those to be used on site.*

#### *18.4.1.2 Additional tests for load-bearing masonry*

- Compressive strength (BS EN 1926). *The specimens should be cut so that the load is as in the proposed use.*

#### *18.4.1.3 Additional tests for paving and flooring*

- Flexural strength (BS EN 12372 or BS EN 13161) *The specimens should be cut so that the load is as in the proposed use.*
- Slip resistance (BS EN 1341 or BS EN 14231).
- Abrasion resistance (BS EN 1341 or BS EN 14157).

#### *18.4.1.4 Additional tests for locations subject to freeze/thaw action*

- Freeze/Thaw resistance (BS EN 12371). The resistance to freeze/thaw cycles is assessed by recording changes in visual

**Table 18.1** Tests appropriate for different uses of natural stone in buildings

| *Test method* | *Use* | | | | | | |
|---|---|---|---|---|---|---|---|
| | *Cladding* | *Non-load-bearing facing units* | *Copings* | *Sills/ Lintels* | *Load-bearing units* | *Roofing* | *Paving*[1] |
| Petrographic description | ✓ | ✓ | ✓ | ✓ | ✓ | ✓ | ✓ |
| Density | ✓ | ✓ | ✓ | ✓ | ✓ | ✓ | |
| Compressive strength | | | | | ✓ | | ✓[2] |
| Flexural strength | ✓ | | | ✓ | | ✓ | ✓[3] |
| Slip resistance | | | | | | | ✓ |
| Abrasion resistance | | | | | | | ✓ |
| Durability | ✓ | ✓ | ✓ | ✓ | ✓ | ✓ | ✓[4] |

[1]Including setts and kerbs. [2]Applies only to setts. [3]Excluding setts. [4]Does not apply to internal paving.

appearance and changes in elasticity and flexural strength. A more detailed description of this test can be found in Annex B.

*18.4.1.5 Additional tests for locations subject to salts*

- Resistance to salt crystallization (BS EN 12370). A more detailed description of this test can be found in Annex C.

*18.4.1.6 Additional tests for sandstone or slate*

- Acid immersion (BRE BR 141). The resistance to natural acidity from polluted atmospheres is determined using the acid immersion test.

A full list of European test methods for natural stone is given in Annex A.

## 18.5 Properties of stone

The increasing requirement to test stone is in part being driven by an increased requirement for engineering design calculations to accompany proposals to use certain building elements made of stone, particularly in cases where the design is innovative or where the specification is in terms of performance rather than as a prescription. The elements can range from paving and cladding to cantilevered steps, lintels and balustrades. At present there is no single source of data that can provide a comprehensive range of properties of stone, and so testing is often required. But as stone suppliers gain more experience of testing stone, an increasing amount of the data being asked for (for example tensile strength, modulus of rupture, or compressive strengths) has probably been measured for many of the more commonly specified stones from the EU, and so suppliers of building stone are able to provide data for their own stones at the design stage. There are also a number of web or CD databases, for example "Natural Stones Worldwide" (www.natural-stone-database.com).

## 18.6 Maintenance

Planned and routine maintenance is the simplest and most cost-effective form of intervention in ensuring the required performance. Without such maintenance, repairs and major refurbishment become an increasing necessity.

## 18.7 Cleaning

The revised British Standard, BS8221 (in two parts), provides practical advice for contractors in using different cleaning methods and their suitability to different surfaces. However, clients, specifiers and project managers may not know what surface or soiling type is present, or what cleaning methods are available. BRE Digest 418, Part 2 (Murray et al., 2000b) sets out the main cleaning methods, although the categories used may represent a wide range of techniques and products. The choice of cleaning method will reflect the purpose and extent of the cleaning, for example routine maintenance or refurbishment, and the historical importance of the building.

Once the nature of the soiling is known, a cleaning method must be chosen that will break the bond between the soiling or coating and the masonry surface while leaving the surface undamaged. This is usually achieved by softening, dissolving or physically abrading the soiling using one or more cleaning methods. In some cases the masonry surface beneath the soiling becomes destabilized by the cleaning method and it is only when the soiling is removed (by, say, water washing) that the damage is seen. For this and many other reasons, it is very important that any contract for cleaning masonry surfaces includes trial cleaning before full cleaning operations start.

Whatever type of stone is used in a building, it will at some stage begin to appear dirty and so may need cleaning. Deposits on buildings are typically carbonaceous materials, silts and clays from arable land, and soluble salts in the form of aerosols. Staining and darkening of the stone in buildings can be caused by a variety of agents such as general urban grime and run-off from metals, but biological growths have become very prominent as the concentrations of acidic gases have reduced. Biological growths are found on building materials in conditions where their needs of moisture, food and light are met. While the acid metabolism product from organisms can etch limestone, lichens can obscure carvings and inscriptions and algae may produce disfiguring stains on masonry, the most commonly observed effect is the darkening of the surface. This darkening can be very prominent on ferruginous sandstones where there are complex interactions between iron, manganese and the acidic metabolism products from biological growths. The method of removing such growths by cleaning and treatments with biocidal washes such as quaternary ammonium compounds is well documented. Both this and a detailed review of the effect of biological growths on sandstones can be found in Cameron et al. (1997).

Paint and graffiti are particularly difficult to remove from porous masonry because they can be absorbed into the surface pores. If removal with water is unsuccessful then more aggressive materials will be needed. If paint removers are to be used, it is preferable to use water-based ones that comply with BS3761 (Specification for solvent-based paint remover). Paint strippers based on alkali materials can leave harmful deposits in the stone and so should be used with great care, and only as a last resort. Often paint can only be removed by the use of a grinding disk or grit blasting equipment; these are likely to cause physical damage to the surface of the stone and represent a hazard to the operatives, and should only be used with extreme care – predominantly they should be used as either wet grit blast or a vacuum recovery system in addition to essential personal protection systems.

### 18.7.1 General guidance

If possible, the cleaning of a building surface should be carried out as part of a planned maintenance programme. This is more likely to be a feasible proposition for modern stone-clad buildings than for ancient monuments where interventions will always be very limited in extent and frequency. A survey of the building and areas for cleaning is a prerequisite to the actual operation.

The specification for cleaning a particular building will depend on a number of factors such as stone type, the type of deposit to be removed, and whether a particular cleaning method would create a hazard to the fabric of the building or people. The techniques used in cleaning building surfaces are based on various types of material or method of application: grit, abrasives, chemicals, water washing and laser cleaning. Generally speaking, the simpler and less aggressive methods should be used first, followed by more aggressive methods only when necessary. Cleaning methods for various types of stone are described below.

Further guidance on the relevant health and safety requirements can be found in relevant HSE documents and in BRE Digest 448. A summary of the methods used for cleaning different stone types is given in Table 18.2.

### 18.7.2 Cleaning limestones and marble

Water washing is the conventional method for cleaning calcareous stones such as marble and limestones. Water is applied to the area to be cleaned to soften the dirt, which is then removed by light abrasion using a nylon brush or other non-ferrous tool. Heavily soiled areas where large sulphate "crusts" have built up may have the bulk of the crusts removed by mechanical means prior to final cleaning with water, and sometimes it may be necessary to use abrasive or alkaline cleaners as a last resort in localized areas.

## 18.7.3 Cleaning sandstone and unpolished granite

Although these materials can sometimes be cleaned with water, it is usually necessary that they be cleaned by either abrasive methods or chemical cleaning. Dry abrasive methods can generate free-flying silica dust that can cause silicosis, so operatives and members of the public should be protected. The problem is worse with dry grit blasting than with wet abrasive blasting (with or without grit), but the water used in the latter does not remove all the active silica. The water flows will also need to be collected and properly disposed of. After cleaning with dilute acid some iron-containing rocks can develop a brown stain due to the iron coming to the surface. For this type of stone very specialist advice is needed to select the most appropriate cleaner composition and health and safety measures.

## 18.7.4 Cleaning slates and polished granites

For these materials all that is required is that they are washed with clean water with a mild detergent and finished off with a chamois leather. On no account should acid be used to clean polished granite, as this will remove the polish completely.

## 18.7.5 Cleaning techniques

### *18.7.5.1 Water washing*

There many variations in the way in which water washing can be carried out including water jets, persistent wet mists, intermittent or pulse washing, water lances and steam cleaning. There are two inherent dangers associated with water washing, both of which are easily overcome with a little care. The first danger is the risk of frost damage to stonework saturated with water if cleaning takes place in cold weather, and for this reason, washing should not take place if there is the likelihood of frost.

The second danger is that of water penetration of the building leading to damage to internal finishes. The likelihood of this happening can be substantially reduced by using as little water as possible. It is not necessary or desirable to have water flooding down the surface of a building during cleaning; rather it is only necessary to keep the surface moist. This will soften the dirt just as quickly, enabling mild abrasion with bristle or nylon brushes to loosen the dirt, allowing it to be rinsed away. Water lances, which are of limited use alone, can be used to rinse away loosened dirt, often without the need of abrasion.

### *18.7.5.2 Grit blasting*

There are two methods of grit blasting: wet and dry. In both methods air is used to blow an abrasive grit into the dirt which is

**Table 18.2** Stone cleaning methods

| *Method* | *Technique* | *Similar techniques or phrases* | *Surface type* | | | | | |
|---|---|---|---|---|---|---|---|---|
| | | | *Limestone lime mortar joints* | *Sandstone* | *Brick ceramic and terracotta* | *Concrete hard cement mortars* | *Polished* | *Granite* |
| Water washing | Fine sprays | Mists, nebular mist | b | | | | | |
| | Sponging | | r | r | r | r | | |
| | Low pressure rinsing | Water jets, water lance | r,o | r,o | r,o | r,o | | |
| Chemical cleaning | Acids | Hydrofluoric, hydrochloric, phosphotic acid neutralizer | | b,s | b,s | b,s | | b,s |
| | Alkalis and bleaches | Caustic chemicals, amines, ammonia, hyprochlorites | b,c,s,o | b,c,s,o | b,c,s,o | b,c,s,o | b,c,s,o | b,s,o |
| | Solvents | | c,s | c,s | c,s | c,s | c,s | c,s |
| Soaps and detergents | Soap, ionic detergents | | s,o | s,o | s,o | s,o | s,o | s,o |
| | Non-ionic detergents | | s,o | s,o | s,o | s,o | s,o | s,o |
| Air abrasives | Micro-air abrasives | | b,c,s | b,c,s | b,c,s | b,c,s | | b,c,s |
| | Fine air abrasives | Soft media | b,c | b,c | b,c | b,c | | b,c |
| | Fine air abrasives and water | Soft media, vortex | b,c | b,c | b,c | b,c | | b,c |
| | Air blown grit abrasives | Grit blasting sand blasting | b,c | b,c | b,c | b,c | | b,c |
| | Air blown frit abrasives and water | Grit slurry | b,c | b,c | b,c | b,c | | b,c |
| Brushes | Brushes | Brush-down | b,o,r | b,o,r | b,o,r,s | b,o,r | b,o,r | b,o,r |
| Water as abrasive | Cold pressurized jets | Water lance | o,r | o,r | o,r | o,r | o,r | o,r |
| | Hot pressurized jets | | o,r | o,r | o,r | o,r | o,r | o,r |
| | Micro-steam cleaning | | b,c,s | b,c,s | b,c,s | b,c,s | b,c,s | b,c,s |
| | Steam cleaning | | b,c,s | b,c,s | b,c,s | b,c,s | c,s | |

Prior to cleaning it is important to take in to account the variable standard of surface weathering shown by individual materials (e.g. linestone, sandstone and terracotta). Further guidance is given in BS8221-1 and BS8221-2. b = black soiling, c = coatings, s = residual stains, o = algae, plants and organics, r = residues from light dust or abrasive particle removal scoured away. In the case of wet grit blasting water is introduced into the air/abrasive stream. Both the size of the gun nozzle and the grit size can be varied according to the delicacy to the work and the degree of soiling. Extreme care must be taken in all cases, however, because a great deal of damage can be done if the operatives are inexperienced. Air pressures and abrasive feed to dry air abrasion nozzles should be carefully controlled. The lowest pressure practicable for the cleaning operation should be used.

Traditionally, the grits employed for abrasive cleaning have been sand and flint, but these are now known to be a health hazard due to their free silica content and so should not be used. The hazard is not eliminated with wet grit blasting, and the Health & Safety Executive advises that non-silica grits should be used wherever possible. Even if the non-silica grits are used, free silica can be released from sandstone, granites, and brick during abrasive cleaning and adequate safety precautions (for the protection of the operator and the public) are essential. After dry grit blasting, residual dust will remain on the building, and likewise a residual slurry will cover parts of the building after wet grit blasting. In both cases the residue can be washed off using a pressure water lance, provided risk of excessive mechanical damage and water ingress are acceptable. Failure to do so will cause staining because the residues contain dirt that can be driven back into the pore structure of the stone by rain.

*18.7.5.3 Mechanical cleaning*

Mechanical methods of cleaning include brushing, scraping and rubbing with carborundum stones. The use of power tools with carborundum heads to "spin off" the dirt is no longer considered acceptable except in very limited and specific cases where the process can be controlled, because considerable damage can be caused to soft stones and stone mouldings. The technique produces dust and if siliceous materials are being worked the dust can be dangerous to health. Again, adequate safety precautions are essential.

*18.7.5.4 Chemical cleaning*

Chemical cleaners fall into two main classes, acids and alkalis. The main danger in using chemical cleaners, apart from H&S issues on site, is contamination of the stone with soluble salts which can cause crystallization damage. These arise either from the cleaning material itself or from the reaction of the cleaner with the fabric of the stone. With acid cleaner, soluble salts can arise from the use of neutralizing agents which are said to deactivate the residual acid. If the acids are used correctly neutralizing agents are not necessary. Alkaline cleaners can react with acidic species in the atmosphere to produce salts; once in the stone, soluble salts are extremely difficult to remove and can cause progressive damage for a long time.

## 18.8 Repair of decayed stonework

Probably the most important consideration in the durability of stone masonry is the condition of the mortar joints. If a wall is showing distress – even if nothing else is done – the pointing should be made good (see below) often, but more work will need to be done and this should always be carried out before re-pointing. The extent of any repair work is usually determined by a combination of the conservation needs and the cost. There are four basic methods for repair, each of which will usually be accompanied by re-pointing:

1 de-scale, i.e. remove any loose material
2 dress back the stonework to a sound surface
3 replace in new stone
4 dress back the stonework and rebuild in mortar (this process is known as plastic repair).

Methods (1) and (2) are often quite adequate from the point of view of durability, and are particularly suited to situations where the decay is fairly uniform. Redressing stone is removal of loose pieces from the face by tooling rather than brushing. The procedure is rarely used on historic masonry because it removes the face of the stone and therefore the original tooling or other finishes. Redressing should be used in cases of severe surface deterioration within otherwise sound blocks. The work should be carried out by skilled stone masons using hand techniques (BS8221 Part 1)

Methods (3) and (4) are more labour-intensive and, therefore, more expensive, but their relative cost depends on the particular job to be done. Plastic repair is sometimes regarded as a lower cost option; however, good plastic repairs are often more expensive than replacement but do allow retention of a greater amount of existing stonework. For example, to repair an ashlar block in mortar will probably be more expensive than replacing in new stone when labour is taken into account, but repair of carved stone such as string courses may be more cost-effective than replacement.

Plastic repair to coping and other details that frequently become saturated is not recommended unless the repaired coping is to be further protected by lead or similar durable sheet.

### 18.8.1 Re-pointing

Great care should be taken when specifying re-pointing because much damage can be done both in preparation of the joint before re-pointing and by the use of inappropriate mortars. Removal of the old mortar can be achieved in a variety of ways, and the method chosen will depend on the condition of the joint. Those joints where the existing mortar is badly decayed can easily be raked out with a bent spike, whereas those that are less decayed may need to be cut out with a hammer and plugging chisel. In extreme cases when very hard mortar must be removed, it may be necessary to use power-driven abrasive disks, but this requires great care to avoid damaging the edges of the stones. Whatever method is used, care should always be taken to avoid injury to operatives and damage to the stones.

Joints should be cut out square and to a depth of at least 25 mm or 1.5 times the thickness of the joint, whichever is the greater. After cutting out, all the dust should be removed from the joint and the joint should be rinsed with water. When the joint is pointed the stone should be damp, but not saturated.

The durability of the pointing will depend on two factors: the mortar used and the finish profile given to the face of the joint. The mortar used for re-pointing should be no stronger than the stone that will surround it, and should be tamped into the joint such that is well compacted and there are no air voids left. BS8221 Part 1 gives tables of mortar mixes for use with various types of stone. Finishes to the joint that have mortar projecting forward of the joint (e.g. ribbon pointing) are to be avoided, as are joints where the face of the mortar meets the stone at a shallow angle (feather edges). Where the arrises of the stone have been lost, the joint should be recessed to avoid feather edges, or profiled so that the mortar face meets the stone at right angles or nearly so.

### 18.8.2 Plastic repair

The term "plastic repair" refers to the building up of decayed stonework with a mortar. The word "plastic" refers to the consistency of the mortar mix, and does not imply that it contains any plastic (i.e. polymeric) material. When this type of repair is employed is essential that the area to be repaired is properly prepared. Any decayed stone should be cut away and the edges of the repairs should be undercut to provide a dovetail key. Feather edges to repairs should never be used; for larger areas a mechanical key in the form of ragged non-ferrous dowels should be used. Non-ferrous screws, or non-ferrous wire firmly fixed to the back ground, should be employed.

Plastic repairs should never continue across mortar joints. Rather each block of stone should be treated independently of its neighbours and the repaired block pointed afterwards in the normal manner when the repairs have cured. Mortar should be mixed to match the colour and texture of the stone being repaired. Ashurst and Ashhurst (1988) give some examples of repair mixes, which should never be stronger than the stone. The repair is built up in layers up to 10 mm in thickness, each layer being allowed to cure to allow the initial shrinkage to occur. In some cases it may be necessary to use stainless steel armatures to provide support for the repair. The final layer is put on proud of the finish level and worked back when green to reproduce the texture and finish of the surrounding stonework. Iron tools should not be used in the final working of the surface.

### 18.8.3 Repairs in new stone – piecing-in

Repairs by this method can be quite small, down to 5 $cm^2$ on the face. The thickness of the repair will depend on the method of fixing and the size of the repair, but will not normally be less than 2.5 cm. Large pieces of stone can be fixed by normal masonry techniques using cramps and ties where appropriate. Smaller pieces can be grouted in or fixed with non-ferrous dowels and resin, and grouted after the resin has set.

## 18.9 Stone treatments

Stone treatments are materials that, when applied to friable stonework, are intended to stabilize decaying stonework and slow down or stop further decay. Although treatments have been in use for more than 100 years, their acceptance is still very limited in the UK partly because of past experiences where after the application of some "wonder treatment" the surface of the stone has spalled, but also because most treatments are irreversible and so are inconsistent with current conservation practices. In mainland Europe there has been a wider acceptance of chemical treatments, and the application of consolidants in conservation work and in new build is much more common. There are "consolidants" that essentially deposit solid material in pores – silicates, silicofluorides etc. or organic resins. They try to "improve" by reducing porosity/absorption, increasing density and providing greater strength in friable material. Then there are pore-lining, absorption-reducing, water-repellent materials that do not reduce porosity – e.g. silane, silicone, siloxane. The two approaches are very different in character, behaviour and performance. To many people "treatment" means "silane" but over the years several materials have been tried, ranging from early work with shellac and linseed oil to the current use of (among others) silicon-based products and acidic esters.

### 18.9.1 Performance requirements of a consolidant

What is required of a consolidant or repellent depends on the environment in which the stone is expected to survive. If the stone is expected to be exposed in a controlled environment such as a museum, the approach to consolidation will be different to that for material to be exposed outdoors. For building exterior stone a treatment should achieve the following.

- A consolidant should penetrate deeply into the stone. This is a most important feature. Treatments that penetrate only a few millimetres can produce a surface skin of consolidate material that has markedly different mechanical and physical properties to the underlying stone. This can lead to the surface of the stone blistering off.
- If a water-repellent surface is required then the treatment may remain on the surface of stone or only penetrate by a few millimetres. It is important that the appearance of the surface is not altered by the application of the treatment.
- In both cases, the treated part of the stone should remain sufficiently permeable to water vapour in order to avoid an accumulation of moisture and/or salts behind the treated stone.

Further information on the use of silanes can be found in Wheeler (2005) and Martin *et al* (2002).

## 18.10 References and bibliography


Ashurst, J. and Ashurst, N. (1988). *Stone Masonry: Practical Building Conservation*, Volume 1. Gower Publishing, London, UK.

Ashurst, N. (1994a). *Cleaning Historic Buildings. Volume 1: Substrates, Soiling and Investigations*. Donhead Publishing, Shaftesbury, UK.

Ashurst, N. (1994b).*Cleaning Historic Buildings. Volume 2: Cleaning Materials and Processes*. Donhead Publishing, Shaftesbury, UK.

BRE Information Papers IP 9/99 (1999). *Cleaning exterior masonry. Pre-treatment assessment of a stone building*. BRE, Watford, UK.

BRE Information Papers IP 10/99 (1999). *Cleaning exterior masonry. Assessing cleaning of a brick-built church*. BRE, Watford, UK.

BRE Good Repair Guides GRG 24 (1999). *Re-pointing external brickwork walls*, BRE, Watford, UK.

BRE Good Repair Guides GRG 27 (2000a). *Cleaning external walls of buildings. Part 1. Cleaning methods*. BRE, Watford, UK.

BRE Good Repair Guides GRG 27 (2000b). *Cleaning external walls of buildings. Part 2. Removing dirt and stains*. BRE, Watford, UK.

BS 8221-1:2000 (2000a). *Code of practice for cleaning and surface repair of buildings. Cleaning of natural stones, brick, terracotta and concrete*. British Standards Institution, London, UK.

BS 8221-2:2000 (2000b). *Code of practice for cleaning and surface repair of buildings. Surface repair of natural stones, brick and terracotta*. British Standards Institution, London, UK.

Cameron, S., Urquhart, D.C.M. and Young, M.E. (1997). *Biological growths on sandstone buildings: Control & treatment*. Historic Scotland Technical Advice Note 10, Historic Scotland, Edinburgh, UK.

English Heritage (2003). *Stone roofing*. English Heritage Research Transactions Volume 9, James and James, London, UK.

Henry, A. (ed.) (2006). *Stone Conservation: Principles and Practice*. Donhead Publishing, Shaftsbury, UK.

Leary, E. (1983).*The Building Limestone of the British Isles*. HMSO Report SO 36, HMSO, London, UK.

Martin, W., Mason, D., Teutonico, J.-M., Chapman, S., Butlin, R.N., and Yates, T.J.S. (2002). Stone consolidants: Brethane™. Report of an 18-year review of Brethane™-treated sites. In Fidler, J. (ed.) *Stone: English Heritage Research Transactions Volume 2*. James and James, London, UK.

Murray, M., Pryke, S. and Yates, T.J.S. (2000a). *Cleaning exterior masonry, Part 1 Developing and implementing a strategy*. BRE Digest 418. BRE Press, London, UK.

Murray, M., Pryke, S. and Yates, T.J.S. (2000b). *Cleaning exterior masonry: Methods and materials*. BRE Digest 418 Part 2. BRE Press, London, UK.

Pryke, S. (2000). *Cleaning buildings. Legislation and good practice*. BRE Digest 448. BRE Press, London, UK.

Stone Federation (2005). *Natural stone flooring: Code of Practice for the design and installation of internal flooring*. Publication FC02, SFGB, Folkestone, UK.

Wheeler, G.E. (2005). *Alkoxysilanes and the Consolidation of Stone*. Getty Trust Publications, Santa Monica, CA.


## Annex A: List of European specifications and test methods for natural stone

### Specifications

- Semi-finished products
  - EN 1467 Specification for Rough Blocks
  - EN 1468 Specification for Rough Slabs
- Internal and external cladding
  - EN 1469 Specification for Cladding
- Internal flooring
  - EN 12057 Specification for Modular Tiles
  - EN 12058 Specification for Slabs for Floors and Stairs

- Slate roofing
  - EN 12326–1 Specification
  - EN 12326–2 Test methods
- Natural stone masonry
  - EN 771-6 Specification for Masonry Units. Natural Stone Masonry Units
- Paving, setts and kerbs
  - EN 1341 Slabs of natural stone for external paving. Requirements and test methods
  - EN 1342 Setts of natural stone for external paving. Requirements and test methods
  - EN 1343 Kerbs of natural stone for external paving. Requirements and test methods

### Test methods

- EN 1925 Natural stone test methods – Determination of water absorption coefficient by capillarity
- EN 1926 Natural stone test methods – Determination of compressive strength
- EN 1936 Natural stone test methods – Determination of real density and apparent density and of total porosity and open porosity
- EN 12370 Natural stone test methods – Determination of resistance to crystallisation
- EN 12371 Natural stone test methods – Determination of frost resistance
- EN 12372 Natural stone test methods – Determination of flexural strength under concentrated load
- EN 12407 Natural stone test methods – Petrographic description
- EN 12440 Denomination of natural stones
- EN 12670 Terminology of natural stones
- EN 13161 Natural stone test methods – Determination of flexural strength under constant momentum
- EN 13364 Natural stone test methods – Determination of breaking load around a fixing point
- EN 13373 Natural stone test methods – Determination of geometric characteristics on units
- EN 13755 Natural stone test methods – Determination of water absorption at atmospheric pressure
- EN 13919 Natural stone test methods – Determination of resistance to ageing by humidity, temperature, $SO_2$ action
- EN 14066 Natural stone test methods – Determination of thermal shock resistance
- EN 14147 Natural stone test methods – Determination of ageing by salt mist
- EN 14157 Natural stone test methods – Determination of abrasion resistance
- EN 14158 Natural stone test methods – Determination of rupture energy
- EN 14205 Natural stone test methods – Determination of Knoop hardness
- EN 14231 Natural stone test methods – Determination of slip resistance by means of the pendulum tester
- EN 14579 Natural stone test methods – Determination of sound speed propagation
- EN 14580 Natural stone test methods – Determination of static elastic modulus
- EN 14581 Natural stone test methods – Determination of linear thermal expansion coefficient
- EN1341 Annex E Determination of unpolished slip resistance
- EN 772-4 Determination of real density and apparent density and of total porosity and open porosity
- EN 772-1 Determination of compressive strength

## Annex B: Freeze–thaw test - Determination of resistance to freeze–thaw cycles

[Note: Full details can be found in European Standard BS EN 12371.] A representative number of test specimens, but not fewer than five (usually 280 mm × 50 mm × 50 mm with any anisotropic feature parallel to the long axis) are prepared from the material under test and dried in an oven at 70°C ± 5°C to constant weight. After cooling the specimens are immersed in water at 20°C ± 5°C until constant mass is reached. Initial measurements of apparent volume and the natural fundamental resonant frequency are then made. The specimens are then placed vertically in a tank inside a freezing chamber capable of reducing the specimen temperature to < –10°C in 6 hours and then thawing the sample by the addition of water at 5°C for 6 hours. This 12 hour cycle is repeated for up to 240 times with repeat measurements of apparent volume and natural frequency made at least every sixth cycle up to the 48th cycle, and then at least every 12 cycles thereafter.

### Assessment of results

The specimens are classified as failed when there are specified changes in the natural frequency of the stone or in the apparent volume. The mean number of cycles before failure occurred is recorded.

## Annex C: Sodium sulphate crystallization test - Resistance to salt crystallization cycles

[Note: Full details can be found in European Standard BS EN 12370.] A representative number of test cubes (40 × 40 × 40 mm) are prepared from the material under test and also from samples of stone of known durability, and dried in an oven at 103 + 2°C to constant weight. When the variability of the test material is unknown, six test pieces are taken. After cooling, the cubes are completely immersed in individual containers in a solution containing the equivalent of 14 g of sodium sulphate decahydrate in every 100 g of solution (the specific gravity of the solution at 20°C is 1.055). The immersion is continued for two hours during which time the temperature of the solution is kept at 20 ± 0.5°C. After immersion the test cubes are dried in an oven arranged to provide a high relative humidity in the early stages of the drying and to raise the temperature of the cubes to 103 ± 2°C in not less than 10 and not more than 15 hours. The initial high relative humidity may be obtained by placing a tray of water in the cold oven and switching on the heater for half an hour before putting in the cubes. About 300 ml of water has been found to be adequate in an oven capable of taking 48 cubes. The cubes are left in the oven for at least 16 hours and are then cooled to room temperature before they are re-immersed in fresh sodium sulphate solution. This cycle of operations is carried out 15 times and the cubes are washed in deionized water, and then weighed after drying if they are still coherent.

### Assessment of results

The results are usually reported in terms of loss of weight, expressed as a percentage of the initial dry weight, or as the number of cycles required to induce failure if the test cube is too shattered for weighing before the end of the 15th cycle. A photographic record of the final condition is often informative. Conclusions about the durability of the stone when subjected to salt, for example in a marine environment or where subject to de-icing salts, are drawn by comparing the results obtained with the test pieces with those obtained with the cubes of stone of known resistance.

19

# Polymers in Construction
## An Overview

**Arthur Lyons** MA MSc PhD Dip Arch Hon LRSA
De Montfort University

### Contents

## 19.1 Introduction

Polymers, at the molecular level, are chain structures composed of many hundreds or thousands of smaller repeat monomer units. Naturally occurring polymers include wood, which is based on cellulose, rubber (polyisoprene) and many traditional adhesives such as casein and animal glues. The synthesised polymers derived from the chemical and petrochemical industries are now the larger sector. The UK plastics industry has an annual turnover of £17.5 billion. Worldwide, approximately 4% of crude oil production is converted into plastics products, while consuming a further 3–4% of all oil production as fuel in the process. Less than 10% of plastic material is currently recycled in the UK.

## 19.2 Polymerisation processes

Polymerisation is the joining together of small monomer units into long-chain molecules. In the simplest case, that of the polymerisation of ethylene gas into polyethylene (PE), the individual small monomer units are repeatedly joined end to end to produce the long carbon backbone chain. A similar process converts vinyl chloride into polyvinyl chloride (PVC), styrene into polystyrene and tetrafluoroethylene into polytetrafluoroethylene (PTFE). This type of reaction is addition polymerisation, in which no by-products are eliminated in the process.

Addition polymerisation requires an initiator to start the process. These are usually free radical initiators, but ionic polymerisation is more specific for some monomers. Free radical initiators such as benzoyl peroxide split into two free radicals – highly reactive but neutral species, which then set off the chain addition polymerisation reaction. Ionic polymerisation may be anionic or cationic, depending whether the initiating species is negatively or positively charged. In all cases a chain reaction commences, linking the small monomer units end to end into longer and longer molecular units, i.e. the polymer. Anionic addition polymerisation is initiated by agents such as n-butyl lithium and is used to polymerise certain vinyls and polyethers. Cationic polymerisation is initiated by agents such as acids, boron trifluoride, aluminium trichloride or titanium tetrachloride and is used in the manufacture of some polyethers and butyl rubber. The condensation polymerisation process involves the elimination of water between adjacent monomer units and the incorporation of atoms such as oxygen or nitrogen into the backbone of the macromolecular chains as in polyesters and polyamides respectively.

Depending on the type of initiator and the presence of any special catalyst, the polymer chains produced may be linear or branched. In the case of polyethylene, a highly branched low-density soft and waxy product (LDPE) is obtained under standard free radical polymerisation, but in the presence of a stereo-specific catalyst (Ziegler–Natta catalyst) a high-density, more rigid linear polymer (HDPE) is produced. The degree of branching significantly affects the packing of adjacent polymer chains.

When two or more different monomers are polymerised together, the product is a copolymer. Depending on the affinity of the monomers they will copolymerise randomly, alternately or in block sequences. Block sequences that will have some characteristics of the individual homopolymers can be made by mixing oligomers (slightly prepolymerised) components. Thus the copolymer may be at the sub-microscopic nanoscale, spheres of one homopolymer within a matrix of the other polymer if the individual monomer units lack mutual affinity.

When produced, most polymers consist of randomly oriented molecular chains. However, if the solid material is stretched in one direction, there is a tendency for the chains to become aligned, turning an amorphous material into a partially crystalline polymer. This modifies the physical properties and leads to anisotropy. In certain cases local crystalline regions may be produced on solidification from the melt, but their extents are limited by the entanglement of the molecular chains. In the molten state the individual molecular chains of a thermoplastic polymer move freely relative to each other, allowing the polymer to be moulded. As the material cools the freedom of movement is more restricted down to the point of solidification. However, even in the solid state, the molecular chains of thermoplastics have some movement, giving flexibility. Only when the temperature is lowered below the 'glass transition temperature' does the polymer become rigid and brittle. This particular temperature varies with different polymers or copolymers and is significantly modified by the addition of plasticisers. Generally the softening of thermoplastics on heating is fully reversible, provided that excessive temperatures which would cause degradation are avoided. Thermoplastics tend to swell with organic solvents as these force the molecular chains apart, which is the action of plasticisers in materials such as PVC.

Thermosetting plastics have a three-dimensional cross-linked structure, formed by the linkage of adjacent molecular chains. This structure prevents the melting and re-forming of thermosets, as the molecular chains are not free to move relative to each other on heating. Thermosetting plastics are therefore made and cured in their required form, from liquid resins, powders, partially cured polymers or two-component systems composed of a resin and a cross-linking hardener. Because of the rigid three-dimensional structure, most thermosetting plastics are resistant to swelling by solvents and are harder materials than thermoplastic products. Thermosets can only be recycled as solid plastic aggregates.

Elastomers are long-chain polymers in which the naturally helical or zig-zag structure of the molecular chains is free to uncoil with applied loads and recover when the load is removed. The extent of the extensibility depends on the nature of the polymer chains and the degree of cross-linking between adjacent chains. Some degree of cross-linking, as in the vulcanisation of rubber, is usually required to ensure the material returns to its original form when the load is released, rather than exhibiting plastic flow.

Most commercially produced plastics for specific markets contain small proportions of additives. Plasticisers are added to PVC to increase flexibility for floor coverings, electrical insulation and roof membranes. Fillers such as chalk, sand and carbon black are added to reduce costs, improve fire resistance or increase opacity. For example, titanium dioxide is added to PVC to produce a shiny white surface. Dyes and pigments are blended to give colour; stabilisers are incorporated to give appropriate resistance to UV degradation.

## 19.3 Plastics forming processes

Thermoplastic polymers are usually supplied to fabricators as sheets or granules for forming. Continuous extrusion through a die is the normal process for manufacturing profile sections such as pipes, glazing extrusions and rainwater goods. For producing sheet material, such as damp-proof membranes, a continuous molten plastic tube is extruded which is then inflated to expand it into a cylindrical plastic sheet. This is then rolled flat and trimmed. Laminated sheets can be manufactured by calendering different layers of polymers together between heated rollers.

Injection moulding is the standard process for producing individual components. Thermoplastic granules are heated and then rammed between the two matched halves of the mould. When the plastic is cooled, the mould is opened and the component removed. For thermosetting resins the resin mixture is injected at a low temperature and the mould heated to activate the cross-linking process. An alternative process for thermosetting resins involves simultaneously injecting two polymer precursors into the mould where they mix, polymerise and cross-link.

Hollow thermoplastic components are manufactured by blow moulding. A molten tube of plastic is extruded into the die followed by an air blast which inflates the tube to the internal shape of the mould. Thermoforming requires thermoplastic sheet material to be heated over a mould which is then evacuated to suck the plastic into the mould form. The process may be supported by the application of air pressure on the other face of the plastic sheet, as this gives an enhanced fit to the mould form. Compression moulding is used for thermosetting polymers. A measured quantity of the resin precursor is injected into the mould, which is then closed off. Heat and pressure are applied so the material flows into all the void spaces where it polymerises and hardens to the required form.

Pultrusion is a process used for manufacturing reinforced constant cross-section profiles. Reinforcing strands, usually glass fibres, are pulled through a bath of uncrosslinked thermosetting resin, then through a heated die of the appropriate cross-section which initiates the cross-linking and solidification of the thermosetting resin to the required form, which is then pulled through and cut to appropriate lengths.

## 19.4 Development of natural product plastics and synthetic thermosetting resins

Early plastics were formulated from natural products. Natural rubber from the *Hevea brasiliensis* tree had been known for many centuries, but its hardening by the addition of sulfur was developed by Charles Goodyear in the USA in 1839, and solid vulcanite was first patented in 1843 by Thomas Hancock in the UK. Ebonite is the product containing 30–40% of sulfur and is currently used for making bowling balls. Natural rubber, predominantly cis-isoprene, is now normally reinforced with carbon and treated with anti-oxidants to prevent degradation. It is currently used for flooring and laminated rubber–metal anti-vibration and anti-earthquake bearings for buildings and structures. Its use for electrical cables has been largely superseded by the use of PVC. Most of the natural rubber is currently produced in Thailand, Malaysia and Indonesia, but it also grown in Africa, India, China and South America. Natural rubber production was approximately 11 million tonnes per year in 2011.

Cellulose nitrate, the first semi-synthetic polymer, was made by Alexander Parkes from cellulose fibres and nitric acid in 1856 and named Parkesine. It could be formed into small objects such as cutlery handles and ornaments. Parkesine was rebranded as Xylonite in 1869. The material was thermoplastic as it could be moulded when heated, and then retained its shape on cooling to room temperature. In a further development in 1870, the American John Wesley Hyatt added camphor as a plasticiser to the cellulose nitrate formulation and produced Celluloid, which was used in the production of billiard balls and photographic film. Cellulose nitrate and acetate were the basis of lacquers used in wood finishing.

The first fully synthetic polymer was Bakelite, discovered by the Belgian-born chemist Leo Hendrik Baekeland in 1907. Bakelite, a dark brown coloured phenol-formaldehyde resin, was manufactured from phenol (carbolic acid form coal tar) and formaldehyde under pressure at an elevated temperature. Leo Baekeland is quoted as writing in his diary at the time, 'I know this will be an important invention.' The material, similar in appearance to wood and with good heat resistance and electrical insulating properties, was quickly taken up for the manufacture of small components including radio cabinets and ashtrays. Within construction the main uses were for electrical components and secondary components such as toilet seats and door knobs.

The 1920s saw the development of urea formaldehyde resin, manufactured from carbon dioxide, ammonia and formaldehyde. This was the first light-coloured thermoplastic to which pigments could be added, giving pastel shades and coloured patterns like marble. The material became popular for the production of tableware and for laminates and adhesives. Melamine formaldehyde resins followed in the 1930s, giving a more heat-resistant product. Melamine then became the more favoured material for the production of tableware and domestic items despite its slightly higher cost than urea formaldehyde. Melamine thermoset laminates continue to be used extensively for the production of hygienic, coloured or patterned work surfaces and wall linings.

## 19.5 Development of thermoplastics

In 1933, Eric Fawcett and Reginald Gibson, while experimenting with high-pressure ethylene, produced the waxy solid low-density polyethylene, which went into production with ICI in 1939. Developments in the following decade led to the widespread use of polythene for a range of domestic outlets. Production of polythene is currently approximately 70 million tonnes per year, which includes high-density polythene developed in 1951 by Paul Hogan and Robert Banks. Currently grades range from very low density, through medium to high density with molecular weights ranging from low to high, combining to give a wide range of physical properties. High-density polythene (HDPE), which has a low proportion of branching, is used in water pipes, while medium-density polythene is used in gas pipes and fittings. Cross-linked polyethylene (PEX), which has some elastomeric properties, is used for certain potable water pipes and underfloor heating pipework as it can withstand temperatures up to 90°C. Low-density polyethylene (LDPE) is widely used for damp-proof membranes.

Around the 1930s the petrochemical industry developed techniques for the manufacture of monomers and their polymers from petroleum. These included polymethyl methacrylate (acrylic PMMA), polystyrene (PS) and polyvinyl chloride (PVC). Polymethyl methacrylate was commercially produced by Rohm and Haas in 1927 and by ICI under the trade name Perspex in 1934. It is extensively used for light fittings and signage; also for glazing due to its greater resistance to breakage and greater light transmission than standard glass. Acrylic is widely used in sanitary ware, particularly for baths, where it is usually reinforced on the underside with glass-fibre reinforced polyester (GRP).

Although polyvinyl chloride was discovered by accident in the nineteenth century through experiments with vinyl chloride, it was not until 1926, when Walter Semon invented a plasticiser, that the material became commercially usable. It was developed as an insulator for electric cables during the Second World War, and this has continued to be a major use of PVC. Globally, the construction industry consumes over 50% of the world production of PVC. In addition to electrical insulation, plasticised PVC is widely used for floor finishes, single-layer roofing systems and many small components produced by injection moulding techniques which were significantly developed in the 1930s. Unplasticised polyvinyl chloride (PVC-U) was developed in the 1940s and has since made a major contribution within the construction industry. Rigid PVC has replaced a large proportion of the heavier cast-iron rainwater goods and soil, drainage and water supply pipes. Production techniques offering continuous lengths, coloured as required and stabilised to prevent degradation by weathering and sunlight, are now used to manufacture hollow sections for double glazing frames, cladding, fascias, bargeboards, etc. The co-extrusion of high-impact PVC-U with a core of PVC foam increases the strength of units such as sills, soffits and cladding. Clear extruded PVC-U profile sheeting offers an alternative to glass-fibre reinforced cement for agricultural units, and PVC plastisol is the most widely used finish to sheet steel cladding and roofing on commercial buildings. Chlorinated polyvinyl chloride (CPVC), a derivative of PVC with a higher softening temperature, can be used for hot water systems.

Although initially discovered in the nineteenth century, polystyrene (PS) was not produced commercially until 1937, when the accidental polymerisation of the monomer in storage could be prevented. Expanded polystyrene was developed by Ray McIntire of Dow Chemicals in 1954 and is widely used in construction for insulation including within structural insulated panels (SIPs) and in permanent insulating formwork. Expanded polystyrene is also used for creating voids in precast and *in-situ* concrete. Extruded polystyrene, although slightly denser than expanded polystyrene, has a lower thermal conductivity and is therefore also used for floor, wall and roof insulation.

Polycarbonates were discovered independently by Daniel Fox of General Electric and Hermann Schnell of Bayer in 1955. Commercial production began in 1960. Polycarbonates are widely used as vandal- and bullet-resistant glazing as they are virtually unbreakable, with a higher impact resistance than the alternative polymethyl methacrylate.

Polypropylene (PP) was discovered by Paul Hogan and Robert Banks of Phillips Petroleum in 1951 when they were attempting to convert ethylene and propylene into petroleum. A significant development in 1954 was the formulation by Guilio Natta of specific reagents for catalysis of the polymerisation reaction. Polypropylene is stiffer than polyethylene and is used for manufacturing pipes, water tanks, WC cisterns, etc. Certain breather membranes are multi-layer systems incorporating a combination of polypropylene, polythene and glass-fibre reinforcement; also many geotextiles for soil stabilisation are manufactured from non-woven, heat-bonded, continuous polypropylene fibres.

Polytetrafluoroethylene (PTFE) was discovered in 1938 by Roy Plunkett at DuPont's laboratories. The accidental polymerisation of tetrafluoroethylene had produced a white waxy solid. PTFE is widely used where low friction is required – for example, plumbing tape – and also in PTFE-coated glass-fibre tensile structure fabrics as a more durable alternative to PVC-coated polyester. Closely related is polyvinylidene fluoride (PVDF), which was discovered by Pennsalt in 1960 and is used as a durable smooth coating to sheet steel. Ethylene tetrafluoroethylene (ETFE), a copolymer of ethylene and tetrafluoroethylene developed by DuPont in 1970, is now the standard membrane material for inflated structures such as the Leicester Space Centre and the Eden Project in Cornwall.

Acrylonitrile butadiene styrene (ABS) is a two-phase material composed of a copolymer of styrene and acrylonitrile (SAN) and some terpolymer, with butadiene dispersed in the SAN matrix. The ratios of the three monomers can be varied to change the physical properties of the resultant ABS. Reinforcement or polymer blending with poly methyl methacrylate (MABS) or polyamides (ABS/PA) also enhances the range of physical properties available. ABS was a natural development in the 1950s of the styrene-acrylonitrile copolymers produced in the 1940s. ABS is extensively used in construction for rainwater and drainage goods because it retains strength at low temperatures.

Other thermoplastic polymers used within construction include nylon, developed by Wallace Carothers and patented by DuPont in 1935. Nylon is used for small low-friction components such as hinges, but tends to degrade under prolonged exposure to sunlight. Nylon carpets are durable and hard wearing.

## 19.6 Development of elastomers

Currently 60% of the rubber market is taken up by the synthetic elastomers. In the late nineteenth century it was discovered that natural rubber is a polymer of isoprene, and this monomer was first polymerised by Bouchardt around 1880. However, it took shortages of natural rubber in two world wars to accelerate the commercial development of synthetic elastomers. Several products were then developed including styrene-butadiene, polybutadiene, neoprene, EPDM (ethylene propylene diene monomer), butyl rubber and nitrile rubber; of these, styrene-butadiene, neoprene, EPDM and butyl rubber are used extensively in construction. Production of polyisoprene with the same chemical composition and physical properties as natural rubber was finally achieved in 1954 using the stereo-specific Ziegler–Natta catalysts. It was then discovered by Shell in the 1960s that the addition of styrene monomer to the anionic polymerisation of isoprene in block form led to the production of a rubber that required no vulcanisation. These thermoplastic elastomers were named Kraton and are the basis of certain adhesives and sealants.

Styrene-butadiene styrene block copolymers developed in 1965 are a key component in the SBS polymer modified bitumen roofing sheets, which are more elastic and durable than the earlier standard oxidised bitumen sheets. They withstand the higher temperatures experienced by well-insulated roofs.

Neoprene (polychloroprene) was the first synthetic rubber developed by Wallace Carothers' research group at DuPont in the 1930s. It is resistant to chemical attack, heat and also oils, which tend to swell other elastomers. It is therefore used extensively in construction for seals and gasket systems. It can only be obtained in black.

EPDM (ethylene propylene diene monomer) is a terpolymer typically consisting of 60% ethylene with propylene and a small proportion of a diene such as ethylidene norbornene or 1,4-hexadiene. The material has good resistance to sunlight and ozone as the residual double bonds are in the side chains, and therefore when oxidised they do not cause degradation of the main polymer chains. EPDM can be obtained in any colour and is used widely for roof membranes, weather stripping, gaskets and cable insulation.

Butyl rubber was developed by William Sparks and Robert Thomas at Standard Oil in 1937. It is a copolymer of typically 98% isobutylene with isoprene. It is impermeable to air, with good chemical and weathering resistance. Within construction butyl rubber is used as liners to landfill sites and decorative water features. It is also the basis of certain adhesives.

## 19.7 Fibre-reinforced polymers

Glass-fibre-reinforced plastics (GRP) were developed during the Second World War, and have become a significant material for the manufacture of many architectural components such as cladding panels, barge boards, dormer windows, entrance canopies and classical columns where multiple production of the product from an individual mould is possible. GRP is also used for custom-built structures such as the replacement of church spires, where a traditional replacement would be cost-prohibitive. The hand lay-up moulding process for making GRP components lends itself to design freedom and the ability to thicken or reinforce the material as appropriate.

The standard matrix material for GRP is polyester resin, although other thermosets including phenolic, epoxy and polyurethane resins are also used. Glass fibres are continuous rovings, chopped strand or woven mat, but may be replaced by Kevlar or carbon fibres for added strength at an additional cost.

## 19.8 Inorganic polymers

Polysiloxanes or silicones are polymers with backbones consisting of alternating silicon and oxygen atoms rather than mainly carbon atoms as in organic polymers. Siloxanes were discovered in 1927 by Frederic Kipping and developed into a patent for the General Electric Company in 1943. The standard polymer is polydimethyl siloxane produced from the monomer dichlorodimethyl siloxane. Siloxanes are heat-resistant and less readily attacked by oxygen or ozone due to the inherent strength of the silicon–oxygen bond. They also can be used to low temperatures as they have the lowest glass transition temperature of any elastomer. The wide range of manufactured products is associated with varied side groups (methyl, vinyl, phenyl, etc.) attached to the backbone silicon atoms. Silicones are extensively used within construction as low- to high-modulus sealants and gaskets. Partially polymerised silicones in hydrocarbon solvent can be used as water repellents for concrete and masonry. The typically 5% polymerised material penetrates the concrete pores where it then becomes fully polymerised to form a water-repellent silicone-resin lining.

## 19.9 Sealants and adhesives

Although used in relatively small quantities compared with the major construction materials, sealants and adhesives play a significant role in ensuring the successful working of buildings. Sealants are classified as plastic, elastoplastic or elastic depending on their key physical properties and appropriate application.

The development of sealants and adhesive mirrors the development of plastics generally. Until the twentieth century adhesives and sealants were based on natural materials such as beeswax, tar and various derived animal, vegetable and mineral products. From about 1910 phenol-formaldehyde was used in the plywood industry. This was followed by urea-and melamine-formaldehyde, neoprene and polyvinyl acetate (PVA) wood glue by the 1930s. Polyurethanes were developed by Otto Bayer in 1937 and as well as adhesives and sealants are used extensively in rigid and flexible foams and surface coatings. Acrylic resins (polyacrylates) similarly evolved during this period and are available as both water-based and solvent-based products, mainly used for internal sealing such as around new windows.

In the 1950s a whole new range of products evolved including the epoxies, cyanoacrylates and the anaerobic adhesives. The epoxy resins were developed from the early work of Castan and Greenlee in 1936. They are structural adhesives that will successfully bond timber, metal, ceramics, stone, glass and rigid plastics. The cyanoacrylates, or instant adhesives, were discovered by Harry Coover of Eastman Chemical Company and then patented in 1956. The stabilised monomer immediately polymerises in contact with surface moisture, forming a strong polymer bond between the two surfaces. A range of cyanoacrylate adhesives specifically target different substrates including metals, rubber, plastics, wood and ceramics. Anaerobic adhesives derived from methacrylates were developed in 1953 by Vernon Krieble of Loctite Corporation. They will seal threaded pipe fittings by curing between two metallic surfaces while out of contact with atmospheric oxygen. Polysulfides were also developed in this period as adhesives and sealants, as either one- or two-component systems. They have good resistance to ozone and sunlight and are widely used in construction to seal floor and pavement joints and in civil engineering projects to create waterproof joints in reservoirs and dams.

Hot-melt adhesives, applied through a glue-gun, were developed in the 1960s. They are usually based on the copolymer ethylene vinyl acetate (EVA). The 1970s saw the development of pressure-sensitive and contact adhesives based on the acrylics, polyurethane and non-crosslinked rubber. Hot-melt pressure-sensitive adhesives are frequently based on ethylene ethyl acrylate (EEA) or ethylene methyl acrylate (EMA).

More recent developments have been related to modifications to the previous range of adhesives, particularly the production of waterborne materials to reduce the use of ozone-depleting solvents escaping into the atmosphere. Many such adhesive products are polymer dispersions characterised by a two-phase system consisting of fine solid particles within the liquid water phase. The dispersion typically contains 40–60% polymer, stabilised to prevent coagulation during shelf-life. Products include solvent-free gap-filling adhesives that can replace traditional nail fixings. Curable hot melt systems are available based on both rubber and synthetic polymers; typical uses include the manufacture of laminated timber panels and post-formed work surfaces. A further advantage of hot melt adhesives over solvent-based products is no loss of thickness on setting, compared to a typical 50% reduction in thickness as solvents evaporate.

## 19.10 Polymer-modified cement mortar and concrete

Polymer latexes (aqueous colloidal dispersions of polymers) incorporated into standard concrete mixes enhance the water and frost resistance of the subsequent hardened concrete. The replacement of typically 10–15% cement binder by latex also increases the strength of bonding to any steel reinforcement and to adjacent existing concrete; the latter is particularly useful in the case of remedial work on damaged or spalled fairfaced concrete. Standard polymers are acrylic (polymethyl methacrylate) latex, epoxy resin and styrene-butadiene rubber. Typical uses include large structures and floor surfaces that are non-dusting.

Polymer-modified cement is used for thin-joint mortar masonry systems, where the standard 10 mm joint is reduced to 2–3 mm. The advantages include rapid setting of the mortar, with full strength achieved after 2 hours, permitting more courses to be laid each day, which is particularly relevant for the use of large-format blocks.

Polymer concrete consists only of aggregate, and polymer latex as a total replacement for the standard Portland cement binder. Resin concrete has greater frost, chemical and abrasion resistance than standard Portland cement concrete, but it is more expensive. Binders are usually acrylic, epoxy, polyester or polyvinyl ester resins. Fillers may be sand, crushed stone or any non-absorbent material. Fibres may be incorporated for additional strength properties. Typical uses are for drainage components where the low water absorption, chemical resistance and low permeability are advantageous.

## 19.11 Polymer surface finishes

Polymer-based paints, varnishes and wood stains have taken over from the early products based on linseed oil, shellac and creosote. In paints, the binder, which is the component that solidifies to produce the surface film, is now usually based on alkyd, vinyl or acrylic resins. Alkyd resins oxidise and polymerise in air whereas vinyl and acrylic emulsions lose water by evaporation, causing the dispersed polymer to coalesce into a continuous layer. Alkyd resins are manufactured from polyhydric alcohols and dibasic organic acids and are generally dispersed within a solvent system, whereas vinyl and acrylic resins are water-based emulsions with only a small VOC (volatile organic compound) content. Environmental concerns over the release of VOCs into the atmosphere have led to the development of 'high solids' paints with considerably reduced levels of organic solvents, and a switch towards predominantly water-based products. However, not all solvent-based paints can yet be replaced successfully by water-borne alternatives. The range of specialist polymer paints is extensive, including acrylated rubber paints for chemical resistance, silicone water-repellent masonry paint, and epoxy paint for resistance to abrasion and chemicals.

The standard wood stains are alkyd resins in organic solvent (e.g. white spirit), and water-borne acrylic or alkyd/acrylic blends. Polymer varnishes, as water- or solvent-based systems, are mostly based on polyurethane or modified alkyd resins. Most manufacturers are now offering water-based dispersion products as viable alternatives to solvent-based products to reduce their environmental impact.

## 19.12 Engineering properties

The physical and mechanical properties of most traditional construction materials such as steel, concrete, masonry and timber are comparatively constant over the typical working temperature range of –10°C to +30°C. However, properties, for example elastic modulus, of many plastics including thermosetting resins may vary considerably over this range of temperature. In most cases this is attributed to the 'glass transition temperature'. Above this point plastics tend to more rubbery and prone to creep, whereas below this temperature they are hard and ultimately fail in a brittle manner. Thus for any specific construction outlet, the variation of physical properties within the anticipated working temperature range should be fully characterised.

## 19.13 Degradation of plastics

The degradation of plastics by heat, UV or ozone is most frequently attributed to the breakdown of a proportion of the polymer macromolecules into shorter chains. Discolouration, frequently yellowing, is caused by the production of molecular units with double bonds, and surface crazing may develop where degradation has caused cross-linking and embrittlement of the surface. Plastics that initially contain plasticiser, such as PVC, can lose the plasticiser by its migration, leaving a more brittle product prone to cracking.

## 19.14 Plastics in fire

All plastics are organic materials and are therefore, to differing degrees, combustible. Burning organic materials generally produce carbon monoxide and carbon dioxide, but in addition some plastics containing nitrogen, such as polyurethane foam, produce hydrogen cyanide on combustion, while PVC evolves hydrochloric acid. Some plastics such as acrylics and polystyrene add significantly to the heat of a fire and may produce burning droplets and/or smoke.

The European classification of building materials with respect to fire is defined in BS EN 13501-1: 2007. The classification describes seven Euroclasses from A1, A2, B through to F. A1 Euroclass materials do not contribute to the fire even during the fully developed fire stage. A2 Euroclass materials do not contribute significantly to the fire load and fire growth. The Euroclasses B down to F have decreasingly stringent criteria for resistance to flame when exposed to thermal attack by a single burning item under the strict test conditions specified in the standard BS EN 13823: 2010. There are additional classifications relating to smoke production, from s1 which is the most stringent to s3 which sets no limits; also on flaming droplets from d0 requiring no flaming droplets/particles to d2 which sets no limits. A separate parallel classification applies to floorings, with categories $A1_{fl}$ Euroclass materials making no contribution to the fire at any stage of its development, down to $F_{fl}$ which describes materials for which no fire performance is determined or for which the material cannot be classified to the more stringent ratings. An additional classification relates to smoke production, with s1 relating to limited and s2 to no limitation of smoke production respectively. The reaction to the standard fire test for linear pipe thermal insulation products is similar to that for general construction products with ratings from $A1_L$ to $E_L$, categories s1 to s3 for smoke production and d0 to d2 for flaming droplets/particles.

The British Standard BS476 Part 7: 1997 also remains current as the method of testing for the surface spread of flame, but eventually the European classification will become the norm. Specifiers must ensure that they have accurate information from the manufacturer on the precise fire class for materials to be used in construction. Fire classes within one polymer type may differ considerably according to constituents, and applied treatments also may be subject to change as the manufacturer modifies the product. Ignitabilty tests are described in BS ISO 11925 and BS EN ISO 11925: 2010.

## 19.15 Recycling of plastics

Just under a quarter of all plastics used in the UK are consumed by the construction industry. Of all plastics waste, including domestic waste, less than 10% is currently recycled, although most industrial process scrap is reprocessed as it is clean and of known composition. In principle the majority of thermoplastics, because they are re-meltable and re-formable, could be recycled. Most post-consumer plastic waste can be separated into types by a complex series of froth flotation techniques, and other electronic plastic-type detection devices are being further developed. In general the recycling of plastics uses considerably less energy than is required to manufacture the virgin material, but the economies vary significantly depending on the source and condition of the recycling materials. Plastic components are gradually being identified by a recycling coding in seven categories: polyethylene terephthalate (PET – 1), high-density polythene (HDPE – 2), polyvinyl chloride (PVC – 3), low-density polythene (LDPE – 4), polypropylene (PP – 5), polystyrene (PS – 6), and others (7). However, this categorisation is currently used more for domestic plastic recycling than for construction materials.

Some plastics, after sorting, can be regranulated for direct remoulding. Recycled PVC can be used as the core of PVC pipes by co-extrusion surrounded by new material. Recycled PVC is used for floor tiles, decking, guttering, panelling and electric cable covers. Polystyrene can be recycled by solvent extraction for re-use in insulation products. High-density polythene may be recycled into simulated timber (plastic lumber) for decking and garden furniture, also for floor tiles. Similarly recycled low-density polythene is converted into landscape timber, lumber and panelling. Polyethylene terephthalate (used for making bottles) can be recycled into polyester carpets. Polythene, polypropylene and PVC are blended with saw dust and other wood waste to produce wood–plastic composites for architectural mouldings, decking and window profiles. Other products, some of which currently use recycled plastics in their manufacture, include drainage channels and membrane sheeting. Chopped and ground plastics can be used directly as thermal insulation and as a padding filler, for example under sprung gymnasium floors.

The development of corn-based plastics, such as polylactic acid (PLA), will require the use of more agricultural land and with biodegradability will reduce the landfill or recycling burden. However, these products are relevant largely only to the packaging of construction materials.

## 19.16 Current and future developments

Over the next few years, a key thrust for development of plastics within construction will be towards more eco-friendly products requiring the use of less energy in production and fabrication. There will also be a continuing trend to reduce the quantities of volatile organic compounds (VOCs) in their production, particularly in the case of certain adhesives, foamed plastics and surface finishes. The clear marking of plastic products with their compositions will promote recycling after their useful life in buildings. Electronic techniques for the recognition of different plastic waste products will make their separation and recycling more efficient, aiming towards government recycling targets.

An area of further development is likely to be associated with the various optical properties of plastic sheet and film. Materials that appear to change colour are of interest to architects and interior designers who wish to create dynamic visual effects on the external facades or interiors of buildings. Dichroic plastic films cause the observed colour and opacity to change depending on the view point and the direction and intensity of any light sources. For example, the colour can change between green, gold and orange or between purple and blue. Thermochromic pigments in fibre-reinforced plastic cladding will make the building appear to respond to the external temperature. Light-conducting plastic channels embedded in concrete create rippling effects, responding to movement of the observer or changes in the lighting. Liquid crystal films create rainbow colour effects similar to that observed with oil and water, while phosphorescent plastics will glow for up to 20 hours.

The drive to higher levels of thermal insulation in construction led to the production of polypropylene fibre-reinforced aerated concrete, which offers a combination of strength and insulation properties for use in floors, walls and roofing panels. Further developments in insulation materials will increase the interest in aerogel, a silica gel containing up to 99.8% air, the lightest known solid material, having a density of only 3 mg/l (ρ air = 1.2 mg/l). It is currently used to manufacture translucent polycarbonate glazing units with very low U-values.

It is anticipated that future developments will involve the production of nanocomposite plastic materials. The addition of nano-sized particles to plastics at rates of typically 5%, as an alternative to conventional additives such as glass fibres, generates composites with significantly different properties including enhanced durability, strength, and high-temperature, UV and fire resistance. Current developments relate mainly to thermosetting polymers, but it is anticipated that the standard thermoplastics such as polyethylene, polypropylene, PVC and polyethylene terephthalate (PET) will eventually dominate the nanocomposite market. Current nanomaterials are clay minerals (surface modified montmorillonite), carbon black and carbon nanotubes. As costs of these new materials fall in real terms, it can be predicted that they will become incorporated into more standard products, particularly as the required loading of nanomaterials in polymers is only a few per cent, in contrast to glass fibres within glass-fibre reinforced polyester at typically 20–30% content.

## 19.17 British Standards

- *BS476 Part 7: 1997.* Fire tests on building materials and structures. Method of test to determine the classification of the surface spread of flame of products.
- *BS ISO 11925.* Reaction to fire tests. Ignitability of building products subjected to direct impingement of flame: Part 1: 1999. Guidance on ignitability. Part 3: 1997. Multi source test.
- *BS EN ISO 11925.* Reaction to fire tests. Ignitability of products subjected to direct impingement of flame: Part 2: 2010 Single flame source test.
- *BS EN 13501-1: 2007.* Fire classification of construction products and building elements. Classification using test data from reaction to fire tests.
- *BS EN 13823: 2010.* Reaction to fire tests for building products. Building products excluding floorings exposed to the thermal attack by a single burning item.

## 19.18 Bibliography


Akovali, G, (2005) *Handbook of polymers in construction.* Shrewsbury, UK: Smithers RAPRA Press.

Engelsmann, S., Spalding, V. and Peters, S. (2010) *Plastics in architecture and construction.* Basle, Switzerland: Birkhäuser.

Hall, C., Chen, J.-F. and Hollaway, L.C. (2010) *Polymers and polymer fibre composites.* London, UK: Thomas Telford.

Links International (2012) *Architecture and construction in plastic.* Barcelona, Spain: Leading International Key Services.

Lyons, A. (2010) *Materials for architects and builders*, 4th edition. Oxford,UK: Butterworth-Heinemann.

Van Uffelen, C. (2008) *Pure plastic: New materials for today's architecture.* Berlin, Germany: Verlagshaus Braun.

# 20 Acrylic Plastics

**Arthur Lyons** MA MSc PhD Dip Arch Hon LRSA
De Montfort University

## Contents

## 20.1 Introduction

Acrylic plastics form a group of thermoplastic polymers with a wide range of composition, characteristics, application and performance. In addition to being thermo-formed, many 'acrylic' compounds are also soluble in organic solvents or dispersible as aqueous emulsions. They include polymethyl methacrylate (PMMA), polyacrylonitrile (PAN), polyacrylates and cyano-acrylates (CA). The materials are polymerised from their parent monomers using free-radical initiators in bulk, suspension, solution or emulsion.

Polymethyl acrylate was first produced in 1880 by the Swiss chemist Georg Kahlbaum, and commercially produced by Rohm and Haas in 1927. Polymethyl methacrylate was first produced by ICI under the trade name Perspex® in 1934 and subsequently by DuPont as Lucite®. It is available in sheet form for thermoforming and granules for injection moulding.

Polyacrylonitrile, commonly used as acrylic textile fibres, is also the precursor for carbon fibres, which are manufactured by oxidation of fibres at 250°C followed by carbonisation in an inert atmosphere at 2500°C to produce 92–100% carbon. Cyanoacrylates are the basis of contact adhesives known as super-glues, while poly methyl and ethyl acrylates are the film forming components of acrylic paints. Acrylic-modified concretes have greater resistance to water penetration and enhanced bonding to reinforcement; also acrylic-modified cements have good adhesive properties for concrete surface repairs, for example where reinforcement corrosion has caused surface spalling.

## 20.2 Manufacture

All acrylic polymers are products of the chemical and petrochemical industries. Acrylic acid and acrylic esters are manufactured by primary chemical producers and inhibited against polymerisation for transportation to the polymer producers. Methyl methacrylate is produced in very large quantities (estimated 3.2 Mt/annum), mainly from acetone and hydrogen cyanide, for conversion to polymethyl methacrylate sheeting and injection moulding material. Cast polymethyl methacrylate (CS) is formed between two sheets of glass and polymerised under controlled water bath or oven conditions. Extruded polymethyl methacrylate (XT) is manufactured from the polymer in granular form. The two basic types have some differences in their properties due to variations in their molecular structures resulting from the alternative fabrication processes.

Cyanoacrylates are a family of adhesives. The standard domestic and industrial grades are based on methyl and ethyl cyanoacrylate, whereas the specialist medical adhesives are the octyl and n-butyl cyanoacrylates.

## 20.3 Chemical composition

Figure 20.1 illustrates the basic chemical repeat unit for polymethyl methacrylate, but molecular weight distributions will depend upon polymerisation conditions and the required properties for specific end uses.

Polymethyl methacrylate is soluble in most aromatic and chlorinated hydrocarbons, esters, ketones and tetrahydrofuan. dichloromethane or trichloromethane can be used as a solvent adhesive giving a clear joint. It is largely resistant to aliphatic hydrocarbons (e.g. hexane and octane), inorganic acids and alkalis but is attacked by organic acids such as acetic acid and phenols.

Crosslinked acrylics are insoluble but swell in chlorinated hydrocarbons. However, debonding solvents for cyanoacrylate adhesives based on acetone, nitromethane or methylene chloride are commercially available.

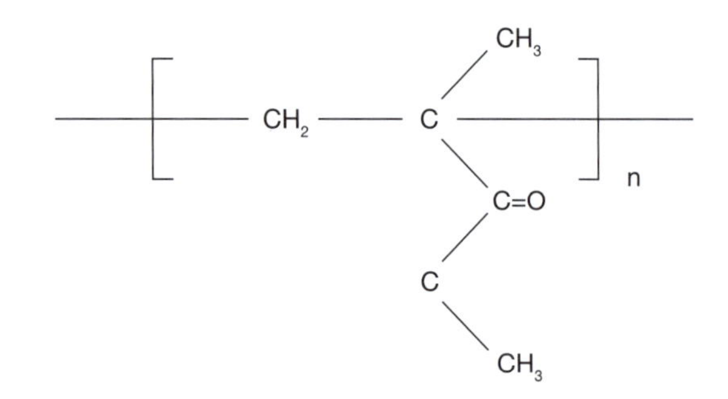


**Figure 20.1** Polymethyl methacrylate structure

## 20.4 Standard forms of polymethyl methacrylate

Clear cast acrylic sheet is typically available in a wide range of thicknesses from 2 mm to 50 mm and blocks up to 200 mm in sheet sizes up to 3050 × 2030 mm. Standard stock coloured sheets are normally available in 3 and 5 mm with greater thicknesses to order. Extruded sheet, which is cheaper than cast sheet, is manufactured to 2050 mm width and can be available in lengths longer than 3050 mm, subject to transportation constraints. It is manufactured to closer thickness tolerances, but with a narrower colour range. Cast sheet is available to an extensive colour range including solid, translucent, transparent, opal, fluorescent, anti-reflective, mirror and various striated, crystal and stone effects. Additional forms of acrylic sheet include high molecular weight UV filtering resin for specific glazing applications and clear prismatic sheet for specialist glazing and fluorescent light fitting diffusers. An impact-modified grade of polymethyl methacrylate gives enhanced impact resistance to BS6262-4: 2005. A black product allows only the transmission of infrared radiation. Textured finishes in a range of classical and modern designs are produced to give requisite levels of obscuration.

Acrylic polymer granules are available for injection moulding applications, extrusion, co-extrusion and the manufacture of composites incorporating inert fillers. Typical injection moulding temperatures are within the range from 220°C to 240°C.

The copolymer methyl methacrylate-acrylonitrile-butadiene-styrene (MABS) is available in powder and pellet form, also modified or unmodified by the addition of additives, fillers and colourants. The properties of the material combine the toughness and impact strength of acrylonitrile-butadiene-styrene (ABS) copolymer with the transparency of polymethyl methacrylate (PMMA) (see Table 20.1). It is formed by extrusion or injection moulding.

Methyl and ethyl cyanoacrylates contact adhesives form colourless transparent bonds. Selection of an appropriate grade depends on the substrates to be bonded, the speed of adhesion required and the anticipated service conditions. Different viscosity grades are produced to ensure appropriate wetting and/or gap filling of the contact surfaces. Acrylic resins are used in coatings and adhesives.

## 20.5 Specification of polymethyl methacrylate

The standards BS ISO 8257-1: 2006 and BS EN ISO 10366-1: 2002 define the designation system for polymethyl methacrylate and the copolymer methyl methacrylate-acrylonitrile-butadiene-styrene (MABS) respectively. The designation, in a set of data blocks, includes name (PMMA or MABS), application and method of processing, critical physical properties, content of

**Table 20.1** Typical physical properties of standard grade clear polymethyl methacrylate

| *Property* | *Cast sheet* | *Extruded sheet* |
|---|---|---|
| Relative density | 1.19 | 1.19 |
| Water absorption, % | 0.2 | 0.2 |
| Light transmission, % (3 mm clear) | 92 | 92 |
| Refractive index | 1.49 | 1.49 |
| Thermal conductivity, W/m K | 0.189 | 0.189 |
| Softening point, °C | >110 | >105 |
| Maximum service temperature, °C | 80 | 95 |
| Thermoforming temperature, °C | 170 | 155 |
| Tensile strength, MPa | 75 | 70 |
| Flexural strength, MPa | 116 | 107 |
| Flexural modulus, GPa | 3.21 | 3.03 |
| Elongation at break, % | 4 | 4 |
| Charpy un-notched impact strength, kJ $m^{-2}$ | 12 | 10 |
| Coefficient of linear expansion, $K^{-1}$ | $7.7 \times 10^{-5}$ | $7.8 \times 10^{-5}$ |

The standards BS EN ISO 7823 Parts 1–3 give minimum values for certain physical properties that are lower than these typical commercial values.

fillers and any required supplementary information. The standards BS EN ISO 7823 Parts 1–3 define the requirements, including appearance, colour, dimensions and physical properties, for cast, extruded and continuous cast sheets of polymethyl methacrylate.

Polymethyl methacrylate has good impact resistance. The standard 3 mm material is rated to Class C (lowest), and 8 mm to Class A (highest) standard to BS6206: 1981, which requires safe breaking without sharp pointed protrusions, under controlled impact testing. The impact modified grade of polymethyl methacrylate is appropriate for use to BS6262-4: 2005, in critical building locations where safety glass is required.

## 20.6 Dimensional stability

Allowances must be made for the large thermal movement of polymethyl methacrylate. For example, when used as glazing, 5 mm per metre run should be allowed for thermal movement in both dimensions to prevent the build-up of stresses which may lead to crazing. A temperature range of –20°C to 70°C should be anticipated within the UK. Rubber and other profile sealing strips must be compatible with the material. The rebate depth must be sufficient to allow for thermal movement, including contraction at lower temperatures. The thickness of glazing should be calculated on impact strength and for external windows on predicted wind loading according to Eurocode 1 BS EN 1991-1-4: 2005.

Extruded sheet flows on heating rather than deforms, so it will not return to flat sheet on reheating.

## 20.7 Durability

Polymethyl methacrylate can be scratched, but acrylic polishes are available to restore the surface finish. The satin (light-diffusing) finishes are less sensitive to surface scratching. The material has good weathering properties as it is transparent to near-UV, and only absorbs radiation from the middle- and far-UV wavelengths. Loss of physical properties is normally gradual, but the material may suffer from surface crazing if stresses generated in the fabricating or thermoforming processes are not fully eliminated by subsequent annealing between 70 and 90°C. Extruded sheet is more susceptible to stress crazing or cracking than cast sheet, and crazing may be accelerated by exposure to solvents or high-energy radiation. Plasticised PVC sealing strips should not be used in contact with the material, as they cause stress cracking. However, butyl rubber, neoprene or polysulfide sealing strips are appropriate, as are some EPDM and silicone sealant systems. Polymethyl methacrylate exhibits creep related to applied load and increased temperature.

## 20.8 Uses of polymethyl methacrylate in construction

Polymethyl methacrylate can be worked with standard workshop tools including laser engraving and cutting. However, as a brittle material, machining speeds must be kept low and the sheet temperatures not allowed to rise in the process. Fine-tooth saws should be used, and routing processes should be cooled with an air jet. Flame polishing of cut edges produces a good finish but tends to produce discolouration with the deeply coloured products.

A minimum cold bending radius of 200 times the thickness is manageable for cast sheet and 300 times the thickness for extruded sheet. Draping, vacuum forming, air-pressure forming, press forming and ram moulding are all techniques used for forming acrylic components.

Polymethyl methacrylate is frequently used as a glazing material, particularly for moulded rooflights and barrel vaulting. Corrugated and multiskin sheets are used as roofing products. A modified surface finish can give a reduced condensation effect internally or a self-cleaning effect externally. The impact modified product is used for balcony in-fill panels where all edges should be fixed to prevent the glazing being sprung from the frame. Load-spreading devices should be used at fixing points to avoid the build-up of stresses which could lead to stress crazing of the material.

Polymethyl methacrylate may be used in thick section or laminated to other glazing materials to provide appropriate protection against vandalism and firearms. Decorative illuminated floors, made of sufficiently thick units to restrict deflection, can be an architectural feature. Polymethyl methacrylate can be used for acoustic barriers; 3 mm produces a sound reduction of 26 dB, and 12 mm a reduction of 35 dB. Road noise barriers may incorporate polyamide fibres to prevent sheet fragments becoming detached on vehicle impact.

Some furniture and much interior and exterior shop signage are made with acrylic plastic due to its ease of forming, machining, fixing and its inherent optical properties. Polymethyl methacrylate is used extensively for moulding baths, spas and shower trays, which are usually reinforced with sprayed GRP on the underside. Polymethyl methacrylate composites incorporating pigments and quartz fillers are used to manufacture durable rigid components such as kitchen sinks.

Co-extrusion is used for the production of composite polymethyl methacrylate coated PVC for linear components such as barge boards and soffit units where advantage is taken of the durable finish.

Light fittings are frequently produced from injection moulded heat-tolerant grades of polymethyl methacrylate, and prismatic material is the norm for fluorescent light diffusers.

The work surface material Corian® is composed of one-third polymethyl methacrylate and two-thirds minerals based on aluminium trihydrate from bauxite. It is manufactured in a wide range of coloured finishes in sheets of dimensions 6, 12.3 and 19 mm to 930 × 3680 mm. Corian is a non-porous, durable finish that is affected only by strong acids, ketones, chlorinated solvents and equivalent chemicals. The standard product is rated with respect to fire to Euroclass C –s1, d0 and to BS476 Part 7: 1997 as Class1. Corian is widely used for worktops, sinks, basins, and hospital and laboratory work surfaces.

## 20.9 Proprietary brands of polymethyl methacrylate

The more common trading names for polymethyl methacrylate are: Perspex, Plexiglas, Lucite, Acrylite, Acrylplast, Altuglas, Limacryl, Plazcryl and Polycast.

## 20.10 Hazards in use

Eye protection should be worn while working on polymethyl methacrylate to prevent the risk of eye damage from swarf and chips. The performance in fire should always be considered in relation to its appropriate use in construction. Polymethyl methacrylate is a combustible material.

## 20.11 Performance in fire

Polymethyl methacrylate burns, producing carbon monoxide, carbon dioxide and little smoke, but burning droplets, which continue to burn. The extruded product and cast sheets of less than 3 mm are classified to Class 4 to BS476 Part 7: 1997. Cast sheets of thickness 3 mm and above are rated Class 3 to the British Standard. Polymethyl methacrylate is categorised with respect to fire to Euroclass E of BS EN 13501-1: 2007, when tested according to BS EN ISO 11925-2: 2010.

## 20.12 Sustainability and recycling

The vast majority of plastics are currently dependent on the petrochemical industry, and cannot therefore be considered to be sustainable until sufficient alternative sources of organic precursors based on agricultural production are developed. The largest sector of recycled plastics is PVC, and this is normally used for lower grade products. Polymethyl methacrylate from industrial and commercial sources can be fully recyclable when clean and separated from other waste materials.

## 20.13 British Standards

- *BS476 Part 7: 1997.* Fire tests on building materials and structures. Method of test to determine the classification of the surface spread of flame of products.
- *BS6206: 1981.* Specification for impact performance requirements for flat safety glass and safety plastics for use in buildings.
- *BS6262-4: 2005.* Glazing for buildings. Code of practice for safety related to human impact.
- *BS ISO 8257-1: 2006.* Plastics. Poly(methyl methacrylate) (PMMA) moulding and extrusion materials. Designation system and basis for specifications.
- *BS EN 1991-1-4: 2005.* Eurocode 1. Action on structures. General actions.
- *BS EN ISO 7823.* Plastics. Poly(methyl methacrylate) sheets. Types, dimensions and characteristics: Part 1: 2003 Cast sheets. Part 2: 2003 Extruded sheets. Part 3: 2007 Continuous cast sheets.
- *BS EN ISO 10366-1: 2002.* Plastics. Methyl methacrylate-acrylonitrile-butadiene-styrene (MABS) moulding and extrusion materials. Designation system and basis for specifications.
- *BS EN ISO 11925-2: 2010.* Reaction to fire tests. Ignitability of products subjected to direct impingement of flame. Single-flame source test.
- *BS EN 12600: 2002.* Glass in building. Pendulum test. Impact test method and classification for flat glass.
- *BS EN 13501-1: 2007.* Fire classification of construction products and building elements. Classification using data from reaction to fire tests.
- *BS EN 60695-11-10: 1999.* Fire hazard testing. Test flames – 50 W horizontal and vertical flame test method.

## 20.14 Bibliography


Brydson, J.A. (2005) *Plastics materials.* New Delhi, India: CBS Pub & Dist.

Cousins, K. (2002) *Polymers in building and construction. RAPRA market report.* Shrewsbury, UK: RAPRA Technology.

Dufton, P.W. (2002) *Polymers in building and construction. RAPRA market report.* Shrewsbury, UK: RAPRA Technology.

Halliwell, S., (2003) *Polymers in building and construction.* Report 154. Shrewsbury, UK: RAPRA Technology.

Lyons, A. (2010) *Materials for architects and builders*, 4th edition. Oxford, UK: Butterworth-Heinemann.


## 20.15 Websites

- Altuglas International: www.altuglas.com
- Bay Plastics: www.bayplastics.co.uk
- Lucite International: www.lucitesolutions.com
- Plexiglas: www.plexiglas.com
- Polycast: http://spartech.com
- Limacryl: www.limacryl.be

# 21 Polycarbonate Plastics

**Arthur Lyons** MA MSc PhD Dip Arch Hon LRSA
De Montfort University

## Contents

## 21.1 Introduction

Polycarbonates (PC) are a family of mainly thermoplastic polymers characterised by the presence of oxygen linkages in the carbon backbone chain, associated with the carbonate group. Their good UV resistance, robustness and excellent light transmission result in widespread application in glazing. The most common polycarbonate is that produced from the aromatic precursor bisphenol A. However, crosslinked aliphatic polycarbonates, such as poly dialyldiglycol carbonate (DADC), which is the thermosetting resin used in the production of many optical lenses, are also commercially produced.

Polycarbonates were discovered independently by Dr Daniel Fox of General Electric and Dr Hermann Schnell of Bayer in 1955. The two companies filed similar patents for the production of linear polycarbonates of high molecular weight, and began commercial production by 1960. In addition to various grades of the general-purpose material, polycarbonate is available as flame-retardant, impact-modified, high melt strength and glass fibre reinforced products. Blends with ABS (acrylonitrile–butadiene–styrene) and polyesters offer a further increased range of physical properties.

The majority of components manufactured in polycarbonate are formed by extrusion, injection moulding, vacuum forming or blow moulding. Polycarbonate foam is used for its strength in sandwich form components.

## 21.2 Manufacture

Polycarbonate is produced from the reaction of bis-phenol A and carbonyl chloride (phosgene). Bis-phenol A is manufactured by the reaction of phenol with acetone under acidic conditions. World production is expected to reach 4.5 million tonnes by the end of 2016. The compatibility of polycarbonate with a range of other polymers has led to the production of a range of blends targeting specific physical properties. The most significant blended product is with acrylonitrile–butadiene–styrene, as the PC/ABS blend exhibits low-temperature toughness and increased resistance to stress cracking. Blends with aromatic polyesters include polyethylene terephthalate (PET) and polybutylene terephthalate (PBT). The latter blend has the chemical resistance associated with PBT while retaining the low-temperature toughness of polycarbonate. Rubber-modified polycarbonate has enhanced impact resistance.

## 21.3 Chemical composition

Figure 21.1 illustrates the basic chemical repeat unit for bisphenol A polycarbonate, but molecular weight distributions will depend on polymerisation conditions and the required properties for specific end uses.

Polycarbonate is resistant to most acids, but not alkalis. It is attacked by aromatic hydrocarbons, chlorinated hydrocarbons, ethers, aldehydes, ketones, esters and tetrahydrofuran. These agents cause swelling and surface whitening. Acetone and xylene cause stress cracking and crazing. Strong alkalis and amines destroy polycarbonate.

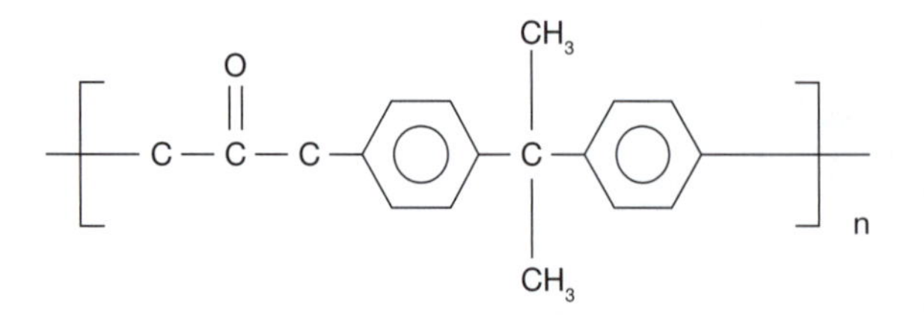


**Figure 21.1** Chemical repeat unit for bis-phenol A polycarbonate

## 21.4 Standard forms of polycarbonate

Polycarbonate is available commercially as granules for thermoforming by injection moulding, by extrusion to form linear components, also for calendering into sheets between 1 mm and 15 mm thickness and films of less than 1 mm. The different grades are differentiated by viscosity, colour and required physical properties. Colours – transparent, translucent or opaque – may be defined by the RAL colour system.

Formed polycarbonate is normally available as thin film and in clear sheet with thicknesses from 0.75 to 13 mm for standard sizes up to 2050 × 3050 mm. A more limited thickness range is produced in transparent bronze, grey, green, red and blue, although virtually any colour can be produced to order. A similar colour range is available in translucent and opaque. Larger flat sections to 50 mm, rod and tubes are available for special applications.

Embossed patterned sheets in the range 2 to 6 mm including leather-grain and prismatic finishes offer controlled light transmission and obscured through vision. Corrugated sheet applications include lightweight hall and stadium roofs, industrial buildings and swimming pools. Sheets are typically available in the thickness range 0.75 to 1.5 mm to widths of 1295 mm and lengths determined by client requirements and transportation limitations. Much extruded polycarbonate is used as extruded twin-wall and triple-wall flat 'sheet' for use in cladding/glazing. These multi-wall sheets achieve higher thermal insulation and better spanning than simple flat or corrugated sheet.

For increased impact properties, glass fibre to 50% fill, carbon fibre, aramid and stainless steel fibre-reinforced products are marketed. For low-friction products polycarbonate is blended with PTFE. Specialist products include nickel-coated carbon fibre blends, also flame-retardant materials and polymers with enhanced heat resistance. Physical properties of sheet polycarbonate are given in Table 21.1.

The toughness of polycarbonates increases with molecular weight. High molecular weight grades remain ductile to lower temperatures, but generally have higher viscosities and are therefore more difficult to process. Special grades with improved viscous flow are available to resolve the balance between impact resistance and viscosity. For an enhanced combination of toughness and stiffness, the addition of 10% glass fibre reinforcement in polycarbonate is effective.

## 21.5 Specification of polycarbonate

The standard BS EN ISO 7391-1: 2006 defines the designation system for polycarbonate and polycarbonate blends. The designation is a set of data blocks that include the name of the plastic, the method of processing, any additives, key physical properties including viscosity, melt flow and impact resistance, and any added reinforcement.

Polycarbonate and the impact-modified grades have excellent impact resistance. The standard BS6262-4: 2005 refers to polycarbonate as being very difficult to break.

## 21.6 Dimensional stability

Polycarbonate has a large coefficient of linear expansion and allowances should be made for significant thermal movements. When used for glazing an expansion of 3 mm per metre should be allowed for thermal movement in both directions, and a minimum of 15 mm of edge cover is required. A temperature range of –20°C to 70°C should be anticipated within the UK. Normally all four

**Table 21.1** Physical properties of sheet polycarbonate

| *Property* | *Extruded sheet* | *Extruded with 10% glass fibre* |
|---|---|---|
| Relative density | 1.18–1.22 | 1.23–1.24 |
| Water absorption, % (equilibrium) | 0.12–0.35 | 0.1–0.3 |
| Light transmission, % (3 mm clear) | 80–91 | |
| Refractive index | 1.59 | |
| Thermal conductivity, W/m K | 0.20 | 0.19–0.26 |
| Softening point, °C | 147–153 | 150 |
| Maximum service temperature, °C | 114–138 | 128–146 |
| Thermoforming temperature, °C | 300–304 | 254–310 |
| Tensile strength, MPa | 58–65 | 55–85 |
| Flexural strength, MPa | 85–105 | 93–124 |
| Flexural modulus, GPa | 2.2–2.6 | 2.3–5.3 |
| Elongation at break, % | ≥50 | 2–14 |
| Charpy un-notched impact strength, K J $m^{-2}$ | No failure | No failure |
| Coefficient of linear expansion, $K^{-1}$ | $65–70 \times 10^{-6}$ | $32–60 \times 10^{-6}$ |

As there are many grades of polycarbonate, ranges of data are listed.

edges should be supported, and the glazing thickness calculated to prevent excessive flexing according to BS5516: 2004 and BS6399-2: 1997.

When the material is cold bent, there is some relaxation after the process and slight overbending is therefore necessary. Cold bending can cause slight colour variation along the bend line and any textured surfaces should be on the compression face. Polycarbonate exhibits creep related to applied load and increased temperature.

## 21.7 Durability

Polycarbonate can be scratched and may be affected by this form of graffiti in public places. However, proprietary scratch removal polish is commercially available to remove minor abrasions. The standard material has a tendency to yellow on long-term exposure to sunlight as it absorbs UV radiation below 400 nm. This causes degradation of the polymer with associated reductions in physical properties. Dependent on the degree of exposure and the climatic conditions, the material will suffer from hazing, dullness of the surface and ultimately crazing. However, UV stabilised grades reduce discolouration, and co-extrusion with a 40 μm surface treatment can also reduce weathering and the effects of UV radiation. The weather-protected surface is indicated on the protective masking sheet to ensure its correct orientation. Contact with PVC, which may contain plasticisers, is not advised. Sealants and gaskets should be compatible with polycarbonate to avoid stress cracking at edges and fixing points. If a sealant primer is used it also should be compatible with polycarbonate. Polycarbonate is highly resistant to accidental impacts and deliberate vandalism.

## 21.8 Uses of polycarbonate in construction

Polycarbonate can be worked using standard fabrication techniques, such as bending, drilling and routing, and does not normally split, chip or break. Cutting should be done with fine-tooth blades under a blast of air to remove swarf. Drilling should be carried out at low speeds, also with an air blast to avoid localised heating. Punching is inadvisable as it leaves microcracks which can cause subsequent failure. Laser cutting also creates internal stresses and leaves carbon deposits at the cut edge. However, limited-speed water-jet cutting is usable. Polycarbonate can be polished to a high gloss finish. It can be joined mechanically, solvent jointed using methylene chloride, or cemented with standard epoxy, polyurethane, hot melt or silicone adhesives. Polycarbonate can also be ultrasonically welded.

Polycarbonate sheet can be formed by a variety of processes including drape forming at 155°C, and mechanical or vacuum forming at temperatures in the range 170–225°C.

Polycarbonate sheet with UV protection is used extensively for roof lights, roof domes, smoking shelters, car ports, road barriers, greenhouses and covered walkways, because of its impact resistance and durability. However, plastic sheet roofing tends to be noisy under heavy rain. The hard coated optical grades are used for standard and security glazing, where they may be subject to frequent cleaning or polishing and occasional graffiti. Bullet-resistant grades are laminates that absorb the impact energy without dangerous spalling. Polycarbonate is also used as an alternative to acrylic for signage.

Polycarbonate is extensively used for double-, triple-, five- and seven-wall transparent and translucent roofing systems for conservatories and garden room roofs and for a wide range of commercial buildings. Thicknesses range from 6 to 40 mm according to the required thermal properties. Widths range typically to 2100 mm and lengths to 6 m or even 9 m in some cases. Polycarbonate has the advantage of high impact resistance, good thermal properties and the choice of clear, opal or bronze for appropriate solar control. A typical 35 mm five-wall system has a U value of 1.25 $W/m^2 K$.

For glazing, polycarbonate sheet thicknesses to resist wind pressures should be calculated according to the standards BS5516 Parts 1 & 2 and BS6399-2: 1997. The mean sound reduction achieved by plastics glazing is typically 18 dB for a single 3 mm sheet and 26 dB for a 10 mm sheet (BS5516-2: 2004).

Polycarbonate units can be cold curved on site to a radius of 100 times that of the panel thickness for the uncoated material and to 175 times for the UV-protection coated product. Tighter curvatures risk stress cracking and damage to the protective UV and weathering coating.

## 21.9 Proprietary brands of polycarbonate

The more common trading names for polycarbonate are Makrolon, Makroclear, Lexan, Xantar, Calibre, Panlite and Zelux.

## 21.10 Coatings

Polycarbonate products may be painted, printed or vacuum metallised. Some paints contain thinners which may cause stress cracking and a reduction in impact strength. Polyurethanes, epoxies, vinyls and acrylic paints are all used, but generally the approved acrylic paints are the most appropriate.

Cleaning agents based on alcohols (methyl, ethyl, butyl or isopropyl alcohol) or aliphatic hydrocarbons (heptane or hexane) should be used to avoid surface damage.

## 21.11 Hazards in use: Performance in fire

The performance in fire should always be considered in relation to the appropriate use of polycarbonate in construction. Standard polycarbonate softens and deforms under fire conditions, with slow burning and some smoke generation. Special flame-retardant grades can resist severe flammability tests.

Most standard non-flame-resistant grades of 3–6 mm polycarbonate sheet sheets achieve a Class 1Y rating to BS476 Part 7: 1997 and produce little smoke. The Y suffix is due to the softening and then melting of the sample under test, with some flaming droplets after 4 minutes. When tested to BS EN 13501-1: 2007, UV-protected polycarbonate 3–5 mm sheet, without flame retardant, typically achieves a Euroclass reaction to fire classification of B-s2, d0. Users must however, verify the particular fire characteristics of their specified product.

## 21.12 Sustainability and recycling

The vast majority of plastics are currently dependent on the chemical and petrochemical industries, and cannot therefore be considered to be sustainable until sufficient alternative sources of organic precursors based on agricultural production are developed. Commercial polycarbonate waste is recycled by grinding and reprocessing into pellets for re-use by the industry.

## 21.13 British Standards

- *BS476 Part 7: 1997.* Fire tests on building materials. Method of test to determine the classification of the surface spread of flame of products.
- *BS5516 Part 1: 2004.* Patent glazing and sloping glazing for buildings: Code of practice for design and installation of sloping and vertical patent glazing.
- *BS5516 Part 2: 2004.* Code of practice for sloping glazing.
- *BS6206: 1981.* Specification for impact performance requirements for flat safety glass and safety plastics for use in buildings.
- *BS6262-4: 2005.* Glazing for buildings. Code of practice for safety related to human impact.
- *BS6399-2: 1997.* Loading for buildings. Code of practice for wind loads.
- *BS EN ISO 7391-1: 2006.* Plastics. Polycarbonate (PC) moulding and extrusion materials. Designation system and basis for specifications.
- *BS EN ISO 11925-2: 2010.* Reaction to fire tests. Ignitability subjected to direct impingement of flame. Single-flame source test.
- *BS EN 12600: 2002.* Glass in building. Pendulum test. Impact test method and classification for flat glass.
- *BS EN 13501-1: 2007.* Fire classification of construction products and building elements. Classification using data from reaction to fire tests.
- *HB10139: 1997.* Wind loading ready reckoner for BS 6399 Part 2: 1997.

## 21.14 Bibliography

Brydson, J.A. (2005) *Plastics materials.* New Delhi, India, CBS Pub & Dist.

Cousins, K. (2002) *Polymers in building and construction. RAPRA market report.* Shrewsbury, UK, RAPRA Technology.

Dufton, P.W. (2002) *Polymers in building and construction. RAPRA market report.* Shrewsbury, UK, RAPRA Technology.

Glass and Glazing Federation (2003) *Glazing with plastics.* London, UK, GGF.

Halliwell, S. (2003) *Polymers in building and construction.* Report 154. Shrewsbury, UK, RAPRA Technology.

Hopkins, C. (2006) *Rain noise from glazed and lightweight roofing.* BRE Information Paper IP 2/06. Watford, UK, Building Research Establishment.

Lyons, A.R. (2010) *Materials for architects and builders.* 4th edition, Oxford, UK, Butterworth-Heinemann.

## 21.15 Websites

- Bay Plastics: www.bayplastics.co.uk
- Makrolon: www.makrolon.de
- Lexan: www.sabic-ip.com

# 22 Polyethylene (Polyethene) Plastics

**Arthur Lyons** MA MSc PhD Dip Arch Hon LRSA
De Montfort University

## Contents

## 22.1 Introduction

Polyethylene is the most widely used plastic material, with extensive uses ranging from domestic and commercial packaging, bags, bottles and films to industrial pipes, containers, machine parts and electrical insulation, depending on the particular grade. The material is correctly termed polyethylene or more recently, polyethene, but the registered trade name 'polythene' is often used as a convenient abbreviation.

Polyethylene is produced by the polymerisation of the monomer ethylene (ethene). The polymerisation conditions of temperature, pressure and catalysis determine the extent and type of chain branching, the degree of crystallinity and the molecular weight distribution of the product.

The synthesis of polyethylene was discovered accidentally in 1933 by Reginald Gibson and Eric Fawcett of ICI, when they produced a white waxy material from the high-pressure reaction of ethylene and benzaldehyde in the presence of traces of oxygen. The technology was refined by Michael Perrin into the industrial process for low density polyethylene (LDPE) in 1935, patented in 1936, and developed to the commercial production named 'polythene' by 1939.

Subsequent advances were associated with the development of specific catalysts that enabled the polymerisation to take place at lower pressures, producing the denser product, high density polyethylene (HDPE). In 1951 Phillips Petroleum developed a catalyst based on chromium trioxide, and this was followed in 1953 by the Ziegler catalysts based on titanium chloride and organo-aluminium compounds. In 1976, the metallocene organo-metallic complexes were developed as specific catalysts by Kaminsky. These can activate the production of very low-density polyethylene (LDPE) and linear low-density polyethylene (LLDPE) as a copolymer incorporating a small proportion of an α-olefin, such as butene or hexene.

Research is continuing to develop further advanced catalysts based on the metallocences, using complex ligands attached to nickel, iron, palladium, zirconium and other metals to enhance specificity of function.

## 22.2 Manufacture

The different grades of polyethylene are produced by variations in the manufacturing processes (Table 22.1). The two distinct manufacturing processes are the high pressure system for low (LDPE) and medium density (MDPE) polyethylene and the low pressure system for high density (HDPE) and linear low density polyethylene (LLDPE).

Low density polyethylene is manufactured from ethylene under a pressure between 1000 and 3000 atmospheres (100–300 MPa) at a temperature between 100 and 300°C, in a stirred autoclave or tubular reactor with benzene or chlorobenzene as a solvent. Recent advances have reduced the required pressure to typically 160 MPa. The stirred autoclave has different zones, operating at different temperatures to control the molecular weight distribution and degree of branching of the product. Tubular reactors permit the feed of ethylene and catalyst at different locations to control the polymerisation process. The free radical addition polymerisation reaction is initiated by oxygen or organic peroxide initiator. Molecular weight distributions depend on the exothermic reaction conditions and the addition of chain transfer agents such as propane. Polymer chains are highly branched, typically at intervals of 10 to 25 monomer units, and polymer yields of 20% are usual with the unreacted ethylene being recycled.

High density polyethylene and linear low density polyethylene are manufactured by various low pressure technologies, as a slurry, in solution, or in the gas phase using a fluidised bed, which is the most common. In the fluidised bed system, compressed ethylene, hexene, hydrogen and nitrogen are fed through the base of a fluidised bed containing the transition metal catalyst. The hexene produces the short chain branches and the hydrogen acts as a chain terminator to control the molecular weight. The powdery polyethylene is then degassed and extruded into pellets. The slurry and solution production processes use the Phillips chromium, Ziegler–Natta aluminium or metallocene catalyst systems to ensure the production of linear polymer chains. The polymer, which is insoluble in the reaction mixture, is filtered off, washed and dried.

Ultra-high molecular weight polyethylene (UHMWPE), also known as polyethylene of raised temperature resistance (PE-RT), is an ethylene-octene copolymer synthesised in a solution process which produces a controlled side chain distribution.

Cross-linked polyethylene (PEX) is produced by treatment of medium or high density polyethylene. The three types PEX-A, PEX-B and PEX-C indicate the processes used to initiate the cross-linking process. PEX-A is manufactured by incorporating a peroxide, such as dicumyl peroxide, into the polymer, which then becomes cross-linked at the elevated temperature of extrusion. The density of the cross-linked product is slightly reduced from the raw material, e.g. from 950 to 939 kg/m$^3$ as the degree of crystallinity is reduced. PEX-B is formed by adding a silane, such as vinyl silane, together with a catalyst to the polyethylene resin. The moisture-cured cross-linking occurs after formation of the product when it is subsequently steamed or immersed in a hot water bath. PEX-C is cross-linked in its final form by irradiation with an electron beam. Care is required to ensure the appropriate level of cross-linking within the bulk of the material without embrittlement of the surface. Silane and radiation cross-linked polyethylenes have similar densities (e.g. 940 kg/m$^3$) to their raw materials as cross-linking takes place in the solid phase.

**Table 22.1** Grades of polyethylene

| | | *Density, kg/m$^3$* |
|---|---|---|
| LLDPE | Linear low density polyethylene | 900–939 |
| VLDPE | Very low density polyethylene | 880–915 |
| LDPE | Low density polyethylene | 916–925 |
| MDPE | Medium density polyethylene | 926–940 |
| HDPE | High density polyethylene | 941–970 |
| HMWPE | High molecular weight polyethylene | 947–950 |
| UHMWPE | Ultra-high molecular weight polyethylene | 930–935 |
| PEX | Crossed-linked polyethylene | 926–970 |

## 22.3 Chemical composition

Figure 22.1 illustrates the basic chemical structure of polyethylene, but molecular weight distributions and extent of branching will depend on polymerisation conditions.

The molecular weight of UHMWPE is usually between three and six million, giving rise to reduced efficiency in the packing of the chains and therefore a lower density. The molecular weight of HDPE is 10 times less, with a low degree of branching giving efficient packing and a resultant high density approaching the theoretical maximum of approximately 1000 kg/m$^3$ for the polymer. Only a few long chain branches can significantly change the rheological properties of the polymer. LDPE has a high degree of both long and short branch chains, whereas the LLDPE only has short branches from its copolymerisation with

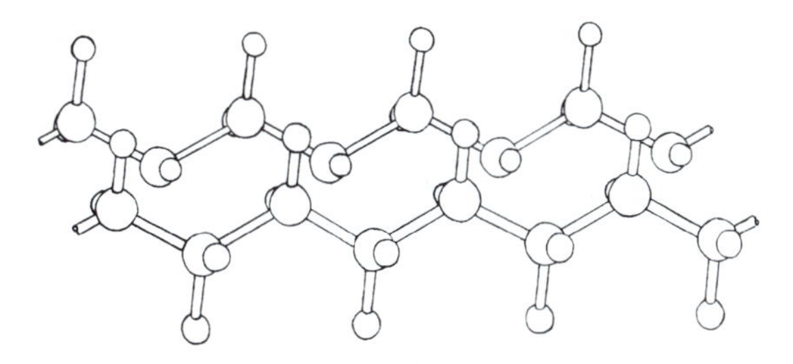

**Figure 22.1** Structure of the polyethylene molecule. Large circles – carbon atoms; small circles – hydrogen atoms

α-olefins, such as 1-hexene. Similarly, VDLPE has a high proportion of short chain branches associated with the copolymerisation of ethylene with α-olefins, including 1-butene, 1-hexene and 1-octene. A typical molecular weight distribution (polydispersity) is illustrated in Figure 22.2.

The three types of PEX have different chemical compositions. In PEX-A, carbon to carbon bonds are formed by free radical initiation, with a high level (usually up to 70%) of cross-linking. For PEX-B, the added silane forms carbon–silicon bonds, and in PEX-C the irradiation splits carbon–hydrogen bonds, allowing the cross-links to form between adjacent chains, but to a lesser degree than in PEX-A. A greater level of cross-linking than 90% would lead to increased brittleness and stress cracking.

Crystallinity in polyethylene, which refers to zones in which the polymer chains are oriented and packed closely together in a regular manner, relates closely to the degree of branching. The balance of the material is amorphous. Long and short side chains in LDPE produce relatively low levels of crystallinity (35–55%); similarly LLDPE with a significant proportion of short side chains has low to moderate crystallinity (35–60%). By contrast, HDPE with few side chains (e.g. 1 per 200 carbon atoms) has a higher degree of crystallinity (85–90%).

In addition to co-polymerisation with α-olefins for controlling molecular structures, ethylene may be co-polymerised with a range of other monomers including vinyl acetate to produce ethylene vinyl acetate (EVA), also acrylic esters to produce ethylene acrylates (e.g. EMA, ethylene methyl acrylate; EBA, ethylene butyl acrylate; 2EHA, ethylene 2-ethyl hexyl acrylate). Certain vinyl acetate and acrylic ester terpolymers additionally incorporate maleic anhydride or glycidyl methacrylate.

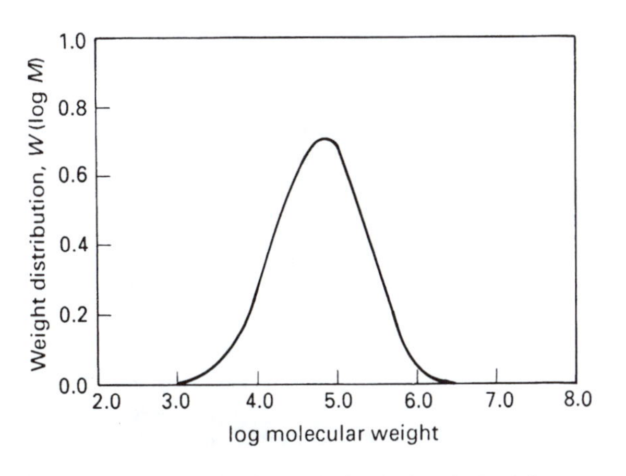


**Figure 22.2** Typical molecular weight distribution of polyethylene

## 22.4 Standard forms of polyethylene

Polyethylene in most grades is manufactured as powder or pellet form for subsequent forming processes. For construction purposes the main forms are extrusions, pipe, conduit, sheet and film. Ultra-low molecular weight polyethylene waxes are used within coatings and adhesives.

## 22.5 Typical physical properties of polyethylene

These are given in Table 22.2. LDPE is flexible but with good fatigue and chemical resistance and low moisture permeability. HDPE has rigidity and a tensile strength higher than that of low and medium density polyethylene but a slightly lower impact strength. Generally HDPE has good chemical resistance, low temperature strength and good processability at low cost.

UHMWPE has chemical inertness and good environmental stress crack resistance to solvents, oils and detergents. It has a low coefficient of friction, with good wear and abrasion resistance. UHMWPE has high impact and fatigue resistance and can be machined as wood. The ultra-high molecular weight distribution leads to a high melt viscosity and a melting point of 130°C; it is therefore mostly processed by compression moulding, ram extrusion or ram injection. It is available at a moderate price compared to low cost HDPE.

The broad molecular weight distribution of LLDPE ensures that is easily extruded for products such as pipes and cable jacketing and has good impact resistance. Polyethylene products with narrow molecular weight distributions generally have better optical properties and resistance to environmental stress cracking.

Melt viscosity is an important property in relation to the manufacture of polyethylene products by the various fabrication techniques. The melt index (MI) is recorded as the flow in g/10 min under standard test conditions of 2.16 or 21.6 kg. Generally the melt index is inversely related to the average molecular weight.

## 22.6 Specification of polyethylene

The key properties for the specification of polyethylene are density, melt index and molecular weight distribution. These require appropriate specification according to the fabrication

**Table 22.2** Typical properties of low, medium and high density polyethylene

| *Property* | *Units* | *Polyethylene* | | |
|---|---|---|---|---|
| | | *Low density (LDPE)* | *Medium density (MDPE)* | *High density (HDPE)* |
| Density | $kg/m^3$ | 910–925 | 925–940 | 940–970 |
| Tensile strength at yield | MPa | 6–26 | 13–28 | 20–34 |
| Elongation at break | % | 120–650 | 50–850 | 10–1200 |
| Flexural modulus | MPa | 50–300 | 150–1100 | 700–1700 |
| Vicat softening point | °C | 72–100 | 78–118 | 80–133 |
| Coefficient of thermal expansion | $\times 10^{-4}/°C$ | 1.3–1.9 | | |
| Thermal conductivity | W/m K | 0.4 | | |

These values are only typical, as there is a wide range of products in each category.

processes and physical requirements of the end product. The standard fabrication processes are extrusion, blown film, blow moulding and injection moulding.

Extrusion is used for continuous pipe and the production of thin film. Blown film involves the blowing of air into the extruded pipe immediately as it emerges from the die. The blown film is then usually flattened to a double sheet and cut along one or both edges. Blow moulding is used for the manufacture of individual components such as tanks and bottles, and in some cases uses blends of different average molecular weights with bi-modal or multi-modal molecular weight distributions which are advantageous to the process. Injection moulding is used for the production of slightly more complex forms that require pressure to force the fluid melt into all corners of the mould. Additionally polyethylene is processed into extruded or cast sheet and laminates.

Additives to polyethylene include UV stabilisers, colour pigments, flame retardants (e.g. antimony bromide or aluminium hydroxide), slip agents (e.g. fatty acid amides, to reduce the surface coefficient of friction), anti-static and carbon filler. Additives to improve processing performance include anti-oxidants to ameliorate the effect of elevated temperatures during forming.

## 22.7 Dimensional stability

The thermal movement of polyethylene is one of the highest in the thermoplastics range at $1.3–1.9 \times 10^{-4}$ °C$^{-1}$, compared to PVC at $0.7 \times 10^{-4}$ °C$^{-1}$ and steel at $0.12 \times 10^{-4}$ °C$^{-1}$. It is therefore necessary to make appropriate allowances for thermal movement where significant components are exposed to temperature variations. However, buried pipes will be restricted in their movement and subjected to only minor temperature variation. Corrugated drain pipes will take up any thermal movement within the corrugations.

## 22.8 Durability

Polyethylene is not resistant to UV without the addition of a stabiliser. Some stabilisers are grafted onto the carbon backbone chain, while others are intimately mixed into the polymer prior to forming. Where HDPE is used as geomembrane liners in landfill sites the service life is controlled by antioxidant depletion and the resistance of the material to subsequent oxidation; also long-term stressing may lead ultimately to environmental stress cracking. Polyethylene materials should not be stored for extended periods in direct sunlight.

## 22.9 Uses of polyethylene in construction

The main uses for polyethylene in construction are for pipes, ducting for services, sheet membranes and components. Blue polyethylene pipes are now standard for mains potable water supply and similarly yellow for gas supply, although this colour coding in not fully applied across Western Europe. Some countries, including the USA, use a continuous colour stripe on black polyethylene to indicate usage. Nominal outside diameters for thermoplastic pipes are listed in BS2782-11: 1997 and wall thicknesses in BS ISO 4065: 1996.

The quality assurance processes for ensuring fitness for purpose of polyethylene pipes include consideration of resistance to internal pressure, the effects of creep deformation and also fracture and rapid crack propagation resistance. The resistance to internal pressure is verified to the standard BS EN ISO 1167 Parts 1 and 2: 2006. Creep is covered by BS4618 Part 1.1: 1970 and brittle or ductile impact failure is described in BS4618 Part 1.2: 1973. The resistance to rapid crack propagation is assessed by the test BS EN ISO 13478: 2007. The standard grades for polyethylene pipes are PE80 (MDPE – medium density polyethylene) and PE100 (HPPE – high performance polyethylene), which is UHMWPE.

### 22.9.1 Gas distribution

The gas supply system in the UK uses orange and yellow (or peelable yellow with brown stripes) polyethylene pipe, although the standard BS EN 1555-2: 2010 allows for black (PE80 or PE100), yellow (PE80) or orange (PE100). Black (PE80) can have yellow stripes, while black (PE100) may have yellow or orange stripes. Peelable layers may have black, yellow or orange stripes. The standard refers to pipes of 16 mm to 630 mm outer diameter in two tolerance levels with respect to outside diameter, also deviations from roundness and wall thicknesses. The minimum markings on the pipe, at least every metre (Figure 22.3) must include the standard, the manufacturer's details, critical dimensions, tolerance grade, material, date of manufacture and the utility 'gas' (as appropriate). The European standard covers pipework for gas supply rated to 10 bar which equates to 1 MPa pressure, using polyethylene with a minimum density of 930 kg/m$^3$. However, Table 22.3 illustrates pressure ratings in relation to the polyethylene grade and the standard dimension ratio (SDR), which approximates to the ratio between the nominal outside pipe diameter and the nominal wall thickness. These are listed in the Gas Industry Standard publications (GIS/PL2 Parts 1 to 8: 2008).

For replacement piping, slip-lining of new pipes into old pipe work is carried out, where possible to reduce trenching costs and disruptions to the highway, as illustrated in Figures 22.4 and 22.5. The renovation of old underground gas supply networks is described in BS EN 14408: 2004. The Part 3: 2004 describes the use of close-fit pipes for renovation work. Joints can be electrofusion sockets or saddles, fusion-butt welded or mechanical. The specification for polyethylene pipes for gas is defined in the

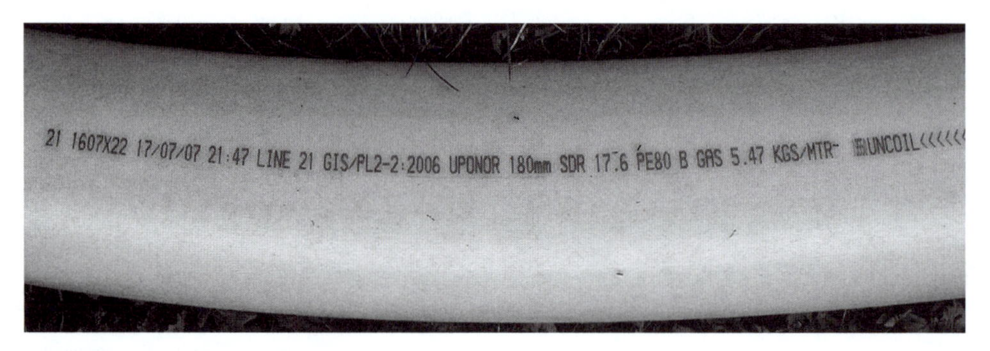


**Figure 22.3** Typical markings on 75 and 180 mm PE80 gas pipe

**Table 22.3** Maximum operating pressure ratings for polyethylene gas pipe at 0°C (GIS/PL2-1: 2008 and GIS/PL2-2: 2008 )

| *Polyethylene grade (colour code)* | *Outside diameter (mm)* | *Standard dimension ratio (SDR)* | *Maximum operating pressure (bar)* |
|---|---|---|---|
| PE80 | 16–140 | 11 | 5.5 |
| (yellow) | 200 | 11 | 4.4 |
| | 250 | 11 | 4.0 |
| | 315 | 11 | 3.4 |
| | 400 | 11 | 2.9 |
| | 450 | 11 | 2.7 |
| | 500 | 11 | 2.5 |
| | 90–250 | 17.6 | 3.0 |
| | 315 | 17.6 | 2.7 |
| | 400 | 17.6 | 2.3 |
| | 500 | 17.6 | 2.0 |
| | 355–500 | 21 | 2.0 |
| | 140–315 | 26 | 2.0 |
| PE100 | 250–630 | 11 | 7.0 |
| (orange) | 55–630 | 21 | 2.0 |

SDR = nominal outside diameter/nominal wall thickness. 10 bar = 1 MPa.

standard BS ISO 4437: 2007. This also includes references to testing samples for slow crack growth (BS EN ISO 13479: 2009); also the electrofusion and butt fusion of joints (BS2782-11: 1997, BS ISO 11413: 2008 and BS2782-11: 1997, BS ISO 11414: 2009 respectively).

In the butt-fusion jointing of pipes, the ends are cleaned and squared off. A flat electric heater is inserted between the pipe ends and pressure applied to hold them to the non-stick heater surface. The heater is then removed and the pipes forced together until cool, when the jointing bead is removed both internally and externally. Electrofusion involves the use of couplers or saddles containing heating elements which can be connected to the appropriate electrical supply for the necessary heating period (Figure 22.6). As for butt-fusion jointing, the pipe ends must be thoroughly cleaned or the removable coatings peeled off before the fittings are attached to the pipes. Couplers are used for jointing pipes and saddles for attaching smaller diameter service pipes.

Polyethylene pipes may be subjected to 'squeeze off' whereby the pipe is compressed with two clamps, at a distance apart of less than twice the wall thickness, to cut off the gas flow for maintenance purposes. The resistance of the pipe to gas pressure after this operation is tested according to the method described in BS EN 12106: 1998. In addition to homogeneous polyethylene pipes, the standard BS ISO 4437: 2007 refers to polyethylene pipes with co-extruded layers, also pipes with an external peelable layer (yellow with brown axial stripes) that requires removal before electrofusion or mechanical jointing to expose the clean white polyethylene. This process eliminates the requirement to prepare the pipe end on site before jointing.

**Figure 22.5** Slip-lining of polyethylene gas pipe into old main

Cross-linked polyethylene pipe systems may also be used for the conveyance of gas (BS ISO 14531). The Standard gives specifications for pipes, mechanical jointing, heat-fusion jointing and system design (BS ISO 14531 Part 1: 2002, Part 2: 2004, Part 3: 2010 and Part 4: 2006 respectively).

### 22.9.2 Water distribution

Blue or blue with brown stripes polyethylene pipes are generally used for potable water supply within the UK, although the standard BS EN 12201-2: 2011 permits either blue or black with blue stripes. Pipes from 16 mm to 1600 mm outer diameter (OD) are described at three tolerance bands relating to deviations

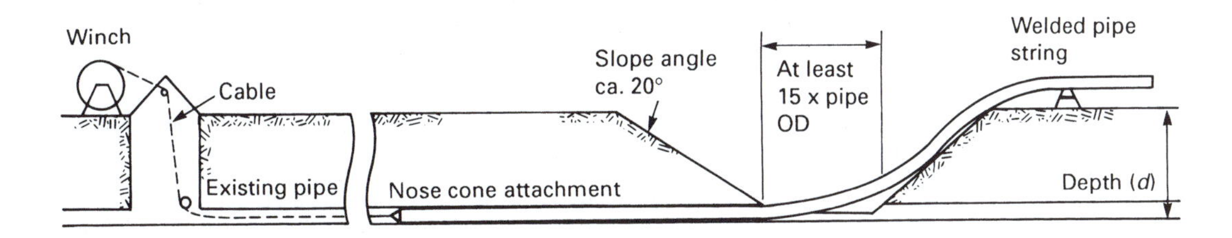


**Figure 22.4** Slip-lining process

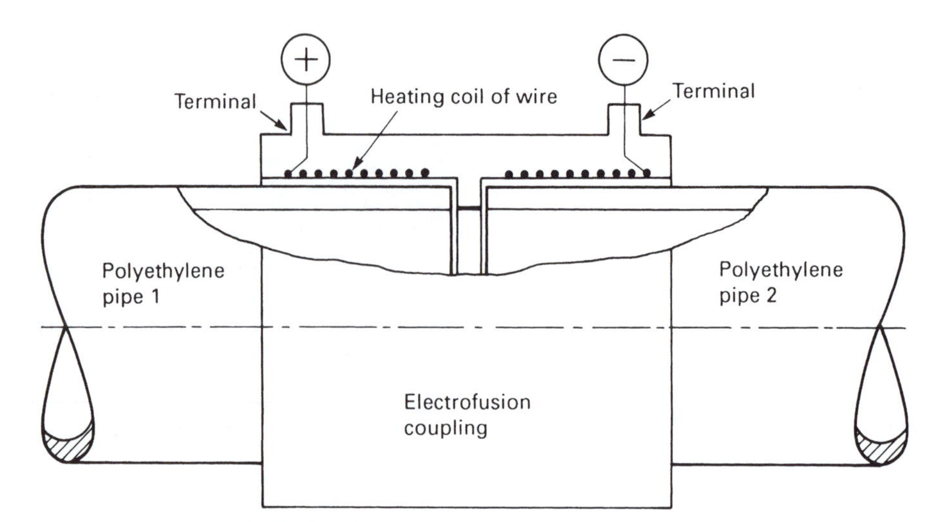


**Figure 22.6** Jointing of polyethylene pipes by electrofusion

**Figure 22.7** Typical markings on 90 mm PE100 water pipe

from roundness and nominal dimensions. Pipes must be marked at intervals, not less than once per metre, indicating the standard number, pressure rating, critical dimensions, material, manufacturer's identification and date of manufacture (Figure 22.7). The European standard covers pipes rated up to 25 bar which equates to 2.5 MPa pressure with a normal operating temperature of 20°C. Whilst the European Standards specify the requirements for pipelines irrespective of materials and then also give performance specifications for appropriate materials, the Water Industry Information and Guidance Notes (IGNs) and Water Industry Specifications (WIS) give the specification of operational systems. Table 22.3 shows the colour coding for potable and non-potable water systems and Table 22.4 (IGN 4-32-18: 2003) details the choice of pressure ratings for polyethylene pipe systems for water supply and sewerage as a function of SDR (outside diameter/wall thickness). The minimum required strength for the material PE100 (HPPE) is 10 MPa and that for PE80 (MDPE) is 8 MPa to fulfil the requirements of minimum design life of 50 years at 20°C. This, with a safety factor of 1.25, gives a design stress of 8 MPa for PE100 and 5 MPa for PE80 pipes. The pressure ratings (Table 22.5) for specific pipes of these materials are then determined from the SDR.

Peelable pipes require stripping back the polypropylene film to solid white polyethylene before electrofusion or butt fusion jointing as shown in Figure 22.8. Pipes incorporating an aluminium barrier must be marked accordingly, Figure 22.9. The brown stripes on a blue background indicate a composite layered pipe.

### 22.9.3 Waste water distribution

Polyethylene is a standard material for non-pressure underground drainage and sewerage pipes (BS EN 12666-1: 2005) and their fittings, such as shallow inspection chambers (BS EN 13598-1: 2010). The preferred colour for pipes is black, although other colours may be used. For ancillary fittings the preferred colours are black, orange/brown or dusty grey, but again other colours may be used. Additionally polyethylene is a standard material for soil and waste discharge pipes within buildings (BS EN 1519-1: 2000), including the discharge of waste at high temperatures. The preferred colour is black although other colours may be used. In all cases the units must be marked with the appropriate standard, the manufacturer's details, material, key dimensions, appropriate application and date of manufacture. The renovation of old underground non-pressure drainage and sewerage networks by lining old pipes with polyethylene piping is detailed in BS EN 13566 and BS EN ISO 11296. The standards describe lining with continuous pipes (BS EN 13566 Part 2: 2005), close fit pipes (BS

**Table 22.4** Polyethylene (PE) potable and non-potable/sewerage water pipe systems

| *Application* | *Material* | *Colour code* | *Nominal size range (mm)* | *Specification* |
|---|---|---|---|---|
| Potable water | PE80 | Light blue | 20–63 | BS EN 12201 Part 2: 2011 |
| Potable water | PE100 | Dark blue | 90–1000 | BS EN 12201 Part 2: 2011 |
| Potable water (barrier pipe) | PE80 and aluminium | Light blue with axial brown stripes | 25–180 | WIS 4-32-19: 2011 |
| Potable and non-potable water | PE100 | Dark blue with axial brown stripes | 90–630 | BS EN 12201 Part 3: 2011 |
| Pipes for above-ground potable, non-potable and sewerage and below-ground non-potable and sewerage | PE100 | Black | 90–1000 | BS EN 12201 Part 2: 2011 |

WIS – Water Industry Specification.

**Table 22.5** Nominal pressure ratings for polyethylene water pipe (IGN 4-32-18: 2003)

| *Polyethylene grade* | *Standard dimension ratio (SDR)* | *Nominal pressure rating (bar)* |
|---|---|---|
| PE80 | 11 | 12.5 |
| | 17 | 8 |
| | 26 | 5 |
| PE100 | 11 | 16 |
| | 17 | 10 |
| | 21 | 8 |
| | 26 | 6 |

Applicable to dimensions of less than 355 mm. Based on a minimum design life of 50 years at 20°C. SDR = nominal outside diameter/nominal wall thickness. 10 bar = 1 MPa.

EN ISO 11296 Part 3: 2011), cured-in-place pipes (BS EN ISO 11296 Part 4: 2011) and spirally wound pipes (BS EN 13566 Part 7: 2007). Cured-in-place sewer repairs are also described in WIS 4-34-06: 2010.

Polyethylene pipe is used for buried and above-ground pressure drainage and sewerage systems (BS EN 12201 and DD CEN/TS 13244-7: 2003). Typical situations are sea outfalls, pipes suspended under bridges, pipes laid in water as well as those laid in ordinary ground. The European standard is valid up to a maximum operating pressure of 25 bar or 2.5 MPa at a reference temperature of 20°C, but individual pipes and components may have lower pressure ratings. The pipes should normally be of a density greater than or equal to 930 kg/m$^3$ and colour coded black or black with brown stripes.

Structured-wall piping systems in polyethylene (BS EN 13476: 2007) for non-pressure underground drainage and sewerage are manufactured to two key types. Type A have smooth internal and external surfaces and are subdivided into Type A1 with a core of axial ribs, foam or intermediate thermoplastic layer, and Type A2 in which the inner and outer layers are connected by a spiral or

**Figure 22.8** Peelable water pipe jointing

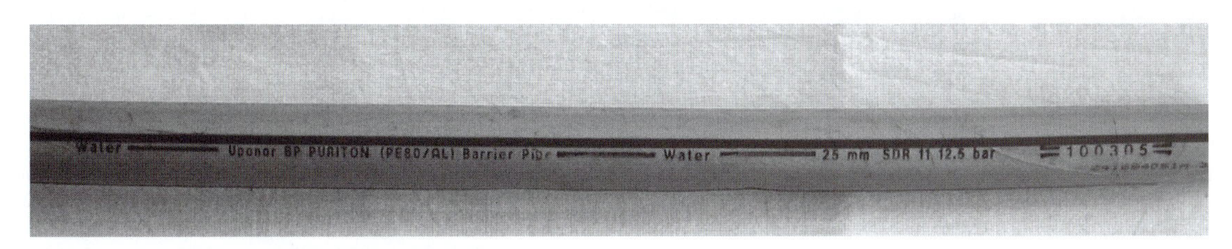


**Figure 22.9** Typical markings on 25mm PE80 barrier water pipe

radial ribs. Type B pipes have a smooth internal surface but a profiled external surface of solid or hollow annular or spiral ribs. The preferred colours for this system are black, orange-brown or dusty grey, although other colours may be used.

### 22.9.4 Hot and cold water systems

Cross-linked polyethylene (PEX) pipes are used for both hot and cold water installations including under-floor heating systems. The cross-linking enhances the physical properties of standard polyethylene sufficiently to withstand elevated domestic temperatures and pressures. The standard BS EN ISO 15875-2: 2003 permits the use of polyethylene cross-linked by any of three methods (PEX-A, PEX-B or PEX-C). The standard BS7291-3: 2010 requires technical information to be marked at not less than metre intervals on the pipe, including the word 'barrier' if the pipe is laminated, and gives the maximum UK service conditions.

There are two distinct types of types of laminated PEX pipes. For certain piped hot and cold water systems, including under-floor heating, an oxygen barrier is required in the pipe to prevent the ingress of oxygen into the water, which might subsequently cause corrosion of any steel components within the system. In one system, the oxygen barrier is a layer of ethyl vinyl alcohol copolymer (EVOH) (typically 0.13 mm) which is fixed to the outer and inner layers of PEX by a thin layer of adhesive, creating a sandwich product. An additional layer preventing UV penetration may also be included. The alternative and more common product uses an interlayer of aluminium foil as the barrier and this composite plastic–aluminium pipe is often designated as PEX-AL-PEX. The composite product is less flexible than PEX tubing, but retains its bent form and has slightly less thermal movement than the cheaper equivalent standard PEX pipe. Composite pipes may be rated up to 95°C.

Polyethylene of raised temperature resistance (PE-RT) offers good stress crack resistance and thermal conductivity, making it a suitable alternative to PEX for hot water systems especially in multi-layer metal and plastic composite pipes. Unlike PEX, it can be easily welded, creating fewer mechanical joints.

Individual components such as static thermoplastic storage tanks may be manufactured from polyethylene by blow or rotational moulding techniques to BS EN 13575: 2012.

### 22.9.5 Conduits

High density polyethylene extrusions are widely used as conduit, particularly in road works, for threading other service cables and pipes. Typical systems are twin-wall extrusions consisting of a corrugated outer wall and a smooth inner wall, which are heat-welded together in the twin extrusion process as illustrated in Figure 22.10. Appropriate jointing systems consisting of seals and coupler prevent the ingress of solid matter, and water if specified, into the conduit. These conduits are generally available in a wide range of colours to suit the relevant colour coding relating to the contained pipes, electricity or telecommunication cabling. Conduit jackets may also incorporate low density PEX foam insulation for reducing heat loss, for example between buildings. Insulated jackets may incorporate several colour coded pipes and a heating cable if frost protection is required.

**Figure 22.10** Flexible services conduit

### 22.9.6 Electrical uses

PEX is used for the insulation of electric cables intended for medium to high voltages, as it has good dielectric properties and can be continuously rated to conductor temperatures up to 90°C. High density polyethylene and EVA copolymer are also used for outer jacketing of electric cables. Also more complex blends of low molecular weight polyethylene, polyethylene-olefin copolymer and UHMWPE are used to produce the required physical properties from multi-modal molecular weight distributions on forming and cross-linking. Additives, such as chlorosulfonated polyethylene and magnesium oxide, may be incorporated to reduce the evolution of smoke in fires.

### 22.9.7 Membranes

Low density polyethylene is a standard material for damp-proof membranes (DPMs) of typically 250, 300 and 500 µm thicknesses. Appropriate grades of polyethylene sheet may be used for radon or carbon dioxide protection, but on more heavily contaminated brownfield or industrial sites where a gas barrier for carbon dioxide, radon and methane protection is required, or for tanking under hydrostatic pressure, high density polyethylene laminated with an aluminium foil core is available. Polyethylene is extensively used for geomembranes on landfill sites to BS EN 14576: 2005 and BS EN 14575: 2005.

Polyethylene is one of several materials approved for use in damp-proof courses (DPCs) to BS6515: 1984. A gas-resistant version, to prevent the transmission of radon, carbon dioxide and methane, incorporates an aluminium foil layer. Elastomeric laminates of modified bitumen on cross-laminated polyethylene sheet form suitable waterproof tanking materials for under-floor slabs and buried walls. For wall tanking an appropriate primer is usually applied to the masonry or concrete prior to the application of the self-adhesive membrane, which requires sheet protection before backfilling to eliminate the risk of puncturing.

Derivatives of polyethylene are standard materials for single-ply roofing systems. Ethylene propylene rubber (thermoplastic polyolefin, TPO) is a fully recyclable, heat weldable product, typically produced as a 1.2 mm sheet up to 3.6 m wide for mechanical or adhesive fixing. Alternative materials for single-ply roofing systems include chlorinated polyethylene (CPE) which can be reinforced with polyester and backed with a polyester fleece, mineral fibre or asphalt saturated felt; similarly, chlorosulfonated polyethylene (CSPE) which has good durability due to its resistance to elevated temperatures and chemical attack.

### 22.9.8 Insulation

Extruded closed-cell non-crosslinked polyethylene foam sheet (e.g. Ethafoam) can be used for thermal and acoustic insulation. It can be laminated with standard polyethylene sheet to provide a DPM or for acoustic insulation in partitioning.

### 22.9.9 Adhesives and additives

Other uses for polyethylene in construction include its use as a tackifier addition to adhesives and sealants, which can fine-tune the physical properties of the product in relation to viscosity, setting time, adhesion, durability and low-temperature performance. Also it is extensively used as EVA copolymer in hot melt adhesives ('glue sticks'). The wide range of properties and uses of EVA depend on the copolymer content and melt flow index. EVA and other ethylene acrylate copolymers (e.g. EMA, EBA and 2-EHA) are used as additives to bitumen, usually in the range 3–10% by weight, to modify mechanical behaviour, particularly to reduce high-temperature deformation and low-temperature cracking.

### 22.9.10 Rope

Polyethylene fibre ropes of three and four strands are described in the standard BS EN ISO 1969: 2004.

## 22.10 Proprietary brands of polyethylene

There are a large number of registered trade names for polyethylene and derivatives used in construction. The list includes Polythene, Alathon, Alkathene, Carlona, Dowlex, Hypalon, Ethafoam, Hypalon, Marlex, Petrothene, Plastazote, Rigidex, Ultrapoly and Visqueen, but it is not exclusive.

## 22.11 Hazards in use

Generally polyethylene is non-toxic; the solid material does not pose any health hazard if used correctly. Polyethylene may produce static electricity sparks. Airborne particles may cause respiratory or eye irritation, and burning material generates fumes. The molten material can cause burns due to its adherence to the skin.

## 22.12 Performance in fire

Polyethylene is combustible and most types start to melt in fire at 105–135°C and decompose by 300°C. On burning it will produce soot and some flaming droplets. Combustion gases are carbon monoxide, carbon dioxide, but also hydrocarbons, aldehydes, acrolein and other organic chemicals, depending on the fire conditions. Fire-retardant grades are produced commercially. Closed-cell polyethylene foam boards are also combustible and should not be exposed to flame. Chlorinated polyethylene has enhanced fire resistance over polyethylene.

## 22.13 Sustainability and recycling

While most polyethylene is currently produced from petroleum products, new developments are leading towards some manufacture from ethanol made from sugar cane as a renewable resource.

Waste material produced during the manufacturing process is usually recycled, as it is free from contamination. Post-consumer waste requires separation, transportation, cleaning and repelletising, and these operations are carried out by specialist companies. Recycled polyethylene can be used in certain specified situations, particularly in co-extrusion of thick section pipes in which virgin material sandwiches the recycled product, ensuring an appropriate surface quality and no contamination to the flowing gas or liquid.

The use of non-virgin polyethylene, including the characteristics of the recycled PE that should be taken into account, is described in the publications BS EN 15344: 2007 and DD CEN/TS 14541: 2007. Certain manufacturers produce polyethylene membranes for construction use from 100% recycled material. As most post-consumer recycled polyethylene is not suitable for food packaging, its use where acceptable in construction is an environmentally appropriate outlet.

Large quantities of domestic polyethylene waste are converted into 'plaswood' for manufacturing durable boardwalks, bollards and assorted street furniture items. While ordinary polyethylene sheet or film may take up to 100 years to degrade fully in the absence of the UV effects of sunlight, biodegradable polyethylene decomposes in approximately 5 years in the presence of sunlight, moisture and micro-organisms, depending on its manufacture.

## 22.14 British Standards

- *BS2782-11 Method 1121B: 1997.* Methods of testing – plastics. Thermoplastic pipes, fittings and valves. Thermoplastic pipes for the conveyance of fluids. Nominal outside diameters and nominal pressures. Metric Series.
- *BS ISO 4065: 1996.* Thermoplastic pipes. Universal wall thickness table.
- *BS ISO 4427.* Plastics piping systems. Polyethylene (PE) pipes and fittings for water supply: Part 1: 2007 General. Part 2: 2007 Pipes. Part 3: 2007 Fittings.
- *BS ISO 4437: 2007.* Buried polyethylene (PE) pipes for the supply of gaseous fuels. Metric series. Specifications.
- *BS4618.* Recommendations for the presentation of plastics design data. Mechanical Properties: Part 1.1: 1970 Creep. Part 1.2: 1973 Impact behaviour.
- *BS5955-8: 2001.* Plastics pipework. Specification for the installation of thermoplastic pipes and associated fittings for use in domestic hot and cold services and heating systems in buildings.
- *BS6076: 1996.* Specification for polymeric film for use as a protective sleeving for buried iron pipes and fittings.
- *BS6515: 1984.* Specification for polyethylene damp-proof courses for masonry.
- *BS7291-3: 2010.* Thermoplastics pipes for hot and cold water for domestic purposes and heating installations in buildings. Specification for cross-linked polyethylene (PE-X) pipes and associated fittings.
- *BS7523: 1991.* Specification for preformed polyethylene (PE) materials for thermal insulation of pipework.
- *BS ISO 11413: 2008.* Plastic pipes and fittings. Preparation of test piece assemblies between a polyethylene pipe and an electrofusion fitting.
- *BS ISO 11414: 2009.* Plastic pipes and fittings. Preparation of polyethylene pipe/pipe or pipe/fitting test piece assemblies by butt fusion.
- *BS ISO 11922-1: 1997.* Thermoplastic pipes for the conveyance of fluids. Dimensions and tolerances. Metric Series.
- *BS ISO 13953: 2001.* Polyethylene pipes and fittings. Determination of tensile strength and failure mode of test pieces from a butt-fused joint.
- *BS ISO 14531.* Plastics pipes and fittings. Cross-linked polyethylene (PE-X) pipe systems for the conveyance of gaseous fuels. Metric series. Specifications: Part 1: 2002 Pipes. Part 2: 2004 Fittings for heat fusion jointing. Part 3: 2010 Fittings for mechanical jointing. Part 4: 2006 System design and installation guidelines.
- *BS ISO 17484-1: 2006.* Plastics piping systems. Multi-layer pipe systems for indoor gas installations.

- *BS EN ISO 1167.* Thermoplastic pipes, fittings and assemblies for the conveyance of fluids. Determination of the resistance to internal pressure: Part 1: 2006 General method. Part 2: 2006 Preparation of pipe test pieces.
- *BS EN 1519-1: 2000.* Plastics piping systems for soil and waste discharge (low and high temperature) within the building structure. Polyethylene (PE).
- *BS EN 1555.* Plastics piping systems for the supply of gaseous fuels. Polyethylene (PE). Part 1: 2010 General. Part 2: 2010 Pipes.
- *BS EN ISO 1969: 2004.* Fibre ropes. Polyethylene 3 and 4 strand ropes.
- *BS EN ISO 11296.* Plastics piping systems for renovation of underground non-pressure drainage and sewerage networks: Part 1: 2011 General. Part 3: 2011 Lining with close fit pipes. Part 4: 2011 Lining with cured-in-place pipes.
- *BS EN ISO 11298.* Plastics piping systems for renovation of underground water supply networks: Part 1: 2011 General. Part 3: 2011 Lining with close fit pipes.
- *BS EN ISO 11542-1: 2001.* Plastics. Ultra high molecular weight polyethylene.
- *BS EN 12106: 1998.* Plastics piping systems. Polyethylene pipes. Test method for resistance to internal pressure after application of squeeze off.
- *BS EN 12201.* Plastic piping systems for water supply, and for drainage and sewerage under pressure. Polyethylene (PE): Part 1: 2011 General. Part 2: 2011 Pipes. Part 3: 2011 Fittings. Part 4: 2012 Valves. Part 5: 2011 Fitness for purpose of the system.
- *BS EN 12666-1: 2005.* Plastics piping systems for non-pressure underground drainage and sewerage. Polyethylene (PE). Specification.
- *DD CEN/TS 13244-7: 2003.* Plastic piping systems for buried and above ground pressure systems for water for general purposes, drainage and sewerage. Polyethylene (PE): Guidance for the assessment of conformity.
- *BS EN 13476.* Plastics piping systems for non-pressure underground drainage and sewerage. Structured wall piping systems: Part 1: 2007 General requirements and performance characteristics. Part 2: 2007 Specification for pipes and fittings with smooth internal and external surface and the system. Type A. Part 3: 2007 Specification for pipes and fittings with smooth internal and profiled external surface and the system. Type B.
- *BS EN ISO 13478: 2007.* Thermoplastic pipes for the conveyance of fluids. Determination of resistance to rapid crack propagation (RCP). Full scale test.
- *BS EN ISO 13479: 2009.* Polyolefin pipes for the conveyance of fluids. Determination of resistance to crack propagation. Test method for slow crack growth on notched pipes.
- *BS EN 13566.* Plastics piping systems for the renovation of underground non-pressure drainage and sewerage networks: Part 2: 2005 Lining with continuous pipes. Part 7: 2007 Lining with spirally-wound pipes.
- *BS EN 13575: 2012.* Static thermoplastic tanks for the above ground storage of chemicals. Blow moulded or rotationally moulded polyethylene tanks.
- *BS EN 13598-1: 2010.* Plastics piping systems for non-pressure underground drainage and sewerage. PVC-U, PP and PE.
- *BS EN 13967: 2012.* Flexible sheets for waterproofing. Plastic and rubber damp proof sheets including plastic and rubber basement tanking sheet. Definitions and characteristics.
- *BS EN 14408.* Plastics piping systems for the renovation of underground gas supply networks: Part 1: 2004 General. Part 3: 2004 Lining with close-fit pipes.
- *BS EN 14575: 2005.* Geosynthetic barriers. Screening test method to determine resistance to oxidation.
- *BS EN 14576: 2005.* Geosynthetics. Test method for determining resistance of polymeric geosynthetic barriers to environmental stress cracking.
- *BS EN ISO 14632: 1999.* Extruded sheets of polyethylene (PE-HD).
- *BS EN 15344: 2007.* Plastics. Recycled plastics – Characterisation of polyethylene (PE) recyclates.
- *BS EN ISO 15875.* Plastics piping systems for hot and cold water installations. Cross-linked polyethylene (PE-X): Part 1: 2003 General. Part 2: 2003 Pipes. Part 3: 2003 Fittings. Part 5: 2003 Fitness for purpose of the system.
- *BS EN 60695-11-10: 1999.* Fire hazard testing. Test flames – 50 W horizontal and vertical flame test method.
- *PAS 25: 1999.* Plastics frames for use in gully tops and manhole tops for pedestrian areas.
- *DD CEN/TS 14541: 2007.* Plastics pipes and fittings for non-pressure applications. Utilisation of non-virgin PVC-U, PP and PE materials.

## 22.15 Bibliography

Brydson, J.A. (2005 ) *Plastics materials.* CBS Pub & Dist, New Delhi, India.

Clough, R. and Martyn, R. (1995) *Environmental impact of building and construction materials. Plastics and elastomers.* Construction Industry Research and Information Association, London, UK.

Cousins, K. (2002) *Polymers in building and construction. RAPRA market report.*, RAPRA Technology, Shrewsbury, UK.

Donskoi, A., Shaskina, M. and Zaikov, G., (2003) *Fire resistant and thermally stable polymers derived from chlorinated polyethylene.* Brill Academic Publishers, Leiden, The Netherlands.

Dufton, P.W. (2002 ) *Polymers in building and construction. RAPRA market report.* RAPRA Technology, Shrewsbury, UK.

Gas Industry Standard GIS/PL2-1: 2008. *Polyethylene pipes and fittings for natural gas and suitable manufactured gas. Part 1 General and polyethylene compounds for use in polyethylene pipes and fittings.* National Grid.

Gas Industry Standard GIS/PL2-2: 2008. *Polyethylene pipes and fittings for natural gas and suitable manufactured gas. Part 2 Pipes for use at pressures up to 5.5 bar.* National Grid.

Gas Industry Standard GIS/PL2-8: 2008. *Polyethylene pipes and fittings for natural gas and suitable manufactured gas. Part 8 Pipes for use at pressures up to 7 bar.* National Grid.

Halliwell, S. (2003) *Polymers in building and construction.* Report 154., RAPRA Technology, Shrewsbury, UK.

Hazlehurst, J. (2009) *Tolley's basic science and practice of gas service.* 5th edition. Newnes-Elsevier, Amsterdam, The Netherlands.

Lefteri, C. (2006) *Plastics 2.* Rotovision, Crans-Près-Céligny, *Switzerland.*

Lefteri, C. (2008) *The plastics handbook.* Rotovision, Crans-Près-Céligny, *Switzerland.*

Lyons, A.R. (2010 *Materials for architects and builders.* 4th edition. Butterworth-Heinemann, Oxford, UK.

Scott, D.V. (2000) *Advanced materials for water handling. Composites and thermoplastics.* Elsevier Advanced Technology, Oxford, UK.

Water Industry Specification WIS 4-32-19 (2011). *Polyethylene pressure pipe systems with an aluminium barrier layer for potable water supply in contaminated land.* Water UK.

Water Industry Specification WIS 4-34-06 (2010) *Specification for localised sewer repairs using cured-in-place systems with or without re-rounding.* Water UK.

Water Industry Information & Guidance Note IGN 4-32-05: 1986 *Specification for polyethylene (PE) pipes for sewer linings (non-pressure applications).* Water Research Commission.

Water Industry Information & Guidance Note IGN 4-32-18: 2003 *The choice of pressure ratings for polyethylene pipe systems for water supply and sewerage duties.* Water Research Commission.

## 22.16 Websites

- British Polythene Industries: www.bpipoly.com
- BP Plastics Holding: www.bpplastics.com

- Dow Chemical Company: www.dow.com
- DuPont: www2.dupont.com
- Exxon Mobil Chemical: www.exxonmobilchemical.com
- Glynwed Pipe Systems: www.glynwedpipesystems-uk.com
- National Grid: www.nationalgrid.com
- Crosslinked polyethylene information: www.pexinfo.com
- Uponor: www.uponor.co.uk
- Visqueen Building Products: http://visqueenbuilding.co.uk
- Water UK: http://water.org.uk
- Water Industry Specifications: www.wis-ign.org
- Water Regulations Advisory Scheme: http://wras.co.uk
- Water Research Commission: www.wrcplc.co.uk
- Wavin Plastics: www.wavin.com

# 23 Polypropylene (Polypropene) and Polybutylene (Polybutene-1) Plastics

**Arthur Lyons** MA MSc PhD Dip Arch Hon LRSA
De Montfort University

## Contents

## 23.1 Introduction

Polypropylene is a stiff, chemically resistant, translucent thermoplastic widely used for domestic packaging, containers and automotive parts, and also as fibres for ropes and carpets. While some physical properties are similar to those of polyethylene, polypropylene has a significantly higher softening point and good stress crack resistance, hence its use in certain engineering applications.

Polypropylene (polypropene) is produced by the polymerisation of the monomer propylene (propene) (Figure 23.1). The polymerisation conditions of temperature, pressure and catalysis have a profound effect on the resulting polymer, particularly in respect of the tacticity of the product, associated with the relative orientation in three dimensions of the bulky methyl groups along the carbon backbone chain. Atactic polypropylene, in which the methyl groups are distributed randomly from side to side along the chain, is amorphous. By contrast, the tactic products (Figure 23.2) in which the methyl groups are all oriented one way (isotactic) or in a regular alternating pattern (syndiotactic) are more crystalline and with higher melting points. The differing structures lead to different properties and applications. Thus atactic polypropylene is used as the modifier in APP (atactic polypropylene) bitumen roofing systems and tactic polypropylene is used in the extrusion of pipes. The relative proportions of tactic and amorphous polymer can be adjusted in the polymerisation process to produce the required physical properties. Additionally copolymerisation increases the product range. Polybutylene (polybutene-1) (Figure 23.1) is produced by the polymerisation of butylene (butene-1).

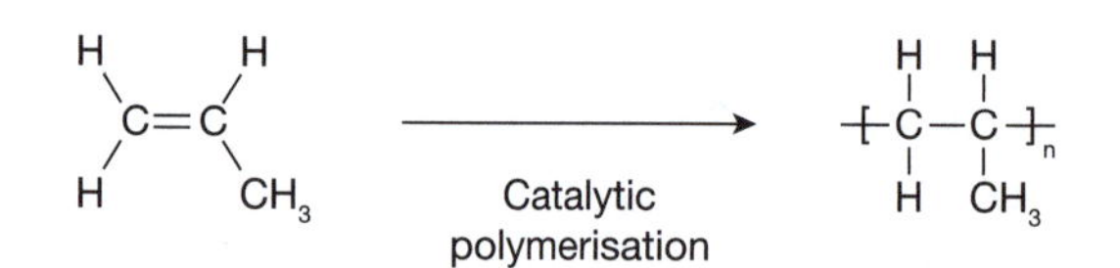


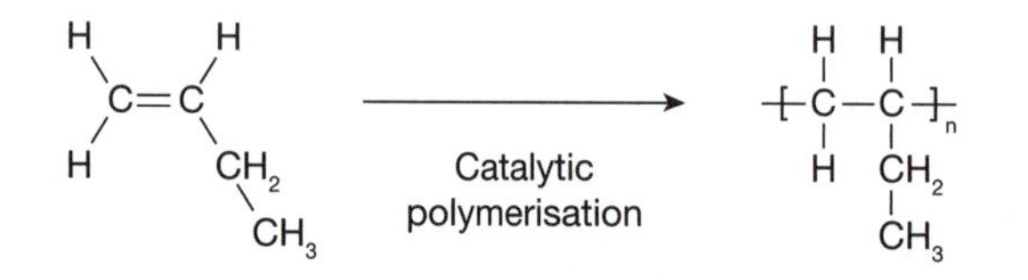


**Figure 23.1** Polymerisation of propylene and butene-1

Polypropylene was first produced from propylene gas by Natta in Spain in 1954, using the titanium chloride/organo-aluminium catalysts and technology developed for polyethylene. This was followed a year later by the first synthesis of polybutylene. Commercial production of polypropylene started in 1957. Initially the polypropylene produced had only a moderate level of crystallinity, but subsequent developments of the sterospecific catalysts by 1975 led to the production of 90% isotactic polypropylene. Industrial production of polybutylene from butene-1 commenced in Germany in 1964 and in the USA in 1968; it is now also produced in Italy, The Netherlands and Japan.

## 23.2 Manufacture

Polypropylene is manufactured from high-pressure propylene in a slurry, a solution or a gas phase process. According to the polymerisation conditions, various grades of polypropylene are produced. Atactic polypropylene, which is amorphous and rubbery, is produced by free radical initiated polymerisation. The stereo-specific Ziegler–Natta catalysts produce largely isotactic polypropylene, with the methyl groups on one side of the chain,

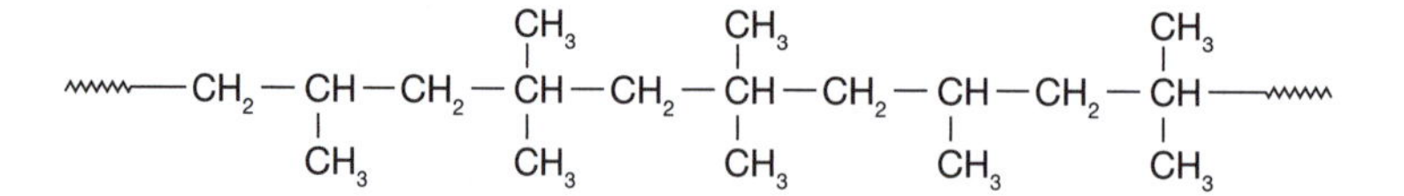

Atactic (random) polypropylene

Isotactic polypropylene

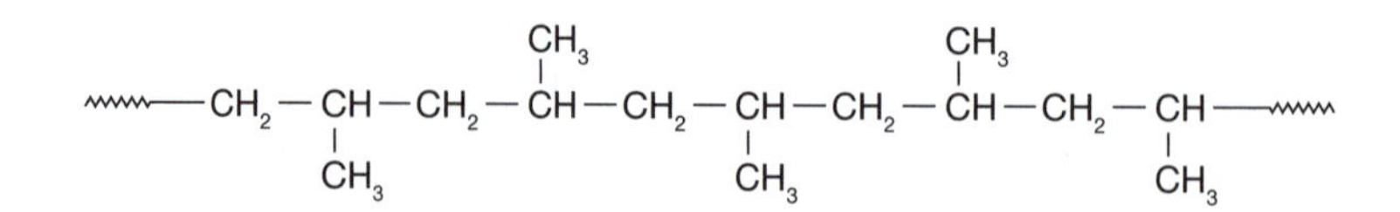

Syndiotactic polypropylene

**Figure 23.2** Atactic, isotactic and syndiotactic polypropylene

causing the macromolecules to coil into helices, which then align in groups to form crystalline zones. The Kaminsky metallocene catalysts are significantly more specific and can produce atactic, isotactic or syndiotactic polymers according to the exact form of the catalyst. A high degree of tacticity can be produced when required as well as high molecular weight distributions. Metallocene catalysts can also produce combinations of tactic and atactic block polymers.

In addition to homopolymers of polypropylene, copolymers with small quantities of polyethylene can be produced to widen the available range of material properties. These may be either random or block copolymers according to the dispersion of the additional polymer units.

Polybutylene (polybutene-1), produced by Ziegler–Natta catalysis, is a linear, isotactic, high molecular weight, semi-crystalline material. It is mainly used as the homopolymer rather than copolymerised with other polyolefins. It can be blended with polypropylene, but it is not compatible with polyethylene.

## 23.3 Chemical composition

Figure 23.1 illustrates the basic chemical repeat units for polypropylene and polybutylene (polybutene-1), but molecular weight distributions will depend on polymerisation conditions and the required properties for specific end uses.

## 23.4 Standard forms of polypropylene and polybutylene

Polypropylene and polybutylene are manufactured in powder or granular form for subsequent forming processes. For construction purposes the main forms of polypropylene are extrusions, pipe, membranes, formwork, fibres and mouldings. Polybutylene is mainly used for pipe extrusions.

## 23.5 Typical properties of polypropylene and polybutylene

Properties are given in Table 23.1. Polypropylene is a translucent stiff polymer with good impact resistance at ambient temperatures. The material has good fatigue resistance and electrical insulating properties. It may be opaque or coloured but not fully transparent. Polypropylene is resistant to stress cracking and to acids and alkalis, but not to hydrocarbons and halogenated hydrocarbons. The material has poor resistance to UV unless stabilised or protected. Additives are therefore added to polypropylene to give UV protection and to enhance other end-use properties. Polypropylene fibre is non-absorbent.

**Table 23.1** Typical properties of polypropylene and polybutylene (polybutene-1)

| *Property* | *Unit* | *Polypropylene* | *Polybutylene* |
|---|---|---|---|
| Density | kg/m$^3$ | 890–910 | 910–940 |
| Tensile strength at yield | MPa | 12–43 | 20 |
| Coefficient of thermal expansion | $\times 10^{-4}$/°C | 1.0–2.0 (typically 1.4–1.5) | 1.3 |
| Maximum constant temperature use | °C | 80 | 95 |
| Melting point | °C | 120–165 | 127–130 |
| Thermal conductivity | W/m K | 0.10–0.22 | 0.19 |

These values are only typical, as there is a wide variation associated with polymer tacticity and molecular weight distributions.

The block copolymer of polypropylene and ethylene has enhanced impact resistance at low temperatures (–20°C). The random copolymer is a clearer and flexible product.

Polybutylene is less stiff than polypropylene; it has good low-temperature impact resistance, also good creep resistance and elastic recovery. Polybutylene does not suffer from environmental stress cracking and it is resistant to dilute acids, alkalis and alcohols. It is, however, affected by aromatic and halogenated hydrocarbons.

## 23.6 Specification of polypropylene and polybutylene

The key property of polypropylene is its tacticity, but in addition, the melt flow index and molecular weight distribution need to be appropriately specified for the particular end-use. Polybutylene is predominantly produced in the isotactic form, although this is available in a range of molecular weight distributions. Polybutylene can take high loadings of filler, which is advantageous for specific uses such as halogen-free flame-retardant composites. Copolymers and block homopolymers of polypropylene offer a wide range of properties from rigid to rubbery elastomeric materials.

The standard fabrication processes for both polypropylene and polybutylene are extrusion, blown film, blow and injection moulding, compression and rotational moulding, calendering and cable coating. The materials are normally supplied to the fabricators as powder, or granules and also fibres for polypropylene. Additives include processing and UV stabilisers, colour pigments, carbon and other fillers, anti-static and fire resisting agents.

The standards BS5139: 1991 and BS EN ISO 1873-1: 1995 detail the designation and specification system for polypropylene. The equivalent standard for polybutylene (polybutene-1) is BS EN ISO 8986-1: 2009. The standards detail the technical information required for full specification as a series of codes within a data block set.

## 23.7 Dimensional stability

The thermal movement of polypropylene is typically $1.5 \times 10^{-4}$ °C$^{-1}$ and that of polybutylene is $1.3 \times 10^{-4}$ °C$^{-1}$, compared to PVC at $0.7 \times 10^{-4}$ °C$^{-1}$ and steel at $0.12 \times 10^{-4}$ °C$^{-1}$. It is therefore necessary to make appropriate allowances for thermal movement where significant components are exposed to temperature variations. However, buried pipes will be restricted in their movement and except for hot water systems, will be subject only to minor temperature fluctuations.

## 23.8 Durability

Polypropylene and polybutylene require the addition of UV stabilisers to prevent degradation when exposed to sunlight. These materials should not be stored for extended periods in direct sunlight prior to use in construction.

Polypropylene does not exhibit stress cracking and is highly fatigue resistant; hence its use in integral hinges, for example, the snap-shut insulating boxes around electrical connectors in domestic light fittings. The high melting point (up to 160°C for certain grades) gives the material an advantage over most standard grades of polyethylene. Polybutylene is unaffected by environmental stress cracking, has low creep and the advantage of low-temperature impact resistance.

## 23.9 Uses of polypropylene and polybutylene in construction

The main uses for polypropylene in construction are for pipes and their fittings, drainage access chambers, tanks, vapour and moisture control membranes, permanent shuttering for concrete, fibre incorporation in concretes, wood plastic lumber and small components. Polybutylene is mainly used for pipework as an alternative to cross-linked polyethylene (PE-X) pipe.

### 23.9.1 Water distribution

Polypropylene water pipes may be manufactured from polypropylene homopolymer (PP-H), polypropylene block copolymer (PP-B) or polypropylene random copolymer (PP-R), as specified in the standard BS EN ISO 15874: 2003. They may be used for hot and cold water (whether for human consumption or not) and for heating systems. Only manufacturers' own recycled polypropylene in addition to virgin material may be used in the production process. However, with the increasing concern for environmental issues the use of recycled polypropylene in non-pressure pipes is considered by the draft paper DD CEN/TS 14541: 2007. Pipes must be opaque with a maximum light transmittance of 0.2% and may be without or with a barrier to diminish the diffusion of gases and the transmission of light. The standard BS EN 15874-1: 2003 details the maximum service conditions for hot water supply, underfloor heating and high-temperature radiators. Preferred sizes in the range 12 to 160 mm (dimension class A) are quoted in addition to sizes based on copper pipes (dimension classes B1 and B2) and non-preferred systems (dimension class C). The standard requires pipes to be marked at not less than metre intervals with the standard number, manufacturer's mark, nominal dimension class and sizes, material, application class and operating pressure.

Polybutylene pipes for hot and cold water systems are specified in the standard BS EN ISO 15876: 2003. Pipes may incorporate an EVOH (ethyl vinyl alcohol copolymer) barrier layer of less than 0.4 mm thickness and may be used for potable water. As for polypropylene, only the manufacturer's own recycled polypropylene in addition to virgin material may be used in the production process and pipes must be opaque with a maximum light transmittance of 0.2%. The recent British Standard BS 7291: 2006 details the maximum service conditions for these pipes including a maximum malfunction temperature for sealed central heating systems of 114°C, which is greater than the 100°C permitted in the earlier standard BS EN ISO 15874-1: 2003. The standard BS7291-2: 2010 requires pipes to be marked at not less than metre intervals with the standard number and date, manufacturer's mark, nominal dimension class and sizes, material and application class. Polybutylene is slightly permeable to oxygen, therefore in heating systems, the primary circuit water should contain a corrosion inhibitor or barrier pipes should be used, in which case the pipes must be marked accordingly. Polybutylene pipes are sufficiently flexible to be coiled, providing the coil diameter is not less than 20 times the nominal pipe diameter. A number of manufacturers offer push-fit plumbing fittings for polybutylene systems.

Polypropylene and polybutylene (polybutene) pipes of nominal diameters from 12 to 1600 mm for industrial applications are described in the standard BS EN ISO 15494: 2003. The standard details tolerances and jointing by electrofusion, butt fusion, socket fusion and mechanical fixings.

### 23.9.2 Waste water distribution

Polypropylene pipework is frequently used for soil and waste discharge within buildings and for non-pressure drainage and sewerage as an alternative to other plastics products. Additionally polypropylene with mineral modifiers is used for underground drainage systems. Pipes coded 'B' are for use inside buildings and outside buildings fixed to the wall. Coding 'BD' refers to inside the building and buried in the ground within the building structure.

Polypropylene soil and waste discharge pipes and fittings for use within the building structure are specified in the standard BS EN 1451-1: 2000. The standard refers to pipes of 32 to 315 mm nominal outside diameter in homopolymer (PP-H) or copolymer (PP) with appropriate additives. The standard colours are grey, black or white but other colours are permitted. Pipes must have their full designations marked at least every metre. Jointing is normally by ring seal or butt fusion.

Polypropylene underground drainage and sewerage pipes are specified in BS EN 1852-1: 2009. Pipes are coded 'D' for use under and within one metre of the building structure and 'UD' for outside the building structure. Pipes should be black, orange-brown or dusty grey. The standard refers to pipes of nominal outside dimension between 110 and 1600 mm. The standard requires pipes to be marked at not less than two metre intervals with the standard number, manufacturer's mark, area application code, nominal dimensions, and the polypropylene material, including either normal elastic modulus (PP) or high elastic modulus (PP-HM).

Structured-wall piping systems in polypropylene (BS EN 13476: 2007) for non-pressure underground drainage and sewerage are manufactured to two key types. Type A have smooth internal and external surfaces and are subdivided into Type A1 with a core of axial ribs, foam or intermediate thermoplastic layer, and Type A2 in which the inner and outer layers are connected by a spiral or radial ribs. Type B pipes have a smooth internal surface but a profiled external surface of solid or hollow annular or spiral ribs. Variations include slotted pipes for land drainage and pipes manufactured with a profiled black polyethylene outer layer joined to a coloured smooth inner pipe. The preferred outer colours for this system are black, orange-brown or dusty grey, although other colours may be used. Structured-wall pipes may be designated for application area codes 'U' and 'D'.

The use of mineral modified polypropylene (PP-MD) is specified in the standard BS EN 14758-1: 2012. The material consists of polypropylene to which either calcium carbonate or talc (magnesium silicate) has been added to less than 50% content by weight. Pipes and fittings are normally coloured as standard polypropylene (black, orange-brown or dusty grey) and standard sizes are in the range 110 to 1000 mm nominal outside diameter. The marking, at a minimum of every two metres, must include the coding, PP-MD for mineral modified polypropylene. Mineral modified polypropylene has a higher density in the range 100 to 140 kg/m$^3$, a reduced coefficient of thermal expansion (0.7–1.2 $\times$ $10^{-4}$/°C) and an increased thermal conductivity (0.2–0.6 W/m K) compared to standard polypropylene. Pipes are coded 'U' or 'UD' as appropriate to the application. Jointing uses the ring seal system.

### 23.9.3 Membranes

A range of construction membranes are manufactured from laminated polypropylene. Polypropylene membranes for pitched roofs include those manufactured from woven or non-woven polypropylene laminates incorporating polyolefin copolymer (e.g. polypropylene/polyethylene) or polyethylene. This combination gives the appropriate resistance to wind-blown dust, rain and snow together with vapour control to prevent condensation. Equivalent products are appropriate as air leakage layers within flat roofing systems. Certain systems incorporate aluminium foil for additional thermal insulation.

Moulded polypropylene sheeting can be used as an internal damp-proofing membrane to moist walls. The three-dimensional moulding permits water to move behind the sheeting which can then be applied with standard internal finishes.

Certain single-ply roofing systems are manufactured from FPA (flexible polypropylene alloy) reinforced with glass-fibre scrim. The water-proofing layer is typically 1.2–2.0 mm in thickness and can be hot-air welded. An underlay of polyester felt together with ballast, adhesive or mechanical fixing are standard systems.

A range of civil engineering geotextiles are manufactured from polypropylene due to its inertness to most ground contaminants including acids, alkalis, chlorides and sulfates. Filtration geotextiles may be woven or non-woven needle-punched membranes giving protection to geomembrane liners.

Standard cast, extruded and biaxially oriented polypropylene films and sheeting are covered by the standards BS ISO 17557: 2003, BS EN ISO 15013: 2007 and BS ISO 17555: 2003 respectively.

### 23.9.4 Polypropylene units

Moulded polypropylene is used not only for pipe manufacture but also for a range of large units such as manhole and access chambers (BS EN 13598: 2010), tanks for rainwater management systems, small wastewater treatment systems (BS EN 12566-3: 2005) and domestic cisterns (BS4213: 2004). In addition small components, such as spacers to ensure a uniform depth of horizontal mortar joint in brickwork, are manufactured from polypropylene.

### 23.9.5 Concrete shuttering

Polypropylene permanent concrete shuttering is commercially available as a wide range of products. These include smooth or fluted sheets, usually bonded as double layers to give strength during the concrete setting and hardening processes. Expanded polystyrene insulation (typically 50–200 mm) may be incorporated between the polypropylene web spacers. Units can offer radon or methane protection if specified. Non-woven polypropylene fabric formwork liners for concrete reduce the production of surface micro-cracks associated with the early drying out of the setting and hardening surface. Variations of the system include multi-layer products allowing for the escape of excess moisture through the plastic fabric or mesh backing.

### 23.9.6 Fibre-reinforced concrete

Polypropylene fibres for incorporation in concrete are manufactured by extrusion and stretching to improve tensile properties, followed by coating with a surfactant to ensure bonding to the cement. Fibres are typically 18–22 μm in diameter and 12 to 19 mm long. Polypropylene fibres can improve the post-cracking impact resistance of concrete. In addition plastic shrinkage and plastic settlement can be reduced. The inclusion of polypropylene fibres in high-quality concrete improves its performance in fire by reducing spalling.

### 23.9.7 Electrical uses

Polypropylene is an alternative to PVC for the insulation of electric cables to BS EN 50290-2-25: 2002, which specifies the solid and cell grades appropriate for this application. Polypropylene has the advantage that it emits less smoke and no toxic halogens in fire. It therefore is used in low-ventilation environments including tunnels. Pressure-sensitive adhesive tapes for electrical purposes may be made from polyethylene or polypropylene to BS EN 60454-3-12: 2006.

### 23.9.8 Rope

Rope is manufactured from mono-filament and multi-filament fibre; also split film polypropylene. Standard ropes are three-, four- and eight-strand to BS EN ISO 1969: 2004 and BS EN ISO 1346: 2004. Polypropylene twine, mainly used for agricultural purposes, is covered by BS EN 12423: 1999.

### 23.9.9 Adhesives

Polybutylene is used in certain hot-melt adhesives.

## 23.10 Proprietary brands of polypropylene

Some of the more common trading names for polypropylene are Fortilene, Olevac, Profax, Propylux, Protec and Tecafine, but the list is not exclusive.

## 23.11 Hazards in use

Generally polypropylene and polybutylene are non-toxic; the solid material does not pose any health hazard if used correctly. The materials may produce static electricity sparks. Airborne particles may cause respiratory or eye irritation.

## 23.12 Performance in fire

Polypropylene and polybutylene will burn in a fire. Melting plastics may spread in a large fire and soot will be produced. Combustion gases will be oxides of carbon, but as the materials are hydrocarbons, no halogen-containing toxic or corrosive fumes are produced. Molten polypropylene is known to cause burns due to its adherence to the skin. Heavily filled polybutylene is more resistant to fire than the unfilled polymer.

## 23.13 Sustainability and recycling

Polypropylene and polybutylene, like most thermoplastics, are products of the petrochemical industry, hence current production is dependent on non-renewable oil resources. Waste material produced during the manufacturing process is usually recycled, but generally only this and virgin material are acceptable to the various British and European Standards for pipe and other production. However, the characteristics of post-consumer polypropylene that may be recycled are described in the publication DD CEN/TS 14541: 2007.

The BRE Digest 480 discusses the opportunities offered by the formation of wood plastic composites (WPCs) from recycled plastics including polypropylene. Wood fibres, fire retardants, UV stabilisers, lubricants and pigments are necessary components within the recycled plastic mix which then can be moulded into piles, beams, joists, decking, etc. The material has the advantage of durability and low maintenance and is considered to be an eco-composite as it uses waste from the timber industry and recycled plastics.

Polypropylene materials for recycling should show the resin identification code 5 (Figure 23.3).

## 23.14 British Standards

- *BS ISO 161: 1996.* Methods of testing – plastics. Thermoplastic pipes, fittings and valves. Thermoplastic pipes for the

**Figure 23.3** Polypropylene resin identification code

conveyance of fluids. Nominal outside diameters and nominal pressures. Metric Series.

- *BS476 Part 7: 1997.* Fire tests on building materials and structures. Method of test to determine the classification of the surface spread of flame of products.
- *BS ISO 1167-1: 2006.* Thermoplastic pipes, fittings and assemblies for the conveyance of fluids. Determination of the resistance to internal pressure. General method.
- *BS ISO 4065: 1996.* Thermoplastic pipes. Universal wall thickness table.
- *BS4213: 2004.* Cisterns for domestic use. Cold water storage and combined feed and expansion (thermoplastic) cisterns up to 500 l. Specification.
- *BS4618.* Recommendations for the presentation of plastics design data. Mechanical Properties: Part 1.1: 1970 Creep. Part 1.2: 1972 Impact behaviour. Part 1.3: 1975 Strength.
- *BS4991: 1974.* Specification for propylene copolymer pressure pipe.
- *BS5139: 1991.* Method of specifying general purpose polypropylene and propylene copolymer materials for moulding and extrusion.
- *BS 5955-8: 2001.* Plastics pipework. Specification for the installation of thermoplastic pipes and associated fittings for use in domestic hot and cold services and heating systems in buildings.
- *BS 6076: 1996.* Specification for polymeric film for use as a protective sleeving for buried iron pipes and fittings.
- *BS 7291.* Thermoplastics pipes for hot and cold water for domestic purposes and heating installations in buildings: Part 1: 2010 General requirements. Part 2: 2010 Specification for polybutylene (PB) pipe and associated fittings.
- *BS ISO 11922-1: 1997.* Thermoplastic pipes for the conveyance of fluids. Dimensions and tolerances. Metric Series.
- *BS ISO 17484-1: 2006.* Plastics piping systems. Multi-layer pipe systems for indoor gas installations.
- *BS ISO 17555: 2003.* Plastics. Film and sheeting. Biaxially orientated polypropylene (PP) films.
- *BS ISO 17557: 2003.* Plastics. Film and sheeting. Cast polypropylene (PP) films.
- *BS EN 728: 1997.* Plastics piping and ducting systems. Polyolefin pipes and fittings. Determination of oxidation induction time.
- *BS EN 1167.* Thermoplastic pipes, fittings and assemblies for the conveyance of fluids. Determination of the resistance to internal pressure: Part 1: 2006 General method. Part 2: 2006 Preparation of pipe test pieces.
- *BS EN ISO 1346: 2004.* Fibre ropes. Polypropylene split film, monofilament and multifilament. 3, 4 and 8 strand ropes.
- *BS EN 1451.* Plastics piping systems for soil and waste discharge (low and high temperature) within the building structure. Polypropylene (PP): Part 1: 2000 Specification for pipes, fittings and the system.
- *PD CEN/TS Part 2: 2012.* Guidance for the assessment of conformity.
- *BS EN 1852.* Plastic piping systems for non-pressure underground drainage and sewerage. Polypropylene: Part 1: 2009 Specifications for pipes, fittings and the system.
- *DD CEN/TS Part 2: 2009.* Guidance for the assessment of conformity.
- *DD CEN/TS Part 3: 2003.* Guidance for installation.
- *BS EN 1873.* Plastics. Polypropylene (PP) moulding and extrusion materials: Part 1: 1995. Designation system and basis for specifications. Part 2; 2007 Preparation of test specimens and determination of properties.
- *BS EN ISO 1969: 2004.* Fibre ropes. Polyethylene 3- and 4-strand ropes.
- *BS EN ISO 8986.* Plastics. Polybutene (PB) moulding and extrusion materials: Part 1: 2009 Designation system and basis for specifications. Part 2: 2009 Preparation of test specimens and determination of properties.
- *BS EN ISO 11296.* Plastic piping systems for renovation of underground non-pressure drainage and sewerage networks: Part 1: 2011 General. Part 3: 2011 Lining with close fit pipes. Part 4: 2011 Lining with cured-in-place pipes.
- *BS EN ISO 11298.* Plastic piping systems for renovation of underground water supply networks: Part 1: 2011 General. Part 3: 2011 Lining with close fit pipes.
- *BS EN ISO 11925-2: 2010.* Reaction to fire tests. Ignitability subjected to direct impingement of flame. Single-flame source test.
- *BS EN 12056-1: 2000.* Gravity drainage systems inside buildings. General and performance requirements.
- *PD CEN/TR 12108: 2012.* Plastics piping systems. Guidance for the installation inside buildings of pressure piping systems for hot and cold water intended for human consumption.
- *BS EN 12423: 1999.* Polypropylene twines.
- *BS EN 12566-3: 2005.* Small wastewater treatment systems for up to 50PT. Packaged and/or site assembled domestic waste water treatment plants.
- *BS EN 12828: 2003.* Heating systems in buildings. Design for water-based heating systems.
- *BS EN 13476.* Plastics piping systems for non-pressure underground drainage and sewerage. Structured wall piping systems of PVC-U, PP and PE: Part 1: 2007 General requirements and performance characteristics. Part 2: 2007 Specification for pipes and fittings with smooth internal and external surface and the system. Type A. Part 3: 2007 Specification for pipes and fittings with smooth internal and profiled external surface and the system. Type B.
- *BS EN ISO 13478: 2007.* Thermoplastic pipes for the conveyance of fluids. Determination of resistance to rapid crack propagation (RCP). Full scale test.
- *BS EN ISO 13479: 2009.* Polyolefin pipes for the conveyance of fluids. Determination of resistance to crack propagation. Test method for slow crack growth on notched pipes.
- *BS EN 13501-1: 2007.* Fire classification of construction products and building elements. Classification using data from reaction to fire tests.
- *BS EN 13566.* Plastics piping systems for the renovation of underground non-pressure drainage and sewerage networks: Part 2: 2005 Lining with continuous pipes. Part 7: 2007 Lining with spirally-wound pipes.
- *BS EN 13598-1: 2010.* Plastics piping systems for non-pressure underground drainage and sewerage. PVC-U, PP and PE. Specification for ancillary fittings.
- *BS EN 13967: 2012.* Flexible sheets for waterproofing. Plastic and rubber damp proof sheets including plastic and rubber basement tanking sheet. Definitions and characteristics.
- *BS EN 14408.* Plastics piping systems for the renovation of underground gas supply networks: Part 1: 2004 General. Part 3: 2004 Lining with close-fit pipes.

- *DD CEN/TS 14541: 2007.* Plastics pipes and fittings for non-pressure applications. Utilisation of non-virgin PVC-U, PP and PE materials.
- *BS EN 14575: 2005.* Geosynthetic barriers. Screening test method to determine resistance to oxidation.
- *BS EN 14576: 2005.* Geosynthetics. Test method for determining resistance of polymeric geosynthetic barriers to environmental stress cracking.
- *BS EN 14758.* Plastics piping systems for non-pressure underground drainage and sewerage. Polypropylene with mineral modifiers (PP-MD). Part 1: 2012 Specifications for pipes, fittings and the system.
- *DD CEN/TS Part 2: 2007.* Guidance for the assessment of conformity.
- *DD CEN/TS Part 3: 2006.* Guidance for installation.
- *BS EN ISO 15013: 2007.* Plastics. Extruded sheets of polypropylene (PP). Requirements and test methods.
- *BS EN ISO 15494: 2003.* Plastics piping systems for industrial applications. Polybutene (PB), polyethylene (PE) and polypropylene (PP). Specifications for components and the system. Metric Series.
- *BS EN ISO 15874.* Plastics piping systems for hot and cold water installations. Polypropylene (PP): Part 1: 2003 General. Part 2: 2003 Pipes. Part 3: 2003 Fittings. Part 5: 2003 Fitness for purpose of the system.
- *DD CEN ISO/TS Part 7: 2003.* Guidance for the assessment of conformity.
- *BS EN ISO 15876.* Plastics piping systems for hot and cold water installations. Polybutylene: Part 1: 2003 General. Part 2: 2003 Pipes. Part 3: 2003 Fittings. Part 5: 2003 Fitness for purpose of the system.
- *DD CEN ISO/TS Part 7: 2003.* Guidance for the assessment of conformity.
- *BS EN 50290-2-25: 2002.* Communication cables. Common design rules and construction. Polypropylene insulation compounds.
- *BS EN 60454.* Pressure sensitive adhesive tapes for electrical purposes: Part 3-12: 2006 Specifications for individual materials.
- *BS EN 60695-11-10: 1999.* Fire hazard testing. Test flames. – 50 W horizontal and vertical flame test method.

## 23.15 Bibliography


BRE Report 384 (2000) *Performance of high grade concrete containing polypropylene fibres for fire resistance: the effect on strength.* Building Research Establishment, Watford, UK.

BRE Report 395 (2000) *Effect of polypropylene fibres on performance in fire of high grade concrete.* Building Research Establishment, Watford, UK.

BRE Digest 480 (2004) *Wood plastic composites and plastic lumber.* Building Research Establishment, Watford, UK.

Brydson, J.A. (2005) *Plastics materials.* CBS Pub & Dist, New Delhi, India.

Cousins, K. (2002) *Polymers in building and construction. RAPRA market report.* RAPRA Technology, Shrewsbury, UK.

Dufton, P.W. (2002) *Polymers in building and construction. RAPRA market report.* RAPRA Technology, Shrewsbury, UK.

Halliwell, S. (2003) *Polymers in building and construction.* Report 154. RAPRA Technology, Shrewsbury, UK.

Karger-Kocsis, J. (1994) *Polypropylene. Structure blends and composites: Structure and morphology.* Vol. 1. Chapman & Hall, London, UK.

Karger-Kocsis, J. (1994) *Polypropylene. Structure blends and composites: Copolymers and blends.* Vol. 2. Chapman & Hall, London, UK.

Karger-Kocsis, J. (1994) *Polypropylene. Structure blends and composites: Composites.* Vol. 3. Chapman & Hall, London, UK.

Karger-Kocsis, J. (1998) *Polypropylene. An A–Z reference.* Kluwer Academic Publishers, Dordrecht, The Netherlands.

Lyons, A.R. (2010) *Materials for architects and builders.* 4th edition, Butterworth-Heinemann, Oxford, UK.

Nwabunma, D. & Kyu, T. (2008) *Polyolefin blends.* Wiley-Blackwell, Hoboken, NJ.


## 23.16 Websites

- Polybutene Piping Systems Association: www.pbpsa.com
- British Plastics Federation: www.bpf.co.uk
- European Plastic Pipes and Fittings Association: www.teppfa.org
- Polypipe: www.polypipe.com
- Warmafloor (GB): http://warmafloor.co.uk

# 24 Polystyrene

**David Thompsett**
MA (Cantab)
Independent Consultant, Technical Advisor to British Plastics Federation
Formerly Jablite Ltd

## Contents

## 24.1 History

Synthetic polystyrene was first made over 135 years ago. However, it was not until 1936 that the first commercial production became available. In that year, some 800 tonnes was produced in Germany, increasing to 5000 tonnes by 1942.

In its solid state, polystyrene is a clear, non-crystalline, plastic material which is brittle. Improvements to its impact strength can be made by compounding with more rubbery polymers. These high-impact grades can be extruded or injection moulded into products for use inside buildings; for example, in light fittings (luminaires) when adequate ultra-violet light resistance is needed to cover fluorescent tubes or as a lightweight substitute for glass in shower cubicles or screens.

It is, however, as a foamed plastic that polystyrene is mainly used in construction. Techniques for producing suitable foams were not available until 1952, when expanded polystyrene was first exhibited at the Dusseldorf Trade Fair by BASF, a Germany company, although a patent had been granted in 1949 in the name of Dr F. Stastny. By the mid-1960s the benefits of the material were more widely recognized and applications began to be developed. Today, polystyrene foams, which are extremely versatile in manufacturing techniques, are widely used for thermal insulation.

## 24.2 Manufacture

Two basic processes are available to produce foams from polystyrene. In order to simplify the nomenclature we will use the following abbreviations:

- *PS* – a polystyrene foam produced by either an expansion process or an extrusion process
- *EPS* – a polystyrene foam manufactured by moulding blocks, boards or shapes from expandable beads, sometimes referred to as bead board
- *XPS* – an extruded polystyrene foam manufactured into boards.

In any given application EPS or XPS may equally well be used and, in these cases, reference will be made to PS foams. Where either one material or the other is more suitable for technical reasons this will be pointed out in the text by specific reference to EPS or XPS.

Foams are produced by dissolving a volatile gas into the polymer and allowing subsequent controlled expansion of the polymer by application of heat, which softens the polymer resulting in a cellular structure. As the two materials are derived from the same chemical building block, shown in Figure 24.1, they share several attributes such as similar softening points and chemical resistance. Differences that occur are due largely to the processes by which the boards are manufactured.

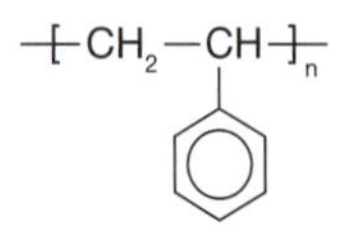


**Figure 24.1** EPS/XPS monomer

### 24.2.1 Manufacture of EPS

Suspension polymerization of the styrene building blocks is carried out in the presence of pentane to produce long polymer chains incorporating pentane in solution. Granules from this process are then coated and graded by sieving. Coatings are used to improve subsequent processing. Various sizes of expandable bead are found to be suitable for different applications, many outside the building industry. For example, small beads are used to manufacture disposable cups and other grades may be preferred for moulded packaging applications. Incorporation of flame retardants, where required, takes place in the polymerization process. Consequently, flame-retardant and normal grades are manufactured in different production batches.

These beads are then expanded to about 50 times their volume, under controlled steam pressure, either through continuous or batch pre-foamers. It is at this point that the final density of the product is largely controlled. The beads soften at 100°C, allowing expansion of the pentane and the polymer. The beads are cooled, creating a partial vacuum within the closed cells, which is corrected by migration of air into the cells over a period of hours.

To produce EPS boards, the beads are usually then transferred to a mould, e.g. a box, typically of up to 7.5 m × 1 m × 1 m. Once the lid is closed, steam is introduced by holes in the walls of the mould, which causes the thermoplastic beads to soften once more and attempt to expand further. Now constrained by the mould, the beads fill the interstices and fuse together to form a block.

After cooling, the block is de-moulded for subsequent cutting into boards using hot wires. Alternative machines are produced to allow the manufacture of continuously moulded boards and semi-continuously moulded blocks. In both these cases there is great flexibility in product length. Shape-moulded products are increasingly manufactured for specific building applications. We are all familiar with this type of product when used as a packaging medium for electrical goods, such as refrigerators or televisions.

The resulting products are air-filled foams and usually range in density from 8 to 50 kg $m^{-3}$, or even higher. Traditionally, EPS is white in colour, although special coloured varieties are available. In recent years development to improve the thermal conductivity has taken place and products with this enhanced performance are grey in colour; see Section 24.4.

### 24.2.2 Manufacture of XPS

Starting from polystyrene granules containing no blowing agents, XPS is manufactured by heating in an extruder the polymer blended with flame retardants, various additives to control nucleation of the cells and a blowing agent, the choice of which affects the thermal conductivity.

As the polymer melts, the chosen blowing agent is introduced under pressure to dissolve in the polystyrene melt. At the end of the extruder this melt of blended polystyrene and blowing agent passes through a dye into atmospheric pressure and, consequently, expands due to the drop in pressure. Control and sizing of this expanded material determines the density and properties of the boards produced. It is normal for each manufacturer to colour-code his own production. Boards can be produced either singly, in which case a hard skin surface results, or as blocks for subsequent cutting into boards with hot wires. Blocks up to approximately 150 mm thick are manufactured in widths of up to 1.2 m. Densities produced by this method range from about 28 kg $m^{-3}$ upwards.

A variation of this process allows, by application of a vacuum to the material after it exits the dye, lower densities to be produced. Values down to 20 kg $m^{-3}$ are claimed.

Until recently, chlorofluorocarbon blowing agents (CFCs) have been used for their combination of low boiling point and non-flammability to produce XPS. This also brought benefits in thermal conductivity performance, even after allowing for some outward migration of the gas. Due to the current concern over the long-term effect of CFCs on the ozone layer, an international agreement, the Montreal Protocol, has phased out CFCs and HCFCs, and manufacturers of XPS now use HFA and carbon dioxide ($CO_2$) blowing agents, which have a zero ozone depletion value. The choice of blowing agent used will have an effect on the thermal conductivity.

## 24.3 Physical properties

### 24.3.1 Introduction

In its solid state, polystyrene has a density of 1050 kg $m^{-3}$. Its physical properties are well known. For building use the polymer is mostly used in expanded form as a cellular foam. To a very large extent, the solid content of the matrix determines its physical properties, although the distribution of material and imperfections in the cellular structure also play a part. It follows that the density of the foam is the most significant factor in determining the physical properties.

The physical properties also vary according to the method of manufacture. Moulding of EPS will give slightly different properties to XPS material. This is mainly due to density differences, but also because the cellular structure of XPS is more uniform and has 100% closed cells. EPS, being made up of many discrete beads physically bonded together, has a labyrinth around the beads because total fusion of the bead surface area is not possible. The individual beads are 100% closed cell and the foam is considered 'substantially' closed cell.

Manufacturing variations will result in a normal distribution of a physical property around the mean at any given density. Each physical property described below has a high confidence level.

### 24.3.2 European Standards (BS ENs)

Standards covering all insulation products are now produced by CEN, the European Committee for Standardisation, in Technical Committee 88. These are mandated by the European Commission to produce common standards for all construction products to fulfil the requirements of the Construction Products Directive, to be superseded in July 2013 by the Construction Regulations. These standards will be reviewed at 5 year intervals and updated when necessary. Input from the UK is via the British Standards Institution (BSI).

The effects of these new standards, the harmonized part of which is the mandated standard, are quite radical. Previous British Standards, now withdrawn, were relatively simple documents relying on other test methods to measure the less common properties as required. The EN approach is to make, as far as possible, the format of all insulation standards the same. This includes common test methods for physical properties and common ways to measure fire properties of all construction products. The most rigorous change has been a philosophical one: that of not allowing standards to impose levels of performance for individual properties or combinations of two or more properties to generate a product type. Instead levels are set from which the manufacturer can choose to target his product performance. The reasoning is that manufacturers should be allowed to innovate and obtain the benefits without the need to await a standard review. The result of this approach is that in theory, many different products could be produced. In practice this is not feasible commercially or technically, and a more restricted range of products is available in the market.

The standards now include references to all required test methods which have been rewritten and also define a rigorous statistical approach to the declaration of thermal performance, whether it be thermal conductivity ($\lambda$) or thermal resistance ($R$). These declared values (so called 90/90 values) must now be representative of the values achieved by 90% of the production with a confidence level of 90%.

A separate standard, BS EN 13172, covers the evaluation of conformity of production and details different requirements dependent on both reaction to fire characteristics and other characteristics.

### 24.3.3 Compressive strength

This property is in many ways the most significant for both PS products in that the decision as to which grade to specify centres on the design loadings anticipated.

The cellular structure of expanded polystyrene gives it a complicated mechanism for resisting compression forces. On one hand there is a series of struts and ties with associated cell faces that make up the cell structure and, on the other, the gas within the cell also resists compression in the short term; long-term pressure above the elastic limit will allow creep to occur. In the totally homogeneous structure of XPS the stress–strain curve follows the classic pattern, with a yield occurring at between 5% and 10% strain. EPS has a more complicated structure because of the existence of the inter-cellular labyrinth and, in fact, does not yield in the true sense of the term.

At levels of compression above 2% strain, both materials will exhibit long-term creep deformation caused by the pressure of the gas in the cells reducing to near atmospheric and thereby increasing the stress taken on the cell walls. As a general rule, as the density increases the compressive strength increases and the propensity to creep under a given load reduces. It should be noted that the EPS is described as carrying a compressive stress at a given strain, rather than a compressive strength. For normal purposes the strain is taken as 10%. This level of strain is chosen so that it can be measured consistently during quality-control testing.

The elastic limit for PS lies at approximately 1.5–2% strain. Values of stress quoted at a nominal 1% compression can, therefore, be used for design purposes.

Both EPS and XPS use the declared compressive stress value in the product designations. Thus EPS 100 indicates a value of 100 kPa at 10% compression.

### 24.3.4 Bending strength

Bending strength is a function of how well EPS foam has been made and is a measure of the fusion between the individual beads, the bending strength increasing with density. Some properties such as tensile strength can be correlated with bending strength. See EPS White Book published online by Eumeps, which represents the EPS industry (www.eumeps.org). This test is not cited in the XPS standard.

### 24.3.5 Dimensional stability

The manufacture of PS foam relies on the thermoplastic nature of the material. Heat is supplied during the manufacturing processes to produce a foam. On exiting the machinery, the material is substantially above room temperature and tends to 'grow' slightly. A conditioning period is therefore required for the material to stabilize before cutting to the finished dimensions. In the case of EPS, the post-moulding conditioning period is usually satisfied within 24–48 h of moulding. Because XPS is subjected to a higher temperature and becomes molten and strained during the extrusion process, the conditioning period is substantially longer.

Once conditioned, PS boards and products can be evaluated for dimensional stability at room temperature or under specific conditions according to BS EN 1603 or BS EN 1604.

### 24.3.6 Moisture absorption

PS foams are essentially hydrophobic and will not readily accept moisture into the cell structure. Where the surface is cut, rather than moulded, there is a layer of open cells and moisture may ingress into this area. Moisture may be forced into the cell structure by pressure over a given period of time, the rate of absorption being roughly proportional to the pressure and logarithmically proportional to the time period.

Under identical conditions EPS will generally have a higher moisture uptake than XPS because the labyrinth surrounding the bead structure provides an additional path of ingress. Moisture absorption is also dependent on the density and the processing conditions, especially for EPS. The important point to remember is that neither material will readily absorb moisture from its surroundings and that pressure is required to force moisture into the cellular structure. Because of the inert nature of the material it will not support capillarity and can, therefore, be used in a situation where it bridges moisture barriers.

The difference in water uptake between EPS and XPS is, in most applications, unimportant. However, the better performance of XPS does mean that some applications, for example inverted roof insulation, remain the prerogative of XPS. Relevant tests are BS EN 12087 for long-term water absorption by immersion and BS EN12088 for long-term water absorption by diffusion.

### 24.3.7 Moisture vapour transmission

Moisture vapour acts as a gas in its mechanisms of transfer through materials. PS foams allow the transmittance of moisture vapour through the matrix, but severely retard the rate of transfer. The labyrinth around interconnecting beads of EPS allows a tortuous path through the material but does not provide free passage. In common with most properties of PS foams, an increase in density results in a reduction in water vapour transmission.

In some circumstances XPS is used with one or both skins intact and this markedly reduces the water vapour transmission rate. The same is true, although to a lesser extent, with individually moulded EPS units. In the materials used commercially the water vapour transmission rate will determine the need for a vapour control layer. Most insultants, including EPS, will need a vapour barrier in flat-roof-insulation applications or a vapour check for internal drylining of walls. XPS, however, with a better vapour resistivity, can often be used without a vapour control layer. It is important to treat all applications and the need for vapour barriers on their merits. BS EN 12086 details the test method to determine water vapour transmission properties.

### 24.3.8 Tensile strength

In the context of normal insulation boards, the tensile strength is not important, although it provides a natural resistance to accidental damage during transportation. Tensile strength becomes important when the PS foam is bonded to another material with adhesives to form a composite panel or is a part of external insulation system (ETICS), for example. The tensile strength is largely governed by the density of the foam, although the dispersion of material throughout the matrix and the cellular structure also plays a part. This is particularly true for XPS, where the method of manufacture allows the cell orientation to be adjusted. EPS can have its tensile strength adjusted within limits by the degree of fusion between the beads, which is a function of the processing conditions. However, limits are imposed by the natural tensile strength of the solid polymer (test method BS EN 1207 is used to measure tensile strength perpendicular to the board faces).

### 24.3.9 Shear strength

The stress placed on PS foam is rarely of a shear nature unless the material is being used as the structural core to a sandwich panel. In these circumstances, the surface area at any cross-section determines the shear strength and, consequently, the density is a limiting factor under the normal condition of uniform material dispersion. This is true for XPS, but only true for EPS if the matrix is sufficiently well fused, which is usually the case, to transfer stress without cracks starting to form. PS foams are elastic up to a maximum of 2% strain, at which point plastic deformation occurs.

To take account of this, the shear modulus is used in structural calculations, since the shear deformation contributes to the overall deflection of the panel. The shear modulus is proportional to the density of the material, within its elastic limit.

### 24.3.10 Young's modulus of elasticity

XPS acts in a traditional manner when stressed and has a yield point, after which the material is subject to strain with very little increase in stress. EPS exhibits a different stress–strain relationship to most standard building materials. To use a measurement of the stress–strain relationship with a high degree of confidence it is necessary to remain within the elastic limit of the material. The elastic limit is certainly no greater than 2% and it is usual to use a modulus of elasticity derived from the stress–strain relationship at 1% deformation. Provided that the stress limits on the material for any given property remain within the 1–2%, region they can be safely used.

### 24.3.11 Creep

When a material is placed under load it shows two reactions to stress. The first is instantaneous and the second is time-dependent. The difference between the initial strain and the strain at any future time period, under constant load, is described as the long-term-strain deformation, or creep.

Creep is a phenomenon that occurs in all materials and under any conditions of stress but is more pronounced with thermoplastics. It is generally immeasurable when the stress–strain relationship remains within the elastic limits of a material. For low-density materials, such as plastic foams, where the elastic limit is quite low, i.e. typically 1–2% strain, creep can be considered not to occur at a stress that does not cause a strain greater than 1%. Tests are long-term and therefore quite expensive (BS EN 1606), with results predicted to a 50 year term given sufficient data. Figure 24.2 shows a creep prediction for EPS.

### 24.3.12 Poisson's ratio

Poisson's ratio gives a measure of the relationship between the lateral strain and the longitudinal strain under a given load. EPS foam has a Poisson's ratio of between 0.1 and 0.2. The actual value will depend more on the orientation of the cell structure than on the density of the foam. This low Poisson's ratio allows fill behind bridge abutments of lightweight PS foam blocks which then transmit a minimum force on the abutment, thus reducing the strength needed in the concrete abutment.

## 24.4 Thermal conductivity

Thermal conductivity, expressed as a lambda value ($\lambda$) in units of W/m K, of a foam product is a combination of three factors:

1 conduction through the matrix
2 conduction through the gas contained in the cells
3 radiation across the cells.

For EPS this gives a curve of decreasing lambda against density which flattens out above about 25 kg/m$^3$ as the balance between these effects changes. Figure 24.3 illustrates this relationship for white EPS and is based on a total of nearly 4000 measurements.

The recent development of a low lambda product improves the performance at the lower density range in particular by reducing the conduction due to radiation. As the product in its end use condition is air filled, the pentane blowing agent having dissipated

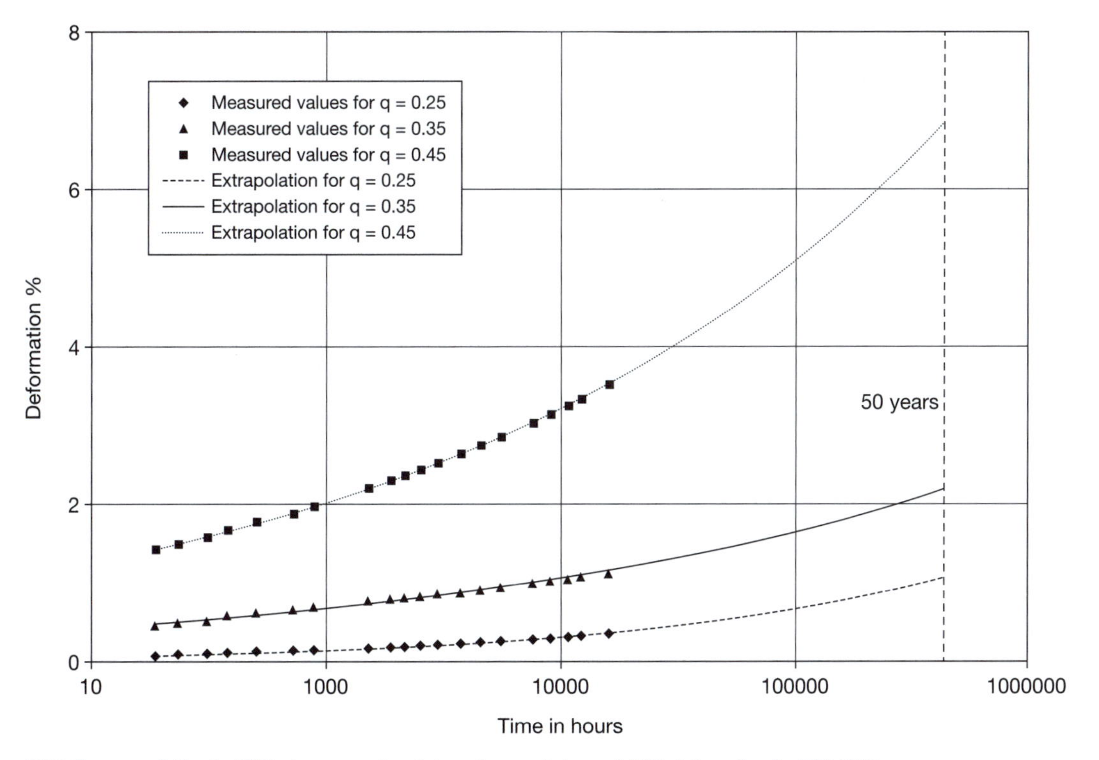


**Figure 24.2** Creep prediction for EPS where $q$ = ratio of stress/imposed stress at 10% deformation for EPS 1000

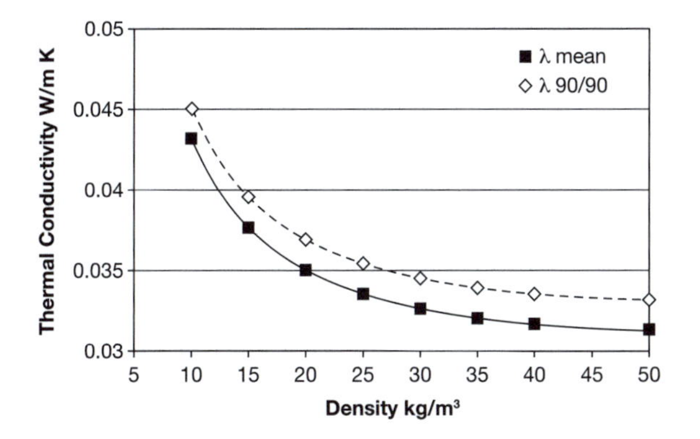


**Figure 24.3** Mean and 90/90 curves for white EPS

from the cells during manufacture and subsequent ageing, the thermal conductivity remains stable.

XPS, on the other hand, has taken advantage historically of the use of low-conductivity blowing agents such as CFCs and HCFs, both of which, now banned under the Montreal Protocol to protect the ozone layer, are no longer in use. HFA blowing agents have replaced these to give a low thermal conductivity product, and themselves may be replaced by other more benign gases, as they are developed, with lower greenhouse warming effects. Indeed, products are now available using carbon dioxide but these forgo the advantage of low-conductivity gases and have a higher lambda value.

All insulation standards declare a thermal conductivity according to the so-called 90/90 value, which represents 90% of the product with a statistical confidence of 90%. There are procedures in place for products using enhancing blowing agents to test for the ageing effects, i.e. a decline in lambda over time. Any such effect has to be reflected in the declared value and is important for polyurethane (PUR and PIR) and phenolic foams and also for XPS.

## 24.5 Fire properties

Polystyrene is a combustible material. As a low-density foam, the material is easily ignitable. For this reason it is recommended that PS foams should always be protected by a fire-resistant material when used in building applications. Certain decorative products, e.g. tiles and veneers, are permitted to remain uncovered, but only in conditions where they are fully adhered to a fire-resisting substrate.

Two types of EPS raw material are produced by bead manufacturers, one of which contains a flame-retardant additive that increases the resistance to ignition. This addition will improve the European Reaction to Fire Classification from Class F to Class E when tested according to the small flame test BS EN 11925-2. All XPS products contain a flame retardant and therefore achieve a Class E

When a high-temperature heat source or an open flame is placed into contact with Class F material it will burn fairly rapidly, whereas Class E material will shrink away from the ignition source. Prolonged exposure of Class E material to an ignition source may cause some combustion, but this will cease if the ignition source is removed. Tests for the presence of a flame-retardant additive are determined by the test referred to above.

The presence of a flame-retardant additive does not materially affect the behaviour of polystyrene once a fire has become established, the overall performance being determined by the protective layer. Once the maximum continuous-operating temperature of the foam has been exceeded, the material will soften and shrink. This becomes more evident as the temperature increases above 100°C. The material will also start to form a gaseous vapour once it is molten. Sustained flaming can be initiated by a spark or flame provided that there is adequate ventilation and the surface temperature is maintained above about 350°C. Spontaneous ignition can occur in the gaseous layer above the surface at temperatures above about 490°C, provided that the gas–air mixture is within flammability limits.

Once the foam has commenced combustion, it requires approximately 160 times its own volume of air to burn fully. Volumes of air less than this will prevent full combustion, reduce the calorific contribution of the material to a fire and will, consequently, significantly increase the density of the smoke evolved. The combustion of foam is difficult to sustain in narrow cavities, such as those in brick cavity walls, due to the lack of air flow through the cavity. If the cavity were to be open at the top and bottom, allowing cross-ventilation, combustion of the PS foam would be possible.

The calorific value of PS is 34 MJ/kg, which is nearly three times the calorific value of softwood (12 MJ/kg). However, the commercial densities of PS foam range from 10 to 50 kg/m$^3$, while softwoods have densities of 500 kg/m$^3$ upwards. A comparison between the calorific value of PS at 20 kg/m$^3$ and softwood would give values of 680 MJ/m$^3$ and 6000 MJ/m$^3$ respectively. The maximum heat liberated with full combustion of PS foam would be around one-ninth that released by an equal volume of softwood under the same conditions.

Test methods for reaction to fire have been harmonized and products classified according to BS EN 13501-1 into A1, A2, B, C, D, E and F according to a series of test methods. The highest level, class A1, covers non-combustible products down to classes that have a higher tendency to burn, to class F, which can signify a failure to reach class E or 'no performance determined'.

For naked PS products containing a flame retardant, class E is obtained though class D can be reached for thinner and low-density products. The test involved is a small flame test, BS EN ISO 11925-2.

To achieve higher classification, products must be tested in the Single Burning Item (SBI) test (BS EN 13823), which consists of an open corner with no ceiling and a gas burner. This test is not relevant to naked PS foams, which melt quickly in the presence of the gas burner. However, as mentioned earlier, PS foams are used with a protective facing such as plasterboard and can the achieve B classification. In full, a result B-$s_1d_0$ indicates that smoke is the best classification with little contribution and $d_1$ indicates no flaming droplets/particles within 600 s from the start of the SBI test. This result is also achieved by other insulation products, including mineral fibre when installed behind plasterboard.

### 24.5.1 Toxicity

As with any carbonaceous material, PS foam will evolve hot gases and smoke during the combustion process. The evolution process will depend on factors such as temperature, ventilation rate, available oxygen, local concentration and position of the materials involved. From these factors alone it can be seen that combustion is a complex process. PS foams will break down to carbon dioxide ($CO_2$), carbon monoxide (CO), water and traces of styrene. Traces of other, more complex, compounds may occur, but these can be effectively ignored. In a situation where the air flow is sufficient to allow full combustion, the two main compounds given off will be $CO_2$ and water. As the air flow is reduced the proportion of CO to $CO_2$ increases.

The toxicity of these compounds is relatively low compared with some other construction materials and is a function of oxygen deprivation, rather than an attack on the human organism. CO has a higher affinity for the oxyhaemoglobin in the blood stream than does oxygen and can be fatal if inhaled for 1–3 min at 10,000–15,000 parts per million (ppm). The inhalation of oxygen prior to fatality will usually reverse the effects of CO and no lasting damage will occur if the oxygen is applied sufficiently quickly. Carbon dioxide is toxic only in the sense that it prevents oxygen reaching the lungs. The temperature at which the gas is inhaled is more likely to cause damage than the gas itself.

BS6203 Guide to fire characteristics and fire performance of expanded polystyrene materials (EPS and XPS) used in building applications gives a great deal of information on this topic and includes an annex of examples where PS foams are used.

## 24.6 Chemical properties

Polystyrene is essentially an inert material. It is unaffected by the normal range of building materials that it comes into contact with, i.e. cement, plaster, concrete, etc. It is acceptable to use the material in ground that harbours sulphates. There are certain materials that PS foam should be separated from, notably silicon injected damp-proof courses (DPCs), plasticized plastics (particularly polyvinyl chloride (PVC)), sealants and adhesives containing volatile solvents and coal-tar–pitch-based damp-proof membranes.

## 24.7 Applications

Applications are generally not covered by product standards, which are concerned with the production of products and the definition of the range of properties delivered. Annex ZA, included in all building product standards, provides a link between the Mandate M103 given under the Construction Products Directive (89/106/EEC) and the intended use of the product. In the case of all thermal insulation products the intended use is defined merely as thermal insulation for buildings, which is not helpful in defining the methods of utilizing the insulation in a way to ensure that designed performance is achieved. Most manufacturers therefore continue to hold Agrément certification for specific applications in which the construction details are defined and limitations imposed on product use (see www.bbacerts.co.uk). Other independent certification bodies can fulfil the same function.

### 24.7.1 Wall insulation

The insulation of walls can be carried out on three effective interfaces: the internal face, the external face and within the wall sandwich. Each system has its own advantages and is appropriate in the right circumstances.

#### *24.7.1.1 Internal lining*

PS foam is placed behind a suitable fire-resistant covering to provide the required fire resistance. The exposed face may be plasterboard or mineral-based boards or, in exceptional circumstances, wet plasters. This latter method is now uncommon, except for small areas, because of the difficulty of providing a mechanical fix between the plaster and the wall behind the PS foam. Modern methods of construction emphasize the need to provide mechanical fixings to dry-lined systems.

Laminates of PS foam and plasterboard are readily available and speed installation. The normal method of installation for laminates is to use a heavy adhesive compound as a gap-filling agent to bond directly to the blockwork or mechanically fixed to metal firings. The use of adhesive provides a method for levelling the surface of the internal wall finish and its adhesion qualities are secondary. Mechanical fixings must be used to provide long-term support to the laminate, particularly to provide stability in the event of a fire.

*24.7.1.2 Cavity walls*

PS foam historically had a large market share in partial fill cavity wall insulation where a cavity wall is insulated against the inner leaf and a cavity retained to prevent moisture ingress.

The first row of wall ties is placed at 600 mm centres, usually below the DPC, with the bottom edge of the boards resting on them. Because the PS foam does not support capillarity it is permitted for the boards to bridge the DPC. Insulation boards are attached to the constructed leaf after all excess mortar has been removed and the wall ties brushed clean. A lift of no more than 1350 mm should be constructed before attaching the insulation and bringing up the other leaf.

The maintained cavity should never be less that 25 mm, although a recommended 50 mm retained cavity to allow for workmanship tolerances and imperfections is now normal. The wider the maintained cavity the easier it is to keep clean, using both cavity battens and post-cleaning methods. Since the purpose of the cavity is to avoid moisture penetration through the wall, it is imperative that the cavity is left clear and clean. More stringent insulation requirements have favoured the use of lower thermal conductivity insulants such as polyurethanes, though developments in EPS feedstock has produced a grey material with a 20% improvement in thermal conductivity.

A full fill product has been designed with tongue and grooved joints moulded into the board. This ensures that there will be no rain penetration across the cavity. Installation is similar to other built-in full-fill cavity batts, with the grey products offering the lower lambda value of 0.030 W/m K. Figure 24.4 illustrates a full-fill product in EPS. Figure 24.5 Illustrates a full-fill product using low lambda EPS.

For some domestic and industrial buildings a system of EPS beads or granules may be blown into the cavity after it has been fully constructed. This has the advantage of allowing the cavity wall to be built in a traditional manner and inspected to ensure its cleanliness before the cavity is closed and the bead pumped in under low air pressure. It is usual to provide an aqueous adhesive to the beads during the installation (the 'bonded' process) to prevent any settlement. The improved lambda products now give EPS an edge in upgrading existing buildings and for fully filling a cavity in new build where this approach is chosen. Figure 24.6 shows the blown-bead system.

*24.7.1.3 External insulation*

External insulation can be carried out by fixing or applying a weatherproof layer over the face of the insulation. Two basic methods are available – wet or dry – defined by the method of applying the weather-proofing. Dry systems use a mechanically fixed cladding material, usually fixed onto timber battens, such as tiles, slates, timber or plastic boarding. PS foam is placed either beneath or between the battens and the cladding is then attached. Care must be taken to ensure that the cladding is properly installed with correct detailing and that it is properly maintained.

In common with all external-insulation methods, it is vital to ensure that if any vapour-control layer or damp-proof membranes are used they are correctly positioned. This must be on the warm side of the insulant. Beneath slates or tiles a 'breathable' sarking felt should be used to avoid condensation damage to timbers.

Wet systems of external insulation, externally insulated composite systems (ETICS) have been used in the UK mainly for

**Figure 24.4** A full-fill product in EPS

**Figure 24.5** A full-fill product using low-lambda EPS

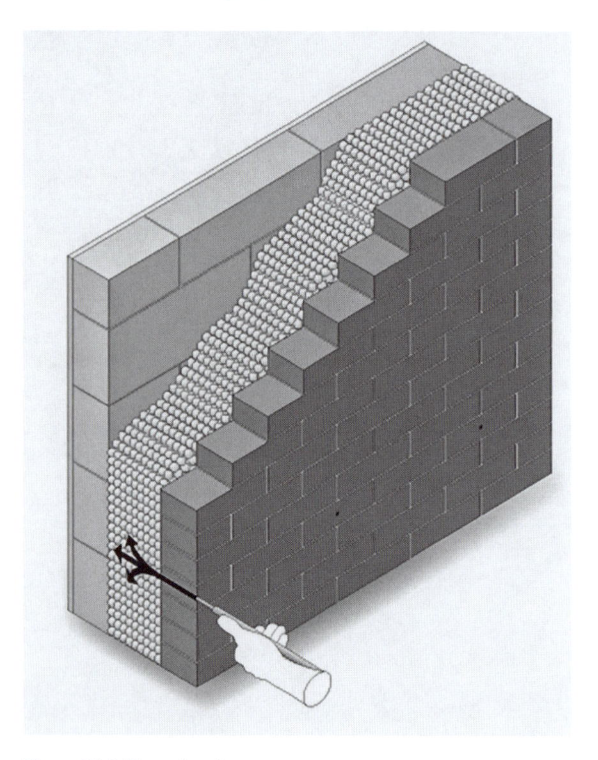

**Figure 24.6** Blown bead

upgrading of existing buildings but are very popular in parts of Europe for new buildings. The UK has been largely wedded to the tradition of brickwork. With increasing bricklaying skill shortages and the pressure to improve thermal performance, alternative constructions are more widely used including ETICS systems to provide insulation over solid walls and for use with EPS formwork systems. EPS foam in its expanded form (but not extruded) is the most popular material for this application throughout Europe and North America, although systems based on XPS are available. It is essential to use a rigid, low-weight, stable material as the substrate for the render or cracking of the render will result. To ensure the stability of the foam a period of ageing, usually six to eight weeks, is required. This allows natural shrinkage of the material to take place before use.

By placing the insulation on the outside of the structure the position of potential interstitial condensation is moved to the outside of the wall where it can do the least damage and allows the wall to breathe naturally.

The common method of construction is to use a thick adhesive to attach the insulation to the structure. This needs to be sound and free of loose material and dirt to ensure good adhesion. The adhesive also acts as a gap filler and allows the insulation to be laid flat and provide a level base for the finishes.

Reinforcement meshes are bonded to the insulation and supplementary fixings used to ensure that the render system remains in place in the event of fire. It is recommended that some of the fixings should be metal. Each different system has its own particular render which is a proprietary mixture. This is usually applied in two coats, with the base and finish coat being of slightly different formulations. Both thick (20 mm) and thin (6 mm) render systems claim advantages, and both systems have been used successfully for many years through Europe and North America.

External wall insulation in the UK has, primarily been used as a refurbishment method, although the basic advantage of providing thermal storage in the wall structure would make it attractive for low-energy housing if the aesthetics were acceptable. It provides a solution to the twin problems of rain penetration and condensation in existing buildings and many of the 1960s system-built houses have benefited from a new outer skin. More recently it has been used more and more on new build.

Due to concern over their performance in a fire, BRE has produced a report on the performance of various proprietary external-wall-insulation systems. To overcome the potential, but minimal, risk of fire spread via the insulation material in multistorey buildings, cavity barriers of glass or rock fibre should be installed at appropriate centres, e.g. 10 m, in the insulation layer.

The Loss Prevention Certification Board (LPCB) evaluates ETICS systems using a large-scale test and results can be seen on www.redbooklive.com. A system based on EPS has obtained an ETIC1 classification, the same as for mineral fibre.

While there is at present no harmonized EN standard for ETICS, a standard (not harmonized) BSEN 13499 for EPS has been produced and work is ongoing for other insulants including XPS.

*24.7.1.4 Timber-framed walls*

While PS foam is used extensively in masonry constructions, it is less common in timber-frame construction. In Europe and North America PS foam is often used as a sarking board on the outside of the structural sheathing. Common practice in the UK is to use a mineral-fibre quilt between the timber studwork or low thermal conductivity polyurethanes with a sheathing board applied externally to resist racking of the frame. There is no reason for not using PS foam, particularly as factory-finished components allow accurate placing of materials.

In all circumstances where PS foam is used in composite wall construction, PVC electrical cables must be isolated from the insulation materials. If they are maintained in contact the plasticizer in the PVC will migrate into the PS foam. This has the effect of softening the PS foam and causing embrittlement of the PVC-sheathed cable. The embrittlement improves the electrical-insulation properties of the cable but makes it susceptible to damage if it is subsequently moved. There is no danger of an electrical fire caused by the loss of electrical insulation but, if fully surrounded by any insulation product, the cables should be derated by a factor of 75%.

A rigid cable ducting, which can be rigid PVC, or a reliable air gap is sufficient to eliminate this effect.

*24.7.1.5 Structural insulated panel systems (SIPS)*

SIPS panels consist of a core of lightweight insulation, typically polystyrene (EPS or XPS) bonded to rigid timber facings, typically oriented strandboard (OSB). Widely used in the USA, these products are now manufactured by several companies in the UK (Figures 24.7 and 24.8).

Their advantages are to give a quick method of construction using well-insulated panels that are manufactured under factory controlled conditions. The continuous insulation layer largely eliminates cold bridges. The walls constructed with these products can be designed with suitable facings internally to provide the required level of fire resistance, and a wide range of external finishes can be also be accommodated. Products can achieve a one hour fire resistance under loads found in domestic construction and will need independent certification for Building Control purposes. For example, the British Board of Agrément has certified SIPS systems.

**Figure 24.7** Cross-section of an SIP panel

*24.7.1.6 Insulating concrete formwork*

Insulating concrete formwork (ICF), otherwise known as permanently insulated formwork (PIF), is an insulated in-situ concrete system of building which is quick to construct and offers a level of performance significantly better than that available from more traditional approaches to building. Popular in Germany, where it was introduced in the early 1950s, it has grown rapidly in North America, initially as a basement construction, and now has an 8% share of the low-rise housing market in Canada and the USA.

It has been growing in the UK self-build sector for some years, with a performance in excess of the Building Regulations' requirements for insulation. With increasing demands for higher insulation levels, these systems should take an increasing share of the construction market.

With accompanying focus on ECO-homes and the Code for Sustainable Homes, its use is becoming more widespread and broadening into mainstream construction, particularly with UK Housing Associations. It is also one of just four approved walling systems for domestic basements (see www.basements.org.uk).

There are several different designs of ICFs, which are usually made from insulating foam, particularly EPS, and are manufactured as separate panels that are connected with plastic connectors or ties; pre-formed interlocking hollow blocks are another form of IFC (Figure 24.9). The ability to mould intricate shapes with EPS is a huge benefit for these systems.

The lightweight insulating components lock together without intermediate bedding materials, such as mortar. After assembling the formwork, concrete is pumped into the formwork system to provide the structure for the building. Depending on the design of the system used, the formwork may be delivered as a hollow moulded block or come flat-packed with sheet insulation or moulded boards that are assembled on-site using proprietary ties or spacers.

Walls up to 3 m high can be cast in one pour of concrete. Excellent U-values, as low as 0.11 $W/m^2$ K, can be achieved.

## 24.8 Roof insulation

Roofs are either pitched using tiles as weather proofing or flat, when a waterproofing membrane is applied. This can be the traditional bituminous felt or mastic asphalt or one of the polymeric single-ply membranes, of which PVC is the most widely available.

Figure 24.8 A large SIP panel being craned into position

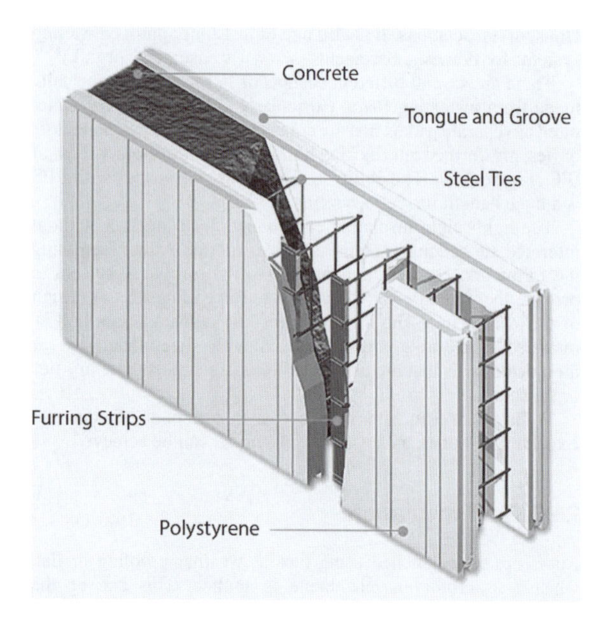


Figure 24.9 An ICF in section

### 24.8.1 Pitched roofs

In pitched-roof construction the roofs are primarily designated as cold-roof constructions. This means that the void created externally to the insulation, whether it be a loft or between rafters, requires ventilation. Ventilation of the space is the prime means of preventing condensation in the timbers and subsequent rot. As high thicknesses of insulation, already 250 mm and increasing, are used the risks for condensation in the loft and for cold water tanks increases. With pressure on land usage there is a tendency to utilize the loft space for accommodation, giving rise to insulation at rafter level and the use of factory-supplied structural roof elements.

A warm roof can be constructed with the insulation placed above the rafters. This approach can be taken with new build or where tiles are being renewed on buildings. Care must be taken to ensure continuity of insulation at the junction with the wall insulation to prevent a cold bridge. Tiling can then be carried out in the normal way.

Composite roof elements consisting of sheet timber facings factory bonded to an insulating core of EPS, and which may incorporate timber edges and/or counter battens, are a standard roof construction in the Netherlands and are being introduced into the UK. They provide an effective way to construct a structural roof system but require a crane to position the large panels which span between eaves and ridge of a pitched roof (Figure 24.11).

Figure 24.10 An ICF product in construction

Figure 24.12 Retrofitting insulation in a pitched roof

For existing buildings insulation can be retro fitted below the sarking felt and protected by plasterboard to provide extra living space. In this case, the PS panels fit between and below the rafters prior to fitting the plasterboard, and care must be taken to ensure that a ventilation gap is retained above the insulation (Figures 24.12 and 24.13).

### 24.8.2 Flat roofs

Flat roofs can be designated as either cold or warm construction related to the position of the insulation and the structural deck. In a cold roof construction where the insulation is below the deck, there is a requirement to ventilate the void formed between timber

Figure 24.11 Roof elements were used to construct this roof

**Figure 24.13** A roof insulated at rafter level

rafters immediately below the timber deck, and to ensure that no dead areas of air are created. In different orientations of the roof this total ventilation may be extremely difficult and, if possible, this type of construction should be avoided.

The warm roof construction, where the insulation is positioned above the structural deck, is therefore increasingly used because a ventilated air space below the deck is not necessary. Condensation risk calculations should always be carried out to ensure no continued build-up of condensation.

*24.8.2.1 Warm-roof construction*

Warm roofs differ in respect of the position of the insulation layer relative to the weatherproofing. Traditionally, insulation is positioned below the weatherproofing. Alternatively, if the insulation layer is positioned above the weatherproofing the roof is termed an 'inverted', 'upside-down' or 'inverted roof membrane' application.

The use of PS in these applications varies, with each type of material taking prominence in a particular field. EPS is used in the standard type of construction where the insulation is below the weatherproofing membrane, while XPS is used in the inverted type of application.

The omission of ventilated void in a warm-roof construction means that an alternative method of condensation prevention has to be adopted. This generally implies a condensation-risk analysis calculation in accordance with BS 5250. The calculation is designed to determine the likely risk of condensation in a roof construction, the amount of condensation over a winter and summer period and the overall net retention figure, if any. It is normally shown that with the inclusion of a vapour control layer between the structural deck and the EPS insulation the condensation risk is low, with no annual net retention of moisture.

A standard warm-roof construction will incorporate the following elements:

- *A structural deck.* This can be of almost any material construction available on the market today: timber, concrete, metal or a hybrid of these materials. Whether the deck forms an internal ceiling finish, or if this needs to be considered separately, it does not influence the insulation application requirements.
- *A vapour control layer.* This is a minimum type 1B felt to BS747 or, in the case of a profiled deck, a reinforced felt type 1F to BS 747. It may be necessary, where high humidity conditions prevail in the area below the roof, to provide an upgraded or double vapour-control layer to prevent interstitial condensation occurring in the roof. An upgraded vapour barrier will normally include a foil interface which gives it a superior vapour resistance.

The vapour-control layer will be bonded to the previously primed deck. All joints in the vapour-control layer must be lapped and sealed. Around the perimeter the vapour-control layer must be taken up in excess of the insulation thickness to be bonded into the weatherproofing membrane.

*24.8.2.2 EPS insulation*

The use of EPS insulation boards, either uniform or tapered, and with a factory bonded felt or fibre board overlay, in warm-roof constructions is covered, and approved where possible, by the major roofing associations in the UK. In practice, the choice between uniform and tapered EPS insulation boards depends on a variety of factors. Basically, these can be divided into two categories: those roofs that have drainage falls incorporated in the structure and those that require the drainage falls to be created above the structural deck. In the case of the use of tapered EPS boards, the prime objective is to create drainage falls on the roof. Due to the thickness of insulation involved, however, the overall thermal transmittance value of the roof will normally comply with, or better, the Building Regulations requirements.

Where the EPS insulation board is used with a pre-felted surface, the application of a minimum 13 mm thick fibre board, perlite or cork board overlay is required. The overlay board acts as a heat sink over the EPS when either built-up roofing or mastic asphalt weatherproofing is applied.

Laminates of EPS and fibre board are available in both uniform and tapered form as an alternative to the site-applied version above.

The EPS boards must have a minimum compressive stress value of 100 kPa where normal maintenance or domestic loads are envisaged. Where there is the likelihood of higher loadings, such as commercial areas, a value of 150 kPa would be used.

EPS boards are applied to the vapour control layer by mopping an area of the vapour control layer with hot bitumen and laying the EPS board onto the surface. The initial high bitumen temperature is absorbed by the vapour control layer. All EPS boards will be laid with tight butted joints and up to all perimeters.

In the case of a mastic asphalt weatherproofing system, a 50 mm wide gap must be left between the EPS boards and all upstands. This is then filled with a rigid infill material, such as earth, damp sand/cement or timber, to act as a support to the final mastic asphalt fillet around the perimeter of the roof.

Pre-felted EPS boards have the advantage at this stage in the roof construction of being able to provide a temporary weatherproofing layer if joints and edges are sealed with felt. Where an EPS/fibre board laminate has been used, the first layer of the weatherproofing must be applied immediately. In addition, taping of the insulation boards joints may be necessary to prevent bitumen ingress where a BUR weatherproofing is applied.

*24.8.2.2.1 Overlay board* The application of an overlay board to pre-felted EPS is carried out by applying a coat of hot bitumen compound to the overlay board and laying this onto the felt surface of the EPS insulation. Overlay boards should be fully butted and, in the case of a mastic asphalt weatherproofing, the overlay board should overlap the rigid infill perimeter detail by 50%.

*24.8.2.2.2 Weatherproofing* On to the overlay surface can be applied either BUR or mastic asphalt specification in accordance with CP144: Parts 3 and 4 and relevant trade associations.

In addition to the standard weatherproofing options, a family of single-layer membranes is now widely available in the UK. These membranes do not generally involve the use of hot bitumen in their application. A felt or fibre board surface to the PS insulation is not normally necessary, though fibrous layers may be specified when PVC single-ply membranes are used to prevent migration of plasticizers from the membranes to the PS foam. Both EPS and XPS are used below single-ply membranes.

#### *24.8.2.3 XPS insulation*

Where XPS is used in an inverted roof system, irrespective of the deck, the insulation is laid over the weatherproofing membrane. Protection of the weatherproofing membrane is the foremost benefit of this system. Degradation from thermal shock and solar exposure is reduced to an absolute minimum. Estimates have been put forward that the temperature fluctuation in the protected weatherproof membrane can be less than 1°C. Installation of the system can be carried out in virtually any weather conditions, as disturbance of the weatherproofing is not necessary.

In the thermal-transmittance calculation for an inverted roof construction, the thickness of the insulation is increased by 20% to compensate for intermittent heat loss. This loss is created by the flow of rainwater below the XPS board and over the weatherproofing.

An inverted roof consists of the following layers:

- *Waterproof membrane over structural deck.* This is followed by a protection sheet or a cushioning layer, as required by some PVC membranes to prevent reaction between XPS and membrane. Alternatively, this is used as a cushioning layer over an irregular surface.
- *XPS insulation.* This is laid loose over the surface of the roof and vertically at upstands to reduce thermal bridging.
- *Filter fabric and ballast.* A filter fabric may be required, again when using plastic membranes, to prevent fines from percolating through the ballast and damaging the membrane. On built-up roofing or mastic asphalt roofs a filter fabric is not necessary if the ballast is 20–30 mm thick and clean. Ballasting of the XPS boards is necessary to prevent their movement by wind uplift pressures. Different thicknesses of XPS board require different depths of ballast to maintain their stability. On roofs subject to traffic the use of paving slabs in lieu of ballast is necessary. The paving slabs are supported on spacer pads to allow drainage below.

## 24.9 Floor insulation

The use of PS in insulated floor construction has been widely accepted since the early 1960s in timber, traditional solid concrete and modern pre-cast concrete or beam and block floor systems. With today's awareness of the need for energy conservation measures and the legislative requirements of Building Regulations, the wide variety of possible floor insulation applications is of major significance. The versatility of PS allows optimum use of a material, available in a range of grades, to meet various thermal and structural requirements in the sphere of floor construction.

PS used in floor construction to provide thermal and/or structural requirements is very versatile because of the performance characteristics of the material. The same characteristics give a range of advantages in use, surpassed in all aspects by no other insulation material – it is lightweight, has good structural strength, is moisture-resistant and durable, is an excellent thermal insulation and yet is easy to install and is compatible with most standard building materials. Insulation board may be positioned vertically on either face of the foundation or horizontally either above or below the slab.

There exist several individually designed systems for floor construction which from the design stage have considered the need to incorporate significant thicknesses of insulation.

The use of PS in floor construction can be grouped in the following application areas:

1. perimeter insulation
2. underslab insulation
3. overslab insulation
4. timber floor insulation
5. high load floors/cold store floor insulation
6. structural floor systems.

### 24.9.1 Perimeter insulation

In the case of ground floors, it is generally considered most economical to insulate the whole floor area. This is especially true in the case of houses and other small buildings where it is not possible to meet insulation requirements by solely using perimeter insulation. In buildings of larger floor area where the intrinsic heat loss from the floor is lower, the major source of heat loss from the slab is the edge, which can be insulated independently of the whole floor using slab-edge or perimeter insulation with either EPS of XPS.

Vertical-perimeter insulation is normally applied to a depth of at least 600 mm.

Installation of the insulation boards against the foundation is carried out prior to backfill around the foundations. The surface to receive the insulation board should be flat and even to allow firm contact between the surfaces. Backfill of the trench should be carried out with care to prevent damage to the insulation board.

### 24.9.2 Underslab insulation

In addition to giving a solid ground-floor slab excellent thermal insulation properties, the use of PS insulation boards below it provides a rigid, robust and easy–to-install base on which to lay a damp-proof membrane in sheet form. Its high water-resistance characteristic makes PS insulation ideal for use above or below the damp-proof membrane under a ground-floor slab.

The range of material grades available in EPS and XPS make possible their use under differing load criteria from domestic to heavy-duty constructions.

The high compressive-strength characteristics of PS foam allows non-load-bearing walls of both masonry and stud partition construction to be built directly off the finished concrete slab without the need for slab strengthening below them.

In preparation for the laying of insulation boards or a damp-proof membrane, the hardcore base must be well compacted and suitably blinded to leave a level and even surface. The insulation boards are then laid over the prepared surface with joints closely butted. Installation of insulation vertically around the perimeter of the slab to prevent cold bridging is then carried out. Where the damp-proof membrane has been laid below the insulation, all board joints should be taped to prevent grout ingress.

The concrete ground floor slab is then laid to the specified thickness and finished as required. Protection to the insulation boards may be required if materials are to be barrowed across the surface.

In Scandinavia high-density products are even used under the areas on the edge of concrete raft foundations where the wall loadings occur. This approach gives complete insulation of the ground floor.

In the UK restrictions on the amount of backfill allowable, to restrict settlement of ground floors over poorly consolidated hardcore, has stimulated the use of thick boards of EPS as an alternative fill material. Boards of 300 mm and greater thicknesses have been used, with the bonus of very low U values.

Figure 24.14 EPS used over a concrete slab

Figure 24.15 Underfloor heating system

### 24.9.3 Overslab insulation

PS insulation boards can be laid directly over a structural floor slab or system to provide a rigid and robust base for a suitable floor finish. Finishes normally applied over the polystyrene are a minimum 18 mm thick flooring-grade chipboard to BS5669 or a minimum 65 mm thick sand–cement screed.

In domestic applications the lower strength PS boards are used and will adequately support the loads associated with this type of occupancy. If the loading conditions exceed domestic equivalents, then higher grades of PS foams may be necessary.

Load-bearing and other heavy partitions must be supported by a treated-timber batten placed below the floor finish in lieu of the polystyrene. Generally this applies only where a chipboard floor finish is to be used (Figure 24.14).

The surfaces onto which the polystyrene insulation boards are to be laid must be flat and even. Minor irregularities over a tamped slab or between the elements of a pre-cast floor system may be overcome by the insulation boards. If significant irregularities occur they must be suitably removed or levelled by use of a proprietary screed, self-levelling or otherwise. A suitable damp-proof membrane may be used directly under the insulation boards, if required. The insulation boards are laid over the prepared surface and tightly butted together.

#### *24.9.3.1 Chipboard floor finish*

If the damp-proof membrane has been laid below the floor slab, or if a screeded pre-cast floor system is considered, a vapour-control layer of minimum 1000 g polyethylene must be laid over the PS prior to the chipboard. The vapour check must be taped at joints and taken up the perimeter walls at least 100 mm to finish behind the skirting boards.

Flooring-grade chipboard, tongued and grooved all round, is then laid with staggered joints over the vapour control layer with all joints fully glued with a standard PVA woodworking adhesive. Temporary wedges are used around the floor perimeter in order to tighten the chipboard joints when the floor is complete.

A 10 mm wide expansion gap must be allowed at all chipboard perimeters. In runs in excess of 10 m, expansion gaps must be provided in the floor and treated-timber battens must be placed below expansion joints.

At positions in the floor where the tongued and grooved joint of the chipboard is lost, a treated-timber batten must be positioned below the boards, lieu of the polystyrene, to support the butt-jointed chipboard. The batten should be firmly fixed to the structural floor. This requirement for a timber batten also applies at external doorways, but not around all room perimeters. Protection of the chipboard from water spillage should be as recommended in BS 5669.

#### *24.9.3.2 Screed-floor finish*

A 65 mm thick sand–cement screed is laid directly onto the polystyrene. The recommendations of BS8204 should be followed, with the screed area not exceeding 15 m$^2$ with a length-to-width ratio not in excess of 1.5:1. If these limits are exceeded, or the screed thickness reduced to a minimum 50 mm, a light-gauge galvanized-metal reinforcement should be placed centrally in the screed.

Underfloor heating systems using warm water piping are conveniently located using designed EPS panels and the covered with a screed (Figure 24.15).

#### *24.9.3.3 Timber-floor insulation*

Insulation of timber ground floors can be achieved by positioning suspended EPS between the joists or XPS over or between the joists. This form of floor insulation is particularly appropriate in renovation work. The under-floor space must be ventilated.

When insulation is placed between joints, a treated-timber batten or galvanized fixing is positioned on the side of the joists to give sufficient height to accommodate the insulation thickness below. An air gap of 25 mm between the insulation and timber-boarded floor finish will improve the overall thermal resistance of the floor.

Insulation boards should be cut to fit tightly between the joists and rest upon the previously positioned supports. Pinning of the top surface with galvanized nails prevents insulation-board movement.

XPS insulation boards can be laid directly over the floor joists of new or existing floors. Once installed, they provide the base for the required timber-board floor finish which is fixed through the insulation to the timber joists.

In refurbishment work, both types of PS can be laid over an existing timber floor prior to laying a suitable floor finish.

### 24.9.4 High-load floors/cold store floors

Higher grades of PS materials make the use of PS in areas of high-load application possible. While the general construction details remain similar to those previously described, the PS strength requirement has to be adjusted, by choice of grade, to suit the situation. A typical example of this area of use is that for cold-store floors.

Polystyrene insulation boards are used to reduce the amount of heat entering the cold store from the surrounding sub-base. Additionally, the ground temperature is maintained and assists in the prevention of possible ground heave. Insulation boards in polystyrene are available in various grades, based on this strength requirement, to suit differing load criteria.

Preparation of the ground to the required level should ensure that it is firm and flat. Placing of a sub-base, clean and free from contaminants, should be in layers and allow adequate compaction of each.

Subsequent to blinding the sub-base, if necessary, a screed is laid over the surface. Screed thickness is dependent on its designated use, either as blinding or as a carrier for heating cables laid to prevent freezing of the soil and subsequent ground heave. The screed must be finished to at least a fine tamp or equivalent finish.

A suitable vapour control layer must be laid over the screed and turned up at all abutments. Insulation boards are then laid, either in two layers breaking joints or in a single layer with suitable joint detail to avoid cold bridging. The vapour control layer is then turned down onto the insulation at perimeters and a 500 g polythene laid over the whole surface of the PS to prevent grout ingress and subsequent cold bridging. A concrete slab to the requisite design is then placed over the insulation and finished, as appropriate. Historically cold store floors have been insulated with XPS due to the high loadings in such buildings, though EPS is also specified.

### 24.9.5 Structural floor systems

EPS has been developed for use in a series of floor systems where it becomes part of the main structure. Used in conjunction with pre-stressed concrete beams, EPS infill panels replace the more traditional block or hollow pot infill. In addition to providing the infill to the beams, the panels give the floor system an extremely high insulation value.

Variation in the EPS panel design allows their use with different pre-stressed concrete-beam sections, which effectively designates the structural strength of the whole system.

The reinforced structural screed topping system allows a higher degree of load capacity than the chipboard system. In both systems the EPS panels are designed to prevent cold bridging of the beams by including an insulation layer above or below the beam, as well as in between. EPS panels used in these systems also provide the benefits of handleability, light weight, speed of construction and consistent quality.

Preparation of the site and substructure to receive the beams is as that for a traditional building. As with a standard beam-and-block system, the beams are set and erected in accordance with the beam manufacturer's layout drawing. Solid and spacer blocks are used to position the beams accurately. The EPS panels are then easily fitted between the beams with closely butted joints (Figure 24.16). Additional details for the system are available from the beam manufacturers.

Finishes on the system are laid in accordance with the respective requirements for chipboard or reinforced screed. With the chipboard system, a polythene vapour-control layer is first positioned over the panels and turned up at perimeters. An 18 mm thick flooring-grade chipboard, with tongued and grooved edges, is then laid over the vapour check and the joints glued. An expansion gap of 10–12 mm must be left at all perimeter edges. Temporary wedging of the chipboard at perimeters allows the

**Figure 24.16** Beam and EPS infill

expansion gap to be maintained when the chipboard is finally tightened after gluing.

To form the structural screed system an A98 steel mesh reinforcement to BS4482 is laid over the concrete beam tops, topped and tied as necessary. A screed, generally grade 35, containing 10 mm aggregate, is then laid. Its minimum thickness over the beam head must be 40 mm. Compaction and finishing of the screed is then carried out as necessary.

### 24.9.6 Clay heave

There is an application in foundations using a lightweight EPS product to accommodate forces due to clay heave. Ring beams, supported on piles, can be protected from the pressures generated when a previously dry clay soil expands as water is taken up. Sites cleared of trees, for example, can exhibit this phenomenon on certain clay types. Careful production of EPS boards will provide the correct balance of properties (i) to provide adequate support to concrete ring beams during concreting, with a minimum amount of compression; and (ii) to allow a significant movement to occur safely as the clay expands.

Clearly, the downward forces of the building must be able to counter the upward transmitted pressure. Typically, EPS provided for this application will compress by more than 50% of its original thickness under loads of 40 kPa. Thus, building foundations can be protected from clay heave, though the technique is not suitable for normal in-situ floor slabs, which do not have adequate strength. A better solution for floors is to provide a suspended ground floor, creating a void into which the soil can move.

### 24.9.7 Ventilation products

Health considerations in housing have stimulated the use of products to remove such gases as radon and methane from below foundations. Without appropriate measures these gases, when present, can permeate the living accommodation. Various proprietary products are available to create voids beneath a concrete floor slab to allow the gases to vent naturally or, if necessary, with the aid of forced ventilation. This solution will involve the use of impermeable membranes below the concrete to protect the occupants fully.

## 24.10 Civil engineering

The use of EPS in civil engineering applications is significant. The properties that have stimulated its use include its light weight, good compressive strength, versatility of shape and the availability

**Figure 24.17** EPS spanning suspended beams

of the material in large blocks. It is the flexibility of sizes, combined with the relative price of EPS compared with XPS, which dictates that very little extruded product is used in civil engineering. It is true that XPS is used in colder parts of Europe for frost protection of roads and railways, and has been used in road construction as a lightweight fill in the UK, but in general the former applications are not relevant to the British climate and, in the latter, EPS is more competitive and convenient because of the availability of larger block sizes. The following applications, therefore, relate only to EPS, except where stated.

### 24.10.1 Shutters for concrete structures

The versatility of shaping EPS blocks has allowed a greater freedom in the design of the external form of poured concrete structures. Simple curves and profiles, difficult or indeed impossible to produce in any other way, can simply be cut from blocks and supported on timber frames to tie in with more conventional formwork. A facing of polypropylene sheet, or a proprietary coating for shapes, other than simple curves, is used to protect the EPS and facilitate release from concrete.

Re-usable demountable shutter systems have been developed from this idea to produce formwork for steel-reinforced cast concrete floors in large buildings. On a smaller scale, a similar technique can be used to produce custom-built arch forms for brickwork or reproduce cut-stone work profiles to effect repairs to existing buildings. EPS cylinders or profiled blocks can be used inside poured concrete to reduce the dead load of a structure, particularly bridges, to maximize design strength and thereby to economize on total construction costs. Large bridge constructions use this method to reduce the mass of poured concrete. It is important that the EPS does not absorb excessive moisture from curing concrete, has good compressive resistance against concrete hydrostatic pressures and adequate cross-breaking and compressive properties – to ensure that it can be properly restrained to resist the flotation forces from wet concrete during pouring and vibration. EPS satisfactorily performs these functions using a grade produced specially for the civil-engineering industry.

More recent road-building schemes have, for environmental reasons, called for sections of road to be built below ground level. A technique called 'cut and cover' is used, in which a channel is excavated, the road built in the bottom and, in effect, a very wide bridge is built across the top of the channel, so that the whole construction can be recovered with soil. Poured concrete, using EPS void forms, lend themselves very well to this technique which has been used, for example, on the A1 to the north of London at Hatfield. The use of EPS voids, in an otherwise solid concrete construction, creates a series of I-beams and optimizes the use of the materials involved.

### 24.10.2 Flotation

Using the same properties of light weight and water resistance, EPS is used to provide buoyancy in a variety of applications, ranging from simple floats for the pens used on commercial fish farms to floating piers and walkways in marina developments. Protection is normally provided to give impact resistance, where required, by encasing the EPS blocks in concrete or with another suitable covering. A low-density product fulfils the requirements for this application.

### 24.10.3 Road construction

Scandinavian countries, particularly Norway, have pioneered the use of large EPS blocks as foundations for road embankments. Increasing interest in the UK has resulted in several projects using EPS and, occasionally, XPS.

The reason for using this apparently expensive material to build up such embankments is to reduce the load of the embankment onto the subsoil. It is then possible to work over poor ground conditions without resorting to inherently expensive solutions, such as piling. It is also possible to save construction time, where long-term preloading might otherwise have been used to stabilize the soil. One additional benefit in built-up areas is that the transport of a high-volume, low-density product has environmental advantages over heavy traditional fills, and EPS has been specified largely for this reason alone.

EPS 100 is normally found to have sufficient load-bearing capacity and, at a nominal density of 20 kg $m^{-3}$, can considerably reduce downward pressures of traditional fill materials with densities in the region of 2000 kg $m^{-3}$. For design purposes, a conservative density of 100 kg $m^{-3}$ can be used to make allowance for long-term water absorption, where continuous or frequent wetting of the base layers occurs. Proper consideration of suitable drainage needs to be taken, as do data on historic water-table levels to ensure that unwanted flotation forces cannot occur.

Vertical loads imposed on EPS blocks are transmitted downwards with very low lateral thrust. This enables design savings to be made when using this technique against structures such as bridges, or retaining walls, which do not then have to withstand the usual lateral forces from granular fills (Figure 24.18).

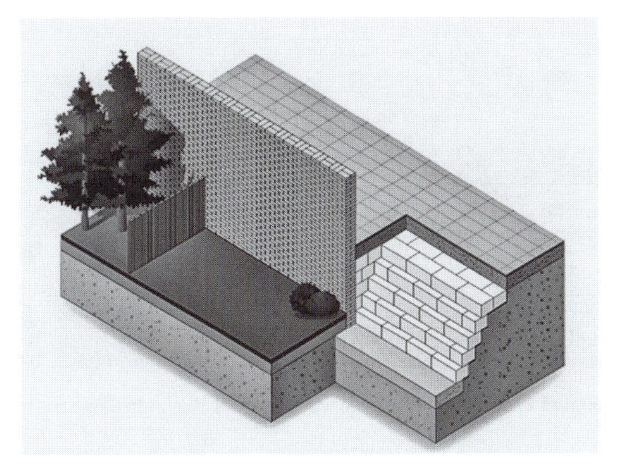

**Figure 24.18** EPS used to reduce thrust against a retaining wall

The Norwegian experience, which is well reported, extends back to the early 1970s and this is now a standard technique for road construction on poor sub-soils in that country and the UK (Figure 24.19).

**Figure 24.19** Placing EPS blocks in a road foundation

**Figure 24.20** Rail embankment during construction with EPS to replace a deteriorating steel bridge

Use of XPS for frost protection under railway tracks is established in Scandinavia, and higher grades of EPS (up to EPS 500) are being used for rail embankments in complex combinations with lower density products (Figure 24.20).

### 24.10.4 Foundations

Large quantities of EPS are used in some continental countries, particularly in Scandinavia where frost protection in important, for construction of raft foundations. This technique has also been used in the UK, with the purpose of economizing on the use of reinforced concrete or building on poor ground. Proprietary systems are available to provide insulated formwork into which reinforcement is placed and concrete is poured to complete the foundation.

## 24.11 Environmental performance

With increasing emphasis on environmental issues from carbon dioxide emissions to pollution to the environment generally, it is unsurprising that building products are being seen as an area where significant improvements can be made. Large volumes of waste building materials have casually been discarded over the years and better choice of products, better design and construction and understanding of the issues involved is only correct.

Insulation products are no less prone to this scrutiny than the more energy-intensive products. The tools being developed to measure environmental impact are becoming established, though there is no doubt that they will be developed further as experience is gained.

Ecoprofiles for individual insulation products are available and cover a list of 13 parameters. The detail can be found on the website www.greenbooklive.com or www.thegreenguide.org.uk. To enable comparison, all insulants are compared on the basis of a common functional unit which is defined as 1 $m^2$ of insulation having a thermal resistance of 3 $m^2$ K/W, roughly equal to a product approximately 100 mm thick with a thermal conductivity of 0.033 W/m K.

The topics covered by this profiling are: climate change, water extraction, mineral resource extraction, stratospheric ozone depletion, human toxicity, ecotoxicity to freshwater, nuclear waste, ecotoxicity to land, waste disposal, fossil fuel depletion, eutrophication, photochemical ozone creation, and acidification. This is not the place to go into detail on each of these topics which are scored from A+ to E, the results from each category being divided into equal band widths. The value of a 'B' for one topic is not comparable with that from another, but an overall rating is also given for each product. Most of the commonly available insulation products achieve an overall rating of A or A+ because of their purpose to save energy and reduce carbon emissions over their lifetime, which can be measured in decades. Where the ozone-depleting gases have been used a poor rating results, and the published summary rating of E for XPS reflects this. No doubt re-evaluation with the newer gases being used will give a higher rating closer to that of EPS – A+.

Care should be taken when choosing between insulants not to favour one aspect unduly; a balanced decision should be made between the requirements of the application and the environmental evaluation. The energy saved by all insulants is many times the energy used in their production, and over their lifetime a product that contains less embodied energy will show only a small percentage advantage over one with a higher value.

The Code for Sustainable Homes is another wider evaluation of our built environment that the construction industry has to address. This Code evaluates nine categories of environmental impact for housing and, using weighting factors on credits given for the categories listed below, awards a percentage score which is then converted into six Code Levels.

The overall category descriptions are: energy and CO2 emissions; water; materials; surface water run-off; waste; pollution; health and wellbeing; and ecology. A high number of available credits and weighting is given to the energy and materials categories, which account for half of the available credits. Insulation performance clearly has a significant part to play in the Code.

The code is linked to the requirements for energy conservation in the Building Regulations (Part L) and improvement over these requirements. A review of the Code in the light of changes to Part L will be needed, and debate over how this should be done is currently under way.

New homes are obliged to have a rating calculated to the Code for Sustainable Homes as a means to move towards zero-carbon homes. This, coupled with a proposed improvement in energy use by 25% in 2010 and 44% in 2013 and then to zero-carbon homes in 2016 compared to today's standards, will increase the use of all insulation products.

## 24.12 Standards and bibliography

BS EN 13163:2001. Thermal insulation products for buildings. Factory made products of expanded polystyrene. Specification.

BS EN 13164:2001. Thermal insulation products for buildings. Factory made products of extruded polystyrene foam (XPS). Specification.

BS EN 13172:2008. Thermal insulating products. Evaluation of conformity.

BS EN 13501-1:2007. Fire classification of construction products and building elements. Classification using data from reaction to fire tests.

BS EN 13823:2002. Reaction to fire tests for building products. Building products excluding floorings exposed to the thermal attack by a single burning item.

BS EN ISO 13370. Thermal performance of buildings – Heat transfer via the ground – Calculation methods.
BS EN 14933:2007. Thermal insulation and light weight fill products for civil engineering applications. Factory made products of expanded polystyrene (EPS). Specification.
BS EN 14934:2007. Thermal insulation and light weight fill products for civil engineering applications. Factory made products of extruded polystyrene foam (XPS). Specification.
BS6203:2003. Guide to fire characteristics and fire performance of expanded polystyrene materials (EPS and XPS) used in building applications.
BS747. Specification for roofing felts.
BS5628. Code of practice for use of masonry: Part 3, material and components, design and workmanship.
BS5534. Code of practice for slating and tiling. Part 1 Design.
BS8204. Screeds, bases and in situ floorings.
BS8217. Code of practice for built-up felt roofing.
BS8218. Code of practice for asphalt roofing.
BRE Report 262. Thermal insulation: avoiding risks – Third edition 2002.
BRE Green Guide to Specification.

## 24.13 Websites

- EPS Eumeps. The White Book: www.eumeps.org
- EPS: www.eps.co.uk
- XPS: www.exiba.org

25

# Polytetrafluoroethylene (PTFE) and Ethylene Tetrafluoroethylene Copolymer (ETFE)

**Arthur Lyons** MA MSc PhD Dip Arch Hon LRSA
De Montfort University

## Contents

## 25.1 Introduction

Polytetrafluoroethylene (PTFE) is well known for its low-friction properties and particularly for its use in domestic non-stick cookware. Its low coefficient of friction and chemical inertness are reflected in its commercial use for pipework, low-friction industrial components and coated tensile fabric structures. Ethylene tetrafluoroethylene copolymer (ETFE) is used mainly as sheet material, either as single layer or as building envelopes formed from inflated foil cushions. ETFE exhibits some of the properties of PTFE, such as chemical inertness and temperature resistance but with the advantages associated with thermoplastic forming properties which are absent in PTFE.

Polytetrafluoroethylene is produced by the polymerisation of the monomer tetrafluoroethylene. It was discovered in 1938 by the DuPont chemist Dr Roy Plunkett, who, while working on refrigerant gases, observed that a frozen compressed sample of tetrafluoroethylene had spontaneously polymerised into a white waxy solid. The manufacturing process and trade mark 'Teflon' were registered, and commercial production commenced in 1946.

Ethylene tetrafluoroethylene is a copolymer of the two monomers ethylene and tetrafluoroethylene. Its discovery in 1970 was registered by DuPont under the trade name 'Tefzel'. ETFE and other fluorinated polymers such as fluorinated ethylene propylene (FEP) were developed through research following the discovery of PTFE.

## 25.2 Manufacture

Polytetrafluoroethylene is produced by free radical initiated aqueous polymerisation of the monomer under pressure. Polymerisation in suspension produces a white granular material, while the dispersion process produces a fine white powder or a milky white dispersion stabilised by wetting agents. As, unlike most thermoplastics, PTFE does not melt when heated, unconventional processing techniques are required.

Granular compression moulding resin is composed of small fibrous particles which can be fabricated by milling and pelletisation. For extrusion into profiles, either a pressed billet is forced through a die under high pressure or PTFE paste is extruded. PTFE tape is manufactured by extrusion followed by sintering if specified. Where solid block material is required, the PTFE pellets are pressed then sintered at 360°C. Granular compression moulding PTFE resin may be blended with inorganic fillers including metal powders for specific applications.

A range of aqueous dispersions is produced with different solids content from 30% to 60% PTFE according to the anticipated processing and application. These are suitable for coating surfaces or impregnating fabric materials. Coating normally involves a sequence of dipping, drying and baking operations to build up the required thickness. Calendering may also be used to ensure an appropriate finish. Finally the laminate is heated to a temperature of approximately 337°C, which is above the crystalline melting point of the resin, to sinter the material. The substrate, such as glass fibre fabric, must withstand this high-temperature process. Where impregnation of a material by unmelted PTFE granules is required, the final high-temperature sintering is omitted.

Modified PTFE polymers incorporate small proportions of fluorinated comonomers such as perfluoropropyl vinyl ether (PPVE) to alter manufacturing processes and product properties (e.g. surface porosity, stiffness and creep resistance).

Ethylene tetrafluoroethylene copolymer is produced by copolymerisation of the two monomers. Unlike PTFE, it is thermoformable by the standard fabrication techniques of injection-, transfer-, compression- and blow-moulding, also extrusion. It is manufactured in powder or pellet form for subsequent processing.

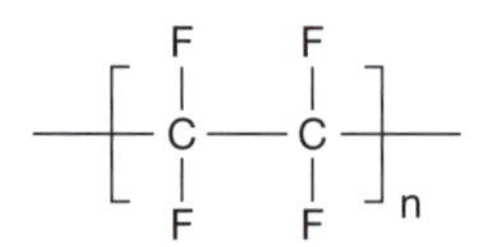


Polytetrafluroethylene (PTFEE)

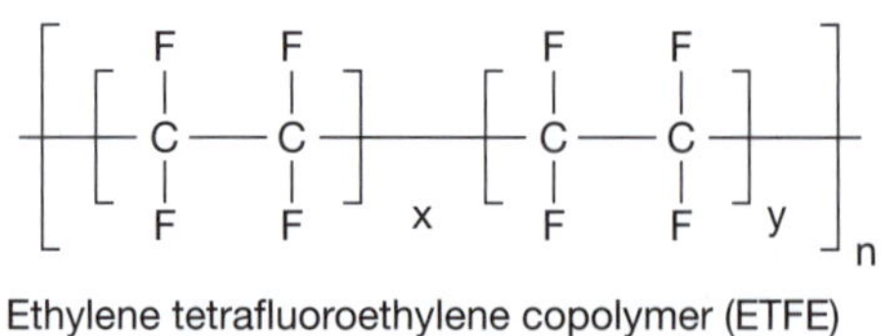


Ethylene tetrafluoroethylene copolymer (ETFE)
(x and y are variable numbers)

**Figure 25.1** Composition of PTFE and ETFE

A variation on ETFE is ethylene chlorotrifluoroethylene copolymer (ECTFE), a tough plastic with the highest abrasion resistance of any fluoropolymer film.

## 25.3 Chemical composition

Figure 25.1 illustrates the structures of linear polytetrafluoroethylene (PTFE) and ethylene tetrafluoroethylene copolymer (ETFE).

## 25.4 Standard forms of polytetrafluoroethylene and ethylene tetrafluoroethylene copolymer

PTFE is produced in granular form for fabrication into film and components, and as solid blocks for machining. The aqueous dispersion grades are used for coating and impregnating sheet materials. The standard forms are tape, sheet, plate, rod and tube. Most PTFE products are white but pigments may be added if required. Fillers include glass or carbon fibres, graphite, molybdenum disulfide and metals including bronze.

ETFE is a thermoplastic resin, available in powder or pellet form for moulding; it may be coloured to specification. The stock material is sold as sheet, rod and tubing. Fillers include glass and carbon fibres.

## 25.5 Typical physical properties of polytetrafluoroethylene and ethylene tetrafluoroethylene copolymer

The main characteristics of PTFE are its chemical inertness, electrical properties and very low coefficient of friction. PTFE is inert to virtually all industrial chemicals and solvents, unaffected by weathering and usable within an operating temperature range of –200°C to 260°C. It has a crystalline melting point of approximately 327°C, but does not become fluid, so it may be formed by sintering at 360°C; the material degrades at temperatures over 400°C, which is much higher than other thermoplastic materials.

PTFE has excellent dielectric properties as both the power factor and dielectric constant are low and do not vary with temperature. It is an excellent electrical insulator, especially as it is not wetted by water and is non-absorbent, thus electrical resistivity remains very high even under humid conditions.

The strength properties of PTFE and its anisotropy depend on its fabrication process. The tensile strength and modulus of elasticity are relatively low, but are retained over a wide temperature range. The impact resistance of PTFE is good even at sub-zero temperatures; however, its resistance to wear and creep is low in relation to other engineering materials. The mechanical properties, including strength and wear resistance, may be enhanced by the addition of fillers including glass fibres, carbon fibres and metals.

PTFE exhibits plastic memory such that, when it is subjected to tensile or compressive stresses below the yield point, some permanent deformation is produced. The original form is recovered by reheating in a stress-free condition. As the material is relatively soft, machine tools must be kept sharp. PTFE exhibits a significant transition point at 19°C, in which the crystalline component of the material is changing its geometric form; this is associated with significant dimensional changes, which may affect components designed to operate around this critical temperature. The greater the proportion of crystallinity, which normally varies from 55% to 95%, the greater the dimensional changes at this transition temperature.

PTFE has a greasy feel and the lowest coefficient of friction of all solid materials, usually within the range 0.04–0.09. The coefficient of friction is independent of temperature but reduces to a stable value with increased load.

Base PTFE is defined by BS EN ISO 12086-1: 2006 as containing no more than 1% of modifying copolymer. Modified PTFE products have enhanced surface finishes with less porosity and micro-voids, increased stiffness and creep resistance, and enhanced electrical resistance properties.

Ethylene tetrafluoroethylene copolymer (ETFE) as a film is a tough, stiff and tear-resistant material with good stress crack and weather resistance. The material work hardens over a 300–400% elongation. It is usable over a wide range of temperatures from –180°C to 165°C. It has slightly lower chemical resistance than the totally fluorinated PTFE, but has the advantage of being thermoformable. The high light transmission of ETFE (approximately 95%), compared to PTFE coated glass fibre tensile fabrics (8–16%) makes the film eminently suitable for architectural applications. ETFE film has a low surface energy appropriate to release applications. Table 25.1 compares the physical properties of PTFE and ETFE.

The closely related product ethylene chlorotrifluoroethylene copolymer (ECTFE) has similar properties to ETFE, but with enhanced chemical resistance. ECTFE has the greatest abrasion resistance and highest dielectric strength of currently available fluoropolymer films.

**Table 25.1** Typical properties of PTFE and ETFE

| *Property* | *Value* | | *Units* |
|---|---|---|---|
| | *PTFE* | *ETFE* | |
| Density | 2.13–2.24 | 1.70–1.76 | $gm/cm^3$ |
| Tensile strength | 20–40 | 40–50 | MPa |
| Elongation at break | 300–400 | 300–400 | % |
| Flexural modulus | 500–520 | 900–1000 | MPa |
| Tensile modulus | 400–550 | 800–965 | MPa |
| Melting point | 327 | 250–275 | °C |
| Maximum service temperature | 260 | 150–165 | °C |
| Minimum service temperature | –200 | –180 | °C |
| Coefficient of friction | 0.04–0.09 | 0.20–0.23 | |
| Water absorption | < 0.01 | < 0.03 | % (24 h at 23°C) |
| Refractive index | | 1.40 | |
| Light transmission | | 92–95 | % |
| Thermal conductivity | 0.25 | 0.24 | W/m K |
| Coefficient of linear expansion | $10 \times 10^{-5}$ | $9.4 \times 10^{-5}$ | °C |
| Dielectric constant, 1 MHz | 2.1 | 2.5–2.6 | |
| Dissipation factor, 1 MHz | < 0.0002 | 0.006 | |
| Volume resistivity at 23°C and 50% relative humidity | $> 10^{18}$ | $10^{17}$ | |
| Breakdown voltage | 9 | 12 | kV/0.1 mm |

Typical values for reference only, not for specification.

## 25.6 Specification of polytetrafluoroethylene and ethylene tetrafluoroethylene copolymer

The Standard BS EN ISO 12086-1: 2006 defines the designation system for the specification of fluoropolymers moulding materials. The five data block system allows for the specification of the material, its application or method of processing, specific properties, fillers and any additional required information. It is applicable to PTFE, ETFE and other fluoropolymers. The Standard BS EN ISO 13000-1: 2005 specifies requirements for PTFE semi-finished products, such as tape, sheets, rods and tubes. The specification for E-glass fibre filled PTFE is described in the Standard BS6564-3: 1990.

The Standard BS7786: 2006 defines the specification for unsintered PTFE tapes for general use and BS EN 751-3: 1997 gives the classification and requirements for unsintered PTFE tape for sealing threaded metal pipe joints for gases and liquids. Requirements for PTFE insulated electric cables are specified in the Aerospace Standard BS 3G 210: 1996. The series of European Standards BS EN 1337 relates to all types of structural bearings. Within this set, Part 2: 2004 refers to sliding elements, Part 3: 2005 to elastomeric bearings and Part 7: 2004 to spherical and cylindrical PTFE bearings.

## 25.7 Dimensional stability

The thermal movement of PTFE is $1.0 \times 10^{-4}$ °C$^{-1}$, and that of ETFE $0.94 \times 10^{-4}$ °C$^{-1}$ compared to PVC at $0.7 \times 10^{-4}$ °C$^{-1}$ and steel at $0.12 \times 10^{-4}$ °C$^{-1}$. It should, however, be noted that the phase change for PTFE at 19°C from helical to hexagonal crystalline form causes an additional expansion of 1.3% to 1.8% depending on the degree of crystallinity. Also, outside the critical range 19–25°C, thermal expansion for PTFE is not linear, and because of its manufacturing processes the material is anisotropic, therefore it exhibits some variation in the coefficient of expansion with direction.

## 25.8 Durability

Polytetrafluoroethylene is an extremely durable material. It is unaffected by chemical reagents and solvents except the most caustic materials such as reactive alkali metals and fluorine under pressure. Fluorinated hydrocarbons cause some reversible swelling. PTFE is unaffected by exposure to the atmosphere and UV light, but it is degraded by high-energy radiation.

Ethylene tetrafluoroethylene copolymer is a durable material with high resistance to weathering and exposure to UV light. It is unaffected by most chemical agents, although ethylene chlorotrifluoroethylene copolymer (ECTFE) has superior chemical resistance. Ethylene tetrafluoroethylene copolymer has slightly enhanced resistance to high energy radiation over PTFE. The ETFE envelope of the Eden Project in Cornwall (Figure 25.6) has a life expectancy of 25 years.

## 25.9 Uses of polytetrafluoroethylene and ethylene tetrafluoroethylene copolymer in construction

A major use for PTFE in construction is for bearings and sliding surfaces where its low coefficient of friction, chemical inertness and resistance to weathering and water absorption make it an ideal material. Typical applications are the supports for major buildings and structures including sliding movement joints in bridges. The low friction of PTFE is also advantageous in pipelines and tapes for sealing threaded joints.

A major use of PTFE in construction is for PTFE/glass fibre tensile roof structures, due to the durability and self-cleaning nature of the material. PTFE coated glass fibre fabric has the advantage of an anticipated life expectancy of 25–30 years compared to the alternative PVC coated polyester with an anticipated life expectancy of 15–20 years.

Other uses for polytetrafluoroethylene in construction include PTFE and expanded PTFE electrical insulation and in composites such as expanded PTFE thermal insulation incorporating low-density silica or organic polyimide aerogels.

Most ETFE in construction is used as foils, either as single layers or more frequently as inflated cushion systems. Foil thicknesses usually vary between 10 and 150 μm. Other applications include bearings, electrical components, insulation and covers to photovoltaic systems.

### 25.9.1 PTFE bearings

Several types of PTFE-based bearings are available. The low-friction surfaces either permit rotation over cylindrical or spherical surfaces, or provide sliding over planar surfaces. Normally plane sliding bearings allow only translation movement, but they can permit some rotation about the axis perpendicular to the sliding plane. Cylindrical and spherical bearings may be used in conjunction with flat sliding elements; the extent of linear displacement is normally limited by a combination of guides and restraints. Figure 25.2 shows a PTFE spherical bearing and Figure 25.3 shows a guided bearing allowing unidirectional translational movement.

**Figure 25.2** Spherical PTFE bearing (courtesy of Ekspan Ltd)

**Figure 25.3** Guided PTFE bearing (courtesy of Ekspan Ltd)

The set of European Standards BS EN 1337 'Structural bearings' describes the full range of bearings including installation and maintenance, as follows:

- Part 1: 2000 General design rules
- Part 2: 2004 Sliding elements
- Part 3: 2005 Elastomeric bearings
- Part 4: 2004 Roller bearings
- Part 5: 2005 Pot bearings
- Part 6: 2004 Rocker bearings
- Part 7: 2004 Spherical and cylindrical PTFE bearings
- Part 8: 2007 Guide bearings and restrain bearings
- Part 9: 1998 Protection
- Part 10: 2003 Inspection and maintenance
- Part 11: 1998 Transport, storage and installation.

In bearings, the PTFE surface is normally the lower surface with the upper mating surface oversailing to prevent the ingress of dirt which would scour the sliding surfaces. For low-pressure applications the upper surface may also be PTFE, but for most structural applications polished stainless steel is used. Backing plates are normally ferrous metals, although for cylindrical and spherical bearings aluminium alloy is an alternative for the convex element to BS EN 1337-2: 2004.

In sliding bearings to BS EN 1337-2: 2004, the PTFE is either recessed or bonded to the backing material to be retained in the correct position. The backing metal plate must be sufficiently rigid to retain its unloaded shape and resist shear forces under all loading conditions. The thickness of the PTFE is related to the maximum plan dimensions to the formula in BS EN 1337-2: 2004. Where pressures in recessed PTFE bearings exceed 5MPa, a uniform pattern of dimples is required to retain the lubricant, typically silicone grease. Silicone grease does not affect the sliding surfaces and is effective down to temperatures as low as –35°C. PTFE sheets may be subdivided into a maximum of four identical parts, subject to a minimum size. In PTFE sheets bonded to elastomeric bearings where undimpled PTFE is used, it must be a minimum of 1.5 mm thick and initially lubricated.

Typical designs for cylindrical and spherical bearings are illustrated in BS EN 1337-7: 2004. Cylindrical bearings may allow displacements in any directions or be restrained by internal or external guides to one linear direction. Spherical bearings similarly may allow translational displacements, be restrained by linear guides or be fixed by a restraining ring. The PTFE may be attached to either the concave or convex backing plate and detailed according to BS EN 1337-2: 2004.

The surface mating with the PTFE must be highly polished, be protected from the risk of corrosion and comply with BS EN

**Figure 25.4** Heavy-duty PTFE skidway system (courtesy of Fluorocarbon Co. Ltd)

1337-2: 2004 in respect of deviations from the plane or curved surface. Mechanically polished stainless steel is the standard material, and this is fixed to the backing material by full surface bonding, continuous fillet welding or, in the case of flat sheets, by screwing or riveting. The minimum thickness of stainless steel is 1.5 mm for flat and cylindrical bearings, but 2.5 mm for spherical bearings and screwed or riveted flat sheet.

The PTFE used for bearings must be pure sintered polytetrafluoroethylene without filler or recycled material. Composite PTFE materials are permitted for sliding guide surfaces. The blends may be PTFE with lead (typically 49–62%) or PTFE with glass fibres and graphite. The lubricant dimple pattern is illustrated in BS EN 1337-2: 2004. For the purposes of calculating the ultimate limit state the contact area is taken as the gross area without deduction for the dimples where the minimum sheet dimension is greater than 100 mm, but the dimple area is deducted from the gross area for PTFE sheet sizes with a dimension less than 100 mm.

## 25.9.2 PTFE skidways

In addition to its use in bearings, PTFE is frequently used in skidways for moving heavy structures such as bridges and oil rigs. The low coefficient of friction and the absence of stick–slip characteristics are advantageous. For high load applications, glass-fibre filled PTFE or sintered bronze impregnated with PTFE is used. The system may be PTFE skates running on an accurately laid stainless steel track, or alternatively polished skates running on PTFE pads located on a prepared track base. A typical heavy-duty system of skidway plates would require 2.5 mm PTFE bonded to 3 mm steel plates. For heavy-duty PTFE skid pads welded to the bottom of a large structure, dimpled and filled PTFE is frequently used for sliding over polished steel or timber. Large structures such as oil rigs, of up to typically 30,000 tonnes, may be launched at sea from PTFE lined skid beams on a cargo barge. Figure 25.4 illustrates a heavy duty skidway system for launching an oil rig.

The static coefficient of friction between PTFE and mating surfaces (e.g. stainless steel) is only slightly higher than the dynamic coefficient. This narrow difference leads to the lack of stick–slip behaviour. The coefficient of friction reduces with reduced speed, also with increased compressive stress (Table 25.2) and temperature (Table 25.3). Long-term test data (BS EN 1337-2: 2004, Table 25.3) shows that the coefficient of friction rises with continuous running. In initial use, PTFE is transferred to the mating surface reducing the coefficient of friction, but this gradually wears to an equilibrium state when the loss due to abrasion equates to the PTFE transferred (TRRL Report LR 491). Therefore, for design purposes, the long-term value of the coefficient of friction is used.

Where a combination of horizontal and vertical forces is encountered, to withstand higher stresses and reduce wear on the slides, filled PTFE is often used. This is normally associated with a higher coefficient of friction as indicated in Table 25.2.

**Table 25.2** Coefficient of friction for dimpled lubricated PTFE against stainless steel (BS EN 1337-2: 2004)

| Contact pressure $\sigma_p$ MPa | ≤5 | 10 | 20 | ≥30 |
|---|---|---|---|---|
| Coefficient of friction $\mu_{max}$ PTFE dimpled/ stainless steel | 0.08 | 0.06 | 0.04 | 0.03 |

The characteristic compressive strength for PTFE for main bearing surfaces up to 30°C is 90 MPa. For guides, the coefficients of friction for plain PTFE and filled composite materials to stainless steel are taken as $\mu_{max} = 0.08$ and $\mu_{max} = 0.20$ respectively.

**Table 25.3** Coefficient of friction for PTFE against stainless steel in long-term tests as a function of temperature (BS EN 1337-2: 2004)

| *Temperature, °C* | *Total slide path* | | | |
|---|---|---|---|---|
| | *5000 m* | | *10,000 m* | |
| | $\mu_{static}$ | $\mu_{dynamic}$ | $\mu_{static}$ | $\mu_{dynamic}$ |
| –35 | 0.030 | 0.025 | 0.050 | 0.040 |
| –20 | 0.025 | 0.020 | 0.040 | 0.030 |
| 0 | 0.020 | 0.015 | 0.025 | 0.020 |
| 21 | 0.015 | 0.010 | 0.020 | 0.015 |

### 25.9.3 PTFE pipes

The low friction and almost universal resistance to solvents and corrosive chemicals over a wide temperature range (typically –30°C to 200°C) makes PTFE an ideal material for industrial pipework. PTFE tubing is a standard stock item, but PTFE is also commonly used for lining metal pipes and fittings to give the combination of metallic strength and PTFE inertness. Where inflammable fluids are to be transported the PTFE liner can incorporate antistatic filler to reduce volume resistivity. Industrial hoses, manufactured from low-crystallinity PTFE to maximise flexibility and permeability resistance, may be encased in braided stainless steel. PTFE is also used extensively for pipe supports and seals.

### 25.9.4 PTFE tape

PTFE tape is commonly used to seal threaded joints in pipework, by acting as a deformable filler, but without causing friction within the joint. There are two standards referring to the specification for unsintered PTFE tapes. BS7786: 2006 refers to three grades for general use, including with common liquids and gases, within the temperature range –270°C to 260°C. The three grades relate to the weights H (300–150 g/m$^2$), M (150–75 g/m$^2$), and L (75–25 g/m$^2$) with thicknesses in the range 0.05 mm to 0.25 mm. No additives other than processing lubricant are permitted within the PTFE so that it can be used in an oxygen-enriched environment. The tape is therefore natural white in colour. PTFE tape likely to come into contact with potable water must satisfy the requirements of the standard BS 6920-1: 2000.

The standard BS EN 751-3: 1997 refers to unsintered PTFE tapes for sealing metallic threaded joints in pipework specifically for use with town, natural and LPG (liquefied petroleum gas) (5 bar/0.5 MPa), also hot water systems (7 bar/0.7 MPa), gas appliances (0.2 bar/0.02 MPa) and LPG gas storage (20 bar/ 2 MPa) within a temperature range from –20°C to 125°C. Two classes of PTFE tapes F and G relate to their application to fine and course threaded joints respectively. Tapes for sealing gas pipe joints are generally a thicker grade than for water and may be tinted yellow. Colour tinted tape should not be used for oxygen or potable water.

### 25.9.5 PTFE roofing

PTFE coated glass fibre fabric for tensile structures is described in Chapter 36.

### 25.9.6 PTFE electrical insulation

PTFE is used for electrical insulators, including high voltage, due to its dielectric properties including high arc resistance. Components may be manufactured by extrusion or from block material by CNC (computer numerical control) engineering techniques. Expanded PTFE is used in the manufacture of coaxial cables. The Standard BS3G 210: 1996 refers to the specification for PTFE insulated wires that are either extruded or spirally lapped and sintered for consolidation. PTFE insulated cables are resistant to significantly higher temperatures and emit less smoke in fire than PVC insulated cables, but they are not categorised as low smoke zero halogen (LSZH) products due to the high fluorine content.

### 25.9.7 ETFE cushion roofing

The standard system for ethylene tetrafluoroethylene copolymer building envelopes consists of inflated double or multilayer ETFE cushions restrained by aluminium extrusions, sealed with an EPDM perimeter strip and supported by a structural steel framework. The cushions are inflated to a low pressure which is constantly maintained through a network of pipes and a computer-controlled system. The cushions in the Leicester Space Centre (Figure 25.5) are formed as horizontal bands. However, more frequently to create the required building geometry, a blend of various polygons is used. Hexagons predominate in the domes for the Eden Centre in Cornwall (Figure 25.6), while a complex pattern of hexagons, pentagons, rectangles and triangles is required to produce the aesthetic and to fit the cuboid form of the Beijing 2008 Olympics National Swimming Centre (Fig. 25.7). In the Eden project the largest foil unit has a surface area of 75 m$^2$ and the maximum span in the Beijing Swimming Centre is approximately 8 m. Foils may be clear or coloured as in the light blue of the Beijing Olympic Swimming Centre. Standard colours are blue, yellow, red and green.

The 300–400% elongation afforded by ETFE allows for considerable deflections of the supporting structure, associated with wind and potential snow loads, although on sloping surfaces the low friction will allow reduce snow build-up and also ensure self-cleaning. Multilayer systems enhance thermal insulation values (typical U-values for horizontal cushions are: two-layer – 2.94 W/m$^2$ K; three-layer – 1.96 W/m$^2$ K; four-layer – 1.47 W/m$^2$ K and five-layer – 1.18 W/m$^2$ K).

Multilayer systems offer the option of modifying the visual transparency and levels of solar gain. Graphics can be printed on the various foils and photosensitive layers can be incorporated to respond to external environmental conditions. Compared to traditional glazing, the membranes have only 1% the weight of equivalent glass panels, significantly modifying the requirements of the structural system.

The material can be punctured with sharp objects, but as a thermoplastic it can be repaired by patching with new material. The external drumming noise from heavy rain makes the system inappropriate for certain building enclosures; however, the problem can be ameliorated by the use of multilayer cushion systems. Conversely, noise created within the enclosure is not

**Figure 25.5** Banded ETFE cushions – National Space Centre, Leicester. Architects: Grimshaw Architects. Photograph: Arthur Lyons. (This image was published in *Materials for Architects and Builders*, Arthur Lyons, 4th edition, p. 315. Copyright: Butterworth Heinemann, 2010)

reverberated at the building envelope but escapes directly through the plastic material.

### 25.9.8 ETFE sheeting

In addition to its standard use for multilayer cushion roofing systems, ETFE is also used as a single layer roof membrane. The material work hardens over a 300–400% elongation to a tough material. It offers an alternative to PTFE coated glass fibre and PVC coated polyester systems, with the advantage of visual clarity and good durability. ETFE has been used internationally for numerous projects from small canopies to major building works, capitalising on its resistance to UV degradation.

ETFE is also used as an alternative to glass as the cover for photovoltaics as it has a low refractive index (approximately 1.4) and high light transmission, permitting efficient photovoltaic power generation.

## 25.10 Proprietary brands of PTFE and ETFE

There are a large number of registered trade names for PTFE used in construction. The list includes Teflon, Daikin-Polyflon, Dyneon, Fluon, Fluorinoid and Halon.

Similarly there are several trade names for ETFE, including Tefzel, Texlon, Norton and Fluon.

## 25.11 Hazards in use

PTFE and ETFE in normal use are inert and non-toxic, but inhaling PTFE dust during machining should be avoided. Vapours associated with high-temperature fabrication processes may cause flu-like symptoms in humans and they should therefore not be inhaled. Tobacco products must be kept clear of potential contamination by PTFE particles.

**Figure 25.6** Polygonal ETFE cushions – Eden Project, Cornwall. Architects: Grimshaw Architects. Photograph: Arthur Lyons (This image was published in *Materials for Architects and Builders*, Arthur Lyons, 4th edition, p. 314. Copyright: Butterworth Heinemann, 2010)

**Figure 25.7** Coloured ETFE cushions – Beijing 2008 Olympics National Swimming Centre. Architects: China State Construction and Engineering Corporation, PTW Architects and Arup. Photographs: Courtesy of Vector Foiltec

## 25.12 Performance in fire

Both PTFE and ETFE resist ignition and do not promote flame spread. The presence of fluorine in the materials makes them self-extinguishing. Within a fire situation the contribution to the heat and smoke is small, but at temperatures over 400°C PTFE and ETFE degrade to carbonyl fluoride and hydrogen fluoride. Additionally between 450°C and 650°C, PTFE may produce toxic fluoro compounds depending on the fire conditions.

Under fire conditions, PTFE/glass fibre membrane roofs withstand relatively high temperatures, and will normally fail only when exposed seams reach approximately 270°C. By contrast, hot gases (above 200°C) impinging on ETFE cushions cause the material to soften and shrink away from the gas plume, allowing the gases to vent to the atmosphere. No burning droplets of molten ETFE fall down, as any small fragments of plastic are carried up into the rising gas plume.

PTFE/glass fibre tensile roofing normally meets the requirements of BS476 Part 7 Class 0 and Part 6 Class 0. ETFE is classified as Euroclass $Bs_1$, $d_0$ to the European Standard BS EN 13823: 2010.

The fire safety of PTFE-based materials used in buildings was the subject of a full report by Purser et al. (1994) for the Fire Research Station of the Building Research Establishment.

## 25.13 Sustainability and recycling

PTFE is an expensive plastic to produce, so waste created in the manufacturing process is usually cleaned and ground down into powder for re-use. Burning waste at high temperatures releases toxic chemicals including trifluroacetate to the atmosphere, and the latter persists for long periods.

ETFE is fully recyclable as is a thermoformable polymer. Current environmental developments in ETFE cushion technology include the incorporation of nanogel to increase thermal insulation and the attachment of photovoltaic panels to capture solar energy.

## 25.14 British Standards

- *BS3G 210: 1996.* Specification for PTFE insulated equipment wires and cables, single- and multi-core, with silver plated copper conductors (190°C) or nickel plated copper conductors (260°C).
- *BS ISO 3547.* Plain bearings. Wrapped bushes: Part 4: 2006 Materials.
- *BS6564.* Polytetrafluoroethylene (PTFE) materials and products: Part 3: 1990 Specification for E glass fibre filled polytetrafluroethylene.
- *BS6920.* Suitability of non-metallic products for use in contact with water intended for human consumption with regard to their effect on the quality of the water: Part 1: 2000 Specification.
- *BS7786: 2006.* Specification for unsintered PTFE tapes for general use.
- *BS7899.* Code of practice for assessment of hazard to life and health from fire: Part 1: 1997 General guidance. Part 2: 1999 Guidance on methods for the quantification of hazards to life and health and estimation of time to incapacitation and death in fires.
- *BS EN 751.* Sealing materials for metallic threaded joints in contact with 1st, 2nd and 3rd family gases and hot water: Part 3: 1997 Unsintered PTFE tapes.
- *BS EN 1337.* Structural bearings: Part 1: 2000 General design rules. Part 2: 2004 Sliding elements. Part 3: 2005 Elastomeric bearings. Part 4: 2004 Roller bearings. Part 5: 2005 Pot bearings. Part 6: 2004 Rocker bearings. Part 7: 2004 Spherical and cylindrical PTFE bearings. Part 8: 2007 Guide bearings and restrain bearings. Part 9: 1998 Protection. Part 10: 2003 Inspection and maintenance. Part 11: 1998 Transport, storage and installation.
- *BS EN 1992.* Eurocode 2. Design of concrete structures: Part 1-1: 2004 General rules and rules for buildings. Part 2: 2005 Concrete bridges. Design and detailing rules.
- *BS EN 1993.* Eurocode 3. Design of steel structures: Part 1-1: 2005 General rules and rules for buildings. Part 2: 2006 Steel bridges.
- *BS EN 1994 Eurocode 4.* Design of composite steel and concrete structures: Part 1-1: 2004 General rules and rules for buildings. Part 2: 2005 General rules and rules for bridges.
- *BS EN ISO 12086.* Plastics. Fluoropolymer dispersions and moulding and extrusion materials: Part 1: 2006 Designation system and basis for specifications. Part 2: 2006 Preparation of test specimens and determination of properties.
- *BS EN ISO 13000.* Plastics. Polytetrafluoroethylene (PTFE) semi-finished products: Part 1: 2005 Requirements and designation. Part 2: 2005 Preparation of test specimens and determination of properties.
- *BS EN 13501.* Fire classification of construction products and building elements: Part 1: 2007 Classification using data from reaction to fire tests. Part 2: 2007 Classification using data from fire resistance tests. Part 5: 2005 Classification using data from external fire exposure to roof tests.
- *BS EN 13823: 2010.* Reaction to fire tests for building products. Building products excluding floorings exposed to thermal attack by a single burning item.
- *BS EN 60695-11-10: 1999.* Fire hazard testing. Test flames – 50W horizontal and vertical flame test method.

## 25.15 Bibliography


*Architects' Journal.* Rocket science. 5 July 2001, p. 27.

*Architects' Journal Specification.* Focus on fire engineering. April 2007, p. 29.

*Architects' Journal Specification.* Patterned envelopes. February 2007, p. 35.

BRE (1994) *Fire safety of PTFE-based materials in buildings.* Report 274, Building Research Establishment, Watford, UK.

Brydson, J.A. (2005) *Plastics materials.* CBS Pub & Dist, New Delhi, India.

Clough, R. and Martyn, R. (1995) *Environmental impact of building and construction materials. Plastics and elastomers.* Construction Industry Research and Information Association, London, UK.

Cousins, K. (2002) *Polymers in building and construction. RAPRA market report.* RAPRA Technology, Shrewsbury, UK.

Dufton, P.W. (2002) *Polymers in building and construction. RAPRA market report.* RAPRA Technology, Shrewsbury, UK.

Halliwell, S. (2003) *Polymers in building and construction.* Report 154. RAPRA Technology, Shrewsbury, UK.

Hollaway, L., Hall, C. and Chen, J-F. (2010) *Polymers and polymer fibre composites.* Thomas Telford, London, UK.

HVCA (2004) *Domestic central heating installation specification.* Heating and Ventilating Contractors' Association , London, UK.

Lefteri, C. (2006) *Plastics 2.* Rotovision, Crans-Près-Céligny, Switzerland.

Lefteri, C. (2008) *The plastics handbook.* Rotovision, Crans-Près-Céligny, Switzerland.

Lyons, A.R. (2010) *Materials for architects and builders.* 4th edition. Butterworth Heinemann, Oxford, UK.

Purser, D.A., Fardell, P.J. and Scott, G.E. (1994) *Fire safety of PTFE-based materials in buildings.* BRE Report 274. Building Research Establishment, Watford, UK.

Transport and Road Research Laboratory (1991) *PTFE in highway bridge bearings.* DOE TRRL Report LR 491. Crowthorne, UK.

Transport and Road Research Laboratory (1991) *Bridge bearing survey and problem assessment.* DOE TRRL Contractor Report 260. Crowthorne, UK.

## 25.16 Websites

- Building Research Establishment: www.bre.co.uk
- DuPont: www2.dupont.com
- Ekspan Structural Products: http://ekspan.co.uk
- Fluorocarbon Group: www.fluorocarbon.co.uk
- Saint-Gobain Performance Plastics Corporation: www.plastics.saint-gobain.com
- Saipem: www.saipem.it
- DuPont Teflon: www.teflon.com
- Vector Foiltec: www.vector-foiltec.com
- Verseidag-Indutex GmbH: www.vsindutex.de

# 26 Polyvinyl Chloride Polymers

**Godfrey Arnold** BSc CEng MIChemE
Godfrey Arnold Associates

**S.R. Tan** MSc, CEng, FIMMM
INEOS Vinyls UK Ltd

## Contents

## 26.1 Introduction

The use of polyvinyl chloride (PVC) in the construction industry has become established around the world, especially over the past 45 years. The amount of PVC used for particular applications varies from country to country for a variety of reasons, which may include the type of building methods used, the protection of vested interests, the availability of traditional materials and many others. However, if the European market for PVC in building is surveyed, it will be seen that this major thermoplastic has achieved an important status and provides extremely useful products for this sector. This chapter seeks to explain some of the technical background of the product and describes the use and properties of PVC in the more important applications.

By far the largest applications of PVC in the construction industry are in the pipe and profile sectors. In the European Union unplasticized PVC pipes and profiles constitute 53% of all PVC applications while cable covering, sheet and flooring account for a further 12% of PVC use. About 1% of PVC is used for plasticized PVC profiles in the construction sector, which therefore constitutes approximately two-thirds of the PVC market.

## 26.2 General description

Polyvinyl chloride is the most widely used thermoplastic in the construction industry. It is a polymer of vinyl chloride, and has a chlorine content of 57%.

The product name is often abbreviated to PVC but this needs to be more carefully qualified to enable the user to understand the nature and properties of the product being used. Other terms that are often encountered in industry relating to PVC are 'vinyl' and 'UPVC' and, although neither of these terms is now officially sanctioned by the International Organisation for Standardisation (ISO) nomenclature, they are widely used, especially where the consumer is concerned and, therefore, need to be understood.

For the specifier and consumer the term 'PVC' often means a material or product made from a PVC composition, i.e. an intimate mixture of a vinyl chloride polymer or copolymer with various additives, some of which may be present (e.g. plasticizer or filler) in very substantial and, occasionally, predominant proportions.[1]

The additives that are used with PVC resins are needed to enable processing to take place and include products such as heat stabilizers, lubricants, fillers and pigments. These additives also have an influence on the final properties of the product. Oher additives are often used, such as processing aids and impact modifiers, which also have a process- or property-enhancing effect. Another important group of additives is plasticizers, because they can impart to the PVC product a very wide range of flexible properties. Thus two types of PVC product are used in the building industry: (i) rigid or unplasticized PVC, which is based on a composition that does not contain plasticizer, and (ii) flexible or plasticized PVC. These types are defined by ISO.[2] Traditionally the rigid forms of PVC have been called UPVC, but ISO nomenclature states that such products should be termed PVC-U and flexible PVC compositions and products should be termed PVC-P. However. it is likely that the term UPVC will still remain in use for some considerable time. Vinyl is a term generally used in the USA (and to a much lesser extent elsewhere) for all PVC applications. However, for certain applications, such as flooring, it is widely used.

## 26.3 Production of PVC

PVC is a polymer of vinyl chloride, which is derived from oil and salt. By means of the electrolysis process the salt forms sodium hydroxide and chlorine and the latter is used in the manufacture of vinyl chloride. The monomer is nowadays commonly produced by the chlorination of ethylene to make ethylene dichloride, followed by pyrolysis and oxychlorination of the EDC to make vinyl chloride.

Although other routes to vinyl chloride are possible, such as by manufacturing vinyl chloride from acetylene using a coal-based route (particularly used in China), the ethylene route accounts for all of the PVC produced in Europe. The polymerization process can be carried out in a number of ways to produce polymers with different processing characteristics.[3] The main types of polymers used in products made for the construction industry are suspension and emulsion homopolymers of vinyl chloride.

The processes used are all designed to produce a white free-flowing powder which for suspension polymers, the most common form used for applications in the construction sector, have an average particle size of 110–170 μm. This powder can then be mixed with the additives referred to earlier to produce the required compounds.

There are a number of variables associated with the detail of the manufacturing methods for producing resins, but the most important of these as far as the product is concerned is the 'K' value. This is a number related to the molecular weight of the resin and has an important effect on the processing behaviour of the material and some of its physical properties. Most of the PVC products used in the construction industry are based on resins with 'K' values in the range 57–70.

In addition to the suspension and emulsion polymers, there are also some speciality polymers that are classified as graft copolymers which are used to impart enhanced properties, especially in impact strength, to the PVC product. Such polymers are sometimes used for products such as window profiles, but similar impact results can be obtained by using standard suspension polymers together with the appropriate additives. A special form of emulsion polymer, designated a paste polymer, is also manufactured; it is very fine and has a low uptake of plasticizer yet forms a stable dispersion when plasticizers are incorporated. Such dispersions are called plastisols. These polymers are used in a number of applications, but as far as the construction industry is concerned the major areas where such products are used are for coating steel and for many types of flooring.

## 26.4 Compounding of PVC

PVC resin cannot be used by itself, but has to he manufactured into compounds before it can be processed into finished articles. These compounds are a blend of PVC polymer with a range of specialized additives. There is a strong link between the formulation and processing of the PVC compound, its conversion into the product and the physical properties and in some cases, the fire performance of the product. There are two basic forms of compound that are used in the products for the building industry and these are differentiated by their physical form. The first is called a dry blend or powder blend and is, as its name suggests, a mixture of all the required ingredients in powder form. The second form of compound is called a granulate or pellet and is produced by melt compounding the dry blend mentioned above and pelletizing the resultant product. Both forms of compound are widely used in industry, although for larger scale operations such as pipe or window profile extrusion there is a predominant use of dry blends. A schematic diagram of the compounding of PVC is shown in Figure 26.1.

The equipment in which dry blends are manufactured is called a high-speed mixer and consists of a vessel in which a blade rotates at high speed, providing an intimate mix of the ingredients. The blade also heats up the ingredients by a frictional mechanism and the blend is discharged at a typical temperature of about 120°C into a cooling blender. In the cooling blender, which is

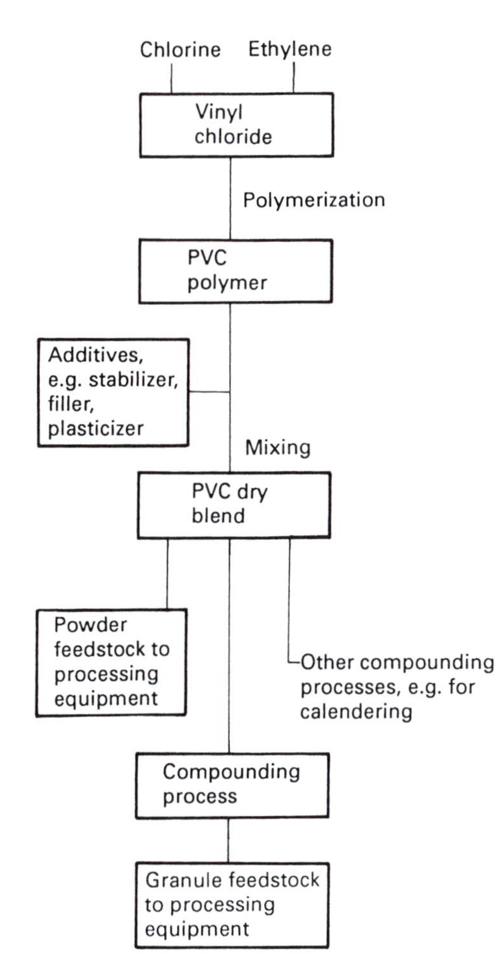


**Figure 26.1** Diagrammatic scheme for the preparation of PVC compound

generally cold-water jacketed, the blend is cooled to a temperature of around 50°C and then discharged.

In the case of granulate, a powder blend, which in many cases has not gone through the full process described above, is fed to various forms of extruder where it is compounded before being extruded through appropriate die plates which cut the resultant melt into cylindrical pellets.

## 26.5 Processing of PVC compositions

In order for any PVC product to be manufactured, appropriate processing equipment must be used. The equipment will vary according to the final product, but virtually all products for the building industry are made by extrusion, injection moulding, or calendering. The main rigid PVC products described, such as pipes, sheet and profiles, are produced by the extrusion process and this also applies to flexible PVC profiles and sheet. A more detailed description of processing techniques is available in the literature.[1,3]

## 26.6 Properties and applications of unplasticized PVC (PVC-U)

In its final product form, PVC-U is stiff, robust and light in weight, with good chemical and weathering resistance. PVC-U compares favourably with most other thermoplastics in standard flammability tests and it does not require the use of flame-retardant additives to achieve this performance. Compared with most other thermoplastics, PVC-U is relatively stiff and strong at room temperature. However, its upper continuous-service temperature is limited to about 60°C since PVC-U is essentially an amorphous material with a glass transition temperature (the temperature at which softening of the material starts to occur) of 80°C. Some important properties of PVC-U are given in Table 26.1.

**Table 26.1** Typical properties for PVC-U compositions

| *Property* | *Test method* | *Value* |
|---|---|---|
| Tensile strength @ 23°C | ISO527 | 45-55MPa |
| Tensile modulus(1% strain), 100 sc, 23°C | ISO R899 | 2200-3000MPa |
| Charpy impact strength | ISO 179 Notch Type A | 6 $Jm^{-1}$ to no break |
| Vicat softening point | ISO 306B | 70-85°C |
| Poisson's ratio | | 0,4 |
| Coefficient of linear expansion | | $6 \times 10^{-5}$°$C^{-1}$ |
| Coefficient of thermal conductivity | | 0.16$Wm^{-1}$°$C^{-1}$ |
| Relative density | ISO1183 | 1.38-1.5 |
| Oxygen index | ISO4589 | 45-50% |

The range of properties shown is dependent on the formulation of the compound.

PVC-U is resistant to a wide range of chemicals and is only readily attacked by some ketones, aromatic and chlorinated hydrocarbons and some nitroparaffins. There are comprehensive lists published on the chemical resistance of PVC and these should be consulted when relevant.[1,4] A summary or the chemical resistance of PVC-U to some common materials is given in Table 26.2.

The long-term properties of PVC-U are important when it is used in engineering applications,[5] and both the creep and fatigue properties of PVC-U have been extensively studied with the objective of allowing good design practice. Typical creep

**Table 26.2** Chemical resistance of PVC-U to some common chemicals

| *Chemical* | *Temperature (°C)* | |
|---|---|---|
| | *20* | *60* |
| Alcohol (40% aqueous) | ++ | + |
| Antifreeze | ++ | ++ |
| Dishwashing liquid | ++ | ++ |
| Detergent (diluted) | ++ | ++ |
| Furniture polishes | ++ | ++ |
| Gas oil | ++ | ++ |
| Lanolin | ++ | ++ |
| Linseed oil | ++ | ++ |
| Mineral oils | ++ | ++ |
| Moist acidic atmosphere | ++ | + |
| Moist alkaline atmosphere | ++ | + |
| Motor oils | ++ | ++ |
| Petrol | ++ | ++ |
| Sea-water | ++ | ++ |
| Vinegar | ++ | ++ |
| Vegetable oils | ++ | ++ |

++, resistant; +, practically resistant.

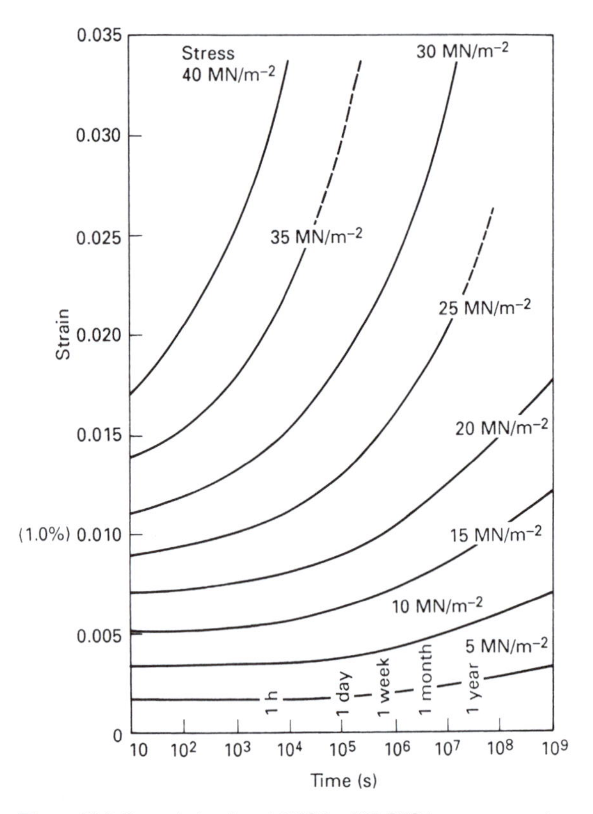


**Figure 26.2** Creep in tension at 20°C for K67 PVC in a pressure pipe formulation

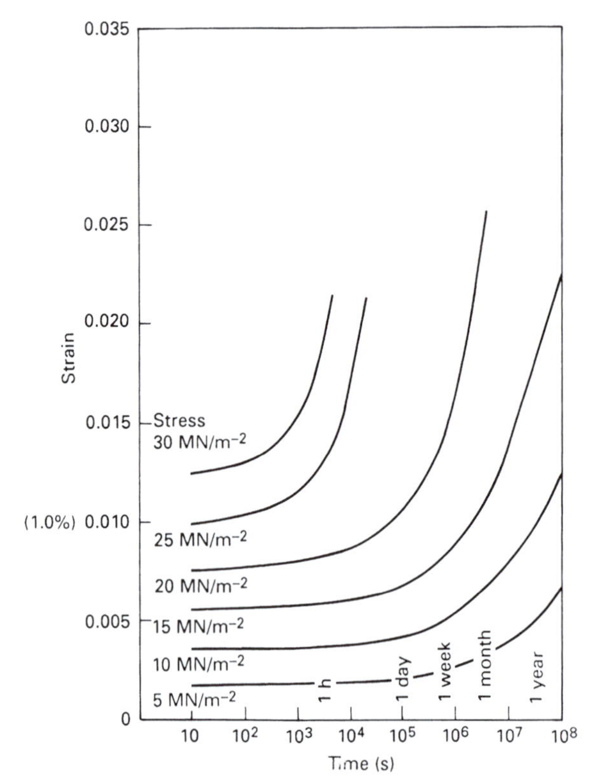


**Figure 26.3** Creep in tension at 50°C for K67 PVC in a pressure pipe formulation. Samples conditioned at 50°C prior to test for 83 days

data are given in Figures 26.2 and 26.3. Similarly, the influence of modulus with time and temperature has been measured since this can have a marked effect on the design of the PVC-U product (Figure 26.4).

The impact strength of PVC-U is often an important factor affecting its use and performance. Although PVC-U is inherently a tough material, its impact strength can be substantially enhanced by the use of special rubbery additives or by using specific copolymers that incorporate a rubbery material such as butylacrylate. Other types of modifier include methacrylate/butadiene/styrene (MBS) and chlorinated polyethylene. The effect of impact modification is shown in Figures 26.5 and 26.6, which show the performance of unmodified and impact-modified PVC using both un-notched and notched specimens. In addition, in Figure 26.6 the notched impact strength of some other thermoplastics is compared with that of PVC-U.

## 26.6.1 PVC-U pipe

PVC-U pipe is the largest single application for PVC worldwide and typically consumes 25% of all the PVC used. This means that in the EU in 2005 nearly 1.34 Mt of PVC was used for pipes of various types.

The general benefits of using PVC pipes in comparison with more conventional materials are:

1 low weight
2 high tensile and pressure resistance
3 high stiffness
4 high durability
5 good creep characteristics
6 excellent hydraulic characteristics
7 good abrasion resistance
8 good resistance against bacterial growth
9 non-flammable properties
10 easy to manufacture joints
11 flexibility – resistance to ground heave.

PVC-U pipes have been accepted for a wide range of applications and these are summarized together with a short description of the essential properties of PVC-U appropriate to that application in Table 26.3.

*26.6.1.1 Major types of pipe*

As can be seen in Table 26.3, there are numerous types of pipe, each of which uses one or more properties of PVC for its effectiveness. It will also be noted that virtually all forms of pipe are subject to British and/or European Standards. The properties and performance of these pipes are thus comprehensively characterized. Tests include a number of standard checks such as those for impact strength and softening point, and most forms of pipe have specialized testing relevant to the application, such as pressure testing for water mains pipes and deformation testing for drainage pipes. The different building methods and availability of competitive materials in each country often dictate the degree of penetration that PVC-U pipe has in each main sector of the pipe industry.

(a) *Pressure pipe.* PVC-U for this application has been widely used globally for over 40 years, mainly as water mains but also as pipe for chemical plants and, with increasing availability of advanced processing technology, has resulted in a very high-quality product. It is generally used in the range of 75–315 mm diameter, but larger sizes up to 630 mm are also installed.

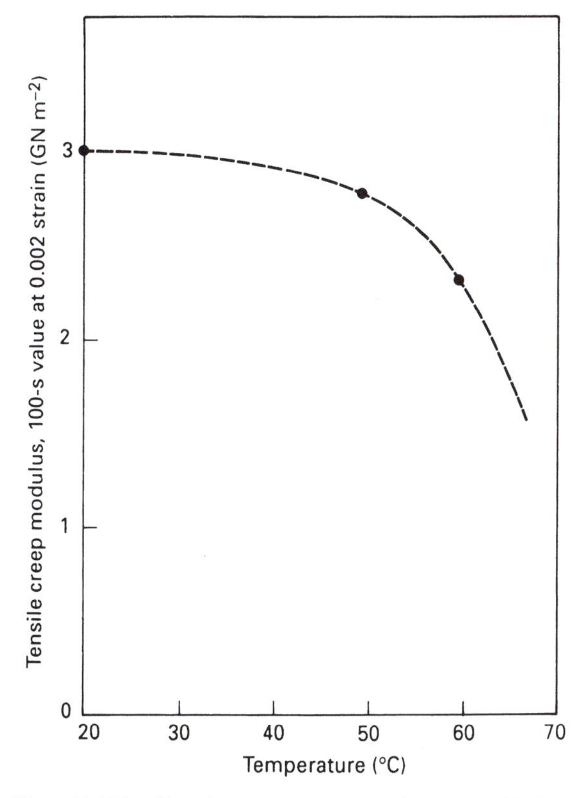


**Figure 26.4** The effect of temperature on the tensile modulus of PVC-U

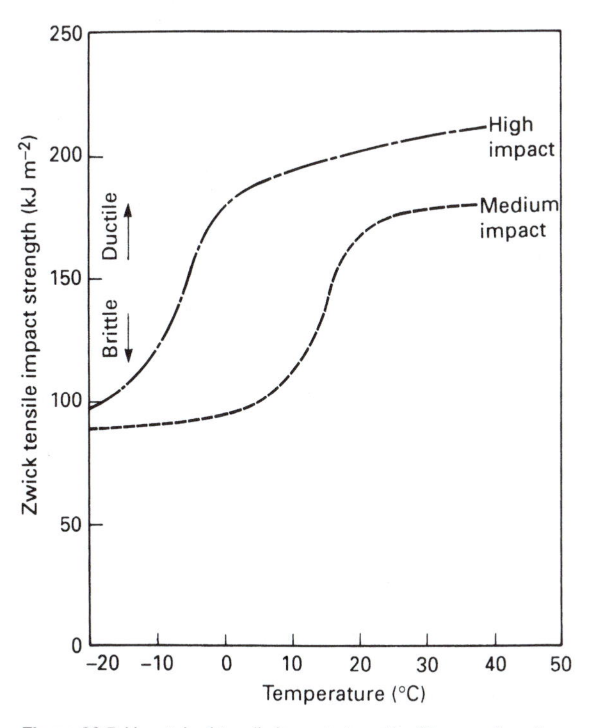


**Figure 26.5** Unnotched tensile impact strength of two grades of PVC-U compound

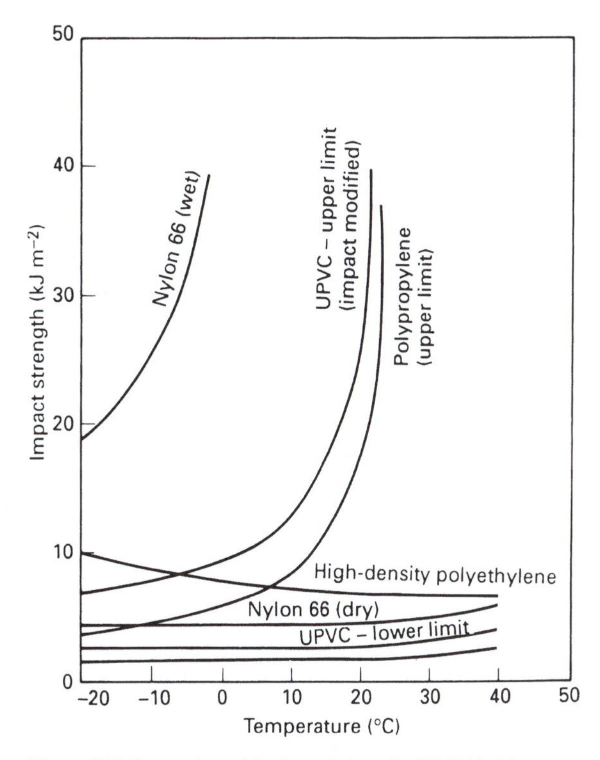


**Figure 26.6** Comparison of the impact strength of PVC-U with some other thermoplastics. Charpy specimens with 250 μm notch tip radius

(b) *Soil pipe and rainwater goods.* Soil pipe made from PVC-U has achieved strong penetration of the market across Europe and other parts of the world and is totally accepted by the users. A number of detailed designs of rainwater goods are available and a number of colours, e.g. black, white, grey and brown, are manufactured. Rainwater goods made from PVC-U are particularly used in the UK, where they have an overwhelming share of this market sector.

(c) *Drain and sewer pipe.* This has been a developing market globally over the past 40 years. Although competing against traditional ceramic pipe, PVC-U drain and sewer systems have strongly established themselves in the market because of their superior properties. Drain and sewer pipes are generally installed in diameters up to 315 mm but larger sizes can be manufactured and are used particularly in Germany.

(d) *Waste pipe.* This is a relatively minor application for PVC. Because of the higher temperatures experienced in such pipes with diameters in the range 32 mm to 50 mm, the basic PVC compound is modified with the addition of a proportion of chlorinated PVC polymer (PVC-C), which has the effect of increasing the softening point – and thus the maximum working temperature – of standard PVC materials. Such pipes can operate satisfactorily for short periods at temperatures approaching 100°C.

(e) *Land-drainage pipe.* This product strongly established itself a number of European markets for use as agricultural drainage pipe in the 1970s and 1980s. The pipe, which is strong but lightweight, competes successfully with porous clay tiles and is made in a corrugated form so that it can be coiled, allowing for easy transportation and installation.

(f) *Ducting.* This product is used for the containment of telephone wiring and other cabling. Telephone companies and cable TV operations are major users and it is also used for street lighting and road and railway signalling cables.

**Table 26.3** Descriptions of types of PVC-U pipe

| *Type of pipe* | *Important properties* | *Relevant Standards* |
|---|---|---|
| Potable water pipe and industrial pressure pipe | High tensile strength, high modulus, toughness, good creep characteristics, good flow characteristics, good chemical resistance, leak-tight joints. | EN1452-2, EN1452-3 (fittings), EN1456, EN15493, BS3505, BS4346-1 (fittings), ISO 4422 |
| Potable water pipe – molecular oriented | As above, but lighter and tougher. | ISO 16422 |
| Soil pipe | High modulus, toughness, good weathering. | EN1329-1 |
| Rainwater goods | High modulus, toughness, good weathering, non-corrosive. | EN607, EN12200-1, EN1462 |
| Waste pipe | Good temperature resistance, tough. | BS5255, EN1566-1 (PVC-C only) |
| Hot-water pipe | High temperature resistance, light weight. | ISO EN 15877-2 and 3, (PVC-C only) |
| Drain, sewer pipe | High stiffness, good creep characteristics, high tensile strength, leak-tight joints. | EN1401-1 |
| Land drainage pipe | High stiffness/unit weight, toughness, good weathering. | BS4962 |
| Twin wall + ribbed pipe | High stiffness, low weight, toughness, good creep characteristics. | EN13476-1, EN13476-3 |
| Foam core sewer pipe | High stiffness, low weight, good creep characteristics | EN13476-1, EN13476-2 |
| Ducting pipe | High stiffness/unit weight, toughness, good creep characteristics | EN50086-2 and -4 |
| Conduit | High toughness, good insulating properties | BS4607 |

(g) *Electrical conduit.* Because PVC is both non-conductive and non-corrosive, electrical arcing cannot occur with PVC conduit. This product is now widely used and has shown itself to be superior to steel conduit in many sectors of the building market. The PVC-U materials used to produce conduit and trunking have been tested for ignitability to BS476 Parts 5 and 12, and for flame spread to BS476 Part 7, when a Class 1 rating was achieved.

(h) *Hot water plumbing pipe.* This type of pipe is only made possible with the use of chlorinated PVC (PVC-C), which has a chlorine content of about 65% compared with 57% for standard PVC. This has the effect of increasing the maximum working temperature of standard PVC by up to 40°C. While it is used to a small extent in Europe, it finds much wider use in the USA.

*26.6.1.2 Special forms of pipe*

A number of special pipes which have been designed for particular applications and are finding their own niche in the market place. Examples of these are described below.

(a) *Ribbed pipe.* In order to maintain cost-effectiveness and competitiveness against traditional sewer-pipe products such as concrete or clay, it has been a requirement to provide a lighter weight PVC pipe which has within its design the desired properties essential for this application, the most important of which are stiffness and resistance to deformation under load. An extruded ribbed pipe where a series of peripheral ribs is produced on the outside of the pipe by a patented sizing technique fulfils this need, and these products are now being manufactured in diameters above 200 mm. The process, which saves about 40% by weight on a conventional pipe, has been developed by Uponor, which now has a number of licensees.

(b) *Foamed pipe.* This is yet another form of pipe mainly developed for the drain and sewer market, although in France such products are also used for soil pipe and ducting. Increasingly popular, the product typically consists of a foam core with co-extruded skins of solid PVC and provides a weight saving of about 30% over conventional pipe. Foam core pipes are now widely used in Europe and especially in Germany, France and the Republic of Ireland.

(c) *Twin-wall pipe.* In another design solution to provide a weight saving for sewer pipe, a manufacturing method has been developed using co-extrusion technology to produce a lightweight product with a smooth inside wall and a corrugated outer wall. Such pipes are increasingly common in the European market. In addition to their use for sewer and waste water drainage applications, such pipes are also being used for highway drainage.

(d) *Hollow-walled pipe.* For the same application as ribbed pipe, a form of pipe is produced by Wavin called Wavihol which consists of a pipe with closely spaced oval, longitudinal channels in the wall. This also gives strength combined with a substantial weight saving. This product is only found in France and only in the larger diameters in the range 500–630 mm.

(e) *Spirally wound pipe.* This type of pipe consists of profile shaped to give ribs when the profile is wound round a mandrel to manufacture a pipe. Although these pipes are used in the USA and Australia for sewer pipes and their relining, they have not yet been widely introduced to the European market.

(f) *High-strength, molecular oriented PVC pipe.* This type of product, produced by uniaxial or biaxial orientation, is designed to offer pressure resistance with lower wall thicknesses and a weight saving of up to 40%. The product also has superior impact and fatigue resistance as compared with its conventional counterpart, and has been successfully used in the water-mains market. In recent years its use has been increasing significantly.

(g) *Impact modified pipe or PVC-A.* There are various types of PVC pipes that contain a proportion of a rubbery material or impact modifier which is designed to enhance significantly the impact strength of the pipe. Such materials will also have improved ductility. The applications where these pipes are used include the potable water pipe sector, gas distribution (especially in the Netherlands), piping for compressed air, chilled water and aggressive slurries in mines, particularly in South Africa and pipeline renovation applications including sewer relining. In this last application the pipe is warmed and prefolded before being inserted into the pipe requiring renovation and, once installed is inflated to its normal diameter by means of steam under pressure.

### 26.6.2 PVC injection-moulded fittings

Most of the pipe applications described in the above sections require fittings to enable the pipe systems to be joined together and operate properly as a system. For example, there is a wide range of fittings associated with rainwater pipes and gutters, soil and drainage systems, to give these PVC applications a great deal of flexibility in use. Such fittings are produced from PVC by injection moulding and the products manufactured have essentially the same properties as the pipe or gutter with which they are associated.

### 26.6.3 Rigid PVC profiles

The generic term 'rigid PVC profile' applies to a very large number of products used in the building industry. Profiles made from PVC-U account for the largest use of PVC in Europe, amounting to around 28% of production or about 1.5 Mt. These profiles have often been developed as substitutes for existing products made from more traditional materials including wood, aluminium or steel. By virtue of its properties, PVC-U is relatively easy to produce in shapes of various sizes and quite high degrees of complexity and, therefore, it is impossible to cover here the full range of manufactured products. The main application areas where significant quantities of PVC are used are in window frames and roller shutters, but there is a very wide range of other outlets. These products are all manufactured from unplasticized PVC in various designs and often incorporate hollow sections to provide lightweight strength and thermal insulation. In addition, some applications such as roofline products, cladding and window sills are also available in foamed PVC (PVC-UE), which has a density of between 0.5 and 0.8 compared with a density of 1.4 to 1.5 for its unfoamed equivalent (see Section 26.6.3.3)

#### *26.6.3.1 Window profiles*

The use of PVC-U for window profiles has become the largest single application for PVC profiles over the past 30–40 years across Europe. The pattern of usage varies from country to country; in total around 1.6 Mt of PVC-U window profile is manufactured across the EU, the CIS and Turkey. This application is also the fastest growing sector for PVC. In the UK alone almost 300 kt of window profile is manufactured, while in Germany the volume is more than double this amount. France is also a major producer of PVC-U window profile, while more recently large markets have developed in Russia, Turkey and to a lesser extent Poland. The benefits of using PVC-U for window frames are summarized in Table 26.4.

The materials used for PVC window frames are predominantly impact-modified PVC where the PVC polymer is either copolymerized with a rubbery additive or the formulation contains a rubbery ingredient such as butylacrylate. Some window frames are also manufactured from special formulations that do not contain impact modifier. Typical properties of an impact modified window-frame material are given in Table 26.5.

One of the main benefits of PVC window frames, especially when compared with those made from aluminium, is that PVC has a very low thermal conductivity of 0.16 W m$^{-1}$ °C$^{-1}$, which is even lower than that of timber (Figure 26.7).

Window frames are fabricated by joining mitred profiles using hot-plate welding. By choosing the correct temperature and pressure, welds can be made that are as strong as the base material. The effects of temperature and pressure are shown in Figure 26.8.

A series of British and European Standards[6–9] relating to this application have been published and these cover the performance of the window as well as the physical, mechanical and durability characteristics of the PVC profiles.

**Table 26.4** Benefits of PVC-U window frames

| *Property* | *Benefit* |
|---|---|
| Low cost | Extremely competitive |
| High rigidity/unit cost | Lower product cost |
| Strong, tough, durable | Longer product life |
| Lightweight | Easy handling |
| Excellent weathering resistance | Long service life |
| Excellent chemical and weathering resistance | No maintenance |
| Self-extinguishing | Will not cause fire or enhance its development |
| Low thermal conductivity | No condensation |

**Table 26.5** Typical properties of a high-impact PVC-U window compound

| *Property* | *Test method* | *Value* |
|---|---|---|
| Tensile strength at 23°C | ISO527 | 44 MPa |
| Tensile modulus (1%) at 23°C | ISO899 | 2250 MPa |
| Flexural modulus at 23°C | ISO EN 178 | 2400 MPa |
| Flexural yield strength at 23°C | ISO178 | 76 MPa |
| Relative density at 23°C | | 1.4–1.5 |
| Charpy impact strength at 23°C | ISO EN179-2 | 14 kJ m$^{-1}$ (V notch) 40 kJm$^{-1}$ (U notch) |
| Retention of impact strength after accelerated weathering to 8 GJ m$^{-2}$ | BS2782 method 359 | >80% |
| Colour fastness | EN513 | Grey scale 4–5 |
| Vicat softening point | ISO EN306 | 81°C |
| Flammability (oxygen index) | ISO 4589 | 45% |
| Fire resistance | BS476-7 | Class 1 (most resistant) |
| Coefficient of linear thermal expansion | | $6 \times 10^{-5}$ °C$^{-1}$ |
| Coefficient of thermal conductivity | | 0.16 W m$^{-1}$ °C$^{-1}$ |
| Reversion | 1 h at 100°C | 1.6% |

#### *26.6.3.2 Other rigid profiles*

The remaining types of rigid PVC profiles all have essentially the same benefits as window frames, although in many cases the performance criteria are somewhat less stringent than for window frames. The major applications for these profiles in the construction industry include cladding products, cable management systems and roller shutters, but other outlets include products

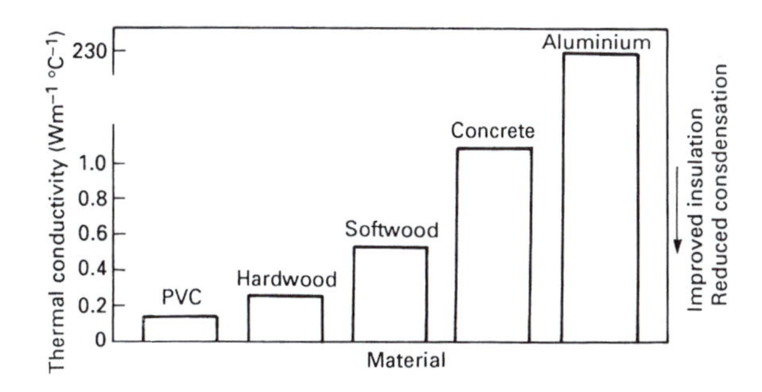


**Figure 26.7** Comparison of the thermal conductivity of PVC-U compound with some other materials

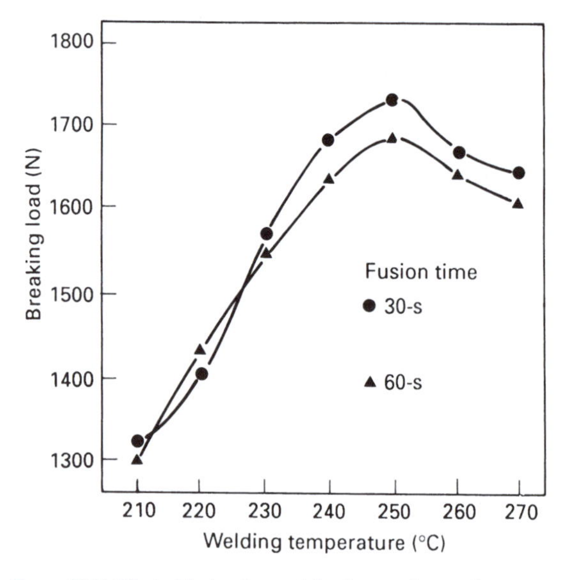


**Figure 26.8** Effect of fusion time welding temperature on the corner strength of PVC-U window profile

such as skirting boards, architraves and curtain rail, which find wide application in buildings. The case of roller shutters is a good example of an application being very successful in one country and not in another: in Germany PVC is used for the majority of roller shutters, while in the UK there is almost no use of PVC for this application.

*26.6.3.3 Cellular (foamed) PVC profiles (PVC-UE)*

There are a number of special properties that make PVC-UE a particularly useful product for some applications, even compared with PVC-U. These are:

1 low volume cost
2 good heat- and sound-insulation properties
3 good dimensional stability
4 low reversion and expansion
5 high stiffness/unit weight
6 availability of easy installation techniques.

These products are typically produced for building industry applications from rigid PVC formulations that have chemical blowing agents incorporated, the most common being sodium bicarbonate. PVC-UE foam profiles possess similar properties to rigid PVC, but the products typically have a density of about one-half to one-third that of its PVC-U equivalent. Cellular PVC foam products used for building applications tend to have a high degree of stiffness because they are generally made by co-extruding a solid PVC skin, which accounts for around 15% of the total linear weight, over a foam PVC core. A British Standard for these products is also available.[10]

Typical applications of PVC-UE profiles include fascias, barge boards, soffits, window boards and sills, window sub-frames, trim sections and cladding.

## 26.6.4 Rigid PVC sheet

Flat or corrugated PVC sheet is used extensively in the building sector for roofing applications and partitioning. The product is available in a range of colours, either clear or opaque and varying in thickness from about 1 to 25 mm. The main service requirements are typical of those that have to be met by other PVC applications used outdoors, i.e. good weathering characteristics and adequate robustness, but in the case of transparent sheet, which is the dominant product sold in the marketplace, it is necessary to ensure a high degree of clarity that is retained over many years. The fire properties of PVC sheet are also of importance, particularly where the product is used for applications such as wall lining. This aspect is dealt with in Section 26.9.

The applications of PVC sheet are numerous and include glazing, lining for tunnels and roofs, roof and dome lights, and roofing, which probably consumes the greater part of the PVC found in this sector of the market.

Rigid PVC sheet for building is produced by one of two methods. Most sheet, whether flat or corrugated, is produced by using a conventional extrusion process fitted with an appropriate die, a polishing stack and corrugating equipment, if required. It is also possible to specify products with either a matt or a gloss finish or embossed with a pattern. There is also a special type of biaxially oriented sheet available which is claimed to impart enhanced impact resistance together with good retention of the property at temperatures down to –40°C. The performance, properties and dimensions of PVC corrugated sheeting are specified in EN1013 Part 3.[11]

Another form of rigid sheet is a double-skin box section of clear rigid PVC which is made from an impact-modified formulation. This product is used for light partitioning in buildings such as sports halls.

Foamed PVC sheet is a growing application and its uses in the construction sector include partitioning, cladding, various forms of panelling and false ceilings, where it replaces traditional products such as plasterboard, hardboard and timber. Such sheet has excellent thermal and acoustic properties.

## 26.7 Flexible PVC products

The number of flexible or plasticized PVC products used in the building industry is considerably less than in the rigid sector because, as their name implies, these products are flexible and therefore do not have the rigidity that makes unplasticized PVC such a useful construction material. Nevertheless, there is still significant usage of flexible PVC in this sector and the more important of these uses are covered in this section. Furnishings such as PVC wall coverings and leather cloth are not included in this chapter.

Because of the influence of plasticizer, which can vary the softness of the product, the properties of plasticized PVC will obviously vary according to both the amount and type of plasticizer present. The typical properties given in Table 26.6 give an indication of the range covered by plasticized PVC products.

### 26.7.1 Vinyl flooring

An important application of flexible PVC in building, accounting for about 5% of PVC usage in the EU, is flooring, which comes

**Table 26.6** Typical properties of plasticized PVC

| *Property* | *Unit* | *Value* |
|---|---|---|
| Relative density | | 1.19–1.68 |
| Tensile strength | MN m$^{-1}$ | 7.5–30 |
| Elongation at break | % | 140–40 |
| Hardness (Rockwell) | | 5–100 |
| Brittleness temperature | °C | –20 to –60 |
| Volume resistivity at 23°C | Ω cm | $10^{10}$–$10^{15}$ |
| Ageing resistance | | Excellent |
| Ozone resistance | | Very good |

basically in two forms. The older and very established use of PVC for floor tiles and solid flooring is made by the calendering of PVC compositions which generally contain significant proportions of mineral fillers. This product is often used in industrial buildings and in hospitals, offices and schools. These products are manufactured in a wide variety of designs, have excellent wear characteristics and can be easily cleaned. Many PVC flooring products are covered by European Standards,[12] while specialist products such as safety flooring are subject to extra specifications.

The more modern types of vinyl flooring widely used in the domestic sector and some specialized industrial and commercial applications are much more complex products which are available in a number of configurations. In its simplest form it is a two-layer product where there is a wear layer made from PVC backed by a thicker foam layer. The wear layer will generally be printed and will incorporate a thin clear PVC coating to protect this. The flooring made from a multi-layer construction generally consists of a clear wear layer, a thin foam layer backed by fabric and there is also occasionally an additional layer between the foam and the fabric. Again a wide range of designs is available together with the added option of embossing, which greatly adds to the types of products available. The main properties of this type of flooring are rapid recovery after compression, good wear and soiling resistance and good dimensional stability, together with a 'feel' that the solid type of PVC flooring does not possess. There are also many specialized uses for vinyl flooring materials which can be designed to provide anti-slip characteristics, low acoustic properties, and electrostatic discharge protection. This range of properties allows the use of vinyl flooring products in hospitals, cleanrooms and wet areas.

### 26.7.2 Flexible tubes and profiles

Although flexible pipes are used to a very minor extent compared with their rigid equivalents, some special products do find their way into building applications. In particular, heavy duty hose, which is reinforced by a synthetic fibre braid and consists of a plasticized PVC outer layer designed to give high abrasion resistance, is found useful in association with various forms of building equipment. Other forms of this reinforced hose are where the rigid support is supplied by a continuous coil of rigid PVC over which the flexible PVC is extruded, giving the product the required flexibility. An important application for plasticized PVC profile used in the construction industry is that of water stop. This is a fairly thick extruded section which, when embedded in concrete, forms a watertight diaphragm that prevents the passage of fluid through the joint. It is used for the leak-proofing of joints in concrete structures such as dams and water and waste treatment plants. There are a number of other important uses for flexible PVC profiles which include an increasing use for gasketing in association with PVC window frames and increasingly as a co-extruded strip together with rigid PVC, draught excluders, the coating of wire used as chain-link fencing and a wide variety of trims and nosings for internal parts of buildings.

### 26.7.3 PVC cable covering

Plasticized PVC has been used for more than 50 years as both insulation and sheathing material on a wide range of cables. It is the most widely used thermoplastic in this application, accounting for about 7–8% of all PVC used in Europe and a quarter of all flexible PVC. While cables are used in a large number of applications in addition to the construction industry, the use of cables for wiring in buildings and for machinery and appliances are important sectors.

Cables can also be covered with a range of other materials including polyethylene, cross-linked polyethylene and thermoplastic rubbers, depending on the performance required and the type of cable. However, PVC is by far the dominant material, with about 55% of the overall market for thermoplastic insulation and sheathing in Europe in 2005.

**Table 26.7** Main areas of use for PVC-P cable covering in the construction sector

| *Sector* | *Energy systems* | | *Housing* |
|---|---|---|---|
| | >1 kV | <1 kV | |
| Use | ++ (1) | ++ (2) | ++ (1,2) |

++ Major use; + minor use. 1 As sheathing; 2 as insulation.

Cables are generally classified as low-voltage (up to 1 kV), medium-voltage (1 to 30 kV), and high-voltage (greater than 30 kV). PVC is ideal for low- and many medium-voltage applications up to about 6 kV. At higher voltages, materials with better dielectric properties, such as polyethylene, are used for insulation, while PVC may still be acceptable for the sheathing (Table 26.7).

The reason why PVC has achieved this success is that it combines a range of excellent properties with low cost. PVC materials can be tailored to meet a range of specific needs such as electrical properties, abrasion resistance, flexibility, tear resistance, fire resistance and cost, and can be formulated to provide satisfactory performance over a wide temperature range. More detailed information can be obtained in PVC for Cables – The Specifiers Choice.[13]

Some other materials, such as polyethylene, have better electrical properties, while others will have better performance at high temperatures. However, there are often installations where a number of materials will be seen as possessing the correct electrical and mechanical properties. In these cases, the cost of the cable becomes a significant factor – an area where PVC has a proven advantage.

Cable design requires the use of an insulation layer together with an external sheathing layer. In some designs, such as armoured cables, bedding or 'filler' layers are also employed to provide special performance. The inherent flexibility of PVC compound formulation provides the cable designer with a comprehensive range of property options, allowing the optimization of the product on both performance and cost criteria.

The PVC formulation can be modified to provide various degrees of softness, tensile strength and elongation, abrasion and tear resistance, with options of glossy or matt surface finish. The versatility of PVC allows for operation under a wide range of temperatures by means of specific formulation design. Very flexible formulations using special plasticizers can be used which enable PVC covered cables to remain flexible down to –65°C. In normal practice a minimum working temperature of –15 to –40°C is selected, which allows the use of low-volatility, heat-stable formulations that can extend the maximum working temperature to a range of 70 to 85°C, and even higher in special cases.

Dielectric performance is the key electrical property of cable insulation. This is defined in terms of volume resistivity, electric strength, dielectric constant and dissipation factor (Table 26.8). PVC insulation maintains its integrity under short circuit conditions below temperatures of about 150°C. This means that it is possible to optimize the compound to provide specific electrical performance. For example, where PVC is being used at higher voltages (when dielectric heating and energy loss become

**Table 26.8** Typical electrical properties for PVC-P cable covering

| *Volume resistivity (Ω cm)* | *Electric strength (kV mm$^{-1}$)* | *Dielectric constant* | *Dissipation factor (IEC250)* |
|---|---|---|---|
| $10^{11}$–$10^{15}$ | 14–20 | 4.5–5.5 | 0.06–0.11 |

important design considerations), the formulation ingredients can be selected to provide specific properties.

The fire hazards associated with cables and cable covering are a complex subject and have been the subject of much research. An overview is provided in Section 26.9.

### 26.7.4 Flexible sheeting

Flexible PVC sheeting, which can be calendared or extruded, is used in the building industry as a liner for canals, water reservoirs, landfill sites, tanks and pools. Another important application is roofing membranes replacing bitumenized felt.

For this application the membrane is often reinforced with polyester or glass fibre to provide enhanced mechanical properties. Its main benefits are that it has good low-temperature flexibility, good weathering performance, a high resistance to puncture and impact and, especially in roofing applications, it has excellent resistance to fire propagation. In some cases a lacquer coating is applied to the top surface to resist staining from airborne dirt and pollutants. The above applications utilize calendared sheeting and in its extruded form flexible PVC sheeting is most often seen as a clear product used for industrial doors. This product can be 5 mm thick or over and up to 1.5 m wide. Sometimes the doors are made up of strips, each 200–300 mm wide.

## 26.8 PVC/metal sheet laminates

PVC/metal sheet laminates find an important application in the construction industry, mainly being used for walls and cladding. PVC-covered steel provides excellent abrasion, weather and chemical resistance while at the same time providing an aesthetic function, being available in a wide range of colours. Such coatings have exhibited lifetimes of up to 40 years. The coating can be made either from a flexible PVC film or by using a coating of PVC plastisol.

## 26.9 Fire performance of PVC

Both rigid and flexible PVC are very widely used in the construction industry and the fire performance of the product is dependent to some extent on the formulation and, in particular, on whether the product being tested is made from rigid or flexible PVC. The chlorine content of PVC is 57% and this contributes towards efficient flame-retarding properties. In PVC-U applications the PVC polymer element is 80% or more, and therefore the fire behaviour of PVC determines the performance of the product. In plasticized PVC applications the presence of the plasticizer increases the flammability of the composition and product. In the case of critical applications it is possible to incorporate special flame-retardant plasticizers, and another option is to incorporate special flame-retardant additives.

The behaviour of plastics in fire tests and fire scenarios is a complex issue and it is not possible in a review of this nature to provide all the details required to develop a comprehensive understanding of the parameters involved and the assessment of suitability of PVC in different applications. A more detailed review of the subject can be found in some PVC industry booklets.[14,15] More detailed sources are provided in these which give extensive detail and test results of PVC fire behaviour in all construction applications.

### 26.9.1 Fire testing

A variety of laboratory tests are available to determine the combustion properties of PVC, and the results of these taken together give an overall perspective of the behaviour of PVC under actual fire conditions.

The key parameters for understanding the way fires grow from ignition to a fully developed fire are ignitability, rate of flame spread, rate of heat release and smoke generation. Fire testing is carried out both on materials and on the final products. Material tests are often small-scale bench tests which bear little or no relationship to the finished product. However, over recent years a great deal of work has been undertaken by international bodies such as the ISO and the IEC (International Electrotechnical Commission) to develop larger scale tests. The data from such testing is increasingly being used in fire safety engineering to evaluate potential fire hazards.

From this work it has become clear that appropriate PVC formulations are able to meet the toughest requirements. In particular there has been a seven year programme on cable covering which demonstrated that fire-retardant PVC insulation and sheathing are securely positioned in the range of materials used in high-risk, high-fire-load installations.

PVC is difficult to ignite using most common ignition sources. The ignitability of unplasticized PVC is one of the lowest among thermoplastics and even the ignitability of plasticized compositions is better than that for wood (Table 26.9). The most important cause of fire death is the rapid release of heat from a burning material, and PVC is among the materials with the lowest rates of heat release.

A burning object will spread fire to nearby products only if it gives off enough heat to ignite them, and this heat release is the most important property in terms of fire hazard. Tests[16] have shown that most materials are more prone to ignite other products than is PVC. PVC also exhibits a very good performance when tested against the various flammability tests involving spread of flame, and many PVC compositions meet the best classifications for combustible building materials.

Although under burning conditions both flexible and rigid PVC will produce more smoke than wood, the smoke densities under non-flaming conditions for these materials are very similar, and tests that have been carried out with PVC-U test panels to simulate real fire conditions have shown that these produce significantly less heat and smoke than plasterboard or wood. The

**Table 26.9** Ignition properties of PVC

| | *Test method* | *PVC-U* | *PVC-P* | *Wood* |
|---|---|---|---|---|
| Flash ignition temperature | ISO871 | 400 | 330–380 | 210–270 |
| Self-ignition temperature | ISO871 | 450 | 420–430 | 400 |
| Oxygen index | ISO4589 | 50 | 23–33 | 21–23 |
| ISO radiant core | ISO5657 | | | |
| Ignition time @ 30 kW $m^{-2}$ | | 112 | 50–75 | 30–90 |
| Ignition time @ 50 kW $m^{-2}$ | | 33 | 17–26 | 4–30 |
| Needle flame test | IEC 60695-11-5 | Non-ignitable | Ignition only at high levels of plasticizer | Ignitable in less than 20 s |

toxicity of PVC smoke is within the normal range of all currently used materials. PVC is unusual in that, when it burns, it releases hydrogen chloride. However, hydrogen chloride odour is easily detectable at less than 1 ppm and, in common with most burning materials, it is highly irritant. However, it does not incapacitate or become dangerous until it reaches concentrations much higher than those measured in real fires. While the presence of HCl in fire effluents does pose a hazard, it is carbon monoxide, emitted by all organic material when burning, that is the most dangerous toxic material produced in fires.[17]

Hydrogen chloride is also unique among fire gases in that its concentration decays by rapidly reacting with many construction surfaces. Where this does not occur, quick and effective neutralization of corrosive damage is a priority and can save electric and electronic devices and structural elements in buildings. Research has shown that the environment of all fires is corrosive, regardless of the fuel or of the metal exposed, and that there is no correlation between the acid gas emissions of materials when they burn and the corrosive potential of their smoke.[18]

A number of other tests have been carried out to show the excellent fire performance of unplasticized and plasticized PVC. For example, the Fire Research Station[19] carried out tests on windows made from wood and PVC and concluded that there was no difference in fire performance under the conditions used, while PVC flooring performed well in a radiant panel test (ASTM E648-78).[20]

In full-scale room–corridor testing, PVC flooring has been found to evolve much less heat than all other types of flooring with the exception of wool carpet, which gave comparable readings to that of the PVC. A wide variety of materials have been tested in these trials and the PVC flooring and wool carpet were the only ones that did not reach the flashover stage and where the fire did not extend to the full length of the corridor. In radiant panel testing for flooring, which forms the basis for the Euro-Classification of floor coverings, PVC achieves the highest level of fire performance and is considerably better than other thermoplastic types of flooring. Fire classification for construction products is designated in BS EN 13501.

The UK Building Regulations[21] allow the use of PVC pipe to pass through the walls of buildings – the only thermoplastic material that has this approval. PVC-U sheeting is also highly rated in these regulations and is the only thermoplastic material to satisfy the TP(a) rigid requirements for all end-use thicknesses. This excellent performance means that PVC-U sheeting can be used in almost all thermoplastic rooflight and lighting diffuser applications as well as for external glazing.

Small-scale ignition tests are not totally appropriate for applications such as cables, where the cables represent a high or concentrated fire loading and where the ignition source may be of a very high energy. Because of this a larger scale, a more stringent flame spread test – IEC 332 Part 3 – has been developed and in this test cables are bunched together and installed on a vertical ladder of the type used in power stations. An extensive research programme[22] has been carried out by the PVC industry to evaluate the performance of PVC cable covering on actual cables. The results of this work showed that both the cable insulation and sheathing materials are key to the results of the test. The results showed that while standard PVC insulation was acceptable in low-intensity situations, as the intensity of the fire increased fire-retardant PVC insulation and sheathing offered the maximum benefit in resisting flame spread. Cables employing these materials are widely used in large-scale power and communication installations.

## 26.10 Future developments of PVC in the building industry

The use of PVC has been established in construction applications for well over 40 years, so it can be regarded in many ways as having reached maturity. However, new ways of using PVC are continually being found.

One example of this is the development over the past 10 years of wood–plastic composites (WPCs). The global market for WPCs now stands at 500,000 tonnes per year,[23] with 85% of this in the USA where the main application is in decking. Growth rates over the past few years have been in double digits and even over the next few years, growth rates of more than 10% are expected. WPCs based on PVC represent about 20% of the WPC market in the USA, and markets have recently emerged in Europe and Japan as the manufacturing technology has spread around the world. Applications in construction include doors, windows, siding and railings, as well as decking; PVC WPCs have advantages of higher stiffness and better fire properties than other WPCs.

Nanotechnology is still in its infancy, but already it has found applications within the plastics packaging and automotive areas. It is only a matter of time before this is extended into construction products. The application of nanotechnology to PVC has come a little later than it did with other thermoplastics, partly because of the higher melt viscosity of PVC, but also because the clays on which the early research work was done were not suitable for PVC, making processing difficult. Alternative nano compounds have been found for PVC and work is going on to look at their application at both the compounding and polymerization stages of production.

So, in spite of more than 40 years of developments in PVC, it seems that the continuing work of the raw material manufacturers, processors and machinery manufacturers is likely to lead to the development of new or improved PVC products for the construction industry.

## 26.11 Sustainability, environmental impact and recyclability

The PVC industry is gaining recognition for its sustainable development progress. In its various building applications PVC is extremely durable, and at the end of life it can be recycled without losing essential qualities.

It is widely used to satisfy people's basic needs in products that make life safer, more comfortable and more enjoyable. In terms of economic and social development, the European PVC industry supports 23,000 businesses and jobs for half a million people.

It is widely accepted that there are three components to sustainability – the so called 'three pillars', which take into account economic, social and environmental aspects.

The economic sustainability parameters have been evaluated by the use of life cycle analysis (LCA) which has included whole-life costs for PVC products. Many of the applications considered in this chapter are designed as long-life applications such as pipes and windows, i.e. in excess of 30 years, while the majority of the other products used in the building industry are medium-life applications such as flooring, with lifetimes estimated at about 20 years. A number of LCAs have been carried out over the past 25 years examining the use of PVC in various building applications such as rainwater pipes, windows, cables and flooring and comparing PVC with alternative materials that could also be used for the relevant application. A UK Government study[24] demonstrated that PVC products have significantly lower whole-life costs than alternative products with equivalent performance. Other studies – such as those carried out by GUA (Gesellschaft fuer umfassende Analysen GmbH)[25] – have demonstrated significant cost and performance advantages for a variety of PVC building products. With energy ratings for products becoming of increasing interest combined with the UK Government's Sustainable Homes initiative, the window profile manufacturers and associated technologies have put considerable effort into ensuring the best energy ratings possible. It is now possible to find a number of PVC window systems in the UK that have the highest possible Grade A rating.

In the flooring sector many vinyl floor coverings are formulated for ultra-low volatile organic compound (VOC) emissions, which means that these are only detected at extremely low levels and fall well within recommended limits for good indoor air quality. Some flooring products in the UK have achieved the highest possible rating over a 60 year service life using an environmental profiling methodology developed by the Building Research Establishment (BRE) and assessed in accordance with its Green Guide to Specifications.

Social sustainability for PVC can be demonstrated by the fact that PVC products combine high performance with affordability, which enables many essential products to be made available for a wide cross-section of the community. A good example of this is the provision of clean water supplies flowing through PVC pipes which offer widespread protection to all the community. PVC is also used for many building applications because of its low maintenance requirements in comparison to traditional materials. PVC pipes and window profiles are good examples.

The main challenge to any examination of sustainability resides in its environmental performance. The EU published a Green Paper on PVC in 2000[26] which identified ecological sustainability issues that the PVC industry should address. The industry has devoted much effort over many years to ensure that its manufacturing processes meet existing legislation and industry standards. To this end all European companies involved in the production of VCM and PVC have drawn up a charter for manufacture which is independently audited on a yearly basis. While all chemical and other industrial processes have an effect on the environment, it is a fact that the emissions to air and water arising from the PVC manufacturing process have decreased markedly over the past 15 years or so. It is the industry's view that further reductions can and will be made over the next few years.

One of the cornerstones of the commitment is protection of the environment, and the PVC resin manufacturers have signed a charter whose aim is to reduce environmental impact and improve eco-efficiency through compliance with the stringent terms of this charter. An independent organization of repute – Det Norske Veritas – monitors compliance with the terms of the charter on a regular basis and publishes its findings, which can be seen on the ECVM website (see Section 26.13).

Two priority areas requiring attention by the industry were: (1) the types of additives used in some PVC formulations (especially some stabilizers based on heavy metals, and some plasticizers used for flexible PVC applications; and (2) improving the waste management of PVC products.

Over the past 25 years there has been an extended debate over the effect that the production and applications of PVC have had on the environment. A variety of concerns have been raised over this period, a number of which achieved a good deal of publicity. The PVC industry, especially in Europe and the USA, was able to show by the use of scientific fact that many of these concerns were misplaced and that in a number of ways PVC exhibited a lesser environmental impact than many other materials.

In 2000 the industry in Europe brought together all the interested organizations involved in the manufacture of PVC and its applications to issue the Vinyl 2010 Voluntary Commitment. This provided a 10 year plan on sustainable development, with defined targets and deadlines to enhance sustainability of PVC products and production over the full lifecycle. The initiative uses a lifecycle approach to the production and use of PVC by seeking to minimize the environmental impact of the production process, using chemicals responsibly and reducing the environmental impact of products once disposed of, such as by recycling.

This voluntary commitment is independently audited on a regular basis and has wide support and encouragement. The Monitoring Committee includes members of the EU Commission and European Parliament and the initiative has received a lot of support and encouragement from this committee, which also includes the chairperson of the Green Grouping in the European Parliament.

While at one time doubts were expressed about the future use of PVC there is now no real pressure against it as a material and in its diverse applications, many of which are essential to the health and safety of the community. A request for a Directive in the EU on PVC was rejected in 2007 by an overwhelming vote in the European Parliament, which has subsequently demonstrated strong support for its continued use.

A core part of policy of Vinyl 2010 has been to promote dialogue with stakeholders including the EU, the European Commission, trade unions and consumer associations, and to promote openness and disclosure and share best practice and good product stewardship. In 2004, Vinyl 2010 became a member of the UN Partnership for Sustainable Development. This initiative creates partnerships with organizations that work to implement the sustainable development goals set out by various UN resolutions such as Agenda 21, Rio +5 and the Johannesburg Plan of Implementation.

The European PVC industry has put in place a major programme to address these two priority areas while also continuing progress on sustainable development related to maximizing the efficiency of the manufacturing processes – including sharing and adopting best practice. This programme, known as Vinyl 2010, includes commitments to (a) minimizing risks to the environment and human health – for example, by setting targets to reduce and finally eliminate the use of lead stabilizers, (b) maximizing recycling and recovery rates for PVC products, and (c) realizing socially responsible best practices.

The recycling initiative has set challenging targets to be met by the pipe, window profile and flooring sectors, and good progress is being made in these applications. PVC cable covering has been recovered for many years because of the need to reclaim the valuable copper conductor wiring, and the recovered material is used in a wide variety of applications. Over 35 kt of PVC flooring products from European sources is now being recycled and being used for the backing of new flooring and carpets. Many thousands of tonnes of post-use pipe and window profile products is also being recycled across Europe and being fed back into the PVC processing chain, and significant progress is being made in other application sectors. The whole process is independently audited and more extensive details on the progress of recycling activities can be obtained from the Vinyl 2010 literature – see Section 26.13.

## 26.12 References


1 Titow, W. Y., *PVC Technology*, 4th edition. Elsevier Applied Science Publishers, Amsterdam, The Netherlands (1984).
2 International Organization for Standardization. ISO 1163: Part 1 1995 Plastics – Unplasticized poly(vinyl chloride) (PVC-U) moulding and extrusion materials – Part 1: Designation system and basis for specifications. ISO2898 Plastics – Plasticized poly(vinyl chloride) (PVC-P) moulding and extrusion materials – Part 1: Designation system and basis for specifications 1996.
3 Burgess, R. H. et al. (Eds) Manufacture and Processing of PVC. Elsevier Applied Science Publishers, Amsterdam, The Netherlands (1982).
4 British Standards Institution. CP 312-1(1973), CP312-2(1973): Code of practice for plastics pipework (thermoplastics material).
5 European Vinyls Corporation. Engineering design with unplasticized polyvinyl chloride. EVC19 (1989).
6 British Standards Institution. BS 7412(2002), BS EN 14351 (2006). Plastics windows made from unplasticized polyvinyl chloride (PVC-U) extruded hollow profiles.
7 British Standards Institution. BS EN12608 (2003). Unplasticized polyvinylchloride (PVC-U) profiles for the fabrication of windows and doors. Classification, requirements and test methods.

8 British Standards Institution. BS7722 (2002) Surface covered PVC-U profiles for windows and doors.
9 BS6375 pts 1, 2, 3 (2004) Performance of windows and doors.
10 BS7619: 1993 Specification for extruded cellular unplasticized PVC (PVC-UE) profiles.
11 British Standards Institution. BS EN 1013-3 (1997): Light-transmitting profiled plastic sheeting for single skin roofing – Part 3: Specific requirements and test methods for sheets of polyvinyl chloride (PVC).
12 British Standards Institution. BS EN 649: 1997. Resilient floor coverings. Homogeneous and heterogeneous polyvinyl chloride floor coverings.
13 PVC for cables – The Specifier's Choice. European Vinyls Corporation brochure EVC48, 1997.
14 British Plastics Federation. PVC in Fires (1996).
15 The Vinyl Institute. USA. Fire Properties of PVC (1996).
16 Smith, E. E. Ignition. Heat release and noncombustibility of materials. ASTM STP 502 (A. F. Robertson, Ed.). American Society for Testing and Materials, Philadelphia, PA. 119 (1972).
17 M. M. Hirschler. Fire Retardance, Smoke Toxicity and Fire Hazard. In *Flame Retardants* 1994, pp. 225–237.
18 J. D. Ryan, V. Brabauskas, T. J. O'Neill, and M. H. Hirschler. Performance Testing for the Corrosivity of Smoke (1990). National Technical Information Service Report.
19 Fardell, P. J., Murrell, J. M. and Rogowski, Z. W. The Performance of UPVC and Wood Double Glazed Windows. Fire Research Station (UK) and British Plastics Federation, London, UK (1984)
20 American Society for Testing and Materials. ASTM E648:78.
21 UK Building Regulations. Approved Document B. Fire Spread. London, UK (2000).
22 M. A. Barnes et al. A comparative study of the fire performance of halogenated and non-halogenated materials for cable applications. Fire and Materials 20, 1–37 (1996).
23 Boyer, C. PVC, the Definitive Choice for Sustainable Natural Fibre Composites. Presented at the 9th International Conference on PVC, Brighton (Institute of Materials), UK, 26–28 April 2005.
24 UK Department of the Environment, Transport and the Regions. Life Cycle Assessment of Polyvinyl Chloride and Alternatives. February 2001.
25 GUA – Gesellschaft für umfassende Analysen. The Contribution of Plastic Products to Resource Efficiency. Vienna, Austria, January 2005.
26 European Union. Commission of the Communities. Green Paper – Environmental Issues of PVC. COM (2000) 469 final.

## 26.13 Websites

Detailed information on PVC and its applications is available from various sources on the Internet. Some of these sites contain a great deal of environmental information and details of what is being done in the PVC field to promote sustainability and recyclability. A few sites offer some studies of technical issues affecting specific applications.

The sites given below, while not exhaustive, will be useful to specifiers who want to know more about PVC materials in construction.

1 www.vinyl2010.com – general information on the environmental issues relating to PVC
2 www.ecvm.org – the major European PVC trade organization
3 www.plasticseurope.org – the major European trade association for all plastics manufacturers – including PVC – in Europe
4 www.recovinyl.org – PVC recycling activities in Europe
5 www.pvc4pipes.org – PVC pipes. Many technical papers
6 www.plasticsconverters.eu – general information on the manufacture and use of PVC products
7 www.eppa-profiles.org – general information on the use of PVC profiles in Europe
8 www.teppfa.org – information on the European plastics pipes sector including PVC
9 www.agpu.de – environmental issues affecting PVC in Germany
10 www.bpf.co.uk – the major UK trade association, with a number of sector groups including a Vinyls group, a Windows group and Cellular Foam group
11 www.ibk-darmstadt.de – (IBK = Institute for Building with Plastics) – the major institute covering the use of plastics in building in Germany
12 www.vinylinfo.org.– The Vinyl Institute website – the major USA PVC trade association
13 www.vinylsum.org.uk – the website of a network of university and industrial partners which is addressing some of the sustainability challenges
14 www.pvcplus.de – a website based in Germany with the objective of promoting PVC applications in Europe
15 www.beamainstallation.org – the UK trade organization for the installation of cable conduit and trunking. The site has a very useful list of FAQs about the fire performance of PVC-U

# 27 Thermosetting Resins

**Shaun A. Hurley** BSc PhD MRSC
Consultant, formerly with Taylor Woodrow Technology Centre

## Contents

## 27.1 General description

### 27.1.1 Thermoplastics, thermosets and elastomers

Organic polymers are formed when relatively small molecules (monomers) are combined to form very large, high molecular weight compounds (macro-molecules) that have repeating structural units. The chemical nature of these units provides the generic description of the polymer (e.g. vinyl chloride leads to poly(vinyl chloride)) and determines many of its fundamental properties.

The chain-like organic polymer molecules are conventionally depicted in an extended linear form (see, for example, Figure 27.3), as this conveniently conveys the chemical composition. However, in reality, the molecules commonly exist in highly irregular, randomly coiled arrangements (Figure 27.1(a)). In some cases, a degree of branching from the main chain may also be present (Figure 27.1(b))—a variation that is responsible, for example, for the marked difference in properties between high- and low-density polyethylene.

Linear and branched polymer chains are physically entwined, either randomly or in some form of alignment (due to intermolecular electrostatic forces). However, there are no fixed chemical linkages between them and each chain is a discrete entity. With sufficient heating, such polymers soften and then form very viscous liquids; on cooling, they re-solidify. This process can be repeated indefinitely without changing the chemical structure (provided that the temperature is limited to avoid thermal decomposition).

Polymers of this type, which can be fabricated or reshaped by heating (and by the application of pressure), are conventionally known as 'thermoplastics' (e.g. polyethylene and polystyrene). They can melt and flow because the polymer chains are relatively unrestrained and are thus able to move past one another. Most thermoplastics can also be dissolved in various solvents, a further demonstration of chain separation.

The linear or branched chains of certain polymers can be irreversibly extended and cross-linked via chemically reactive groups that form network junction points (Figure 27.1(c)). Polymers with this extensive network structure are infusible and insoluble and are known as 'thermosets'. The precursors (or 'pre-polymers') are known as reactive thermosetting polymers, and although conversion to a thermoset is a polymerization reaction, it is more commonly referred to as the curing or hardening process. In some cases, the reaction can be promoted solely by heating but, more generally, the network structure is formed by the addition of a second reactive material, which usually has a low molecular weight.

The shape of a thermoset cannot be usefully changed by the application of heat or pressure once the characteristic network has been formed. Although softening usually occurs with increasing temperature, decomposition takes place prior to melting. Furthermore, although thermosets may be swollen (and, thus, physically degraded) by the absorption of some powerful solvents, the cross-linked structure prevents true dissolution.

The properties of thermosets, for example elastic modulus or chemical resistance, can be varied widely by changing the chemical nature of the chain segments and/or the frequency of the junction points (i.e. the cross-link density). Such control of the molecular structure is achieved by choice of the appropriate thermosetting polymers, the co-reactants and the reaction conditions. The commercial availability of a very wide range of reactive components allows considerable scope for tailoring the properties of these materials for many specific applications.

The microstructure of many elastomers (Figure 27.1(d)) is intermediate between that of thermosets and thermoplastics. Here, the degree of cross-linking is too low to give the characteristic properties of thermosets. However, it is sufficient to restrain long-range movement of the molecular chains and, thus, it controls not only the lack of both solubility and softening with temperature, but also the typical recovery of shape and dimensions on release of an applied stress.

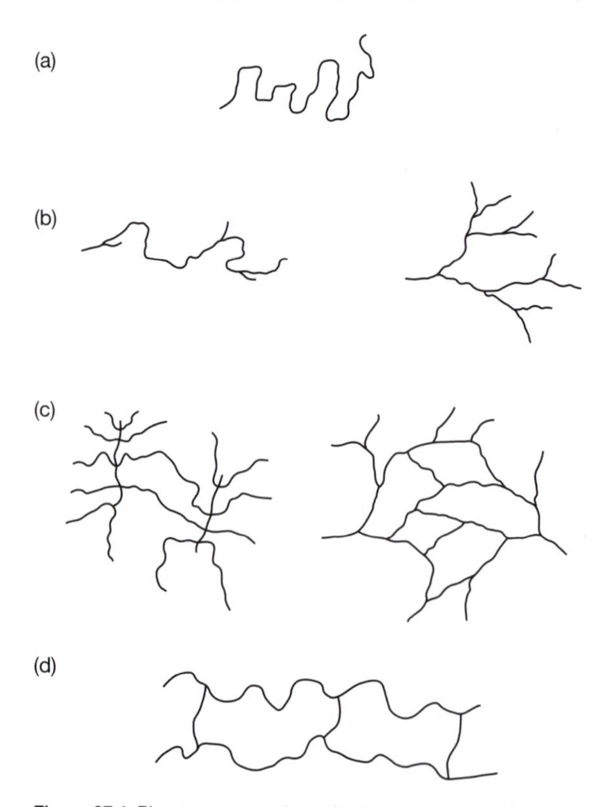


**Figure 27.1** Planar representations of polymer microstructures: (a) randomly coiled linear polymers (thermoplastics); (b) branched polymers (generally thermoplastics); (c) cross-linked polymers (thermosets); (d) elastomeric polymers

### 27.1.2 Thermosetting resins

The term 'synthetic resin' is used to describe man-made, thermosetting pre-polymers because, although some are low melting-point solids, many have the viscous, sticky consistency of naturally occurring resins, e.g. those secreted from coniferous trees.

Conventional terminology can be confusing, as both the cross-linked polymers and the pre-polymers are commonly described as 'resins'. Thus, one component of a two-pack epoxy adhesive is an epoxy resin which, on reaction with the second component (the hardener or curing agent), gives a cured adhesive that is also frequently referred to as an epoxy resin. Strictly, the term 'resin' should be applied only to the precursor—the cured material is simply an 'epoxy'. This dual use is widespread but, fortunately, the specific context usually conveys the exact meaning.

Thermosetting pre-polymers (whether viscous liquids or soluble, fusible solids) generally consist of low/intermediate molecular weight polymers. In both physical forms, the molecular chains possess specific reactive (or 'functional') groups.

The following generic types of thermosetting resins, with a sub-division that reflects significant differences in the cure mechanism, are most commonly used in construction applications:

1 epoxies, furanes and polyurethanes
2 unsaturated polyesters, vinyl esters and certain methacrylates/acrylates
3 phenol-, urea-, resorcinol- and melamine-formaldehydes.

The systems that incorporate formaldehyde are widely employed in building applications; for example, as laminates, mouldings, adhesives, surface coatings, and as binders in the manufacture of chipboard. However, these resins are usually used under factory conditions, as heat is essential for the promotion of the cross-linking reactions and the fabrication process often requires the application of pressure. Consequently, these systems are not dealt with in the present chapter, which discusses only the other resins listed above—materials that are widely used on site, where cure takes place at the point of service and at ambient temperatures.

In the cured state, these resin systems are, in general, relatively rigid, with a high strength. They are formed from proprietary products that may be supplied as unfilled liquids or as systems containing different amounts of mineral fillers, thus giving the consistency of a grout, a paste or a mortar.

The following, chemically associated materials are also excluded from discussion in this chapter: cellular (foamed) products; elastomeric joint sealants based on polyurethanes and silicones; fibre-reinforced polymer composites, precast polymer concrete; and coatings based on polysiloxanes.

### 27.1.3 The cure of thermosetting resins

In the uncured state, thermosetting resins have limited application—the use of higher molecular weight, linear chain, epoxy resins as solvent-borne coatings representing an exception to the general need for cross-linking. The essential curing or hardening process is induced via the reactive, functional groups situated either within or at the ends of the resin molecules. These groups may react together or, more commonly, with another reactive material, thus giving the characteristic cross-linked network structure.

This cross-linking takes place by utilizing reactions that are specifically designed to give the three-dimensional network (rather than simple branching or lengthening of linear polymer chains). Consequently, the number of functional groups present is important, as is their chemical nature and molecular position.

Some cure reactions take place only if heat is applied. However, in many cases, sufficient cross-linking for excellent service properties can be obtained, at normal or even relatively low ambient temperatures, by using appropriately reactive thermosetting resins and co-reactants.

Two common routes to the cross-linked network structures are illustrated in a simplified schematic form in Figure 27.2. The co-reactant may take the form of a separate curing agent (hardener) that becomes chemically linked within the network structure. In this case, provided that the temperature is not too low, reaction between the functional groups on the thermosetting resin and those of the curing agent commences immediately upon mixing, thus leading to a gradual increase in the degree of cross-linking (Figure 27.2 (route a)). In other cases, the thermosetting pre-polymer and a lower molecular weight cross-linking agent may be pre-mixed, as reaction between the two does not occur unless an initiator or catalyst is added to stimulate the process—cross-linking then commences at a very early stage (Figure 27.2 (route b)).

In the type of cross-linking scheme illustrated in Figure 27.2 (route a), a reactive group is situated at each end of the relatively low molecular weight resin molecule. Consequently, only linear polymers would be obtained if the co-reactant also contained only two reactive groups. For cross-linking, one of the reacting materials must have at least three reactive groups in order to generate network junction points and, following initial branching of the chains, a cross-linked final structure.

For the type of reaction shown in Figure 27.2(route b), the network structure is obtained by cross-linking fairly high molecular weight, linear pre-polymers containing reactive groups within the molecular chain. In this case, the reactive groups (denoted by the symbol ']'), are difunctional (i.e. capable of reacting with two additional molecules). In principle, only one of these groups within each pre-polymer chain would be sufficient for the chains to be linked by reaction with a low molecular weight, difunctional co-reactant. In practice, a greater density of cross-linking is achieved by using pre-polymers with a number of reactive groups within the chain structure.

For both of the above cases, strong covalent bonds are formed during the cross-linking process; although the cure reaction may be accelerated, retarded or terminated, it is irreversible once it has taken place.

Applied heat may be used to accelerate the rate of the reaction or, following initial hardening at ambient temperature, it may be used to induce further cross-linking in the solid state. The latter process, known as 'post-curing', can lead to greater rigidity, improved chemical resistance and/or an increase in bond strength. However, it is most frequently used in factory-based operations (e.g. adhesive bonding) rather than on site.

### 27.1.4 Fabrication and physical forms

As noted earlier, cured thermosetting resins cannot be shaped by heat and/or pressure. Consequently, fabrication for service has to be accomplished by placing the system immediately after mixing, e.g. by pouring, pumping, spreading, brushing, or spraying. Conversion to the cross-linked network structure and the required rigid form then follows by irreversible chemical reaction at a rate pre-determined by the specific formulation.

Rather than representing a problem, this inherent processing requirement confers versatility. It allows resins to be fabricated not just under routine factory control, but also on site under ambient conditions; an invaluable property in many civil engineering and construction applications where placing and fixing a preformed rigid shape would be impossible or undesirable.

Thermosetting resins also provide the basis for a very wide range of proprietary products that are developed by:

1. the addition of modifying materials, such as inert fillers and solvents, to obtain a variety of physical forms
2. varying the chemical nature of the resin and the co-reactant to give different properties.

The physical form, prior to cross-linking, can range from a low-viscosity liquid to a non-flowing mortar or concrete, with obvious benefit to specific applications. The rate of the hardening reaction can also be controlled to suit particular requirements; for example, use under different ambient conditions or time to achieve the service properties. Furthermore, as the detailed chemical structures resulting from the cross-linking reaction are virtually unlimited, the service properties of thermosets can vary very widely.

It should be emphasized that this diversity does not give the user complete freedom to interchange generic types, or even specific formulations, as not all resin systems are suited to a given application or prevailing conditions.

## 27.2 The production and curing reactions of thermosetting resins

A description of the manufacturing processes used to produce the main types of thermosetting resins is given below, together with examples of how these resins are cured. Due to the chemical complexities and numerous variations that are possible, this account is necessarily both simplified and illustrative. Furthermore, although discussion of the chemical principles governing various cure reactions is beyond the present scope, it must be emphasized that these principles determine such features as: the need for thorough mixing; the rate of gelation and cure; the ambient conditions under which a reaction will take

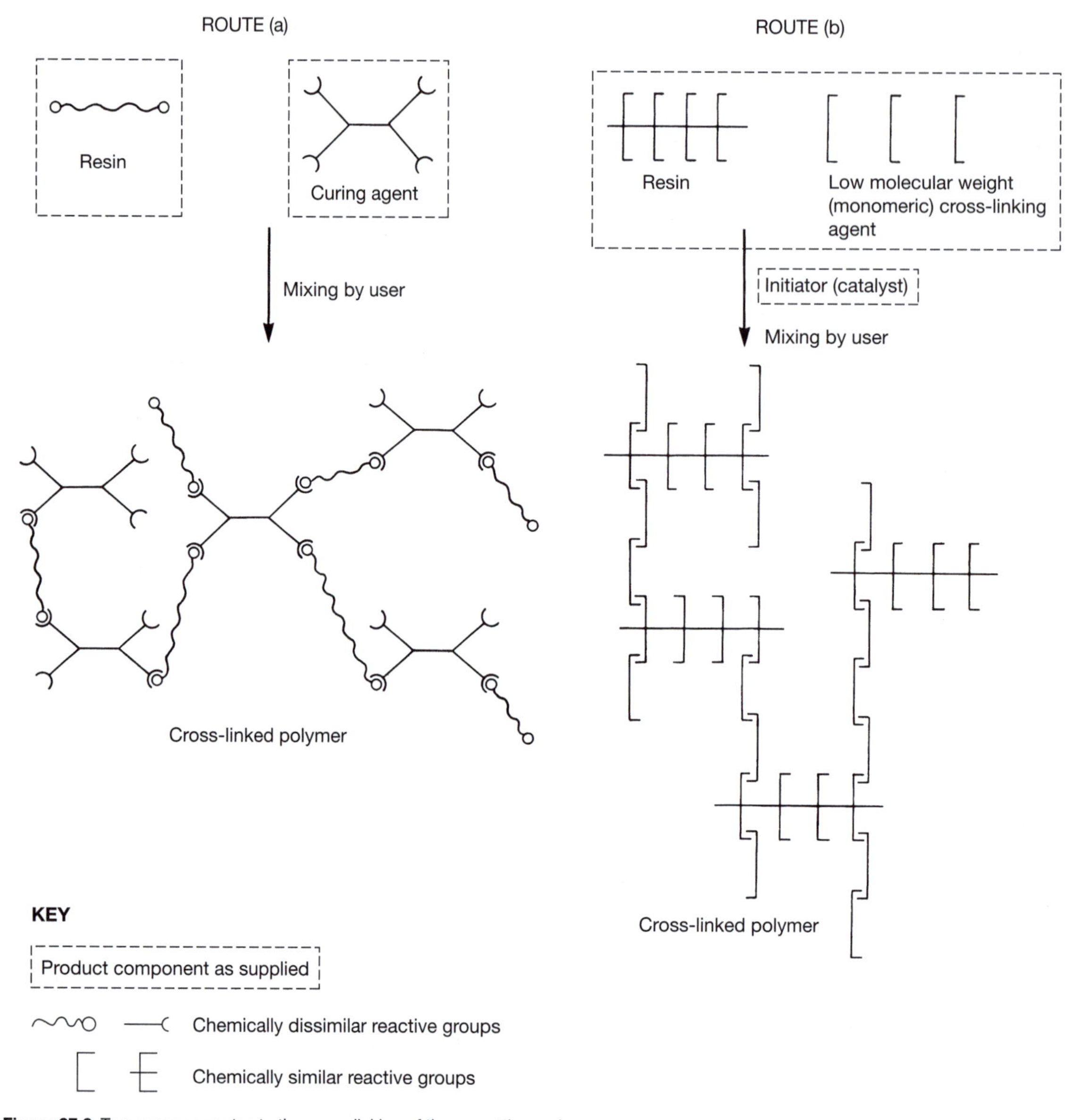

**Figure 27.2** Two common routes to the cross-linking of thermosetting resins

place; and the properties developed. Consequently, they have a significant influence on the manner in which various products must be used and the specific applications that are suitable for a given resin type—versatility and high performance do not imply or impart tolerance of misuse.

### 27.2.1 Epoxy resins

Epoxy resins were first synthesized in the late 1930s, with commercial manufacture by Ciba A.G. of Switzerland and Shell Chemicals commencing in the early 1950s. Epoxies are one of the major thermosetting resins, produced in high volume and with extremely diverse applications due to their excellent adhesion to a wide variety of substrates (including wet surfaces), a high strength-to-weight ratio, outstanding toughness, high electrical resistance and excellent durability and chemical resistance.

Although many special epoxies are available, most formulations used in construction are based on standard liquid resins that are synthesized from epichlorohydrin and bisphenol-A or, following a later development, bisphenol-F. As shown in Figure 27.3, the bisphenols are synthesized from phenol and either acetone (bisphenol-A) or formaldehyde (bisphenol-F); epichlorohydrin is produced from propylene and chlorine.

As the initial resin structure shown in Figure 27.3 contains two glycidyl ether groups per molecule, this relatively low molecular weight liquid resin is often referred to as a diglycidyl ether of bisphenol-A (DGEBA). Various degrees of purity are available, the 99.9% grade being virtually monomeric DGEBA (i.e. $n = 0$), while the 95% 'Technical Grade', which is suitable for most applications, contains some higher molecular weight resins (i.e. $n \geq 1$).

By varying the bisphenol/epichlorohydrin ratio, as well as the reaction conditions, higher molecular weight resins can be

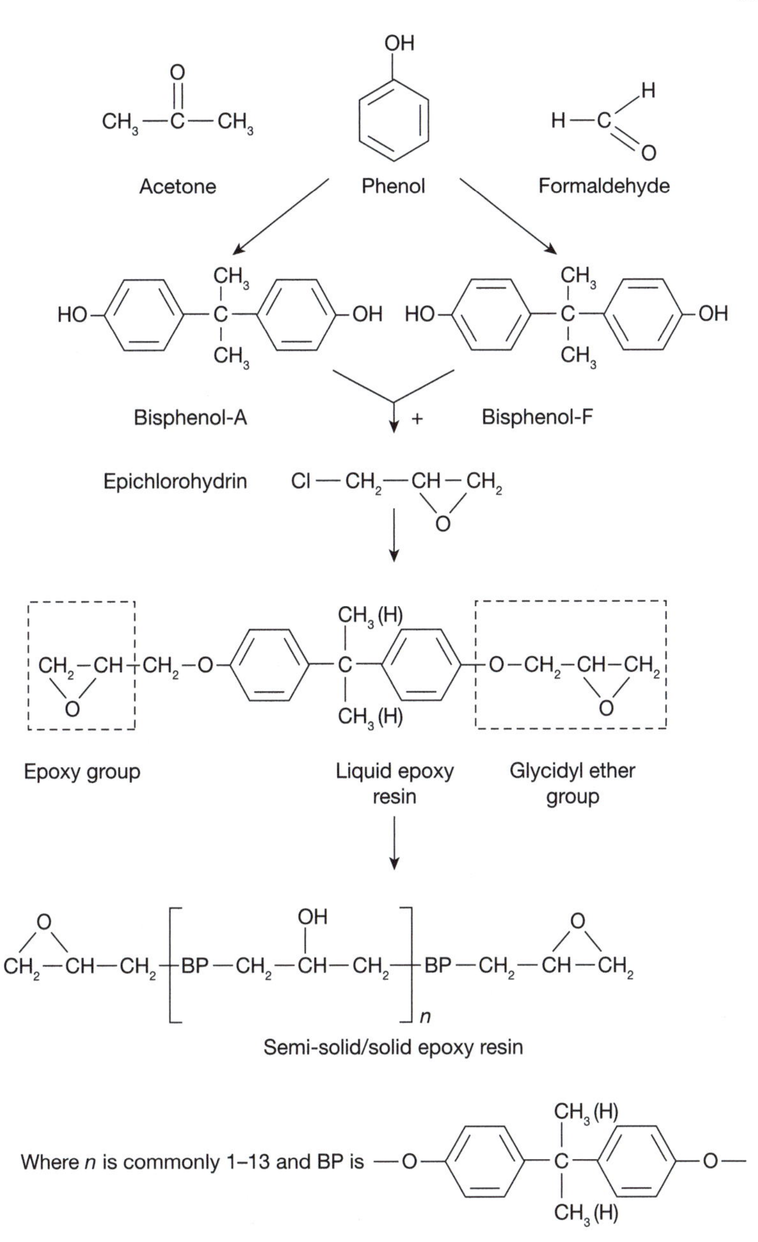

**Figure 27.3** Production of epoxy resins

produced by successive addition of further reactants via the terminal epoxy groups. Thus, semi-solid to solid thermoplastic epoxy resins that contain a range of chain lengths can be obtained. It is then necessary to refer to an average molecular weight and, when this is sufficiently high, the resins can be used, for example, as solvent-based, one-pack coatings that have usable properties without any cross-linking.

The viscosity of the 'Technical Grade' DGEBA is approximately 15 Pa s, which, at 25°C and below, is rather thick for easy handling (golden syrup has an approximate viscosity of only 0.3 Pa s). To obtain resins with a lower viscosity, various modifiers are commonly added in amounts of 5–10 parts or more per 100 parts of resin. These viscosity reducers, or 'diluents', can be reactive materials that become chemically incorporated within the final cured structure (e.g. relatively low molecular weight glycidyl ethers) or non-reactive products that also act as conventional plasticizers in the cured resin, e.g. high boiling point aromatic hydrocarbons and pine oil. Some diluents, for example benzyl and furfuryl alcohols, may also act as cure accelerators.

The incorporation of diluents can have economic benefits, due to both a unit-cost reduction and the possibility of increasing filler loadings when a lower viscosity resin binder is used. Such use should not be viewed as giving inferior, 'cheap' formulations; when chosen appropriately and used at an optimum level, diluents can give a wide range of benefits, e.g. easier handling, more convenient reactivity and usable life, and improved surface wetting, air release, impact resistance and flexibility. An optimized formulation containing such diluents can, in fact, outperform an undiluted system.

The typical epoxy resins described above cannot be cross-linked by heat alone—even heating to 200°C has little effect.

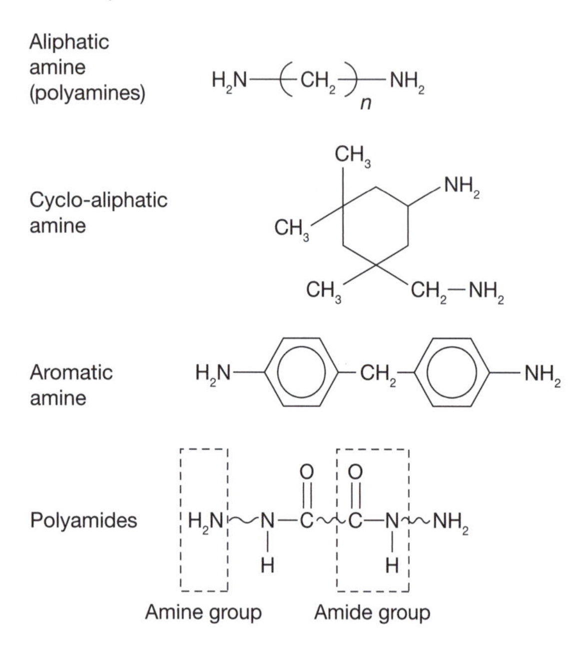


**Figure 27.4** Some epoxy resin curing agents

In order to convert the resins into a network structure, it is necessary to add a co-reactant (i.e. a curing agent). The selection of the specific curing agent has a critical effect on properties such as usable life and cure-rate, the acceptable temperature range during use, viscosity, mechanical properties, chemical resistance, adhesion, initial colour and colour stability. Manufacturers of curing agents for epoxies may offer a range of at least several dozen products. Fortunately, selection is eased as many of these curing agents fall into reasonably well-defined chemical classes that have well-established general characteristics.

The most commonly used epoxy curing agents for typical civil engineering and construction applications are based on aliphatic, cyclo-aliphatic and aromatic amines, and polyamides derived by reaction between vegetable oil carboxylic acids and aliphatic amines (see Figure 27.4). Special amine curing agents that emulsify the hydrophobic liquid epoxy resins are used for water-borne formulations.

In general, these amine curing agents are supplied as formulated, proprietary materials rather than as pure chemicals; for example, as blends which may also contain catalysts and other cure modifiers. Curing agents are also supplied in the form of higher molecular weight 'resinous amines' (adducts) derived from reaction between an epoxy resin and a low molecular weight amine. While the amine functionality and the basic performance are retained, such adducts can impart more convenient mixing ratios with the resins (e.g. 50–70 parts by weight (pbw)/100 pbw, rather than 10–20 pbw/100 pbw), a greater range of pot-lives, less sensitivity to atmospheric conditions and safer handling (due to the lower volatility and toxicity).

The detailed pathway of the cross-linking cure reaction varies with the particular amine used as co-reactant and is too complex to illustrate here. However, a notable feature of the curing reactions is that low molecular weight volatile materials are not evolved, thus avoiding the formation of voids in the solid.

As the amine curing agent reacts by addition to the resin and becomes a permanent part of the cross-linked structure, a certain resin/curing-agent mix ratio is required to obtain an optimum degree of cross-linking and, therefore, a good balance of properties. The amount of curing agent required for a particular epoxy resin type is recommended by the supplier and is often based on experimental evaluation as well as theoretical calculations. Misproportioning of the resin/curing agent, commonly beyond the ±10% level, is likely to lead to inferior properties. However, with some polyamide curing agents a fairly wide variation in mixing ratios may be employed to modify the flexibility of the cured resin without detriment to other properties.

It is essential that the resin and curing agent are thoroughly mixed, so that the reactive groups are brought into close proximity. Poor mixing may allow sufficient reaction for hardening to take place but, as completion of the cure reactions becomes severely retarded in the solid state, the required service properties may not be attained.

The simplest form of an epoxy system (unfilled resin and curing agent) is used for two-pack gap- or crack-filling grouts, adhesives, clear coatings and primers. Such systems also represent the starting point for a range of formulated products that are obtained by the incorporation of pigments and fillers; for example, coatings, pastes, mortars, filled grouts and floor-toppings. Heavily filled systems (e.g. a filler/binder ratio of ≥ 3 pbw/1 pbw) are usually supplied as three-pack products, allowing the resin and curing agent to be well mixed prior to addition of the filler.

### 27.2.2 Unsaturated polyesters, acrylates and vinyl esters

Although the detailed structures differ, these thermosetting resins are related, as cross-linking takes place via the reaction of unsaturated or double carbon-carbon bonds (C=C). This reaction is initiated by free radicals, generated by the interaction between an added initiator (catalyst) and an accelerator in the resin.

#### *27.2.2.1 Unsaturated polyester resins*

Polyesters contain the recurring ester group, shown in Figure 27.5, in the main chain.

A wide variety of polyester-based materials is available commercially, for example: surface coatings (alkyds), plasticizers (phthalates) and fibres (Terylene). Those of interest here are linear polymers containing aliphatic unsaturation within the main chain, i.e. reactive C=C bonds that provide sites for cross-linking. Polymers of this type first became available in the USA in 1927, and by about 1949 their commercial use for glass-fibre-reinforced laminates was firmly established.

Linear unsaturated polyesters are prepared by the reaction between a saturated diol and a mixture of polyfunctional (often dibasic) carboxylic acids (or the corresponding anhydrides). Although one of the acids must contain the aliphatic carbon–carbon double bond, the other (modifying) acid may be saturated. The function of the modifying acid is to reduce the frequency of reactive sites in the polymer chain and, hence, to lower the cross-link density and brittleness of the cured resin.

Many diols and dibasic acids are employed in the production of unsaturated polyesters; those shown in Figure 27.6 are among the most common. Variations can markedly influence the properties of the thermoset; for example: the flexibility and toughness, resistance to thermal degradation and chemical resistance.

At room temperature, the unsaturated polyester itself is a thermoplastic solid of limited use. The molten polymer could be cross-linked directly, but the reaction would be slow and the resultant structure would give poor properties. To avoid these limitations, the polyester is dissolved in an unsaturated, reactive

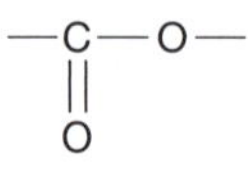

**Figure 27.5** Recurring ester group

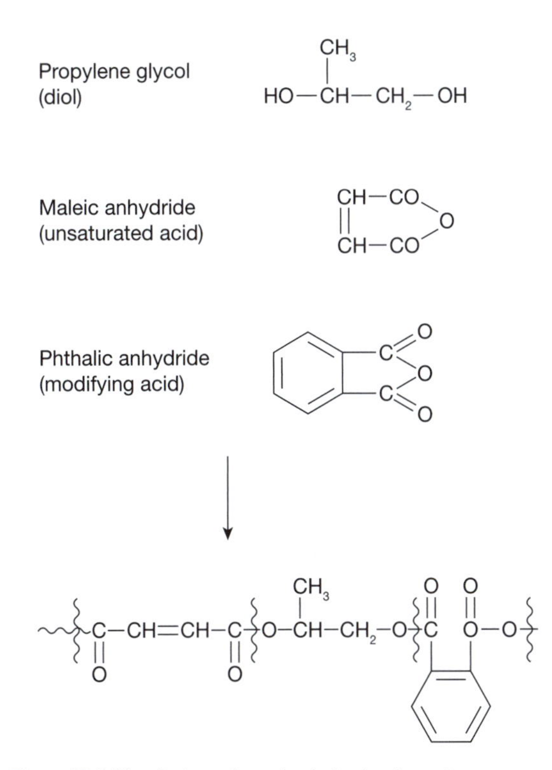


**Figure 27.6** Manufacture of unsaturated polyester resin

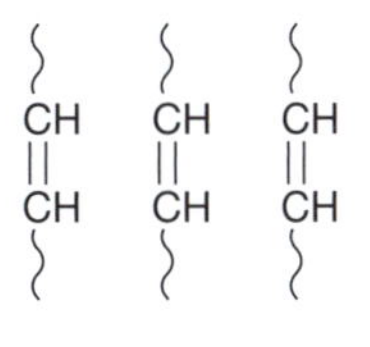


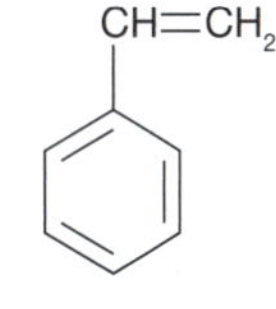


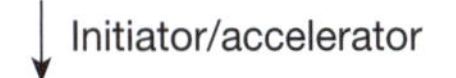


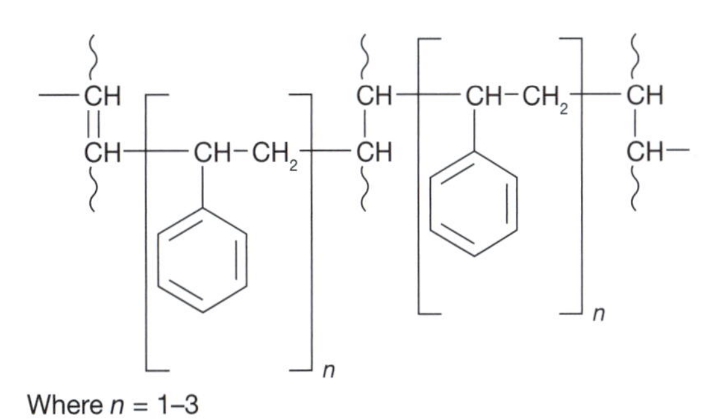


**Figure 27.7** Curing of unsaturated polyester resins

solvent as part of the manufacturing process. This additional component, invariably a vinyl monomer and most commonly styrene, acts as a cross-linking agent while also giving a liquid resin of usable viscosity. The unsaturated polyester resins used to formulate proprietary products are, therefore, solutions of the reactive polyester (65–70%) in a reactive monomer (30–35%). Reaction between the two components gives the cross-linked network structure.

In contrast to epoxy resins, these liquid polyesters contain all of the reactive groups that are essential for cross-linking; formation of the cured network structure is induced by a triggering mechanism. The reaction can be brought about solely by the action of heat, or even light. For this reason, unsaturated polyesters usually have a lower shelf-life than epoxies—even at normal ambient temperatures—although stabilizers/inhibitors are added during manufacture to avoid premature self-gelation. Stability is also enhanced by storage at low ambient temperatures in dark containers.

Heat- or light-promoted cross-linking would be very difficult to control. Hence, the cure of polyesters is brought about by the addition, immediately before use, of a relatively low concentration (up to 5%) of an organic peroxide initiator, e.g. benzoyl peroxide. Such peroxides (actually complex mixtures of peroxides and hydroperoxides) can be decomposed relatively easily to give very reactive species known as free radicals. Once the low stabilizer concentration has been neutralized, these free radicals induce rapid cross-linking of the linear main chains—the link being formed by reaction of the C=C bonds in these chains and those in the monomer/solvent molecules. This type of reaction leads to some cross-linking at a relatively early stage of the overall cure process. In contrast, the epoxy–amine addition reactions result in a more gradual build-up of cross-linking.

When a peroxide initiator is used alone, heat is usually necessary to obtain an acceptably rapid cure (temperatures of about 70–120°C are required). To induce cure at normal and low ambient temperatures, a resin-soluble accelerator (activator or promotor) is also added, usually at a similarly low level to the peroxide. In the presence of such accelerators, commonly metal salts (e.g. cobalt octoate or cobalt naphthenate) or aromatic amines (e.g. dimethylaniline), the peroxide initiators are rapidly decomposed into the cure-promoting free radicals without the application of heat. (The peroxide initiators and the accelerators must always be stored and added to the resin separately, as direct mixing can lead to an explosive reaction.)

Although the peroxide catalyst is always mixed with the resin immediately before use, accelerators can be incorporated during the resin manufacturing process while retaining a satisfactory shelf-life. The use of such pre-accelerated resins is obviously very convenient and, hence, they are readily available.

If the complex reactions between initiator, accelerator and resin are omitted, the cure reaction of polyesters may be represented by the scheme shown in Figure 27.7 where, as with epoxies, void-forming, volatile by-products are not evolved. The effect of the inhibitor, initiator and accelerator on the cure rate is summarized by the gel time data given in Table 27.1.

As pure peroxides are extremely hazardous, the peroxide catalysts are supplied in a diluted form; for example, as pastes

**Table 27.1** Gel times for polyester resins

| *Polyester resin** | *Approximate gel time* | |
|---|---|---|
| | *20°C* | *100°C* |
| Without inhibitor | 2 weeks | 30 min |
| With inhibitor | 1 year | 5 h |
| With inhibitor and catalyst | 1 week | 5 min |
| With inhibitor, catalyst and accelerator | 15 min | 2 min |

Data from British Industrial Plastics Ltd. *Inhibitor, 0.01% hydroquinone; catalyst, 1% benzoyl peroxide; accelerator, 0.5% dimethylaniline.

(diluted with a plasticizer) or powders (dispersed onto calcium sulphate). The former are convenient when the proprietary, resin-based product is a liquid or lightly filled paste. However, for more highly filled two-pack grouts and mortars, the peroxides in powder form can be pre-blended with the fillers. Care in formulation is necessary, as the peroxides are easily decomposed by many materials, e.g. carbon black pigment and iron salts present in fillers. Loss of activity during storage can lead to the delay, or complete absence, of gelation.

In contrast to epoxies, the formulator generally carries out relatively little modification of polyester resins. Once a pre-accelerated resin has been selected, formulation is concerned mainly with the fillers/pigments and with the level of catalyst required for a given usable life.

Within limits, polyesters are generally more tolerant of site misuse than epoxies. Reasonable variation in the catalyst level tends to affect the gel/cure times, rather than the service properties. In addition, as all the reactive sites are intimately pre-mixed in the liquid resin, cross-linking is likely to be satisfactory even when the catalyst and resin are mixed somewhat less than thoroughly.

*27.2.2.2 Acrylate resins (acrylic esters)*

Monomers containing both ester groups and carbon–carbon unsaturation can provide linear thermoplastic polymers with pendant ester side-groups (the ester groups in unsaturated polyesters are *within* the main polymer chain). Two common monomers of this type are methyl acrylate and methyl methacrylate (Figure 27.8), although many other types are available, thus allowing the polymer properties to be varied widely.

The polymerization can be carried out, at ambient temperatures, using initiator/accelerator systems similar to those used to cross-link unsaturated polyesters via a free radical chain reaction. By incorporating an acrylate monomer containing two (or more) carbon–carbon double bonds, a cross-linked network structure can be obtained, as shown in Figure 27.9.

A wide variety of both linear and cross-linked acrylate-based polymers are used as surface coatings, laminates and glass-fibre-reinforced plastics. However, the thermosetting acrylates described above are often used on site as unfilled resins for injection and impregnation and as filled systems that give patching mortars and floor toppings. These products usually take the form of a viscous syrup containing the acrylate monomer(s), cross-linking agent(s) and the accelerator; as for polyesters, the filler/peroxide blend is then added on site.

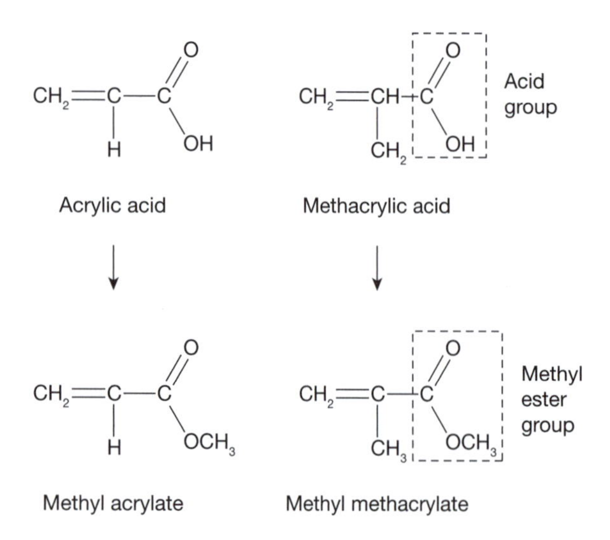


**Figure 27.8** Acrylates

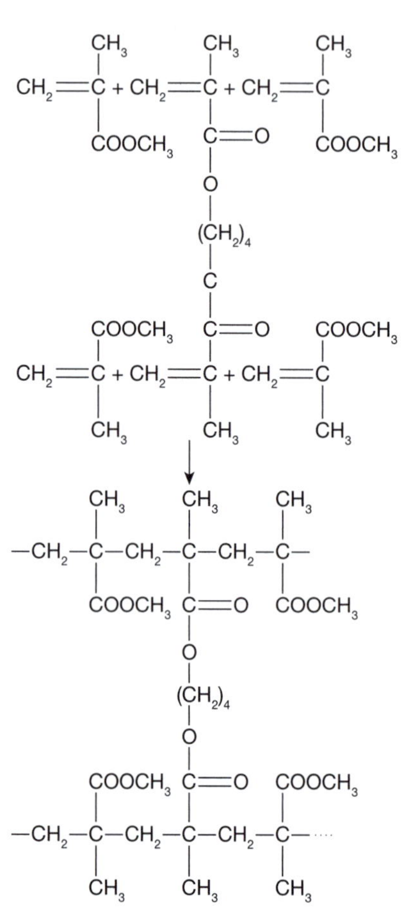

**Figure 27.9** Cross-linking of acrylate resins

*27.2.2.3 Vinyl ester resins*

As discussed in Section 27.2.2.1, unsaturated polyesters contain both repeating ester linkages and unsaturated carbon bonds along the backbone of the main polymer chains (see Figure 27.10 (a). After cross-linking, the ester groups remain unchanged and the network structure may also contain some unreacted carbon double bonds. Both groups are susceptible to chemical attack, such as hydrolysis and oxidation.

Improved chemical resistance can be achieved by modifying the carboxylic acid/diol chemistry, leading to a reduced number of ester groups in the main polymer chain (Figure 27.10(b)). With vinyl ester resins, this modification is taken a step further. Both the ester groups and the carbon double bonds are confined to the chain ends of the polymer, the main segment of which can be based, for example, on an epoxy (Figure 27.10(c)) or a polyurethane (Figure 27.10(d)). In addition to the improved chemical resistance arising from the reduced number of ester groups in the network, the longer chain segments between cross-linking points in vinyl esters usually give greater toughness compared to the more densely cross-linked polyesters.

As with the polyesters, styrene is usually employed as a solvent/cross-linking monomer for the resin and cure is promoted

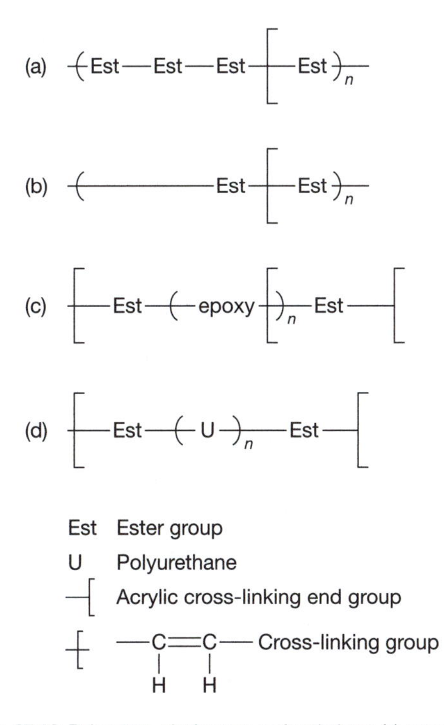


**Figure 27.10** Polyesters, vinyl esters and variations: (a) general-purpose polyester resin; (b) polyester with improved chemical resistance; (c) and (d) vinyl ester variations

at ambient temperatures using a peroxide initiator and a metal salt or amine accelerator.

Vinyl esters are more expensive than conventional polyesters. For many site applications, such as bedding and anchoring grouts, mortars and adhesives, the performance of the latter is quite satisfactory. Consequently, vinyl esters tend to be used only where their superior performance is essential; for example, for service at elevated temperatures and in chemically aggressive conditions. They are frequently used in factory processes—rather than on-site—to produce reinforced castings and laminates. The full potential of the composites can then be realized by carrying out post-curing, at elevated temperatures, in order to further improve the service properties.

### 27.2.3 Furane resins

Although not widely used, furane-based resins are met in construction applications as filled jointing cements, grouts and mortars and, also, as surface coatings/linings and fibre-reinforced laminates. The cured resins have outstanding resistance to heat, abrasion and many very aggressive chemicals.

The dark, free-flowing liquid resin used in these compositions is obtained by the controlled polymerization of furfuryl alcohol, to a predetermined degree of conversion, using an acid catalyst at a high temperature (*c.* 100°C). The resin obtained in this process, occasionally supplied in a ketone solvent, can then be cross-linked at room temperature using an organic acid catalyst such as p-toluenesulphonic acid (the exact mechanism of this reaction is uncertain)—as shown in Figure 27.11.

Unlike the resins considered earlier, which are mostly synthesized from petrochemicals, furane resins are derived from agricultural residues such as corn cobs and rice/oat hulls. These materials yield furfuraldehyde (furfural) which is then converted to furfuryl alcohol.

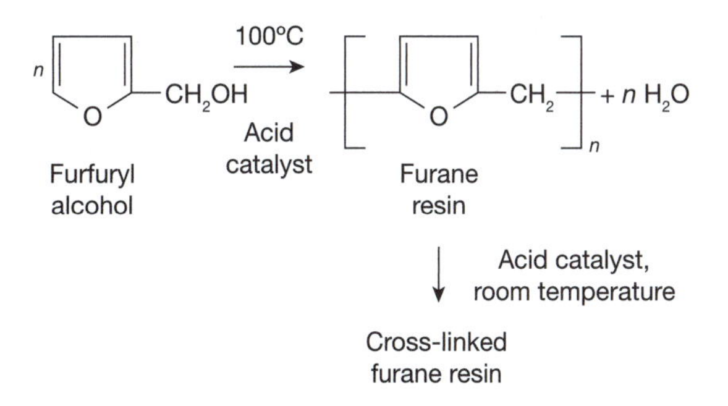


**Figure 27.11** The production of furane resin

The furane resin systems are usually supplied as two-pack products—the liquid resin and a powder blend of the acid catalyst and fillers (silica sands or, where resistance to hydrofluoric acid or hot concentrated alkaline solutions is required, carbon flour).

### 27.2.4 Polyurethanes

The commercial development of polyurethanes dates from the late 1930s following the pioneering work on plastics and fibres by Otto Bayer in Germany. Subsequently, intensive research demonstrated the versatility of polyurethanes and they became used as a basis for many adhesives, rigid foams, surface coatings, elastomers and flexible foams. The established mass production of flexible foam in many countries, by the mid to late 1950s, led to the ready availability of polyurethane raw materials and, consequently, the development of other applications followed readily.

Although they are used in mortars, and more extensively in flooring applications, polyurethanes are perhaps most commonly encountered during construction activities as coatings and elastomeric sealants for movement joints. As such materials are not dealt with in the present chapter, the origins and formation of polyurethanes will be dealt with only briefly.

The characteristic urethane linkage (–NH–CO–O–) is formed by reaction between an isocyanate and a hydroxyl group, as shown in Figure 27.12.

The reaction between a diisocyanate and a diol will produce a linear polyurethane. Cross-linked polyurethanes can be obtained from the reaction between a compound containing three or more isocyanate groups and a diol, but this approach is of limited commercial importance. A more common route to the cross-linked polymers is via the reaction between a diisocyanate and a polyhydric compound (a polyol).

Schematically, this reaction is similar to the cross-linking of epoxy resins because the diisocyanate is analogous to the epoxy resin and the polyol may be equated to a multifunctional amine. In both cases, but particularly in the case of polyurethanes, side-reactions may complicate the simple cross-linking schemes outlined here. For example, isocyanates react with any chemical group containing an active hydrogen atom, including the urethane linkage itself. Such reactions may also contribute to the cross-linking process.

Polyisocyanurates represent a particular modification of polyurethanes in which the diisocyanates groups initially react together to form a ring structure before undergoing further reaction with a polyol. Foam insulation products are thus manufactured for use in wall and roofing panels.

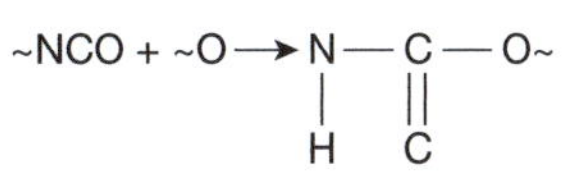

~NCO + ~O ⟶ N — C — O~
(H on N; C double-bonded below C)

**Figure 27.12** Reaction of an isocyanate and a hydroxyl group

$$\sim NCO + H_2O \longrightarrow NH_2 + CO_2$$

$$NH_2 + \sim NCO \longrightarrow \sim NH — CO — NH\sim$$

**Figure 27.13** Reaction scheme for branched, isocyanate terminated prepolymers

Polyureas can be formed if diisocyanates are reacted with compounds containing amine groups (rather than the polyols). These elastomeric materials have been available and used since the 1990s as very tough, sprayed coatings/linings for the protection of steel and concrete.

The most common diisocyanates used to prepare polyurethanes are toluene diisocyanate (TDI), diphenylmethane diisocyanate (MDI) and various isocyanate terminated prepolymers.

Numerous polyols are used in the production of polyurethanes. These range from low molecular weight, aliphatic diols, such as 1,4-butanediol, to polymeric hydroxy compounds (commonly based on polyethers or polyesters in the form of viscous liquids or low melting point solids). Hydroxy terminated polyester resins, for example, may have linear or branched structures, thus giving a wide range of polyurethane-based products with different degrees of cross-link density. Blends of polyols (as with the diisocyanates) are frequently used to obtain the properties required.

The polyurethanes considered above are supplied as two-pack products. However, single-pack polyurethanes are also available. These are based on branched, isocyanate terminated prepolymers that cure on exposure to moisture according to the (simplified) reaction scheme shown in Figure 27.13.

Careful formulation and the avoidance of excessive moisture during application are necessary in order to control the evolution of carbon dioxide (CO2), thus preventing the formation of voids in the cured polymer.

## 27.3 Modification of thermosetting resins

A very wide range of properties can be obtained by appropriate selection of the resin and co-reactant(s) used to produce thermosets. Such variation of the basic elements may be viewed as determining the structural skeleton, or framework, for fully formulated, usable products. The more specific characteristics required for given applications can then be obtained by the incorporation of additional materials. Such components allow the formulator to tailor products to specific requirements, as they provide scope for the control of cost, handling characteristics and rate of cure, in addition to the final properties.

Some materials may modify the resin system merely by their physical presence. Others have a more direct, chemical effect on the cure reaction and, in some cases, they may become part of the cross-linked structure. It is important that the type of modifier is carefully chosen and its level of addition optimized so that, in achieving certain requirements, other properties are not affected adversely; inevitably, this may lead to some compromises in the final formulation.

The materials commonly used to modify thermosetting resin systems may be grouped in the following general categories:

1. reactive diluents and resinous modifiers
2. fillers, pigments and reinforcing agents
3. specific additives that, at low concentrations, can modify such characteristics as cure rate, bulk rheology, surface appearance and texture, air-release, adhesive capabilities, electrical conductivity and fire retardancy.

These components and their effects are discussed below in more detail.

### 27.3.1 Reactive diluents and resinous modifiers

The use of reactive diluents in epoxy resin systems was discussed briefly in Section 27.2.1. In general, such diluents are low molecular weight compounds containing at least one reactive epoxide group. When such diluents have a functionality of at least two, they become fully incorporated into the network structure, in a similar manner to the resin itself. Monofunctional diluents also become part of the polymer, but they act as termination points in the growth of the cross-linked structure, thereby decreasing the cross-link density.

A prime function of such diluents is to lower the viscosity of the resin, thus giving greater ease of handling and the potential for higher filler loadings. However, they may also have a significant effect on other properties; for example: the prevention of resin crystallization at low temperatures, cure rate, mechanical properties and chemical resistance. The surface tension of the system can also be modified, bringing improvement in adhesive properties, the ease of wetting fillers/pigments and the release of air entrained during mixing.

In the case of polyurethane resins, the addition of low molecular weight hydroxy (or isocyanate) compounds can fulfil a number of the functions discussed above. With unsaturated polyesters and acrylates, the level or type of cross-linking monomer may be altered, but property variations are more commonly achieved by changing the basic resin system or by blending several resins.

Various modifiers in a resinous form are available, particularly for epoxy systems, and many take part in the cure reaction. They include epoxy functional polymers with very flexible structures, isocyanates, phenolics, silicones and polysulphides.

Non-reactive materials are also commonly added to resin-based systems to modify such properties as flexibility, toughness, thermal shock resistance and chemical resistance. They include pitch, bitumen and various naturally occurring or synthetic resins.

### 27.3.2 Pigments, reinforcing agents and fillers

Pigments are usually added to resin systems by the formulator although, in some cases, they are made available in a dispersed, concentrated form, allowing selection of colour by the user. Resin-based flooring systems, in particular, are available with a wide range of decorative effects, achieved, for example, by the incorporation of various mineral fillers or coloured thermoplastics in flake form. Micaceous iron oxide is also often used as a pigment in coating systems, as its lamellar structure can increase durability significantly by restricting water ingress.

The mechanical reinforcement of resin systems used in construction is most commonly achieved by the use of glass or carbon fibres, often in mat or woven form. Glass flakes are also employed, particularly in coatings, to increase weather and chemical resistance and, hence, durability. The resilience of some resin systems can also be improved by the addition of crumbed rubber.

In volume terms, inert mineral fillers represent the major method of modifying thermosetting resins (and reducing their cost). While fulfilling the primary function of controlling the consistency and handling properties (producing, for example, a paste, a fluid grout or a mortar), such fillers also have a significant effect on many properties; for example: strength, hardness, impact and abrasion resistance; long-term creep; the thermal coefficient of expansion; and the volumetric shrinkage and evolution of heat that occurs during the cure reaction.

The following materials are widely used as fillers in resin systems: silica flour (particle size $< 100\ \mu m$), sand and, occasionally, larger aggregate; various forms of calcium carbonate (crushed limestone or natural/precipitated chalk); barytes; and silicates, talcs, refined clays, alumina, mica and slate powder. Calcined bauxite, which is extremely resistant to polishing, is used to give surfacing materials with good, long-term skid resistance. Hollow spheres, manufactured from synthetic polymers or glass (or obtained as a by-product in the production of pulverized fuel ash) are used to give lightweight, filled resin systems that can be applied to vertical/overhead surfaces at an appreciable thickness (typically up to approximately 75 mm).

Although fillers may affect the rate of cure, by reducing the heat generated during the exothermic reaction, they are not generally used to exert chemical control over the cross-linking process. Care must be taken, therefore, to avoid undesirable side-reactions between fillers and the other components of the system. For example, fillers containing even traces of iron salts may deactivate the peroxide catalysts used with polyester, vinyl ester and acrylate resins; alkaline fillers may react adversely with the acidic catalysts used to cure furanes.

More obviously, the use of certain fillers must be avoided due to undesirable chemical reactions under service conditions, e.g. calcium carbonates when contact with acids is likely.

In addition to their chemical nature, the shape and particle size distribution of fillers must be given close attention. Whereas an angular shape can give filled systems with good troweling properties, the use of more rounded particles assists packing, thus minimizing the voids content and allowing higher filler loadings to be achieved. Silica sands are usually well graded (although occasionally gap graded) with a particle size distribution ranging from < 100 μm up to 2–3 mm. However, the level of very fine material required to minimize the voids content in the filler is often reduced to maintain acceptable handling properties. The gradings required are usually achieved by blending sands having a narrow particle-size distribution.

Fillers used in resin systems must be very dry. Moisture levels of only 0.1–0.2%, for example, are often sufficient to cause a marked drop in mechanical strength. Manufacturers use kiln-dried sands supplied in bags. Sands commonly found on construction sites, even those appearing to be very dry, must not be added, as the formulated properties of the product are likely to be downgraded significantly.

### 27.3.3 Miscellaneous additives

These additives, incorporated by the formulator, will be discussed only briefly, despite their importance, as the specific types and functions are so extensive.

Numerous organic materials are used to accelerate the cure of epoxy resins, e.g. phenols, carboxylic acids, alcohols and tertiary amines. The cure of polyurethanes is commonly accelerated by compounds of tin.

Various surface active materials, e.g. organo-silicon compounds, are used to promote adhesion by chemically coupling the cured resin to the substrate surface, to assist the release of entrained air and, particularly in coatings, to control the surface appearance.

Organic compounds containing halogens (chlorine and bromine) and inorganic materials such as antimony trioxide are used to achieve fire retardancy.

The addition of electrically conducting graphite to flooring systems can prevent a build-up of static electricity and the risk of sparking in hazardous areas.

Waxes that float to the surface can be incorporated in polyester systems to reduce the emission of vapour from the hazardous styrene monomer and, in acrylate resins, to eliminate cure inhibition at the surface by atmospheric oxygen.

Rheological control aids (thixotropes) are also widely used in resin-based systems. These materials, which include chemically treated clays, castor-oil derivatives, fumed silica and certain polymer-based fibres, impart sag resistance, thus assisting application to inclined, vertical and overhead surfaces. Their use in poured or pumped grout systems can limit, or even eliminate, flow in the absence of applied pressure.

By virtue of molecular attractions, these thixotropes form a gel-like structure within the fluid resin system. This structure does not depend on permanent chemical bonding and, thus, it can be broken down by a mechanical action such as stirring, giving a more fluid consistency. If the system is then left in an undisturbed state, the gel-like structure is reformed at a rate that depends on the particular thixotrope. 'Non-drip' paints also demonstrate this behaviour.

## 27.4 Properties

The variations discussed in Sections 27.2 and 27.3 enable thermosetting resin systems to be specifically formulated for a wide range of construction applications. Thus, the versatility of these products assists in readily meeting the following demands under various site conditions: appropriate handling characteristics for different applications; satisfactory cure; and good durability. Although the proprietary products usually have a high cost in comparison with many other construction materials, resin-based systems have demonstrated over many years that they can provide very cost-effective performance.

While the very wide range of properties that can be obtained with these materials is obviously invaluable to the end-user, it presents some difficulty in summarizing their specific benefits. Consequently, only a brief guide to the main features of each generic resin type is given below. Some relative advantages and disadvantages of each resin type are summarized in Table 27.2,

**Table 27.2** Relative advantages and disadvantages of resin-based systems

| *Resin type* | *Benefits* | *Disadvantages* |
|---|---|---|
| Epoxies | Excellent adhesion, good chemical resistance, little cure shrinkage, relatively tolerant of cold/wet cure conditions, good mechanical properties, wide scope in formulation, wide variation in pot-life and cure time, good shelf-life | High viscosity, health and safety, chalking caused by sunlight, slow cure, limited high-temperature performance, high thermal coefficient of expansion, critical mixing |
| Polyurethanes | Good chemical resistance, wide range of flexibility, good mechanical properties, little cure shrinkage, good adhesion, wide formulation scope | Health and safety, water sensitivity during cure, critical mixing, high thermal coefficient of expansion |
| Polyesters | Fast cure, low-temperature cure, mixing tolerance, chemical resistance, good against acids, good mechanical properties, low viscosity | High cure shrinkage, adhesion limited by shrinkage, health and safety, relatively short shelf-life, chemical resistance against alkalis can be poor, exotherm |
| Vinyl esters | Similar to polyesters but with improved chemical resistance and elevated-temperature performance | Similar to polyesters (but with improved chemical resistance) |
| Acrylates | Similar to polyesters but cure shrinkage has less consequence, good intercoat adhesion | Health and safety, exotherm |
| Furanes | Outstanding acid resistance and high-temperature performance | Sensitive to alkaline conditions/surfaces during cure, brittle, high coefficient of thermal expansion |

**Table 27.3** A comparison of resin and ordinary Portland cement (OPC) based mortars and concretes

| *Property* | *Unit* | *Epoxy system* | *Polyester system* | *OPC system* |
|---|---|---|---|---|
| Compressive strength | N mm$^{-2}$ | 55–110 | 55–110 | 15–70 |
| Flexural strength | N mm$^{-2}$ | 20–50 | 25–30 | 2–5 |
| Tensile strength | N mm$^{-2}$ | 10–25 | 10–25 | 1–5 |
| Compressive modulus | kN mm$^{-2}$ | 0.5–30 | 2–20 | 20–30 |
| Tensile elongation at break | % | 0–15 | 0–5 | 0.1–0.2 |
| Linear coefficient of thermal expansion | $\times 10^{-6}$ °C$^{-1}$ | 25–35 | 25–35 | 7–12 |
| Water absorption, 28 days | % w/w | Up to 5 | Up to 5 | 5–15 |
| Maximum service temperature under load | °C | 70–80 | 70–80 | ≥300 |
| Rate of usable strength development at 20°C | Days or weeks | 1–7 days | <1 day | 1–4 weeks |
| Cure shrinkage | – | Very low/negligible | Significant | Significant |

and Table 27.3 provides a comparison of properties for epoxy, polyester and cementitious mortars/concretes.

For more specific data, publications in the Bibliography should be consulted, together with suppliers' literature. The following points are significant when considering the properties of these systems:

1 The specific properties of seemingly similar products may differ markedly due to differences in the formulations. These variations can also detract from the value of case histories, although proven applications remain useful in a more general sense.
2 Many measured properties of resin systems depend on the specimen size and test method used—test procedures published by different authorities often vary.
3 When properties are determined in the laboratory, both curing and testing are conventionally carried out at constant temperature. The period required to develop the same level of cure may differ significantly under laboratory and site conditions, due to fluctuating ambient temperatures in the latter case. Consequently, it must be ensured that adequate cure has been achieved when, for example, a certain level of strength or chemical resistance is critical to service performance.
4 Care must be taken if elevated temperatures are used to carry out accelerated testing. Post-curing of the resin may lead to properties that are not representative of those found under normal service conditions.
5 Short-term properties are widely quoted, whereas information on longer term performance, e.g. creep, fatigue and thermal compatibility, is often unavailable—although many demanding uses have been satisfactorily demonstrated in the field. For some applications (for example, the structural repair or upgrading of reinforced concrete), the development and publication of new European Standards should assist in the generation of this additional information.

### 27.4.1 Health and safety

Many resin products contain components that, in certain circumstances, can lead to both acute and chronic systemic effects via ingestion or skin absorption, respiratory problems, eye damage and dermatitis. While the suppliers of raw materials and formulators devote considerable effort to the reduction and elimination of such problems, few chemicals of this nature can be made completely safe. Mandatory requirements for product labelling and the documentation of detailed information on health and safety matters should ensure that users of resin systems are fully aware of both the potential risks to health/safety and the appropriate protective measures.

In general, epoxy resins have a low vapour pressure and low odour, although both may be increased by the presence of low molecular weight diluents or solvents. Many curing agents tend to be more volatile, often with a marked ammoniacal, pungent odour. However, specifically formulated, low-odour curing agents are available, enabling the risk of tainting to be eliminated when work has to be carried out in food-processing areas, e.g. coating or the installation of flooring products. Most epoxy resin systems present little, if any, health risk once cured; some coatings, for example, are approved for use in contact with potable water. Both epoxy resins and their curing agents generally have high flash points and, consequently, fire risks are relatively low.

The cross-linking monomers present in polyester, vinyl ester, and acrylate resins are significantly volatile and possess distinctive odours. They also have relatively low flash points and, therefore, must be treated as a significant fire hazard. Although most organic peroxides are explosive, often on impact, the catalysts for these resin systems are supplied in a dilute, dispersed form, thus virtually eliminating this hazard. However, even these proprietary catalysts can react explosively if mixed directly with the cure accelerators.

Following normal cure at ambient temperatures, resins of this type often contain relatively high levels of unreacted monomer. Consequently, special catalyst systems and an elevated temperature post-cure may be necessary where contact with potable water or a foodstuff is likely to occur.

Particular attention must be given to the suppliers' instructions for handling polyurethanes, as isocyanates are hazardous by inhalation. Often this hazard is reduced by using isocyanates that have a relatively low vapour pressure under normal ambient conditions, i.e. polyisocyanate prepolymers.

### 27.4.2 Shelf-life

The product shelf-life stated by the supplier assumes that the storage recommendations will be followed, particularly in the case of polyester, vinyl ester and acrylate resins.

A shelf-life of at least 1 year can be expected for most epoxies, two-part polyurethanes and furane resins. A similar shelf-life can be obtained with the polyester and similar resins, although these systems are inherently less stable as cross-linking can occur under ambient conditions. Consequently, a shelf-life of only 6 months is often given for these products.

### 27.4.3 Viscosity

Thermosetting resin systems may be used as relatively low-viscosity, free-flowing liquids or, in other applications, as heavily filled systems that require heavy-duty, forced action mixers. Between these extremes, proprietary products are available as thixotropic liquids, pastes, putties and flowing grouts containing fillers.

Due to the essential presence of very low viscosity, cross-linking monomers (e.g. styrene and various acrylates), polyesters, vinyl esters and acrylates generally provide the lowest viscosity resins available (< 1 Pa s at 25°C). However, with appropriate diluents and curing agents, epoxy resin systems can have sufficiently low viscosities (< 5 Pa s at 25°C) for deep injection into concrete crack widths of only 0.05–0.1 mm.

For all resin systems, the viscosity changes quite markedly over the temperature range −10°C to +40°C and, consequently, when working at either extreme, handling difficulties can often be avoided or minimized by appropriate storage prior to mixing; an appreciable storage time can be required, as resins respond very slowly to changes in ambient temperature due to their low thermal conductivity.

### 27.4.4 Pot life and cure rate

With suitable formulation, the pot-life (or usable life) of thermosetting resins can be varied from a few minutes up to at least several hours under a range of ambient conditions. To meet some special needs, the gel time of epoxy resins may be extended for many hours—for example, resin-assisted bonding between fresh and hardened concrete where extensive shuttering has to be fixed.

Ambient temperatures have such a marked effect on the pot-life of resin systems (10°C can change pot-life by a factor of 2) that many manufacturers produce specific grades of some products for particular climatic conditions; for example, summer and winter grades.

The addition of fillers to a liquid resin system can lead to either an increase or a decrease in the usable life—a reduction in the exothermic heat of reaction (see Section 27.4.5) may slow the rate of reaction, but an unusable consistency may be reached at a lower degree of cross-linking.

Once a thermosetting resin has set to a hard mass, it does not immediately possess the intended service properties. Initial solidification represents a transition stage that can be developed at a relatively low degree of cross-linking. The reactions giving the network structure continue in the solid phase, although at a much reduced (and continually reducing) rate. As the extent of cross-linking continues to increase, an initial cure stage is reached where some service requirements can be adequately met, e.g. the ability to sustain a high proportion of full loading.

The quoted service properties are then fully developed in the final stages of the cure reaction. Even here, some potential for cross-linking may remain unrealized because the molecular motion required for reaction becomes so restricted. Raising the cure temperature assists this motion; further reaction can then result, for example, in improved chemical resistance or an increased capability to sustain loads at elevated temperatures. Use of heat in this manner is usually referred to as post-curing (a process that can occur naturally, to some extent, with a change in the service conditions).

Depending on the particular formulation, the initial cure of resin systems is commonly attained within 24 h (at 20°C). For some epoxy and polyurethane systems, several hours may be sufficient to achieve this state. With polyesters, vinyl esters and acrylates, initial cure may take place in only 5–10 min.

In general, epoxies and polyurethanes exhibit a gradual approach to the final cured state and full service properties may not be obtained for 7–14 days at 20°C; many polyesters, vinyl esters, acrylates and furanes, however, can reach the fully cured state well within 1 day.

At low temperatures, the effective activity of the chemical groups taking part in the cure reaction can be severely reduced, as molecular mobility in the solid phase, even at a low degree of cure, can become very restricted. Consequently, cure rates at low temperatures are very much reduced and, in some cases, very little, if any, reaction takes place.

The reactions leading to cross-linking of polyesters and similar resins can take place relatively rapidly, even at temperatures down to about −10°C. However, satisfactory cure of epoxies usually requires a temperature of no less than 5°C (10°C for some systems); many polyurethane systems will cure satisfactorily down to about 0°C and furanes may be used at 5°C–10°C.

Many systems may appear to cure well at low temperatures—for example, a high compressive strength may be obtained. However, the system may show this strength because it is, in effect, 'molecularly frozen' and in a glass-like state due to the low temperature. If the strength of a test specimen (e.g. a 40 mm cube) cured at a low temperature is determined after allowing, say, 2–3 h to warm to 20°C, then lack of cure may be shown by a much reduced strength and by considerable deformation prior to failure (in contrast, brittle failure may be observed at the lower cure temperature despite a relatively low degree of cure). Lack of cure at a low temperature is usually a temporary phenomenon, as reaction will continue if the temperature of the resin is increased.

### 27.4.5 Exotherm

The reactions leading to cross-linking evolve heat and are, therefore, termed 'exothermic'. Most resin systems and many substrates are poor conductors of heat. Consequently, if the system is very reactive and/or lightly filled, a considerable rise in temperature (and shortening of the pot-life) may occur, even in a volume of only a few litres or less. In the extreme, with unfilled epoxy resins, the heat evolved may be sufficient to cause boiling and charring. Polyesters tend to solidify prior to reaching the peak temperature and, in this case, an extreme exotherm can lead to expansive cracking of the solid mass. Aside from these extremes, failures can occur due to the contraction on cooling following the cure exotherm—this shrinkage can cause cohesive fracture or stress concentrations at a resin/substrate interface that result in adhesive failure.

Where resins are used in very small volumes, or thin films, initial setting can be delayed, as the exothermic heat of reaction is more effectively dissipated; consequently, the temperature of the material remains close to ambient. This lengthening of the usable life can be of benefit—for example, when resins are used to repair concrete by slowly injecting fine cracks.

Due to the exotherm effect, the pot-life of a resin system should always be referred not only to a given temperature, but also to a given mass of mixed material.

### 27.4.6 Cure shrinkage

In addition to the thermal contraction discussed above, shrinkage also occurs as a result of the chemical and physical changes that take place during cross-linking, both prior to and after gelation.

Epoxies and polyurethanes undergo very little shrinkage of this type, and that which does occur rarely has any effect on the service performance. However, polyesters, vinyl esters and acrylates undergo much higher levels of cure shrinkage (10–20% on a volumetric basis for unfilled resins, and 5–10% for typical filled systems). When a significant proportion of this shrinkage occurs in the solid state, the use of the resin for applications such as gap filling or floor toppings may not be possible, as the shrinkage stresses may lead to debonding and/or cracking. This is often the case for polyesters and vinyl esters, but not so for the acrylate resins that are widely used as flooring products.

### 27.4.7 Mechanical properties

Resin products can give very high-strength solids (compressive strengths up to 100 N mm$^{-2}$), although the intended strengths may be much lower where other requirements take precedence. Similarly, modulus and ultimate strain capacity can be varied widely and thus, while some products are quite brittle (strain at failure < 1%), others can be rigid and tough, or soft and flexible (up to 100% strain at failure).

Excellent impact and abrasion resistance can be achieved, leading to the extensive use of epoxies, acrylates and polyurethanes for industrial flooring where the performance requirements can be very demanding. Some tar-extended epoxies, in combination with a bauxite aggregate, have sufficient resistance to abrasive wear for use as anti-skid road surfaces.

Resistance to dynamic loads and fatigue can also be very good, as demonstrated by the wide use of epoxies and polyesters to bond holding-down bolts for vibrating machinery baseplates fixed to concrete. Epoxy grouts have also been used extensively to fill gaps under heavy-duty crane rails.

### 27.4.8 Creep and stress relaxation

Creep of load-bearing resin systems rarely causes problems, as many applications utilize filled products at normal ambient temperatures and both the stress and aspect ratio are often very low. However, strain may increase unacceptably with time at more elevated temperatures (see Section 27.4.10). Where creep effects are of concern, advice should be obtained from the manufacturer; in some cases, specific tests may need to be carried out. Generally, creep effects will be minimized at low ratios of imposed load to short-term ultimate strength and with low aspect ratios.

The associated property of stress-relaxation is often beneficial and epoxies can be formulated to give this effect in a controlled manner. Thus, differential stresses between a resin and substrate, due, for example, to marked differences in the thermal coefficients of expansion or to cure shrinkage, may be reduced or alleviated.

For some structural applications—for example, anchoring at elevated temperatures—the possibility of creep rupture may cause concern and require investigation (i.e. eventual failure at a lower level of stress than observed under normal short-term loading). This phenomenon may be assessed experimentally, for a given elevated temperature, by recording the time to failure for a range of applied stress levels and thus deriving a limiting stress below which creep rupture should not take place.

### 27.4.9 Chemical resistance

Thermosetting resin systems generally have good resistance to water and dilute mineral acids and alkalis. Depending on the resin type and the specific formulation, good resistance against some strong acids and alkalis may also be obtained. Furanes, vinyl esters and certain polyesters, in particular, may be used in contact with some fairly concentrated acids, even at elevated temperatures. Resistance against solvents and organic materials is more variable; for example, some epoxies are readily attacked by the organic acids contained in fruit concentrates and milk, but most products are very resistant to aliphatic hydrocarbons such as petrol.

### 27.4.10 The effects of temperature and fire

Although elevated temperatures do not cause thermosets to melt and flow, they do induce softening and possibly significant changes in such properties as modulus, creep, strength, and chemical resistance.

A unique and fundamental property of organic polymers, arising from their chain-like structure, is the glass transition temperature (*Tg*), which depends on both the detailed formulation and the specific cure conditions. When a polymeric material below its glass transition temperature is heated, it will demonstrate a fairly sharp transition from a hard solid to a softer, more flexible, rubber-like material, as illustrated in Figure 27.14.

The glass/rubber transition is solely a physical effect, reversible indefinitely without detriment to the material. For thermosetting resins (in contrast to thermoplastics and as shown in Figure 27.14), eventual liquid flow above *Tg* is prevented by the chain cross-linking.

The glass transition point corresponds to the onset of significant motion of quite long segments of molecules, thus characterizing a rubbery state where viscoelastic effects are more pronounced. Below *Tg*, this motion is, in effect, frozen, and only relatively short parts of the molecules are able to move against local restraints. The consequence of cooling below *Tg* is demonstrated by the effect of impact on rubber cooled in liquid nitrogen: it is very brittle, or glass-like, and shatters easily (note: some polymers, e.g. polycarbonates, remain tough below *Tg*).

*Tg* values for the thermosetting resins cured at ambient temperature, as discussed here, are typically in the range 30–70°C (furanes, exceptionally, have values > 70°C). For comparison, the *Tg* of natural rubber is about –70°C and that for rigid PVC is about 87°C; epoxy systems requiring elevated temperatures for cure will commonly have a *Tg* above 100°C.

Two particular effects on service performance stem from the phenomenon of a transition temperature:

1. It is often necessary to limit the upper service temperature of a resin system, particularly in structural applications.
2. Where a low modulus, rubber-like flexibility is essential, then the polymer selected should have a *Tg* below the service temperature, e.g. for movement joint sealants.

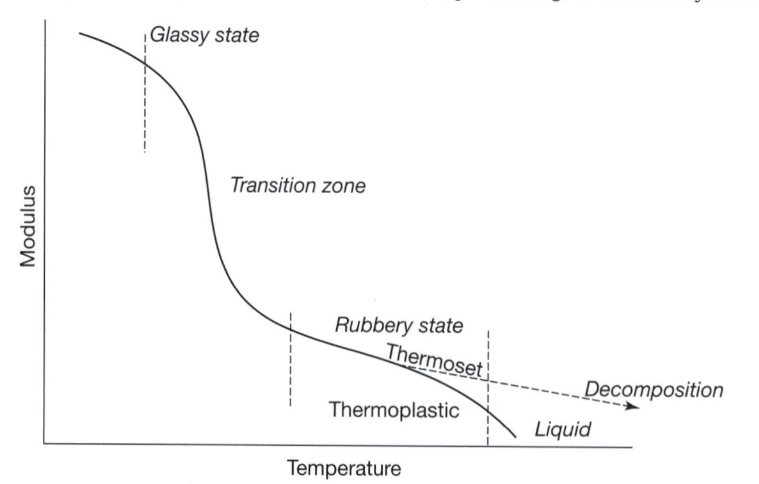


**Figure 27.14** Polymer modulus as a function of temperature

Within limits, the formulator can adjust the specific components used in a resin product in order to control the *Tg* for given application/service conditions.

While many systems can be used non-structurally at temperatures up to at least 70°C, certain load-bearing applications may need to be considered more carefully if service temperatures above 40–50°C are anticipated. However, changes in structure at these temperatures are more likely to be due to beneficial post-cure reactions, rather than to permanent degradation. Where service temperatures are in excess of about 70°C, particularly in non-structural applications or uses where chemical resistance is important, then furanes, vinyl esters and certain polyesters are often suitable.

Resin systems may also be formulated to have good resistance to thermal shock, thus allowing epoxies and polyurethanes, for example, to be used as steam-cleanable floor toppings.

The effects of fire on any material are complex and, consequently, this aspect of resin performance is not considered here in any detail.

Cross-linked resins are not highly combustible but, as with any organic material, the effects of fire must be assessed in certain applications. The surface spread of flame and rate of fire penetration may be reduced by certain fillers, particularly at high filler/binder ratios, and thus many systems are self-extinguishing. However, the effect of fire does need to be considered carefully at the design/selection stage, particularly in structural applications where a sudden loss of properties with a rapidly increasing temperature could have serious consequences. The evolution of toxic materials from burning resins should also be considered.

### 27.4.11 Adhesion

Most thermosetting resin systems adhere well to a variety of substrates, provided that suitable surface preparation has been carried out. Polyurethanes, and particularly epoxies, are perhaps outstanding in this respect, being assisted by the very low level of cure shrinkage and consequent absence of stress concentrations at the bonded interface.

Epoxy systems, if appropriately formulated, can give excellent long-term adhesion even when applied to wet surfaces. It is necessary, however, to distinguish between good curing and good adhesion in wet conditions, as the latter does not necessarily accompany the former. Once beyond the initial stages of cure, epoxy resin surfaces are not particularly receptive to adhesive bonding, a characteristic that can cause problems with multi-coat or multi-layer applications.

## 27.5 Applications

It is not possible in a chapter of this length to discuss in detail the wide range of applications of resin-based products in construction and civil engineering. However, the following subsection gives a brief overview of some common uses that are carried out on site—uses that demonstrate the relative importance of different properties and, also, the versatility of these materials in meeting various demands for high performance.

Many of these applications have become areas of specialist knowledge and activity. The publications in the Bibliography should be consulted for more detailed consideration of specific applications. In addition, the extensive commercial/technical literature, which is readily available from product suppliers, will give an excellent overview of many uses.

### 27.5.1 Typical uses

Examples of typical site applications of resin based products are summarized in Table 27.4.

Epoxy resins are most commonly used for adhesive applications, mainly due to their inherently good adhesion to

**Table 27.4** On-site uses of resin systems in construction and civil engineering

| *Type of use* | *Examples of typical applications* |
|---|---|
| Adhesive bonding | Anchoring of rock bolts, fixings and cavity wall ties<br>Ground anchors<br>Segmental construction of bridges and other structures<br>Steel or GRP reinforcement bonded externally to concrete<br>Brick slips and tiles<br>Connectors for steel reinforcement in concrete<br>Rubber strip/sheeting for water-proofing<br>Fresh to hardened concrete<br>Relatively rigid joint sealants (e.g. some floor joints)<br>Miscellaneous applications with concrete, steel, stone, ceramics, GRP composites and glass |
| Grouting and gap/void filling | Waterproof grouting of tiles and brick linings<br>Support of machinery base-plates, crane rails and bridge bearing pads<br>Bedding of manhole frames, kerbstones and runway lights<br>Sleeve/jacket encasement of columns and piles |
| Repair | Patching of concrete<br>Rapid repair of airfield runways and aprons<br>Resin injection of cracked concrete and delaminated renders, tiles and floor screeds<br>Timber repair<br>Consolidation of masonry/brickwork by injection or impregnation |
| Surface protection and waterproofing | Surface coatings, linings and renders<br>Impregnation of porous substrates such as concrete and timber<br>Bridge deck overlays |
| Uses connected with wearing surfaces | Floor screeds<br>Paving formed from resin-bound large aggregate<br>Skid and slip-resistant surfacing<br>Waterproof membranes for bridge, car parking and podium decks, roofs and balconies/walk-ways<br>Expansion joint nosings for roads, bridges and floors |

many substrate types, their negligible cure shrinkage and their ready availability in a variety of physical forms, thus providing good gap filling properties and enabling the thickness of the bond-line to be varied over a wide range. When formulated appropriately, they can bond and cure well in wet conditions; in some cases, underwater.

The use of polyester resins for anchoring steel bars, dowels, studs or bolts into a base material—for example, concrete or masonry—is a very well established application (some vinyl esters are also used). In this case, advantage is taken of the ability of these resins to cure extremely rapidly, even at very low temperatures. Consequently, a service load may be applied to the fixing/anchor after a very short cure period (< 1 day). The potentially detrimental effect of cure shrinkage is generally offset by using filled products and by limiting the width of the annular gap. Mechanical 'keying' of the resin to both the steel and the substrate material is also routinely recommended.

The installation of bonded fixings has been assisted for many years by pre-packaging the resin system in a glass capsule or a 'sausage' type of cartridge. Mixing is then carried out in situ by driving the steel bar/bolt into the capsule/cartridge with a drill. Polyesters and vinyl esters are particularly suited to this method of use, as the reactive resin and the cross-linking agent are pre-mixed; consequently, this very simple type of mixing can be effective (see Section 27.2.2.1).

Segmental construction utilizes epoxy resins semi-structurally, as the bonded layer, formed between post-tensioned precast units, is subjected predominantly to compressive stresses. In addition to acting as a uniform stress-distributing layer, the resin also forms a waterproof joint. The use of epoxy adhesives in this application dates from the construction of the Choisy-le-Roi Bridge over the River Seine in 1963. Since then, the technique has been used to construct bridges and viaducts in many parts of the world. The reinforced concrete superstructures for Sydney Opera House, the Parc des Princes Stadium and the Montreal Olympic Stadium were also constructed segmentally using epoxy adhesives.

The need to strengthen (or stiffen) load-bearing elements in existing concrete structures can arise for several reasons; for example: upgrading to accommodate changes in use, such as higher traffic loading on bridges or increased loading on floors, or as compensation for damaged or deficient reinforcement. The (epoxy) adhesive bonding of steel plates (or now more commonly, glass or carbon fibre-reinforced composite plates) to an external face of the element, thus increasing the flexural capacity, has developed since the early 1960s into a well-established, routine and very cost-effective technique. This method has also been applied to cast iron structures, which cannot be welded easily, and to timber beams, masonry walls and reinforced concrete columns.

The use of flowing, resin-based grouts for load-bearing applications represents one of the earliest applications of these materials in construction. Epoxies and polyurethanes (and, for some uses, polyesters) containing mineral fillers/aggregates are still used extensively, although a wide range of cementitious systems, including expansive or shrinkage compensated products, is now available. High early strength, excellent resistance to dynamic stresses and the negligible cure shrinkage of epoxies/polyurethanes provide obvious benefits, although the thickness may have to be limited to avoid detrimental effects due to the reaction exotherm.

Resin-based products were used extensively in the past for the patch repair of concrete, following the spalling and delamination induced by corrosion of the mild steel reinforcement. However, these repairs are now more usually carried out using products from the wide range of proprietary cementitious systems that have become available over the past 25 years or so. In addition to lower cost, these products, which are often modified with polymer dispersions, provide a much closer match with the physical and chemical properties of the parent concrete, for example modulus and coefficient of thermal expansion. It is now universally accepted that this compatibility is more critical for good long-term performance than the high strength of resin systems that was formerly seen as a major benefit.

Resin-based products, particularly unfilled epoxies, continue to be used for repairs requiring injection under a controlled pressure into cracks in concrete and small gaps resulting from the delamination of tiles, renders and floor screeds. Although a wider variety of materials may be used where cracks have to be sealed just to prevent water ingress, resins retain a unique position where a structural repair is required.

Waterproofing and protective coatings are produced from many polymers, including thermoplastic and elastomeric types. Epoxies and polyurethanes contribute very significantly to the major and worldwide coatings industry. The performance benefits that they bring include: excellent adhesion to many types of substrate (which can include damp/wet surfaces), good durability under many exposure conditions, resistance to a wide range of aggressive chemicals, abrasion/impact resistance and impermeability.

In addition to these properties, epoxy and polyurethane resin systems allow coating formulations to be varied without difficulty in order to obtain specific characteristics. They can also be used as solvent or water-borne products or as solvent-free, high-build coatings.

Polyesters and vinyl esters are often used as thick, fibre-reinforced protective linings, particularly where resistance to elevated temperatures and very aggressive chemicals is essential; furanes are also suitable for this application, although they are used less commonly.

Flooring applications are a very significant market for resin products based on epoxies, polyurethanes and acrylates. The range of products, which are usually applied to concrete bases to give a wearing finish, includes: penetrating sealers, coatings with a thickness of up to 1 mm (a fine aggregate may also be encapsulated to provide slip/skid resistance), flow applied systems with a thickness of 2–3 mm (often referred to as self-levelling/smoothing systems) and heavily filled, trowel-finished products with a thickness of 6 mm or more.

The thicker, seamless systems in particular have excellent resistance to impact, abrasion, aggressive chemicals and the ingress of liquids. They can also provide rapid installation and curing without cracking, good resistance to slip in wet conditions and, with some systems, a wide range of decorative effects associated with the inclusion of many different types of aggregate. Depending on the formulation, steam-cleaning is possible.

Waterproofing exposed, concrete car parking decks is a particularly demanding application, requiring cohesive and adhesive tolerance of trafficking stresses, impact and abrasion resistance, the accommodation of thermal movement (and, possibly, concrete cracking) and good resistance to skidding/slipping, fuel/oil spillage, weathering and the ingress of deicing salts. Suitable waterproofing systems are usually formed from a primer (often an epoxy), a crack-resistant membrane (a flexible polyurethane or a fibre-reinforced acrylate or polyester) and finishing coats encapsulating an anti-skid/slip aggregate (polyurethanes or acrylates).

The use of resin-bound aggregate to form highly permeable paving (on a permeable base) is a relatively recent development, contributing to sustainable urban drainage systems. Rainwater is filtered and then passes directly into the ground, thus assisting the maintenance of natural water levels.

Given their versatility, thermosetting resins will undoubtedly continue to be used as widely in the future as they have been in the years since their applications in construction were initiated in the 1950s.

## 27.6 Responsibilities of the formulator and the end-user

The formulators of proprietary resin-based products act as a link between the end-user and the chemical companies that manufacture a very wide variety of resins, curing agents, catalysts and additives. In general, the user has little scope to produce or modify resin systems to suit his particular needs and, consequently, must select a proprietary product from the range tailored by the formulator for specific applications. This supply chain, which also exists with other site-applied products such as surface coatings and joint sealants, is viable if all parties accept certain responsibilities.

Usually, the end-user or specifier (contractor, engineer or consultant) does not possess sufficient specialized knowledge and experience to deal with the complexities involved in the formulation of an acceptable resin product. Thus, the formulator is not merely a blender of product components—assisted by experience and testing, these components must be selected to give a suitable balance of properties. Often, this involves some compromise in order to optimize the following requirements:

1 appropriate handling properties
2 satisfactory curing under a wide variety of site conditions
3 suitable properties for the service conditions envisaged.

A difficulty frequently confronting the user or specifier is that of appropriate selection from an extensive number of products for a given application, particularly as technical data sheets vary in their content and do not necessarily provide all the detail that is required. In this case, discussions should be held with the manufacturers, possibly assisted by advice from independent experts. However, the benefits of such contacts depend on the end-user supplying as much information as possible regarding the intended application. To give just one example: to assess the effect of spillage onto a resin flooring product, it is necessary to know not only the type of chemicals, but also their concentrations, temperature and the maximum contact time.

For some applications, the detailed property requirements may be difficult to quantify and assess. The comparison of short-term data for different products can also be difficult, as test procedures can vary, although the publication of many new European Standards in recent years has resulted in agreed, standard methods being used more widely. Documented case histories are often used to compensate for the absence of long-term testing. However, for commercial or technical reasons, formulations may be changed from time to time, thus limiting the usefulness of some case histories and existing test data. For some critical applications, special test programmes may be carried out or be required for product approval to a given specification.

Once a product has been selected, attention must be given to the supplier's recommendations for the methods of storage, mixing and placement. Manufacturers provide comprehensive instructions for the use of their products and these must be read carefully and understood before commencing work. The applicator must accept that a chemical reaction is being carried out at the point of use and that all such reactions are affected by the prevailing conditions, particularly temperature. Use at unacceptably low temperatures (outside the formulator's recommendations) is most likely to result in a very slow rate of curing and a significant delay in the attainment of the properties required for satisfactory service, e.g. loading or exposure to aggressive chemicals. At unduly high temperatures, there may be insufficient time to place the material satisfactorily prior to hardening. Such temperature limits will vary with the formulation as well as the resin type. Consequently, due attention must be given to storage prior to use and the likely weather conditions at the time of application and cure. Rheological characteristics at either temperature extreme can easily lead to poor adhesion if the material is too viscous to wet the surface thoroughly.

With virtually all resin systems, care must be taken to mix the components thoroughly in the correct proportion. Pre-weighed packs ensure correct proportioning, provided the packs are used in their entirety—as usually required, although there are some exceptions. Power-driven paddles or, for large packs of heavily filled products, forced action mixers ensure that mixing can be achieved satisfactorily.

Consideration must be given to the pack sizes that are best suited to a specific use in order to avoid an excessive exotherm and hardening prior to placement. The purchase of large packs, with a lower unit cost, can be false economy if material is wasted or poorly placed. The supplier's recommendations concerning surface preparation and the use of primers must always be followed, as of course should all requirements concerning health and safety matters.

Both suppliers and users play key roles in achieving satisfactory service properties and the successful, cost-effective use anticipated by specifiers and engineers. When these responsibilities are fully accepted, with due co-operation, the great potential of resin-based systems can be fully realized.

## 27.7 Bibliography

*Note (i):* The publication of European Standards (ENs) and Technical Approval Guidelines ((ETAGs) has proliferated since this chapter was first published in 1992. The ENs/ETAGs, and also the Standards published in the USA (ASTMs), that are relevant to thermosetting resins and their applications are too numerous to list here. However, details can be found on the websites of various National/International Standards Organizations.

*Note (ii):* A number of publications from the 1970s and 1980s, referred to in the earlier edition of this chapter, have been omitted here, although they provide some useful commentary on the early developments in this area and, furthermore, they also contain information and guidance that remains useful today. These publications, and many others, are referenced in Volumes 1–3 of the 'state-of-the-art' reports published by CIRIA in 2000 and detailed below (CIRIA Report C537 and CIRIA Project Reports 78 and 79).

### Conferences and symposia

*American Concrete Institute Symposium Publications:* SP-21 (1968); SP-40 (1973); SP-58 (1978); SP-69 (1981); SP-89 (1985); SP-99 (1987); SP-116 (1989); SP-137 (1993); SP-165 (1996); SP-166 (1996); SP-169 (1997); SP-214 (2003). Details of the papers given at these symposia can be found at: www.concrete.org/pubs/journals/sp-home.asp.

*International Congress on Polymers in Concrete:* London, UK (1975); Austin, TX (1978); Koriyama, Japan (1981); Darmstadt, Germany (1984); Brighton, UK (1987); Shanghai, China (1990); Moscow, Russia (1992); Oostende, Belgium (1995); Bologna, Italy (1998); Honolulu, HI (2001); Berlin, Germany (2004); Chuncheon, Korea (2007). See also: www.icpic-community.org.

Ohama, Y., Bibliography on polymers in concrete, *International Congress on Polymers in Concrete* (1990).

### Books and articles

American Concrete Institute, *ACI manual of concrete practice.* This manual, which is regularly updated, contains many guidance documents that are relevant to various applications of thermosetting resins. See also: www.concrete.org/pubs/newpubs/mcp2012.htm.

Cadei, J. M. C., Stratford, T. J., Hollaway, L. C. and Duckett, W. G., *Strengthening metallic structures using externally bonded fibre-reinforced polymers*, CIRIA Report C595, CIRIA, London, UK (2004).

*Encyclopedia of Polymer Science and Technology*, Wiley, Hoboken, NJ. The following articles can be obtained from the online edition of this publication: (i) Ulrich, H., *Polyurethanes*, Volume 4, 26–72 (2001); (ii) Pham, Ha. Q. and Marks, M. J., *Epoxy Resins*, Volume 9, 678–804

(2004); (iii) Nava, H., *Unsaturated polyesters*, Volume 11, 41–64 (2004); (iv) Gotro, J., and Prime, R. B., *Thermosets*, Volume 12, 207–260 (2004). See: http://mrw.interscience.wiley.com/emrw/9780471440260/home.

Feldman, D., *Polymeric building materials*, Elsevier/Applied Science, London, UK (1989).

Gatfield, M. J., *Screeds and floorings: a guide to their selection, construction, testing and maintenance*, CIRIA Report R184, CIRIA, London, UK (1997).

Hare, C. H., *Protective coatings – fundamentals of chemistry and composition*, Technology Publishing Company, Pittsburgh, PA (1994).

Hollaway, L. (ed.), *Polymers and polymer composites in construction*, Thomas Telford, London, UK (1990).

Hollaway, L. and Leeming, M. B. (eds), *Strengthening of reinforced concrete structures*, Woodhead Publishing, Cambridge, UK (1999).

Hurley, S. A., *The use of epoxy, polyester and similar reactive polymers in construction, Volume 1: The materials and their practical applications*, CIRIA Report C537, CIRIA, London, UK (2000).

Hurley, S. A., *The use of epoxy, polyester and similar reactive polymers in construction, Volume 2: Specification and use of the materials*, CIRIA Project Report 78, CIRIA, London, UK (2000).

Hurley, S. A., *The use of epoxy, polyester and similar reactive polymers in construction, Volume 3: Materials technology*, CIRIA Project Report 79, CIRIA, London, UK (2000).

Hutchinson, A. R., *Joining of fibre-reinforced polymer composite structures*, CIRIA Project Report 27, CIRIA, London, UK (1997).

Lee, H. and Neville, K., *Handbook of epoxy resins*, McGraw-Hill, New York, NY (1982).

Mays, G. C. and Hutchinson, A. R., 'Engineering property requirements for structural adhesives', *Proceedings of the Institution of Civil Engineers*, Volume 85, Part 2, Issue 3, 485–501 (1988).

Mays, G. C. and Hutchinson, A. R., *Adhesives in civil engineering*, Cambridge University Press, Cambridge, UK (1992).

May, C. A., *Epoxy resins: chemistry and technology*, 2nd edition, CRC Press, Boca Raton, FL (1988).

Mays, G. C., 'Structural applications of adhesives in civil engineering', *Materials Science and Technology*, Volume 1 (Nov. 1985).

## Trade associations

*Adhesives and sealants:* The British Adhesives and Sealants Association: www.basa.uk.com/Home.aspx

*Concrete repair:* The Concrete Repair Association: www.cra.org.uk

*Fixings and anchors:* The Construction Fixings Association: www.fixingscfa.co.uk

*Flooring:* The Resin Flooring Association: www.ferfa.org.uk

*Waterproofing membranes:* The Liquid Roofing and Waterproofing Association: www.lrwa.org.uk

# 28 Natural and Synthetic Rubbers

**Vince Coveney** PhD
University of the West of England, Bristol (Faculty of Environment and Technology)

## Contents

## 28.1 Introduction

Rubbers or elastomers or elastic hydrocarbon polymers – the terms are widely used interchangeably – are a large class of materials which, because of their properties, are of strategic importance. They are soft (Young's modulus $E \approx 1$ N/mm$^2$ yet with bulk modulus $K \approx 1000$ N/mm$^2$ so that Poisson's ratio $\nu \approx 0.5$) and can undergo near elastic extensions of ~100% and more: i.e. they are approximately hyperelastic at room temperature.[1,2] The foregoing mechanical characteristics of rubbers mean that particular techniques are often required in the stress analysis of these materials.[3] Natural rubber (NR) was the first elastomer and is still one of the most heavily used. Its mechanical properties epitomize those of a fully rubbery material. Ethylene propylene rubbers (EPM and EPDM) are synthetic elastomers extensively used in construction because of their excellent resistance to weathering. The strength, and related properties, of NR and of EPDM are compared with those of other materials in Table 28.1.

In some, especially dynamic, applications full rubbery behaviour offers great advantages. In other cases, such as window-frame seals and roofing membranes, only a limited degree of elasticity is required. In the latter category there is considerable overlap between elastomers, which are sometimes modified to make them more like thermoplastics, and thermoplastics, which have been modified to make them more rubbery.

At the opposite end of the spectrum from (normally vulcanized) natural rubber lie materials that are closely related chemically to rubbers but are not physically rubbery. Examples include the ebonites, which are hard materials obtained by the vulcanization (heating) of natural or certain other rubbers containing very high levels of sulphur. The latter class of materials lies outside the scope of this chapter.

Additives make an important contribution in the performance of finished elastomers. Without the reinforcing action of fine particles of carbon (carbon black, used in rubber since *c.* 1904) or silica, some synthetic rubbers would be so weak as to be of little use. Without the protective action of added antioxidants, antiozonants and carbon black, the surface of natural rubber and other rubbers based on unsaturated hydrocarbon chains would deteriorate within 1 year or so in many environments. Nevertheless, the base polymer of an elastomer remains of key importance.

The symbols and abbreviations used in this chapter are, as far as possible, taken from the International Standard ISO 1629.[8]

## 28.2 Production of base elastomers

Elastomers are generally composed of long chains of repeated molecular units. The (small) chemical compounds which, when joined, make up the polymer chains are called 'monomers'. A co-polymer is a polymer formed from two types of monomer.

The elastomers covered in this section are limited to major types and those finding widespread use in civil engineering.

### 28.2.1 Natural rubber

Although there are other substances present in raw natural rubber that have an effect on the final properties of the final material, the key component is the long-chain polymer (chemically, *cis*-1,4-polyisoprene[9]) which is formed from *c.* 10, 000 repeat units giving an average molecular weight of ~1 million times that of a hydrogen atom (Figure 28.1). Large numbers of plant species produce milky aqueous suspensions (latices) of *cis*-polyisoprenes, but only two species are of commercial interest. The principal rubber-producing plant is the tree *Hevea brasiliensis,* a native of Brazil, which thrives in warm climates with high rainfall. Substantial quantities of natural rubber are produced in Malaysia, Indonesia, Thailand, China, Sri Lanka, tropical America and tropical Africa.

Rubber is tapped from incisions made in the bark of the *Hevea* tree; there is, though, intermittent interest in a second rubber-producing plant which has to be crushed to yield its rubber latex: the guayule bush found in Mexico and Arizona.[10]

As is the case for many other agricultural products, the consistency and yield of *Hevea* natural rubber have been progressively increased through selective breeding: yields in excess of 2000 kg per hectare per year are now obtained.

Some tapped latex is concentrated to about 60% rubber content and stabilized with ammonia and other preservatives, but most is converted directly into dry rubber. The latex from the trees is typically diluted to about 12% rubber content. Acetic or formic acid is then added to coagulate the rubber which is then washed, dried and formed into solid bales. An important additional source of base rubber is 'cup lump' which comes from latex that has coagulated naturally in the tapping vessel. Although rubber from the tree is subject to natural variation, schemes for classification and guaranteed technical specifications have been available to ensure a consistent product since the mid-1960s.

A small proportion of natural rubber is modified chemically into methylmethyacrylate grafted natural rubber (MG rubber) or epoxidized rubber (ENR). (See TARRC, Section 28.11.)[11]

### 28.2.2 Synthetic *cis*-polyisoprenes

These polymers, the nearest synthetic equivalent to natural rubber, are manufactured from isoprene monomer which is available in substantial quantities from petroleum refinery operations. In order to achieve degrees of *cis*-1,4 content approaching the near 100% present in the natural rubber molecule, one of two catalyst systems is normally used: an alkyllithium system (giving approximately 93% *cis* content) or a Ziegler–Natta system (giving approximately 97% *cis* content). These small percentage differences in *cis* content can lead to significant differences in processing and properties of the end product.

### 28.2.3 Butyl (isobutene–isoprene)

The base polymers for these synthetic rubbers are produced by co-polymerizing isobutylene with a small proportion of isoprene monomer dissolved in methyl chloride or, alternatively, in the presence of catalysts such as $BF_3$ or $AlCl_3$. The temperature, which is kept low, controls the molecular weight of the polymer. Like isoprene, and most other 'feedstock' chemicals used in the manufacture of synthetic elastomers, isobutylene monomer is available in substantial quantities from petroleum refinery processes. The nearness to saturation (of the carbon chain with hydrogen atoms and corresponding lack of double bonds) has a major effect on the properties obtained with the material. The small amount of isoprene facilitates sulphur vulcanization.

### 28.2.4 Styrene–butadiene

As the name implies, styrene–butadiene rubbers (SBRs) are co-polymers of styrene and butadiene monomers—either as random or as block co-polymers. The proportion of styrene can vary between 20% and 40% in the rubber grades, but can be over 60% in the case of high styrene resins.

Styrene can be produced from the petroleum derivative ethylbenzene by a range of processes. Butadiene monomer can be formed by dehydrogenation or oxidative dehydrogenation of butanes or butenes or by cracking of petroleum fractions.

Much SBR is still manufactured by polymerization in an aqueous emulsion of styrene and butadiene monomers. This produces essentially random polymers. However, since the 1960s, solution-polymerized SBRs have been produced commercially.

**Table 28.1** Comparison of material properties (ranges are given where appropriate)

| Property | Unit | Natural rubber | | | | EPDM filled[e] | Polyethylene (high density) | Nylon 66 | Aluminium | Glass (common) | Concrete | Wood[f] (oak) | Water |
|---|---|---|---|---|---|---|---|---|---|---|---|---|---|
| | | Unfilled[a] | Filled[b] | Ebonite[c] | Foam[d] | | | | | | | | |
| Density | Mg m$^{-3}$ | 0.95 | 1.12[f] | 1.08–1.10 | 0.1–0.6 | 1.2 | 0.94–0.97 | 1.14 | 2.70 | 2.4–2.7 | 2.1 | 0.74–0.77 | 1.00 |
| ***Thermal properties*** | | | | | | | | | | | | | |
| Specific heat | J g$^{-1}$ °C$^{-1}$ | 1.83 | 1.50 | 1.38 | 1.89 | 2.09[g] | 2.3 | 1.7 | 0.95 | 0.84 | 1.89 | 1.37 | 4.19 |
| Thermal conductivity | W m$^{-1}$ °C$^{-1}$ | 0.15 | 0.28 | 0.16 | 0.045 | 0.39 | 0.45–0.52 | 0.25 | 230 | 1.1 | 1.2 | 0.38 | 0.65 |
| Coefficient of linear expansion (× 10$^5$) | °C | 22 | 18 | 6 | 5–10 | 1.89 | 11–13 | 8 | 2.3 | 0.9–1.1 | 1.0–1.5 | 0.5[i] | 7 |
| ***Mechanical properties*** | | | | | | | | | | | | | |
| Hardness | | | | | | | | | | | | | |
| Brinell | kgf mm$^{-2}$ | ‘1 | ‘1 | 11–12 | – | – | 1 | – | 22 | ~340 | – | – | 0 |
| International rubber scale[j] | IRHD | 45 | 65 | 100 | <20 | 65 | 98 | 100 | 100 | 100 | 100 | 100 | 0 |
| Tensile strength | MPa | 24[k] | 24[k] | 40–60 | 0.2–1.5 | 17 | 20–35 | 60–80 | 70–100 | 30–90 | 20–50 | 40–120 | 0 |
| Elongation at break | % | 700 | 550 | 3–8 | 200–300 | 450 | 20–600 | 60–100 | 5 | 3 | – | – | 0 |
| Young's modulus, $E$[l] | MPa | 1.9 | 5.9 | 3000 | 0.05–0.8 | 5.9[g] | ≈1000 | 1000–3000 | 70 000 | 60 000 | 25 000–35 | 9000 | 0 |
| Bulk modulus, $K$ | MPa | 2000 | 2000 | 4170 | – | – | 3300 | – | 70 000 | 37 000 | – | – | 2050 |
| Poisson's ratio, $\nu$ | – | 0.499 | 0.499 | 0.2 | 0.33 | 0.499 | 0.38 | 0.38 | 0.345 | 0.22 | 0.2 | – | 0.5 |
| ***Acoustic properties*** | | | | | | | | | | | | | |
| Velocity of sound[m] | m s$^{-1}$ | 1600 | 1700[g] | – | – | – | – | – | 6400 | 5000–6000 | 4300–5300 | 4000[g] | 1500 |
| ***Electrical properties*** | | | | | | | | | | | | | |
| Volume resistivity | Ω m | $10^{14}$ | $10^{10}$[n] | $10^{14}$ | $10^{14}$ | $10^{10}$[g,n] | $10^{15}$ | $10^{12}$ | $10^{-8}$ | $10^9$ $10^{11}$ | $10^7$ $10^{10}$ | $10^8$ $10^{11}$[o] | $10^2$ $10^5$[p] |
| Dielectric constant (relative permittivity) | – | 2.5–3.0 | 15 | 2.8 | – | – | 2.25–2.35 | 4.0–4.6 | – | 7 | – | 3–6 | ≈80 |
| Dissipation factor | – | 0.002–0.04 | 0.006–0.14 | 0.004–0.009 | – | – | <0.005 | 0.01–0.02 | – | 0.005 | – | 0.04–0.05 | – |

[a]Gum vulcanizate containing vulcanizing ingredients only. [b]Vulcanizate containing ~50 parts of a reinforcing carbon black per 100 parts, by weight, of natural rubber. [c]Unfilled ebonite containing ~40 parts of sulphur per 100 parts of industrial rubber. Ebonite, which is also known as 'hard rubber' in the USA, has a maximum glass transition temperature of ~80°C, whereas unmodified natural rubber has a transition temperature as low as –70°C. For further information on natural rubber ebonite, see Refs 4 and 5. [d]Properties of latex foam and other cellular products from natural rubber will vary with extent of foaming or expansion. Young's modulus of open-cell foam can be estimated from Young's modulus of the solid rubber component and the density of the foam (Ref. 6). [e]Typical formulation (120 parts of black, 95 parts of oil) for roof sheeting given by Easterbrook and Allen (Ref. 7). [f]Wood is strongly anisotropic. Where appropriate, properties refer to tests along grain. [g]Estimated value. [h]Across grain, ~0.2 W m$^{-1}$ °C$^{-1}$. [i]Across grain, ~5 × 10$^{-5}$. [j]Hardness scale range is 0 (e.g. liquid) to 100 (e.g. glass). [k]Calculated on the original unstrained cross-sectional area; when calculated using on the cross-section just before break, the value can be as high as 200 MPa. [l]For isotropic materials $E$ is related to $G$ and $K$ as follows: $E = 2G(1 + \nu)$ and $E = 3K(1 - 2\nu)$. [m]Compression varies in medium or large extent. [n]Depends on filler type and concentration. High values are obtained by replacing carbon black with silica. Using specially conductive carbon black, values under 10 Ω m can be achieved. [o]The higher value applies to paraffinic timber. [p]For distilled water.

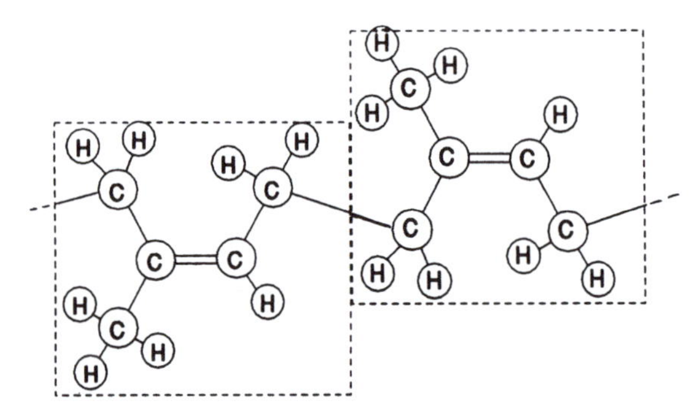


**Figure 28.1** *cis*-1,4-polyisoprene

The solution polymerization process allows more scope for control of the final polymer structure.

### 28.2.5 Butadiene (polybutadiene)

Polybutadiene (BR) is generally produced by polymerization of butadiene monomers in solution. The reaction can take place in a number of organic solvents in the presence of a variety of catalysts. The choice of reaction conditions, including catalyst type, has a major effect on the structure of the polymer.

### 28.2.6 Nitrile

The less specialized grades of acrylonitrile–butadiene rubbers—popularly known as nitrile rubbers (NBR)—are prepared by emulsion polymerization of acrylonitrile and butadiene monomers at low to moderate temperatures. Proportions of acrylonitrile in the final polymer are varied from below 20% to over 50% to produce materials with a wide range of properties.

There are a number of ways in which acrylonitrile monomer is produced commercially, ethylene being the usual starting point. The production of butadiene monomer is described in Section 28.2.5.

### 28.2.7 Ethylene-propylene

Ethylene–propylene (EPM) rubbers are the copolymers of ethylene and propylene alone; in EPDM rubbers a third monomer is also present in small quantities to introduce the unsaturation needed for sulphur vulcanization (see Section 28.4). By varying the proportions of ethylene, propylene and the type and amount of the third monomer, significant changes in properties may be achieved.

Production of ethylene-propylene rubbers involves a polymerization process using Ziegler–Natta-type catalysts.[12] Polymerization is frequently performed in organic solution, but sometimes the reactions take place in suspension.

### 28.2.8 Chloroprene (polychlorophene)

Polychloroprene (CR) has a structure related to that of polyisoprene, the methyl group in the isoprene repeat unit being replaced by a chlorine atom. Chloroprene monomer is now predominantly produced from butadiene. Polymerization is performed in emulsion with suitable catalysts, a latex of raw chloroprene rubber being produced.

Chloroprene rubbers are available with varying chemical compositions and this can have a marked effect on the tendency to crystallize—an important parameter in categorizing these materials. Commercial chloroprene rubber consists largely of *trans*-polychloroprene—in general, the closer the proportion of this constituent is to 100%, the greater the tendency of the material to crystallize rapidly. Co-polymers of chloroprene and 2, 3-dichlorobutadiene are resistant to low-temperature crystallization; chloroprene–sulphur co-polymers (the DuPont G types, for example) have intermediate crystallization properties.

### 28.2.9 Polyurethane

The term 'polyurethene' covers a wide range of polymeric materials with urethane links. The materials are formed by the reactions of polyols with diisocyanates. Polyurethanes can be divided into polyester (AU) and polyether (EU) based materials. Polyesters are manufactured by reaction of dibasic acids with diols. (Ethylene glycol, 1, 2-propylene glycol, and diethylene glycol are examples of diols used when linear polymer segments are required.) For chain branching or ultimate cross-linking, triols are used. Propylene and appropriate glycols are the common starting points or 'feedstocks' for the manufacture of polyurethanes. The two types of polyether commonly used in the production of polyurethane elastomers are polypropylene glycols and polytetramethylene glycols.

### 28.2.10 Silicone

These differ from other elastomers in that the main chain molecular structure consists of silicon and oxygen rather than carbon atoms—as is the case for natural and most other rubbers. As a consequence, and in marked contrast to most other synthetic elastomers, the production of silicone rubber is not heavily reliant on petroleum products. Production of silicone polymers commonly involves the reaction of elementary silicon with an alkyl halide (frequently methyl chloride). The resulting chlorosilanes subsequently react with water to give hydroxyl compounds which then condense to give a polymer structure. In the case of silicone rubbers containing vinyl groups (i.e. PVMQ and VMQ), intermediate reactions between acetylene and chlorosilanes are used. Other substituent groups are methyl (e.g. MQ), phenyl (e.g. PMQ) and fluorine (e.g. FMQ).

### 28.2.11 Chlorosulphonated polyethylene

Chlorosulphonated polyethylene (CSM) is produced by the random substitution of sulphonyl chloride groups and chlorine onto preformed polyethylene molecules. This is achieved by reacting a solution of polyethylene in boiling carbon tetrachloride with chlorine and with sulphur dioxide. The process converts polyethylene into a rubbery material and introduces alternative sites for cross-linking.

## 28.3 Available forms

Elastomers are available in bale (nominally solid), particulate or liquid forms prior to vulcanization (see Section 28.4). An indication of the forms available for each common elastomer type is given in Table 28.2.

## 28.4 Vulcanization and 'compounding'

Vulcanization, known also as curing, is the process whereby the polymer chains of a base elastomer are chemically cross-linked (Figure 28.2). The process was first applied to natural rubber in 1839 by Charles Goodyear. Previously, unvulcanized natural rubber had been used for rubber balls, erasers and mackintoshes, but vulcanization resulted in an altogether stronger, more environmentally resistant material and paved the way for the use of rubber in mechanically demanding applications.

**Table 28.2** Available forms of common elastomers

| *Elastomer* | *Bale/slab* | *Chip/pellet* | *Particulate* | *Latex* | *Liquid/paste* | *Other* |
|---|---|---|---|---|---|---|
| Natural (NR) | **[a] | | * | * | * | |
| Synthetic *cis*-polyisoprene (IR) | **[a] | | | * | * | |
| Butyl (IIR) | ** | | | * | * | |
| Styrene–butadiene (SBR) | **[b] | | | * | | |
| Butadiene (BR) | **[a] | | | * | * | |
| Nitrile (NBR) | ** | * | * | * | * | |
| Ethylene–propylene (EPM/EPDM) | **[a] | * | * | | | *[c] |
| Chloroprene (CR) | | ** | * | * | | |
| Polyurethane (AU/EU) | * | | | | ** | |
| Silicone (MQ, etc.) | *[d] | | | | * | |
| Chlorosulphonated polyethylene (CSM) | | * | | | | |
| Ethylene acrylic (AEM) | *[d] | | | | | |
| Polyacrylic (ACM/ANM) | * | * | * | ** | | |
| Fluorinated hydrocarbon (FPM or FKM) | | ** | | | * | |
| Polysulphide (T) | ** | | | | * | |
| Epichlorohydrin (CO/ECO) | * | | | | | |
| Polypropylene oxide (GPO) | * | | | | | |
| Polynorbornene | *[a] | | * | | | *[d] |

*Available form; **very common form. [a]Oil-extended grades are available. [b]Oil-extended SBR is common. Carbon black filled master-batches with and without oil are also available. [c]Crumb and friable bale forms also available. [d]Often partly compounded.

The original vulcanization process used sulphur and high temperatures (hence the name 'vulcanization', from Vulcan, the Roman god of fire). The majority of cross-linking processes for natural and other rubbers still involve the use of sulphur and elevated temperatures. However, vulcanization conditions can vary a great deal from high-temperature (up to 200°C) cures lasting only a few seconds, through 'normal' cures typically lasting around 15 min at about 150°C, to room-temperature vulcanization taking place over days or longer. The properties of the vulcanized rubber or, as it is often called, the 'vulcanizate', are influenced by the number of cross-links and their type; the properties can also be affected by other molecular modifications to the elastomer that occur during the vulcanization process.

The term 'compounding' in rubber technology refers to the mixing of various ingredients into rubber. Mixing of nominally solid rubber is performed on open two-roll mills and in closed mixers; the high shear forces cause the rubber to flow and disperse ingredients and can also be used to progressively break down the molecular weight or 'masticate' certain rubbers. 'Mastication' reduces the elasticity of the unvulcanized material and facilitates shaping.[13] In addition to the vulcanizing chemicals, the ingredients usually include particulate fillers. Altering the quantity and type of filler is a principal method of adjusting the modulus, strength, fatigue and damping properties of natural and many other rubbers. The fillers can be non-reinforcing (clay, chalk and some carbon black powders fall into this category) or reinforcing (these can be sub-μm powders of carbon black or silica–carbon black or silica). Research and development is being performed on the possible use in rubbery materials of truly nanometric fillers such as intercalcated or exfoliated clay, polyhedral oligomeric silsesquioxane or carbon nanotubes. Non-reinforcing fillers, added to cheapen or harden a rubber, generally have a neutral or negative effect on strength properties. Reinforcing fillers, in contrast, tend to have a positive effect on strength properties—this is particularly true for some of the synthetics which are weak without reinforcing fillers. Silicone rubber is an example of an elastomer whose strength properties benefit greatly from the presence of appropriate silica filler.

Frequently, rubber compounds include some oil or other plasticizer added to aid processing of the rubber prior to vulcanization and/or to influence the properties of the vulcanizate—notably stiffness or low-temperature resistance. The most commonly used classes of oils are naphthenic, aromatic and paraffinic; all three are products of the petrochemical industry. For silicone rubber, silicone oil is generally used.

Many elastomers have antidegradants of various kinds added during compounding to improve the resistance of the final vulcanizate to attack by oxygen or ozone, for example. Natural, butadiene and styrene-butadiene rubbers are examples of elastomers whose environmental resistance is greatly improved by the presence of appropriate antidegradants.

Vulcanized elastomeric and rubber products can be given their final shape in a number of ways. Machining of vulcanized rubber

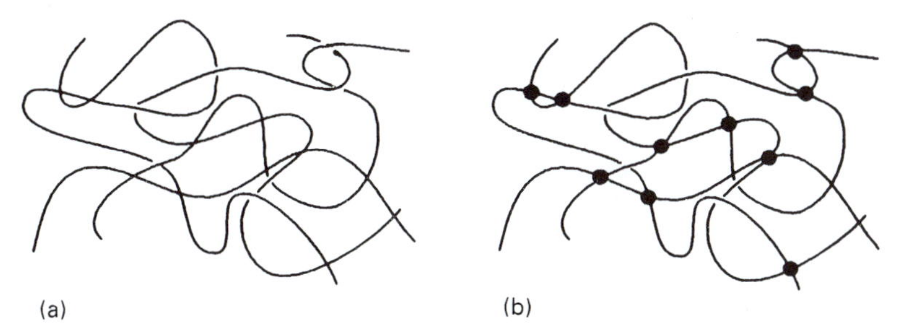


**Figure 28.2** Elastomer chains and cross-linking: (a) elastomer chains coiled, kinked and entangled but not chemically cross-linked; (b) elastomer chains after vulcanization, cross-linked (•) together

is relatively uncommon, except, perhaps, for trimming, but calendering of unvulcanized sheet, coating of fabric from rubber solution, extrusion and moulding are all common.

Broadly speaking, there are three types of moulding. In compression moulding a quantity of unvulcanized material is placed in a single-cavity mould with a closing lid or plunger which is then pressed between heated platens of a press, whereupon the elastomer is shaped and vulcanized. Transfer moulding is also widely practised and here the plunger is located in the first cavity; when the mould is under the press the rubber is forced through transfer ports into the second, shaping, cavity. Compression or transfer moulding time is normally not less than 5 min, although there is considerable variation depending on size of component, rubber formulation and temperature.

Injection moulding of a thin-walled elastomer component typically takes less than 1 min for conventional elastomers. For thermoplastic elastomers (TPE), however, chemical cross-linking does not take place during moulding; instead, within the TPE, thermoplastic areas, molten during and solid after moulding, take the place of chemical cross-links. Because there is no cure time very rapid injection moulding cycles are possible for TPEs.

With conventionally moulded elastomers it is not unusual for products to be heated in an oven after removal from a hot mould; this process, termed 'post-cure', is particularly common for certain speciality rubbers, e.g. fluorinated types. Post-cure can improve elastic recovery in the vulcanizate and reduce tendency to creep without the need for long and expensive moulding cycles.

Steam autoclaves and molten-salt baths are just two other ways in which rubber articles are cured; further methods are described by Crowther.[14]

### 28.4.1 Natural rubber

Vulcanization of natural rubber is normally carried out at temperatures in the range 120–180°C, although in some processes cross-linking at room temperature is used. In general the higher the temperature, the more rapid is the cure. Most natural rubber (NR) is cross-linked with sulphur-based vulcanizing systems. However, high levels of sulphur and long cure times would be required without additional ingredients. Accelerators speed up the cross-linking processes. Vulcanizing systems with relatively high ratios of sulphur to accelerator are called 'conventionally accelerated'; these systems give rise to cross-links most of which are formed by many sulphur atoms (polysulphidic). The lower the ratio of sulphur to accelerator, the fewer sulphur atoms per cross-link. Examples of conventionally accelerated, efficient or low-sulphur vulcanization (EV) and semi-EV formulations are given in Table 28.3; a wide variety of natural rubber formulations may be found in the Bibliography and References (e.g. Malaysian Rubber Producers' Research Association,1979[15]). EV formulations give improved long-term high-temperature properties at the expense of strength and fatigue properties. The improved 'reversion' resistance (stability of properties at high temperature) of efficient sulphur vulcanizates also enables products with thick cross-sections to be moulded more quickly through an increase in cure temperature. Examples of sulphur-free vulcanizing systems are those based on (dicumyl) peroxide and 'Novor' (urethane), the latter giving similar results to an EV system. Normal sulphur level formulations are more resistant to low-temperature crystallization than are EV or dicumyl peroxide formulations (see Section 28.6).

Natural rubber vulcanizates of different moduli can be obtained by varying the degree of cross-linking; however, this method is limited since too few cross-links result in a material with rather high creep, while too many cross-links limit the extensibility of the material. Oils can be added to reduce the modulus of the final vulcanizate, but more commonly are added to modify processing behaviour; high-viscosity oils are sometimes used in natural rubber to increase the damping of the final vulcanizate (similar comments apply to many other elastomers and notably to the speciality material polynorbornene).

**Table 28.3** Natural rubber formulations

| *Formulation* | *Parts by weight* | |
|---|---|---|
| *EDS24* | | |
| Natural rubber, SMR CV | 100 | |
| Zinc oxide | 5 | |
| Stearic acid | 2 | |
| Carbon black, GPF (ASTM N-660) | 20 | |
| Process oil | 2 | |
| Antioxidant/antiozonant, HPPD | 3 | |
| Antiozonant wax | 2 | |
| CBS (accelerator) | 0.6 | Conventionally accelerated normal sulphur level, vulcanizing system |
| Sulphur | 2.5 | |
| *EDS4* | | |
| Natural rubber, SMRL | 100 | |
| Zinc oxide | 5 | |
| Zinc 2-ethylhexanoate | 1 | |
| Carbon black, SRF (ASTM N-762) | 30 | |
| Antioxidant, TMQ | 2 | |
| Antiozonant, DOPD | 4 | |
| MBS (accelerator) | 1.7 | Semi-EV, moderately low-sulphur system |
| TBTD (accelerator) | 0.7 | |
| Sulphur | 0.7 | |
| *EDS42* | | |
| Natural rubber, SMR CV | 100 | |
| Zinc oxide | 5 | |
| Stearic acid | 2 | |
| Carbon black, FEF (ASTM N-550) | 20 | |
| Process oil | 2 | |
| Antioxidant/antiozonant, HPPD | 3 | |
| Antiozonant wax | 2 | |
| MBS (accelerator) | 2.1 | EV, low-sulphur system |
| Sulphur | 0.25 | |
| TMTD (accelerator) | 1.0 | |

### 28.4.2 Synthetic *cis*-polyisoprene

Synthetic *cis*-polyisoprene, the artificial analogue of natural rubber, is generally a softer material to process and care can be needed to obtain good dispersion of fillers on the open mill. The base polymer contains less, and different, non-rubber material than does natural rubber: for similar additions of vulcanizing agent this can result in the cure being somewhat slower for the synthetic material. Synthetic *cis*-polyisoprene is sometimes blended with NR to 'fine-tune' the processing characteristics.

### 28.4.3 Butyl

There are three broad types of butyl rubber base polymer: unmodified butyl rubbers (IIR), chlorobutyl rubbers (CIIR), and bromobutyl rubbers (BIIR).

Cross-linking of IIR may be carried out with sulphur-based or non-sulphur-based vulcanizing systems (quinoid or phenolic resin, for example), but cure is sluggish, particularly for variants with few chemical double bonds (low unsaturation); this sluggishness necessitates the use of powerful accelerators and/or high temperatures and long cure times. Chlorobutyl and bromobutyl rubbers are less sluggish to cure and can be successfully co-vulcanized with rubbers such as NR and SBRs. Carbon black is routinely used as a reinforcing filler, and diluent oils are added to aid processing.

### 28.4.4 Styrene–butadiene

Styrene–butadiene rubber (SBR) is a generic term for a wide variety of rubbers. Factors affecting final properties of the vulcanizate and processing include: the relative numbers of styrene and butadiene units and their sequence distribution in the polymer chains; molecular weight (distribution) of the polymer chains; and molecular branching. Much SBR continues to be produced by emulsion, rather than solution, processes; the range of properties for emulsion SBRs is more restricted (see Section 28.2.4); comments in this and subsequent sections are directed primarily towards this more restricted class of SBRs.

SBRs are processed in a conventional manner, but do not break down on milling (see earlier in Section 28.4); they are normally vulcanized with sulphur-based systems. EV systems offer the same kinds of advantages over higher sulphur level vulcanizing systems for SBR as for NR (flat cure characteristic and thermal stability of vulcanizate), without the prime disadvantage of worsened fatigue properties.

SBRs require quite high (*c.* 50 pphr (parts per hundredweight of base elastomer)) quantities of reinforcing filler (usually carbon black, but sometimes silica) to achieve good strength properties. Oils are frequently included in SBR formulations—naphthenic or aromatic types are normally used.

### 28.4.5 Butadiene

The comments on compounding made above for SBR generally apply to butadiene rubbers (BRs) also. Although it is sometimes used alone, polybutadiene is generally used in blends with NR, SBR, and IR or with (the thermoplastic) polystyrene, in this last case to provide impact resistance.

### 28.4.6 Nitrile

Nitrile (acrylonitrile butadiene rubbers, NBR) vary in their acrylonitrile content; this is a key parameter affecting the properties of the final vulcanizate: the higher the acrylonitrile content, the better the oil resistance and the less rubbery the material—particularly at low temperatures. NBR is frequently blended with polyvinylchloride (PVC) to improve the ability of NBR to withstand attack by ozone, etc.

Similar comments on processing, vulcanization and reinforcement apply to nitrile rubber as to SBR, except that for optimum high-temperature endurance silica fillers are sometimes favoured over carbon black. The liquid diluents used are dissimilar to those used for SBR—phthalate plasticizers and aliphatic esters being commonly used to improve the low-temperature flexibility of NBR.

Carboxylated and hydrogenated nitrile rubbers are both elastomers in their own right and have their own processing requirements.

### 28.4.7 Ethylene–propylene

Reinforcement by carbon black, or sometimes silica, is often required for ethylene–propylene rubbers (EPDM and EPM); particular care needs to be taken with dispersion of the filler. Cross-linking of EPM rubbers is achieved with peroxide systems. For EPDM cross-linking is possible with a vulcanizing system based on sulphur and supported by powerful accelerators; however, peroxide vulcanizates are also used, especially when non-staining and improved high-temperature properties are required. Large quantities of carbon black filler and diluent oil are often incorporated into ethylene–propylene rubbers. Since chemical reactions may take place between peroxides and naphthenic or aromatic oils, paraffinic oils are the commonly used diluent for EPMs; for sulphur cured EPDMs paraffinic and naphthenic oils are used.

### 28.4.8 Chloroprene

The different classes of chloroprene rubber (CR, also known as polychloroprene) exhibit rather different processing behaviour: sulphur modified grades can be permanently softened by chemical peptizers and by heat before vulcanization. Vulcanization is generally carried out with sulphur-free systems for all classes of polychloroprene at temperatures in the range 150–205°C. The vulcanizing system is usually based on metal oxides, typically of zinc and magnesium; red lead oxide cures can be used to give improved water resistance. CR can also be vulcanized with sulphur-based systems.

Polychloroprene possesses some inherent resistance to fire but its performance, and that of many other rubbers, can be improved by avoiding the use of flammable additives and by incorporating combustion-inhibiting additives such as antimony trioxide, which acts synergistically with CR, zinc borate and hydrated alumina.

Inclusion of carbon black filler in CR formulations is standard, although the intrinsic strength of the material means that filler reinforcement is not obligatory. Naphthenic and aromatic oils are used as diluents. Ester plasticizers can also be used to soften the final vulcanizate at room and especially low temperatures; however, their incorporation can facilitate (time-dependent) low-temperature crystallization and under some circumstances can, paradoxically, lead to the material being stiffer (pre-yield) after long times at low temperature. Use of sulphur in the vulcanizing system and the presence of certain resinous plasticizers inhibit low-temperature crystallization.

### 28.4.9 Polyurethane

The term 'polyurethane rubber' applies to a broad class of linear block co-polymer materials: alternating molecular blocks of isocyanates, polyols and other substances.

An elastomeric product can be produced from a 'castable' polyurethane in a 'one-shot' process by reacting polyol, diisocyanate and chain extender. More commonly, the polyol and diisocyanate are reacted together to form a prepolymer. The finished elastomer is formed by the prepolymer being 'cross-linked' by a diol or diamine chain extender. A considerable amount of cross-linking often occurs after removal of the product from the mould.

More 'conventional' (in rubber terms) processing and vulcanization can be applied to the 'millable' polyurethanes. In general, these are of relatively low molecular weight and are essentially linear polymers. Cross-linking is achieved by a sulphur, peroxide or diisocyanate cure system, depending on the type of millable polyurethane. Fillers are sometimes used but are not obligatory in polyurethanes, which often have intrinsic strength.

Thermoplastic polyurethane rubbers (TPU) were among the first commercial thermoplastic elastomers. They include truly elastomeric materials but are processed and formed in ways more akin to plastics than to conventional rubbers. Elastic recovery tends to be poorer for TPUs than for the castable grades.

### 28.4.10 Silicone

Many silicone rubbers are not cured by any process resembling conventional vulcanization, and cure conditions are extremely varied for these materials.

The familiar room temperature vulcanizing (RTV) one-pack silicone rubbers (for bath sealant etc.) are cured by atmospheric humidity—limiting the thickness practicable. Some two-pack RTV silicone rubbers (for flexible moulds etc.) are cured by addition (A + B → C) and others by condensation (A + B → C + $H_2O$) cross-linking reactions.

Neither methyl-containing (MQ) nor phenyl-containing (PMQ) rubbers can be vulcanized with sulphur: peroxide cross-linking is generally used. Vinyl-containing silicone rubbers (VMQ and PVMQ) are also normally cross-linked by peroxide. An alternative method of cross-linking is 'hydrosilation', which involves the reaction between a polymer containing vinyl groups; and a material containing hydrosilane (Si–H) groups; this is the type of reaction used in the cure of liquid silicone rubbers for example.

Silicone rubbers tend to be intrinsically weak; the use of reinforcing fillers is therefore of crucial importance: silica filler with a 'high structure' (strong tendency to cluster together) is most commonly used.

### 28.4.11 Chlorosulphonated polyethylene

A significant proportion of chlorosulphonated polyethylene (CSM) rubber is used in sheet form without curing. A wide range of cross-linking systems are available, each giving particular advantages and disadvantages. Carbon black may be used as a filler, although other fillers such as silica enable bright colours, including brilliant white, to be achieved. Liquid diluents are sometimes used, but for high-temperature applications these need to be chosen with care.

## 28.5 Moulding of thermoplastic rubber and elastomer

Thermoplastic elastomers (TPEs), like other thermoplastic materials, are set into shape by cooling of the hot melt. This contrasts with 'conventional' (thermoset) rubbers whose final shape is fixed by vulcanization. Moulding can, therefore, be much faster for TPEs than for thermoset elastomers; furthermore, it is more practicable to reuse TPE scrap. The foregoing two features together with the ability to cover a wide range of moduli have led to a significant increase in the use of TPEs in recent years. The properties of moulded TPEs are, however, more sensitive to details of flow in the mould than is the case for mould-vulcanizing elastomers. This, and the fact that most TPEs show poorer elastic recovery than do high-quality cross-linking elastomers, is not a great disadvantage in a large number of applications. Whereas for conventional elastomers compounding is generally carried out by the component manufacturer, for thermoplastic elastomers pre-compounded grades are supplied, normally in particulate form.

### 28.5.1 Block co-polymer thermoplastic elastomer

Block co-polymer TPEs are composed of blocks of hard polymer (amorphous with high glass transition temperature or crystalline) alternating on a molecular scale with soft rubbery polymer (low glass transition temperature). At high temperatures the hard segments soften enough to allow the material to flow into shape. However, up to moderate operating temperatures the hard segments link the soft segments together and have a similar function to chemical cross-links and reinforcing filler in a conventional vulcanized and filled rubber. Styrene–butadiene–styrene (SBS) polymers are an early example: the glassy polystyrene blocks link the rubbery polybutadiene blocks. Some ethylene–vinyl acetate (EVA) materials can also be regarded as TPEs.

Polyether–polyester block co-polymer rubbers first became commercially available a decade after SBS, but now, together with polyurethane and polyether–polyamide materials, epitomize the high mechanical performance TPEs. In polyether–polyester TPEs most of the polyester (hard) material is crystalline at operating temperature but some is in the amorphous phase with the polyether; this morphology makes for a strong material.

Polyurethane TPEs consist of urethane-sparse (soft) and urethane-rich (hard) segments; their morphology is broadly similar to that of polyether–polyester TPEs.

Polyether–polyamide block co-polymer TPEs consist of soft (polyether) and hard (polyamide) segments.

### 28.5.2 Thermoplastic polyolefin elastomer

These materials (sometimes known as TPOs) are essentially physical blends between rubbers such as ethylene–propylene rubber (EPM, EPDM), natural rubber (NR) or nitrile rubber (NBR) and polyolefin thermoplastics—typically polypropylene (PP) but sometimes polyethylene.[16] Either the rubber or the thermoplastic may be the continuous phase, depending on the proportions present, method of preparation and other factors. More rubbery materials can be obtained by vulcanization of the rubbery phase during blending with the thermoplastic (rather than prior to blending); these materials are sometimes called 'thermoplastic elastomer alloys'. Further cross-linking is sometimes carried out after the formation of the final product, thereby blurring the demarcation between these materials and conventional elastomers. Fillers and liquid diluents are sometimes incorporated into TPEs, although such additions are not normally made by the moulder as they can adversely affect the morphology of the TPO.

## 28.6 Properties of elastomers

The properties of an elastomer are normally categorized into mechanical aspects such as strength, modulus or damping, and into the elastomer's tolerance to 'environmental' factors such as high or low temperature, and exposure to oils, ozone or ultraviolet radiation. The balance between the importance of mechanical properties and the degree of hostility of the environment will help determine whether an elastomer is suitable for a particular function. Simplified tables and guides to the performance of different elastomers may sometimes be misleading, as environmental factors can interact with each other and with mechanical conditions. Although indications are given here of some of the complications, the primary concern in the present section is with broad indications of elastomer behaviour.

Certain elastomers, notably natural rubber (NR) and chloroprene rubber (CR), undergo partial conversion from an amorphous to crystalline structure on stretching to high strains, thereby giving these materials inherent advantages in strength and resistance to fatigue. Both materials partially crystallize, and thus harden, with time at low temperatures—but temperatures well above the glass transition. Non-strain-crystallizing elastomers such as silicone rubber and most SBRs require reinforcing fillers such as carbon black or silica to give them the strength required for many applications. Some rubbers, such as ethylene–propylene and silicone, have high innate resistance to degradation by ozone and by diatomic oxygen in the atmosphere, whereas others, notably nitrile (NBR) and fluorinated hydrocarbon rubbers, are relatively unaffected by many mineral (petroleum-based) oils.

The design of a component can have a profound influence on the performance with regard to fatigue or environmental attack; fatigue life can be greatly improved by the avoidance of stress concentrations, while thick sections of 'non-resistant' rubber can often function adequately despite oil splashes or ozone attack.

Many elastomers capable of withstanding high temperatures for long periods without deterioration but may have significantly lower typical strength at the elevated temperatures compared to more normal temperatures. Other elastomers can sometimes be used successfully above their normal operating limit for short exposures. What constitutes an acceptably long life varies according to the application from hours, or even less, to many decades.

It must also be emphasized that the properties for a given elastomer type can vary widely depending on 'compounding', ingredients, and processing details (see Section 28.4). Furthermore, elastomers are frequently blended with each other.

### 28.6.1 Natural rubber

Natural rubber's (NR) combination of inherent strength and high elasticity means that, mechanically, it sets the standard for other elastomers. The damping of NR vulcanizates is generally low. It is unsuitable for heavy exposure to mineral (petroleum-based) oils; because of this and its inability to tolerate temperatures much above 100°C, NR is classified as a 'general-purpose' rubber. It is, however, resistant to a surprisingly wide range of liquids. NR's vulnerability (and that of other elastomers such as SBR) to oxidative ageing and ozone attack can be overcome in many circumstances by incorporating protective agents and avoiding high stresses at the surface of the component; in others the bulk of a component means that some surface damage is acceptable. Protection against ultraviolet radiation is achieved by incorporating some carbon black filler in the formulation.

A typical operational temperature range is –60 to +80°C (continuous). Time-dependent stiffening due to crystallization can occur at temperatures well above –60°C; the maximum crystallization rate occurs in the region of –25°C. Although such stiffening is reversed by moderate strains or warming, the use of crystallization-resistant (e.g. conventionally sulphur vulcanized) formulations may be advisable in some applications.

### 28.6.2 Synthetic *cis*-polyisoprene

Isoprene rubbers (IR) are chemically the closest of the synthetic rubbers to NR, although they normally contain a slightly lower percentage of *cis*-1, 4-polyisoprene and have significantly less molecular branching. The vulcanizates are typically somewhat weaker than NR, especially at elevated temperatures.[17] A typical operational temperature range for IR is –60 to +80°C.

### 28.6.3 Butyl

The damping of butyl rubber (IIR) is typically much higher than that of NR, and this attribute is exploited in some applications.

Among the elastomers IIR has one of the lowest permeabilities to gases, although the permeabilities of some epichlorohydrin rubbers and NBRs can be even lower. (It should be noted that the gas permeability depends on the temperature.) IIR has good resistance to oxidation and ozone but not to many mineral oils. Resistance to damage by flexing, abrasion and tearing can, with appropriate compounding, come close to that of NR, but IIR does have somewhat higher creep, and lower tensile strength. The typical operational temperature range is –60 to +120°C.

### 28.6.4 Styrene-butadiene

Like NR, SBRs are vulnerable to mineral oils and often require protection against oxidation and ozone attack. Tensile strength can be increased by an order of magnitude (and extension to break increased somewhat) by appropriate fillers. Filled SBR has good abrasion resistance but is weaker than NR in other respects. Mechanical damping is typically higher than for NR. A representative operational temperature range for SBRs is –40 to +80°C.

### 28.6.5 Butadiene

Butadiene rubbers (BRs) are commonly available with low, medium, high and very high (> 96%) contents of *cis*-butadiene monomeric units. (Materials such as *trans*-polybutadiene rubber are not dealt with here.) Very high *cis* content BRs strain crystallize and this improves the otherwise low strength somewhat.

BRs, like NR and SBR, frequently require protection against oxygen and are vulnerable to mineral oils. Mechanical damping can be even lower than for NR. However, other strength properties are acceptably high only for filled vulcanizates. BR is frequently used in blends with NR and SBR to enhance abrasion resistance. The normal operational temperature range is –70 to +80°C.

### 28.6.6 Nitrile rubbers

Nitrile rubbers (acrylonitrile butadiene, NBR) were some of the first oil-resistant elastomers and continue to be popular for this reason. Oil-resistant grades have better resistance than does CR and, when specially compounded, are more resistant to heat ageing, but are significantly weaker and less rubbery than CR. Oil resistance and minimum usable temperature both increase with acrylonitrile content. NBRs can be attacked by ozone and are sometimes blended with polyvinyl chloride (PVC) to increase their resistance.

Carboxylated nitrile rubber (XNBR) has significantly improved strength and abrasion properties, and hydrogenated nitrile rubber (HNBR) has excellent performance at high temperatures in chemically difficult environments.

The operational temperature range is –20 to + 110°C.

### 28.6.7 Ethylene–propylene rubbers

Ethylene–propylene rubbers are a range of materials essentially immune to ozone cracking and attack by diatomic oxygen. Other useful attributes include the ability to accept high loadings of filler. These rubbers have good tolerance to water, but not to mineral oils. Inherently less rubbery and much weaker than NR, their strength properties can be fairly good if appropriate fillers are used. Although good bonding to a more rigid substrate can be achieved, it is potentially more difficult than for many other elastomers. In general EPM has better high-temperature performance than does EPDM.

The operational temperature range of ethylene–propylene rubbers is –40 to +130°C.

### 28.6.8 Chloroprene

Chloroprene rubbers (CRs) have greater tolerance than NR to high temperatures, mineral oils and oxygen, while their strength properties approach those of NR. Chlorine compounds evolved on combustion give CR a level of fire retardancy but, in some circumstances, can pose problems themselves because of toxicity and corrosion, for example. Improved fire resistance can be obtained with appropriate additives. Special formulations may be required for prolonged contact with water. In common with other elastomers where halogens forms part of the polymer structure, resistance to high-energy ionizing radiation is poor.

Crystallization-resistant grades are available, but in other grades quite rapid crystallization can take place especially at temperatures around –10°C (see Section 28.6.1).

A normal operational temperature range for CR is –30 to +100°C.

### 28.6.9 Polyurethane

Urethane rubbers comprise a particularly wide range of materials and thus it is difficult to make general statements on performance. Many are strong materials, very much stiffer, less elastic and possessing higher mechanical damping than NR; several types have very good resistance to abrasion. Urethane elastomers are generally tolerant of oxygen and mineral oils, but polyester type urethane (AU) rubbers are vulnerable to degradation by moisture or immersion in aqueous fluids (hydrolysis), especially in warm conditions, and to consequent microbial attack. Polyether type urethane (EU) rubbers are very much more resistant to aqueous fluids, although these too can be attacked at high temperatures. On combustion, polyurethanes can produce highly toxic substances.

A typical operational temperature range of urethane rubbers is –30 to +80°C.

### 28.6.10 Silicone

Silicone rubbers are notable for their wide operational temperature range and their excellent resistance to oxygen. Their already good fire resistance can be enhanced by appropriate additives. The permeability of silicones to gases is higher than for many elastomers; they are inherently weak materials, although significant improvements in strength properties can be obtained by using appropriate fillers. Apart from fluorosilicones, the tolerance of silicone rubbers to mineral oils is not good.

The operational temperature range of silicone rubbers is typically –90 to +230°C.

### 28.6.11 Chlorosulphonated polyethylene

Chlorosulphonated polyethylene (CSM) rubbers are notable for their ability to produce brilliant white yet durable vulcanizates. They are more resistant to fire than are many other elastomers and are more resistant to water and to oxygen. They are moderately resistant to mineral oils (depending on the chlorine content of the CSM). The strength properties of these generally rather stiff elastomers are fair.

The operational temperature range of CSM rubbers is –20 to +120°C.

### 28.6.12 Chlorinated polyethylene

Chlorinated polyethylene (CM) rubbers are resistant to oxygen and water. They have fair tolerance to mineral oils. CM rubbers are less rubbery than CSM rubbers, themselves not very rubbery elastomers.

A typical operational temperature range of CM rubbers is –20 to +100°C.

### 28.6.13 Ethylene–methylacrylate

Ethylene-methylacrylate rubber (AEM) is notable for its ability to give high mechanical damping over a wide temperature range. It is also capable of accepting substantial amounts of filler (fire-retardants, for example). Vulcanizates of this rubber have generally good resistance to oxygen and to water, are moderately tolerant of mineral oils and have fair strength properties.

A representative operational temperature range of AEM is –40 to +150°C.

### 28.6.14 Fluorinated hydrocarbon

Fluorinated hydrocarbon rubbers (FPM or FKM) have excellent resistance to mineral oils, oxygen and high temperatures and are, therefore, classed as speciality elastomers. They are also tolerant of water, but their tolerance to high-energy ionizing radiation is poor. Strength properties of the vulcanizates are only moderate and tend to be stiffer than for many other elastomers. The density of FKM is *c.* 50% higher than that of natural rubber.

The operational temperature range of FKM is typically –20 to +230°C.

### 28.6.15 Polysulphide

Polysulphide rubbers are noted for their resistance to a wide range of fluids including mineral oils and water; they tolerate oxygen well. Polysulphides have low gas permeability but are mechanically rather weak.

A normal operational temperature range of polysulphide rubbers is –40 to +80°C.

### 28.6.16 Epichlorohydrin

Epichlorohydrin rubbers have good resistance to oxygen and many mineral oils. ECO (the co-polymer of epichlorohydrin and ethylene oxide) can give low mechanical damping. Gas permeability for CO (the homopolymer) can be significantly lower than for butyl rubber. However, epichlorohydrin rubbers can show some vulnerability to water; their strength properties are fair.

A typical operational temperature range is –10 to +130°C for CO and –20 to +130°C for ECO.

### 28.6.17 Polypropylene oxide

Polypropylene oxide rubbers (GPO) are resistant to attack by oxygen and are fairly tolerant of water and of mineral oils. Vulcanizates of the material can give low mechanical damping but are not particularly strong.

The operational temperature range of GPO is typically –50 to +130°C.

### 28.6.18 Ethylene–vinyl acetate

The properties of these materials depend markedly on composition. Ethylene–vinyl acetate (EVA) rubbers have good resistance to oxygen and to water. They have some tolerance to petroleum-based oils. Their strength properties are fair.

A normal operational temperature range of EVA rubbers is –40 to +120°C.

### 28.6.19 Polynorbornene

Polynorbornene rubbers (PNRs) are highly plasticized polynorbornene and the properties of the vulcanizates can depend strongly on the plasticizing oil. PNRs are notable for the fact that very soft materials can be obtained with reasonable strength properties. These rubbers are generally tolerant of water but are vulnerable to many mineral oils and to oxygen.

A representative operational temperature range of PNRs is –40 to +80°C.

### 28.6.20 Properties of thermoplastic elastomers

In general, thermoplastic elastomers (TPEs) show less complete immediate recovery after deformation than do many conventional elastomers, and their creep rates also tend to be higher. Harder materials can be obtained with TPEs than is normal for most conventional elastomers (even polyurethanes); however, for these high-modulus TPEs yield strains may be only *c.* 10%. Block copolymer types (e.g. polyether–polyamide, polyether–polyester and polyurethane TPEs) can have Young's moduli of hundreds of MPa—i.e. beyond the range of what is normally thought of as rubbery or elastomeric.

#### *28.6.20.1 Styrene–butadiene–styrene block copolymers and related materials*

The resistance to the environment of styrene–butadiene–styrene (SBS) block co-polymers is similar to that of SBR, but is improved in hydrogenated grades. A typical operational temperature range is –60 to +80°C, although strength can be significantly reduced at temperatures below the upper limit.

*28.6.20.2 Polyether–polyester block co-polymers*

Polyether–polyester block co-polymers (YPBOs) are high-strength materials which are resistant to ozone and also to mineral oils and many other fluids. At high temperatures, additives may be required to give protection against attack by water.

The operational temperature range of YPBOs is typically –40 to +120°C.

*28.6.20.3 Polyurethane thermoplastic elastomers*

Thermoplastic polyurethane (TPU) elastomers are generally strong materials which have good resistance to tearing and abrasion. TPUs, particularly polyester types, tolerate ozone and many mineral oils well, but have some susceptibility to hydrolysis (attack by water). Resistance to water is better for ether-based types and for harder materials (see Section 28.6.9).

Operational temperatures of TPUs typically range from –40 to +110°C.

*28.6.20.4 Ethylene–vinyl acetate co-polymers*

Thermoplastic EVAs have good resistance to ozone but are vulnerable to mineral oils. Although not inherently flame-resistant they are capable of accepting large amounts of fire retardant filler; the resulting materials are not readily flammable and give off only minimal levels of noxious gases and smoke. Thermoplastic EVAs are not normally classed as high-strength elastomers.

The operational temperature range of thermoplastic EVAs is –30 to +80°C.

*28.6.20.5 Polyether–polyamide block co-polymers*

These materials are resistant to ozone and a wide range of fluids, including petroleum-based oils. They have good strength and fatigue properties. They have lower densities (by *c.* 20%) than TPUs.

The operational temperature range of polyether–polyamide block co-polymers is –40 to +140°C.

*28.6.20.6 Thermoplastic polyolefin rubbers*

The properties of thermoplastic polyolefin (TPO) rubbers vary very widely according to the type of rubber and thermoplastic and the method of preparation of the TPO. However, many TPOs have high resistance to ozone and general weathering. Resistance to mineral oils is generally not good unless an oil-resistant rubber is used. Many TPOs cannot withstand such high stresses as the block co-polymer polyether–polyester, polyether–polyamide or the polyurethane TPEs, and they are less resistant to abrasion. Some soft TPOs can, however, withstand and recover from high strains. The operational temperature range depends on the polymers used, but for EPDM/PP TPOs is –50 to +120°C.

## 28.7 Example applications

Although not used in such quantities as some other classes of materials, elastomers perform important functions in many present-day structures, especially where seals or flexible joints are required. Elastomers are used extensively to protect other materials from abrasion or corrosive fluids; they are also used to protect structures from vibration, shock or even earthquakes.

Elastomers can be: unvulcanized or partially vulcanized; unreinforced or reinforced with fabric or with steel plates; monolithic or cellular.

### 28.7.1 Applications in buildings

Elastomers are used in many buildings for effective and durable weatherproofing of roofs. Elastomers with good environmental resistance such as ethylene–propylene rubber (EPM, EPDM), chlorosulphonated polyethylene (CSM), butyl rubber and plasticized polyvinylchloride (PVC), which can be regarded as a thermoplastic elastomer, are popular in single-ply roofing. Sometimes the elastomers are not vulcanized. CSM is particularly suitable where a brilliant-white surface is required for thermal or aesthetic reasons.

Elastomer composites are also used for roofing: polyester-reinforced EPDM or glass-fibre-reinforced chloroprene rubber (CR) coated with CSM, or IIR-based materials are examples. Although bituminous material is commonly used for roof membranes, its temperature-dependent mechanical properties are not always adequate; these can be improved by additions of elastomeric material; styrenic block co-polymers and natural rubber are two of the elastomers employed for this purpose. EPDM, silicones, CR and now thermoplastic elastomers (thermoplastic polyolefin (TPO) alloy) provide preformed joints for curtain walling and other types of exterior cladding panels. Gaskets are often made from expanded elastomer—frequently CR is used.

For *in situ* formed expansion joints, polysulphide, silicone, acrylic and polyurethane elastomers are used—either uncured or partially cured. Seals for windows are commonly made from EPDM or thermoplastic elastomers. Again excellent environmental resistance is a key factor.

Damp-proof membranes for floors and also walls can be made by applying a coat of elastomer-containing material before the cementitious layer.

Elastomers are widely exploited in intumescent and other fire seals and coatings, especially where flexibility is required. Depending on the application, systems such as epoxy/polysulphide elastomer or rubbers, such as ethylene–propylene, capable of accepting high loadings of appropriate fillers are used. Latices of rubbery polymers, added to cementatious surfaces, can improve crack resistance and durability in some circumstances.

Whole buildings, or parts of them, can be isolated against ground-borne vibration by supporting the structure on appropriately designed bearing pads. Good strength properties and proven durability are of paramount importance here and natural (NR) or polychloroprene rubbers (CR) are the elastomers of choice. The bulk of the components and the low thermal conductivity of rubber give a good margin of safety in the case of a fire.[18] If further fire resistance is required, the bearings may be covered in fire-retardant material.

An extension of this application has been to protect buildings and their contents against earthquakes by means of elastomeric bearings. Ability to withstand large strains, relatively low modulus and, above all, stability of bulk properties over many decades are all requirements of the vulcanizate, which is often based on NR.

Rubbery expansion joints and NR or CR bearing pads can also be fitted to allow for (differential) thermal expansion in buildings, parts of buildings or other large structures. Cellular elastomers are frequently used, in conjunction with other means, for sound deadening within buildings.

### 28.7.2 Fixtures, fittings and services

Elastomeric flooring in the workplace or in public places (e.g. stations and airports) performs specific functions. Ethylene–propylene rubber (EPM, EPDM), NR or combinations of elastomers provide non-slip surfaces. NR, SBR or EPDM incorporating specialized carbon black filler produce electrically conductive and anti-static flooring. More basic impact- and chemical-resistant floor coatings can be made from epoxy/polysulphide rubber. EPDM is a popular material used for draught-proofing strip because of its environmental resistance.

Because of its good all-round tolerance to aqueous environments, EPDM is also widely used in pipe joint seals for potable water and waste-water pipelines; the use of NR in these applications dates back over 100 years. Seals for oil and gas pipelines are generally based on nitrile (NBR) and epichlorohydrin (CO, ECO) rubbers because these have resistance to the substances concerned (see also Section 28.7.5).

Adhesives and mastics based on rubbery polymers are used in many areas of construction: adhering wall panels and plywood subflooring are two examples. Seals for insulating ('double') glazing require good environmental resistance: polysulphide, EPDM, urethane, butyl, silicone and thermoplastic (e.g. TPO alloy type) elastomers are used according to the balance of properties required and cost. In thermal, solar panels, EPDM, nitrile, silicone and thermoplastic elastomers are all used.

Heavy-duty conveyor belting frequently has a composite structure, with NR or SBR normally forming the carcass in general-purpose belting. Vulcanizates based on EPDM often form the cover if heat resistance is required, while CR, NBR or NBR/PVC elastomers are used for oil resistance. CR and PVC are also employed for their superior fire resistance—in mining, for example. Belting can also be made from butyl or urethane rubber. Moving walkways are composite structures, but have NR as a major component because of its strength properties.

Elastomeric seals are likely to be used whenever fluids (e.g. water or air) are piped or ducted within buildings. For example, EPDM seals are widely used in central heating systems.

A further major application of elastomers in construction is in electrical cable and flex. Plasticized PVC is used for general wiring (50 Hz, < 250 V) but the polarity of the material rules it out as an insulator for high-voltage or high-frequency applications. Furthermore, the physical properties of plasticized PVC are inadequate for mechanically arduous conditions. In high-voltage and high-frequency communications cables insulation can be provided by a non-polar elastomer such as EPDM or by cross-linked polyethylene; thermoplastic elastomers (TPO type) can also be used in these areas. The choice of cable jacketing material depends on the mechanical and chemical operating environment. Thus for good mechanical properties, NR is often chosen. Where good resistance to abrasion is required, urethane rubber is an option. Chlorosulphonated polyethylene (CSM) cable jacketing has a good balance of oil, heat and general environmental resistance characteristics and can be brightly coloured. When fire resistance is a particular requirement, halogenated elastomers, e.g. CR or plasticized PVC, may be used. For cable jacketing with low toxicity/smoke emission, elastomers such as EPDM, EVA, or ethylene–methylacrylate are heavily loaded with fire-retardant additives.

### 28.7.3 Piping, hose, reservoirs and tanks

Where abrasive particles are present in a fast-flowing aqueous environment, channels and tanks are frequently lined with NR or polyurethane to protect the underlying material (e.g. steel). Where the mechanical conditions are less arduous, butyl and sometimes CSM rubbers are used.

Butyl rubber (sometimes blended with EPDM) finds frequent application as a waterproofing membrane for ponds and reservoirs, while NR or CR is often employed for waterstops for leak protection at joints in water containment structures. CSM rubber, with a suitable reinforcing fabric, has been used as a cover material for potable water reservoirs. Flexible piping and hose are two further areas that exploit the particular characteristics of elastomers.

### 28.7.4 Roads, railways and bridges

In many bridges provision for thermal expansion of the deck is made by supporting it on compliant bearings. Such bearings can be plain elastomer pads, laminated rubber/steel pads or 'pot bearings' (confined elastomer units combined with sliders). The elastomers of choice for use in bridge bearings are generally NR and CR; unconfined rubber discs in combined elastomer/slider units are occasionally used—these exploit the high-modulus, high-strength characteristics of urethane elastomer.

Protective, waterproof, elastomeric coatings for bridge abutments are frequently NR or urethane rubber based. While preformed joint seals for roads, road bridges and runways are predominantly made from CR in the USA, EPDM is used extensively in northern Europe and SBR in Holland. The housings for 'cat's eyes' (the reflective lane markers ubiquitous in the UK and widely used elsewhere) are made from NR.

Scientific investigation of the modification of bituminous road materials by elastomers dates back some 50 years.[19] Elastomers can be incorporated into the bituminous road materials for two main reasons: to facilitate road making at lower temperatures and to improve the performance of the road surface. Particular applications that benefit from elastomer modification include: overseal banding for crack repair, stress-relieving membrane to avoid surface cracking, bridge deck membranes, and surface dressing binder for improved chipping retention on 'highly stressed' sites (e.g. inclines and roundabouts). Benefits of elastomeric modification on longer stretches of road can include greatly improved durability and improved grip. Depending on the detailed requirements, climate and cost factors, the elastomer may be a thermoplastic elastomer, EVA or SBS for example, natural rubber or reclaimed rubber.

Elastomers are frequently employed to improve the performance of rail systems where concrete sleepers are used. Rail pads are made from NR but EVA or polyurethane is sometimes used, while (under the) ballast mats made from NR are used in Germany and the UK for example to reduce vibration to the surroundings. Other systems of vibration isolation by means of elastomers have been implemented where ground-borne vibration from trains would be a particular problem—as was the case in the London–Heathrow underground link.

### 28.7.5 Oil, gas, chemical engineering and mining

Vibration isolation is a particular requirement on oil rigs and production platforms, where accommodation modules and noisy plant can be in close proximity. NR and CR are often the elastomers used for the construction of vibration isolation bearings, although other elastomers may be used if there are special requirements; frequently the bearings double as compliant elements to allow relative thermal expansion between supported and supporting structures.

Highly stressed flexible joints for tethering systems in offshore oil platforms have been made from NR and NBR, while polyurethane is employed as a coating in many situations, e.g. mooring cable buoys in flotation units for oil-spill containment barriers. Urethane rubber has also been used to make pipeline cleaning 'pigs'.

Numerous oil-well parts are made from NBR since high acrylonitrile content NBR has very good oil resistance. Particular circumstances may require the use of other elastomers, for example carboxylated nitrile rubber (XNBR) for improved abrasion resistance. Under conditions requiring improved resistance to temperature, oil and other media, fluorinated hydrocarbon elastomers are used. Examples of elastomeric oil well/pipeline parts are: seals, O-rings and blowout preventers.

In piping of gas, elastomers find widespread applications for seals, valves, etc. Typically, NBR and epichlorohydrin (CO, ECO) rubbers are used.

Abrasion-resistant ball-mill linings can be made from NR; where oils are involved in the separation process XNBR can be used. NR, urethane rubber or XNBR/PVC blends form abrasion-resistant chute linings and screens.

Elastomers are employed in chemical engineering for a wide range of applications including expansion bellows and vibration dampers; the choice of elastomer type depends on the operating temperature and chemical environment.

### 28.7.6 Miscellaneous

One of the ways in which steep slopes of loose earth can be stabilized is by applying suitable rubber latex loaded with grass seed.

Foamed (cellular) rubbers can have much lower stiffness than the solid material, especially to volume changes; this is especially true for open cell foams. Furthermore, foamed rubbers can have even lower thermal conductivities than the solid material—especially for closed cell foams. These two characteristics lead to a wide range of applications. Latex foam (often NR or SBR) is widely used as a cushioning material for carpets etc., while upholstery, mattresses, etc. can be made from latex foam or expanded urethane rubber. Expanded CR is widely used in external gaskets in construction (see Section 28.7.1).

The use of impact-absorbing mats to avoid injury in gymnasiums and playgrounds is extensive. The mats can be made of NR, SBR or butyl cellular rubber. Other types of sports surfaces, e.g. high-quality running tracks, exploit the properties of elastomers to advantage.

## 28.8 Elastomers—appropriate and inappropriate use

Elastomers are a diverse family of materials, and the distinction between them and other engineering polymers is not always a clear one. Nevertheless, designers frequently need to decide whether to use an elastomer and, if so, which one. As illustrated by the foregoing example applications, elastomers are candidate materials wherever flexibility, a high degree of elastic energy absorption, grip/adhesion characteristics or a high Poisson ratio is called for. Elastomers, if appropriately chosen, can also protect the underlying material from abrasive or chemical attack.

### 28.8.1 Strength, fatigue, creep and stress relaxation

One of the first questions to ask in design with an elastomer is whether it can sustain the stresses and strains envisaged in the application. In this context it should be remembered that some rubbers are capable of storing considerably more energy elastically than can, for example, mild steel: NR can store one order of magnitude more energy per unit volume and some two orders of magnitude more energy per unit mass.

In some situations fatigue is the most likely cause of failure. In some elastomers that do not non-strain-crystallize, cracks can propagate in a time-dependent manner, so it can be inadvisable to subject them to steady high stresses. In strain-crystallizing elastomers such as NR and CR, fatigue cracks generally propagate only when there are cyclic changes in stress; the rate of crack propagation is greater when the cyclic stress passes through zero and so, to optimize fatigue life, NR or CR components are often designed to be 'preloaded'.

The weak point of a rubber component is frequently the bond between elastomer and metal or other substrate. The problem can be avoided by not using a bond at all, but bonds are often necessary or desirable. Strict manufacturing controls must be observed to ensure bond reliability, and particular attention needs to be paid in component design to stresses at the bond.

Other long-term considerations include age hardening/softening and creep. All elastomers have the tendency for their elastic moduli to change with time. In many cases these changes are small, and in others, sulphur vulcanized NR for example, hardening processes (cross-linking) are counterbalanced by softening processes.[20,21]

In general, articles should be well vulcanized at the time of moulding, or of post-cure if applicable, to minimize the risk of significant further cross-linking taking place during service. This can be particularly important for certain combinations of elastomer and vulcanizing systems which show a 'marching modulus' effect: the modulus continues to rise virtually indefinitely as cure proceeds. Prolonged vulcanization can sometimes alleviate the problem.

All elastomers to some extent creep under constant load and relax under constant displacement. The two phenomena are related to one another and to damping.[22] 'Physical' (readily reversible) creep and stress relaxation proceed approximately logarithmically with time, and creep rates are often expressed as percentage increase in initial deflection (or percentage decrease in initial stress) per 'decade' (i.e. per 10-fold increase of time). Physical creep and stress relaxation rates can be kept low for unfilled elastomers: in the region of 1.5% per decade for unfilled natural rubber vulcanizates, although for filled vulcanizates the rates are somewhat higher. For high-damping elastomers greater rates of physical creep can be expected. Creep and stress relaxation can also occur through chemical changes taking place in the elastomer. Chemical creep has an approximately linear dependence on time: a typical value for a natural rubber vulcanizate is 0.5% of initial deflection per annum at room temperature for anaerobic conditions.[23,24] Chemical stress relaxation proceeds at a similar rate.[25]

### 28.8.2 Stiffness and dimensional tolerances

Even for a given elastomer an enormous range of stiffness is possible for a component of given overall size and shape. Whereas the objective with steel springs is generally to increase the compliance of steel by design, sometimes to the detriment of fatigue life, this is not generally the case with elastomers; here the aim is often to increase the stiffness in a particular direction, for example by bonding alternating layers of elastomer and much stiffer material—such as steel. The amplification of the material Young's modulus ($E$) to a modified compression modulus ($E_{c1}$) by the shape-factor $S$ (Figure 28.3) effect is given by:

$$E_{c1} \approx E(1 + 2S^2)$$

where $E_{c1} << K$, the bulk modulus. More generally the modified compression modulus is given by

$$\frac{1}{E_c} \approx \frac{1}{E_{c1}} + \frac{1}{K + \dfrac{4G}{3}}.$$

[The shear modulus $G = \dfrac{E}{2(1+\nu)}$;

Poisson's ratio $\nu = \dfrac{3K - 2G}{2(3K+G)}$; $K = \dfrac{2(1+\nu)}{3(1-2\nu)} G$]

The shape-factor effect occurs because the Young's modulus of elastomers is ~3 MPa, whereas the bulk modulus is ~2000 MPa. This difference in moduli can also be exploited in design by confining the rubber; as in a 'pot bearing' for example.

All elastomeric materials possess some damping but the degree varies widely between polymer types and within each polymer type depending on details of compounding as well as operational conditions such as temperature, frequency, pressure level, strain amplitude and mode of deformation. In many cases of dynamic isolation the selected level of damping for a rubber component represents a compromise between requirements for moderate frequencies well above resonance (where low damping

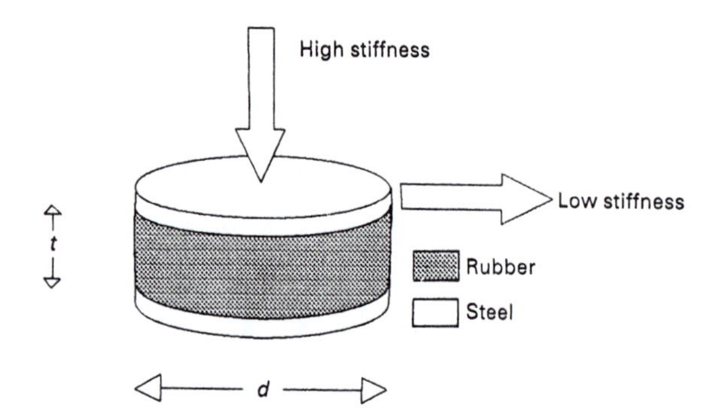


**Figure 28.3** The shape factor for rubber/steel bonded units. $A1 = (\pi d^2)/4$, $A2 = \pi dt$, shape factor = A1/A2

is often desirable), and through-resonance and also higher frequency requirements (for which higher levels of damping are generally desirable).[26,27] Resonant frequencies, which are a key feature in vibration isolation, depend on the ratios of relevant masses to the stiffnesses of spring elements.

Although very precise modelling of practical elastomers remains a challenging area, in many situations analytical and, increasingly, numerical methods provide predictions to a reasonable degree of accuracy;[3] accuracies of ~20% are normal. As far as the reproducibility of vulcanized rubber springs is concerned, tolerances of 10–15% on stiffness values are routine, with much tighter tolerances being readily achievable at extra cost.

The vulcanized elastomer to be used in a component is frequently specified in terms of its indentation hardness, usually in terms of the international rubber hardness degrees (IRHD) scale. Since IRHD measurements are subject to an uncertainty of ±2°, this implicitly creates an imprecision in the specification. The relationship between hardness and elastic modulus is non-linear and complex, particularly for filled elastomers.[28] However, a change in hardness from 60 to 62 IRHD, for example, generally corresponds to a ~12% increase in modulus.

Virtually all elastomers undergo significant shrinkage on cooling after moulding, ~3% (linear) is typical, and this is one of the factors setting normal limits to dimensional tolerances of moulded rubber items. Much finer tolerances can be achieved, and some manufacturers specialize in this area of work.

### 28.8.3 Environmental factors

One of the most important environmental factors for elastomers is temperature: materials that are rubbery at 20°C (at a given time-scale and above) can immediately become brittle at low temperatures (and/or short times) and lose much of their strength at sufficiently high temperatures. Moreover, high temperatures accelerate the deterioration of properties, pyrolysis being the extreme case.

Design must take account of brittleness and, when appropriate, melting or pyrolysis temperatures. It is clear, for example, that care needs to be exercised over the use of high acrylonitrile content nitrile rubber (NBR) in Alaskan winters, but there are more subtle points too. The high temperature resistance of an elastomer is often gauged by its ability to resist time-dependent degradation at the elevated temperature rather than by its strength properties at that temperature, which may be more appropriate. Many temperature-resistant elastomers show a marked instantaneous deterioration in creep and strength properties well below the upper limit of their operating range, and this must be allowed for.

Brittleness marks the end-point of a phenomenon sometimes referred to as 'glass hardening', which results from reductions in mobility (at a given timescale and shorter timescales) of the elastomer molecular chains; glass hardening is essentially instantaneous on the elastomer reaching a given temperature. In contrast to glass hardening, low-temperature crystallization stiffening in affected elastomers, notably NR and CR, can take place over short periods or over weeks or months. The low-temperature-crystallization rate is greatest for NR near –25°C, whereas for CR the rate is greatest near –10°C. At lower temperatures crystallization rate can decrease significantly. Although the effects of this type of stiffening can be diminished or eliminated by warming or strain cycling, it may be advisable in some applications to use crystallization-resistant 'compounds' of NR or CR for some components exposed to low temperatures for long periods.

High or low temperatures are likely to affect a large proportion of the rubber in a product and in so doing can seriously affect its performance; some forms of high energy ionizing radiation also fall into this category. Other potentially harmful influences such as atmospheric gases (e.g. oxygen and ozone), ultraviolet radiation, and liquids (e.g. oils and aqueous fluids) have their main effects at or near the surface; such effects can be serious for small or thin products of vulnerable elastomer but can be less so for large components. Factors such as intensity of exposure, the importance of aesthetic considerations, the mode of deformation, and the bulk and life expectancy of the component need to be allowed for. An elastomer can usually be found that is extremely resistant to a particular kind of chemical attack, but this may not always be an appropriate choice when strength properties or cost considerations are taken into account. Usually a balance must be struck.

Elastomers such as NR and SBR are not normally thought of as having good weathering resistance, but in many cases the resistance is more than adequate provided that the components are sufficiently thick. NR bridge bearings are an example; in many cases these can be expected to outlast the bridge itself. Complications can arise when there are conflicting requirements. Urethane rubber and NR have good resistance to penetrating ionizing radiation but not to high temperatures: the reverse is true of fluorinated elastomers.

It might be thought that urethane rubbers, being resistant to mineral oil, should be widely used in oil seals, but their limited high-temperature range and vulnerability of some grades to contaminant water counts against them. At the same time ethylene–propylene (EPM and EPDM) rubbers, NRs and SBRs are often loosely referred to as being non-oil-resistant; in fact, some such rubbery materials do have good tolerance to many non-mineral (non-petroleum-based) oils.

## 28.9 Standards and specifications

Standards and specifications relating to elastomers exist to provide safeguards and some guidance; however, they cannot take the place of experience and expert advice. Given below is a list of some of the more important standards for elastomers and their uses, together with brief comments on the contents of the standard.

### 28.9.1 General

- BS6716 (1986) Guide to properties and types of rubber.
- BS374-1 (1997) – ISO 3302-1:1996 Rubber tolerance for products – dimensional tolerances.
- BS ISO 9631:2004 Rubber seals. Joint rings for pipelines for hot water supply, drainage and sewerage pipelines up to 110 degrees C. Specifications for the material – ISO 2230:2002 Rubber products – Guidelines for storage.
- BS3558-2:2001 Glossary of rubber terms. Additional British terms – ISO 1382:2002 Rubber vocabulary.

### 28.9.2 Test methods for rubber

The following standards are examples of those dealing with small-scale laboratory tests for quality control, material specification, design parameters and assessment of environmental stability.

- BS903 Methods for physical testing of rubber (Parts A1–A59). *See standards for current and ISO equivalents.*
- Annual Book of ASTM Standards Volume 09.01 Rubber, Natural and Synthetic—General Test methods; Carbon Black. Various International Standard test methods produced by Joint Technical Committee ISO/TC45 (Rubber and Rubber Products); in most cases these have been adopted as National Standard methods

### 28.9.3 Material specifications

*28.9.3.1 General classifications*

- BS5176:2003 Classification system for vulcanized rubbers. Specification.
- ASTM D2000-06ae1 Standard classification system for rubber products in automotive applications. (Although intended for the automotive industry, ASTM D2000 is used for many applications.)
- ISO4632 (1982) Rubber, vulcanized—classification—Part I Description of the classification system.

*28.9.3.2 Individual rubber specifications*

Note that the word 'compound' in rubber technology refers to a formulation, i.e. the ingredients, rather than a chemical compound.

- BS1154:2003 Natural rubber compounds. Specification.
- BS1155:2003 Natural rubber compounds for extrusion. Specification.
- BS2751:2001 General purpose acrylonitrile–butadiene rubber compounds. Specification.
- BS2752:2003 Chloroprene rubber compounds. Specification.
- BS3227:2003 Butyl rubber compounds (including halobutyl compounds). Specification.
- BS6014:2003 Ethylene–propylene rubber compounds.
- BS EN 3207:1998 Rubber compounds. Technical specification.
- BS EN 2259:1995 Silicone rubber (VMQ). Hardness 50 IRHD.
- BS EN 2260:1995 Silicone rubber (VMQ). Hardness 60 IRHD.
- BS EN 2262:1995 Silicone rubber (VMQ/PCMQ) with high tear strength. Hardness 50 IRHD.
- BS EN 60684-3-121 t0 124:2001, IEC 60684-3-121 to 124:2001 Specification for flexible insulating sleeving. Specification requirements for individual types of sleeving. Specifications for individual types of sleeving. Extruded silicone sleeving.

*28.9.3.3 Mastics*

- BS 8218:1998 Code of practice for mastic asphalt roofing.
- BS EN 13108-6:2006 Bituminous mixtures. Material specifications. Mastic asphalt.
- BS EN 12970:2000 Mastic asphalt for waterproofing. Definitions, requirements and test methods.
- BS 8204-5:2004 Screeds, bases and in situ floorings. Mastic asphalt underlays and wearing surfaces. Code of practice.
- BS EN 14693:2006 Flexible sheets for waterproofing. Waterproofing of concrete bridge decks and other concrete surfaces trafficable by vehicles. Determination of the behaviour of bitumen sheets during application of mastic asphalt.

### 28.9.4 Product specifications

*28.9.4.1 Bridge bearings*

- BS5400 9.2 (1983) Steel, concrete and composite bridges. Bridge bearings. Specification for materials, manufacture and installation of bridge bearings.
- BS EN 1337-3:2005 Structural bearings. Elastomeric bearings.
- BS EN 1337-5-2005 Structural bearings. Pot bearings.
- BS ISO 6446:1994 Rubber products. Bridge bearings. Specification for rubber materials .
- ASTM D4014-03 Feb-2003 Specification for plain and steel-laminated elastomeric bearings for bridges.

*28.9.4.2 Vibration isolation of buildings*

- BS6177 (1982) Guide to selection and use of elastomer bearings for vibration isolation of buildings.

*28.9.4.3 Rail pads*

- International Union of Railways, UIC 864–5 Technical specification for supply of rail seat pads (4th ed. 1.1.86).

*28.9.4.4 Seals and gaskets*

*Pipe sealing rings*

- BS EN 681-2-4 (2000) Elastomeric seals. Material requirements for pipe joint seals used in water and drainage applications (Various).
- BS EN 682:(2002) Elastomeric seals. Materials requirements for seals used in pipes and fittings carrying gas and hydrocarbon fluids.
- ASTM C564-03a Specification for rubber gaskets for cast iron soil pipe and fittings.
- ASTM D1869–95 (2005) Specification for rubber rings for asbestos cement pipes.
- ASTM C443M-0 5a Standard specification for joints for concrete pipe and manholes, using rubber gaskets [Metric].
- ASTM C505M-05a Standard Specification for non-reinforced concrete irrigation pipe with rubber gasket joints [Metric].
- ISO4633 (2002) Rubber seals—Joint rings for water supply drainage and sewerage pipelines—Specification for materials.
- ISO6447 (1983) Rubber seals—Joint rings used for gas supply pipes and fittings—Specification for material.
- ISO6448 (1985) Rubber seals—Joint rings used for petroleum product supply pipes and fittings—Specification for material.

*Building gaskets*

- BS4255 (1986) Rubber used in preformed gaskets for weather exclusion from buildings—Part I Specification for non-cellular gaskets.
- BS6093 (2006) Design of joints and jointing in building construction. Guide.
- ASTM C509-06 (2006) Specification for cellular elastomeric preformed gasket and sealing material.
- ASTM C542-05 (2005) Standard specification for lock-strip gaskets.
- ASTM C864-05 (2005) Specification for dense elastomeric compression seal gaskets, setting blocks and spacers.
- ISO3934 (2002) Rubber, vulcanized and thermoplastic – Preformed gaskets used in buildings – Classification, specifications and test methods.
- ISO5892 (1981) Rubber building gaskets—Materials for preformed solid vulcanized structural gaskets—Specification.

*Seals for solar energy systems*

- ISO9553 (1997) Solar energy – Methods of testing preformed rubber seals and sealing compounds used in collectors.
- BS EN 12975-1 (2006) Thermal solar systems and components. Solar collectors. General requirements.
- BS7431 (1991), ISO9808 (1990) Method for assessing solar water heaters. Elastomeric materials for absorbers, connecting pipes and fittings.
- ASTM D3667-05 (2005) Specification for rubber seals used in flat-plate solar collectors.
- ASTM D3771-03 (2003) Specification for rubber seals used in concentrating solar collectors.
- ASTM D3832-79 (2006) Specification for rubber seals containing liquid in solar energy systems.
- ASTM D3903-03 (2003) Specification for rubber seals used in air-heat transport of solar energy systems.

*Highway joints: Paving*

- ASTM D1752-04a Specification for preformed sponge rubber and cork expansion joint fillers for concrete paving and structural construction.
- ASTM D2628-91 (2005) Specification for preformed polychloroprene elastomeric joint seals for concrete pavements.
- ASTM D3542-92 (2003) Specification for preformed polychloroprene elastomeric joint seals for bridges.
- ISO4635 (1982) Rubber, vulcanized—Preformed compression seals for use between concrete motorway paving sections—Specification for material.

#### *28.9.4.5 Reservoir linings*

- BS ISO 13226:2005 Jan-2006 Rubber. Standard reference elastomers (SREs) for characterizing the effect of liquids on vulcanized rubbers.
- ASTM D2643-04 Standard specification for prefabricated asphalt reservoir, pond, canal, and ditch liner (exposed type).

#### *28.9.4.6 Chemical plant linings*

- BS 6374 Lining of equipment with polymeric material for the process industries—Part 5 (1985) Specification for lining with rubbers.

#### *28.9.4.7 Flooring*

- BS1711 (1975) Specification for solid rubber flooring.
- BS3187 (1978) Specification for electrically conducting rubber flooring.
- BS8203 (2001) Code of practice for installation of resilient floor coverings.
- ASTM C1059-99 (1999) Standard specification for latex agents for bonding fresh to hardened concrete.
- BS8204-3 (2004) Screeds, bases and in situ floorings. Polymer modified cementitious levelling screeds and wearing screeds. Code of practice.

#### *28.9.4.8 Electrical*

- BS6195 (2006) Electric cables. Rubber or silicone rubber insulated flexible cables and cords for coil end leads

#### *28.9.4.9 Elastomeric roofing*

- ASTM D3105-99 (1999) Index of methods for testing elastomeric and plastomeric roofing and waterproofing material.
- ASTM D4637-04 (2004) Specification for EPDM sheet used in single-ply roof membrane.
- ASTM D5019-07 (2007) Standard specification for reinforced CSM (chlorosulfonated polyethylene) sheet used in single-ply roof membrane.
- Rubber Manufacturers Association (USA), RMA RP-1-90 (1990) Minimum requirements for non-reinforced black EPDM rubber sheets.

## 28.10 Conclusions

A century ago elastomer selection posed few problems since there was effectively just one elastomer: natural rubber (NR). Today the situation is very different, with well over 30 elastomer types available. Some of these materials have opened new possibilities in architecture through their excellent resistance to weathering; others function effectively for years in aggressive fluids at temperatures that would be unthinkable for the natural material. Yet NR remains the world's most heavily used elastomer because of technical merit and price.

Elastomers make a special contribution to the range of construction materials available because of their elastic deformability. Nevertheless, they remain minority materials poorly understood by many of the technical people who do or could benefit from their use. These circumstances can and do result in costly mistakes being made, sometimes visibly with the failure of a component, on other occasions with the hidden costs of over-specification.

Present-day structures make extensive use of elastomers in roofing, curtain walling and in numerous other less visible sealing and jointing applications. Large numbers of bridges and flyovers are supported on longlife reliable rubber bearings to allow for thermal expansion.

There has been a noticeable increase in the use of elastomeric pads to reduce ground-borne vibration from rail and other traffic. The reasons are clear: economic pressures for the utilization of sites near railways, coupled with demand for lower noise levels in offices and dwellings. These trends seem certain to be maintained. Noise-exclusion and energy-conservation requirements will mean that tightly fitting, multiply glazed windows will become even more commonplace; elastomers will continue to fulfil important functions in these units.

After an earthquake it is vital that emergency services remain functional and co-ordinated. Earthquake base isolation by rubber bearings of communication and co-ordination centres has begun. The number of buildings and their contents protected in this way is set to increase.

Modification of bituminous road materials by elastomers can result in greatly improved life and performance. However, the high initial cost has been a deterrent. Increasing levels of traffic congestion, combined with a reluctance to see ever more roads built, point to a reassessment of the overall economic costs of road damage and repair. Against this background increased use of elastomers in road materials is a distinct possibility.

Because there are already so many base elastomers available, it appears unlikely that there will be any major expansion in their number in the near future. New blends are, however, sure to appear and some existing ones will gain in popularity. Already thermoplastic elastomer (TPE) blends have begun to be used in architectural aspects of construction, a tendency that is certain to continue. Block co-polymer TPEs are already used in asphalt modification, as are other elastomers, but the newer high-performance TPEs now seem poised to take over some infrastructural roles from speciality elastomers.

During the first half of the 21st century there may well be major changes in the pattern of supply of base elastomers driven by shortages in petroleum, which is the origin of most synthetic rubbers. There is likely to be a blurring of the boundaries between chemically modified NR and synthetic rubbers, since eventually the feedstocks for synthetic rubbers may be derived from materials that were living months rather than millions of years previously.

## 28.11 Research organizations, trade associations and professional bodies

- ARTIS, Hampton Park West, Semington Road, Melksham, Wiltshire SN12 6NB, UK.
- British Rubber & Polyurethane Products Association Ltd (BRPPA), 6 Bath Place, Rivington Street,London EC2A 3JE, UK.
- TARRC/Malaysian Rubber, Brickendonbury,Hertford SG13 8NL, UK.
- Rubber in Engineering Committee, Institute of Materials, Minerals & Mining, 1 Carlton House Terrace,London, SW1Y 5DB, UK.
- RAPRA Technology Ltd, Shawbury, Shrewsbury SY4 4NR, UK.

The author wishes to acknowledge the advice and help from colleagues, notably those at Avon Materials Development Centre (now ARTIS), Westbury and at MRPRA (now TARRC), Brickendonbury.

## 28.12 References


1 Treloar, L. R. G., *The Physics of Rubber Elasticity* 3rd Edition, Clarendon Press, Oxford, UK (1975).
2 Coveney, V. A. & Muhr, A. H., 'Elastomer-Based Engineering Products: Design', in *Encyclopedia of Materials Science and Engineering* (ed. R. W. Cahn), Pergamon, Oxford, UK (1993).
3 Boast, D. & Coveney, V. A. (eds.) *Finite Element Analysis of Elastomers,* Professional Engineering Publishing (IMech E), London, UK (1999).
4 Malaysian Rubber Producers' Research Association, *Natural Rubber Technical Information Sheet D18*, MRPRA, Brickendonbury, UK (1976).
5 Malaysian Rubber Producers' Research Association, *Natural Rubber Technical Information Sheet D19*, MRPRA, Brickendonbury, UK (1976).
6 Gent, A. N. and Thomas, A. G., 'Mechanics of foamed elastic materials', *Rubb. Chem. Technol.*, 36, 597–610 (1963).
7 Easterbrook, E. K. and Allen, R. D., 'Ethylene-propylene rubber', in *Rubber Technology* (ed. Morton, M.), 3rd edn. pp. 274–275, Van Nostrand, New York, NY (1987).
8 International Organization for Standardization, *ISO 1629*, Rubber and lattices—Nomenclature (1987).
9 Cunneen, J. I. and Higgins, G. M. C., 'Cis-trans isomerism in natural polyisoprenes', in *The Chemistry and Physics of Rubber-like Substances—Studies of Natural Rubber Producers' Research Association 2* (ed. Bateman, L.), pp. 19–40, MacLaren, London, UK (1963).
10 McIntyre, D., Stephens, H. L. and Bhowmick, A. K., 'Guayule rubber', in *Handbook of Elastomers—New Developments and Technology* (ed. Bhowmick, A. K. and Stephens, H. L.), pp. 1–29, Marcel Dekker, New York, NY (1988).
11 Baker, C. S. L., 'Modified natural rubber', in *Handbook of Elastomers—New Developments and Technology* (ed. Bhowmick, A. K. and Stephens, H. L.), pp. 31–74, Marcel Dekker, New York, NY (1988).
12 Corbelli, L., 'Ethylene propylene rubbers', in *Developments in Rubber Technology—2 Synthetic Rubbers* (ed. Whelan, A. and Lee, K. S.), pp. 57–129, Applied Science, London, UK (1981).
13 Brydson, J. A., *Rubbery Materials and Their Compounds*, Elsevier Applied Science, London, UK (1988).
14 Crowther, B. G., 'Vulcanization by methods other than moulding', in *Rubber Technology and Manufacture* (ed. Blow, C. M. and Hepburn, C.), 2nd edn., pp. 351–356, Butterworths, Oxford, UK (1987).
15 Malaysian Rubber Producers' Research Association, *Natural Rubber Engineering Data Sheets*, MRPRA, Brickendonbury, UK (1979–).
16 De, S. K. and Bhowmick, A. K. (eds), *Thermoplastic Elastomers from Rubber–Plastic Blends*, Ellis Horwood, Chichester, UK (1990).
17 Elliott, D. J., 'Comparative properties of natural rubber and synthetic isoprene rubbers', *NR Technol.*, 18, 69–74 (1987).
18 Derham, C. J. and Plunkett, A. P., 'Fire resistance of steel laminated natural rubber bearings', *Natural Rubber Technol.*, 7 (2), 29–37 (1976).
19 Thompson, P. D., 'The use of rubber in road materials', *Civil Eng.*, 57, 1163–1166 (1962).
20 Barnard, D. and Lewis, P. M., 'Oxidative ageing', in *Natural Rubber Science and Technology* (ed. A. D. Roberts), pp. 621–678, Oxford University Press, Oxford, UK (1988).
21 Azura, A. R. & Thomas, A. G., 'Effect of heat ageing on crosslinking, seision and mechanical properties', Chapter 2 in *Elastomers and Components: Service Life Prediction – Progress & Challenges* (ed. V. A. Coveney), Woodhead Publishing, Cambridge, UK, pp. 27–38 (2006).
22 Gent, A. N., 'Relaxation processes in vulcanized rubber. I Relation among stress, relaxation, recovery and hysteresis', *J. Appl. Polym. Sci.*, 6, 433–441 (1962).
23 Derham, C. J., 'Creep and stress relaxation and their relevance to engineering applications', *Proc. Rubber in Engineering Conference*, Imperial College, London, UK, NRPRA (1973).
24 Stevenson, A., 'Longevity of natural rubber in structural bearings', *Plastics Rubber Process. Appl.*, 5, 253–258 (1985).
25 Albihn, P., 'The 5-year accelerated ageing project for thermoset and thermoplastic elastomeric materials', Chapter 1 in *Elastomers and Components: Service Life Prediction – Progress & Challenges* (ed. V. A. Coveney), Woodhead Publishing, Cambridge, UK, pp. 3–25 (2006).
26 Muhr, A. H., 'Transmission of noise through rubber-metal composite springs', *Proc. Inst. Acoust.*, 11, 627–634 (1989).
27 Muhr, A. H., 'The use of rubber noise stop pads in vibration isolation systems', *Proc. Inst. Acoust.*, 12, 417–427 (1990).
28 Muhr, A. H., Tan, G. H. and Thomas, A. G., 'A method of allowing for non-linearity of filled rubber in force-deformation calculations', *J. Natural Rubber Res.*, 3, 261–276 (1988).


## 28.13 Bibliography


Allen, P. W., Lindley, P. B. and Payne, A. R., *Use of Rubber in Engineering*, Natural Rubber Producers' Research Association, MacLaren, London, UK (1967).
Bateman, L., Moore, C. G., Porter, M. and Saville, B., Chemistry of vulcanization, in *The Chemistry and Physics of Rubber-like Substances* (ed. Bateman, L.), pp. 449–561, MacLaren, London, UK (1963).
Bayer, *Bayer Polyurethanes Handbook*, English edn, Bayer, Leverkusen, Germany (1979).
Bhowmick, A. K. and Stephens, H. L. (eds.), *Handbook of Elastomers—New Developments and Technology*, Marcel Dekker, New York, NY (1988).
Black, R. M., 'Recent developments in cable technology', *Proc. Rubber Plastics Technol.*, 2 (4), 25–37 (1986).
Blackley, D. C., *Synthetic Rubbers: Their Chemistry and Technology*, Applied Science, London, UK (1983).
Blow, C. M. and Hepburn, C. (eds.), *Rubber Technology and Manufacture*, Butterworths, Guildford, UK (1987).
*Blue Book—Materials, Compounding Ingredients, Machinery and Services for the Rubber Industry*, Lippincott & Peto, Akron, OH (1990).
Brown, R. P., *Physical Testing of Rubber*, 2nd edn, Elsevier Applied Science, London, UK (1986).
Buchan, S., *Rubber in Chemical Engineering*, Federation of British Rubber and Allied Manufacturers and NRPRA, London, UK (1965).
Coveney, V. A. (ed.), *Elastomers and Components: Service Life Prediction – Progress & Challenges*, Woodhead Publishing, Cambridge, UK (2006).
Coveney, V. A. and Thomas, A. G., 'The role of natural rubber in seismic isolation—A perspective', *Kautshuk and Gummi Kunststoffe*, 44 (9), 861–865 (1991)
Damusis, A. (ed.), *Sealants*, Rheinhold, New York, NY (1967).
Davies, B. J., *The Longevity of Natural Rubber in Engineering Application*, MRPRA, Brickendonbury, UK (1988).
Dupont, *Elastomer Notebook* (serial), Wilmington, DE.
Feldman, D., *Polymeric Building Materials*, Elsevier Applied Science, London, UK (1989).

Fernando, M. J. and Guirguis, H. R., 'Natural rubber for improved surfacing', 12th Australian Road Research Board Conference, Hobart, Tasmania, August 1984, *ARRB Proc.*, 12 (2) (1984).

Freakley, P. K. and Payne, A. R., *Theory and Practice of Engineering with Rubber*, Applied Science, London, UK (1978).

Göbel, E. F., *Rubber Springs Design*, Newnes-Butterworths, Guildford, UK (1974).

*Green Book—International Standards for Quality and Packing in Natural Rubber*, The International Rubber Quality and Packaging Conferences, Washington, DC.

Grunan, E., *Life Expectancy of Sealants in Building Construction*, Institute for Construction Research, Erstadt, Germany (1974).

Hepburn, C., *Polyurethane Elastomers*, Applied Science, London, UK (1982).

Hepburn, C. and Reynolds, R. J. W. (eds.), *Elastomers: Criteria for Engineering Design*, Applied Science, London, UK (1979).

Hofmann, W., *Rubber Technology Handbook*, Hanser/Oxford University Press, Oxford, UK (1989).

International Institute of Synthetic Rubber Producers, *The Synthetic Rubber Manual*, 10th edn, Houston, TX (1986).

International Organization for Standardization, *ISO Technical Report 7620 Rubber materials—chemical resistance*, Geneva, Switzerland (1986).

Legge, N. R., Holden, G. and Schroeder, H. E. (eds.), *Thermoplastic Elastomers–a Comprehensive Review*, Hauser, Munich, Germany (1987).

Lindley, P. B., Fuller K. N. G., Muhr A. H., *Engineering Design with Natural Rubber*, MRPRA, Brickendonbury, UK (1992).

Loveday, C. A., 'Performance with workability from modified binders', *J. Inst. Highways Transport* (June 1986).

Mernagh, L. R. (ed.), *Rubbers Handbook*, Morgan-Grampian/Design Engineering, West Wickham, UK (1969).

Malaysian Rubber Producers' Research Association, *Rubber Developments* (serial), MRPRA, Brickendonbury, UK.

Malaysian Rubber Producers' Research Association, *Natural Rubber Formulary and Property Index*, MRPRA, Brickendonbury, UK (1984).

Morton, M. (ed.), *Rubber Technology*, 3rd edn, Van Nostrand, New York, NY (1987).

Murray, R. M. and Thompson, D. C., *The Neoprenes*, international edn, E.I. DuPont de Nemours & Co., Wilmington, DE (1963).

Plastics and Rubber Institute, *Offshore Engineering with Elastomers*, Proc. Conference, Discussion Forum and Exhibition, Plastics and Rubber Institute, London, UK, 5–6 June (1985).

Rader, C. P. and Stemper, J., 'Thermoplastic elastomers—a major innovation in rubber', *Progr. Rubber Plastics Technol.*, 6 (1), 50–99 (1990).

*Rubber Red Book*, Communication Channels Inc., Atlanta, GA (1990).

Roberts, A. D. (ed.), *Natural Rubber Science and Technology*, Oxford University Press, Oxford, UK (1988).

Roff, W. J. and Scott, J. R., *Fibres, Films, Plastics and Rubbers—A Handbook of Common Polymers*, Butterworths, Guildford, UK (1971).

Rapra Technology, *Rubbicana Europe*, Shrewsbury, UK (1990).

Smith, D. R. (ed.), *Rubber World Magazine's Blue Book*, Lippincott and Peto, Akron, OH (1990).

Stevenson, A. (ed.), *Rubber in Offshore Engineering*, Adam Hilger, Bristol, UK (1985).

Swedish Institution of Rubber Technology, *Rubber Handbook*, Gothenburg, Sweden (1990).

*Thermoplastic Elastomers—An Introduction for Engineers*, Mechanical Engineering Publications, London, UK (1987).

Whelan, A. and Lee, K. S. (eds), *Developments in Rubber Technology—1, Improving Product Performance*, Applied Science, London, UK (1979).

Whelan, A. and Lee, K. S. (eds), *Developments in Rubber Tehnology—2, Synthetic Rubbers*, Applied Science, London, UK (1981).

Whelan, A. and Lee, K. S. (eds), *Developments in Rubber Technology—3, Thermoplastic Rubbers*, Applied Science, London, UK (1982).

29

# Polymer Dispersions and Redispersible Powders

**Robert Viles** CChem MRSC
Chief Technologist, Fosroc International

**Contents**

## 29.1 Introduction

### 29.1.1 Definitions

Polymers are organic (based on carbon) or inorganic (for example based on silicon). This chapter refers to organic polymers only and to dispersions that are *thermoplastic*, meaning that they soften and melt when heated, and their use as binders for other particulate matter or modifiers for cement, mortars or concrete. Polymers are large molecules formed by chemical bonding of many simple small molecules called monomers. That process is called *polymerisation*.

When the monomer is of one type the polymers are known as homopolymers. An example is polyethylene, made from one monomer, ethylene (Figure 29.1). Polymers are frequently represented by a sequence of letters, although such abbreviations are not systematic. Polyethylene is shortened to PE and a particular (linear high molecular weight) form of it is high-density polyethylene (HDPE)

The chain shown in Figure 29.1 could be branched (with side chains), and that is then a low-density polyethylene (LDPE). There are many combination results of this simple polymer due to conformation, chain length and additives. A range of characteristics may be achieved such as different hardness, rigidity, chemical resistance, and flexural and tensile strength.

Monomers that are readily polymerised by simple addition often contain a carbon double bond, C=C, and addition occurs through that opening.

Where two monomers are used they are called *copolymers*. An example is the use of styrene and butadiene monomers to form SBR, where R denotes "rubber". In its water-containing form the white milky liquid has the appearance and some properties of natural rubber. Such dispersions are called *latices* (or *latexes*). Polymer dispersions are also erroneously called "emulsions". An emulsion is a dispersion of a liquid in another liquid. Most of the polymer dispersions used in construction are tiny spheres of solid dispersed in water. Dispersions are however produced by a process known as emulsion polymerisation where the monomer is emulsified to effect polymerisation (Section 29.2). The formed solid particles are often of the order $10^{-7}$ m (0.1 μm) in diameter, although finer dispersions approaching nanometre sizes ($5–50 \times 10^{-9}$ m) can be produced, derived from ultra-fine emulsion polymerisation [1].

Several monomers can be used, and it is fairly common to have three monomers, resulting in a *terpolymer*, e.g. vinyl acetate–vinyl versatate–ethylene (Vac/VeoVa/E). VEOVA is a registered trademark in many countries.

Typically stable dispersions are copolymers or terpolymers, produced as 40% to 60% active matter in water. On drying above a certain temperature, called the *minimum film forming temperature* (MFFT), the particles from dispersion will coalesce to form a more or less coherent and homogeneous film.

MFFT is a key performance parameter and is related to another property of the base polymer called the *glass transition temperature* (*Tg*), the temperature at which a solid polymer will transform from a hard glassy state to a soft rubber-like state. A low *Tg* will give rise to a softer, more flexible or possibly stickier polymer. This relationship can be different for *hydrophobic* (water hating) polymers and *hydrophilic* (water seeking) polymers; suffice it to say that suitable polymer dispersions must form a film to effect adhesion or cohesion.

Dispersions may be formulated with additives such as stabilisers, antifoams, rheology control agents, coalescing agents, biocides and typical concrete or cement admixtures such as water reducing admixtures. They may be further diluted to provide a suitable primer or ready to use component in a two-pack formulated polymer/cement product.

Details of suppliers and relevant associations are given in the Bibliography.

### 29.1.2 History of polymer dispersions

Crude natural rubber latex is cis-1, 4 polyisoprene (Figure 29.2) – a soft, sticky substance, despite having a relatively high molecular weight (around 1 million).

Reportedly [2] this was first used with Portland cement in England in the early 1920s, to fortify mortar in sea walls to prevent deterioration due to marine exposure and to reduce salt corrosion of reinforcement. This natural product has unique properties but unfortunately tends to coagulate in a highly alkaline

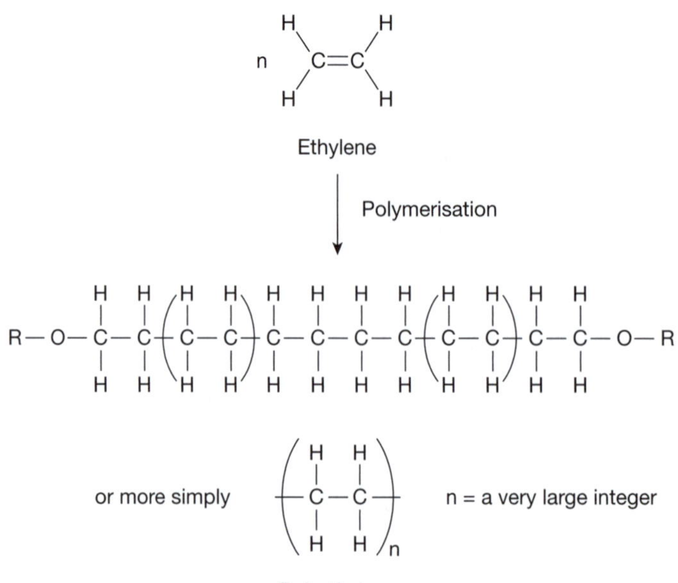


**Figure 29.1** Polymerisation of ethylene

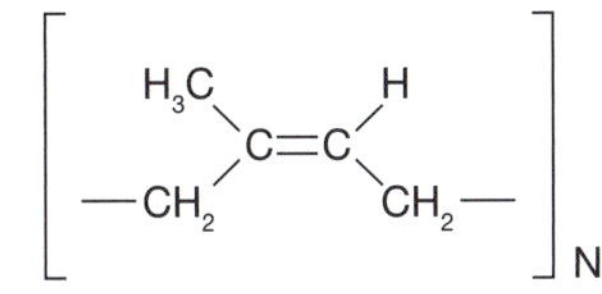


**Figure 29.2** Polyisoprene, cis conformation

calcareous environment and is thus not ideally suited to be formulated with Portland cement. Wider use was made in conjunction with calcium aluminate cements, and adhesive compositions combining the two are still commercially available. This combination was used in early floor screeds, to improve chemical resistance of mortars and in coatings. Unfortunately many solvents and oils attack natural rubber, although it is quite water-resistant and formulations have good adherence to various substrates.

Polyvinyl acetate (PVAc or PVA) was first produced as early as 1912 by Dr Fritz Klatte in Germany and was one of the earliest synthetic dispersions to become widely used, although significant manufacture only began around the mid-1930s. It is even today a familiar adhesive. Dispersions are stable by virtue of pH adjustment and partial hydrolysis which, when more complete, yields polyvinyl alcohol (PVOH or – again – PVA). Polyvinyl alcohol is water-soluble. Both the alcohol and acetate polymers are reasonably compatible with cements or plaster and these combinations work well when dry. Such formulations are however liable to deterioration when wet, since even the high acetate form is susceptible to hydrolysis by the alkali from cement and the alcohol form cannot resist water.

A good dispersion for most construction uses should dry irreversibly.

It was the Second World War and disruption of supply that brought forward the search for alternatives to natural rubber. From the United States came styrene–butadiene rubber (SBR) latex, although it was not until the late 1940s to 1950s that the use of this new material began to appear in cement and concrete.

The second half of the 20th century saw developments of many different types of dispersion such as polyacrylic esters (PAE) or "pure acrylics", styrene–acrylics (S/A) and vinyl acetate–ethylene (Vac/E), with products being synthesised to meet particular application needs. Tile adhesives, screeds, self-levelling floors, repair mortars and protective polymer–cement coatings have all benefited from targeted polymer additive developments.

Polymers that are chemically reactive with another component are not within the scope of this work, although aqueous dispersions containing epoxy resins are widely used in flooring, coatings and adhesives. Some acrylic dispersions are also reactively modified such that the division between chemically crosslinking polymers and simply coalescing dispersions is becoming less distinct. Many polyurethane (PU) dispersions also are capable of crosslinking, although some require heat to "unblock" the crosslinking reaction.

Not discussed here in detail are rheology modifiers such as clay, modified minerals or organic polymers (e.g. cellulose ethers, derivatives or polycarboxylates). Polycarboxylates are admixtures used for water reduction and/or workability retention. These low-level additives contribute significantly to performance but not to the fundamental physical effects of adding polymer to a concrete, mortar, adhesive or coating.

## 29.2 Redispersible powders

Polymers are now increasingly available and used as redispersible powders. This enables a complete single-component powder product to be provided with simply water added at the point of use. The result is a reduction of packaging, reduced transport and certainty of correct polymer dosage.

Many powder polymers are based on ethylene–vinyl acetate (E/Vac or EVA) but copolymers and terpolymers comprising vinyl versatate (VeoVa), vinyl laurate (VL) or vinyl chloride (VC), usually with ethylene (E) and/or vinyl acetate (Vac), are available according to desired properties. In order to facilitate spray-drying, various other additives are often included and typically some 10% of the dry powder is inorganic such as calcium carbonate to avoid "caking" or sintering of the polymer powder.

Traditionally the water resistance of redispersible polymers was partially compromised, but a wide range of applications can now be met. Some popular dispersions such as SBR have been very difficult to spray dry. There are a few examples of highly carboxylated styrene–butadiene polymers/rubbers (CSBR) that are available as powders. These disperse or are solubilised in alkaline media such as Portland cement mixes, by the reaction of the attached carboxylic acid groups with alkali.

## 29.3 Manufacture and design variables

### 29.3.1 Manufacture of polymer dispersions

The process of emulsion polymerisation often produces the polymers used to create dispersions in water. This is a substantial branch of polymer chemistry and technology [3,4]. Specific detail may be found in depth, for example in relation to vinyl acetate polymerisation [5].

Liquid monomer(s) are contained in a reactor and are emulsified in water, as the continuous phase, using emulsifiers (surfactants). The reaction requires an initiator and is conducted in an inert atmosphere. Three basic types of reactor may be used – batch, semi-batch and continuous. For production of final polymer dispersions, a relatively simple batch reactor system is often used. For economic production, capacities of 15 to 100 $m^3$ are deployed. The reaction may be at atmospheric pressure or even up to around 100 bar. A temperature range of 60–100°C is usual, which is controlled by heating and cooling, since the reaction is exothermic.

Emulsion polymerisation has been the subject of much investigation. The basic mechanism was described by Harkins [6,7]. To summarise, the emulsified droplets of monomer are substantially of the order of over 10 to 100 μm in diameter. Most monomers are sparingly soluble in water and the initiator is usually water-soluble. It is thought that initiation may commence in the aqueous phase but that the sub-micrometre droplets of monomer, called micelles, play the critical role in nucleating and developing the polymer. These micelles are initially only around 4 nm in size and are also stabilised by the surfactant and then grow into the polymer dispersion. "Seeding" the reaction with small particles of preformed polymer also facilitates nucleation, and that is an often-used practice. Various other agents may be added to control the polymerisation reaction.

The final dispersion is obtained after removing any residual monomer – "stripping" – and further processing known as "compounding". At that stage a range of additives may be included in the final product, some of which are shown in Table 29.1.

**Table 29.1** Compounding dispersions

| *Agent* | *Function* |
|---|---|
| Water | Adjusts polymer content and viscosity |
| Thickener | Adjusts viscosity |
| Stabiliser | Maintains dispersion stability |
| Antifoam | Prevents excessive air entrainment |
| Anti-oxidant | Prevents discolouration and degradation |
| Biocide | Prevents bacterial growth |
| Buffer | Controls pH and dispersion stability |
| Tackifier | Promotes adhesion |
| Coalescent | Promotes film formation on drying |

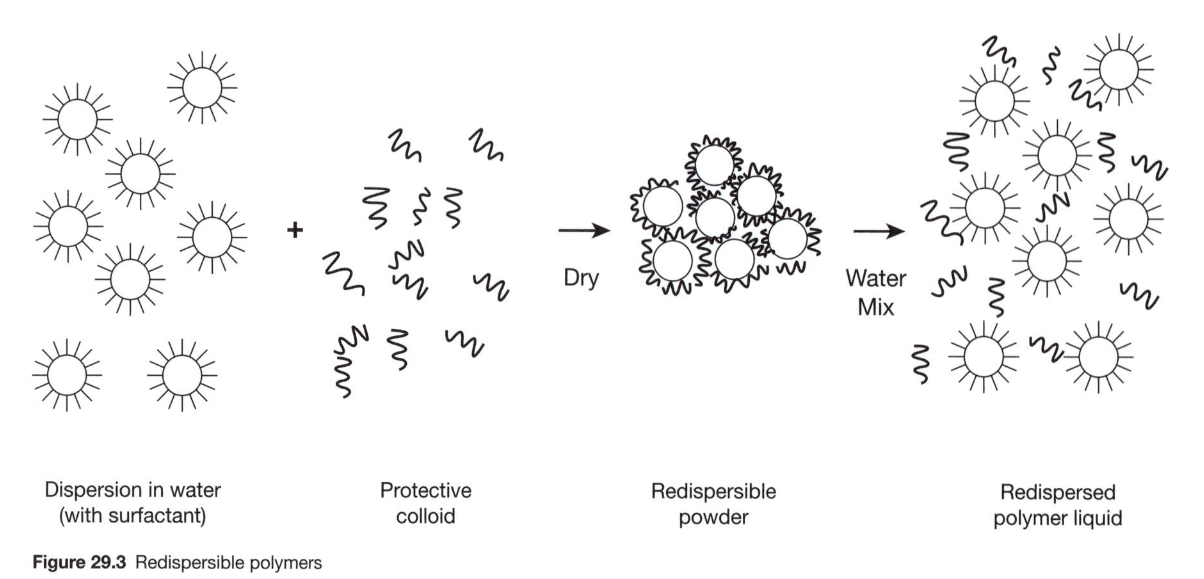


**Figure 29.3** Redispersible polymers

### 29.3.2 Manufacture of redispersible polymers

Since a polymer film for many applications should be irreversibly formed on drying in use, there is quite a technical challenge in spray drying a compound such that it may be reconstituted as an aqueous dispersion with suitable properties. The problem was addressed by surrounding the dispersed particles with a protective colloid to prevent the irreversible coalescence of the dispersed particles on spray drying, shown schematically in Figure 29.3.

The protective colloid is usually a water-soluble polymer such as polyvinyl alcohol. Spray drying is another detailed technology in its own right. Typical polymer dispersions can be spray dried through a rotary atomiser system using a high-speed spinning disc to produce small droplets at the edge of the disc. Air is forced through the drying chamber to remove the water. The powder is recovered either directly through the drying chamber or through air separators/bag filters and may be subjected to further post-drying and cooling. Typically solid spray dried particles are agglomerates of about 30–130 μm diameter and are produced at a rate of about 5 t per hour.

## 29.4 Design variables

### 29.4.1 Monomer type and ratio

The chemical nature of the monomer and resultant structure of the polymer will lead to a variety of properties, particularly with regard to film-forming capability, hardness/softness, adhesion, elasticity and chemical resistance. Table 29.2 gives the general properties assumed by certain homopolymers.

Properties vary considerably with molecular weight, process conditions and any compounded materials. It is possible to approximately calculate end-product properties of heteropolymers, such as glass transition temperature *Tg*. Varying the proportions of monomers may produce desirable combinations or optimise properties. For instance, a high styrene content in a butadiene or acrylic copolymer may make a harder film or composite – but it may not so effectively film form, particularly under low-temperature conditions. Styrene is also an aromatic molecule and its polymers tend to yellow (by oxidation) more readily than a pure aliphatic, although its general chemical resistance is good.

### 29.4.2 Functional monomers

Various "functional" side chains may be added to the polymer structural "backbone". That may be achieved, for example, by taking an unsaturated monomer molecule such as acrylic acid to introduce a free carboxylic acid (–COOH) group into a polymer. Such carboxylation is helpful both to stabilise an SBR dispersion and to improve its compatibility with alkaline materials such as Portland cement. Excessive carboxylation may greatly influence water sensitivity, however, and can, beneficially or otherwise, increase viscosity depending on the presence of alkali and/or polyvalent ions. Calcium, magnesium, aluminium or zinc, carefully used, can improve the final film properties of carboxylated polymers.

**Table 29.2** Properties of homopolymers

| *Monomer* | *Monomer chemical structure* | *Tg/°C* | *Relative hardness* | *Relative elasticity* | *Water resistance* | *Chemical resistance* |
|---|---|---|---|---|---|---|
| 1,3 butadiene | $CH_2$=CH–CH=$CH_2$ | –85 | Soft | High | Good | Good |
| n-butyl acrylate | $CH_2$=CH–C(O)O–$(CH_2)_3$–$CH_3$ | –54 | Soft | High | Good | Fair |
| Methyl acrylate | $CH_2$=CH–C(O)O– $CH_3$ | 10 | Hard | Low | Fair | Fair |
| Vinyl acetate | $CH_2$=CH–O–C(O)– $CH_3$ | 32 | Medium to soft | Medium | Fair | Fair/poor in alkali |
| Vinyl chloride | $CH_2$=CH–Cl | 81 | Medium to hard | Medium | Good | Good |
| acrylonitrile | $CH_2$=CH–CN | 77 | Medium | Medium | Good | Very good |
| styrene | $CH_2$=CH–$(C_6H_5)$ | 145 | Hard | Low | Good | Good |

### 29.4.3 Surfactant type and surface charge on dispersed particles

Different surfactants and functional groups within the polymer give rise to different charges on the particle surface. These may be:

(a) anionic – a negative charge
(b) cationic – a positive charge
(c) non-ionic – a neutral charge.

Surfactants perform an important role in the emulsion polymerisation process and more than one may be deployed. They and/or protective colloids such as hydrophilic polymer dispersions stabilise the final dispersion.

Most dispersions are anionic and/or non-ionic and those are often best compatible with cement. Cationic stabilised dispersions are commonly used with bitumen emulsions (dispersions if the bitumen is a solid).

Cement "plasticisers" are surfactants and may be polymers. Polymer dispersions that do not adversely affect rheology can behave as water reducing or workability extension agents.

### 29.4.4 Particle size

Natural rubber has a mean particle size of about 1 μm while synthetic SBR is 5 to 10 times smaller. The particle size from synthetic dispersions is controlled by the process conditions, notably the amount and type of surfactant.

Very fine particle sizes, sometimes described as "ultra-fine dispersions" may lead to better distribution in the final product or better penetration into a substrate but exposure to coalescing agents, either in formulating or through in-situ use, may reduce some of that advantage. In the ultimate case exposure to a change in pH, an increase in electrolyte, such as a high level of dissolved salts, or exposure to extremes in temperature, particularly freezing, can result in coagulation.

### 29.4.5 Molecular weight and crosslinking

The molecular weight and crosslinking of linear chains will influence final film properties. These parameters are determined to some extent by the polymer type and reactor conditions, but the use of "other" agents as mentioned in Section 29.2.1 are also important. Chain transfer agents, such as mercaptans, are used to terminate a chain while crosslinking agents, such as unsaturated methylol compounds, may be deployed to crosslink chains via methylol group condensation (on heating). The reaction may be halted before completion to avoid excessive crosslinking (e.g. SBR.) Excessive crosslinking and/or too high a molecular weight can lead to partial coagulation or the formation of "gels" which can be removed in the process by filtration.

### 29.4.6 Other process variables

"Seeding" controlled additions during the process and the hydrophobic nature of the developing polymer can lead to the core of a dispersed polymer particle being different to the outer shell. For instance, the outer shell could be, and often is, significantly more hydrophilic than the inner core.

## 29.5 Uses and properties of polymer dispersions

Polymer dispersions used in construction applications are frequently produced from two or more monomers. A seemingly bewilderingly wide range of products is available to the formulator or end user. Terms such as "polymer modified mortar" are meaningless, while specifications citing "acrylic" or "SBR" are wholly inadequate without clear reference to product performance parameters.

### 29.5.1 Polymer dispersion as the only binding agent

For many coatings, adhesives, primers and mastics the polymer content from the dispersion is the only binder. Such materials often contain fillers, e.g. silica sands or calcium carbonate, pigments and other modifiers such as thickeners and coalescing agents to promote film formation.

Coatings are used primarily to protect but also to enhance the appearance of a substrate. The technology is too vast to be even summarised here, but when using polymer dispersions the key elements to consider are as follows:

1 *Is a primer needed?* Most frequently the adhesion of a coating with a substrate relies on the use of an effective primer, although some primer-less systems are available.
2 *What protection is needed?* Resistance to salts and other environmental exposure must be ascertained.
3 *What appearance is satisfactory?* A finish may be textured, glossy, matt, requiring close colour matching or the appearance may be of minor importance to functionality.
4 *How will the system be applied?* Different formulations are best suited to either spray or brush application.
5 *Durability?* Compatibility with the substrate and environmental exposure are the most important properties.

Useful reviews of the state of the art in coatings and "emulsion" polymer technologies can be obtained from the PRA (Bibliography).

### 29.5.2 Dispersions used in and with hydraulic cementitious materials

Cement hardens by simple hydration with water and forms a strong binder well known in concrete and mortar technology. One may ask, "Why use a polymer in cement and concrete?" Some advantages are listed below:

1 *Adhesion promotion.* Bond or "pull off" strengths may increase significantly using an appropriate primer.
2 *Increase in flexural and tensile strength of polymer-modified mortars.* No improvement occurs in compressive strength, but the ratio of compressive/flexural strength for example can be transformed from about 7 to 8:1 for a non-modified concrete to about 4:1 (Figure 29.4).
3 *Increase in elastomeric properties such as crack bridging and movement capacity for coatings and adhesives.*
4 *Improvement in curing.* Cementitious materials less than about 50 mm in thickness may readily lose water. Polymer aids water retention, sometimes by "skinning" on exposure to air.
5 *Improvement in wear rate of flooring.*
6 *Reductions in permeability to aggressive elements* such as chloride or carbon dioxide that may in time corrode (reinforcing) steel. That is evidenced by diffusion data.
7 *Reduced water adsorption and water sensitivity.* That may be useful for mortars and flooring and is essential for adhesives, such as tile adhesives, used in wet conditions.
8 *Improvement in application properties.* Cohesion, trowelling/finishing, spray application and pumping are often improved by using carefully formulated mixes containing polymer.
9 *Economy in use.* Thinner applications are needed either for bedding tiles or for providing effective cover to reinforcement in concrete.

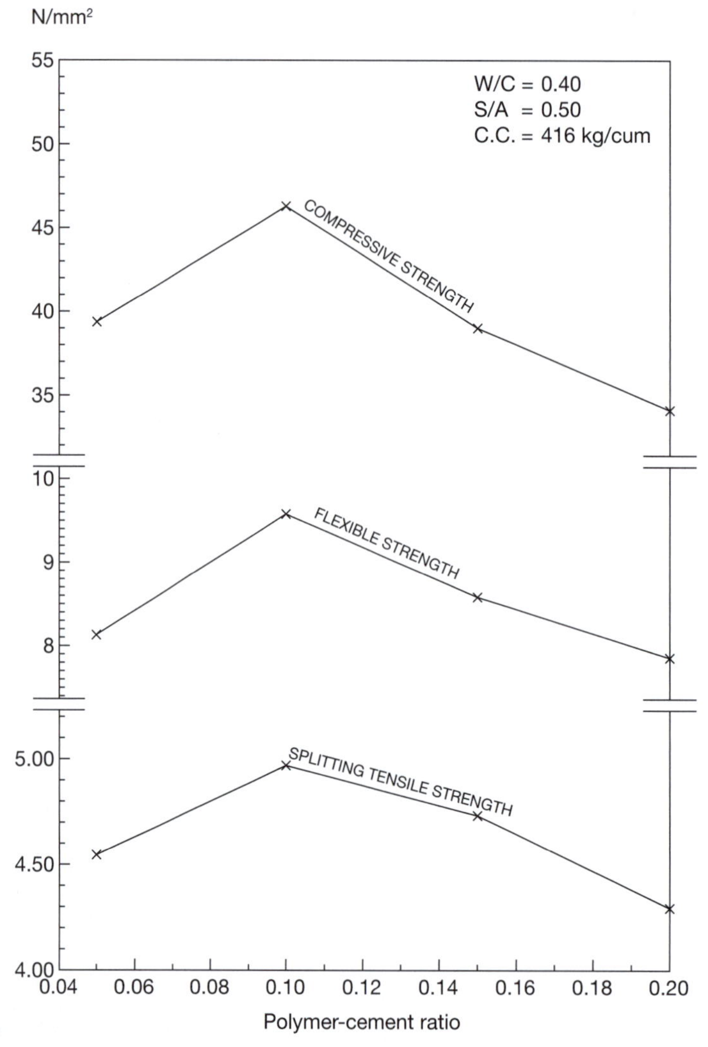


**Figure 29.4** The effect of polymer: cement (dry weight) ratio on mechanical properties of mortars

Such benefits may also be attributable to parameters such as particle size distribution of the filler and other admixtures used in formulated products. It is not always appropriate to use polymer, and certainly in large quantities there may be disadvantages. For instance, a concrete repair mortar may benefit from a low-level addition but excessive amounts of polymer will reduce modulus, increase creep and the product may not be suitable for load transfer when used as large-scale concrete replacement in reinforced structures. A high electrical resistivity (many polymers are intrinsically non conductive) is not desirable for an overlay to be used in cathodic protection applications. Compatibility of each component in the system is the key.

### 29.5.3 Applications summary

Applications range from new build/remediation to repair/refurbishment and comprise:

1 tile adhesives and "thin bed" mortars
2 primers and bonding agents
3 brick and block mortars
4 external, decorative and mortars applied over thermal insulation
5 self-levelling polymer-modified cement flooring and screeds
6 concrete repair and protection mortars and concrete
7 sprayed protective and crack bridging coatings
8 water-proofing, low-permeability and protective coatings for concrete roofs, roof tiles, potable water basins and basement/cellars.
9 architectural and decorative coatings.

## 29.6 Guidelines for polymer selection and practical site use

### 29.6.1 Selection

It is essential to consult the supplier of polymer dispersion or the manufacturer of a ready-to-use product for specific application advice. Beware of similar sounding names, since the substitution of a "butyl" by a "methyl" can give very different character (hardness, flexibility, water resistance), and even the same generic type can develop a range of application properties (flow, slump,

**Table 29.3** Polymer selection

| *Application* | *Environment* | *Monomer/Polymer type* |
|---|---|---|
| Adhesives | Dry | VAc |
| | Wet | VC/E/A or S/B or S/A |
| Screeds/Flooring | Dry | VAc/E or Vac/E/MMA |
| | Wet | S/B or S/A or VAc/VeoVa |
| Repair mortars | Wet & Dry | VAc/VeoVa/E or VAc/VeoVa |
| Flexible cement coatings | Wet & Dry | A or S/B |
| Bitumen modification | | S/B or Cl |
| External mortar coatings | Wet & Dry | A or S/A or VAc/E |
| Elastomeric | Wet & Dry | PU |

A = acrylate (acrylic ester, e.g. butyl acrylate). B = butadiene. Cl = chloroprene. E = ethylene. MMA = methyl methacrylate. PU = polyurethane. S = styrene. VAc = vinylacetate. VC = vinyl chloride. VeoVa = vinyl versatate.

adhesion). Some limited general examples are indicated in Table 29.3.

### 29.6.2 Primers

Ideally a cementitious overlay could be effectively placed on concrete by just saturating concrete with water in place of a primer. Practically that is a difficult if not impossible state to attain. Primers also control "pin holing", a result of air release from the substrate through self-levelling flooring.

The selection of correct primer is critical. For a flooring or mortar application the primer should ideally be just "tacky" prior to the overlayment. Best results are obtained with a material that has no tendency to re-emulsify. Equally it should not dry to a high level of hydrophobicity, since full drying before final application can result in a "disbonding" layer.

Some protective elastomeric coatings used over concrete owe part of their resistance to the primer.

The primer composition and application method needs to be carefully defined. The whole system must be evaluated with specific substrates and overlayments.

### 29.6.3 Polymer-modified cements, mortars and concretes

0.2–5% polymer solids by weight of total may typically modify performance with regard to permeability, adhesion, cohesion, application and pumpability. Such levels of addition are thus adequate for mortars, tile adhesives, cementitious flooring and the like. Many high-performance dry powder products will contain redispersible powders, often of the ethylene–vinyl acetate or vinyl versatate type.

At about 10–15% by weight, the polymer may become the continuous phase when the properties of a "filled polymer" rather than modified cement will be realised. Higher polymer levels are more appropriate for thin layer protective applications where some elastomeric behaviour is required. Note that "curing" of such compositions may mean *drying* and the cement hydration may be purposefully limited within a hydrophobic polymer to just facilitate "drying". A butyl acrylate polymer may be suitable.

Consider the application conditions and requirements and match those to the flexibility, hardness, tackiness and environmental resistance properties of the polymer when choosing a dispersion or redispersible polymer.

Good mixing is always required. At least 3 min mixing in a purpose-designed mechanical forced-action batch mixer should be stipulated. The aggregates and cement should be well mixed first if a multi-component mix is being used. Correct formulation and mixing is necessary to avoid excessive air entrainment and proper development of properties.

Continuous mixers such as progressive cavity pumps may be used for "wet spray" mortars or plasters, thin layer cement-based flooring and "dry mix" screed or concrete. Mixing time may be shorter than in a simple batch mixer but effective mixing is ensured by virtue of higher shear, not just through the pump, but also via the length of hose on the outlet. This is typically over 25 m length and 25 mm diameter hose or less for mortars, with larger proportions for concrete depending on aggregate size. For fluid self-levelling compositions the choice of polymer is one determinant of good flow properties.

"Dry spray" is possible with some redispersible polymers used in powder mixes. although the full benefit of the polymer is possibly best realised in wet spray. The merits of wet spray have been described [8], although dry spray is often convenient.

The main benefits of a formulated dry mix are convenience of use and quality.

Surface preparation is probably the most important step in any application. Consider shot or hydro-blasting and even substantial removal to reveal a sound clean substrate.

Skilled finishing (roller, trowel, sponge according to application) will yield a pleasing appearance for open-faced placements. Overworking should be avoided.

Wet or protective curing during the first day or two is usually beneficial. After that, dry curing may be acceptable according to the environment and specific compositions deployed.

### 29.6.4 Polymer-modified bitumen

The basic properties of bitumen can be improved by polymer dispersions. A common polymer type is a cationic SBR dispersion that is compatible with cationic asphalt emulsions or dispersions. Both hot and cold asphalt systems can be created, with the main benefits of:

1 increased toughness at high temperature
2 greater flexibility at low temperature
3 improved adhesion, water resistance and ageing characteristics.

Such compounds, used mainly for roads and sealants, are best produced in purpose-designed plants. The key product issues are determining compatibility of polymer dispersions with the required bitumen to ensure stability. Care should be taken with equipment, since cationic systems can be destabilised by various metal ions.

Useful cold spray-on, seamless waterproofing membranes can also be produced from two-component systems where a "catalyst" is used to remove water from the polymer-modified bitumen dispersion. This can be applied onto damp surfaces and green (partly cured) concrete.

### 29.6.5 Coatings

Two-component polymer coatings such as epoxy, polyurethane and the rapid action polyureas are not necessarily dispersions, although increasing numbers of water-borne dispersions are available.

A single-component polymer dispersion in water is convenient and safe to use on site by spray or brush application onto many substrates such as wooden flooring, concrete and steel. The choice of polymer binder depends on appearance required or the protection demanded. Often full and effective use demands the application of an appropriate primer, and good surface preparation cannot be over-emphasised.

Elastomeric acrylic dispersions may be a good choice for concrete protection.

Steel protection is afforded by polyurethane–polyester hybrid dispersions and may contain zinc to improve salt resistance, for example.

Wood can be treated by single-component (for example UV curing) polyurethane dispersions. These offer low volatile organic levels with good transparency and excellent water resistance.

Coating formulations can often be complex, with large numbers of added components. They should have good application properties, covering and brushing out well with little drag. That is quite dependent on rheology of the mix. The polymer and most importantly the filler type often control the surface finish.

For polymer cement coating materials where the continuous phase is polymer, dry curing may be required as described in Section 29.6.3.

## 29.7 Health, safety and environment

### 29.7.1 Health and safety

Clean tools and mixers immediately and thoroughly in water where possible after use. If hardened, polymer-modified materials may be susceptible to solvent cleaning (e.g. white spirit) but that should be avoided on health, safety and environmental (HSE) grounds. Polymer dispersions are often preferred for HSE reasons, being essentially free of volatile, toxic and sometimes flammable organic solvents.

Waste and spillages should be cleaned up and properly disposed of, according to regulations.

Many types of cement are alkaline, will defat the skin and may give rise to irritation or dermatitis. Polymer dispersions will usually contain surfactants that may also contribute to skin or eye irritation. At all times during application suitable protective clothing, gloves, eye protection and dust masks should be worn. Maintain good personal hygiene and follow suppliers' safety (MSDS) advice.

### 29.7.2 Fire

Compositions containing less than 1% by weight polymer may usually be considered to be non-combustible. Good fire ratings can be achieved at loadings of 3%. At high polymer contents, about 12–15% by weight, the polymer forms the continuous phase and the behaviour is that of a filled polymer. The contribution to fire and products of combustion or smoke should be carefully considered and assessed by appropriate testing according to regulations and codes (e.g. ASTM E 136, EN1182, EN13501.) Suitability will further depend on the polymer type, the degree of dispersion, the environment and requirements in use.

### 29.7.3 Environment and sustainability

Polymer (dispersion) modified thin layer cementitious mortars can contribute most positively to sustainable concrete systems. They offer a high degree of protection and durability in comparison with simple polymer coatings or plain mortars. This is a result of low permeability to aggressive elements, offering an order of magnitude greater protection than an equal thickness of plain mortar or concrete. Thus 1–2 mm of polymer-modified mortar can have an equivalent cover thickness of 10–20 mm of concrete which is most useful in remediation, refurbishment or repair, using a much lower quantity of material. At the same time such mortars may be fairly permeable to water vapour and have greater compatibility with concrete, ensuring that they remain bonded to the substrate.

The effective use of powdered redispersible polymers enables thin bed adhesives to be prepared and obviates the transport of water.

## 29.8 References


1 Antonietti, M., Basten, R. and Lohmann, S. (1995). Polymerization in microemulsions – a new approach to ultrafine, highly functionalized polymer dispersions. *Macromolecular Chemistry and Physics*, 196(2), 441–466.

2 Laticrete. *History of Latex in Portland Cement Mortars*. www.laticrete.com/portals/0/tds/tds107.pdf.

3 *Encyclopaedia of Polymer Science & Engineering*, Volume 6, *Emulsion Polymerization* (1986). Chichester, UK: Wiley.

4 Lovell, P. A. El-Aasser, M. S. (Eds) (1997). *Emulsion Polymerization and Emulsion Polymers*. Chichester, UK: Wiley.

5 Erbil, H. Y. (2000). *Vinyl Acetate Emulsion Polymerization and Copolymerization with Acrylic Monomers*. Boca Raton, FL: CRC Press.

6 Harkins, W. D. (1947). A general theory of the mechanism of emulsion polymerization. *Journal of the American Chemical Society*, 69, 1428.

7 Harkins, W. D. (1950). General theory of mechanism of emulsion polymerization. II. *Journal of Polymer Science*, 5, 217–251.

8 Austin, S. A, Robins, P. J. and Goodier, C. A. (2002). *Construction and Repair with Wet-Process Sprayed Concrete and Mortar*. Concrete Society Technical Report No. 56. Camberley, UK: Concrete Society.


## 29.9 Bibliography

### Associations

- PRA Coatings Technology Centre, Hampton, UK. Publications such as several reviews on coatings and "emulsion polymer technologies". www.pra-world.com
- The Concrete Society. Camberley, UK. Technical Report TR39, *Polymers in Concrete*, 2nd edition 1994. www.concrete.org.uk
- American Concrete Institute, Farmington Hills, MI. 548.3R-03. Polymer-Modified Concrete. ACI committee 548. www.concrete.org
- The Concrete Repair Association, Farnham, UK. Standards and codes of practice for concrete repair. www.concreterepair.org.uk.
- International Concrete Repair Institute, Des Plaines, IL. Standards and codes of practice for concrete repair. www.icri.org.
- FeRFA, the Resin Flooring Association, Farnham, UK. www.ferfa.org.uk.
- RILEM – International Union of Laboratories and Experts in Construction Materials, Systems and Structures, Bagneux, France. www.rilem.org.

### International Congress on Polymers in Concrete

1 ICPIC London, UK, 1975 (The Concrete Society)
2 ICPIC Austin, USA, 1978 (Univ. of Texas at Austin)
3 ICPIC Koriyama, Japan, 1981 (Nihon University)
4 ICPIC Darmstadt, Germany, 1984 (Technische Hochschule Darmstadt)
5 ICPIC Brighton, UK, 1987 (Brighton Polytechnic)
6 ICPIC Shanghai, China, 1990 (Tongji University)
7 ICPIC Moscow, Russia, 1992 (NIIZhB – Scientific Institute for Concrete and Reinforced Concrete)
8 ICPIC Oostende, Belgium, 1995 (Katholieke Universiteit Leuven, Royal Flemish Soc Engrs)
9 ICPIC Bologna, Italy, 1998 (University of Bologna)
10 ICPIC Honolulu, Hawaii, May 2001 (ICPIC)
11 ICPIC Berlin, Germany, June 2004 (BAM)

### Websites

- Construction Chemistry Directory – www.baustoffchemie.de/en/db/polymerdispersions
- Suppliers of Raw Materials: Polymer Dispersions – www.rhodia-ppmc.com/uk/markets_const.shtm, www.hexionchem.com/english
- Crompton Corporation – www.cromptoncorp.com
- Dairen Chemical Corporation (DCC) – www.dcc.com.tw
- Industrial Copolymers Ltd – www.incorez.com/poly_disp.htm
- Hüttenes-Albertus Lackrohstoff GmbH – www.ha-lack.de
- Paramelt B.V. www.paramelt.com/markets.aspx?tID=3&mID=45&smID=55
- F.A.R. Fabbrica Adesivi Resine S.p.A. – www.gruppofar.it/far/eng
- Prochem AG – www.prochem.ch

- Polymer Latex GmbH & Co. KG – www.polymerlatex.de
- Vinamul Polymers – www.vinamulpolymers.com
- Elotex AG – www.elotex.com
- BCD Rohstoffe für Bauchemie Handels GmbH – www.bcd.at/englisch/eprodukte.htm
- Dynea – www.dynea.com
- Acquos – www.acquos.com
- Dow Reichhold Specialty Latex LLC – www.dowreichhold.com
- Celanese AG – www.celanese.com
- BASF Aktiengesellschaft – www.basf.de/de/dispersionen/products
- Jesons Industries Ltd – www.jesons.net
- CSC Jäklechemie GmbH & Co. KG – www.csc-jaekle.de
- Synthomer GmbH – www.synthomer.com
- Wacker-Chemie GmbH – www.wacker.com
- C.H. Erbslöh KG – www.cherbsloeh.de
- Rohm and Haas – www.rohmandhaas.com
- Johnson Polymer – www.johnsonpolymer.com

## Some global suppliers of polymer (dispersion and/or redispersible powder) modified ready-to-use cementitious products

- Ardex – flooring, tile adhesives – www.ardex.com
- Degussa Construction Chemicals – flooring, coatings, concrete repair systems – www.degussa.com/en/home/construction_chemicals.html
- Fosroc – concrete repair systems, flooring, coatings, waterproofing – www.fosroc.com
- Mapei – tile adhesives, flooring, waterproofing, concrete repair systems – www.mapei.com
- Sika – flooring, coatings, concrete repair systems – www.sika.com

# 30 Silicones, Silanes and Siloxanes

**Arthur Lyons** MA MSc PhD Dip Arch Hon LRSA
De Montfort University

## Contents

## 30.1 Introduction

Silanes, siloxanes and silicones are compounds based on silicon. This significant factor imparts a range of physical and chemical properties quite distinct from the analogous carbon-based organic compounds and polymers. Typical uses for silicon-based compounds and polymers in construction include sealants and water-repellent agents.

## 30.2 Chemical composition

### 30.2.1 Silanes

Silanes are a group of chemical agents based on chains of silicon atoms, analogous to organic materials based on chains of carbon atoms (Figure 30.1). The general formula for silanes is $Si_nH_{2n+2}$. Branched and ring silanes are equivalent to branched and cyclic alkanes. Silanes containing at least one carbon–silicon bond are known as organosilanes.

Attachment of other atoms or groups such as nitrogen, alkyl, methoxy or acetoxy yields silylamines, alkylsilanes, alkoxysilanes and acetoxysilanes respectively (Figure 30.2). These agents react readily with water to form silanols (–SiOH), leading to the stable siloxane (–Si–O–Si–) bond, or –Si–O– bonding to other material surfaces. This reactivity is the basis for the use of silanes in water-repellent surface treatments and as coupling agents for bonding two dissimilar materials.

### 30.2.2 Siloxanes

Siloxanes (Figure 30.3) are chemical compounds with a backbone of alternating silicon and oxygen atoms. They may be linear, branched or cyclic also as short chain oligomers or longer chain polymers.

The R and R′ units in siloxanes may be hydrogen or hydrocarbon groups such as methyl, phenyl or vinyl. Polysiloxanes, also known as silicones, are effectively hybrid inorganic and organic compounds, with an inorganic –Si–O–Si–O– backbone and organic functionality derived from the side chains. The most common example is poly(dimethylsiloxane) or PDMS (Figure 30.4).

The low surface tension of silicones, together with their high temperature and durability, makes them suitable for a range of

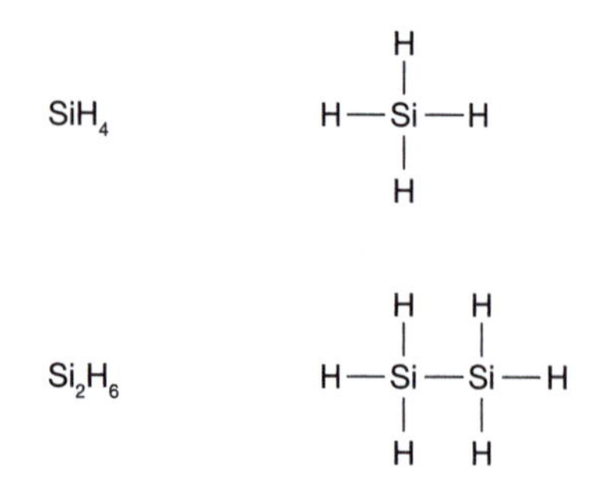


**Figure 30.1** Silane (top) and disilane (bottom)

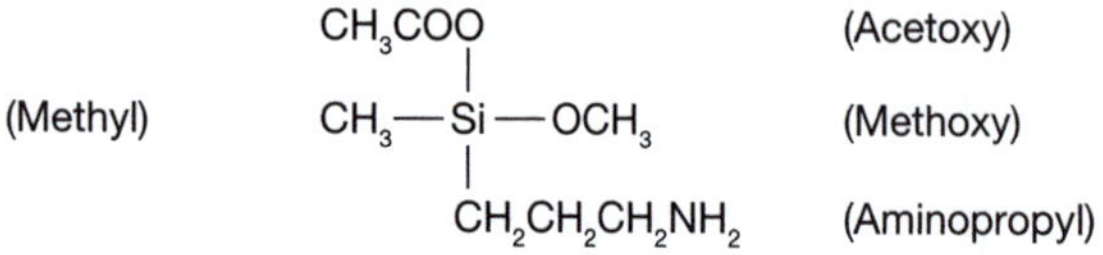

(Acetoxy) $CH_3COO$

(Methyl) $CH_3$—Si—$OCH_3$ (Methoxy)

(Aminopropyl) $CH_2CH_2CH_2NH_2$

**Figure 30.2** Attachment of other atoms or groups to silanes

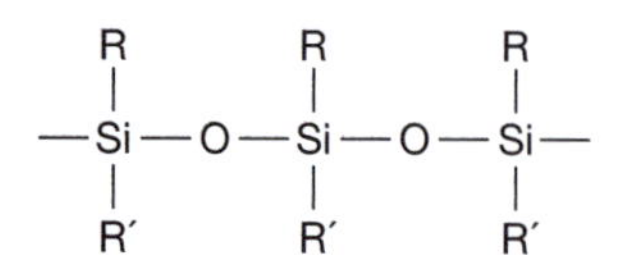


**Figure 30.3** Siloxane oligomer

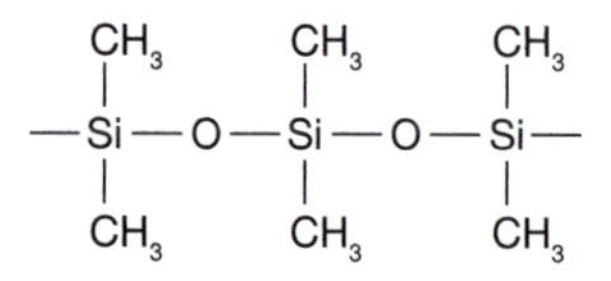


**Figure 30.4** Poly(dimethylsiloxane)

applications in construction. These include silicone rubber gaskets, silicone grease for O-rings, water-repellents, joint sealants and bonding agents. The heat resistance of silicones is important in their function as firestop sealants.

## 30.3 Manufacture

The starting material for silane, siloxanes and silicones is high-grade silicon, which is produced by the thermal smelting of silica sand with carbon.

$$SiO_2 + 2C \rightarrow Si + 2CO$$

Silane is manufactured in a two-stage process. Initially powdered silicon reacts with hydrogen chloride to produce trichlorosilane. This is then catalytically split over aluminium chloride into a mixture of silane and silicon tetrachloride.

$$Si + 3HCl \rightarrow HSiCl_3 + H_2 \text{ (trichlorosilane)}$$

$$4HSiCl_3 \rightarrow SiH_4 + 3SiCl_4 \text{ (silane silicon tetrachloride)}$$

Organosilanes, siloxanes and silicones are produced commercially from the direct reaction of silicon and organo-chlorides. Methyl chloride reacts with pure silicon in a fluid bed reactor to produce a complex mixture of methyl chlorosilanes which is then distilled to extract the primary product, dimethyl dichlorosilane.

$$Si + CH_3Cl \rightarrow (CH_3)_2SiCl_2 \text{ (dimethyl dichlorosilane)}$$

For the production of siloxanes and silicones, the dimethyl dichlorosilane is then hydrolysed to poly(dimethylsiloxane) oligomers. A mixture of linear and cyclic siloxane oligomers are produced, which form the raw materials for a wide range of silicone products.

$$n[(CH_3)_2SiCl_2] + n[H_2O] \rightarrow [Si(CH_3)_2O]n + 2nHCl$$

During the polymerisation of dimethyl dichlorosilane, hydrogen chloride is evolved. This is, however, avoided when chlorine in the precursor is replaced by acetate groups, in which case acetic acid is eliminated on polymerisation. This effect is noticeable when using certain silicone sealants.

Branched chain and cross-linked polymers are produced when silane precursors with three chlorine atoms or reactive groups are

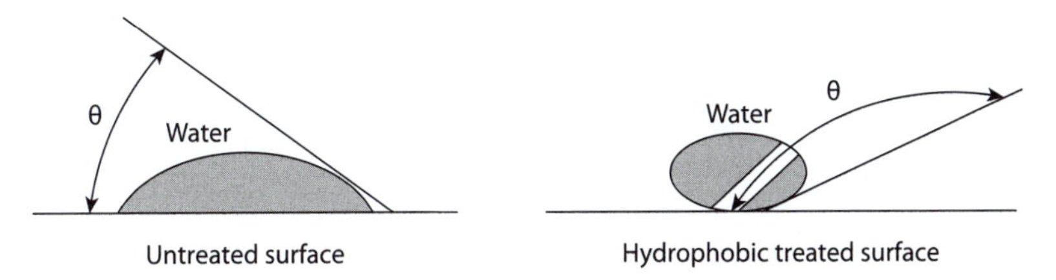


**Figure 30.5** Change in contact angle between water and substrate after treatment with hydrophobic silanes/siloxanes

introduced into the polymerisation process. Conversely, if silanes with three methyl groups are added, these act as siloxane chain terminators to limit molecular weights. This control of molecular weight, together with the wide range of organic reactive groups that may be attached to the inorganic backbone, (typically alkyl, aryl, methoxy, ethoxy, acetoxy or amino), gives rise to the diverse range of physical and chemical properties associated with silicones.

## 30.4 Product range

Silanes and siloxanes are extensively used in the surface protection and waterproofing of concrete. The hydrophobic qualities of the organic functional groups attached to the inorganic backbone chain, change the contact angle between the substrate and water (Figure 30.5).

Sodium and potassium methyl siliconates, dissolved in water or water/alcohol mixtures, are also manufactured as water repellents. Whereas moisture, activated by alkalis, converts silanes and siloxanes to silicones, the reaction of siliconates is induced by atmospheric carbon dioxide. Siliconates are not recommended for use on concrete, particularly when new, but may be applied to brick, certain sedimentary stones and ceramics. Siliconates have a low resistance to alkalis and may form a white residue of sodium or potassium carbonate on the surface if used externally and subsequently affected by rain before complete curing. Siliconates may be applied by dipping, spraying or brushing techniques.

Silanes with diverse functionality are used as coupling agents; for example, to ensure that glass fibres adhere to the polymer matrix within glass-fibre reinforced plastics.

The diversity of physical properties of siloxanes offers a further wide range of applications. Siloxane polymers may be fluids, gels, elastomers or resins. Silicone fluids are essentially linear chains of poly(dimethylsiloxane) (PDMS). Silicone gels are lightly cross-linked, from the incorporation of a small proportion of trifunctional silanes based on either methyl trichlorosilane or vinyl silanes. Silicone elastomers are moderately cross-linked, and may incorporate fillers. Silicone resins are extensively cross-linked by the incorporation of a large proportion of tri- and tetra-functional monomers, such as methyl trichlorosilane, producing a more rigid three-dimensional structure. The range of physical properties can be adjusted to suit the required application.

Other typical uses within construction include adhesives and sealants, and paints, particularly where ageing or heat resistance is important.

## 30.5 Concrete surface treatments and protection

The distinct types of surface treatment to concrete are defined in the Standards BS EN 1504 Part 1: 2005, Part 2: 2004 and Part 10: 2003.

Hydrophobic impregnation produces a water-repellent coating on the inside of the capillary pores but does not prevent transmission of water vapour or ingress of water through hydrostatic pressure. There is no surface film on the concrete and no significant change to its appearance. The standard materials used are silanes and siloxanes.

Impregnation with sealers blocks, or partially blocks, the pores, leading also to a thin discontinuous surface film. Typically organic polymers are used for this process. For a full surface coating of typically 0.1 to 5.0 mm thickness, organic polymers, modified polymers or hydraulic cement modified with polymer dispersions are normally used.

The hydrophobic water-repellent penetrants based on silicon, are normally supplied as single-pack products. Silanes and siloxanes are supplied in organic solvents or dispersions in water, which chemically bond to the concrete pore surfaces and polymerise in-situ.

Standard silanes include iso-butyl triethoxysilane, iso-octyl trimethoxysilane and iso-butyl trimethoxysilane – the last of these being used at a minimum 92% concentration to prevent aqueous chloride ingress into highway structures. These materials are slow to polymerise, and under hot or windy conditions significant quantities of the volatile silanes may be lost, so they are normally used at high concentrations or undiluted.

Siloxanes, which are less volatile, are generally used diluted and have the advantage of faster reaction. In most cases, an appropriate siloxane/silane blend is used. For dense substrates with low permeability a high silane content is used, whereas for absorbent materials, such as natural stone, a high siloxane content is more appropriate. Where the substrate has a high pH, it is usual to use a higher silane content. (Guidance on the site testing of concrete permeability is given the Concrete Society Technical Report No. 31.)

The success of the impregnation is largely determined by the surface condition of the concrete, which must be dry and clear of all loose material, also free from release and graffiti removal agents. If the moisture content of the concrete is 10% or more then penetration is very limited. Impregnation of new concrete must not be effected less than 7 days after placing, or 3 days after remedial repairs. The ambient temperature should be between 5 and 25°C, with wind speeds of no more than 8 km/h.

Generally the depth of penetration for alkylalkoxy silanes and oligomeric siloxanes in organic solvents is determined by a combination of viscosity and surface tension. However, with two-phase emulsions, the viscosity of the oily phase – which relates to molecular weight – is the critical factor, so alkyltrialkoxysilanes penetrate better than oligomeric siloxanes. Micro-emulsion mixtures of silane and oligomeric alkoxysiloxanes may be applied using low-pressure spray techniques.

Silicone resins, typically 5% by weight in organic solvents such as white spirit, may also be used on dry surfaces for immediate water-repellency. However, their molecular weights prevent full penetration into the very fine pores of good-quality concrete, compared to the more fluid silane/siloxane products which polymerise *in situ*.

The water-repellent treatments to concrete do not affect permeability to water vapour or carbon dioxide, so concrete carbonation rates are unaffected. However, corrosion of steel reinforcement

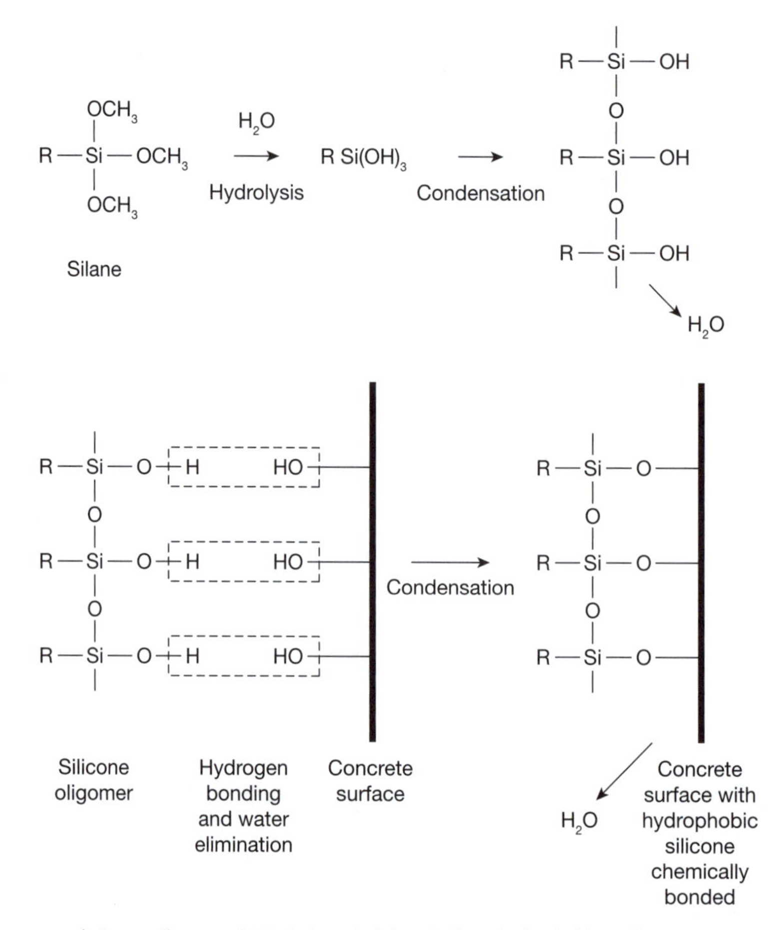


**Figure 30.6** Reaction sequence between silanes and substrate material producing a hydrophobic surface

from the ingress of chloride ions is reduced and in certain cases prevented. Also, because capillary water absorption is reduced, frost resistance is enhanced.

Calculations based on the contact angle of water within silicone coated capillaries show that pores of less than 37 μm diameter will not permit the ingress of moisture under a head of less than 200 mm water pressure. As most pores in concrete are less than this diameter, more typically around 0.5 μm, this explains the effective action of the hydrophobic coating.

## 30.6 Silicone masonry paints

Silicone resin emulsion paints are micro-porous, with low VOC (volatile organic compound) emissions and good weathering resistance. They are suitable for application to brick, stone, lime plaster and render.

## 30.7 Silicone adhesives and sealants

Room temperature vulcanising (RTV) silicone adhesives and sealants are durable and can be used with glass, wood, many plastics, metals and painted surfaces. When cured they are tough, but retain flexibility. Curing is activated by atmospheric moisture and acetic acid is eliminated as the oligomers and acetoxysilanes cross-link to form a three-dimensional matrix. Within a few minutes the surface cures to produce a skin, followed by full curing over a period of hours.

Hot-melt silicone adhesives and sealants offer instant adhesion and cure with low VOC emissions into flexible, weather-resistant cross-linked silicones. Various products give resistance to high temperatures (up to 350°C), UV, ozone, oils and most chemicals, due to stability of the silicon to oxygen bond at a molecular level.

Silicone grease acts as an O-ring lubricant and water-repellent to prevent leaks of moving parts in hot and cold potable water systems. Appropriate products are approved by the UK Water Regulations Advisory Scheme (WRSA Approved materials – lubricants – Section 5160) and by-laws. Some products are approved for use up to a maximum temperature of 85°C.

## 30.8 Chemical damp-proof courses

Chemical damp-proof courses are defined by the British Standard BS6576: 2005. The standard specifies installation methods, including low-pressure injection for solutions of siliconates and microemulsions of silanes or alkylalkoxysiloxanes, and gravity feed injection for aqueous siliconates. Typically for manual insertion, a concentrated viscous silane/siloxane emulsion is injected through 9–16 mm holes at 100–120 mm centres to a depth appropriate to the wall thickness. Injected silicone damp-proofing should be effective for at least 12 years, depending upon local environmental conditions.

## 30.9 Silicone firestop foam

Two-part silicon room temperature vulcanising foam is used for firestopping around pipes and cables that penetrate fire compartments. The material expands to fill voids and therefore must be contained during the curing process. Up to a 4 hour rating can be obtained to BS476 Part 22, depending on the void configuration. Manufacturer's information must be checked to confirm the absence of incompatible materials. Some smoke is generated by the material within a fire situation.

## 30.10 Silane coupling agents

Silane coupling agents capitalise on the combination of inorganic and organic reactivity within the same silicon-based molecule. The silanes comprise a hydrolysable group which bonds to the inorganic substrate and an organofuntional group which attaches to the applied organic material. Therefore the coupling agent can act as the interface between an inorganic material such as glass or metal and an organic coating or adhesive.

The silane bond to the inorganic substrate is generated by the hydrolysis of the reactive group (methoxy, ethoxy or acetoxy) on the silicon, to form an oxygen chemical bond to the substrate (Figure 30.7). To produce good bonding between the coupling agent and the organic coating or adhesive, the materials must be appropriately matched. For example, an epoxysilane will bond to an epoxy resin. In the case of thermoplastic coatings some interdiffusion creates an inter-penetrating network to form the physical bond. A wide range of products is available to match the appropriate silane to the chemical characteristics of the two materials to be bonded.

## 30.11 Stone consolidation

The three major causes of natural stone decay are pollutants, frost and the crystallisation of soluble salts. All are associated with water penetration. While the complete blocking of surface pores increases the rate of stone decay, silanes may be used for stone consolidation which stabilises the friable material, while permitting the continuation of natural weathering. Some colour changes may be observed initially, but these tend to fade after about 18 months. Tetraalkoxysilanes, which form strongly cross-linked polymers, stabilise the stone but with little water-repellency. Alkyl trialkoxysilanes produce less consolidation but more water repellency. Polysiloxanes are flexible but offer greater water repellency. Tests must be carried out on any individual stonework, before application, as natural stones and local environmental conditions vary considerably; also a wide variety of other treatments

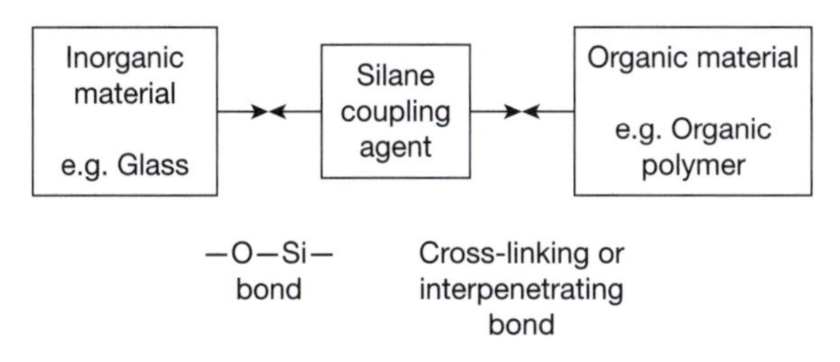


**Figure 30.7** Action of silanes as a coupling agent between inorganic and organic materials

as well as silane-based products are commercially available. Masonry consolidation should only be used under clear guidance from expert conservators and the appropriate authorities where historic buildings are involved.

## 30.12 Durabilty

In impregnated concrete, the cross-linked silicone resin networks are generally durable. However, siloxane bonds in the network may be hydrolysed within alkaline environments, leading to some degradation. If the siliconates produced are water-soluble then they are slowly leached out of the substrate. Silane/siloxane penetrants may not be effective in carbonated concrete as alkalinity is required to activate the in-situ polymerisation reaction. However, the Highways Agency confirms that well-executed hydrophobic impregnation of concrete structures should be effective for at least 15 years, with reapplication appropriate after 20 years.

Silicone adhesives and sealants retain their flexibility and certain grades are resistant to UV, ozone, oils and most chemicals.

## 30.13 Proprietary brands

Dow Corning pioneered the development of silanes and siloxanes in the mid-20th century, and it is currently the largest manufacturer in the field.

## 30.14 Hazards in use

- Solvent-based silicone fluids may be classed as flammable, and may produce some persistent odour after application.
- Silanes are classed as toxic materials, and compliance with current health and safety legislation and manufacturer's recommendations is essential.
- Siliconate fluids may be irritant or corrosive, and precautions are required to prevent skin or eye contact.
- Micro-emulsions may be flammable in concentrated form before dilution, and an irritant or corrosive during application, so suitable precautions are required to prevent skin or eye contact.
- Silicone firestops are known to develop smoke within a fire situation.

## 30.15 Bibliography

Bamforth, P.B., 2005. *Enhancing reinforced concrete durability*. Concrete Society Technical Report No. 61, Part 1. Camberley, UK, Concrete Society.

Bassi, R. and Roy, S.K. (eds.), 2002. *Handbook of coatings for concrete*. Caithness, UK, Whittles Publishing.

Brook, M.A., 2000. *Silicon in organic, organometallic and polymer chemistry*. New York, NY, Wiley Interscience.

Clarson, S.J. and Semlyen, E.R., 1993. *Siloxane polymers*. Englewood Cliffs, NJ, Prentice Hall.

Concrete Society, 1997. *Guide to surface treatments for protection and enhancement of concrete*. Technical Report No. 50. Slough, UK, Concrete Society.

Concrete Society, 2008. *Permeability testing of site concrete. A review of methods and experience*. Technical Report No. 31. Camberley, UK, Concrete Society.

English Heritage, 2012. *Stone (Practical building conservation)*. Farnham, UK, Ashgate Publishing.

Fidler, J. and Teutonico, J.M., 2002. *Stone. Stone building materials, construction and associated component systems. Their decay and treatment*. London, UK James & James Science Publishers.

Garrod, E., 2001. *Stone consolidation halts decay and prolongs life*. The Building Conservation Directory. Tisbury, UK, Cathedral Communications.

Highways Agency, 2003. *Impregnation of reinforced and prestressed concrete highway structures using hydrophobic pore-lining impregnants.* Design Manual for Roads and Bridges BD43/03. Norwich, UK, HMSO.

Horie, C.V., 2010. *Materials for conservation. Organic consolidants, adhesives and coatings.* 2nd edition. Oxford, UK, Butterworth-Heinemann.

Kubal, M.T., 2008. *Construction waterproofing handbook*, 2nd edition. New York, NY, McGraw-Hill.

Mittal, K.L. (ed.) 2009. *Silanes and other coupling agents.* Vol. 5. Boca Raton, FL, CRC Press.

Rochow, E. 2011. *An introduction to the chemistry of the silicones.* New York, NY, Wiley.

## 30.16 Standards

BS476. Fire tests on building materials and structures: Part 22: 1987. Methods for determination of the fire resistance of non-loadbearing elements of construction.

BS8221. Code of practice for cleaning and surface repair of buildings: Part 2: 2000. Surface repair of natural stones, brick and terracotta.

BS6576: 2005. Code of practice for diagnosis of rising damp in walls of buildings and installation of chemical damp-proof courses.

BS EN 1504. Products and systems for the protection and repair of concrete structures. Definitions, requirements, quality control and evaluation of conformity: Part 1: 2005. Definitions. Part 2: 2004. Surface protection systems for concrete. Part 10: 2003. Site application of products and systems and quality control of the works.

## 30.17 Websites

- www.dowcorning.com – Dow Corning Corporation
- www.wrcnsf.com/wras.htm – Water Regulations Advisory Scheme (WRAS) (see 'Water Fittings and Materials Directory' section)

# 31 Adhesives

**Allan R. Hutchinson** BSc PhD CEng MICE FIMMM FIMI
Head of Sustainable Vehicle Engineering Centre, Oxford Brookes University

**James Broughton** BEng (Hons) PhD
Departmental Head of Joining Technology Research Centre, Oxford Brookes University

## Contents

## 31.1 Definitions and terms

An adhesive may be defined as a material that, when applied to surfaces in liquid form, can join them together and resist their separation when cured. The general term 'adhesive' therefore includes all substances capable of holding materials together by surface attachment and includes cement, glue, paste, etc. A structural adhesive may be regarded as a cross-linked polymer for uniting relatively rigid adherends to form a primary structural connection. A semi-structural adhesive would be used chiefly to distribute loads, with the main load-carrying being by some other mechanism in a structure. Non-structural adhesives are required to resist little or no stress and may fulfil a largely sealing role. The consequences of failure in the bondline would be structurally insignificant (IStructE, 1999) although there may be implications for public safety and serviceability.

The term 'adhesion' refers to the attraction between substances whereby when they are brought into contact, work must be done in order to separate them. Adhesion is associated with molecular forces acting across an interface, and involves a consideration of surface energies and interfacial tensions. Adhesion is a vital concept for adhesive and primer materials.

The materials being joined are often referred to as the adherends or substrates. The properties of an adhesive-bonded joint are a function of the bonding, the properties of the materials being joined and the joint design.

## 31.2 Overview

An adhesive must fulfil two fundamental requirements. First it must spread and wet out a surface; intimate contact is required between the molecules of the adhesive and the atoms and molecules in the surface. To do this the adhesive must initially be a liquid of relatively low viscosity. Second, the adhesive must then harden to a cohesively strong solid by one of the following processes:

- loss of solvent or water
- cooling from above melting point
- chemical reaction.

The exception to this is pressure-sensitive adhesive tape. These remain sticky in the form of a very viscous fluid.

In structural adhesives, hardening or curing is always achieved by chemical reaction that involves polymerisation and cross-linking. Polymers are formed by taking many small groups, or repeating units, of atoms and joining them together to form a large molecule. The process may join up groups into a long chain-like molecule or into a branched or three-dimensional cross-linked structure. The two basic processes by which polymers are formed are condensation polymerisation and addition polymerisation. In condensation polymerisation, molecules react with one another because they contain chemical groups that are mutually reactive. There are two such groups per molecule, and if some tri-functional molecules are present then branching or cross-linking will result. Compounds containing double bonds or rings can be polymerised by addition polymerisation, which involves opening the double bonds or rings and joining them together to make a chain. Addition polymerisation is a chain reaction involving the sequential steps of initiation, propagation and termination. In bonded joints, adhesives are applied in an unpolymerised or partially polymerised state; the hardening process that follows is condensation or addition polymerisation. Since cross-linked polymers are both insoluble and infusible, it is essential that the chemical reactions that cause cross-linking occur within the joint.

All adhesives either contain polymers, or polymers are formed within the adhesive bond, and most adhesives are organic in nature. Naturally derived adhesives such as starch, animal glues and plant resins are long-chain molecules, soluble in water, and are either proteins or polysaccharides (cellulose). They have been used for centuries and are still used widely today, and they are a good option for packaging and for joining wood.

Since the 1950s the naturally derived adhesives have been improved and there has been an intense development of synthetic polymers to meet more technically demanding applications. These synthetic polymers and ancillary products, which include thermoplastic and thermosetting types, have been developed to possess a balance of properties that enables them to adhere to a wide variety of surfaces, to possess an adequate cohesive strength and appropriate mechanical properties when cured, to possess good durability, and to meet various application and manufacturing requirements in use.

Thermoplastic adhesives are not cross-linked polymers; they may therefore be softened on heating and rehardened on cooling. Included in this group of adhesives are vinyl polymers and copolymers, such as polyvinyl acetates (PVACs) and ethylene vinyl acetates (EVAs), acrylic polymers such as polymethacrylates, saturated polyesters and polyurethanes. PVACs are used extensively as general-purpose adhesives for bonding slightly porous materials, from floor screeds to timber; they are, however, generally susceptible to the effects of moisture and prolonged stress. Poly(cyano methacrylate), or 'super-glue', is capable of providing strong bonds between metals, glass, ceramics and certain plastics. Polyurethane technology is very versatile, leading to a wide variety of one-part and two-part products that can bond to metals, plastics, timber and glass. One-part formulations often appear as flexible polymers such as building sealants, while two-part adhesives are used where high cohesive strength and improved heat resistance are required.

Thermosetting materials are so called because, when cured, the molecular chains form large three-dimensional cross-linked structures. All structural adhesives are cross-linked because such polymers must exhibit minimal creep. Unlike thermoplastics, they do not melt or flow when heated, but lose strength and become more flexible with increasing temperature. Phenolics, epoxides, acrylics, unsaturated polyesters and some polyurethanes belong to this group and they are all used extensively in building and construction applications. Phenolic resins are the starting point for a wide variety of traditional timber adhesives (see below). Unsaturated polyesters are often used as binders in glass fibre-reinforced plastics (GFRPs), as adhesives for rock bolts and as mortars in conjunction with stone and cementitious materials. However, high shrinkage on cure, poor resistance to creep and sensitivity to damp conditions restricts their use as adhesives significantly. Polyurethanes bond to a wide variety of materials and, when cured, they are tough but flexible. They have been used with success in timber structures and for joining dissimilar material combinations such as sandwich panels and cladding panels. However, like unsaturated polyesters, they are sensitive to the effects of moisture, elevated temperature and creep. Structural acrylics are generally supplied in two parts, the resin and catalyst. Most conveniently, the resin can be spread on one surface and the catalyst on the other, although it is also possible to premix the components. Acrylic adhesives bond to a wide variety of surfaces such as metals and plastics and they cure rapidly, but they are prone to high shrinkage on cure. Like unsaturated polyesters and polyurethanes, they are sensitive to the effects of moisture, elevated temperature and creep. Epoxides, on the other hand, are generally tolerant of many surface and environmental conditions, possess relatively high strength, and shrink very little on curing. The general term 'epoxy' may include materials that vary from flexible semi-elastic coatings and sealants to epoxy resin-based concretes to structural adhesives.

### 31.2.1 Formation of adhesive bonded joints

Five factors of equal importance are involved in the formation of reliable and durable bonded joints. These are:

- adhesive selection
- surface preparation
- joint design
- controlled fabrication
- testing, inspection and quality control.

The purpose of this chapter is to provide information on these factors sufficient to provide a qualitatively correct overall view of the science and technology of adhesive bonding.

## 31.3 Applications

Adhesives have a long history of use in the aerospace, automotive, marine, general engineering and construction industries. Much research has been carried out on the science and technology of adhesion and adhesives relevant to applications in those industries (e.g. Adams et al., 1997), and a great deal can be learnt by transferring appropriate technologies and experiences into the construction sector (Hutchinson and Hurley, 2001). The relative importance of particular properties such as strength, modulus, viscosity, creep and temperature resistance varies from one application to another. Some categories of use are shown in Table 31.1.

**Table 31.1** Structural and semi-structural adhesive applications

| *Application type* | *Description* | *Materials involved* | *Adhesive type* | *Principal application considerations* |
|---|---|---|---|---|
| Segmental construction | Precast concrete elements | Concrete | Epoxy | High compressive strength, durable and gap filling, no need for strong adhesion |
| Anchoring | Rock bolts and ground anchors | Rock, steel and GRP | Polyester | Low cost and rapid cure |
| | | | Vinyl ester | Improved adhesion in wet conditions and higher temperature resistance |
| | | | Epoxy | Improved adhesion in wet conditions, high strength and gap filling |
| | | | Cement | Low cost, good adhesion in wet conditions |
| | Fixings and holding-down bolts | Concrete, masonry, and steel | Polyester | Low cost, rapid cure and low temperature resistance |
| | | | Epoxy | Increased strength and durable |
| | | | Cement | Low cost, low creep, durable |
| | Replacement cavity wall ties | Brickwork, masonry, steel and GFRP (used with epoxy only) | Polyester | Low cost, fast cure and tolerance to low on-site temperature |
| | | | Epoxy | Gap filling and improved adhesion in wet conditions |
| Repair, strengthening and protection | Externally bonded, or near-surface mounted, reinforcement | Concrete, steel, FRP and timber | Epoxy | Thixotropic, high strength, durable and gap filling |
| Miscellaneous | Tiling and brick slips | Ceramics, brick, masonry | Epoxy | Good chemical resistance, gap filling |
| | | | Polyester | High chemical and low temperature resistance |
| | | | SBR | Low cost, instant tack, durable |
| | Support/bearing | Concrete | Epoxy | Gap filling, low shrinkage, high compressive strength |
| | Anti-skid surfacing | Bitumen, concrete, steel, zinc-coated steel, bauxite aggregates | Polyurethane | Large volumes, low cost, and rapid cure |
| | | | Epoxy | Good adhesion |
| | Bedding | Concrete and steel | Polyester | Rapid cure and tolerance to low site temperature |
| | | | Epoxy | Higher compressive strength |
| | | | Polymer-modified cementitious | Low cost, high strength in compression and tension |
| | Grouts/resin injection | Concrete, masonry, timber and steel | Epoxy | Gap filling, can be thixotropic |
| | | | Polyurethane | Flexible joints, and very low temperature resistance |
| | | | Polyester | Low cost |
| | | | (meth) Acrylate | Rapid cure, can provide good flexibility and peel strength |
| | Glulam/structural composite lumber/ plywood | Softwood and hardwood | Phenol formaldehyde | Good moisture and biodegradation resistance, low cost |
| | | | Resorcinol formaldehyde | Good durability, solvent and biodegradation resistance, low cost |
| | | | Phenol resorcinol formaldehyde | Good durability, solvent and biodegradation resistance, low cost |
| | | | Urea-formaldehyde | Protected indoor use only, low cost |

*(Continued)*

**Table 31.1** *(Continued)*

| *Application type* | *Description* | *Materials involved* | *Adhesive type* | *Principal application considerations* |
|---|---|---|---|---|
| | | | Melamine-urea-formaldehyde | Moderately higher temperature resistance compared with urea formaldehyde, also lower cost |
| | Various fixings into timber | Timber | PVA | Biodegradation resistance, indoor use only |
| | | | Caesin | Resistance to organic solvents, low cost, indoor use only |
| | Bridge nosings | Concrete, steel | Epoxy | Good adhesion in wet conditions, resists indirect tensile stresses |
| | Sandwich panels | GFRP, metal, stone and laminated skins Polymeric foam or balsa wood cores | Polyurethane | Low cost, good peel performance |
| | Structural glazing | Aluminium and glass | Silicone | Thick bondlines, flexible, creep-resistant, good adhesion in wet conditions |
| | Joining of structural FRP elements | GFRP primarily. Limited applications with AFRP and CFRP shells and profiles | Epoxy | High strength and gap filling |

Most of the adhesives used in the construction industry are for essentially non-structural applications such as tile fixings, floor coverings and lamination. Examples of such adhesives include ethylene vinyl acetates (EVAs), polyvinyl acetates (PVAs), styrene butadiene rubbers (SBRs) and urea-formaldehydes. However, there are applications in which adhesives are used in a truly structural sense where the consequence of failure is significant. In new build, segmental precast prestressed concrete structures, such as bridges, employ adhesives as a stress distributing waterproof medium in the joints. Other applications include dowels, rock anchors and joints in GFRP structures (Hutchinson, 1997). The use of structural adhesives is more typically associated with repair rather than with new build, and their repair applications include crack injection of concrete, repairs to timber structures, and externally bonded, and near surface mounted, reinforcement of concrete, metallic and timber structures.

In some applications, such as segmental construction, the joint is in compression and adhesion of the adhesive to the concrete is of low importance. In the majority of anchoring applications the joint strength is developed largely as a result of mechanical keying by the adhesive hardening to fill both an irregular (or undercut) hole and the profile of a deformed, threaded, or textured bar. However the development of intimate contact between the adhesive and all of the materials involved provides both a safety margin and enhances reliability. The need to ensure that all materials and installation practices are observed properly was underlined by the collapse of a suspended walkway in the USA in 2006 that was attributed to resin anchor failure (Wikipedia, n.d.).

Further general and detailed information on applications is given by Mays and Hutchinson (1992), IStructE (1999) and Hutchinson and Hurley (2001).

## 31.4 Typical strengthening applications

Epoxy adhesives have been used for the bonding of steel plate reinforcement to the external surfaces of concrete structures for strengthening purposes since the late 1960s. The adhesive provides a shear connection between the dissimilar materials that enables the whole to act as a composite structural unit. The requirements of the adhesive for this application were detailed by Mays and Hutchinson (1988, 1992), and design guidance has been available formally in the UK since 1994 for the strengthening of concrete highway bridges using this technique (Highways Agency, 1994). The 1990s witnessed an increased use of carbon fibre reinforced plastic (CFRP) to replace steel because of advantages associated with improved installation and potentially enhanced long-term durability of the connection. The CFRP can even be pre-stressed prior to bonding and adhesives are used to transfer very large loads through carefully designed bonded steel end-tabs (Hollaway and Leeming, 1999, and Hutchinson, 1997). For this technique, the mechanical requirements of the adhesive for bonding the steel end-tabs to the CFRP are more onerous than they are for bonding steel to concrete because of the large stress transfer involved.

Many different types (aramid, carbon and glass) and forms (pultruded plates, bars and rods; pre-cured shells; on-site laminated materials) of polymer composite material have since been developed to meet the requirements of strengthening a variety of parent materials such as concrete, masonry, metals and timber. The characteristic of most of the design concepts is that structural adhesive bonding is required to effect the transfer in shear of structural loads between the different materials.

### 31.4.1 Concrete structures

In buildings it is sometimes desirable to remove supporting walls or individual supports, leading to the need for local strengthening. A common requirement is the need to strengthen the floor/beam connection in order to increase the capacity for increased loading, or to compensate for inadequate original design by facilitating alternative load paths. Examples include the installation of lift shafts in existing buildings, adding further floors to office blocks and hospitals, enhancing the integrity of power station structures and limiting the deflection of slabs in multi-storey car parks. With bridges and culverts, strengthening may become necessary as a result of damage, deterioration or design code changes. In most cases the materials involved are flat pultrusions bonded to concrete surfaces (Figure 31.1), but bars may also be inserted into pre-cut slots. Bridge deck (and other) columns can be wrapped with reinforcement to preserve integrity, to mitigate the effects of impact damage and to reduce seismic damage. These applications

**Figure 31.1** Application of an FRP strengthening plate to concrete (courtesy Sika Ltd)

can involve pre-cured shells and pultruded sheets, site-laminated fabric materials and automated winding techniques. With slabs, strengthening may be in flexure, shear, or both; with columns, the intention is confinement. Relevant design guidance is given by ACI440 (2002), Concrete Society (2004), Highways Agency (1994) and FIB (2001).

### 31.4.2 Metallic structures

The main applications to date have involved cast and wrought iron structures, often of historical importance. Pultrusions are easily attached to the relatively flat surfaces of beams while fabrics are more suited to curved members and intricate shapes. The purpose of strengthening may be to reduce the effective stresses in the original structure caused by loading or thermal effects, and the former is seen in an innovative use of prestressed CFRP pultrusions in Oxford, UK (Darby et al., 1999) (Figure 31.2). Steel offshore structures have been strengthened in flexure using a special flexible resin infusion process involving 'dry prepreg' carbon fibre fabric layers (Hutchinson, 1997). Relevant design guidance is given by CIRIA C595 (2004).

### 31.4.3 Timber structures

Common resin-bonded methods of repair involve the removal and reinstatement of beam-ends, parts of trussed rafters, and the ends of columns (Broughton and Hutchinson, 2001a, 2001b, 2001c; Mettem and Milner, 2000). Combinations of resin grouts, steel bars, pultruded FRP rods and structural adhesives are used. Strengthening of beams may involve the attachment, or insertion into slots, of steel or pultruded FRP bars or plates using a structural adhesive (Figure 31.3). Many hardwood buildings and bridges of historical importance have been repaired and strengthened with minimum intervention (e.g. Hutchinson, 1997 and www.licons.org). The techniques have been so successful and aesthetically pleasing (i.e. invisible) that there is much interest in making reinforced bonded connections in new build and in enhancing the flexural performance of glued laminated timber beams significantly by including FRP laminations in the section. Design guidance is provided by STEP 1 (1995), STEP2 (1995) and TRADA (1995).

## 31.5 Adhesive types and characteristics

Reviews of adhesive material types relevant to construction are given by Mays and Hutchinson (1992), IStructE (1999) and Hurley (2000). A detailed overview of the science and technology of adhesives for general engineering is given by Adams et al. (1997); some basic chemistry of adhesion and adhesives is provided succinctly by Comyn (1997). This section is concerned largely with structural adhesives.

### 31.5.1 Epoxy adhesives

Epoxies are a very important class of cross-linked adhesives that have been available commercially since the 1940s. As structural adhesives, they are the most widely used and accepted. Epoxy resins are generally supplied in two-component form although one-part heat-cured variants are available to provide a broad range of application characteristics, and mechanical properties when

**Figure 31.2** Hythe Bridge strengthening, Oxford, using prestressed CFRP plates (courtesy Mouchel Parkman) *(Continued)*

**Figure 31.2** *(Continued)*

**Figure 31.3** Timber beam-end repairs using FRP rods (courtesy Rotafix Ltd)

cured. The resin contains epoxide rings, a three-membered ring with two carbon atoms singly bonded to an oxygen atom. The second component is a hardener or curing agent, and hardening is by chemical reaction that produces a cross-linked polymer. The most commonly used epoxy resin is the diglycidyl ether of bisphenol-A (DGEBA). The principal hardeners include aliphatic amines, aromatic amines and acid anhydrides. The choice of hardener results in the distinction between adhesives that cure either at ambient or at elevated temperature. Aromatic amines and acid anhydrides generally require elevated temperatures of 150°C or more to effect cure. However, as a rule of thumb, the rate of chemical reaction is doubled for every 10°C rise in temperature. Commercial formulations are, in general, complex and sophisticated blends of many components, the most important being the resin and hardener. To these components are added fillers, toughening agents, plasticisers, diluents, surfactants, anti-oxidants, and other materials. The blending of the base resin with a variety of materials indicates that the possibilities for altering the properties and characteristics of an epoxy adhesive are numerous.

Epoxy resins have many attributes for use as adhesive agents in construction, namely:

- high surface activity and good wetting properties for a variety of substrates (see section 31.6)
- may be formulated to have a long open time (the time between application and closing of the joint)
- high cured cohesive strength; joint failure may be dictated by adherend strength (particularly with concrete substrates)
- may be toughened, by the inclusion of a dispersed rubbery phase, to improve impact and low-temperature performance
- lack of by-products from curing reaction minimises shrinkage and allows the bonding of large areas with only contact pressure
- low shrinkage compared with polyesters, acrylics and vinyl types; hence, residual bondline strain in cured joints is reduced
- low creep and superior strength retention under sustained load with most formulations
- can be made thixotropic for application to vertical surfaces (i.e. a shearing force is required to make the adhesive flow)
- able to accommodate irregular or thick bondlines (e.g. concrete adherends)
- may be modified by (a) selection of base resin and hardener, (b) addition of other polymers, (c) addition of surfactants, fillers and other modifiers
- normally suitable for service operating temperatures in the range –30°C to +60°C. A higher upper limit may be achieved by using warm- or heat-cured formulations that possess a higher glass transition temperature, $T_g$ (see Section 31.5.10).

A disadvantage of epoxy adhesives is that direct contact with the liquid hardeners can result in skin irritation.

### 31.5.2 Polyurethane adhesives

Polyurethane adhesives are very versatile, with many possible applications. They are generally weaker and more flexible than epoxies, and they are more susceptible to creep and moisture effects. Their operation temperature, up to say 60°C for some formulations, makes them suitable for many structural and semi-structural applications in which the bonded materials are kept reasonably dry. Structural materials for which polyurethane adhesives are suitable include timber, natural stone, glass, aluminium and many plastics. Polyurethane adhesives are generally supplied in a two-component form although single-part formulations, which rely on moisture as the catalyst, are available.

Polyurethane adhesives are made by reacting a low molecular weight polymer containing at least two hydroxyl (OH) end-groups with a diisocyanate. The starting polymers can be polyethers, aliphatic polyesters or polybutadiene. In two-component polyurethane adhesives the polymer and isocyanate are mixed and then applied to the surfaces of a joint. Any hydroxyl groups on the surfaces may react with isocyanate to form covalent bonds between the adhesive and the surface. One-part adhesives consist of low molecular weight, linear molecules that have isocyanate (–NCO) end-groups. Water vapour from the atmosphere diffuses into the adhesive and causes the molecules to join together to form larger linear molecules. Further reaction with urea units results in a cross-linked structure.

One-part adhesives are used extensively for bonding elastomers, fabrics and thermoplastics. They are also used to join combinations of metals, plastics and timber in sandwich panel applications. One-part sealants are very popular for many exterior weather-sealing applications around doors and windows, and in façade engineering. They bond very well to most surfaces and they can be formulated to provide a wide range of properties when fully cured. Two-part adhesives are used in timber joints and repairs, but are sensitive to the moisture content of the timber (e.g. Wheeler and Hutchinson, 1998). In Scandinavia, polyurethanes are favoured over epoxies for perceived health and safety reasons. Because they are not resistant to attack by alkalis, polyurethane adhesives and sealants are not generally suitable for use in direct contact with concrete unless specifically formulated or unless the concrete is primed.

### 31.5.3 Acrylic adhesives

Acrylic adhesives cover a range of materials with a variety of curing mechanisms. For structural applications toughened acrylics are generally used that bond readily to various adherends with minimal surface preparation.

Structural adhesives contain one or more acrylic monomers that may be cured by free radical addition polymerisation at ambient temperature. The principal monomer is an alkyl methacrylate such as methyl methacrylate (MMA) but other monomers may be present, such as methacrylic acid, to confer improved adhesion to metals or improved heat resistance, and ethylene glycol dimethacrylate for cross-linking. Poly methyl methacrylate is often present to increase viscosity and reduce odour, while chlorosulfonated polyethylene is added as a rubber-toughening agent. Reaction is catalysed by components of the initiator, cumene hydroperoxide and N,N-dimethylaniline (Comyn, 1997). Full cure can take place in a matter of minutes. However, there is a large volume decrease when MMA is polymerised, necessitating the inclusion of filler to offset shrinkage. Shrinkage is also the reason why adhesives tend to have poor gap-filling properties. Oxygen can act as an inhibitor of polymerisation so that the exposed edges of a joint may never fully cure and may lead to concerns over durability along the joint perimeter. A further source of concern is that methyl methacrylate monomers are highly inflammable with a flash point of about 10°C.

The adhesives are generally supplied in two-component parts but are susceptible to incorrect proportioning because the initiator is used as a small volume of low viscosity liquid in comparison to the resin. The ratio can vary between 5% and 50% for some commercial products. This means that great care has to be taken in metering and mixing. Acrylics are liable to exhibit significant creep, particularly at high temperatures. Applications for which acrylic adhesives are suitable include the bonding of non-porous thin sheet materials as found in sandwich and cladding panels. They are particularly suitable for applications in which uniformly thin bondlines are achievable, e.g. metal to metal or with plastics, or where rapid hardening is essential.

Cyanoacrylate adhesives ('super-glues') are suitable for use only for non-structural applications involving relatively non-porous materials. Polymerisation is initiated by water, which is adsorbed on all surfaces in the atmosphere, and is completed in seconds. Again, the curing process is inhibited by oxygen, so that the reaction does not take place until the joint is closed and the supply of oxygen is cut off.

### 31.5.4 Silicone adhesives

One-part silicone adhesives are often termed room-temperature vulcanising materials. The principal component is polydimethyl siloxane with a variety of end-groups. These are hydrolysed by moisture from the atmosphere to form hydroxyl groups. The resultant flexible polymer is often used as a sealant material and silicone sealants are popular in many construction applications for performing a sealing function, such as in façade engineering. They bond readily to a wide variety of materials because they possess inherently low surface tension, enabling them to spread and wet out most surfaces. Commercial formulations may contain a wide variety of additives such as fillers and extenders.

Two-part silicones are used for thick sections, such as in structural glazing for bonding glazed units to aluminium profiles. They are catalysed with stannous octoate for a fast cure, or dibutlytindilaurate for a slower cure. Silicone adhesives are soft and compliant and they possess good chemical and environmental resistance. Joints with silicone adhesives can operate over a wide temperature range, from –60°C to 200°C. Their selection for structural glazing applications relates to their stability in fire conditions; they do not burn or support combustion.

### 31.5.5 Polyester adhesives

Polyester adhesives are used in applications in which a rapid gain in strength is required, such as in resin anchors. Formulations are available that allow curing to take place at sub-zero temperatures. The addition of a curing agent or catalyst based on organic peroxide initiates the chemical reaction and promotes cross-linking within the resin. Unsaturated acids are thus cross-linked by inter-polymerisation with styrene to produce the so-called unsaturated polyesters, used as the basis for composite matrix resins and structural adhesives. Contraction during cure can be as high as 10% by volume as a result of thermal contraction, volumetric change during the transition from liquid to hardened solid and, in some cases, solvent loss. Thus, there are usually limitations on the volume of materials that can be mixed and applied at any one time.

The use of polyesters as structural adhesives in many situations is limited by their poor creep resistance under sustained load, their lack of tolerance to damp or wet conditions, particularly to alkaline substrates, and their relatively high intrinsic cure shrinkage.

### 31.5.6 Traditional timber adhesives

The reactions of urea, melamine, phenol or resorcinol with formaldehyde are the basis of aminoresin adhesives for wood (Pizzi, 1983). The adhesives are water-based and contain low molecular weight precursors that cross-link within the bonded joint. The adhesives generally cure at room temperature after the addition of a hardener. The water that is liberated as a by-product of the reaction is absorbed by the wood, such that these adhesives are only suitable for porous adherends. They generally work only in bondlines < 1 mm thick, although fillers can be used to increase this to up to 2 mm.

European Standard EN301 covers the classification and performance requirements of structural wood adhesives; EN302 is the corresponding test standard. These two Standards are applicable to phenolic and aminoplastic adhesives only that may be classified as:

- Type-1 adhesives, which can withstand full outdoor exposure and temperatures above 50°C
- Type-2 adhesives, which may be used in heated and ventilated buildings, and must be protected from the weather. They will withstand outdoor exposure for short periods but are unable to withstand prolonged exposure to weather or temperatures in excess of 50°C.

#### *31.5.6.1 Resorcinol-formaldehyde (RF) and phenol-resorcinol-formaldehyde (PRF)*

Pure resorcinols are made by reacting resorcinol with formaldehyde to create a liquid resin. This resin is completely reacted subsequently with a hardener containing formaldehyde to make a strong joint. Curing may be at room temperature or at elevated temperatures. RF and PRF adhesives are Type-1 adhesives, being very strong and durable. Carbon–carbon bonds are formed in the reaction between resorcinol and other phenols with formaldehyde. These bonds are very strong and durable, and not susceptible to hydrolysis. Thus the adhesives are fully water-, boil- and weather-resistant and will withstand salt-water exposure, making then suitable for marine applications. Members assembled with these adhesives will not delaminate in a fire. These traditional adhesives are widely used in laminated timber, finger jointing of members, I-beams, box beams, etc., both indoors and outside. Best performance is obtained with thin (up to 1 mm) adhesive layers and therefore the materials are not suitable for gap-filling applications.

#### *31.5.6.2 Phenol-formaldehyde (PF)*

Heat-cured adhesives are made by reacting phenol with formaldehyde under alkaline conditions. The adhesive may be supplied as a liquid, powder or film. They are used typically in hot-press fabrication of structural plywood and similar materials; moisture is liberated as a by-product of the reaction. This Type-1 adhesive has the same durability properties as the other phenolic-type adhesives, i.e. fully water-, boil- and weather resistant. In order to make a PF cure at room temperature it must be made acidic. The hardener however is so strongly acidic that it is liable to damage wood, which limits the use of PF adhesives for structural purposes. The extent of such damage would vary between timber species so that trials would need to be conducted to ascertain the severity of any potential problem. Cold-curing PFs were used to a certain extent in the 1950s and 1960s in glulam production. Some buildings made with the glulam actually collapsed many years later, and there is reason to believe that acid damage from the adhesive was the cause (Raknes, 1997).

#### *31.5.6.3 Urea-formaldehyde (UF) and melamine-urea-formaldehyde (MUF)*

Urea-formaldehyde (UF) adhesives are the most important type of the so-called aminoresin adhesives for wood, with a global production of around 6 billion tonnes. UF adhesives are made by reacting urea with formaldehyde; the reaction is accelerated by acid and heat. They may be supplied as liquids or powders, and cure takes place from 10°C upwards. The hot-press types are used for non-structural plywood and chipboard, and they are broken down fairly quickly by combined heat and high relative humidity. Melamine-urea-formaldehyde (MUF) adhesives are closely related to UF adhesives, but some of the urea is replaced by melamine in order to increase resistance against hydrolysis and thereby improve weather resistance. UF and MUF adhesives are Type-2 adhesives and hence are mainly suitable for protected conditions, or where the member is protected from the worst weather conditions. They may be cured from a range of temperatures from 10°C upwards. Unless suitably filled, they should be used in applications in which the bondline is thinner than 0.1 mm. Some MUF adhesives are supplied as hot-press adhesives for plywood.

There is a move towards the formulation of UF resins with lower formaldehyde emissions across Europe, and Japan intends to ban all formaldehyde adhesives for panel products in 2008. Research is in progress on reducing the content of melamine in MUF adhesives while maintaining overall performance (Pizzi, 2004).

#### *31.5.6.4 Casein*

The primary constituent of these adhesives is the milk protein, casein. The adhesive is supplied as a powder consisting of casein and various inorganic salts. When mixed with water, a series of chemical reactions occur. Caseins are less water-resistant, but more resistant to combined heat and high humidity, than urea-formaldehydes, and they have been found to be suitable for load-bearing use in fully protected indoor timber structures. They are only ever used in conjunction with a very thin bondline, and are susceptible to bacteriological and fungal attack. Caseins are probably the oldest type of structural adhesive and have been used for industrial glulam production since the early 1900s. Indeed, a casein-adhesive bonded arch made in 1910 supports the main roof structure of Oslo railway station to this day. Nevertheless, caseins do not meet the requirements of EN301.

#### *31.5.6.5 Polyisocyanates*

Polyisocyanates such as poly 4,4-diphenylmethane-diisocyanate (PMDI) are related to polyurethane adhesives. They react with water and also with wood and lignocellulosic materials. Polyisocyanates are used for the production of water-resistant wood-based panels. Both one- and two-component systems are available; in the former case, moisture from the wood or from a high-humidity atmosphere is required to promote hardening of the adhesive. PMDI is used extensively in the production of exterior-grade oriented strand board, laminated veneer lumber and particleboard.

### 31.5.7 Other adhesives and bonding aids

Materials that belong to this category are adhesives that harden by loss of solvent, adhesives that harden by loss of water, adhesives that harden by cooling, and pressure-sensitive adhesives. An adhesive that has already been polymerised can be dissolved in a liquid or suspended in water as a paste or emulsion. Acrylics can be formulated as emulsions, lattices or solvent-based.

#### *31.5.7.1 Adhesives that harden by loss of solvent*

Contact adhesives are probably the best-known solvent-based adhesives. These are solutions of a polymer in organic solvents that are applied to both surfaces to be bonded. Some time is allowed for the solvents to evaporate and the surfaces are then pressed together. Some interdiffusion of the polymer chains will then occur. The surfaces can also be heated to increase tack. Clear solvent-based adhesives are often solutions of nitrile rubber (a copolymer or butadiene and acrylonitrile) in organic solvents (Comyn, 1997). They have obvious applications for bonding rubber and for bituminous substrates such as in roofing applications. Other contact adhesives are based on neoprene (polychloroprene). Neoprene adhesives exhibit good tack, fast bond strength development and are resistant to oils and chemicals. They are used for bonding rubber and for rubber-to-metal bonding, and for panelling, gypsum board and roofing. SBR adhesives are used for concrete, tiles and roofing.

#### *31.5.7.2 Adhesives that harden by loss of water*

There are environmental pressures to reduce or eliminate the use of organic solvents in adhesives. Industry has responded by developing water-based systems to replace them but there are problems, such as the sensitivity of bonded joints to water. Water-soluble materials are essential to stabilise emulsions but these remain in the adhesive after hardening, so increasing subsequent water absorption of the adhesive.

Polymer lattices are aqueous dispersions of small discrete polymer particles to which are added additives such as coalescents, anti-foaming agents, bacteriocides and anti-oxidants, to improve shelf-life and properties related to end-use. Detailed considerations of the formulation and stabilisation of aqueous emulsions is given by Comyn (1997). Developments in formulation have greatly reduced the sensitivity of these polymers to water and involve some degree of cross-linking of the polymer chains. Adhesives are fast-setting and they stick to a wide variety of materials such as metal, concrete and timber. However, these adhesives will not sustain prolonged stress, particularly at elevated temperatures.

Polyvinyl alcohol and polyvinyl acetate (PVAC) are examples of low-cost adhesives for use as tile cements and for bonding wood. Do-it-yourself wood adhesive is a PVAC latex; the adhesives harden by water migrating into the pores of the wood. Polymers supplied for use as concrete bonding aids include PVAC, acrylics, and styrene acrylates. They usually take the form of a polymer–cement slurry that is applied to a dampened and prepared concrete surface. Many cementitious repair mortars themselves are modified by polymers of similar generic type to those used in bonding aids.

#### *31.5.7.3 Adhesives that harden by cooling*

Hot-melt adhesives are one-part systems that are applied to surfaces as a hot liquid (typically around 170°C) and form a bonded joint by rapid cooling. Substrates that possess inherently high thermal conductivity, such as metals, are not suitable because wetting of the surface by the adhesive is restricted. The polymers used tend to incorporate a variety of monomers and additives, such as anti-oxidants, fillers and waxes. Common hot-melt adhesives are based on EVA, polyamide and polyester.

#### *31.5.7.4 Pressure-sensitive adhesives*

These adhesives remain permanently sticky and find wide application as tapes. They rely on the user pushing the adhesive against a surface to obtain intimate contact with it. The adhesives are based primarily on natural rubber, modified forms of styrene butadiene and acrylics, and they remain as a very viscous liquid throughout the life of the tape. The carrier tape itself is commonly PVC or polyethylene. One face is coated with primer or tie-layer so that the adhesive will stick to it; the other has a release coat so that the tape can be unrolled. Blu-Tack, a product of Bostik, is a linear rubbery polymer, such as polyisobutene with particulate fillers. It behaves as a pressure-sensitive adhesive (Comyn, 1997).

### 31.5.8 Requirements for structural adhesive bonding

Mays and Hutchinson (1988, 1992) identified the principal requirements of structural adhesives for bonding steel to concrete; the performance requirements in relation to concrete strengthening specifically are reviewed by Mays (2001). These requirements are very similar for the cases of bonding most combinations of materials:

- the adhesive should exhibit adequate adhesion to the materials involved and should not be unduly sensitive to variations in the quality or moisture contents of prepared surfaces, or to the cure environment
- the modulus of the material should be > 1 GPa, and the shear

and tensile strengths should be > 12 MPa, at 20°C

- the glass transition temperature, $T_g$, should be > 50°C (see below)
- it should be resistant to the effects of moisture and sustained loading
- it should possess suitable working characteristics for the application temperature
- it may need to be gap-filling and/or thixotropic, depending on the application
- for applications involving the attachment of thin FRP materials overhead, the adhesive should exhibit tack or 'grab' to avoid the need for temporary fixings while it cures.

## 31.5.9 Properties of adhesives

The designer or user is concerned with the behaviour and performance of the selected adhesive from the time it is purchased from the manufacturer, through the mixing, application and curing phases, to its properties in the hardened state within a joint over the intended design life. The adhesive selection process for any particular design is facilitated by reference to performance requirements that embrace three issues:

- working characteristics
- mechanical/physical properties of the adhesive
- bond characteristics and retention.

### *31.5.9.1 Working characteristics*

- Shelf life
- Rheology and sensitivity to temperature
- Ease of mixing and dispensing
- Pot life
- Open time
- Rate of cure development/heating requirement
- Shrinkage
- Ability to fill gaps

### *31.5.9.2 Mechanical performance and bulk property behaviour*

- Mechanical properties (strength, modulus, strain to failure, fracture toughness, temperature sensitivity)
- Heat resistance
- Fire resistance
- Water resistance

### *31.5.9.3 Adhesion and durability of the materials system*

- Requirements for surface preparation/adherend compatibility/adherend moisture content
- Long-term adhesion under a variety of environmental conditions
- Creep resistance (including creep rupture)
- Fatigue resistance
- Moisture resistance
- Moisture-induced dimensional stability
- Temperature sensitivity

## 31.5.10 Mechanical characteristics of engineering adhesives

The properties of adhesive polymers are generally determined by their internal chemistry and structure. The blend of components in many commercial formulations prevents simple relationships from being drawn, although the earlier commentary on adhesive types provides a general basis for expectation. It should be noted that essentially all commercial polymers (plastics, resins, adhesives, etc.) contain additives such as fillers and modifiers to control properties related to processing, end-use, storage, cost, and so on. A comparison of the typical properties and characteristics of some commercial two-part cold-curing engineering adhesives is given in Table 31.2. The range of properties and characteristics relates to formulation. For example, very high modulus values (> 5 GPa) and lower coefficients of thermal expansion are associated with a large amount of coarse particulate filler.

Adhesives possess a low modulus relative to most conventional materials used in construction, and the modulus decreases with increasing temperature. Variations in temperature can transform materials that are tough and strong at 20°C to ones that are soft and weak at 100°C. The glass transition temperature *Tg* denotes a marked change in the mechanical properties of a polymer; structural adhesives are designed to be used below the *Tg*, perhaps up to 20°C below, depending on the requirements of the application. The *Tg* values of typical ambient temperature curing epoxies are around 50 to 60°C (Figure 31.4) but this can be increased by increasing the initial cure temperature by, say, 20°C.

Almost all adhesives absorb moisture, leading to swelling and plasticisation of the adhesive itself. This modifies the response of an

**Table 31.2** Properties and characteristics of some construction adhesives

| *Property (at 20°C)* | *Epoxy* | *Polyurethane* | *Acrylic (toughened)* |
|---|---|---|---|
| Shear strength (MPa) | 15–35 | 5–25 | 15–25 |
| Tensile strength (MPa) | 15–40 | 15–25 | 10–25 |
| Tensile modulus (GPa) | 1–10 | 0.2–2 | 0.1–0.3 |
| Tensile strain to failure (%) | 1–4 | 3–10 | 10 – 30 |
| Fracture energy (kJ m$^{-2}$) | 0.2–1.0 | >5 | >5 |
| Glass transition temperature (°C) | 50–70 | 30–35 | ~100 |
| Coefficient of thermal expansion ($10^{-6}$/°C) | 30–50 | 40–60 | |
| *Typical characteristics* | | | |
| Creep resistance | Generally excellent | Medium | Poor |
| Moisture resistance | Excellent | Fair | Good |
| Heat resistance | Fairly insensitive | Fair | Fair |
| Cure temperature | Between 5 and 50°C | Ambient | Ambient |
| Cure time | 1 to 24 hours, depending on temperature | 1 to 24 hours, depending on temperature | 5 to 60 min, depending on volume |
| Gap filling | Yes | Yes | No |

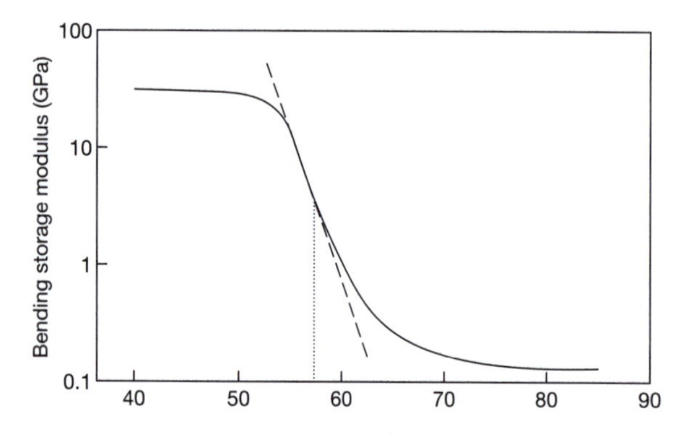


**Figure 31.4** The glass transition temperature for a typical epoxy adhesive, taken from a dynamic mechanical analysis plot

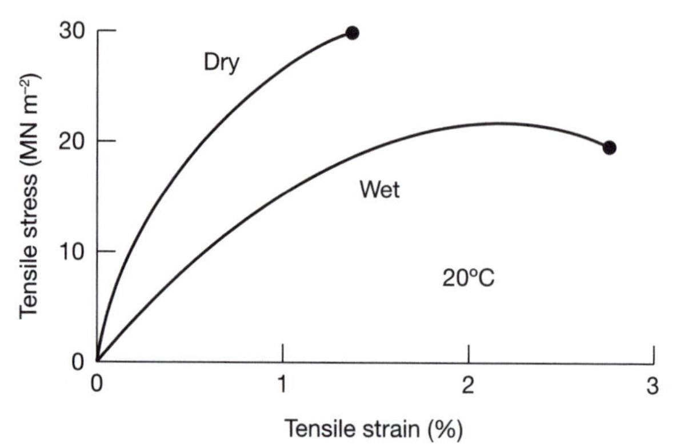


**Figure 31.6** Typical effects of moisture on the mechanical behaviour of an epoxy adhesive in tension at 20°C

adhesive to mechanical deformation in a manner analogous to a lowering of the *Tg*. Such effects are reversible, depending on proximity to moisture in liquid or vapour form. Water and heat thus have similar consequences for the response of an adhesive. Figure 31.5 illustrates the shear response of a typical stiff cold-cure epoxy with respect to temperature when 'dry' and, after a period of moisture absorption, 'wet'. Figure 31.6 indicates the effect of moisture on the tensile response of the same material at 20°C.

### 31.5.11 Classification and testing of structural adhesives

The working characteristics of an adhesive are critical to successful application on site, and these include storage life, usable life and curing requirements. The importance of these aspects is frequently overlooked. The mechanical properties currently form part of the design requirement and their importance should be self-evident. Both of these aspects are handled currently by minimum performance characteristics defined in BS EN 1504 (2004), which covers the use of repair materials for concrete structures. The third aspect, adhesion, is the least well understood of the requirements and yet is crucial for ensuring bond integrity and long-term durability. At the present time there is no reliable direct method for assessing adhesion strength non-destructively, either in the short term or after ageing. The only available approach is to make and test representative bonded coupons.

The performance characteristics defined in BS EN 1504 are limited in their general applicability to adhesive bonding in construction. Crucially, in current practice, no quality control tests are performed that provide complete assurance that the adhesive has stuck sufficiently well to the relevant materials initially. Furthermore, there is no direct assurance through testing of the installed system that a bonded joint will be durable throughout its lifetime.

Appropriate test protocols to assess adhesion properly are therefore required that are linked to quality control testing, so that a rapid and direct comparison with design data can be made as an assessment of performance and quality of workmanship.

A classification scheme was devised in the Compclass project (www.compclass.org.uk) for adhesive systems that embrace the three essential aspects of working characteristics, mechanical/physical properties of adhesive, and adhesion and durability. The scheme, the first of its kind, was developed specifically for strengthening existing structures using FRP materials, but it does have some wider applicability. Adhesive systems are classified using a suite of test procedures for specific application categories (with given loading conditions, parent material type and service environment), though not all material properties may be required for a particular application (see Table 31.3).

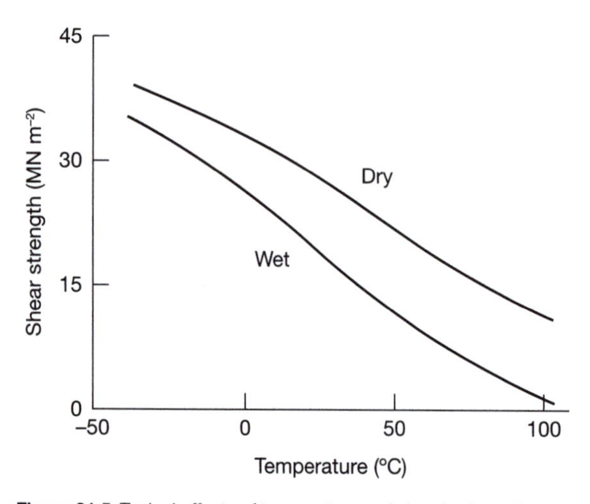


**Figure 31.5** Typical effects of temperature and absorbed moisture on the behaviour of an epoxy adhesive in shear

## 31.6 Adhesion, surface preparation and priming

Adhesives join materials primarily by attaching to their surfaces within a layer of molecular dimensions, i.e. of the order of 0.1–0.5 nm. The term 'adhesion' is associated with intermolecular forces acting across an interface and involves a consideration of surface energies and interfacial tensions. Being liquid, adhesives flow over and into the irregularities of a solid surface, coming into contact with it and, as a result, interact with its molecular forces. The adhesive then solidifies to form the joint. The basic requirements for good adhesion are therefore intimate contact between the adhesive and substrates, and an absence of weak layers or contamination at the interface.

There are six theories of adhesion: physical adsorption, chemical bonding, diffusion, electrostatic, mechanical interlocking and weak boundary layer theories. Because all adhesive bonds involve molecules in intimate contact, the weak attractive forces acting across interfaces associated with physical adsorption must always contribute. A detailed discussion of interfacial forces and adhesion mechanisms is provided by many authors (e.g. Adams et al., 1997; Comyn, 1997; Mays and Hutchinson,

**Table 31.3** Performance requirements for the classification of adhesives and resins for flexural strengthening using FRP materials

| *Performance group* | *Performance characteristic* | *Test method* | *Description* | *Classification value* | |
|---|---|---|---|---|---|
| | | | | *Short term* | *Long term* |
| Working characteristics | Application to vertical surfaces | EN 1799 | Maximum sag flow | 2 | |
| | Application to horizontal surfaces | EN 1799 | Maximum sag flow | 2 | |
| | Storage life | – | *Maximum storage time | 1 | |
| | Workable (pot) life | EN 29514 | *Maximum time from mixing | 1 | |
| | Open time | EN 12189 | *Maximum time for use | 1 | |
| | Cure | BS EN 59 | Time to cure (Hardness) | 1 | |
| Cured bulk properties | Moisture resistance | ISO 62 | % water uptake | 1 | |
| | Temperature dependence | ISO 6721 | Glass transition temperature | 1 | 2 |
| | | ISO 11359-2 | Coeff. of thermal expansion | 1 | 2 |
| | Shrinkage (on curing) | EN 12617-1 and 3 | % shrinkage | 1 | |
| | Bulk performance | ISO 527-1 | Tensile | 1 | 2 |
| | | ISO 178 | Flexural | 1 | 2 |
| Adhesion and durability | Adhesion to FRP reinforcing material | ASTM D 5868 and EN 1542 | Tensile lap shear and Pull-off | 1 | 1 |
| | Adhesion to parent material | ASTM D3762 | Fracture toughness | 1 | 1 |
| | | ASTM D3807 and EN 1542 | Fracture toughness and Pull-off | | |

1. A material characteristic that shall be considered for all intended uses.
2. A material characteristic that shall be considered for certain intended uses.
* Provided by manufacturer.

1992).

One of the most important aspects in adhesive bonding is surface preparation. It is a critical topic that is not often given the attention by designers, operatives and their supervisors that it demands. When surface preparation is undertaken as part of the fabrication of bonded joints, whether in factory or on site, it is vital that the principles behind obtaining satisfactory short- and long-term adhesion are understood by all concerned. Adhesives are frequently blamed for 'not sticking', but the source of the trouble generally lies with the surface preparation. Some particular lessons that have been learnt over many years of structural adhesive bonding in construction are as follows:

- Most adherends need to be prepared properly, especially to secure long-term bond durability. Metals and thermoplastics present the greatest challenges.
- Some metals and FRP materials can be procured in a pretreated and pre-primed state.
- Greater use of primers and coupling agents (see below) formulated specifically for structural adhesive bonding should be used in conjunction with metallic adherends.
- Minor changes in preparation techniques can result in significant differences in the surface condition of materials such as metals. For example, a change to grinding from grit blasting could lead to bonded joints of significantly lower strength and durability.
- Treated surfaces must be bonded immediately, or coated with a primer or chemical coupling agent.

The adhesion mechanism is primarily chemical, but usually involves an element of mechanical keying. Put simplistically, the liquid adhesive will bond to the first material with which it comes into contact: if this is dust, loosely adhering paint, cement laitance or mould releases on plastic surfaces, the joint will clearly be inadequate. The adhesive should also displace air when spreading over the surface so that there are no voids to act as flaws.

The purpose of surface preparation is to remove contamination and weak surface layers. It is not a prerequisite of adhesion that the surface actually needs to be rough. However, the generation of micro-roughness ($< 0.1\ \mu m$) is beneficial, as is the introduction of new chemical groups onto the surface to link with the adhesive (or primer). The key stages involved in achieving this purpose are:

- cleaning
- material removal and surface modification
- further cleaning (to remove contamination introduced by treatments, such as oil-mist, dust or chemical residues).

It should be noted that solvent wiping alone may simply spread oils and greases without removing them from the surface. The use of solvents in any process also has health and safety implications.

### 31.6.1 Primers and coupling agents

Primers, being low-viscosity materials compared to most adhesives, assist adhesion either by partially penetrating the pores of a porous surface or by forming a chemical link between the surface and a relatively high-viscosity adhesive. Primers may also bind and reinforce weak surface layers of certain substrates such as concrete or stone. Improvement in adhesion relates directly to an improvement in bond durability. It is clear that adhesion failure in many real bonded assemblies could have been prevented by priming the substrates. The use of adhesive primers may be more critical in some instances than in others, for instance in association with joints involving metallic substrates or with porous surfaces such as concrete and masonry. In some circumstances, primers are used as a short-term coating to provide temporary protection to metallic surfaces prior to their removal before the application of structural adhesive. This may be done when time delays between the preparation and bonding stages are imposed by working/access constraints.

In structural adhesive bonding, any primers used must possess cohesive strength and not merely provide corrosion inhibition. The danger of film-forming primers, such as micaceous iron oxide ones, is that they generally become the weak link in a joint because they are themselves mechanically weak. The alternative approach utilises a very thin layer of a chemically reactive system based on silane coupling agents, applied as a dilute solution (0.1 to 5%). Plueddemann (1982) and Wake (1987) review such materials in detail. Such adhesion promoters have been shown to work well under controlled conditions on mild steel (Mays and Hutchinson, 1992) and on stainless steel (DTI, 1993), but care is required for the use of these systems on site.

### 31.6.2 Metallic substrates

All metallic surfaces are oxidised such that adhesives actually 'stick' to the metal surface oxide layer and not to the base metal itself. Such joints can therefore cause problems in service because oxide structures, and bonds to them, are susceptible to interfacial degradation by moisture ingress. It can be quite difficult to obtain durable joints that involve high-alloy metals that oxidise rapidly, such as stainless steels and aluminium alloys, unless complex etching and anodising procedures are adopted.

Steel surface treatments vary depending on type – carbon, stainless, and zinc-coated. However, several general points are summarised below.

Dry grit blasting has been applied successfully to carbon steels and stainless steels, and is particularly effective in combination with primers. Hard, sharp, media should be used and dust must be removed after the blasting operation by vacuuming or by brushing with clean brushes. With dry blasting systems, oil and water traps should be used to prevent contamination of the surface.

Wet abrasive blasting can be used in some circumstances and has the advantage that all dust and any residual salts are carried off the surface automatically. Mechanical abrasion such as grinding is often used initially to remove gross contamination and surface irregularities; however, it should be followed by blasting because it does not generally provide a clean and uniform energetic final surface suitable for adhesive bonding. Chemical treatment is not recommended because of the disposal problems associated with residues, but it may currently be the only alternative for metals such as zinc and copper. Commercially available etch primers are available to promote adhesion to some forms of zinc-coated steel, although some types (e.g. EZ, IZ) are more amenable to bonding than others. Hot-dipped galvanised surfaces are very variable in chemistry because of the variations in the processes used. However, reasonable bonds can be obtained by removing most of the zinc with a light grit blast, possibly followed by the application of a primer, depending on the adhesive type. It may seem curious to have to remove most of the zinc, but the level of adhesion that would otherwise be obtained would normally be minimal.

Cast (and wrought) iron surfaces may contain numerous cracks, fissures and voids. These can provide zones of surface weakness or channels through which moisture may gain access to bonded interfaces by capillary action. The strength of wrought iron perpendicular to the surface may also be a design constraint. However, these surfaces respond well to grit blasting and appear to provide similar adhesion characteristics to those associated with grit-blasted carbon steel. Further guidance on the preparation of metallic substrates is presented in CIRIA C595 (2004).

ISO8504 provides guidance on the preparation methods that are available for cleaning metallic substrates, and ISO8501 incorporates the well-established Sa-series for blast cleaning. The most reliable way of demonstrating good bonding is to perform pull-off tests to demonstrate that the adhesive bond is either of sufficient strength or at least stronger than the parent material or other components of the materials system. EN1542 and ISO4624 describe the procedure, but it is recommended to use a 'dolly' 25 mm in

**Figure 31.7** Pull-off test on metallic substrate using 25 mm diameter dolly

diameter (CIRIA C595, 2004) as shown in Figure 31.7. A variety of NDT techniques could theoretically be used but the equipment is expensive and is not generally available to the construction sector for use on site. There are also concerns that NDT techniques are not yet sufficiently reliable.

### 31.6.3 Concrete substrates

The purpose of surface preparation is to remove the outer, weak and potentially contaminated skin together with poorly bound material, in order ideally to expose small to medium-sized pieces of aggregate. This must be achieved without causing micro-cracks or other damage in the layer behind that would lead to a plane of weakness and hence a reduction in strength of the adhesive connection. Any large depressions, blow-holes and cracks must be filled with suitable mortars prior to the application of a structural adhesive. This will ensure a relatively uniform bondline thickness in order to maximise the efficiency of shear stress transfer. The basic sequence of steps in the process of surface preparation should be as follows:

- Remove any damaged or sub-standard concrete and reinstate with good-quality material.
- Remove laitance, preferably by grit-blasting. Techniques such as wire brushing and grinding may be satisfactory for non-structural applications. Other methods such as bush-hammering, acid-etching, water-jetting, or flame-blasting are not recommended for structural bonding applications because the surface is not left in a uniform condition and the sub-surface layers may be damaged. In particular, the use of pneumatic tools can cause significant damage to the underlying concrete (DTI, 1993).
- Remove dust and debris by brushing, air-blast or vacuum.

Additional steps could include:

- further cleaning, with water and detergent, to remove any remaining contaminants
- drying the surface to be bonded
- application of an adhesive-compatible primer, if necessary.

The recommended method of laitance removal, by blasting, is fast, plant-intensive and operator-dependent. Generally a fairly micro-rough but macro-smooth surface is generated, together with a lot of dust; particles of blast media may also be left embedded in the surface. This dust and debris must be removed prior to bonding. Wet blasting is another option that overcomes some of the problems associated with the dust. However, this may create a water disposal problem and the concrete surface must be allowed to dry out to a degree that is suited to the intended adhesive system prior to bonding. Further guidance on the preparation of concrete substrates is provided by ASTM D4258 (1983), Concrete Society (2004), Gaul (1984) and Sasse and Fiebrich (1983).

After preparation, the suitability of the surface should be checked using the partially cored pull-off test (Figure 31.8) as described in EN 1542. The time lapse between preparation and bonding should be minimised in order to prevent any further contamination of the surface. Adhesion to damp and wet concrete surfaces with epoxy adhesives was investigated by Hutchinson and Hollaway (1999) and no effect on bond strength was detected.

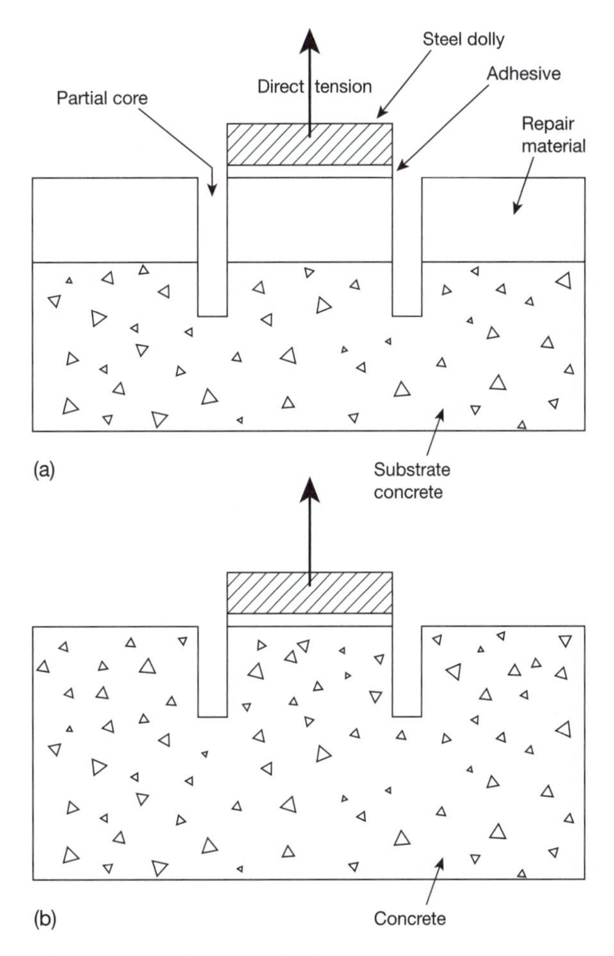


**Figure 31.8** Partially cored pull-off tests on concrete with and without a surface repair coating

### 31.6.4 Timber substrates

There is a wide variety of different timber types, each with its own characteristics; they respond differently to treatments and to being bonded. One factor that affects adhesion is the amount and type of extraneous components present in the timber. Extractives that exist in heartwood may interfere with the spread and penetrability of fluid adhesives and hence with the formation of molecular bonds. Generally, hardwoods are more difficult to bond than softwoods because of their higher densities and the extractives that they contain (such as oils and tannins). Removal of extractives is the simplest way of minimising their negative effects on bonding, but methods such as solvent extraction are expensive. More realistic approaches involve chemical treatment and surface activation. Whatever the species, surface moisture contents at the time of bonding should generally be less than 20 per cent to allow adhesives to flow into the pores and for adhesion to take place: this can be achieved by localised drying. Clearly there will be differences between species and between different adhesive systems, and limits can be established experimentally for different combinations of materials.

The surface layers of timber become oxidised and degraded over time. It is therefore essential to remove material and then bond the surfaces within a day for optimum adhesion. There is general agreement that sound, freshly cut surfaces are ideal for painting and bonding. River et al. (1991) and Davis (1997) provide good accounts of the nature of timber surfaces and their treatment. The general treatments are:

- cutting with a plane, saw, auger, chisel or similar sharp tools
- removal of dust
- localised drying, if necessary
- application of an adhesive-compatible primer, if necessary.

Surfaces that have been sanded or sawn should be cleaned very carefully to remove dust which will be loosely bound or else clog up any cells that have been cut open.

More specialist treatments include the addition of reactive chemicals or the application of radiation energy. Lithium and sodium hydroxides have been shown to improve surface activation and wettability markedly. Oxidative activation by the use of flame, corona or plasma have also been shown to improve wettability for many species of timber, although the treatment levels and effects vary from species to species.

Pull-off, pull-out and compressive block shear tests can be used to demonstrate the adequacy of prepared surfaces. The development and validation of suitable test protocols was the subject of intense development in Europe by CEN/TC193 working group 11 in 2006. It should be noted that the natural variability of timber substrates means that the scatter in data is generally rather large. Broughton and Hutchinson (2001a, 2001b) have reported on the pull-out behaviour of steel rods bonded into timber with epoxy adhesives, and also on the effect of timber moisture content at the time of bonding (Broughton and Hutchinson, 2001c). Guidance on resin-bonded repair systems for structural timber is available from TRADA (1995) and from the LICONS project (www.licons.org).

### 31.6.5 FRP substrates

Fibre-reinforced polymer composite materials are usually based on epoxy, vinylester or polyester resins; they are compatible with commonly used adhesives as well as being readily bondable. The surface of composite materials may be contaminated with mould release agents, lubricants and fingerprints as a result of the production process. Further, the matrix resin may include waxes, flow agents and mould release agents that can be left on the surface of a cured composite. Surface preparation not only serves to remove contamination such as fluorocarbon release agents, but may also increase the surface area for bonding, promote micromechanical interlocking and/or chemically modify a surface. Care must be taken to ensure that only the chemistry and morphology of a *thin* surface layer are modified. It is important not to break reinforcing fibres, nor to affect the bulk properties of the composite.

The main methods of surface preparation for composites are solvent degreasing, mechanical abrasion and use of the peel-ply technique, and these methods are often used in combination (ASTM D2903, 1984; Hutchinson, 1997). The basic sequence of steps in the process of surface preparation should be to *either*

- remove grease and dust with a suitable solvent such as acetone
- remove release agents and resin-rich surface layers by abrasion. This can be accomplished by abrasive papers, light grit blasting using very fine alumina, or cryo-blasting with solid carbon dioxide pellets – the resultant surface should be textured, dull and lustreless
- remove dust and debris by solvent wiping

*or*

- strip off a peel-ply layer.

A peel-ply is a sacrificial layer, normally a polyester or nylon fabric, about 0.2 mm thick that is laid-up on the outermost surfaces of a composite material and co-cured with it. During cure, the peel-ply layer becomes consolidated into the surface of the composite. This provides a uniquely practical surface preparation method such that unskilled operators can achieve a very clean and bondable surface on site. Special considerations associated with peel-ply technology are outside the scope of this chapter, but the subject is discussed by Hutchinson (1997) and by Hutchinson and Quinn (1999).

Combinations of composite resin matrix, peel-ply materials and composite processing conditions must be selected and be assessed carefully by both the specifier and the manufacturer. When the dry peel-ply material is pulled off, only the top layer of resin on the FRP component should be removed, leaving behind a clean, rough, surface for bonding. The resultant surface topography is essentially an imprint of the peel-ply fabric weave pattern.

Lap shear, wedge cleavage and pull-off tests can be used to assess the adequacy of adhesion to FRP materials.

## 31.7 Joint design

Structural adhesive joints must be designed properly for maximum efficiency. Joint behaviour is influenced by many factors including the nature of the applied loading, the geometry of the bonded area, the bulk and surface properties of the adherends, and the properties of the adhesive itself. The guiding principles that should be adopted by all designers of bonded joints are to:

- provide the maximum bond area
- stress the adhesive in the direction of maximum strength (i.e. in shear or compression), and avoid the build-up of secondary tensile stresses
- avoid large stress concentrations which are associated with relatively high-modulus adhesives and flexible adherends
- make the adhesive layer as uniform, void-free and as thin as practically possible
- maintain a continuous bondline.

Structural load-bearing joints in engineering disciplines outside civil, structural and marine engineering tend to be formed with thin (< 1 mm) bondlines, whereas thick bondlines are commonly found in construction applications of adhesives. This is an important difference because it has been shown experimentally that thick bondlines generally result in lower joint strengths whereas theoretical work indicates the opposite. Joint performance is largely dictated by the properties of the adherends. Peel stresses are particularly damaging because adhesive bonds possess less resistance to peel than to any other mode of loading. Peel stresses can also lead to premature failure or delamination of materials such as concrete, timber, FRP and wrought iron. General guidance on engineering bonded joint design is given by Adams et al. (1997). Design guidance is available for the strengthening of concrete (ACI 440, 2002; Concrete Society, 2004; FIB, 2001), metallic structures (CIRIA C595, 2004; Moy, 2001) and timber (TRADA, 1995; STEP 1, 1995; STEP 2, 1995). Design guidance on joining composite materials is given by Clarke (1996).

## 31.8 Joint fabrication

The satisfactory fabrication of joints relies on the provision of suitably qualified materials, detailed designs, and appropriate method statements containing fabrication processes. Training of site operatives and their supervisors is essential to ensure quality of workmanship (CSWIP, 2006). The main topics for consideration are:

- storage of materials, including solvents, primers and adhesives
- protection of the working environment
- control of adherend moisture contents (e.g. concrete, timber and FRP)
- surface preparation of adherends
- priming of adherends (if applicable), including considerations of timing
- adhesive mixing, dispensing and application
- component fit-up, including temporary works and fixtures
- control of bondline thickness and adhesive fillets at joint

edges
- curing
- overcoating and painting.

Detailed general commentaries on these aspects are provided in a variety of sources (e.g. Mays and Hutchinson, 1992; Adams et al., 1997; IStructE, 1999; Lees, 2000). The fabrication of joints involving FRP adherends specifically is covered by Clarke (1996) and by Hutchinson (1997). The fabrication of concrete–composite joints in externally bonded reinforcement is detailed by Hollaway and Leeming (1999) and the Concrete Society (2003, 2004). Fabrication issues concerning FRP strengthening of steel structures are detailed in CIRIA C595 (2004).

## 31.9 Bonded joint durability

Uncertainties concerning long-term performance are still significant, despite more than 30 years' proven adhesives use in construction. It is well known that the most important factor in joint durability is the environmental stability of the adhesive–adherend interface. This is dictated by the type of adhesive, the nature of the adherend and its surface. The experience of all industries is that the mechanical properties of bonded joints may deteriorate under warm and wet conditions (Kinloch, 1983). This is particularly so if the joints comprise substrates that attract water, such as metals or glass, or else permeable substrates such as concrete and timber. It follows that the use of qualified materials, optimisation of surface conditions and pretreatments and high-quality workmanship are the keys to maximising joint durability.

## 31.10 Quality control

Quality control covers many aspects. In the use of adhesives there are three aspects to consider: state of cure, adhesion, and the mechanical properties of the adhesive. The parameters and test methods that govern performance classification are outlined in Section 31.5.11, and a list of test methods for adhesives is summarised in the Appendix to this chapter.

The final mechanical properties of the adhesive layer are strongly affected by the degree of cure. To ensure that the design bonded joint properties are reproduced on-site, test samples of the adhesive should be made in exactly the same manner as the applied adhesive, ideally at the same time and under the same site conditions.

Measurement of the tensile modulus and strength of the adhesive (ISO527) is recommended. Further details of the recommended QA tests, acceptance criteria, sample preparation and dimensions are given by the Concrete Society (2003).

A summary of the recommended QA tests for the application of adhesives to externally bonded reinforcement is presented in Table 31.4, including the number of tests and measurement location (www.compclass.org.uk). Off-site testing should be performed by an experienced independent test house. Where testing involves a bonded joint, it is essential to record the locus of failure as well as the failure load. Some possible failure modes for a concrete/adhesive/FRP system are shown in Figure 31.9.

Non-destructive evaluation (NDE) of bonded joints involves two aspects: the integrity of the bond between the adhesive and substrate material(s), and an assessment of the quality of the cured adhesive layer(s). In general, the integrity of the interface between the adhesive and substrate is of greater importance, particularly in the long term. Visual inspection and simple acoustic tap testing are often sufficient when undertaken by experienced operatives. More sophisticated NDE techniques include laser shearography, thermography and ultrasound (Moy, 2001; Concrete Society, 2003), but the applicability of the methods is cost-, materials- and operator-dependent.

## 31.11 Education and training

Personnel involved in structural adhesive bonding should have had appropriate training and be qualified in the application to be undertaken. This should include the handling of adhesive materials, surface preparation, quality control procedures, and health and safety issues. It is important that the training given provides sufficient technical background to allow personnel to obtain a good understanding as to why key operations such as surface preparation are so important.

It should be noted that training is an essential element of a successful bonded application. Personnel should be the subject of a continuing review of competency with a log kept of experience in the application of bonding systems. This is important as the

**Table 31.4** Summary of QA tests for composites applied on-site

| *Measurement location* | *Property* | *Test method* | *Number of tests (per batch)* | *Comments* |
|---|---|---|---|---|
| On-site | Hardness (state-of-cure of adhesive or laminating resin used in wet lay-up) | BS EN 59 | At least 10 tests, evenly distributed over the composite | On-site hand held test method (Shore D hardness) to assess degree of cure |
| | Adhesion (between parent and FRP reinforcement materials) | EN1542 | At least 5 tests | Pull-off tests using steel dollies: 25 mm diameter for metals, 50 mm diameter for concrete |
| Sample prepared on-site, measured off-site | Glass transition temperature (adhesive or laminating resin) | EN 12614 or ISO 11357-2 or ISO 6721 | At least 3 tests | Same method for $T_g$ measurement should be used throughout the qualification scheme |
| | Tensile modulus and strength (adhesive or laminating resin) | ISO 527 | At least 3 tests | Dumbbell specimen |
| | Adhesion (between adhesive and composite) | Lap shear joint performance ASTM D 5868 | At least 3 tests | Ideal thickness of composite: <5 mm<br>Ideal bondline thickness ~0.5 mm |

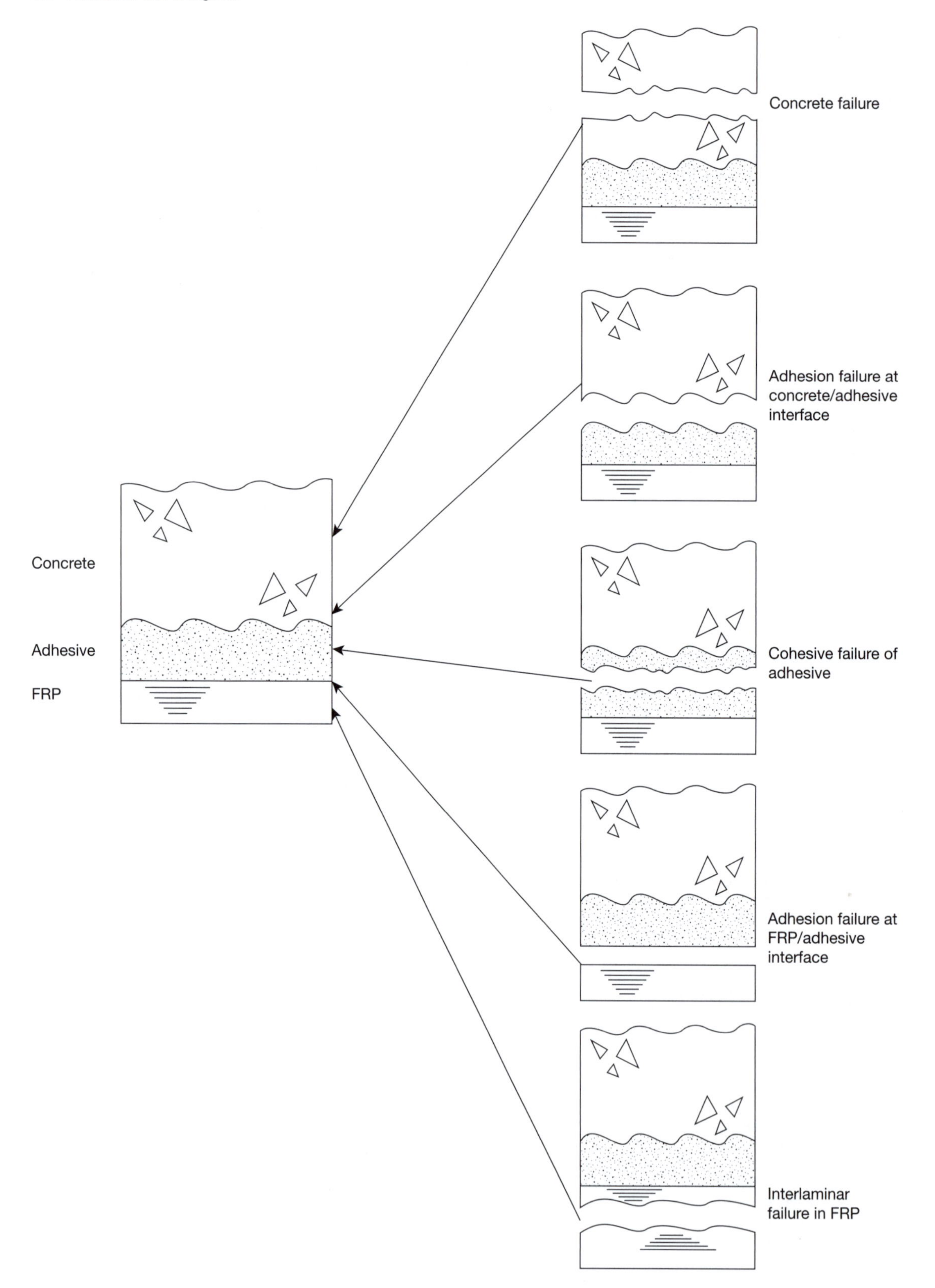


**Figure 31.9** Possible modes of failure associated with joints comprising FRP bonded to concrete

levels of competence and experience achieved by individual applicators should also be considered in the context of particular bonding activities. For example, working in confined spaces or applying material around complicated geometries can pose additional difficulties that should be taken into account.

Supervisory personnel should be trained in the interpretation of method statements and the relevant fabrication techniques and, ideally, should have had a period during which they were engaged in the application of adhesive material systems. Supervisors should also be the subject of a continuing review of competency.

The Concrete Repair Association (CRA) and the Composite Processors Association (CPA) have influenced the Construction Industry Training Board's (CITB) activities on this issue in the UK. NVQ curricula exist for operative and supervisor training in adhesives and composite materials, and a certification scheme for composite materials system application was in development in 2006 (CSWIP, 2006).

## 31.12 Health and safety

Adhesives, primers and solvents will have quoted threshold limit values (TLVs) for short- and long-term exposures. The manufacturer's health and safety data sheets must always be obtained and read before use. They should be used as the basis of any COSHH (Control of Substances Hazardous to Health) assessment. Adhesives should always be used in accordance with the manufacturer's recommendations, taking the necessary precautions such as the use of protective clothing during mixing and application. Similarly, when preparing the surface prior to the application of some adhesives, provision must be made for adequate extraction of noxious fumes, dust, etc. and a suitable supply of fresh air.

Mixing should always be in accordance with the manufacturer's instructions, particularly with regard to the quantity of the adhesive mixed at any one time, as exothermic reactions can lead to excessive temperature rises.

In addition to considering the health and safety aspects of fabricating the adhesive connection, it is necessary to consider the long-term behaviour including the possibility and consequences of failure.

## 31.13 References


ACI440.2R (2002). *Guide for the design and construction of externally bonded FRP systems for strengthening structures*. Farmington Hills, MI: American Concrete Institute.

Adams, R. D., Comyn, J. and Wake, W.C. (1997). *Structural adhesive joints in engineering*. London, UK: Chapman and Hall.

ASTM D2903 (1984). *Standard practice for preparation of surfaces of plastics prior to adhesive bonding*. Philadelphia, PA: American Society for Testing and Materials.

ASTM D4258 (1983). *Standard practice for abrading concrete*. Philadelphia, PA: American Society for Testing and Materials.

Broughton, J.G. and Hutchinson, A.R. (2001a). Adhesive systems for structural connections in timber. *Int. J. Adhesion and Adhesives*, 21, 3, 177–186.

Broughton, J.G. and Hutchinson, A.R. (2001b). Pull-out behaviour of steel rods bonded into timber. *RILEM Materials and Structures*, 34, 236, 100–109.

Broughton, J.G. and Hutchinson, A.R. (2001c). Effect of timber moisture content on bonded-in rods. *Construction and Building Materials*, 15, 17–25.

BS1204 (1993). *Specification for type MR phenolic and aminoplastic synthetic resin adhesives for wood*. London, UK: British Standards Institution.

BS EN 1504 (2004). *Products and systems for the protection and repair of concrete structures*. London, UK: British Standards Institution.

CIRIA C595 (2004). *Strengthening of metallic structures using externally bonded fibre-reinforced polymers*. London, UK: Construction Industry Research and Information Association.

Clarke, J.L. (ed.) (1996). *Structural design of polymer composites – EUROCOMP design code and handbook*. London, UK: E & F N Spon.

Comyn, J. (1997). *Adhesion science*. Cambridge, UK: The Royal Society of Chemistry.

Concrete Society (2003). *Strengthening concrete structures using fibre composite materials: acceptance, inspection and monitoring*. Camberley, UK: TR57.

Concrete Society (2004). *Design guidance for strengthening concrete structures using fibre composite materials*. Camberley, UK: TR55, second edition.

CSWIP (2006). *Installer qualification document*. Cambridge, UK: 1st draft, December, TWI. Cambridge, UK.

Darby, J.J., Skwarski, A., Brown, P. and Haynes, M. (1999). Pre-stressed advanced composite plates for the repair and strengthening of structures. In *Proc. 8th Int. Conf. on Structural Faults and Repair* (M.C. Forde, ed.). Edinburgh, UK: Engineering Technics Press.

Davis, G. (1997). The performance of adhesive systems for structural timbers. *Int. J. Adhesion and Adhesives*, 17, 3, 247–255.

DTI (1993). *Review of substrate surface treatments*. (MTS Project 4, Measurement and Standards Programme on the Performance of Adhesive Joints, 1993–1996, Report No 2), UK Department of Trade and Industry.

EN 301 (1992). *Adhesives, phenolic and aminoplastic, for load-bearing timber structures: classification and performance requirements*. London, UK: British Standards Institution.

EN 1542 (1999). *Measurement of bond strength by pull-off. Products and systems for the protection and repair of concrete structures. Test methods*. Brussels, Belgium: European Committee for Standardisation.

FIB (2001). *Externally bonded FRP reinforcement for RC structures*. CEB-FIP Technical Report. Lausanne, Switzerland: International Federation for Structural Concrete.

Gaul, R.W. (1984). Preparing concrete surfaces for coatings. *Concrete International*, 6, 7, 17–22.

Highways Agency (1994). *Strengthening of concrete highway structures using externally bonded plates, Design manual for roads and bridges, BA 30/94*. Vol. 13, Section 3, Repair, Part 1. London, UK: HMSO.

Hollaway, L.C. and Leeming, M.B. (eds.) (1999). *Strengthening of Reinforced Concrete Structures*. Cambridge, UK: Woodhead Publishing.

Hurley, S. A. (2000). *The use of epoxy, polyester and similar reactive polymers in construction*. London, UK: Construction Industry Research and Information Association.

Hutchinson, A.R. (1993). Review of methods for the surface treatment of concrete. In *Characterisation of surface condition*, MTS Project 4, Report No. 2 on the Performance of Adhesive Joints. London, UK: Department of Trade and Industry.

Hutchinson, A.R. (1997). *Joining of fibre-reinforced polymer composite materials*, Report 46. London: Construction Industry Research and Information Association.

Hutchinson, A.R. and Hollaway, L. C. (1999). Environmental durability. In *Strengthening of reinforced concrete structures* (L.C. Hollaway and M. B. Leeming, eds). Cambridge, UK: Woodhead Publishing.

Hutchinson, A.R. and Hurley, S.A. (2001). *Transfer of adhesives technology*. Project Report 84. London, UK: Construction Industry Research and Information Association.

Hutchinson, A.R. and Quinn, J. (1999). Materials. In *Strengthening of reinforced concrete structures* (L.C. Hollaway and M.B. Leeming, eds.). London, UK: Woodhead Publishing.

IStructE (1999). *A guide to the structural use of adhesives*. London, UK: The Institution of Structural Engineers.

Kinloch, A.J. (1983). *Durability of structural adhesives*. London, UK: Elsevier Applied Science.

Lees, W.A. (2000). Adhesives for engineering assembly. In *Kemps Directory and Yearbook*. Birmingham, UK: Kemps Publishing.

Mays, G.C. (2001). Performance requirements for structural adhesives in relation to concrete strengthening. *Int. J. Adhesion and Adhesives*, 21, 5, 423-29.

Mays, G.C. and Hutchinson, A.R. (1988). Engineering property requirements for structural adhesives. *Proc. Inst. of Civil Engineers*, Part 2, 85, 485-501.

Mays, G.C. and Hutchinson, A.R. (1992). *Adhesives in civil engineering*. Cambridge, UK: Cambridge University Press.

Mettem, C.J. and Milner, M. (2000). *Resin repairs to timber structures: Volume 1, guidance and selection*. High Wycombe, UK: TRADA Technology.

Moy, S.S.J. (ed.) (2001). *FRP composites: life extension and strengthening of metallic structures. Design and practice guide*. London, UK: The Institution of Civil Engineers.

Pizzi, A. (ed.) (1983). *Wood adhesives, chemistry and technology*. New York, NY: Marcel Dekker.

Pizzi, A. (2004). Some considerations on future trends in wood adhesives. In *Innovations on wood adhesives,* COST E34 Conference (M. Properzi, F. Pichelin and M. Lehman, eds). Block 1, Paper 1.1. Biel, Switzerland: HSB Biel/Bienne.

Plueddemann, E.P. (1982). *Silane coupling agents*. New York, NY: Plenum Press.

Raknes, E. (1997). Durability of structural wood adhesives after 30 years ageing, *European Journal of Wood and Wood Products*, 55, 2–4.

River, B.H., Gillespie, R.H. and Vick, C.B.(1991). Timber. In *Treatise on adhesion and adhesives*, Vol. 7 (J.D. Minford, ed.), Chapter 1. New York, NY: Marcel Dekker.

Sasse, H.R. and Fiebrich, M. (1983). Bonding of polymer materials to concrete. *RILEM Materials and Structures*, 16, 94, 293–301.

STEP 1 (1995). *Timber engineering – basis of design, material properties, structural components and joints*. Technical Report, 1st edition. Almere Buiten, The Netherlands: Centrum Hout.

STEP 2 (1995). *Timber engineering – Design details and structural systems*. Technical Report, 1st edition. Almere Buiten, The Netherlands: Centrum Hout.

TRADA (1995). *Resin-bonded repair systems for structural timber*. Wood Information Sheet 22, Section 4. High Wycombe, UK: TRADA Technology.

Wake, W.C. (1987). Silicone adhesives, sealants and coupling agents. *Critical Reports in Applied Chemistry*, 16, 89.

Wheeler, A. and Hutchinson, A.R. (1998). Resin repairs to timber structures. *Int. J. Adhesion and Adhesives*, 18, 1, 1–13.

Wikipedia (n.d.). Big Dig ceiling collapse. http://en.wikipedia.org/wiki/Big_Dig_ceiling_collapse.

www.compclass.org.uk. *Classification and qualification of composite materials systems for use in the civil infrastructure*. Project website. DTI MMS6 Project 2002–2005. Contract Ref: CHBJ/C/005/00028.

www.licons.org. *Low intrusion conservation systems for timber structures*. Project website. EC CRAFT Project 2003–2005. Contract number: CRAF-1999-71216 LICONS.

## Appendix: Test methods for structural adhesives

| *Parameter* | *Performance characteristic* | *Test method* | *Description* |
|---|---|---|---|
| Adhesive material bulk properties and cure assessment | Cure | EN12614 | Glass transition temperature, $Tg$ |
| | | ISO6721<br>ASTM E1640 | Glass transition temperature, $Tg$ using dynamic mechanical analysis |
| | | ISO11357-2<br>ASTM D3418 | Glass transition temperature, $Tg$ using differential scanning calorimetry |
| | Shrinkage (on curing) | EN12617-1 and 3 | % shrinkage |
| | Temperature dependence | ISO11359-2 | Coeff. of thermal expansion |
| | Moisture resistance | ISO62 | % water uptake |
| | Bulk performance | ISO527-1 | Tensile modulus and strength from dumbbell specimen |
| | | ISO178 – general<br>Highways Agency BA 30/94 – specific | Flexural modulus from prism specimen |
| Adhesion and durability assessment | Adhesion of adhesives and primers to a variety of materials | BS 5350–C5<br>ASTM D 1002 | Single lap shear specimen for metals |
| | | ASTM D 5868 | Single lap shear specimen for FRP materials |
| | | ASTM D 2559 | Compressive lap shear specimen for timber |
| | | EN1542 | Pull-off specimen (25mm dia. dolly for metals; 50 mm dia. dolly for concrete) |
| | | ASTM D3762<br>ASTM D 3807 | Double cantilever beam fracture toughness specimen (generally for metal adherends) |
| | | ASTM D3433 | Tapered double cantilever beam fracture toughness specimen |
| | | EN 12188 | Scarf shear specimen (steel to steel) |

32

# Fibre-Reinforced Polymer Composite Materials

**Stuart Moy** BSc PhD CEng FICE
Southampton University

**Contents**

## 32.1 Introduction

What would be the ideal material for engineering? What do engineers want from their materials? The following are the majority of the attributes that are important in construction:

- high or low strength, as required
- high or low stiffness, as required
- low weight
- the ability to tailor properties as needed
- high durability
- low cost.

Fibre reinforced polymer composites offer the engineer most of the above, although they are not the perfect material. The construction industry has difficulty adopting technologies it perceives as being not tried and trusted. Composites have not been available for as long as the conventional construction materials such as steel, concrete, masonry and timber, but they have a significant track record. Using them in construction needs a different design approach and also a different way of thinking.

Fibre-reinforced polymer composites (FRPs) were used first in aircraft radar covers in the late 1930s. Their use spread rapidly in the marine industry so that today they are the most common material for leisure craft construction as well as much larger vessels. As a result they are also utilised in demanding situations such as pressure vessels, pipes and blast panels on offshore platforms (ICE, 2001). The early FRPs used glass fibres (often called GFRPs) as the reinforcing medium, but the development of carbon (CFRPs) and aramid fibres (AFRPs) in the 1960s lead to wider use in the aerospace, automobile and sports equipment industries.

The FRPs offer considerable flexibility of use to the engineer. Strength and stiffness can be controlled by the choice of the appropriate type and amount of reinforcing material. It is possible to have high strength and stiffness simultaneously but it is also possible to have high strength and low stiffness. The tensile strength of an FRP will be greater than that of high-tensile steel while its density will be only about 20% that of steel. Thus the FRPs satisfy the first three requirements in the list above. The fourth can also be satisfied because the reinforcing fibres can be placed as needed. Thus more fibres can be used in the directions where more strength or stiffness is required.

Durability is an issue for all materials, but the FRPs have demonstrated good durability provided sensible precautions are taken. The last item on the list is cost. On a weight-for-weight basis FRPs are more expensive than traditional materials but that is not the comparison to make. FRPs can be manufactured under factory conditions and delivered to site ready to install. Their low weight means they can be installed using fewer people, less plant and less falsework. Consequently installation cost is a better comparator; whole-life costing that considers maintenance is an even better one.

This brief introduction should have whetted the appetite for FRPs. In subsequent sections the material will be considered in detail, including methods of manufacture. Mechanical properties will then be presented together with methods for predicting FRP properties. Uses of the FRPs in construction will be considered, looking at 'non-structural' and 'structural' applications, including upgrade of existing structures as well as new build.

## 32.2 Fibre-reinforced polymer composites

### 32.2.1 A definition of fibre-reinforced polymer composites

Engineering composites are usually formed by intimately mixing two distinct material phases: a fibrous reinforcement and a continuous medium. Using this definition, reinforced concrete is a composite of steel bars and concrete. In FRPs the reinforcement has high strength or stiffness and a low density; the continuous medium, called the matrix, has lower strength and stiffness.

There are three reinforcing fibres in general use: glass, aramid and carbon. Aramid is better known by the trade names Kevlar® or Twaron®. The polymer matrix is a chemical supplied in liquid form, which will undergo a polymerisation reaction and harden under the correct conditions. In the final state the matrix completely encloses the reinforcing fibres, sharing load between the fibres, protecting their surface from damage and inhibiting fracture of the fibres. In order to understand their properties it is necessary to examine these components in more detail.

### 32.2.2 The reinforcing fibres

As stated above, three types of fibre are available; the general properties of the dry fibres are given in Table 32.1. The fibres are brittle, with linear stress–strain characteristics up to failure, as shown in Figure 32.1 for various fibres. It can be seen from the table and figure that there are wide variations in properties, especially for the carbon fibres. A closer look at each type of fibre will explain why this variation occurs.

#### *32.2.2.1 Glass fibres*

As their name implies, these are produced from a bath of molten glass by rapid drawing of fine filaments. The filaments are given a surface treatment, called sizing, which binds and protects them so that they can be wound onto a drum. Incorporating different metallic oxides into the molten silica alters the properties of the resulting fibres. The most common fibres in construction are E-glass, followed by R-glass and S-glass. The last two are more expensive but have higher strength and chemical resistance. Glass fibres are the cheapest and most commonly used fibres in construction applications, particularly in structures that can be designed to accommodate their relatively low Young's modulus.

#### *32.2.2.2 Aramid fibres*

These are manufactured from a synthetic organic compound (aromatic polyamide) by an extrusion and spinning process followed by stretching and drawing. The fibres are produced in low- and high-modulus grades (60 kN/mm$^2$ or 160 kN/mm$^2$). They have high tensile strengths, moderate Young's modulus and low density. Since they produce FRPs with low compressive and shear strengths, their main application in construction has been in the form of cables or prestressing strand.

#### *32.2.2.3 Carbon fibres*

Carbon fibres are produced from three different organic materials, called precursors: rayon, polyacrylonitrile (PAN) and pitch. The manufacturing processes vary from precursor to precursor and are complex; details of each can be found in Hollaway and Head (2001). Each precursor produces fibres with different properties and in different grades: PAN fibres are generally the highest quality and are used extensively in aerospace, fibres from pitch

**Table 32.1** Typical properties of dry reinforcing fibres (ICE, 2001)

| *Fibre* | *Tensile strength (N/mm$^2$)* | *Young's modulus (kN/mm$^2$)* | *Density (kg/m$^3$)* | *Cost (£/kg)* |
|---|---|---|---|---|
| Glass | 3445–4890 | 72–87 | 2460–2580 | 2.5 |
| Aramid | 3150–3600 | 60–160 | 1390–1470 | 20 |
| Carbon | 2100–7100 | 220–900 | 1740–2200 | 10–200 |

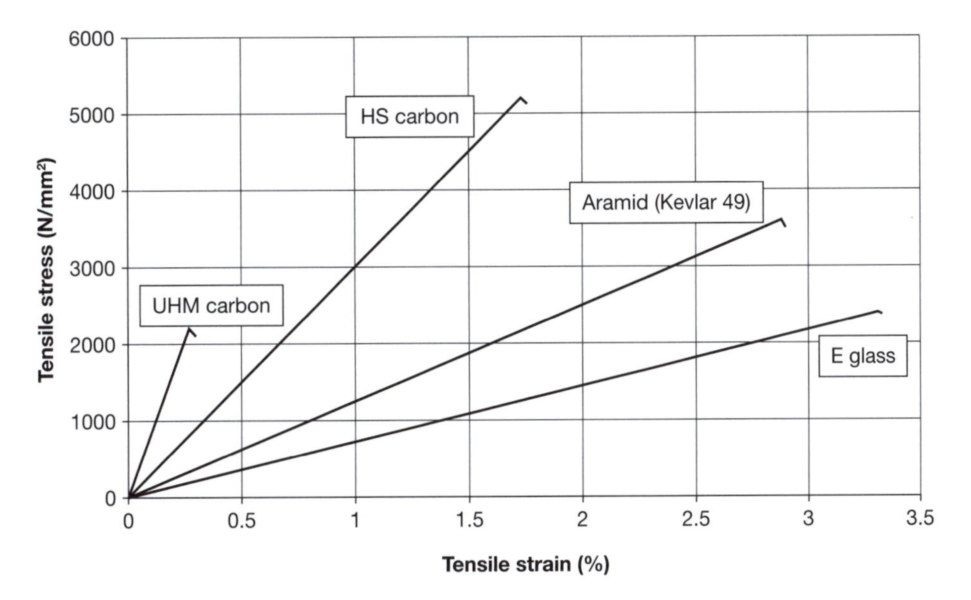


**Figure 32.1** Fibre stress–strain characteristics (by permission of Professor L. Hollaway, University of Surrey)

precursor are cheaper but more variable in consistency. The general characteristics of the carbon fibres are their high strength and Young's modulus and low density. There is a play-off between strength and stiffness: the higher the strength, the lower is the Young's modulus and vice versa. Thus carbon fibres can be found in grades ranging from high strength (HS) to ultra-high modulus (UHM), and provide the best possibility of matching properties to a particular application. Carbon fibres are used when there is a requirement for high strength and stiffness.

#### *32.2.2.4 Some general points about fibres*

A single fibre, often called a filament, has a very small diameter and very high mechanical properties since it is almost flawless. The filaments are grouped together to form strands or tows. A glass or aramid strand will contain several hundred filaments while carbon fibres are grouped into tows. A 12K tow, for example, has 12,000 filaments. Bundles of strands or tows are twisted together to form yarns which can be combined in various ways to produce useful reinforcement fabrics. When fibres are combined together there is a degradation in mechanical properties due to the presence of flaws, thus the practical properties of reinforcement are lower than those of the individual fibres.

The simplest form of reinforcement is the *mat*. Chopped strand mat consists of randomly distributed fibres cut to lengths of around 50 mm and held together by a chemical binder. In continuous strand mat long fibres are swirled together and held in place by a chemical binder. *Unidirectional* (*UD*) reinforcement has a minimum of 95% of the fibres running in one (the 0°) direction. The remaining fibres are woven or stitched at right angles to the main fibres to create a fabric that can be cut to the desired width but remains in a single piece. *Biaxial fabrics* have equal numbers of fibres in the 0° and 90° directions. The two sets of fibres can be woven together using conventional weaving processes or they can be stitched together. *Quadraxial fabrics* have equal amounts of fibres in the 0°, +45°, –45° and 90° directions, and are usually stitched. In woven fabrics the 0° direction is called the *warp* and the 90° direction is called the *weft*. The weaving process necessarily bends the fibres causing stress concentrations in use which further reduce the effective strength of the reinforcement. Stitching produces very little distortion to the fibres and is becoming more popular as a result. A single layer of fabric is called a *ply*. Generally several *plies* are required in the final product and these can be arranged in different directions. The arrangement of all the plies is called the *lay-up*. The reinforcing fibres are chemically resistant and, apart from the aramid fibres, unaffected by UV light. The fibres, particularly carbon, are, however, brittle and susceptible to damage until they are embedded in the matrix.

### 32.2.3 The resin matrix

A *polymer* is an organic compound composed of long molecules which are formed from repeats of a short molecule called a *mer*. As outlined in Hollaway and Head (2001) it is desirable for the resin matrix (polymer) to fulfil the following functions:

- wet out the fibres and cure satisfactorily
- bind and protect the fibres
- disperse and separate the fibres to prevent catastrophic crack propagation
- transfer stresses to the fibres by adhesion and/or friction
- be chemically and thermally compatible with the fibres
- have appropriate fire resistance with limited smoke propagation.

There are many resins but they can be placed into two categories: *thermosetting* and *thermoplastic*. When a thermosetting resin hardens there is considerable cross-linking of parts of the molecule so that the hardening process is irreversible. The main thermosetting polymers used in construction are *polyesters*, *vinylesters*, *epoxies* and *phenolics*; Table 32.2 shows their typical properties. Thermoplastic resins form long-chain molecules without cross-linking when they harden, and heating causes the resin to soften and eventually liquefy. Hence thermoplastic resins can be recovered from the composite and reused. The thermoplastic resins are not used at present in structural applications but are used in geosynthetics.

#### *32.2.3.1 Unsaturated polyester resins*

These are the most common resins in general use. They consist of unsaturated polyester dissolved in styrene. Curing occurs by the

**Table 32.2** Typical manufacturer-quoted properties of the common thermosetting resins

| *Resin* | *Tensile strength (N/mm²)* | *Young's modulus (kN/mm²)* | *Strain at failure (%)* | *Density (kg/m³)* | *Cost (£/kg)* |
|---|---|---|---|---|---|
| Polyester | 50–75 | 3.1–4.6 | 1.0–2.5 | 1110–1250 | ~2.5 |
| Vinylester | 90 | 4.0 | 2.5 | 1070 | |
| Epoxy | 60–85 | 2.6–3.8 | 1.5–8.0 | 1110–1200 | ~5–10 |
| Phenolic | 60–80 | 3.0–4.0 | 1.0–1.8 | 1000–1250 | |

polymerisation of the styrene which forms cross-links to unsaturated sites in the polyester molecules. These resins are relatively inexpensive, easy to process and can cure at room temperature. They offer a good balance of mechanical properties with environmental and chemical resistance.

They do have some drawbacks; particularly:

- moderate adhesive properties
- styrene vapour release at cure; a health risk and subject to emission regulations
- curing is strongly exothermic and can be dangerous without careful control
- the resin shrinks on cure by up to 8%.

*32.2.3.2 Vinylester resins*

These are unsaturated esters of epoxy resins, which means they have similar mechanical properties to the epoxies but require similar processing techniques to the polyesters. Their advantage over the polyesters is reduced water absorption and shrinkage together with better chemical, especially alkali, resistance.

*32.2.3.3 Epoxy resins*

Epoxy resins are used in the majority of high-performance composite structures. They are usually supplied as two-part systems consisting of the resin and a hardener. The different epoxies available offer a wide range of properties. They can be cured at room temperature but the higher performance resins have to be cured at elevated temperature. The processing needs more care than with the polyesters but the epoxies have better mechanical properties, much lower shrinkage during cure, and perform better at elevated temperature. They also have excellent environmental and chemical resistance.

*32.2.3.4 Phenolic resins*

The phenolic resins have useful flame-retardant properties, low smoke generation and high temperature resistance. They are expensive compared to the other resins and are used only in specialised applications.

*32.2.3.5 The effects of temperature*

In general the resins used on site cure at room temperature, although better properties are achieved by curing at elevated temperature, especially under factory conditions. Curing at low temperature is not recommended because it slows the cure rate (many epoxy resins will not cure below 5°C) and also produces a final product with reduced properties since the polymerisation is less complete.

Potentially there is a problem at elevated temperature. When a cured thermosetting resin is heated it displays a mechanical property (tensile strength, for example) against temperature relationship as shown in Figure 32.2. The mechanical property reduces dramatically within a narrow temperature band which corresponds to the resin changing from a glassy solid into a soft viscous material. As shown in the figure the middle of the

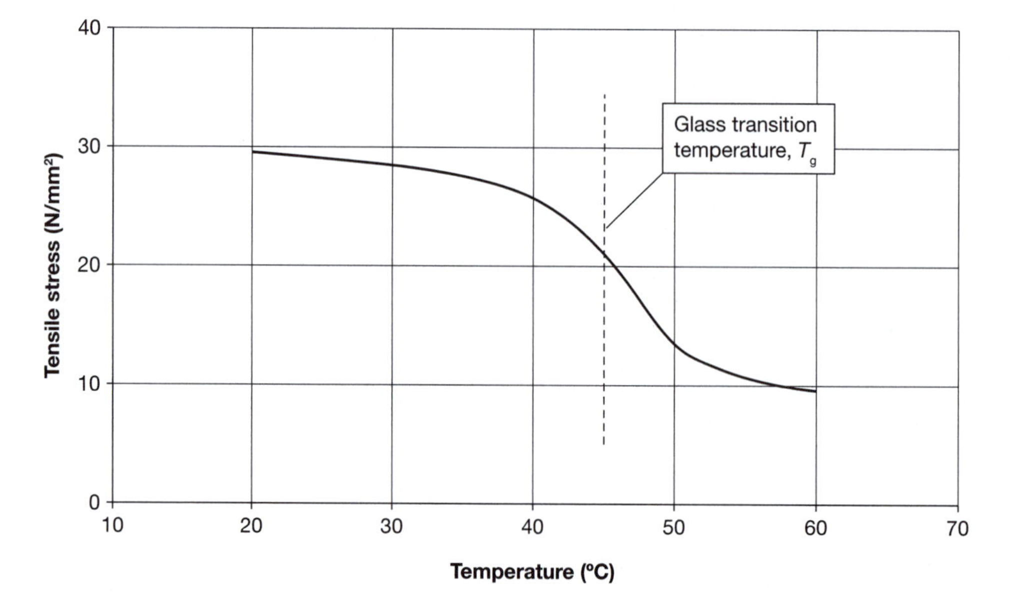


**Figure 32.2** Illustration of the glass transition temperature

temperature range during which this occurs is called the *glass transition temperature*, *Tg*. The ambient curing temperature affects the *Tg* of the cured adhesive for the reasons described in the previous paragraph. Thus a resin cured at an ambient of 10°C will have a lower *Tg* than the same resin cured at an elevated temperature of 40°C (assuming that the resin is formulated for curing at these temperatures). The operating temperature of the composite structure must be carefully assessed because it needs to be significantly lower than the *Tg* of the resin system, a difference of 20°C between the two is generally adequate.

## 32.3 Manufacture of fibre-reinforced composites

The method of manufacture has an effect on the properties of the composite because the process used dictates the proportion of fibres to resin (called the *fibre volume fraction*, $V_f$), the amount of air voids in the material and the overall consistency. A brief overview of the manufacturing processes is needed.

### 32.3.1 Hand lay-up

This is the simplest method. A rigid mould is coated with liquid resin, a layer of dry reinforcement is placed on the resin and then a roller is used to press the reinforcement into the resin until it is soaked (*wetted out*) with resin. This process is repeated until the required number of reinforcement layers is achieved. Even with a formal quality control regime, overall quality depends on the expertise of the operatives.

### 32.3.2 Vacuum bag moulding

This is similar to hand lay-up but with an extra stage. After all layers of reinforcement have been laid down, the component is sealed with a vacuum bag and a vacuum is drawn through the bag. The pressure difference compresses the uncured resin, squeezing out excess resin and entrapped air, producing a component with a higher proportion of reinforcement and better consistency.

### 32.3.3 Resin infusion

Dry reinforcement layers are placed in a rigid mould and covered with a vacuum bag, which is sealed around the edges. A vacuum pump is used to suck the air out of the system, compressing the reinforcement, and then to suck resin through the reinforcement. The resulting component has a low void content and a high fibre volume fraction. The process can be used to produce very large components, using multiple resin inlets and outlets. Since it is a sealed system, chemical emissions are greatly reduced. The RIFT (resin infusion under flexible tooling) process is a commercially proven system for the construction industry.

### 32.3.4 Resin transfer moulding (RTM)

This is similar to resin infusion but is more suitable for large numbers of smaller components. A closed metal mould is used and the dry reinforcement is placed inside. Resin is injected under positive pressure into the mould, sometimes with vacuum assistance (VARTM). The mould can be heated to reduce cure time.

### 32.3.5 Pre-preg

Unidirectional fibres are pre-impregnated with a resin (giving the process its name), usually in a sheet form. The resin cure can be delayed by storing the partially cured sheets in a freezer. When used the individual sheets are laid up in the required orientations as needed. The component is then cured in an autoclave. The process produces the highest quality composites. A variation used on site is the REPLARK system which has been likened to wallpapering. Pre-preg carbon fibres are applied to an existing structure using epoxy resin and the heat given off as the epoxy cures is enough to cure the resin in the pre-preg.

Pre-preg produces custom-made plates and components which can have tapered layers of reinforcement at each end of the plate. This has benefits in reducing adhesive stresses when the plates are bonded to existing structures. Pre-preg is the preferred method for the more expensive carbon fibres.

### 32.3.6 Pultrusion

This is an automated process which produces components of constant cross-section and lay-up. Fibres are passed through a resin bath and then into a heated die where the resin cures. The solidifying product is forced through the die by a mixture of pulling and extrusion. The fibre distribution cannot be changed and the majority of fibres have to run along the length of the component. However, the latest production lines allow more complex fibre distributions. The quality of the composite produced is high.

Since pultrusion is a continuous process it is more suited to mass production rather than custom-made components and it is not possible to taper plate ends.

### 32.3.7 Fibre winding

Another automated process which produces items with rotational symmetry. The fibres are impregnated with resin and wound onto a rotating mandrel. Fibre placement is controlled by a computer-controlled arm which moves along the rotating mandrel. In this way the fibre angle can be controlled. Once the full lay-up has been achieved the component is cured in an oven.

The manufacturing process to be used will depend on the application. In construction a major application of composites is upgrading of existing structures using plate bonding. In general the composite plates can be preformed using the pre-preg or pultrusion processes, but if the surface of the existing structure is very poor the resin infusion method would be better. For new components the appropriate method would depend on the size and shape of the component. Typical fibre volume fraction, $V_f$, and void content for the different methods are given in Table 32.3.

## 32.4 Properties of resin fibre composites

### 32.4.1 Short-term physical properties

The best way to find the mechanical properties of a composite is by testing representative samples, but for unidirectional fibres the properties can be estimated using the *Rule of Mixtures* (Callister, 1999). The resulting equation for Young's modulus in the longitudinal (0° or fibre) direction is:

$$E_{cl} = E_m(1 - V_f) + E_f V_f \qquad (1)$$

**Table 32.3** Fibre volume fraction, $V_f$, and void content for various manufacturing methods

| *Method* | $V_f$ | *Void content* |
|---|---|---|
| Hand lay-up | 0.40 | Up to 5% |
| Vacuum bag moulding | 0.45 | |
| Resin infusion | 0.54 | <1% |
| Pre-preg | 0.60 | 1–2% |
| Pultrusion | 0.60 | |

**Table 32.4** Typical properties of composites produced by vacuum infusion (ICE, 2001)

| *Fibre* | *Reinforcement* | *Longitudinal (0°)* | | | *Transverse (90°)* | | | |
|---|---|---|---|---|---|---|---|---|
| | | *TS (N/mm²)* | *CS (N/mm²)* | *Ect (kN/mm²)* | *TS (N/mm²)* | *CS (N/mm²)* | *$E_{ct}$ (kN/mm²)* | *Density (kg/m³)* |
| Aramid | Unidirectional | 1280 | 290 | 70 | 39 | 150 | 6 | 1370 |
| E-glass | Unidirectional | 700 | 580 | 39 | 72 | 85 | 9 | 1920 |
| E-glass | Quadraxial | 380 | 312 | 23 | 383 | 324 | 23 | 1920 |
| HS-carbon | Unidirectional | 2003 | 797 | 112 | 50 | 85 | 7 | 1510 |
| HS-carbon | Quadraxial | 596 | 420 | 48 | 632 | 404 | 49 | 1510 |
| UHM-carbon | Unidirectional | 1157 | 346 | 310 | 24 | 75 | 6 | 1660 |

TS, tensile strength; CS, compressive strength; HS, high strength; UHM, ultra-high modulus.

and in the transverse (90°) direction:

$$\frac{1}{E_{ct}} = \frac{(1-V_f)}{E_m} + \frac{V_f}{E_f} \qquad (2)$$

where $E$ is Young's modulus, subscripts $c$, $m$ and $f$ refer to the composite, matrix and fibre respectively, and subscripts $l$ and $t$ refer to the longitudinal and transverse directions. There are similar expressions for tensile strength. Table 32.4, based on test results, gives some typical properties. It can be seen that compressive strength is lower than tensile strength and that there is a big difference in longitudinal and transverse properties with unidirectional reinforcement. Figure 32.3 shows elastic modulus values in different directions for various types of reinforcement. It is obvious why the quadraxial fabrics are often called 'quasi-isotropic'.

Since it is not possible or desirable to always obtain properties from an extensive test programme, it is possible to predict properties in different directions using the rule of mixtures and *classical laminate theory* (CLT) (Jones, 1998).

The thermal properties of composite materials can be unusual. Table 32.5 shows the *coefficient of thermal expansion* (CTE) for various fibres, in the longitudinal or warp (0°) and transverse or weft (90°) directions. It can be seen that unusually UHM carbon fibres shorten when heated due to the highly directional properties of the fibres. The fibres are very soft in the radial direction and expand a great amount radially when heated, so that to preserve constant volume they have to contract in the longitudinal direction. The CTE for epoxy resin is typically $60 \times 10^{-6}$ °C$^{-1}$. The CTE for the composite can be calculated from equations based on the rule of mixtures (ICE, 2001). The combination of the fibres and reinforcement mitigates the unusual thermal properties of the carbon fibres, but differential thermal effects in composites need to be considered carefully at the design stage.

### 32.4.2 Long-term behaviour

*32.4.1.1 Durability*

Durability is the ability of the material to meet its performance specification over time and will be affected by the environments, such as water or chemicals, dynamic loading and fire, to which the material is exposed.

*32.4.1.2 Chemical exposure*

Water is probably the most aggressive environment for composites. Its greatest effect is on the interface between the fibres and matrix because it can be absorbed along the interface, reducing the bond

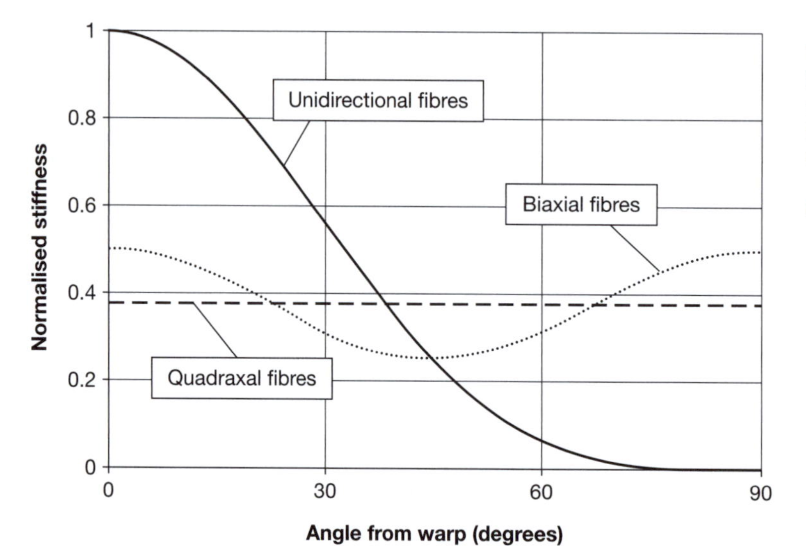


The curves are based on the same fibre volume fraction in each lay-up and have been normalised with respect to the longitudinal stiffness of the UD fibres.

After CIRIA (2004)

**Figure 32.3** Variation of Young's modulus with direction

**Table 32.5** Coefficient of thermal expansion for various fibres (ICE, 2001)

| *Fibre* | *Fabric* | *Fibre angles* | *CTE at 0° (warp) (× $10^{-6}$ °$C^{-1}$)* | *CTE at 90° (weft) (× $10^{-6}$ °$C^{-1}$)* |
|---|---|---|---|---|
| E-glass | Unidirectional | 0° (warp direction) | 8.4 | 20.91 |
| | Quadraxial | 0°, +45°, –45°, 90° | 11.0 | 11.0 |
| HS carbon | Unidirectional | 0° (warp direction) | 0.52 | 25.5 |
| | Quadraxial | 0°, +45°, –45°, 90° | 9.76 | 9.76 |
| UHM carbon | Unidirectional | 0° (warp direction) | –0.07 | 25.68 |
| | Quadraxial | 0°, +45°, –45°, 90° | 9.6 | 9.6 |

between the fibres and the resin. Bulk resins absorb only small amounts of moisture. The interface degradation reduces the strength of the composite but has negligible effect on the stiffness, which is a fibre-dominated property. GRPs show a loss of strength on exposure to water (up to 50% in particularly aggressive situations) but the rate of loss depends on the absorption of moisture, which is affected by temperature, relative humidity and area of material exposed. (Surprisingly this is not a problem in the maritime industry, because stiffness rather than strength drives the design of GRP structures.) CFRPs and AFRPs retain their properties far better during service, usually above 80% of short-term properties, even in the most aggressive environments.

Certain glass fibres are affected by alkaline conditions (pH greater than 11), but not acid. Carbon and aramid fibres are very resistant to both acid and alkali. Diesel and many complex organic pollutants do not affect the composites; their molecules are too large to penetrate the composite. The solution with glass fibres in alkaline conditions, for example reinforcement in concrete, is to use the specially formulated alkali-resistant fibres.

#### *32.4.1.3 Ultraviolet light*

Glass and carbon fibres are unaffected by ultraviolet light but aramid fibres change colour and lose strength. However, when the aramid fibres are embedded in a resin matrix the degradation only occurs close to the exposed surface and the overall mechanical properties are almost unaffected. Direct exposure to sunlight can cause embrittlement of all the resins but this can be prevented by applying a protective coat of paint to the exposed resin.

#### *32.4.1.4 Fatigue resistance*

FRPs are generally more resistant to fatigue loading than metals, but fatigue damage does occur. The mechanisms of fatigue damage are complex in composites but they involve fibre breakage (when the loading is in the direction of the fibres), matrix cracking and shear failure at the fibre/matrix interface. When there are fibres inclined to the fatigue loading the damage is diffused over a large area. The fatigue mechanism involves the accumulation of non-critical cracks within the material, effectively the fibres act as 'crack stoppers' delaying the development of critical cracks. Carbon fibre composites show the greatest resistance to fatigue. As a rule of thumb, if the fatigue loads in carbon fibre composites are less than 50% of the long-term static strength fatigue is unlikely to be a problem, while glass and aramid composites show a greater reduction in properties

#### *32.4.1.5 Fire*

Carbon fibres oxidise in air above 650°C, while glass fibres retain strength up to high temperatures (over 850°C), but neither supports combustion. Aramid fibres cannot be used at temperatures above 200°C. Epoxy- and polyester-based resin materials are flammable, but glass–polyester composites are commonly used in the offshore industry as dedicated fire protection. The degradation of the composite starts at the exposed surface and then slowly proceeds through the thickness. The charred combustion products slow down the spread of damage, providing excellent fire protection to the rear face of the composite. However, the polyesters do emit noxious fumes during the combustion process. Phenolic resins emit less toxic and much smaller amounts of smoke and also display slower flame spread; consequently they are commonly selected in fire-critical situations. Conventional fire protection systems for fibre composites, such as intumescent coatings or insulated panels, are commercially available.

## 32.5 Design considerations

There are design standards, such as the British Standards and the Eurocodes for steel, concrete and timber structures. Similar standards do not exist at present for the FRPs but design guides have been produced: Eurocomp (1996), by the Concrete Society (2003, 2004), the Institution of Civil Engineers (ICE) (2001) and CIRIA (2004). Eurocomp is applicable to new structures, but the others have concentrated on the strengthening of existing structures. Similar guides have been produced in Sweden (Tälsten, 2003), Australia (Oehlers et al., 2005) and the USA (ASCE, 1984).

The UK guides are based on the limit state design philosophy, which is now commonly accepted in the UK, Europe and to some extent the USA. Limit state design represents the uncertainties in loading and material properties by statistical normal distributions from which characteristic loads and properties are calculated; the characteristic loads are multiplied by loading partial safety factors and the characteristic properties are divided by material partial safety factors and then used to find the resistance of the structure. The design is made such that the resistance of the structure is greater than the factored characteristic loads. Statistically this is equivalent to the overlapping region of the two normal curves, which represents when failure would occur, being so small that the probability of failure is acceptably remote, as shown in Figure 32.4.

The approach to loading is identical to that for other materials and does not require elaboration. The suggested approach to material properties is based on the real behaviour of fibre-reinforced composites and allows the designer to use informed engineering judgements about the material. Characteristic properties can be found in two ways. The simple approach would be to use manufacturers' values, which are likely to be conservative; the more sophisticated and usually adopted approach is to conduct a test programme using possible combinations of fibres, resins and the appropriate method of fabrication. This has the added benefit of confirming the suitability of the constituent materials and the fabrication route.

The final material partial safety factor is found by combining a series of factors; these allow for the variability of the material (GFRP is inherently more variable than CFRP), the quality resulting from the method of fabrication, and to account for long-term

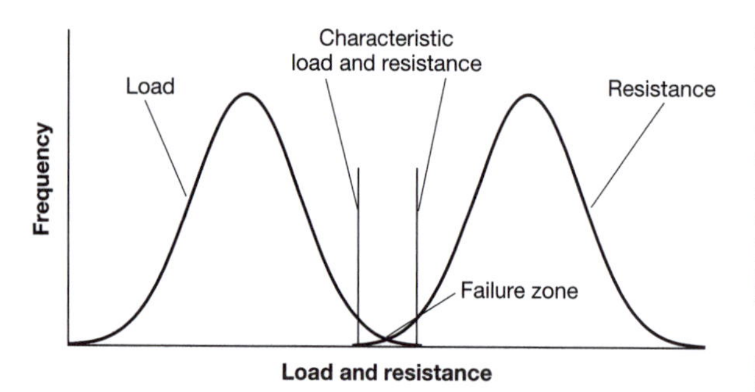


**Figure 32.4** Illustration of the basis of limit state design

degradation from moisture, chemical exposure, UV light, fatigue and creep. Engineering judgements can be made about the levels of long-term exposure, so that a designer confident in the use of FRPs can be less conservative than a designer using them for the first time. At present the final material partial safety factors for composites are quite large, certainly greater than those for steel and concrete. This is because there is only 20 to 30 years' experience of using composites in construction, and much of that with non-structural or non-demanding structural application; as they are used more and in more onerous situations and experience is gained of use over longer periods of time, these factors will come down.

## 32.6 Uses of fibre-reinforced composites in construction

FRPs have proved to be versatile materials in construction. General applications include roofing, cladding, tanks, pipes, trench covers, walkways and access ladders. Strengthening and rehabilitation of existing structures, reinforcement for concrete, cables and ropes and geotechnical applications are all relatively common, and now they are being used for complete structures. Recent developments include motorway gantries and structures for wind generation of electricity. Each will be considered in turn.

### 32.6.1 General applications

Glass-fibre reinforced polymer composites have been used for a wide variety of applications within the construction industry. Cladding and roofing panels are widely available and have shown excellent durability. Examples (Halliwell, 2004), include a naval base in Cornwall, dating from the mid-1970s; Mondial House, London, completed in 1974 and clad with GFRP sandwich panels; a storage building completed in 1968, with a folded plate design and using GFRP wall units and GFRP translucent roof units which are still performing well; and a swimming pool at Southampton University dating from 1970 that has a GFRP roof, which is performing well in contrast to its steel frame, which is corroding badly. GFRP translucent roof panels, with UV protecting additives, are used widely in sports stadia, cutting glare from the sun. GFRP cladding with special surface finishes that can be cleaned easily is used in areas where cleanliness and hygiene are vital, such as operating theatres and cold storage facilities.

GFRP pipes are used widely in the water and wastewater industries and in gas pipelines. They are produced by the filament winding process using double helix filaments to produce strong and durable products. The helix angle can be varied from layer to layer to produce pipes for specific applications. Standard internal diameters up to 3 m and pressure ratings up to 25 bar are readily available. The use of chemically resistant gel-coats ensures excellent durability even in aggressive environments; indeed, the earliest applications of GFRP pipework were in chemical plants. A further application in this area is in storage tanks. Repairs to existing pipes can be carried out using GFRP liners that are placed in the pipe, expanded to fit and cured using hot water pumped through the pipe. This trenchless technology avoids widespread disruption.

Open GFRP tanks, consisting of approximately 1 m square panels bolted together, are used in building and industrial applications. They are delivered piecemeal to site and erected in situ. Flat GFRP panels, capable of withstanding heavy vehicle loads, are used as trench covers by the utility companies. Handrail, ladder and safety cage systems, often manufactured by pultrusion, are found in corrosive environments such as sewage works. If used as escape routes they can be treated to give enhanced fire resistance.

### 32.6.2 Strengthening and rehabilitation of existing structures

Much of the UK's infrastructure was constructed in the Victorian era, and it was completed, along with that of many other developed countries, in the first half of the twentieth century. After the Second World War the need for rapid reconstruction often produced structures with limited durability and, as recent events have shown, there are still many inadequate structures in the world's seismic zones. There is an urgent need for strengthening, repair and retrofit for much of this infrastructure. Incidentally, strengthening, repair and retrofit are different remedial activities; strengthening means an enhancement of structural performance above that originally designed, repair indicates fixing a structural or functional deficiency, and retrofit originally referred to seismic upgrading, but it is often used to cover all three activities.

Although FRP composites are more expensive on a purely weight-for-weight comparison than conventional materials, they have advantages that make them the material of choice in the right circumstances. These arise from the very high strength and low weight of the composites, so that significantly less weight of material is needed to achieve the same effect. The knock-on effect is that installation is easier, requiring less falsework, smaller plant and fewer people to carry out an installation. The FRP is adhesively bonded to the existing structure, the low weight of the composite needing minimal propping during adhesive cure. This usually requires the composite to be prefabricated, but if necessary the fabrication can be carried out in situ using the adhesive properties of the resin to give the bond and easing problems of access. It is important to emphasise that composites are not the universal solution and costings for different situations need to be examined. There are particular considerations for strengthening different materials and these will be examined separately.

A range of adhesives is available for use with FRPs, and the properties of each will be unique. The usual adhesive for strengthening would be a two-part epoxy specially formulated to have adequate shear strength, the ability to fill gaps resulting from imperfections, ambient temperature cure and a glass transition temperature suitable for the structural application (IStructE, 1999). A practical consideration is the need to protect the composite repair from vandalism, whether the deliberate, mischievous variety or the unintentional vandalism resulting from poorly planned work on the structure.

#### *32.6.2.1 Concrete structures*

Flexural strengthening of reinforced concrete is achieved by bonding FRP laminates to the tension face of the member, which can be the top or bottom surface depending on whether it is sagging or hogging bending. The FRP is preformed as a plate and bonded directly to the tension face or the wet lay-up process is used to build up the FRP on the concrete surface. A recent development is near-surface mounted (NSM) reinforcement, which consists of rectangular (preferred) or circular rods glued into grooves cut into the tension face. NSM reinforcement provides better bond between the composite and the concrete, and is more vandal-proof.

Flexural strengthening is designed on the basis of full bond (strain compatibility) between the FRP and the concrete. Care needs to be taken to ensure that the existing steel reinforcement will yield before the concrete fails in compression and the FRP ruptures in tension. Since FRP is linear elastic almost to failure, the moment capacity calculations are based on linear elastic methods – a marked difference from conventional reinforced concrete design. As illustrated in Figure 32.5, there are four possible mechanisms of separation of the FRP from the concrete and each must be checked. Details of the recommended design procedures are given in the Concrete Society Design Guide (2004). Similar recommendations are also given for NSM reinforcement, but there are extra bond failure mechanisms to check. An alternative is to prestress the FRP plate prior to bonding. This has the benefit of relieving existing stresses in the concrete and steel reinforcement when the prestress is transferred to the concrete. Details of the process are given in Hollaway and Head (2001).

It is desirable to use FRP laminates that are as wide and thin as possible. This gives the largest interface between the FRP and the concrete, minimising the bond stress. Since the FRP strengthening systems are generally available as pultruded plates less than 2 mm thick or fabrics with effective thickness up to 0.3 mm, thicker laminates have to be made up by stacking. Inter-laminar shear stresses, which have the potential to cause failure within the FRP, will develop in the adhesive between the layers.

Vertical shear distress in steel reinforced concrete beams is characterised by inclined cracks in the vertical sides of the member. These cracks will form when the principal tensile stress in the concrete becomes too large. Unidirectional FRP placed at right angles to the line of the cracks would be the ideal way to implement shear strengthening, except that it is not possible to predict exactly the crack position and inclination and it is not practical to place FRP in this way. The simplest approach is to apply FRP layers only to the vertical sides of the member, an improved result is obtained by carrying the FRP layers across the tension face (U-wrapping), but fully wrapping the member has the greatest effect. The spacing between the strips has to be carefully controlled. Figure 32.6 shows schematic views of the various approaches to flexural and shear strengthening.

When a concrete cylinder is compressed it shortens but its circumference expands due to the Poisson effect; the compressive strength is determined by the bursting strength of the concrete. Wrapping FRP around the cylinder limits the hoop expansion because the elastic modulus and tensile strength of the FRP are significantly greater than those of the concrete. This confinement of the concrete delays the onset of bursting, producing significant increases in axial load capacity and large increases in deformation capacity (ductility). This is the principle behind FRP wrapping of columns. The wrapping also increases the shear capacity of the member, and if axial fibres are bonded to the concrete inside the FRP wrapping there will also be an increase in flexural strength. The ductility increase is particularly useful in seismic retrofit, which is often done by placing a preformed FRP shell around the concrete and grouting the annulus between the shell and the concrete. These benefits are not limited to circular sections. Research has shown that the core of rectangular sections is confined by compressive forces developed along the diagonals.

#### *32.6.2.2 Masonry structures*

Masonry infill panels are often used to provide horizontal stiffness to structures and they often need to be retrofitted when the structure is assessed for seismic response. FRP composite strips applied diagonally to the surface of the masonry provide additional shear resistance to the infill panel – the truss analogy can be used to assess the amount of FRP required. Both GFRP and CFRP can be used in this manner.

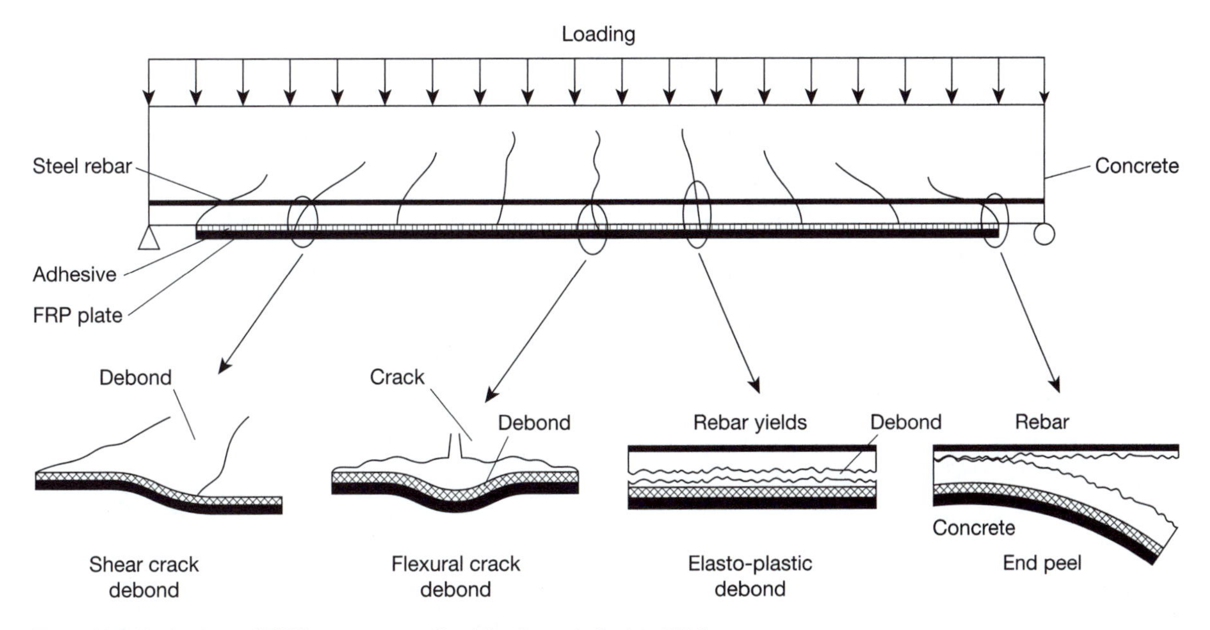


**Figure 32.5** Mechanisms of FRP/concrete separation (after Concrete Society, 2004)

**Figure 32.6** Schematic views of reinforced concrete strengthening (by permission of Sika Ltd)

*32.6.2.3 Metallic structures*

Modern metallic structures are made from steel, generally the standard rolled sections. However, the UK infrastructure, and particularly the railways, has significant numbers of cast and wrought iron structures. FRPs are being used to reinforce steel and cast iron structures and research is being carried out into strengthening wrought iron. Metals are stiff materials and economically ultra-high modulus (UHM) carbon fibre reinforcement in an epoxy resin is the best choice of strengthening material, because it has an elastic modulus at least 50% greater than that of the metal.

Structural upgrade is achieved by bonding CFRP plates to the metallic substrate or by using the resin infusion technique and relying on the adhesive properties of the epoxy resin. Plate bonding has been used to upgrade steel bridges although the rolling imperfections require a relatively thick adhesive layer. Resin infusion has been used on cast iron struts where the poor surface and cooling imperfections made plate bonding impossible.

Wrought iron has not to date been strengthened with fibre composites. It has a lamellar structure because of the production process and there have been concerns that delamination may occur in the iron. Recent research (Moy and Clarke, 2005) has confirmed that delamination of the wrought iron does occur at stresses above the yield stress, especially if the metal is in compression. However, it is felt that FRP strengthening could be carried out subject to certain precautions because the interlaminar shear strength of the wrought iron is greater than the shear strength of the adhesives used.

The FRP is usually applied to the tension face of the metallic structure and has a significant effect on the strength and stiffness. For example, a 7.6 mm thick × 76 mm wide UHM CFRP plate bonded to the tension flange of a 127 × 76 UB 13 would increase the bending strength by 30% and the bending stiffness by 56%. These properties can be calculated using the effective section (modular ratio) method. However, the CFRP would pull the neutral axis of the reinforced section down towards the tension flange, making the compression flange more susceptible to buckling. It is essential that the full effects of the strengthening are checked at the design stage.

The stresses in the adhesive layer can be significant when metallic structures are strengthened. At the ends of the CFRP plate there are high shear and normal (or peel) stresses in the adhesive due to the complex behaviour as the CFRP plate takes up stress. During the construction stage it is important to ensure that the best possible bond is developed between the CFRP and the metal. It is essential that the metal substrate is prepared carefully; the surface must be grit blasted (the SA2½ industry standard is usually specified), pits must be filled with epoxy putty and imperfections ground flat. The blasted surface is chemically active and needs either to be protected using a primer or to have the CFRP applied within four hours. The high adhesive stresses are reduced by tapering the ends of the CFRP plates and by using spew fillets in the adhesive. CFRP, unlike the other FRPs, conducts electricity so there is a theoretical possibility of galvanic corrosion occurring. The adhesive probably provides sufficient insulation but layers of glass fibre have been used as further protection.

### 32.6.3 FRP reinforcement for concrete

In conventional reinforced concrete the bars are steel. The concrete alkalinity usually protects the steel from corrosion, but in aggressive environments where carbonation of the concrete occurs or chloride ions penetrate the concrete this protection will break down. Steel corrosion products produce a volume expansion that eventually spalls the concrete cover off the structure, making it unserviceable. An alternative to steel is to use FRP bars, which have high strength and durability and do not corrode.

FRP bars are manufactured by pultrusion, ideally suited to the unidirectional fibres needed. Alkali-resistant glass fibres are normally used, although hybrid bars, which are a mix of carbon or aramid with glass fibres, have certain advantages. An early problem was developing adequate bond with the concrete, but this was overcome by bonding a layer of sand to the rebar at the end of the pultrusion process and also by producing deformed bars. It is impossible to give definitive properties for the GFRP bars because they vary with manufacturer and also with the size of the rebar. Further details about the use of FRP bars are given in Hollaway and Head (2001). Real applications have been limited by the extra cost of the material compared to steel, but in very aggressive environments the FRP bars offer considerable advantages and overall cost savings. A specialised application is in the foundations of magnetic resonance imaging (MRI) scanners and other magnetic energy instruments because of the magnetic transparency of the GFRP. Similarly GFRP does not conduct electricity and the bars can be used safely in situations where electrical currents could be induced in steel rebar. It is unlikely that GFRP rebar will be used other than in the specialised applications described.

### 32.6.4 FRP ropes and cables

FRPs have been used very successfully in ropes and cables and have significant advantages over steel. As well as the usual advantages in durability and low weight, their excellent fatigue properties are particularly important.

CFRP cables have been used in cable stay bridges and as prestressing tendons. They are fabricated as wires of unidirectional carbon fibres embedded in a polymer matrix and then as assemblies of these wires. It is important that the fibres are kept straight during fabrication to prevent stress concentrations which would reduce the wire strength. Special anchorages have been developed at EMPA (Swiss Federal Laboratories for Materials Testing and Research; Meier and Farshad, 1996), which avoid stress concentrations where the cables enter the anchor and prevent galvanic corrosion.

AFRP rope systems have been around since the 1970s, when they were developed for mooring offshore platforms and for stays to large antennae. These systems carry the trade name PARAFIL™ and are based on Dupont's Kevlar™ fibres. The fibres are unbonded and require purpose-built end terminations. More recent applications have included prestressing cables for concrete and cable stays (Burgoyne, 1987, 1991). They were also used for guys of the Foster-designed Barcelona communication tower.

### 32.6.5 All-FRP composite structures

For many years significant parts of the structure of commercial and military aircraft have been fabricated entirely in composite materials: their low weight, high strength and fatigue resistance and mouldability make them ideal for this application. Similarly, all-GFRP boats and ships have been in service for at least 40 years. The magnetic transparency of the material made it ideal for the mine-hunter *HMS Wilton*, built in 1973.

The construction industry has been slower to realise the potential of the material, but structures made entirely of FRP are now found all over the world. In the UK Maunsell Structural Plastics (sadly no longer in existence) developed the Advanced Composite Structural System (ACCS), which was based on pultruded units commonly called Maunsell planks. These could be simply connected using adhesive to produce a variety of structures. The most striking examples of their application were the Aberfeldy footbridge, Bonds Mill Lift Bridge and the Visitors Centre at the Second Severn Crossing. Maunsell Structural Plastics also developed GFRP bridge enclosure systems which protect the steel girders of steel/concrete composite bridges and also improve their aerodynamic shape. Nine bridges on the approach to the Second Severn Crossing have been treated in this manner.

The consulting engineers Mouchel Parkman (UK) managed a European research consortium that developed the ASSET highway bridge deck, capable of carrying 40 tonne vehicles (Canning et al., 2004). The GRP deck units are pultruded with a complex geometry and sophisticated fibre architecture. They were designed to act compositely with main girders of any material and have been used in an all-FRP West Mill bridge in Oxfordshire. The opening ceremony featured a Sherman tank being repeatedly driven across the bridge. Other deck systems have been developed in the USA (where generous applications of road salt in winter have resulted in serious corrosion problems in conventional decks), Switzerland and Australia.

The low weight of composite material has been used to great advantage in remote areas, utilising helicopters with enough lifting capacity to place a complete bridge. A 17 m span Maunsell ACCS footbridge weighing 1 tonne was lifted into position in Wales. Two bridges were placed across a river at a ski resort near St Moritz, Switzerland. These are removed by helicopter each spring when snow melt carries detritus downstream, and then replaced for the new skiing season.

Composites are now being used in complete buildings such as the Eyecatcher Building in Basle, Switzerland, and in various unusual structures such as the walkway along the Brisbane River in Australia (Van Erp and Ayer, 2004), gantries carrying information signs above roads, windmill blades (up to 60 m long) and other components for wind generation of electricity and beach groynes (Jolly, 2004). The last of these is particularly interesting because the composite material used recycled existing plastic material. On a more mundane level GFRP accommodation units are widely used in industrialised construction of hotels and student accommodation.

### 32.6.6 Geotechnical applications

The geosynthetic materials (such as geotextiles and geomembranes) are FRP composites, with fibres that are often made from polymers. Ground anchors can be made from GFRP, especially for applications in contaminated ground. These materials are used in a wide range of applications such as soil or embankment stabilisation, prevention of erosion and soil consolidation. They are currently being used in tsunami-devastated areas to recover coastal erosion. Some of these specialised applications are discussed further in Hollaway and Head (2001).

## 32.7 Conclusions

This chapter has considered the fibre-reinforced polymer composite materials used in construction. It was shown that these materials consist of two phases: the matrix (resin), which is an organic compound that will polymerise and solidify under appropriate conditions, and the fibre reinforcement that gives the composite its strength. The common resins in construction are polyesters, vinylesters, epoxies and possibly phenolics. The common reinforcements are fibres of glass, aramid or carbon, which can be manufactured in a variety of configurations, such as chopped strand mat, unidirectional fibres and woven or stitched fabrics. Other types of fibre are used in specialised applications. The main features of the materials are high strength (particularly tensile strength), high stiffness of the carbon fibres, low weight and good durability.

There are several manufacturing techniques ranging from hand lay-up to the mechanised processes of vacuum infusion, pultrusion and fibre winding. The manufacturing route influences the properties of the composite formed, the important factors being the volume fraction of the reinforcement and the quantity of entrained air. Also the lay-up affects the mechanical properties; unidirectional fibres result in highly anisotropic behaviour whereas quadraxial fabrics and chopped strand mat produce almost isotropic properties.

Design of FRP composite materials is based on familiar techniques, and design guides for the repair of existing structures are now available. Care needs to be taken with the operating temperature of the structures, thermal stresses (particularly with CFRP), interlaminar shear stresses within the composite and high adhesive stresses at the ends of bonded plates. The limit state philosophy is used with special attention to the material partial safety factors. The construction stage can be a surprise for construction engineers; cleanliness is vital as are very high standards of supervision to ensure the quality of the works. It is recommended that specialist consulting engineers and contractors are employed for this type of work. It is worth mentioning that because the use of FRPs is still in its infancy, careful consideration of in-service monitoring is essential and the guide produced by the Concrete Society (2003) is very helpful in this context.

The uses of FRP composites in construction are manifold: pipelines; storage tanks; cladding and roofing; ladders and walkways; trench covers; the repair, strengthening or retrofit of existing concrete, masonry and metallic structures; new construction ranging from bridge decks and complete bridges to buildings and river walkways. Specialised applications include geotechnics and trenchless repair of pipelines. The examples quoted are only a small selection of the uses of composites in construction. In the total business of the construction industry the FRP composite materials are small players and they will never completely replace steel, concrete and masonry. However, they are important materials and in the right situations they are the best materials to use, and thus should not be ignored.

## 32.8 References


ASCE Manuals and Reports on Engineering Practice No. 63 (1984). *Structural Plastics Design Manual.* Reston, VA: American Society of Civil Engineers.

Burgoyne, C.J. (1987). Structural uses of polyaramid ropes. *Construction and Building Materials*, Vol. 1, 3–13.

Burgoyne, C.J. (1991), *PARAFIL ropes – from development to application.* IABSE Colloquium on New Materials, Cambridge, UK, July.

Callister, W.D. (1999). *Materials Science and Engineering: An Introduction*. Chichester, UK: Wiley.

Canning, L., Luke, S., Täljsten, B. and Brown, P. (2004). *Field Testing and Long Term Monitoring of West Mill Bridge*. Advanced Polymer Composites for Structural Applications in Construction, ACIC 2004, Guildford, UK, April 20–22.

CIRIA Design Guide C595, Cadei, J.M.C., Stratford, T.J., Hollaway, L.C. and Duckett, W.G. (2004). *Strengthening metallic structures using externally bonded fibre-reinforced polymers*. London, UK: CIRIA.

Concrete Society (2003). *Strengthening concrete structures with fibre composite materials: acceptance, inspection and monitoring*. Technical Report 57, Concrete Society, Crowthorne, UK.

Concrete Society (2004). *Design guidance for strengthening concrete structures using fibre composite materials*. Technical Report 55, 2nd edition. Concrete Society, Crowthorne, UK.

Eurocomp Design Code and Handbook, Clarke, J.L. ed. (1996). *Structural Design of Polymer Composites*. London, UK: E. & F.N. Spon.

Halliwell, S. (2004). *Field In-service performance of FRP structures*. Advanced Polymer Composites for Structural Applications in Construction, ACIC 2004, Guildford, UK, April 20–22.

Hollaway, L.C. and Head, P.R. (2001). *Advanced polymer composites and polymers in the civil infrastructure*. Oxford, UK: Elsevier Science.

ICE Design and Practice Guide, Moy, S.S.J. ed. (2001). *FRP Composites: Life Extension and Strengthening of Metallic Structures*. London, UK: Thomas Telford.

IStructE (1996). *A Guide to the Structural Use of Adhesives*. London, UK: SETO.

Jolly, C.K. (2004). *Fibre Composites in Coastal and River Defences*. Advanced Polymer Composites for Structural Applications in Construction, ACIC 2004, Guildford, UK, April 20–22.

Jones, R.M. (1998). *Mechanics of Composite Materials*. 2nd edition. London, UK: Taylor & Francis.

Meier, U. and Farshad, M. (1996). Connecting high performance carbon fibre reinforced polymer cables of suspension and cable stayed bridges through the use of gradient materials. *Journal of Computer Aided Material Design*, 3, 379–384.

Moy, S.S.J. and Clarke, H. (2005). *Assessment of the suitability of wrought iron for strengthening using carbon fibre reinforced polymer composites*. Composites in Construction 2005 – Third International Conference, Lyon, France, July.

Oehlers, D., Seracino, R. and Smith, S. (2005). *Australian guideline for RC structures retrofitted with FRP and metal plates: Beams and slabs*. Composites in Construction 2005 – Third International Conference, Lyon, France, July.

Täljsten, B. (2003). *Strengthening of Existing Concrete Structures: Design Guidelines, 2nd edition*. Division of Structural Engineering, Luleå University of Technology, Sweden.

Van Erp, G. and Ayer, S. (2004). *A Fair Dinkum Approach to Fibre Composites in Civil Engineering*. Advanced Polymer Composites for Structural Applications in Construction, ACIC 2004, Guildford, UK, April 20–22.

33

# Timber and Timber Products

**Andrew Lawrence** MA (Cantab) CEng
MICE MIStructE
Associate Director, Arup

## Contents

## 33.1 Introduction

Timber is a remarkably versatile material. Naturally forming long slender elements ideal for framing, it represents one of our oldest building materials. Originally limited to the size of timber that could be cut from a log, relatively recent developments in processing and adhesive technologies have provided ways to slice, peel or chip the timber and then glue it back together to form larger structural members as well as wood-based panels for a wide range of applications.

In addition to its warm appearance and texture (Figures 33.1, 33.2), timber has many potential advantages. Properly sourced, timber not only is sustainable but is also renewable and has the lowest energy cost of production of any of the building materials. It also locks in carbon dioxide. Timber's cellular nature gives it a strength–weight ratio that is better than mild steel (at least when loaded in its strong direction, parallel to the grain), as well as making it a reasonably good thermal insulant. It is easy to cut or fix on site, when used with a clear finish is cheaper to maintain than white painted surfaces, and it performs well in potentially corrosive environments such as swimming pools.

The purpose of this chapter is to help simplify the potentially daunting field of timber and timber-based products – to explain what is available and its typical characteristics and uses. For more detailed information on a particular subject, the reader is directed to the references at the end of the chapter.

**Figure 33.1** Haberdashers' Hall (London) has consistency because both roof trusses and finishes are surfaced in American white oak (Hopkins Architects, Engineer: Arup). © Arup

**Figure 33.2** Laminated veneer lumber was used to achieve the 600 mm deep members for the 2005 Serpentine Pavilion in London (Architect: Alvaro Siza, Engineer: Arup). © Arup

## 33.2 Properties

### 33.2.1 Basic cell structure

Unlike metals, timber is an anisotropic material, having different properties in different directions. In a very simplistic way, the trunk of a tree is rather like a bundle of drinking straws (Figure 33.3) – cells mainly run parallel to the trunk to support the weight of the tree as well as to carry sap to the leaves. Just as a bundle of straws is strong along its length, but easy to crush or pull apart, so timber is five to ten times stronger parallel to the cells or grain than perpendicular, where it is only weakly held together by the relatively small number of radial cells or rays. It is this anisotropic structure that explains why timber – in its strong direction – has a strength–weight ratio better than mild steel. Perpendicular to the grain, in both compression and tension, timber is notoriously weak. Tension perpendicular to the grain (due, say, to drying shrinkage) can easily lead to fissures.

Growth takes place while environmental conditions are suitable. In softwoods, following the dormant winter season, the *earlywood* is characterised by relatively rapid growth, of a more open texture than the *latewood*. This annual cycle produces a distinctive pattern of growth rings which can be seen most clearly in a transverse section. By comparison in the temperate hardwoods there is often little or no variation between the earlywood and latewood. In tropical areas, growth is more or less continuous but seasonal variations in rainfall may produce a similar, although less pronounced, effect.

The distinction between sapwood and heartwood is discussed in Section 33.2.6.3.

### 33.2.2 Relationship between strength and density

All timber, from whatever species, has the same basic composition – cellulose, hemi-cellulose and lignin. The main variation between species is the density or packing of the cells. There is therefore a reasonably good correlation between density and strength, independent of species. Softwoods or conifers for example generally have a lower density and therefore lower strength than many of the hardwoods.

### 33.2.3 The effect of natural defects on strength

The bending strength of a small piece of timber, free of knots and perfectly aligned with the grain can, in a few species, be as high

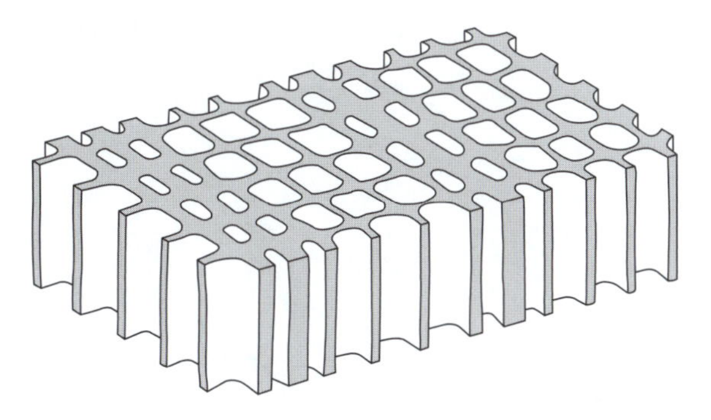

**Figure 33.3** Close up, the cells of a tree resemble a bundle of drinking straws.

as 100 N/mm$^2$. However, in design we are forced to use much lower values to account for the effect of natural *defects.*

The most important defects are knots and slope of grain (Figure 33.4). Knots are the remains of side-branches and disturb the line of the grain, rather as if there were a hole through the timber. A knot cutting through the highly stressed outer tension fibres of a beam could bring a significant reduction in strength. Slope of grain results from cutting a straight member from a bowed trunk and leads to a strength reduction because of the lower strength of the timber perpendicular to the grain. This effect was obviated in medieval cruck frames (Figure 33.5) where the timber was cut from a trunk and main branch, following the line of the grain.

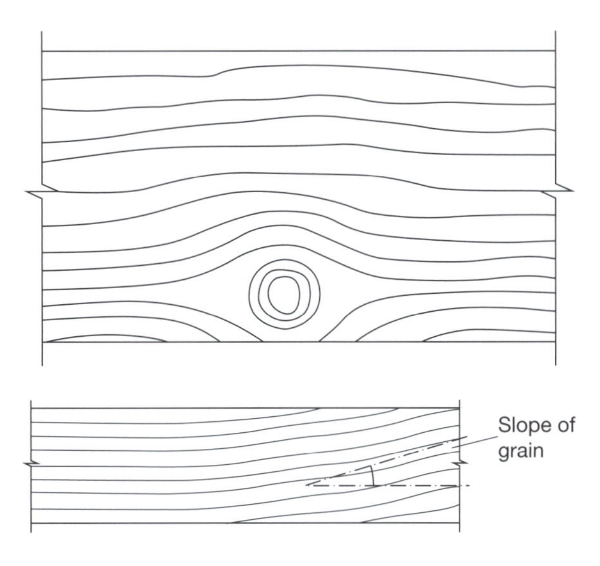


**Figure 33.4** (a) Knots are the remains of branches. *Marginal knots*, which cut through the most highly stressed bending fibres, will have a major effect on bending capacity. (b) Because timber is weaker across the grain, any slope of grain will considerably reduce the axial strength of a member.

### 33.2.4 Drying shrinkage

Not only is timber weaker across the grain, but it also tends to shrink and swell with varying moisture content. A living tree contains nearly 50% water, held partly as sap within the cell cavities and partly in the cell walls. After cutting, water is lost first from the cavities and then from the walls. As the walls dry they shrink in diameter, leading to shrinkage across the grain of up to 7% depending on species. Shrinkage parallel to the grain is almost entirely restrained by the cell fibres. A small number of radial cells also help to reduce the radial shrinkage to about half the tangential value.

**Figure 33.5** Nearly all our surviving medieval frames are in oak. The *cruck* frame generally comprised a trunk and main branch, halved and propped together. © Arup

This non-uniform shrinkage causes distortion on drying (Figure 33.6): the distortion of the square piece into a diamond demonstrates the different rates of radial and tangential shrinkage; floor boards that are quarter- or rift-sawn will stay flatter than boards that are plain-sawn. As an aid to predicting the distortion, imagine that as a piece dries 'the growth rings tend to straighten out'. This can sometimes lead to deep fissures, particularly in larger sections containing the boxed heart of the tree (Figure 33.7). However, the only case when drying fissures will occur through the full depth of the piece (when they are known as splits) is at each end – due to the more rapid loss of moisture from the end-grain, the whole cross-section will try to shrink against the restraint provided by the body of the piece.

To limit shrinkage after installation, we try to use/specify dry timber, with a *moisture content* as close as possible to the in-service moisture content. Moisture content is defined as the weight of the contained water divided by the weight of the dry timber, and can vary from 10–12% in an internal heated environment to 16–18% at an external covered location or above 20% if the timber is exposed to direct wetting. In the past, timber was left to dry naturally. This process, called *seasoning*, could take several years depending on the thickness of the piece. Today, most softwood is kiln-dried in a matter of days, depending on the density and thickness of the member. Hardwoods are generally denser and can take considerably longer to dry, to the point where this dominates the supply cost of dry (i.e. kiln-dried or seasoned) timber. Thus *green* or unseasoned oak might cost only 30–50% of the seasoned material.

### 33.2.5 Seasonal movement

Shrinkage is partly reversible, so that timbers will tend to swell in humid winter conditions and shrink back in the summer. This effect, called seasonal *movement*, can cause the timber to shrink or swell by up to 1%, depending on species and location. Though small, this is enough to cause doors to stick or glued members to delaminate (because of the differential movements of the timber either side of the glue line). Thus for exterior joinery, say, it is important to select a *small movement* timber (1% change in section for every 5% change in moisture content) such as a softwood, rather than say oak, which is *medium movement* (1% change in section for every 4% change in moisture content). To allow for movement in a solid timber floor, intermittent or edge expansion gaps should be provided.

By comparison with its movement characteristics, the coefficient of *thermal expansion* of timber is only one third that of steel or concrete, and is reduced by the drying shrinkage that the increase in temperature will cause. For all intents and purposes thermal expansion of timber may therefore be ignored.

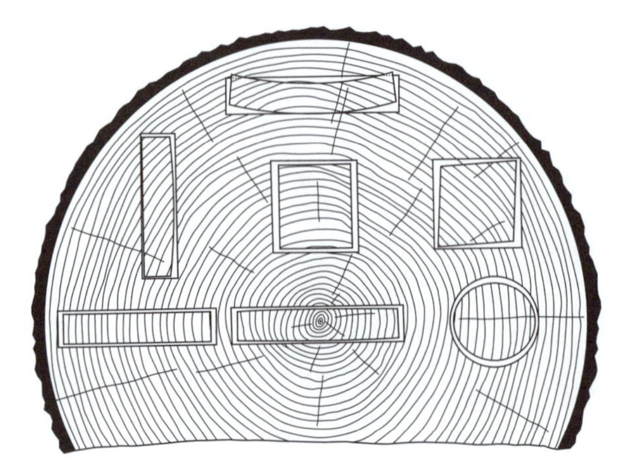

**Figure 33.6** Non-uniform shrinkage across the grain leads to distortion on drying.

**Figure 33.7** At Darwin College, Cambridge, the oak columns were installed green and allowed to season in situ. The fissures are the result of the greater tangential shrinkage (for oak 8%, compared to only 4% radially) (Architect: Dixon Jones, Engineer: Arup). © Arup

### 33.2.6 Durability

#### *33.2.6.1 The risk of fungal decay*

Fungal attack (rot) can only occur in wet timber where the moisture content exceeds about 20%. Even outside, so long as the timber is under cover, the moisture content will not exceed about 16%, so rot can only occur where the timber is directly exposed to rain or condensation, or where it is in contact with wet ground. Dry rot can admittedly be more of a problem – although it can only take hold on wet timber, it is then able to conduct its own water through long shoots to other, drier, parts of the structure. Very limited areas of the UK, as identified in the Building Regulations, are also at risk from house longhorn beetle.

Timbers liable to remain wet for long periods are the most prone to decay. Outside, this might imply shaded horizontal surfaces or timber that has been bedded into the ground, rather than raised up on a plinth. The end-grain, which soaks up water like a straw (the drinking straw analogy again), stays wet longer and is always particularly vulnerable. Inside, the most common causes of rot are leaking roofs, often compounded by condensation within the roof space, particularly where there is poor ventilation, or where the timbers are built into a damp external wall. So long as these risks are avoided, any timber can be used inside without

**Figure 33.8** Timber is ideal for a pool environment, since the moisture content of the timber will be below the threshold for decay. Whereas steel would rust, timber can offer a durable and maintenance-free solution. © Arup

risk of decay. In a swimming pool environment, for example, the combination of temperature and humidity required to avoid condensation will also mean that the moisture content of the timber never exceeds about 16%, usually making timber a more durable option (Figure 33.8) than steel or concrete in this potentially corrosive environment.

### *33.2.6.2 The risk of insect attack*

In temperate climates, where the timber remains in a dry condition, insect attack is generally not significant. The main exception is house longhorn beetle, which can survive even within relatively dry internal building timbers. Fortunately it is only found in limited parts of southern England, as specified in the UK Building Regulations, and preservative treatments of roof timbers in these areas is mandatory.

### *33.2.6.3 Natural durability*

For timber, the term *durability* denotes resistance to fungal or insect attack.

Resistance to fungal attack of various timber species has been measured by 'graveyard' tests of 50 mm square posts driven into the ground. The timber near the ground surface, where there is an adequate supply of both water from the ground and oxygen from the air, is particularly vulnerable to decay, making this a particularly onerous test.

The trunk of a tree consists of a central core of dead material, the *heartwood*, surrounded by the living *sapwood*. The heartwood accumulates extractives, which are toxic to fungal species. The toxicity and thus degree of durability varies between species. The most resistant are some of the tropical hardwoods, many of which are classed (in, for example, EN350 *Durability of wood and wood-based products – Natural durability of solid wood*) as *Very Durable* and even bedded into the ground should last 40 years or more. European oak and Western red cedar are classed as *Durable*. Most of our surviving medieval frames are in oak (Figure 33.5) and both oak and the tropical hardwoods are commonly used for footbridges (Figure 33.9) and lock gates, balconies and decking (Figure 33.10). Western red cedar is chiefly restricted to cladding (Figure 33.11), in view of its softness and relatively low strength. European larch and Douglas fir are both *Moderately Durable*, and therefore suitable for cladding and other locations with only intermittent wetting, but where the water is free to run off allowing the timber to dry fairly rapidly. Most other timbers including the common softwoods (pine and spruce) and most temperate hardwoods (with the exception of oak and chestnut) have little or no natural durability and are not suitable for exterior use without treatment.

When using a naturally durable timber, it is important to specify that *sapwood is excluded* since this contains no extractives and therefore has little or no natural durability (Figure 33.12).

Where there is a risk of decay due to wetting, rather than using a naturally durable timber, it is possible with certain limitations to add artificial toxins called *preservatives*. This is discussed in more detail in Section 33.7. However, the weathering of preserved timber is generally less attractive than that of untreated, naturally durable timbers.

**Figure 33.9** Left untreated, timber that is *fully* exposed to sun and rain will eventually weather to a silver-grey. This bridge is about 15 years old. © Arup

**Figure 33.10** The tropical hardwoods are generally strong and durable, making them ideal for decking. The deck at Chiswick Park uses a timber called Vitex that was available with FSC certification (Architect: Richard Rogers Partnership, Engineer: Arup). © Arup

**Figure 33.11** Western red cedar is a durable timber often used for cladding, and seen here at the new headquarters for a prominent Polish newspaper (Architect: JEMS, Engineer: Arup). © Arup

**Figure 33.12** Even in timbers such as oak with durable heartwood, the sapwood has no resistance to decay, as evidenced by the fungal attack round this oak log. © Arup

Certain durable timbers, notably oak and Western red cedar, are highly acidic, and in wet conditions this will accelerate the corrosion of metal fasteners. In these cases stainless steel should be used in preference to galvanised fixings. Oak and Western red cedar also contain large amounts of tannin, which will slowly wash out, staining adjacent surfaces.

A relatively recent innovation is heat-treated wood, or 'Thermowood'. This is softwood that has been heated to very high temperatures to remove the 'edible' resin and some of the starches, making the timber darker, more dimensionally stable and more durable. However, the timber also becomes more brittle and therefore while it is ideal for cladding, it would not be suitable for structural applications.

#### *33.2.6.4 Design for durability*

Where there is a risk of decay, then as well as using a naturally durable or preserved timber, measures to limit the uptake of water and promote rapid drying will prolong the life of a member. Such measures might include the following.

(a) Roof overhangs to shield vertical surfaces from rain, being aware that this could result in an uneven appearance, due to the uneven exposure to sun and rain. Similarly doors and windows should be set back from the face of the building.
(b) Lifting timber a good 200–250 mm off the ground to prevent 'splash-back'; for example by lifting timber columns up on steel brackets (Figure 33.13).
(c) Protecting exposed end-grain on beams and posts with an end-grain sealer, or better still a sacrificial timber or metal cap.
(d) Providing a slight slope to horizontal surfaces to promote run-off, especially away from joints where water might become trapped. Better still, provide a metal cap to the exposed upper surfaces of members with an air gap between for ventilation. For decking, an adequate slope can be provided by ensuring that the timber cups downwards.
(e) Providing drip grooves under horizontal surfaces and anti-capillary grooves between close surfaces to prevent water being drawn in by capillary action.
(f) By allowing good ventilation to promote drying, especially where there is a risk of condensation such as in roof spaces and floor voids or behind cladding.

**Figure 33.13** External timbers should be lifted well off the ground to help keep them dry and thus prolong their life. © Arup

### 33.2.7 Fire issues

#### *33.2.7.1 Fire resistance*

Timber has a degree of natural resistance to fire. As timber burns it chars and swells. Charcoal is a poor conductor and so insulates the timber behind. Combined with the high thermal insulation characteristics of timber, this ensures that the interior of a member exposed to fire remains cool and structurally sound. The rate of charring is about 0.6 mm/minute (slightly lower for denser timbers), and so to provide 30 minutes' fire resistance, a member would need to be about 20 mm larger on all the exposed faces; in practice the need for an increased member size is often obviated by the lower imposed loads and higher timber strengths that can be taken in the fire condition. Metal fixings also need to be protected, usually with a timber cap. Clearly over-sizing members in this way would be uneconomic for thin domestic joists and wall studs, and so they are instead protected behind a layer of plasterboard.

#### *33.2.7.2 Surface spread of flame*

In larger spaces, exposed timbers are often painted or impregnated with a flame-retardant treatment to provide Class 1 or 0 surface spread of flame resistance (as defined in the UK Building Regulations and BS476, *Fire tests on building materials and structures*). Since only the surface or outer shell of the timber is protected, treatment must be carried out after the members have been cut to final size.

Flame-retardant paints and varnishes often need to be site-applied once the building is weather-tight, since they can be sensitive to moisture. Compatibility with particular timber species and any stains should also be checked, since there is sometimes a risk of discoloration. While some intumesce (i.e. expand under heat, providing an insulating layer), they have no significant effect on the actual fire resistance of a member.

The kiln-drying, required following pressure impregnation with inorganic salt flame retardants, can cause a strength loss of up to 20% and can affect the colour of the timber. Compatibility with any adhesives should be checked with the coating and adhesive manufacturer.

## 33.3 Sources of timber

### 33.3.1 Specifying timber from sustainable sources

Much of Western Europe, including the UK, had already been deforested over 1000 years ago and most timber is therefore imported, although the UK, for example, does now have limited low-grade supplies of plantation-grown softwood timbers – chiefly spruce – as well as oak and some other hardwoods.

Trees can be divided into three distinct families:

- the *softwoods* or conifers such as pine (Figure 33.14) and spruce, which chiefly grow in Northern Europe and Canada, and represent about half the world's forests
- the *temperate* hardwoods of Europe and North America, such as oak and beech
- the diverse family of *tropical hardwoods* found in the rainforests.

Overall the area of softwood and temperate hardwood forest (both predominantly plantation rather than natural forest) is increasing, mainly due to replanting because of the demand for these timbers for paper making and construction. However, the total global forest area is decreasing, due to clearance of the tropical rain-forests for agriculture as well as logging. It is this loss of the tropical forests, bringing with it a spectacular loss of biodiversity, that has focused attention on the sustainability of timber as a resource.

**Figure 33.14** A softwood plantation in Finland. © Arup

It is of course perfectly possible to harvest timber in a sustainable manner, which minimises harm to the forest ecosystem while helping to sustain the local economy. In terms of the rainforests this might imply *selected felling* rather than *clear-felling* of a limited number of older trees, at a rate at which they can be replenished by natural regrowth. Alternatively the tropical hardwoods can be plantation grown, like most of our softwood and temperate hardwood. Organisations such as the Forest Stewardship Council (FSC) have laid down principles for good forest management, against which forests can be independently measured and certified.

In order to ensure that a particular piece of sawn timber has been sourced from a sustainably managed forest, the timber needs to be tracked through the supply chain, referred to as a *chain of custody*. This requires the certified timber to be kept separate from other timber at *every* stage in the chain and again, the FSC and others have laid down principles for supply chain management.

While there are a series of certification schemes for forest management, the principles vary, partly because there is no commonly agreed definition of sustainability; also they do not always incorporate a chain of custody. In response, the UK Government has established the Central Point of Expertise on Timber (CPET), which continuously reviews and compares the various certification schemes for their ability to provide both legal and sustainable supplies of timber. While this is principally intended to help specifiers of timber for public projects, it will help all specifiers reach a view on the effectiveness of the various schemes. While certified supplies of softwoods and temperate hardwoods are relatively easy to obtain, certified supplies of tropical hardwoods remain limited. This will change as developing countries impose tighter forest management practices, but in the meantime specifiers need to allow sufficient lead times and be prepared to accept alternative species depending on availability.

Specifying *clear timber* (free of knots), of whatever species, will generally be a less sustainable as well as an impractical option, since it either entails use of older growth trees that have lost their lower side branches or generates significant wastage unless an alternative use can be found for the large proportion of the timber from a particular tree that will not be knot-free.

Brief mention should be made of the adhesives used in glulam (glue laminated) production. While these are generally derived from petrochemical processes, they make up much less than 1% of the volume of the fabricated element and enable large members to be fabricated from small-diameter plantation-grown timber, avoiding the need to cut large old-growth trees. Although plywood and some other board materials use larger amounts of adhesive, this needs to be weighed against the sustainability of the alternatives such as plastic or metal-based panels, which may well contain higher amounts of petrochemical-derived products or entail larger amounts of energy in their production. Similarly the use of adhesives for glulam beams also enables very large members up to 40 m or more to be fabricated which can compete with steel in terms of spanning capability but entail far less energy in their manufacture.

It is sometimes suggested that reclaimed timber should be specified. While this can offer a more sustainable option in theory, in practice it would be difficult to obtain sufficient material for a contract of any scale, and for the present, it is likely to remain a niche market for floorboards and other domestic elements. Future reuse would at least be aided by using screws rather than nails, to limit damage during dismantling.

### 33.3.2 Softwoods

Nearly all the timber used for construction and joinery is softwood – in the UK, this is principally pine or spruce from Scandinavia and Russia, although a limited amount of fast-grown and therefore relatively low-grade UK timber is also available. Softwood is generally pale and instantly recognisable from the regular pattern of knots (Figure 33.15), remains of the numerous thin side branches. The softwoods or *conifers*, as they are commonly called, grow fast and straight (Figure 33.14), making them relatively cheap and also ideal for cutting into long beams. Although of only modest strength thanks to their low density and

**Figure 33.15** A glulam beam before it has been planed to remove the excess glue. Note how the timbers in each layer are joined end to end using *finger joints*. © Arup

frequent knots, they are adequate for most structural applications (Figure 33.16). European larch, Douglas fir and Western red cedar are sometimes used, instead of pine and spruce, for their resistance to decay, although Western red cedar is really too soft and weak for anything other than cladding (Figure 33.11).

**Figure 33.16** This 100 m span dome (shown here during construction) was built for the Sydney Olympics. Axial compression in the elements is transferred in end-bearing onto steel end plates (Architect: Ancher Mortlock Woolley, Engineer: Arup). © Arup

### 33.3.3 Temperate hardwoods

The *temperate hardwoods* or *deciduous* trees of Europe and North America, such as oak and maple, show much more variation in colour between species, but are generally darker, denser and stronger than the softwoods. They are often beautifully figured and are prized for joinery work (Figure 33.17). This, combined with their relatively slow growth, lack of the conifer's straight trunk, and longer time to kiln-dry, makes them several times more expensive than the softwoods. Correspondingly, they find more applications in joinery and furniture than carpentry, although oak is sometimes used structurally for the combination of its strength, figure and natural resistance to decay (Figure 33.7).

### 33.3.4 Tropical hardwoods

The *tropical hardwoods* are a very diverse grouping. They are generally denser and darker than the softwoods and some, such as teak, have a very beautiful figure as well as a fine grain, making them ideal for high-quality joinery. In consequence they have generally been priced out of the building field, being mainly used for joinery and furniture. However, the tropical hardwoods include some of the strongest and most durable timbers, and they are therefore used for external and particularly marine structures where their natural resistance to decay can be of most value (Figures 13.9 and 13.10).

### 33.3.5 Timber board materials

These are sheet materials manufactured from wood veneers, strands, chips or fibres. Except where noted below, they are generally derived from softwood bonded with non-moisture-resistant glues under heat and pressure. Sheet sizes vary, but are typically 1.2 × 2.4 m.

Plywood and oriented strand board that have been manufactured according to suitable quality control procedures to ensure a consistent strength are suitable for structural applications. Chipboard and fibreboard are relatively weak materials because of the short length and random orientation of the wood fibres; for this reason their use is generally limited to furniture and joinery.

#### *33.3.5.1 Plywood*

Plywoods are manufactured from rotary peeled veneers, with successive veneers oriented at right angles to provide a sheet with dimensional stability and bidirectional bending strength, albeit stronger parallel to the *face grain*.

Structural plywoods are generally manufactured from softwood, although birch and birch-faced plywoods are also available. Although they are bonded with moisture-resistant glues (traditionally referred to as WBP or 'water and boil proof'), the timber itself is not resistant to decay and therefore not suitable for wet exposure for long periods. Unlike non-structural plywoods, *structural plywoods* are manufactured from structurally graded veneers, bonded under rigorous factory conditions, giving them reliable strength properties, derived from testing.

Non-structural plywoods (which may be manufactured from softwood or tropical hardwood veneers) are intended for furniture and joinery applications. These include marine plywoods (comprising durable tropical hardwood veneers bonded with a moisture-resistant glue); veneer plywoods (in which all the veneers are oriented with their plane parallel to the surface of the panel); decorative plywoods (with better grade face veneers sometimes coated with a phenolic film for wear and moisture resistance); and low-cost blockboards (with a central core of wood strips bonded between the outer veneers).

**Figure 33.17** The courtyard roof at Portcullis House (London) uses American white oak, selected for its strength and appearance (Hopkins Architects, Engineer: Arup). © Arup

*33.3.5.2 Oriented strand board (OSB)*

In the UK this is often referred to as 'Sterling board', really a trade name. It is made from large wood flakes arranged in layers, alternately parallel and perpendicular to the face layer to give strength and stiffness in two directions, similar to plywood, but of course weaker because of the shorter fibres.

Structural grades of OSB use moisture-resistant glues and have reliable, albeit limited strengths.

*33.3.5.3 Chipboard*

Like the fibreboards discussed below, chipboards (often called *particleboard*) were originally developed to add value to forest thinnings and sawmill waste. They comprise small wood chips with a resin binder. The resin is generally not moisture-resistant.

*33.3.5.4 Fibreboard*

These are reconstituted from the individual wood fibres. Although they may be bonded using adhesives, often they rely on the inherent adhesive qualities of the fibres themselves, although in this case the lower adhesive content inevitably makes them softer and weaker.

Fibreboards include (a) low-density *softboard* used for thermal and sound insulation, (b) *medium density fibreboard* (MDF) which has a density similar to the softwood from which it was derived, and (c) the higher density and generally thinner *hardboards*.

## 33.4 Structural use of timber

### 33.4.1 Size

*33.4.1.1 Softwood*

As discussed above, most timber used for construction and joinery in the UK is softwood (generally including sap and with mixed species). The predominant species, pine and spruce, grow to a modest size and the dimensions of sawn softwood (kiln-dried to below 20% moisture content) are therefore limited by the 300–600 mm diameter and the taper of a typical log (Figure 33.14). The largest pieces generally available are about 5 m long and 225 mm wide. Larger cross-sections of (French/English grown) oak and (Canadian) Douglas fir can be obtained, perhaps up to 300 or 400 mm square, but only in very limited quantities and these necessarily only partly seasoned. For kiln-dried timber, the maximum thickness is about 75 mm – if much thicker it would split or fissure on drying.

Within the above limitations, kiln-dried sawn timber is available to standard cross-section dimensions. Before use, the timber will generally need to be planed to *finished* sizes, reducing the overall width by about 5 mm for structural timber and 10 mm for joinery, compared with the original sawn dimensions.

*33.4.1.2 Softwood glulam and LVL*

To make larger sections, dried softwood boards can be glued together to make glued laminated (*glulam)* beams (Figure 33.15) or wide boards. Alternatively, the logs can be peeled into thin

**Figure 33.18** Laminated veneer lumber is made from 3 mm thick veneers. © Arup

veneers which are then glued back together to make *laminated veneer lumber* (LVL, Figures 33.2 and 33.18). The additional costs and material wastage associated with their manufacture inevitably double or triple the cost compared with the parent timber.

The size of glulams is limited only by transport, and lengths up to 40 m have been used (Figure 33.8). Widths above 220 mm will rarely be economic, since these will require more than one board to make up the width of any one layer. Additionally the laminates can be easily curved before gluing to create curved members – these will necessarily be fabricated to order and therefore come at a cost premium, particularly for smaller radii which will demand thinner laminates – a radius of about 3 m being the economic limit.

LVL is similar to structural plywood, except that the veneers are generally aligned in one direction to maximise strength as a beam material. In LVL, the short lengths of veneer in each layer are also end-jointed (with a scarf joint), enabling very large and strong beams to be made – currently up to about 2 m wide × 26 m long if required (Figure 33.18).

For most glulam applications, it is usual to specify a planed surface free from glue stains, with exposed knot holes and fissures filled with glued inserts. Where close visual contact is possible, a sanded finish can be specified, usually at additional cost with outer laminations carefully selected to match grain and colour at end joints where practicable, free from loose knots and open knot holes and with reasonable care exercised in matching the direction of grain and colour of glued inserts. Glulam beams under 90 mm wide are usually re-sawn from larger sizes and may show glue stains and minor defects on one side.

#### *33.4.1.3 Hardwood*

Since the volume of hardwood used in the UK is relatively small, merchants only hold limited stocks and availability will depend on species. For the more popular species, seasoned or kiln-dried timber may be available in limited quantities up to 75 × about 200 mm and up to 4 m long. Larger sizes or less popular species may only be available *green* or unseasoned, and only to order. Green timber can also be considerably cheaper because of the time and energy required for kiln-drying of the dense hardwood species.

Although hardwood glulams are occasionally used (Figure 33.17) where requirements of strength, appearance or durability demand, many dense close-grained hardwood species cannot be used because the wood will not 'absorb' the adhesive.

### 33.4.2 Grades

#### *33.4.2.1 The need for grading*

In order to provide a reliable measure of strength, all structural timber must be graded (and marked) to account for strength-reducing defects such as knots and slope of grain. The strength of specific grades is then determined by testing on a representative sample of timbers.

Most grading is still done by eye and referred to as *visual grading*, based on the number and distribution of knots and the slope of grain. Inevitably the grading rules are relatively coarse and can generally only distinguish a high and a low grade, while the remainder of the timber is rejected. Increasingly, for softwoods, *machine grading* is also used, generally relying on the correlation between bending stiffness and strength. This is more accurate than visual grading and can therefore be used to justify higher strengths.

Since, as discussed below, strength is related to moisture content, timber must be graded under similar conditions (kiln-dried to below 20% for internal timber) to those under which it will be used.

#### *33.4.2.2 Softwood grades*

Softwood is usually separated into a high and a low grade, with the remainder being put to non-structural use. For most softwood species, these correspond to the generic *strength classes* C24 and C16 – now common European classifications (as defined in EN338) in which C denotes *conife*r and 16 the ultimate or *characteristic* bending strength in N/mm$^2$. Higher grades up to C30 or C35 can be justified by machine grading, but entail necessarily higher reject rates and therefore a potential cost premium.

Since the common softwoods, such as pine and spruce, have a relatively similar appearance and properties, they are usually specified by grade without reference to a specific species, except where this would be required for durability – such as larch or Douglas fir.

#### *33.4.2.3 Softwood glulam grades*

The great advantage of glulam is that any defects in one laminate will not pass all the way through, creating a member that is about 40% stronger than the parent timber. In LVL, where the laminates are much thinner than in glulam, this averaging effect is even greater and the strength enhancement can be as much as 100%.

The more laminates in a glulam for a given depth, the higher the strength. However, 35–45 mm thick laminates are typically used for reasons of economy, although thinner laminates will be required for tightly curved members.

Glulam can be either *homogeneous* or, again for reasons of economy, *combined*, where a higher grade timber is used for the more highly stressed outer laminates (one-sixth of the depth on both faces). Common European grades (as defined in EN1194) are GL24h and GL32c, where the number again indicates the characteristic bending strength and c/h denote combined/ homogeneous.

Since sawn softwood boards are limited to about 5 m in length, they will need to be finger-jointed (Figure 33.15) to suit the length of the glulam. Where high-grade boards such as C30 or C35 are used for the laminates, the tensile strength of the finger-joints can start to dominate the strength of the member.

Since a moisture content below about 16% is required to ensure penetration of the adhesive, glulams tend to be drier than solid timber at time of installation. While this will limit subsequent shrinkage of the glulam section, large members are still prone to partial depth fissures, as the outside of the member dries and shrinks against the inside.

#### *33.4.2.4 Hardwood grades*

Temperate hardwoods can be obtained in a high and a low grade, while only a single grade is defined for the tropical hardwoods. Depending on species these will correspond to a generic strength class from about D30 (e.g. oak) to D70 (some of the stronger tropical species). Here the D denotes *deciduous,* but is taken to include all hardwoods. Again the number denotes the characteristic bending strength and it can therefore be seen that the hardwoods are in general stronger than the softwoods, due to their higher density and fewer knots. Since most hardwood is destined for non-structural applications, it will usually be structurally graded to order.

### 33.4.3 Member design

Traditional timber design codes, such as BS5268, have been based on permissible rather than limit state design, since the latter is more of an advantage for more plastic materials such as steel and reinforced concrete where a degree of plasticity can be tolerated under ultimate loads. However, newer timber design codes, such as Eurocode 5, also tend to be based on limit state principles.

The grade stresses quoted above are characteristic bending strengths for use in conjunction with Eurocode 5 (EN1995). They indicate the strength of the timber in dry (internal) conditions under short-term loads. To obtain design strengths they need to be divided by a material factor (typically about 1.3 for timber to reflect its greater variability compared with steel), and reduced to account for exposure conditions and load duration.

The trunk of a tree is predominantly in compression, but it has evolved to resist high short-term bending loads generated by the wind. In the same way, a timber beam is strongest and stiffest under shorter term loads such as snow or wind, whereas under sustained loads it will not only tend to creep, but can be up to 40% weaker. The physical reasons for this behaviour are complex and partly related to sliding between the cell fibres.

In strength design, this is accounted for by a *load duration* factor. Other factors account for exposure (in external and therefore damp conditions, timber being 20–40% weaker) and size effects (smaller members being up to 20% stronger in tension). In load-sharing systems, such as floor joists, it is permissible to use a slightly higher stress to account for averaging out of the natural defects in individual members.

In checking deflection, creep under sustained load has conventionally been ignored, on the basis that much of the design load on a domestic joist is only present for part of the time. However, where the dead load is high relative to the live load, or where the floor is used for storage, creep will be more significant and Eurocode 5 provides a way for these effects to be quantified. Creep is only a minor problem for solid timber and glulam, but for composites such as chipboard, the effects can be more pronounced because of creep in the resin.

Other effects that the designer should be aware of are shear deflection (which starts to be significant for beams with plywood webs and for solid members with a span–depth ratio less than about 12) and joint slip (due partly to local crushing of the timber under the fixing, and important in calculating the deflection of assemblies such as trusses).

While detailed member design is beyond the scope of this chapter, it is worth mentioning that the size of a timber member will often be governed by the connections, since it is rarely possible to fit in enough fixings to mobilise the full strength of a member, particularly in bending or tension. Since timber is a relatively low-cost material, it should also be noted that the costs of a timber structure tend to be dominated by the fabrication and assembly cost of the connections, particularly in exposed timber structures incorporating bespoke fabricated steelwork connections.

### 33.4.4 Connection design

Connection detailing can be the most challenging part of timber design and requires a completely different philosophy from other structural materials. Whereas the local stress concentrations in a bolted steelwork connection will be relieved by local plasticity of the metal, timber is more likely to crush or split, particularly when loaded in its weak direction, perpendicular to the grain.

Generically, there are only about four ways of connecting or supporting the end of a piece of timber – direct bearing, dowels (including screws, nails and bolts), timber connectors (split rings, shear plates and toothed-plate connectors, all of which rely on the greater contact area of a ring of steel cut into the face of the timber) and glued joints (scarf joints, finger-joints and very occasionally glued-in fin-plates).

Direct bearing is obviously the most efficient way of resisting the axial load or shear in a member – either bearing directly onto the end of a member in axial compression (Figure 33.16), or under the ends of a beam or joist.

#### *33.4.4.1 Mechanical joints*

Mechanical joints generally comprise *dowel-type connectors* loaded in shear. These include nails, screws, metal dowels and bolts. Strength is usually governed by local crushing of the timber, particularly when loaded perpendicular to the grain. When joining thicker members, bending in the dowel itself can start to become significant. Bolts are stronger than dowels of a similar diameter, because the washers at either end provide some end-fixity, changing the stress distribution along the fixing. The number of fixings that can be 'packed' into a timber is limited by the risk of splitting, and design codes place limits on minimum spacings and edge distances.

Timber connectors (Figure 33.19) include split rings, shear plates and punched metal plate fasteners. They rely on having a large area of steel in contact with the timber, to compensate for timber's relatively low strength.

#### *33.4.4.2 Glued joints*

Glued joints can make stiff and strong connections, limited only by the strength of the timber in shear parallel to the grain. Indeed, they should only be loaded in shear, since direct tension can lead to delamination. They are most commonly used to bond the laminates in glulam and plywood, and in finger-joints to end-joint the glulam laminates (Figure 33.15). Finger-joints are similar to scarf joints, but with less wastage of timber. The efficiency of a finger-joint drops as the slope of the fingers increase, and is defined as the strength of the jointed to the unjointed clear timber.

Because glued joints are stiff, they should never be combined with metal fixings in a single connection, as the stiffer glue will attract the entire load, until it eventually fails and transfers *all* the load onto the fixing. Glues are generally unsuitable in conditions of varying moisture content, since this can lead to differential

**Figure 33.19** For the green oak structure at York Minster, shear plate connectors were used instead of split rings to accommodate the cross-grain drying shrinkage. © Arup

movement of the glued laminates generating tension on the glue line.

Surfaces to be glued need to be flat, clean and dry, while an adequate pressure and temperature needs to be maintained until the adhesive has cured. For this reason gluing needs to be carried out under factory conditions and not on site, unless very tight controls are in place such as heating and a temporary roof. Even in a factory, quality control on the gluing is of vital importance.

*33.4.4.2.1 Types of adhesive* Structural joints need to be creep-free and in practice only a limited number of thermosetting synthetic resins are used, formed from a reaction between formaldehyde and urea, phenol, resorcinol or melamine. Epoxy resins are also used and, while more forgiving in terms of the temperature and pressure required during curing, are relatively expensive. They therefore tend to be limited to repair work because of their ability to bond timber to steel plates/rods and glass fibre.

Phenol and resorcinol formaldehyde (PF, RF or PRF) which form a dark brown glue line are the most widely used, because of their suitability for damp external conditions. Melamine (MF) and melamine urea formaldehyde (MUF) are colourless and also tend to be cheaper, but are only suitable for semi-exterior conditions with occasional wetting (where manufacturers generally recommend that they are over-painted). Urea formaldehyde (UF) is only suitable internally. As an alternative to the formaldehydes, polyurethane-based glues are becoming increasingly common on the Continent; they have a relatively limited track record but do have the advantage of being colourless.

*33.4.4.2.2 Compatibility* Compatibility of adhesives with flame retardants or preservatives should always be checked with the manufacturer. For example, preservatives that incorporate a water repellent, or flame retardants based on inorganic salts, will generally prevent the wood from 'absorbing' the adhesive.

## 33.5 Timber for use in joinery

Detailed guidance on the specification of timber for use in joinery (such as doors and windows) is given in EN942: *Timber in joinery – General classification of timber quality*. This includes guidance on suitable species and (where necessary) preservative treatments, as well as the appearance of the *visible* surfaces of the finished component.

Before specifying, it would be prudent to contact suppliers to determine the quality of timber generally available from a particular species. The temptation to over-specify in terms of, for example, acceptable knot sizes should be avoided since this could lead to significant wastage of timber and expense for the client.

### 33.5.1 Species

For general-purpose *internal* joinery that is to be painted, it will generally not be necessary to specify a particular species. However, for external joinery, a timber with small movement characteristics as well as suitable natural durability (or at least a timber that is amenable to preservative treatment) should be specified.

Other considerations affecting choice of species might include general appearance, strength and workability.

### 33.5.2 Appearance

EN942 defines five classes of timber in terms of allowable knot size, as well as the inclusion of finger joints and laminations, shakes, resin and bark pockets, loose knots, discoloured sapwood, exposed pith and beetle damage. Classes J40 and J50 are intended for general-purpose joinery and could include knots up to 40 or 50 mm diameter respectively, whereas class J30 will require special selection of the timber and therefore command a cost premium. Classes J2 and J10 will command a significant cost premium (due to the high reject rates) and may only be available in limited species.

The choice of finish for joinery may influence the class of timber required. It may be acceptable for joinery which is to be covered with an opaque finish to have features that have been made good using plugs and fillers that would not be acceptable on a clear-finished product.

When the precise appearance of the timber is important to the finished project, the only totally reliable solution may be to inspect the timber prior to use or to require a sample for general approval.

### 33.5.3 Moisture content

It is important to ensure that the component is supplied, stored and installed at a suitable moisture content, relative to the moisture content it will experience in service.

## 33.6 Preservatives and finishes

### 33.6.1 Finishes for exterior timber

Without protection, all timber fully exposed to UV light will eventually weather to a silver grey (Figure 33.9). Exterior finishes can be used to prevent such bleaching and particularly the uneven weathering that might result from, for example, the uneven protection provided by a roof overhang, as well as to prevent the surface checking and splitting that results from alternate swelling and drying.

It is also wise to use a primer or clear woodstain to protect internally exposed members from damp or contamination with dirt, during construction.

#### *33.6.1.1 Paints and varnishes*

Paints are opaque and should offer a life of several years, although they are relatively expensive to maintain. On resinous timbers such as pine and Douglas fir there is a risk of unsightly resin exuding from the knots. It is therefore important to seal the knots, particularly where dark coatings are used on southern aspects, leading to higher heat gains.

Varnishes are similar to paints but without pigments. Because they are colourless, they rapidly embrittle under ultra-violet radiation, causing premature flaking of the varnish and surface degradation of the timber below. They are therefore not recommended for exterior use unless the client is prepared to accept a very expensive, typically annual maintenance regime.

#### *33.6.1.2 Woodstains*

Woodstains can be gloss or matt, and vary from almost colourless to almost opaque. They are much thinner than paints and varnishes, and therefore allow the timber to breathe, while acting as a water repellent.

Their main advantage is ease of maintenance, which usually consists of washing down to remove the products of surface weathering and any loose particles, followed by reapplication. Darker stains require less maintenance because they screen the surface of the timber from harmful ultra-violet radiation.

Some hardwoods can be difficult to coat with low-build stains, because their density and the presence of extractives inhibit adhesion of the stain.

### 33.6.2 Preservatives

Preservatives are used to improve the durability of non-durable timbers in damp (generally external) conditions where there is a risk of decay, by preventing the growth of timber-destroying fungi, as well as to guard against surface mould growth and insect attack.

By definition, preservatives are poisonous not only to fungi but also to humans and wildlife and are therefore becoming increasingly restricted on health and environmental grounds. This is encouraging a gradual move towards the use of naturally durable timbers where there is a risk of decay. For example, the use of CCA (copper chrome arsenic) and creosote is now heavily restricted, and even though theoretically permitted for limited applications they are almost impossible to obtain; modern alternatives are discussed in the next section.

Brush-applied preservatives barely penetrate the timber; they need to be regularly reapplied and are of limited value except in stopping surface mould growth. To be effective, a preservative should be applied in a pressure vessel, so that it penetrates into the surface of the timber forming a protective shell. Better protection is achieved with through thickness penetration, but this is inevitably slower and more expensive. Only some timbers, including most softwoods, are sufficiently permeable to accept treatment. Since only the sapwood is permeable, it is important to limit the amount of heartwood in each piece of timber being treated. The timber must be dry to aid penetration of the preservative, and to prevent subsequent drying fissures from cutting through the protective shell. Similarly site cutting should be avoided or at least followed by a liberal brush-applied application of suitable preservative, or better still dipping the end of the timber directly into the preservative.

There are currently three main types of preservative in general use (water-borne, light organic solvent-borne and micro-emulsions); however, the market is continuing to evolve, partly in response to the requirements of environmental regulations.

#### *33.6.2.1 Water-borne preservatives*

These generally rely on copper (which imparts a green colour to the timber) combined with, for example, boron (such preservatives are termed CCB, and include chromium which acts as a fixative) or organic fungicides and insecticides. Both are believed to be suitable for timber in ground contact, although their track records are still relatively short. Since the preservatives are water-borne, the timber swells and has to be dried after application, which can lead to a degree of distortion and raised grain. They are therefore limited to structural timbers and are not suitable for joinery.

Except where the preservative treatment is through thickness, glulam is best treated after fabrication (subject to available tank size), since planing of the laminates before gluing and of the member after gluing would otherwise remove much of the preservative shell; in the latter instance compatibility of the adhesive with the preservative would also have to be checked. Through-thickness treatment of LVL and plywood is easier to achieve thanks to the micro-fissures generated by flattening of the peeled veneers. It should be noted that the chromium in CCB is likely to become restricted on environmental grounds.

There are also water-borne preservatives based on borates; however, these remain water soluble and so need to be over-painted to prevent long-term leaching and are also not suitable for ground contact. There is also a risk of efflorescence on the surface of the timber.

The fact that boron easily diffuses means that it is commonly used for in-situ remedial work. Boron plugs inserted into holes drilled in timber with fungal damage will diffuse throughout the timber if the timber again becomes subject to moisture, but the plugs will need to be regularly replaced if the wet conditions persist.

#### *33.6.2.2 Light organic solvent-borne preservatives*

These are based on organic fungicides and insecticides, but being borne by light organic solvents (such as white spirit) rather than water, they do not cause wetting distortion of the timber, making them ideal for joinery. They do, however, need a protective coat of paint to be maintained to prevent leaching. There is also pressure from legislation to limit solvent emissions, and they are slowly being replaced by the micro-emulsions described below.

#### *33.6.2.3 Micro-emulsions*

These rely on the same active ingredients as the light organic solvent-borne preservatives discussed above, but now dissolved in a small quantity of strong solvent, which is then mixed with emulsifiers, allowing subsequent dilution with 20 times the volume of water. Although some swelling of the joinery occurs, this is much less than with the water-borne preservatives discussed above.

## 33.7 Bibliography

TRADA, the Timber Research and Development Association, publishes a wide range of excellent guidance, and its notes on, for example, joinery and exterior finishes provide useful reference to the growing number of relevant design, product, manufacturing and testing standards.

Structural design was covered in BS5268: *Structural use of timber*. This has been replaced by EN1995 Eurocode 5: *Design of timber structures* together with a large suite of supporting material and product standards. However, all the key information has been brought together in the Institution of Structural Engineers' *Manual for the design of timber building structures to Eurocode 5* (2007). Other useful references for structural design are the Timber Designers' Manual (Blackwell, 2002) and the guidance notes accompanying the timber clauses of the National Building Specification.

Selection of a suitable species for a particular application is a complex task. The aim of this chapter has been to provide the reader with an adequate understanding of the issues, to enable them to research suitable species with specialists and suppliers.

# 34 Bituminous Materials

**Geoffrey Griffiths** FICE, FIAT
Associate Director UK Infrastructure, Arup

**Robert Langridge** BSc MIAT
Asphalt Quality Advisory Service

**Bob Cather** BSc CEng FIMMM
Consultant, formerly Director of Arup Materials Consulting

**David Doran** FCGI BSc(Eng) DIC CEng FICE FIStructE
Consultant, formerly Chief Engineer with Wimpey plc

## Contents

## 34.1 History and background

Bitumen is one of the oldest known engineering materials and has been used from the earliest times as an adhesive, sealant and waterproofing agent. As long ago as 6000 BC, ship-builders in Sumeria used naturally occurring bitumen found in surface seepages in the area.

In the Indus valley, now in Pakistan, there is a well-preserved water tank that dates back to around 3000 BC. The stone blocks in the walls of the tank are bonded with naturally sourced bitumen and there is a vertical bituminous core in the centre of the wall. This principle is still used today in the design of dams and other water-retaining structures. It is believed that Nebuchadnezzar was one of the early exponents of bitumen, as there is evidence that he used the material for water-proofing the masonry in the construction of Babylon and as grout for stone roads.

In the 19th century the refining of bitumen from crude petroleum oils began, and today most of the bitumen used in construction is derived from crude oil. At refineries crude oil is heated and passed into a fractionating column at atmospheric pressure, where the lighter fractions of the oil, such as kerosene, are removed. The residue from the first fractioning column is heated further and passed into a vacuum distillation column. Steam is injected into this column and other useful materials, such as lubricating oils, are removed. The residue from the bottom of the vacuum distillation column is bitumen.

Bitumen is a versatile material. Its uses now include the construction and maintenance of roads, airfields and roofing; damp-proofing; dam, reservoir and pool linings; sound-proofing; pipe coatings; paints; and many other applications. The American Asphalt Institute has identified over 200 uses of bitumen and the world market for the material is estimated to be in excess of 100 million tonnes per annum, with the vast bulk being used in construction applications.

This chapter presents key information on the nature and sources of bituminous materials. It then gives illustrations, examples and information on the use of bitumen-based materials in building applications and then presents design approaches and materials specification for bitumen-based materials in trafficked pavements – overwhelmingly the largest single application for bituminous materials.

## 34.2 Bituminous materials: Types and definitions

The UK Refined Bitumen Association (www.bitumenuk.com) and the European bitumen producers' organisation Eurobitume (www.eurobitume.eu) report that 85% to 90% of the bitumen supply is used in the construction and maintenance of roads. Of the remaining production, 10% is used in roofing with 5% going to a very wide range of adhesive, coating, sealing and protection applications.

There appears to remain some confusion among designers and constructors as to the distinction between bitumen and other 'black' binding or coating materials also used in construction and related applications. In particular this occurs with coal tars and pitches. There are also geographically initiated differences in terminology such as between bitumen and asphalt.

Bitumen, as described above, is essentially the residue from the refining of crude oil; crude oil being the result of deep below ground pressure and heat conversion of the remains of marine organisms and vegetable matter. There is considerable variation in the nature and properties of the different crude oil sources, and not all are suitable for producing bitumen. Some sources of 'natural' bitumen still exist but are little used compared to the more controlled and consistent refined bitumens.

Coal tar is obtained by the high-temperature carbonisation of bituminous coals and coal tar pitch is extracted as the residue from the partial evaporation or distillation of coal tar. Both materials have previously been widely used in similar water-proofing applications as bitumen, but are much less so now. Although sharing some common outward features with bitumen, these materials have substantially different composition and properties.

In the UK and the rest of Europe, 'asphalt' is a term for a material comprising a bitumen binder with mineral fillers. Typical resulting materials of these types are mastic asphalts and hot rolled asphalt. In North America, however, the term 'asphalt' is used for the material binder in place of bitumen. Thus an asphalt road paving would be termed an asphalt concrete.

There are sources of 'natural' asphalt where the naturally found bitumen binder has become mixed with mineral sources to create the asphalt. Perhaps the most well known source is Trinidad lake asphalt. This has been known to, and used by, Europeans since the 16th century. Extraction of the required quality material requires heating to release trapped water and sieving to remove unwanted solid detritus. The resulting asphalt has a binder content of approximately 54%. For many years it was considered good practice to blend manufactured asphalt with lake asphalt to improve properties and performance. With improved manufactured bitumen quality now available, this blending has become less necessary.

Rock asphalt is another 'natural' asphalt where porous rocks – typically limestone – are impregnated with natural bitumen seepages. Bitumen contents may be 12% by weight. Prominent historic sources of rock asphalt were south-eastern France, Switzerland and Italy.

## 34.3 Modified bitumens

Bitumen is not a single material but is available in a range of properties to meet the end-use requirements. These different bitumens may arise from differences in the feedstock crude oil and from the processing conditions encountered within the refinery. Further property differences can be achieved by blending different bitumen grades and sources or by introducing rheology modifying solvents, or by addition of polymers and particulate or fibrous fillers.

### 34.3.1 Cut-back and fluxed bitumen

Cut-back bitumens are manufactured to reduce the viscosity of the binder in a system using the incorporation of an organic volatile solvent, typically kerosene or white spirit. The reduction in viscosity is useful for application of coatings or sealers by brush or by spraying. For fluxed bitumens similar viscosity reduction is achieved, but the solvents used are essentially non-volatile.

### 34.3.2 Bitumen emulsions

Emulsions (or dispersions) are created by fully dispersing one material in another without chemically combining them. For bitumen emulsions tiny droplets or globules of a bitumen are mechanically dispersed in a water phase. For these emulsions it is common to require a dispersing agent that modifies the droplet surface activation or charge, to enhance deflocculation of the fine droplets in the water phase.

The bitumen content of the typical emulsion products will differ quite markedly and will be determined by the type of bitumen, the nature of the dispersing agent used, the processing plant and the end use.

Bitumen emulsions have particular application in surfacing and coatings where low viscosity and thinner films are required, and the lesser toxicity and fire risk compared to organic solvent systems is advantageous.

## 34.4 Classification of bitumens

With the complexity of the composition of bitumens, the designation of grade tends to be achieved by reference to performance in standard physical property tests. These tests might relate to performance in safety, physical and mechanical loading and durability demands. Classification of bitumen grades is often based on performance in one of two tests; penetration and softening point. 'Penetration' testing, commonly shortened to 'pen' test, is used to assess temperature sensitivity and is taken as the depth of penetration of a standard size and profile needle under a specified load (100 g) at 25°C. Specifications typically require reporting of a range for the sample under test, e.g. 200 ± 20 or 180/220.

Softening point, the beginning of flow or fluidity, is determined in standard tests (ring and ball softening) by constructing a bitumen sample in a steel ring on which a steel ball is placed. The temperature of the test arrangement, immersed in a fluid bath, is slowly increased until the ball deforms the test sample by a specified amount – usually 25 mm. The temperature at which the specified deformation occurs is recorded as the softening point (°C).

For other types of application, different grading systems may be used. One such is viscosity graded bitumens. Various forms of test equipment might be used but will use a similar approach of measuring the flow time for a standard volume of bitumen to flow from an orifice under standard conditions.

These test values cannot be used to calculate fundamental performance of the bitumen but are useful to give indication of likely behaviour, drawing from previous experience with similar materials.

Other standard tests used to control the performance and quality of bitumens include:

- *solubility* – dissolution of the bitumen in an organic solvent to determine mass of insoluble contamination
- *resistance to ageing (hardening)* – determined by loss of weight on heating under standard conditions.

### 34.4.1 Oxidised bitumen

These materials, widely used in building product applications, have useful modification of physical and mechanical properties by, essentially, blowing air through the hot bitumen. This oxidation process changes the chemical and molecular make-up and structure of the bitumen. The degree or extent of blowing can be varied to give 'semi-blown' or 'fully blown' materials. 'Semi-blown' bitumens may also be termed 'air-rectified'. Oxidised bitumens will tend to have tougher, more rubbery, less temperature-sensitive characteristics compared to the 'unblown' materials.

The grading system for oxidised bitumens combines the ring and ball softening value and the standard penetration value. Hence an 85/40 bitumen would have a softening point of 85 ± 5°C and a penetration of 40 ± 5 dmm (dmm = 'decimillimetre', i.e. one-tenth of a millimetre).

## 34.5 Health and safety

Bitumen-based products have been widely used in construction applications for very many years, with a largely good record on health and safety for the public and for construction operatives. Guidance and expert reports on health issues have been produced by the two organisations referenced above (Refined Bitumen Association and Eurobitume) and these are available on their respective websites.

Perhaps the most significant safety risks in producing and constructing with bituminous materials stem from the most common practice of the bitumen being 'hot' during the operations, temperatures of between 100 and 200°C being typical. Hot working brings risks of skin burns – from the plant and equipment or from bitumen 'splashes'. Further risks are from the emission of fumes from the hot bitumen that can cause respiratory tract and eye irritation. Adequate ventilation, personal protection and good work practices are essential components of an overall health and safety risk assessment, required for all construction operations, to minimise risk to people.

In extreme cases of poor control, overheating of the bitumen can lead to ignition and significant combustion of the bitumen and surrounding construction. Although testing for 'flash point' is a routine procedure for bitumens, the behaviour in such tests relates only to laboratory conditions and should not be used as part of a combustion prevention strategy.

## 34.6 Environmental impact of bitumens

Interest in the potential impact on the environment of the extraction, use and disposal of materials is now widespread. There is ever increasing pressure and requirement to analyse and understand these impacts so that minimisation of impacts may be optimised. Such assessments are for most materials complex without a unique basis suitable for all materials.

In the early stages of extraction and processing the issues and assessment of bitumen materials will have some commonality with other oil-derived products, such as petroleum and fuel oils, but with the quantities produced and the very diverse range of products and applications for bitumen-based products the assessments become more difficult. Further bitumen product specificity arises from detail of incorporation with other materials and hence the ability to extract and recycle the bitumen. Recycling of bitumen-based traffic pavements systems is quite widespread, aided by the relatively large volumes available and the relative ease of extraction and incorporation into other products. Such practice is much less readily achieved with small-volume, complex combination situations such as with waterproofing, sealing and grouting applications.

Bitumens are themselves quite durable – dependent on application exposure – which can give benefit in environmental impact studies. Further positive contribution, but perhaps more difficult to quantify, stems from the benefits in using bitumen as a waterproofing and protection system, generating enhanced durability and benefit for the overall construction.

On behalf of the bitumen industry, Eurobitume (www.eurobitume.eu) has commissioned and published a 'life cycle inventory' for bitumen use in many common bitumen based systems such as paving grade bitumens, polymer modified bitumen and bitumen emulsions. It is available from the Eurobitume website.

## 34.7 Further technical advice

The technical data and information presented above represent an introduction and base explanation of bituminous materials. Bitumens are however complex and widely varied materials. Considerable further technical information is available in standards and other texts. Readers of this chapter are recommended to refer to relevant UK and international standards, and for comprehensive technical information to *The Shell Bitumen Industrial Handbook*.

## 34.8 Bituminous materials in building applications

### 34.8.1 Roofing and waterproofing

Overall this is the second biggest market for bitumens after trafficked pavements. The use of bitumen-based products in roof waterproofing is long established and embodies a range of materials types and application principles. Bitumen roofing is most widely associated with so-called 'flat' roofs, but there are application requirements in pitched roof construction too.

#### *34.8.1.1 Built-up roofing*

The term 'built-up' relates to the construction of the complete waterproofing system by in-situ laying-up of several layers (typically two or three layers) with hot bitumen adhesive to form a robust effective waterproofing barrier. Previously termed 'bitumen felt roofing', the term 'built-up roofing' has become preferred, in part to reflect the significantly better performance of modern materials.

Most commonly built-up roofing (BUR) uses bitumen impregnated and coated non-woven fibre substrate ('felt'), although 'flat' roofing products have been based on homogeneous sheets compounded from bitumen and a polymer or pitch and polymer. Although still available, these alternative products comprise a very small sector of the total BUR and 'flat' roofing markets

*34.8.1.1.1 Membranes* In earlier experience with BUR the carrier or reinforcing layer would have been based on felted rag waste fibres; subsequently higher quality natural fibres were deployed and later glass fibres. While able to give some good service, these felt membranes were not very robust and tended to tear and split in service, particularly where incorporated in a roof build-up incorporating high-efficiency thermal insulation.

By the early 1970s, 'felt' carrier fabrics constructed from polyester fibres became available that were suitable for impregnating and coating with bitumen. These 'high-performance' felts proved extremely tough and durable and were able to provide service life in excess of 25 years.

Bitumen felts generally require a first bitumen application to penetrate and saturate the fabric layer. The selection of an appropriate bitumen will be influenced by the actual carrier used and the manufacturing plant, but a bitumen with a penetration of 200 dmm is often used. The bitumen coating uses an oxidised bitumen and possibly fillers and other constituents to achieve a good balance between ease of manufacture, handling during construction and service performance. Different manufacturers and local practice in different countries will lead to different material selections. It is reported that in the UK typical coating grades might be 105/35 or 115/15.

Although these polyester-based bitumen-impregnated membranes can be robust and durable, the stiffness of the overall membrane, particularly in cold weather, can make construction, especially of details, difficult to execute. Further developments in bitumen technology saw the availability of polymer- and rubber-modified bitumen impregnation and coating. Prominent in the materials used were atactic polypropylene (APP) and styrene butadiene styrene, with the latter being the major system used. These polymer and rubber modified bitumens demonstrated superior mechanical and physical properties and durability, together with much enhanced flexibility, useful in membrane installation on the roof.

The surface finish on BUR sheets can be varied depending on need and preference. The default option is a sand finish where a fine silica sand is applied primarily to prevent sheet self-sticking when rolled. If in the completed roof the top surface of the membrane is not visible or exposed – perhaps covered by paving – then a sand finish alone may suffice. Where the membrane is exposed, it is usual to add a layer of bonded mineral granules. Typically the granules are permanently coloured grey, green or sand. The benefit of the granules is that they are decorative, give enhanced protection to incident UV, improve surface fire resistance and give improved aging resistance through reduction of surface temperature.

For prestige buildings it may be desired to have a more decorative finish to the membrane. This can be achieved by finishing the membrane with a fully adhered thin metal sheet, typically aluminium or copper. Good durability can be demonstrated, although such membranes are more expensive to produce and require greater care during installation to avoid surface damage.

*34.8.1.1.2 Bitumen bonding* Most BUR waterproofing is installed as a multi-layer system and fully bonded to the underlying roof construction. In some roofs the membrane might be recommended, because of concerns of substrate movement or vapour dispersion, not to be fully bonded throughout but for the lowest layer to be mechanically fixed to the substrate. The preferred and more widely adopted route is to bond the roof layers fully in a hot bitumen adhesive layer.

Historically the bonding method has been by the 'pour and roll' method. This approach uses a fluid hot bitumen (e.g. 95/25 or 105/35) poured from a container onto the roof substrate as the membrane is unrolled into the bitumen: a well-established practice with a good performance record, but with practical difficulties for roofs with more complex geometry. There are also health and safety issues with transporting of hot bitumen from the heated storage at ground level to the working areas on the roof.

An alternative approach to the hot bonding of the membrane was facilitated by the development of 'torch-on' membranes. With the torch-on system a layer of modified bonding bitumen (usually with atactic polypropylene) is factory-spread onto the back of the membrane. When on the roof, the membrane is unrolled and the bonding bitumen heated to melting using a flame gas-torch. Differing views have been expressed about the suitability or shortcomings of both 'pour and roll' and 'torch-on'. In principle, either method should be capable of achieving satisfactory results.

#### *34.8.1.2 Mastic asphalt waterproofing*

Mastic asphalt roof waterproofing of flat roofs is long established and has a reputation for durability and robustness. Its specification and design have long been supported by British Standards and other authoritative sources. Such performance can certainly be demonstrated where the mastic asphalt is installed and supported on stiff, relatively slow-moving supporting construction. Where less stiffly supported, care is required in the detail design to avoid cracking of the rather brittle mastic asphalt. Good guidance on specification and design is available from organisations such as the UK Mastic Asphalt Council (www.masticasphaltcouncil.co.uk).

As described earlier in this chapter, mastic asphalt normally comprises a blend of bitumen binder (10–18%) together with a natural crushed rock aggregate (normally limestone). Sources of naturally found asphalt are known and have been used successfully, typically blended with other manufactured asphalts including, in recent years, thermoplastic rubber or polymer modification of the bituminous binder. With improved control of reined bitumens, the use of natural 'lake asphalt' has diminished.

Mastic asphalt for roofing is mixed, delivered and laid hot, at temperatures in the region of 170–220°C. Typically for flat roofing the mastic asphalt is laid in two layers to a total thickness of 20 mm, finished with a wood float. For some special installations such as a more heavily trafficked roof, greater thickness may be required.

*34.8.1.3 Rubber-modified bitumen: hot applied*

A number of hot-applied rubber-modified bitumen systems have been developed for waterproofing flat roofs, including planters and water features as well as non-roof applications such as tanking of substructures, bund walls, drainage pits and pile caps. These products, a proprietary blend of bitumens, rubber, fillers and extenders, are supplied to site in cake form for heating (170–220°C) in air- or oil-jacketed melters prior to two-layer spreading. The initial layer is normally 3 mm thick followed by a reinforcing layer of polyester sheet lightly brushed into the still warm base layer. A second 3 mm thick layer is then applied.

*34.8.1.4 Other bitumen membrane applications*

In many constructions with ground contact there may be potential for moisture transmission from the ground into the building space. This is most frequently noted with, but is not limited to, masonry in housing construction and ground-bearing floor slabs generally. The usual design solution to this risk is to incorporate horizontal water-impermeable layers in the floor (damp-proof membrane, DPM) or in the masonry walls (damp-proof course, DPC). Bitumen-based products have a substantial record in fulfilling this performance need using either preformed sheeting or liquid-applied coatings or layers.

A bitumen layer as a liquid can be applied to walls or to floors as a layer a few millimetres thick, protected by an overlay of construction such as sand–cement screed or concrete overslab. To optimise the DPM protection the liquid is best applied in two or three layers. The DPM liquid membrane could be based on a solvent-based solution or a bitumen emulsion.

In principle most of the preformed membranes described in the sections on protection coatings or waterproofing for roofs or below ground could be used to achieve the DPM need, but often the extra cost would not be justified.

For DPC application there will often be extra need to consider the load-bearing capacity, short and long term, of the DPC sheet itself. Frequently such DPC products comprise materials very similar to bitumen-impregnated and coated roofing felts. The specific requirements are contained within British Standards. Homogeneous bitumen/polymer blends in sheet form, as sometimes used in flat roofing construction, are also used as a higher performance, robust DPC.

## 34.8.2 Below-ground waterproofing

Below-ground structures are frequently required to resist the ingress of groundwater for reasons of serviceability, quality of indoor environment and/or durability of the structure. Typical of such structures are the basements of buildings and infrastructure tunnels. In the design of such structures, it might be assumed that the structure itself can provide all or some of the protection against ingress or, alternatively, it may be decided that an external waterproofing – 'tanking' – should be adopted.

There are very many options for such waterproofing need, some of which are based on a bitumen-bound system.

Key parameters or design considerations in waterproofing system selection will include durability in the particular environment, resistance to damage during the construction process and ease of application – particularly in confined excavated working spaces and in hot environments.

Solutions to this need can be either by use of preformed waterproofing sheets or by in-situ application of liquid-based systems.

*34.8.2.1 In-situ applied membranes*

Many of the systems and much of the knowledge described in the sections of this chapter on protection coatings and hot-liquid-applied rubber-modified bitumen roof waterproofing would be appropriate for below-ground applications, but with greater emphasis on post-construction protection.

*34.8.2.2 Preformed membranes*

Preformed bitumen based layer combined with carrier/waterproofing membranes – typically of a polyolefin – combine to form a self-adhesive preformed waterproofing system. These are simply adhered to the prepared substrate as they are unrolled from the 'release paper' – with side and head laps. There are numerous proprietary systems, of generally similar type, available in the market each with variations of detail formulation to allow for hot/cold working conditions. In hot conditions the bitumen component of the membrane can become too 'tacky' to achieve controllable installation. Also in hot conditions the bitumen can allow sliding or slippage of the installed membrane. In cold environments there may be difficulty in achieving good contact and adhesion. Being quite thin, the complete membrane being a few millimetres thick, it is normal recommended practice to apply a protection layer of sand cement screed or preformed board to the outside of the membrane.

## 34.8.3 Flooring

Mastic asphalt used as flooring has a long history, although more recently its adoption has been less extensive, possibly for reasons of lesser familiarity than for any technical shortcomings. The most common applications have been in light industrial facilities, specialist chemical resistance such as acid spillage and car-park surfaces.

The materials used in flooring-grade mastic asphalt are similar to those in flat roof waterproofing, although to counter the higher abrasion conditions a blend with higher aggregate proportion is used. Also, if acid resistance is required, the acid-soluble limestone aggregate is replaced by non-soluble types. Detailed requirements for mastic asphalt flooring are given in the British Standard.

Some degree of self-colouring is possible with mastic asphalt, although the inherent blackness of the binder limits this to the darker reds, greens and browns. If brighter colours are required then over-coating with a pigmented epoxy finish is possible.

## 34.8.4 Protection coatings

Construction buried below ground is often subject to attack from water-borne chemically aggressive species. This attack has the potential to shorten unacceptably the service life of the construction, by corrosion of steel elements such as pipes or piling or by chemical attack of concrete by such mechanisms as sulphate attack or by alkali silica reaction. Bitumen-based coatings can act as a significant physical barrier to the passage of water and aggressive chemical species. This protection against deterioration might be achieved by the provision of below-ground water-proofing – see above – or might be required separately. Bituminous coatings have a long track record in this role, being relatively simple to apply – even to complex geometry – and cost-effective. On the negative, such coatings can be softer than other options and potentially more damageable during construction.

For pipes and other buried steel the coating might be used in conjunction with other measures such as cathodic protection.

There are different levels of bitumen character and product type, ranging from the relatively thin coatings of bitumen in an organic solvent – typically two coats at 100 μm dry film thickness (dft) per coat – to thicker coatings using an oxidised bitumen, e.g. 115/15 blended with mineral fillers and perhaps fibres.

Bitumen-based coatings can be beneficially used to modify the physical interaction between the ground and the structure by allowing 'slip' on or within the coating, depending on its thickness and composition. Such slip characteristic can be used to ease the driving of piles into the ground or to reduce the effects of soil heave or settlement on embedded structures. More detail on this mechanism and use is given in *The Shell Bitumen Industrial Handbook*.

### 34.8.5 Joint sealants

Sealants are required to fill various sized joints created in constructions and adhere to the surfaces, providing a seal against ingress while also experiencing cycling movements – typically opening and closing – of the joint. Although seeming simple, this requirement can be quite demanding for the sealant material. Bitumen-based compounds have a long history of joint sealing service, where their essentially viscoelastic character is beneficial in accommodating movements.

The larger area of current applications for bitumen-based sealants is with asphalt-based pavements. In building applications much of the potential market is taken by non-bituminous 'elastomeric' polymer systems such as polysulphide, polyurethane and silicones. These materials have greater overall movement accommodation and speed of response. Bitumen-based sealants still have a role in joint sealing.

Straight penetration grade or oxidised bitumens are unlikely to be suited to joints where anything greater than a nominal movement is expected. Hot-poured and cold-applied bitumen compounds are available for building work, in both cases the desired movement properties being achieved using thermoplastic polymers to modify the parent bitumen properties. The actual constituent materials' grades and proportions are usually proprietary product-specific.

## 34.9 Bituminous materials in trafficked pavements

Bituminous pavements may be divided into a number of different types of construction; each type is used for a specific application. The following basic categories are found in the UK:

1 *domestic drives* and pedestrian footways
2 *private roads*, supermarket car-parks and access roads to private, light industrial sites
3 *county roads*, roads adopted by district councils; roads, maintained at public expense, of local or regional importance
4 *trunk roads* – roads maintained by central government agencies, of national importance. This category of road includes motorways and major inter-urban highways.

Each type of road is built and maintained to different standards, which are in themselves focused to ensure an efficient and cost-effective level of service. It can be seen that a certain level of overlap exists between the standards: a large county road will be very similar to a trunk road, but a supermarket car park need not be constructed to the same high standard as a motorway. It can be seen that if the intended level of maintenance and pavement use is known it is then possible to design, specify and therefore construct an appropriate pavement to match each application.

This section of the chapter is aimed at introducing bituminous pavements to a technically competent construction professional, allowing them firstly to specify simple structures but then to move on and know when specialist advice is required. An architect or structural engineer should be able to read this text and know when specialist design support is required.

The text is restricted to examining highway and road vehicle trafficked bituminous pavements. Concrete pavements and airfield pavements are outside the scope of this chapter. It is noted that a second school of pavement engineering exists in designing civil and military airfield pavements. Separate design methods and specifications can be found which must be employed to control the successful construction of airfield pavements. A bituminous highway specification should not be used for the construction of an airfield pavement; a substantially higher standard of workmanship and materials control is needed. Aircraft wheel loads and tyre pressures will quickly destroy a standard highway pavement.

### 34.9.1 The basic pavement structure

A number of common features are found within each individual pavement structure. Standard sets of terms are used to describe the different features; it is helpful to understand the terms and the engineering functions associated with each issue. Figure 34.1 describes the standard features.

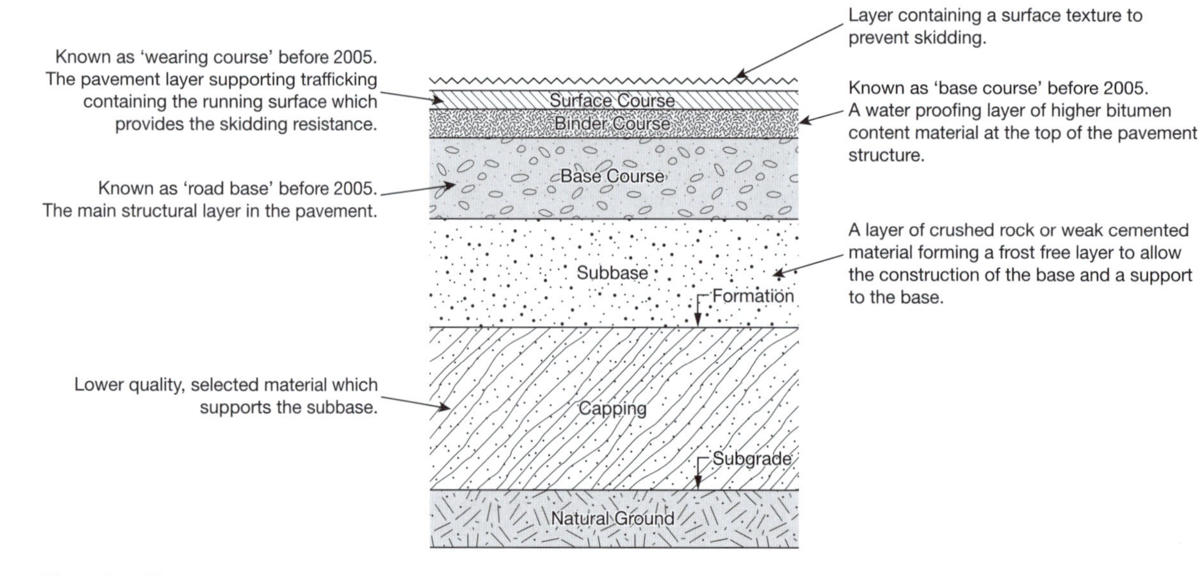


**Figure 34.1** The pavement structure

A European directive complicates an understanding of the common descriptive terms; in January 2005 and again in January 2008 all of the standard, traditional British pavement terms were revised to allow a harmonised understanding of the pavement structure. Figure 34.1 shows both the traditional British terms and the latest EU-approved words. In order to avoid confusion this chapter uses only the standard EU terms, but it should be understood that the term *base* will mean the binder course in pre-2005 texts and the main structural layer in post-2005 and 2008 texts. The 2008 revisions altered the traditional UK pavement materials descriptions, changing the description of bituminous macadam materials to asphaltic concrete. The 2008 revision of the standards radically altered the basis of the specifications by introducing an essentially common set of standards across the EU trading area. The pre-2008 British standards use prescriptive methods and a fixed envelope approach; the current standard uses a declared target approach. The post-2008 approach complies with European Union directives; both systems can be interpreted to provide identical materials but described differently using each set of rules.

### 34.9.2 Design methods

Pavement design is essentially an empirical process; pavement designs are produced from a knowledge of successful practice using a slow process of evolution. Standard successful pavement designs are produced by a process involving the completion and evaluation of trials which then allow new materials and techniques to be adopted within the design of a pavement structure. A number of semi-empirical, rational design methods have evolved but all of the calculation methods require reality checks, obtained from pavement trials, to ensure that the correct calibration factor is employed. Isolated, uncalibrated calculations are likely to produce an unreliable pavement structure.

A number of different pavement design methods can be found; essentially each major country has its own design method. In the UK a technique written by the Transport Research Laboratory is used. The UK method is described in LR1132 [1], TR250 [2] and the most recent document TRL615 [3]. Alternative reliable design methods may be found in South Africa, the USA, Holland, France and Russia. Each design method is intended to be used with an associated construction specification. The specification is an essential element of the design method; it is noted that design methods must not be mixed and matched. The fact that pavement design methods are essentially empirical means that the construction specification is an essential element of the design. The specification is an important part of the recipe.

The UK method describes pavement life by the term 'million standards axles' (msa). The term describes the passage of a certain number of 80 kN axles across a pavement, the pavement deteriorating under the fatigue effect obtained by trafficking using the standard load. Thicker pavements are generally accepted as having more strength, a greater resistance to loading and therefore a higher msa value.

A concept known as a 'long life pavement', described in TR250 and TR615, suggests that a maximum thickness of construction exists beyond which an increased pavement life cannot be extended. The concept is used to limit the cost of new pavement construction, fixing an upper limit on pavement strength. The value is currently fixed at +80 msa in the UK; any pavement with an expected life greater than this value is known as a 'long life pavement'.

### 34.9.3 Standard specifications

The construction specification is a frequently misunderstood and undervalued document. If a pavement is designed in accordance with a specific design method, the pavement then must be constructed to the matching construction specification. In the UK one standard document is accepted and maintained by the central government Highways Agency. The document is known as the Specification for Highway Works [4 & 5]. The document is regularly updated and revised and is the accepted text in England, Scotland, Northern Ireland and Wales. A similar, but slightly, different document is used in the Republic of Ireland. The UK document has developed over the past 50 years from standard Department of Transport specifications; similar documents based on the original document can be found in many former British Empire countries. In the UK some minor alternative, simplified documents can be found but essentially all of the specifications are referenced into the standard Highways Agency document. The standard text contains the following sections that describe pavement construction work:

- Section 600 – Earthworks
- Section 700 – Accuracy of construction
- Section 800 – Sub-base material and cement bound pavement materials
- Section 900 – Bituminous materials
- Section 1100 – Kerbs
- Section 1200 – Footways.

This review uses the standard DMRB [6,7] specification clauses to describe materials wherever practical.

The specification uses two alternative techniques to describe materials and the construction process; the specifications are either method specifications or end product documents. The majority of the traditional clauses use the method or recipe approach to describe pavement construction. An end-product specification is one in which the purchaser defines the final product and then asks the supplier to meet the requirement with the supplier's specific design.

The 2008 British Standard employs a standard four-point system (Table 34.1) to describe bituminous mixtures.

### 34.9.4 Typical pavement types

In order to understand bituminous pavements effectively, it is helpful to know that a number of standard structures may be found. The following details are offered as typical solutions to specific applications. The standard designs can be found in similar forms in many slightly adjusted applications. The structures are all presented over a 3% CBR subgrade strength; materials will achieve this subgrade strength.

#### *34.9.4.1 Domestic drive and footway*

The weakest and lowest standard of construction that is commonly found is the domestic drive or rural footway (Figures 34.2, 34.3 and 34.4). This type of pavement is intended simply to support either a pedestrian or a domestic car. Footways are frequently constructed with the softest available bitumen 160/220 pen with 100 mm of crushed rock sub-base. Drives may be constructed with the slightly harder grade of 100/150 pen material and 200 mm of crushed rock sub-base. The typical materials used in a domestic drive are:

1. 20 mm, Clause 909 [4], AC 6 dense surf 100/150 Dense Surface Course, BS EN 13108 Part 2 [16] minimum PSV 45

**Table 34.1** British Standard description of bituminous mixtures

| *Material description* | *Mix aggregate upper sieve size* | *Material application* | *Bituminous binder grade* |
|---|---|---|---|
| AC/SMA/HRA | D | Surf/bin/base | 60/70 |

AC = the material description; D = the material size description; surf or bin or base = the material application, which is surface course, binder or base.

**Figure 34.2** Footway

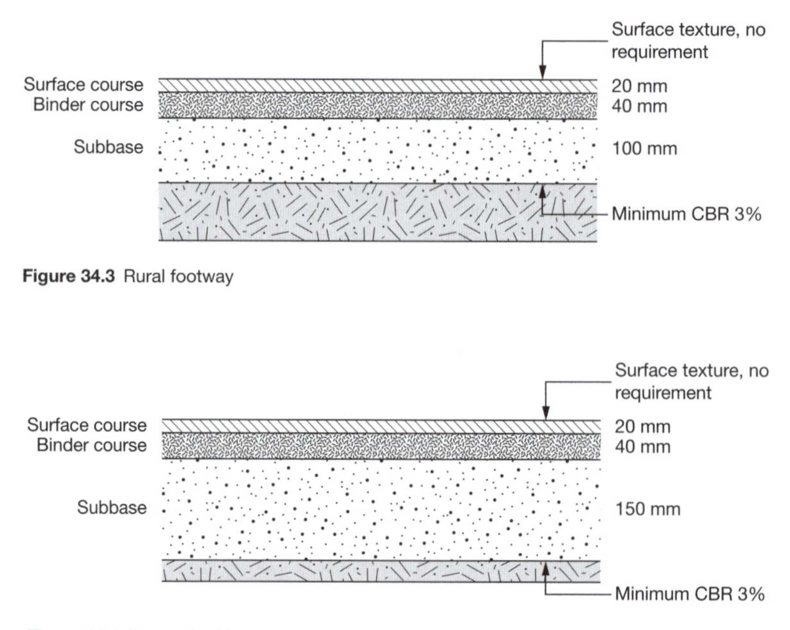


**Figure 34.3** Rural footway

**Figure 34.4** Domestic drive

2 40 mm, Clause 906 [4], AC 20 dense bin 100/150, Dense Binder Course, DBM125, BS EN 13108 Part 1[8]
3 150 mm, crushed rock subbase, Type 1 Clause 803 [4].

Precast concrete edgings are used to form the pavement edges.

*34.9.4.2 Private road or supermarket car park, 0.1 msa*

Domestic supermarket car parks and private unadopted roads are the next step up into the pavement standard (Figure 34.5 and 34.6). A short (10 year) design life is frequently adopted to ensure that the intended works are constructed to the lowest practical cost; this type of pavement must be economically designed and built, to the simplest practical standard. The short design life is adopted as private users frequently require short horizons and do not wish to over-design the pavement. Many private roads will have a change in use after 10 years; the altered client requirement will frequently require the adoption of a different layout and the revision of the pavement.

**Figure 34.5** Private road or supermarket car park

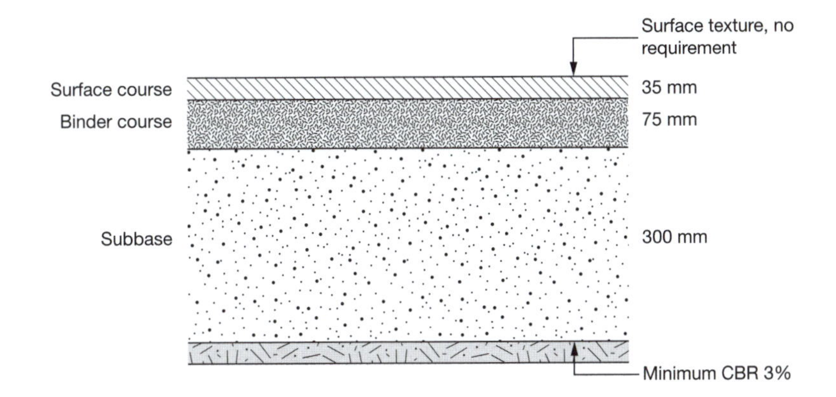


**Figure 34.6** Private road, supermarket car park

**Figure 34.7** Low trafficked county road

The design is constructed to a standard that may become damaged in a hard, cold winter, the owner taking the view that maintenance may be easily completed if the pavement is damaged by frost.

Porous pavements are a recent development for this type of construction; the closed textured, water-proof dense materials may be replaced by open textured material. The open, voided material provides an easy drainage path through the structure, producing an efficient porous pavement. A private road will usually be edged by standard 250 mm precast concrete kerbs, which present a 125 mm high kerb face.

A typical private road can be specified using these materials:

1 35 mm, Clause 912 [4], AC 10 close surf 100/150 close graded Surface Course, BS EN 13108 Part 1 [8], minimum PSV 45, no surface texture requirement
2 75 mm, Clause 906 [4], AC 20 dense bin 100/150 rec, Dense Binder Course, DBM125, Part 1 [8]
3 300 mm, crushed rock sub-base, Type 1 Clause 803 [4].

*34.9.4.3 Adopted, low-speed, housing estate county road, 1 msa*

The housing estate road is the simplest form of adopted road. The design is produced to a standard that is expected to last for 40 years with only a minimum level of maintenance (Figure 34.8). District councils insist on the higher standard of construction as any future maintenance costs are met from the public purse. The pavements are therefore robustly specified to minimise any possible maintenance costs; this type of pavement is frequently maintained at 20 year intervals by the application of a thin, sprayed coat of bitumen which then receives an application of 6 mm stone chippings. The system of maintenance is known as 'surface dressing'. Adopted road edges are supported by standard precast kerbs in a manner similar to private roads.

A low-speed adopted housing estate road may be specified using the following structure and materials:

1 35 mm, Clause 912 [4], AC 10 close surf 100/150, close graded Surface Course, BS EN 13108 Part 1 [8], minimum PSV 45, no surface texture requirement
2 60 mm, Clause 906 [4], AC 20 dense bin 100/150 rec, Dense Binder Course, DBM125, BS EN 13108 Part 1 [8]
3 105 mm, Clause 906 [4], AC 32 dense base 100/150 rec, Dense Base, DBM125, BS EN 13108 Part 1 [8]
4 300 mm, crushed rock sub-base, Type 1 Clause 803 [4].

*34.9.4.4 Major, high-speed, motorway trunk road, +80 msa*

Trunk roads and motorways are constructed to the highest commonly found standard of construction (Figure 34.9). The essential features are as follows:

1 The surface course is a highly engineered material, to produce a skid-resistant surface.

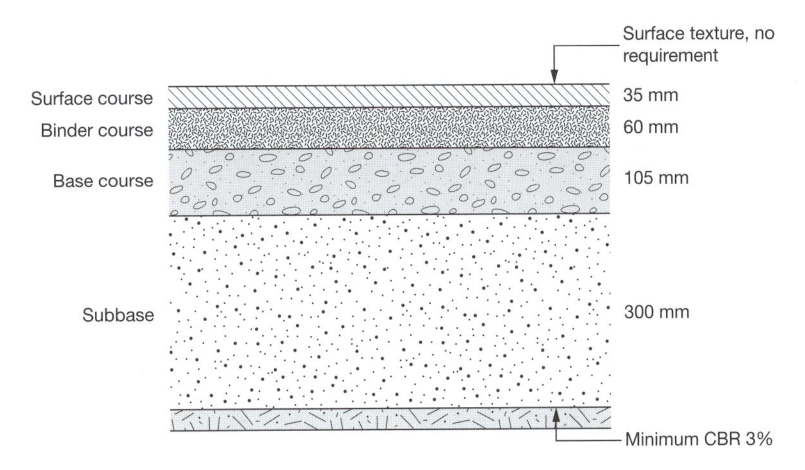


**Figure 34.8** Housing estate road

2 The completed pavement surface is controlled to achieve a good ride quality.
3 The lower pavement subgrade is drained by a system of land drains to ensure a dry lower pavement structure.

Motorway structures are exceptionally difficult to maintain. The motorway must be booked to plan any lane closures; the execution of the works is highly dangerous and is frequently completed in stressful night working conditions. Motorway pavements require careful materials testing and control as well as a superior level of workmanship when compared to the other types of structure. The edges of motorways are drained by open, wide, shallow depressions that act as drains.

**Figure 34.8** Motorway

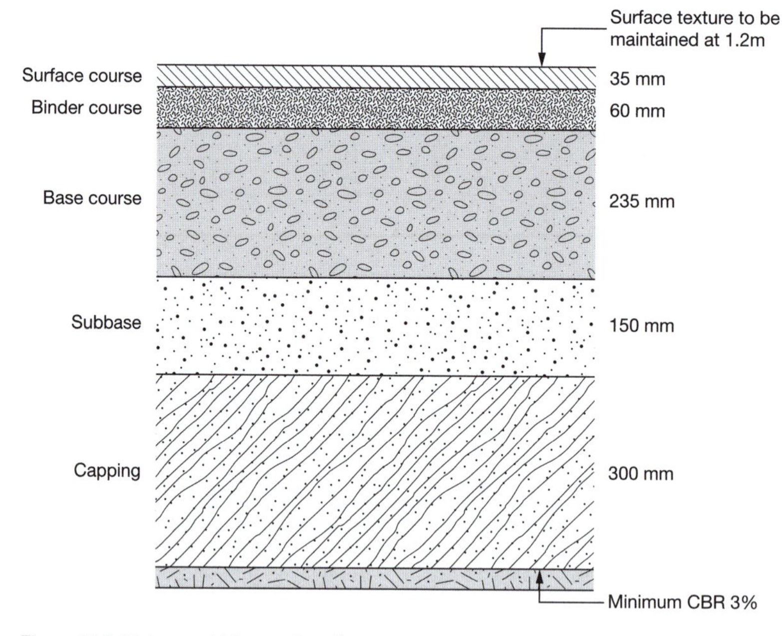


**Figure 34.9** Motorway, high-speed road

A typical motorway will be a long life pavement and may be specified using the following structure and materials.

1 35 mm, Clause 942 [4] TSC with a surface texture to Clause 921, SMA 14 70/100 Thin Surface Course System, Stress Level = 1, minimum PSV 50, minimum surface texture 1.5mm, guarantee period 2 years, level of rutting resistance and noise-reducing qualities are both defined. TSC systems are type-specified using either BS EN 13108-Part 2 or Part 5.
2 60 mm, Clause 906 and 929 [4], AC 20 dense bin 40/60 rec designed binder course, BS EN 13108-Part 1 [8], 40/60pen bitumen.
3 255 mm, Clause 903 and 929 [4], AC 32 dense base 40/60 (Roadbase), laid in two layers typically 120 mm upper and 135 mm lower.
4 150 mm, crushed rock subbase, Type 1 Clause 803 [4], replaced by a cement-bound lean mix concrete in the current UK standards.
5 300 mm, capping as 6F2 [4].

Major motorways require specialist designers and contractors; adopted roads and large private roads will also benefit from specialist engineering assistance.

### 34.9.5 Types of material

Bituminous pavements materials fall into two distinctly different groups, 'hot rolled asphalts' and 'asphaltic concrete', formerly known as 'dense bitumen macadams'. In European terms all bitumen-bound aggregate mixtures are known collectively as asphalts. The 'asphaltic concretes', frequently known as DBM, are a group of materials containing an evenly graded coarse to fine aggregate. The materials are described in BS EN 13108 Part 1 [8], PD 6691:2007 Annex B [15] and contain a bitumen content of between 3.5% and 6.5% with a void content that can vary between 3 and 8%. An extreme open-textured, porous asphalt BS EN 13108 Part 7 [18] version of the material will have a void content of about 10–15%. The AC materials obtain strength from the interlocking aggregate matrix, which must be correctly compacted, at the appropriate rolling temperature, to achieve the required strength. The bitumen acts partly as a glue but also as a lubricant enhancing the compaction process.

ACs are successfully manufactured from any hard, crushed rock aggregate, provided the material is stable and reasonably cubic in shape. Problems will occur if laminar, flaky or non-cubic aggregates are used. Washed river gravels will also prove difficult to use; the smooth washed and polished aggregate surface reduces the stability of the pavement material. Smooth gravel aggregates may be successfully used for pavement materials if cement is added to stiffen the bituminous mix.

'Hot rolled asphalts', frequently known as HRA, are the second important group of materials and are quite different to a DBM in that they consist of a discontinuous mixture of coarse and fine aggregates, known as a 'gap graded' mixture. The materials are described in BS EN 13108 Part 4 [10] and PD 6691:2007 Annex C [15], and are commonly used in areas of the UK where good-quality rock is difficult to obtain. The material gains strength from a combination of the coarse and fine aggregate cemented by a relatively stiff binder, usually 50 pen. A durable HRA will have a void content of between 3% and 6% with a bitumen content that may vary from 5.5% to 9% depending on the required specification. HRA materials tend to be manufactured using the stiffer grades of bitumen but can be more susceptible to rutting when compared to DBM materials.

EME2 materials form a third group of materials which are used in the construction of heavily trafficked motorways and trunk roads. The materials are designed using an extensive system of laboratory trials and obtain much of their strength from the high, very stiff, bitumen content. An EME2 mix will have a bitumen content of typically 5.5% using a very low pen bitumen.

A fourth group of materials are defined using an end-product performance specification. The group are patent surface course and thin surface treatment systems. The principle of performance-based specifications is also available for 'Base and Binder' course materials, but this system has at present proved largely unsuccessful. The complexity involved in defining a suitable performance-based test has made the application of performance specifications impractical in the lower 'base and binder' coarse layers of the pavement.

The major commonly found material types are described in turn within this text.

*34.9.5.1 Surface course materials*

The surface course materials are intended to provide a skidding-resistant, durable finish to the pavement surface. The design of a high-speed road surface course requires careful attention; the following issues are controlled by adjusting the surface course specification:

1 Surface texture, at 1.2 mm, 1.3 mm or 1.5 mm, depending on the road speed and material type.
2 The polishing characteristics of the coarse aggregate: areas that are subjected to heavy braking, at roundabouts and traffic signal approaches, are fixed to the highest level.
3 Rutting resistance: the material is designed to prevent rutting in high-stressed situations.
4 Limiting tyre to road noise: coarse textured materials that have positive uneven textures, e.g. HRA with coated chippings will produce high levels of environmentally damaging tyre-generated road noise.

The design of surface course materials for high-speed roads is outside the scope of this chapter. A specification for a high-speed road surface will require specialist assistance and should not be attempted by untrained designers. Low-speed, low-trafficked roads, drives and footways are more easily designed; the skidding resistance is controlled by simply choosing the correct specification for the coarse aggregates. A minimum polished stone value (PSV) of 50 is generally employed on low-risk sites to prevent simple skidding problems.

The surface course layer is also designed to drain without ponding. Poor surface drainage characteristics can result in aquaplaning problems. A minimum highway cross fall of 2.5% and longfall of 0.5% producing a maximum water film depth of 4 mm is an accepted standard which will prevent ponding or aquaplaning problems.

*34.9.5.1.1 Dense surface course materials to BS EN 13108 Part 1 [8] & BS 594987 2 [ 9]* The UK industry regularly uses nine different surface course materials, each of which is used in slightly different applications. The DBM materials are the preferred solution for providing a suitable surfacing material to most UK low-trafficked and footway pavements. The different materials and the recommended material applications are each described in Table 34.2.

**Figure 34.10** 0/6 Dense Bitumen Macadam to a private low-speed road

The porous asphalt materials are specialist applications. The Clause 8.2 material can be successfully used for the construction of car parks and footways. The Clause 8.1 material was specifically designed to limit tyre noise on trunk roads but has proved difficult to construct successfully.

*34.9.5.2 Hot rolled asphalts, BS EN 13108 Part 4 [10]*

Hot rolled asphalts are divided into two specific groups: the chipped, 30% or 35% stone content and the non-chipped 55% stone content HRA materials. The British Standard contains 15 different surface course specifications that are divided into three groups: design mixes, recipe mixes and performance specification mixes.

*34.9.5.2.1 Chipped, 30% and 35% stone content HRA* The most commonly used material is the 30% and 35% stone content, chipped HRA material. This group of materials is currently found

**Table 34.2** Surface course materials and applications

| *PD 6691 and BS 4987 reference* | *Aggregate grading* | *Material description* | *Application* |
|---|---|---|---|
| Table B.13 Clause 7.1 | 0/14 mm | Open Graded Surface Course Clause 916 [4], AC 14 open surf 100/150 | Low-speed roads |
| Table B.13 Clause 7.2 | 0/10 mm | Open Graded Surface Course Clause 916 [4], AC 10 open surf 100/150 | Low-speed roads |
| Table B.14 Clause 7.3 | 0/14 mm | Close Graded Surface Course Clause 912 [4], AC 14 close surf 100/150 | Low-speed roads |
| Table B.14 Clause 7.4 | 0/10 mm | Close Graded Surface Course Clause 912 [4] AC 10 close surf 100/150 | Low-speed roads and domestic drives |
| Table B.15 Clause 7.5 | 0/6 mm | Dense Surface Course Clause 909 [4], AC 6 dense surf 100/150 | Low-speed roads, domestic drives and footways |
| Table B.16 Clause 7.6 | 0/6 mm | Medium Graded Surface Course AC 6 med surf | Low-speed roads, domestic drives and footways |
| Table B.16 Clause 7.7 | 0/4mm | Fine Graded Surface Course Clause 914 [4], AC 4 fine surf 160/220 | Domestic drives and footways |
| Clause 8.1 | 6/20 mm | Porous Asphalt Surface Course Clause 938 [4] | Specialist application for trunk roads |
| Clause 8.2 | 2/10 mm | Porous Asphalt Surface Course | Porous pavements and a spray-reducing layer for airfield runways |

**Figure 34.11** 20mm Chipped HRA to a high speed county road

**Figure 34.12** 55% stone content Hot Rolled Asphalt to a low speed urban county road

on most major motorways and heavily trafficked highways. The material is a low stone content, dense mixture which receives a spreading of either 14 mm or 20 mm chips (to Clause 915 [4]) rolled into the pavement surface, at the completion of the construction process. The stone chippings provide the skidding resistance; when the chips are correctly applied, the material produces a very successful surface. If the chips are poorly compacted they will be plucked from the pavement surface; when they are over-compacted the pavement will have insufficient surface texture: either case produces a safety hazard. Chipped HRA is accepted as less safe to construct than the more modern material. The application of the chips is a hazard and can be a significant safety risk.

The materials are currently out of fashion for the motorway and trunk road network and have effectively been replaced by thin surfacing. They are used extensively on county roads where their durability is attractive.

Chipped HRA materials are described in [4], BS EN 13108: Part 4 [10] and PD 6691:2007 Annex C [15] (Table 34.3)

*34.9.5.2.2 55% high stone content HRA; known as MTA* The second group of HRA surface course materials are the high stone content HRAs. The material is frequently referred to as MTA (medium temperature asphalt), the name derived from the low pen bitumen materials that are commonly used in the mixtures. It consists of 55% coarse aggregate within the gap graded pavement material. The materials have a stony, textured surface

**Table 34.3** Chipped HRA materials

| | | |
|---|---|---|
| Design mixes | Clause 911 [4] | Type F, Table C.2A Columns 4, 8 and 10<br>HRA 30/10F, 30/14F and 35/14F surf XX/YY des [15]<br>Type C, Table C.2C Column 9 and 11<br>HRA 30/14C and 35/14C surf XX/YY des [15] |
| Recipe mixes | Clause 910 [4] | Type F, Table C.2B Columns 4, 8 and 10<br>HRA 30/10F, 30/14F and 35/14F surf XX/YY rec [15] |
| Performance-based | Clause 943 [4] | Type F Table C.2C Column 9 and 11<br>HRA 30/14C and 35/14C surf XX/YY des and also the required Table C.3 performance requirement |

and do not require the addition of chippings to provide skidding resistance. The coarse aggregate provides the required pavement surface texture. The material can be manufactured with either a sand or crushed rock fine fraction.

MTA is a difficult pavement material to lay and compact successfully. The materials are prone to poor compaction, segregation and binder drainage problems. A well-laid surface will have a regular, even, closed texture with solid, well-formed joints. A badly laid surface may have an open, bony texture with a high void content along the joints. Poorly laid areas may be expected to lack durability and will fret as the pavement ages; eventually the open-textured areas will pothole. The sandy mixtures tend to be more durable than the crushed rock materials; the slightly rounded nature of natural sand particles helps the material to be more cohesive and hence easier to compact, giving a dense impermeable material that can provide a reasonable service life.

MTA should only be laid by an experienced workforce, using a quarry source that is known to produce a durable product.

The high (55%) stone content HRA materials are described in BS EN 13108 Part 4 [10] and PD 6691: 2007 Annex C [15] (Table 34.4).

The Type C mixtures are usually manufactured from crushed rock fine aggregate; the Type F materials from sand. Performance-related high stone content HRA materials are not recommended.

*34.9.5.2.3 Clause 911 [4] design mixes and the Marshall test, BS EN 594987 [12], BS EN 12697-34 [20]* The design mixes referred to within the different HRA surface course materials are designed to fix the stiffness of the intended mix. The Marshall test (Figure 34.15) is an internationally accepted design procedure to

**Table 34.4** High (55%) stone content HRA materials

| | | |
|---|---|---|
| Design mixes | Clause 911 [4] | Type F HRA 55/14F surf XX/YY des [15]<br>Type C Table C.2C Columns 13<br>HRA 55/14F surf XX/YY des [15] Table C.2A Column 12 |
| Recipe mixes | Clause 910 [4] | Type F Table C.2B Column 12<br>HRA 55/14F surf XX/YY rec [15] |
| Performance-based | Clause 943 [4] | Not recommended |

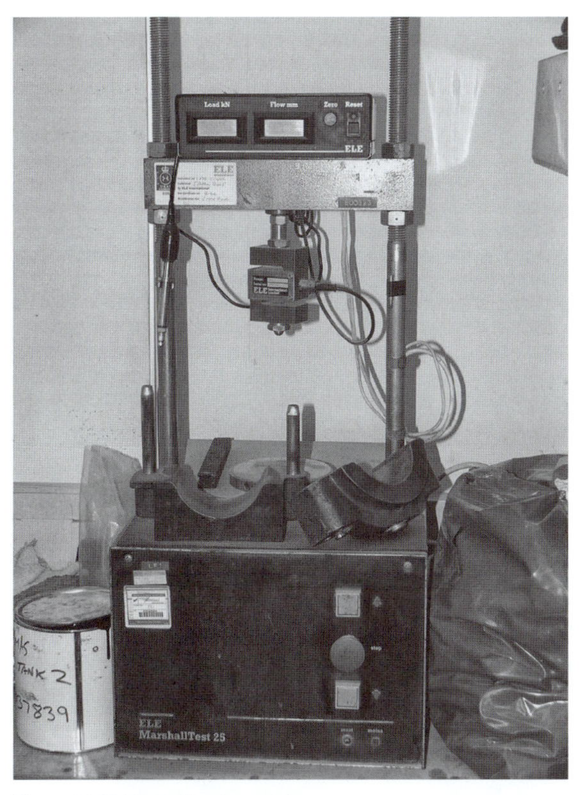


Figure 34.13 Marshall test apparatus

Figure 34.14 Marshall compaction hammer

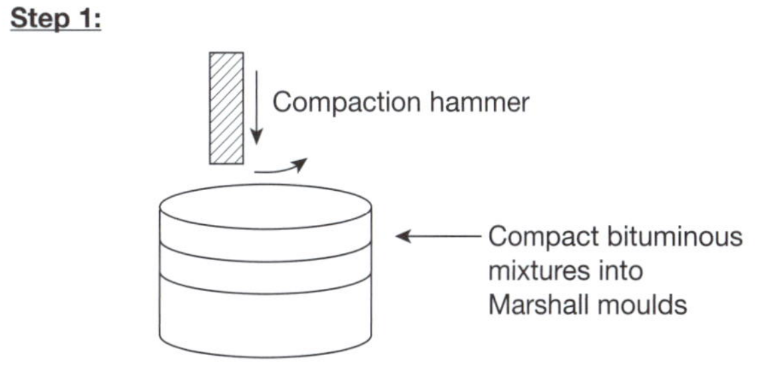


**Step 2:** Age the samples in a hot water bath

**Step 3:**

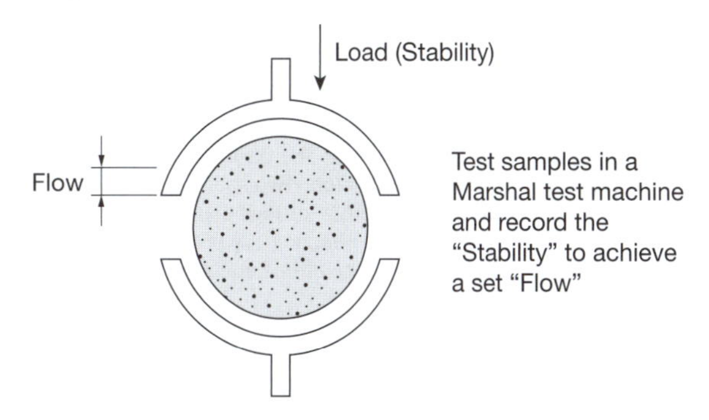


**Step 4:** Complete graphs to optimise the mix design.

**Figure 34.15** The Marshall design method

identify an optimised bitumen content and fix the stiffness of the intended mix. The pavement material mix stiffness is known as the 'stability' and is the failure load (kN) of the mixture when specimens are tested using a 'crushing' test under a set constant load. The amount of strain at peak load (peak stability) is known as the 'flow'. The test is described in BS594987 [12]. The test is completed at the start of the construction process and is part of the materials design process. The higher the planned stability, the stiffer the mixture.

The intended stability values given in Table 34.5 are the required design standard in BS5949 Part 2 [10].

The higher mix stability values of 10 kN produce mixtures that are difficult to compact and can be susceptible to cracking problems. The test involves the production of the following key activities.

1 Prepare a number of aggregate mixtures with different bitumen contents.

**Table 34.5** Required design standard in BS5949 Part 2 [10]

| *Traffic flow, commercial vehicles per lane per day (cvd)* | *Required design stability, kN* |
|---|---|
| Fewer than 1500 | 3 to 8 |
| 1500 to 6000 | 4 to 8 |
| Over 6000 | 6 to 10 |

2 Compact specimens of each mixture into Marshall test moulds using a standard compactive effort and compaction hammer.
3 Extract specimens from moulds and age the samples in a warm water bath.
4 Crush the specimens, recording the mixture 'stability value' at the failure load.
5 Record the strain at 4, above; this is the flow.
6 Plot graphs to calculate the optimised bitumen content and report the test result.

The test is accepted as flawed but is an accepted method for producing a durable HRA surface course material.

*34.9.5.3 End-product, patent surface course materials*

Heavily trafficked, adopted roads are currently surfaced with a group of end-product surface course materials. The materials are a departure from previous practice in that UK pavement specifications have traditionally employed various forms of recipe or precisely specified material. The end-product systems use an alternative specification method – a set of final end-product design requirements to which the material supplier then produces a certified product to meet the specification. The supplier then elects to design and detail the mix; the customer monitors performance to the defined specification. Two specific materials are currently supplied to this method; a surface course material and a high friction resistant surface treatment. A precise description of each method is outside the scope of this document; each specification is briefly described here.

*34.9.5.3.1 Clause 942 [4] thin surface course system (TSC)* The thin wearing course system is the current preferred specification for surfacing high-speed roads. The system is preferred to chipped hot rolled asphalt and is accepted as an improvement on the older materials. A specific problem can exist if the materials are over-compacted or incorrectly manufactured to produce a slippery bitumen film on the pavement surface.

A typical TSC system will consist of a thick bitumen bond coat which is initially sprayed on underlying binder course. A TSC mix will usually consist of a high void content material which is then rolled into the bond coat producing a high void content surface and a thick, dense lower layer. TSC systems can be susceptible to bitumen draining and over-compaction. Bitumen drainage is avoided by either adding fibre filler to the bitumen or using polymer modified binders; over-compaction is avoided by controlling the completed surface texture. A typical TSC system will have an 8% void content controlled using the completed pavement surface texture, which may be controlled to either a 1.2 mm or a 1.5 mm level.

**Figure 34.16** 0/10 Clause 942 TWC in an open-textured bitumen macadam to a low-speed urban road

**Figure 34.17** 0/14 Clause 92 Thin wearing course material to a high-speed urban road

The standard open textured TSC systems are typically either 14 mm or alternatively 10 mm aggregate gradings and are approved and certified by an independent body, the British Board of Agrément, using the HAPAS certification system.

Clause 942 [4] TSC systems currently control the following issues to ensure contract compliance.

1 Site stress level fixed at a five possible levels: Level 0 is the lowest level, Level 4 the highest standard.
2 Road tyre noise is controlled at four levels: Level 0 is the highest noise level, Level 3 the lowest standard.
3 Layer thickness is defined at three levels:
   a Type A, < 18 mm
   b Type B, 18 to 25 mm
   c Type C, 25 to 50 mm.
4 Surface texture; fixed using 'macrotexture' levels at four levels; Level 0, no requirement; Level 3, 1 mm surface texture after 2 years of service.

*34.9.5.3.2 Clause 924 [4] high friction surfacing* A second high-quality pavement treatment is also defined using an end-product test. High friction surface treatments are employed to provide a high-quality skidding resistance at the approaches to roundabouts, traffic signals, pedestrian crossings and any other high-risk sites. The treatment is used as an alternative to providing high-specification PSV aggregate within the surface course material. The system is essentially a high-quality surface dressing system involving the gluing of 4 mm high-quality high-friction aggregate onto a pavement at the approaches to high-risk areas. The length of material application is fixed as a function of the junction approach speed.

Clause 924 HFS are again certified by the British Board of Agrément and are patented systems. Three different standards of materials are defined as Type 1, 2 or 3. Type 1 is the lowest quality; Type 3 the highest.

The systems have a limited life and will quickly abrade off a pavement surface course in highly stressed locations.

*34.9.5.4 Binder course materials*

The layer immediately below the surface course is known as the binder course and is intended to provide a relatively impermeable deformation-resistant layer supporting the surface course.

**Figure 34.18** Traffic signal approach showing Clause 924 High Friction Surfacing

The material is fixed to a thickness of between 40 mm and 60 mm and is designed to be not only rut-resistant but also durable. The layer can be constructed as either a HRA or AC (formally DBM) material. Binder course materials are generally constructed using 0/20mm aggregate, but 0/32 and 0/14 materials can be used.

**Figure 34.19** Clause 924 High Friction Surfacing

Binder course materials can be either recipe mixes or designed mixes; the clause numbers in Table 34.6 are used to describe the standard mixes in the UK standards:

*34.9.5.5 Base (road-base) materials*

The base layer provides the structural support to the pavement system. The layer is usually constructed using a BS13108 Part 1 [8] continuously graded material. The layer can be constructed in either 0/20 mm or 0/32 mm aggregate gradings, which can be specified as either a recipe or design mix layers. The base material design process is described in Clause 929 [4] of the standard specification and involves a process by which a job standard mix is

**Table 34.6** Binder course materials: standard mixes

| | *British Standard* | *Specification [4] reference* |
|---|---|---|
| Design mixes | HRA BS EN 13108 Part 4 [10] | Clause 905/943 |
| | dense/HDM/HMB BS EN 13108 Part1 [8] | Clause 929 |
| | EME2 | Clause 930 |
| Recipe mix materials | HRA BS EN 13108 Part 4 [10] | Clause 905 |
| | Dense BS EN 13108 Part 1 [8] | Clause 906 |
| | HDM BS EN 13108 Part 1 [8] | Clause 931 |

**Table 34.7** Base materials: standard mixes

| | *British Standard* | *Specification [4] reference* |
|---|---|---|
| Design mixes | HRA BS EN 13108 Part 4 [10] | Clause 904 |
| | ACs dense/HDM/HMB Part 1 [8] | Clause 929 |
| | EME2 | Clause 930 |
| Recipe mix materials | HRA BS EN 13108 Part 4 [10] | Clause 904 |
| | ACs BS EN 13108 Part 1 [8] | Clause 906 |
| | HDM BS EN 13108 Part 1 [8] | Clause 931 |

designed and then approved. An alternative EME2 mixture may be used in heavy truck road construction. A successful binder or base material will have a low void content of perhaps 5%, will be well compacted, have an even textured surface and will have a binder content of between 3% and 4.5%. Heavily trafficked base and binder course materials are designed in compliance with Clause 929 [4], which defines the void content and therefore compaction requirements for the completed pavement layer. Base materials follow essentially the same specification standards as the binder course material. The clauses in Table 34.7 fix the materials.

*34.9.5.6 Clause 930 EME2 materials base and binder course materials*

A third group of heavy-duty stiff base and binder course materials are available known as 'Enrobe a Module Eleve 2' materials. This is a group of high bitumen content, stiff low pen materials which can be used as the main pavement layers within a motorway structure. The materials originate from France and require specialist design and construction methods but can produce a durable efficient pavement structure. It is expected that the materials will replace current dense/HDM/HMB and HRA materials within the motorway network. The system is extensively used throughout the European highway network. An EME2 material will have a very low void content, typically 3%, a high bitumen content of around 5% and will be manufactured from relatively small 10 mm or 14 mm aggregate using very stiff, low penetration grade 15 or 20 pen bitumen.

*34.9.5.7 Surface dressing and maintenance of pavement surface course materials*

An accepted maintenance technique for low trafficked bituminous surfaces is to surface-dress the pavement at regular 15-year periods. The technique is used as a method to restore a worn fretting, oxidised or polished pavement surface. The method involves applying a bituminous sprayed bond coat onto the pavement surface; a coating of 4 mm, 6 mm or 10 mm chippings is then spread onto the wet bitumen and rolled into place. A specific design method, Road Note 39 [13], is available to fix the rate of binder application and chipping sizes.

Surface dressed pavements are also used in new pavement constructions. Rural low trafficked roads can be successfully constructed as a combination of crushed rock base layers sealed with a two-coat surface dressing technique. This form of construction is extensively used in Africa and in other arid or semi-arid tropical countries.

**Figure 34.20** 0/4 Surface dressing to a low-trafficked urban low-speed county road

*34.9.5.8 Subbase materials*

The material supporting the bound bituminous material is the sub-base. The standard UK material is Clause 803 [4] Type 1 sub-base. The standard UK material is an even graded crushed rock manufactured from stable but lower quality aggregate compared to the bituminous bound materials. A lower quality Type 2 sub-base material is also available, manufactured from a lower strength rock. Other materials are also available that are waste products from other construction processes, road planning, quarry waste or cement bound materials from good quality as-dug gravel aggregates. Aggregate strength is fixed using the Los Angeles value test. Compaction of both Type 1 and Type 2 material is completed by a method specification linking roller weight, passes and thickness of aggregate layer. Compaction layer thicknesses are restricted to 110 mm, 150 mm or 225 mm; the thicker layers are used with the heavier plant.

Very few construction problems occur with Type 1 or Type 2 material. The materials are relatively user-friendly.

A system of in situ stabilisation materials known as soil stabilisation is also available, but a thorough discussion of this and other hydraulically bound materials is beyond the scope of this text.

*34.9.5.9 Capping materials*

Capping materials are a second group of lower quality, selected as-dug materials that are used in the construction of heavily trafficked roads on relatively weak subgrades. Capping materials can be manufactured from in-situ stabilisation techniques or can be imported selected as dug gravels or stiff clay materials. A thorough discussion of these materials is beyond the scope of this text.

### 34.9.6 Materials definitions and testing methods

*34.9.6.1 Bitumen specifications*

Bitumen forms an essential element of the bituminous pavement. The bitumen provides the lubrication assisting in the compaction of the macadam material and the glue bonding the crushed rock together. Bitumen is specified using a needle penetration test that involves allowing a needle to settle through a sample of bitumen at a defined test temperature of 25°C. The test is known as the 'pen test'; bitumen is supplied in accordance with BS EN 12591. A second test known as the 'softening point test' is used to fix the lower viscosity characteristic of the material. The second test is used to control the compaction temperature of the pavement material (Figure 34.24).

The standard paving grades of bitumen and the applications of the materials are given in Table 34.8.

The pen and softening point tests are used together to identify the bitumen viscosity characteristics.

An alternative viscosity test can be used which directly measures bitumen viscosity at different temperatures, but it is not generally used in the construction industry.

**Figure 34.21** Bitumen pen test apparatus

**Figure 34.22** Nottingham asphalt tester

Bitumen is occasionally modified using the addition of polymers. Polymer modification alters the bitumen viscosity to temperature relationship and other properties such as ductility. Polymer-modified binders are expensive and tend to be used in specialist and proprietary systems where enhanced or guaranteed performance is demanded. A discussion of polymer modification materials and techniques is beyond the scope of this text.

*34.9.6.2 ITSM asphalt stiffness testing using the NAT apparatus BS EN 12697-26*

An important core stiffness test apparatus is available which allows the stiffness and rutting resistance of a cored pavement sample to be recorded. The test is known as the NAT (Nottingham asphalt test), and can be used to measure the completed dynamic stiffness of a cored pavement sample.

Heavily trafficked, adopted roads can be defined using an end-product base and binder course specification. The test apparatus can be used to control an end-product specification but is not regularly used by the construction industry. The standard design mix Clause 929 [4] also includes the test to allow a record of cored stiffness data to be assembled.

*34.9.6.3 Aggregate testing methods*

The coarse and fine aggregates form an important element within the bituminous pavement construction. The aggregate grading is tested using a system of standard sieves. A sample of aggregate is shaken through the pan sieves and the residue held within each layer is weighed. The specification of the coarse aggregate can thus be tested and controlled.

*34.9.6.3.1 Aggregate strength testing* The strength and stability of the aggregate is tested using a number of different tests. The Loss Angeles value (LA) test is the key strength indicator. The test involves placing a sample of aggregate in a drum with fist-sized steel balls. The drum is then rotated to a set time and the mass of fines material knocked off the aggregate is recorded.

The other important aggregate test requirements are as follows:

1 1 *Polished stone value (PSV)* – the ability of an aggregate to withstand polishing from the action of tyres. A high

**Figure 34.23** Typical aggregate grading pans

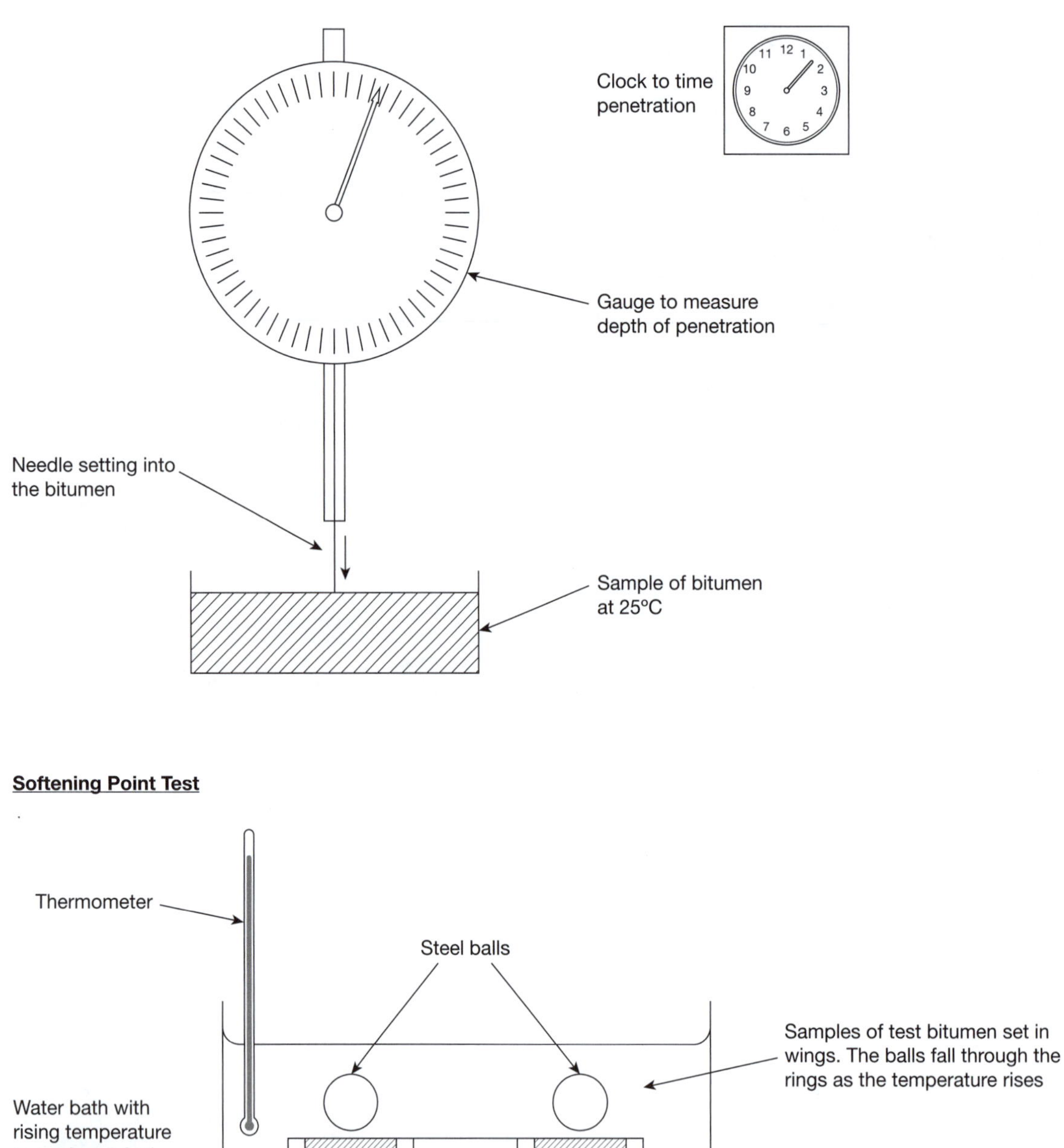


**Figure 34.14** Bitumen material tests

**Table 34.8** Standard paving grades of bitumen and the applications of the materials

| *Bitumen grade* | *Common applications* |
|---|---|
| 160/220 pen | Footways, drives and lightly trafficked pavement areas |
| 100/150 pen | Lightly trafficked roads |
| 40/60 pen | Heavy duty macadams and HRA materials |
| 30/45 pen | Specialist heavy-duty materials |
| 15/25 Pen | EME2 material only |

**Figure 34.25** A set of aggregate grading pans

**Figure 34.26** Polished stone value test equipment

**Figure 34.27** Los Angeles loss test equipment

polishing resistant value is 68; a low value is 45 and will easily polish.

2 *Aggregate abrasion value (AAV)* – the ability of a surface course material to withstand abrasion. A low-value, weak material will have a value of 16; a high-value, strong material will have a maximum value of 11.
3 *Freeze and thaw tests* – measure resistance to freeze–thawing for unbound materials.
4 Cleanliness and general chemical stability.

Specific details of the required aggregate strengths for different materials may be found in the relevant section of the Highways Agency specification Clause 901 [4].

## 34.9.7 Manufacture

### *34.9.7.1 Batching and mixing materials*

Bitumen materials are produced using specialist coating plant of differing configurations (Figure 34.30). Essentially, coarse and fine aggregate is prepared in a quarry by crushing to the required grading. Each element of product is stored in hoppers or silos before being weighed into a mixing drum. The aggregates are then heated and the hot bitumen added. The product is finally loaded into storage silos before loading into lorries to deliver to site.

Two basic types of coating plant exist: drum mixers and pugmill mixers. Drum mixing is a continuous operation; pugmill mixers are used for smaller quantities of material. The plant is controlled electronically to pre-set, designed mix batches. Mobile plants can be used on large projects but the standard plants are fixed quarry-based systems. The expense involved in constructing mobile plants can only be justified on major motorway projects; most bituminous materials are produced from fixed plants.

**Figure 34.28** Mobile drum batching plant

**Figure 34.29** Quarry-based batching

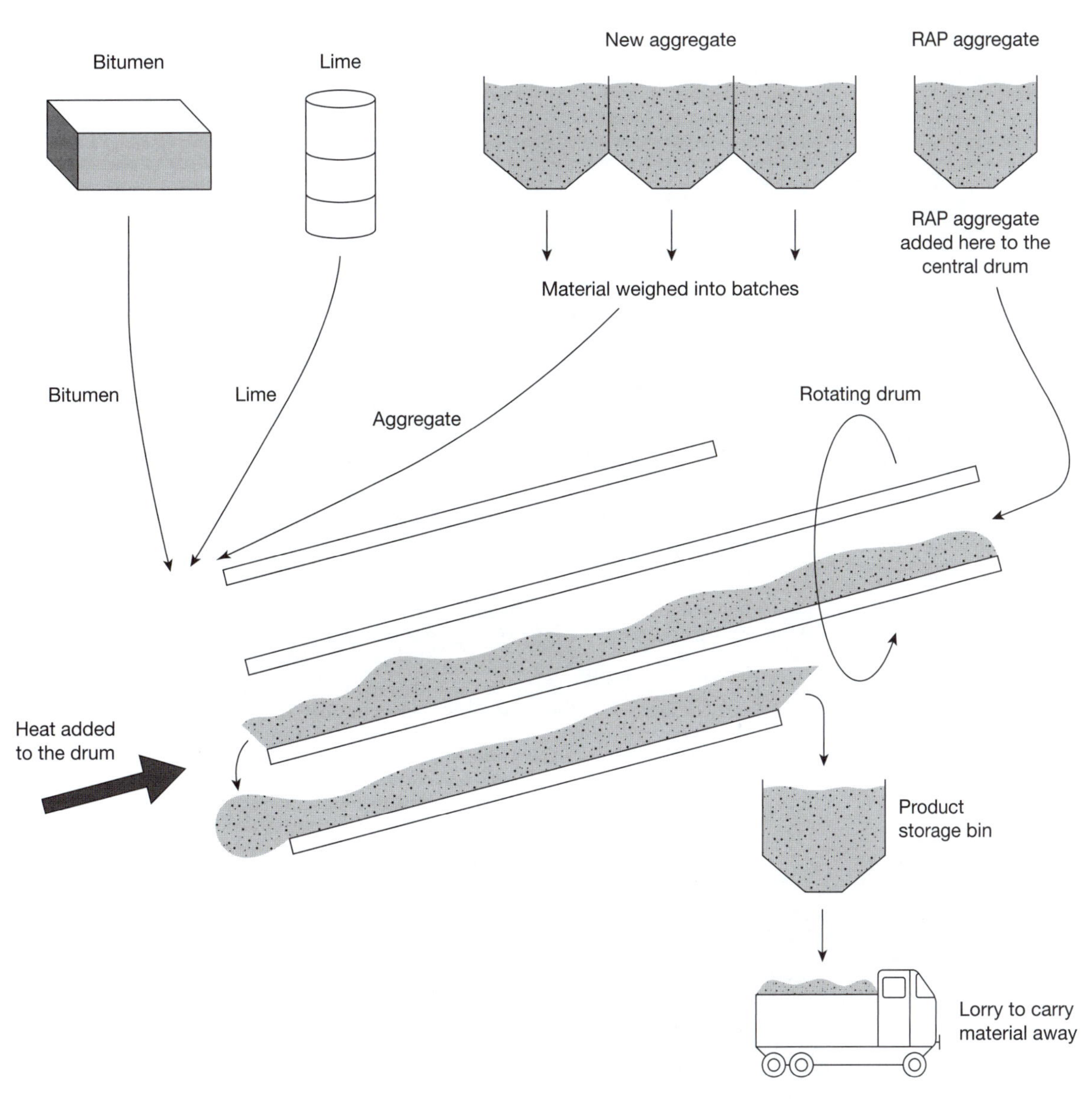


**Figure 34.30** Schematic diagram of a drum mixer continuous coating plant

*34.9.7.2 Constructing the pavement*

Bitumen pavements are constructed using different sets of construction plant to match the required standard of construction. Footways and domestic drives are constructed using hand laying and rolling techniques. Private and small adopted roads use simple paving machines and larger vibrating rollers. Motorway projects require specialist pavers and larger vibrating rollers.

The essential elements involved in constructing a bituminous pavement are as follows.

1 A system to spread the bituminous material into an even, controlled layer thickness. On footways this is completed using hand rakes and shovels; on larger projects a paving machine spreads the material.
2 A system to regulate the layer thickness. Paving machines contain an integral screeding beam which vibrates and tamps the bituminous material into an even layer. The tamping action produces a significant element of compaction.
3 A compaction element. The pavement mat is rolled using either dead weight or vibrating rollers to compact the bituminous material into a stiff durable layer. On large motorway contracts large 17 tonne twin drum vibrating rollers may be used; at the other extreme on footways hand-guided small 0.5 tonne rollers are used.

Each element of the construction process is essential to the final completion of the pavement structure. A successful pavement is produced if each stage is successfully completed.

**Figure 34.31** Road plane removing a fretted pavement surfacing material

**Figure 34.32** An EME2 paving operation

**Figure 34.33** EME2 compaction roller

**Figure 34.34** Footway construction

**Figure 34.35** Pavement cracking in a stiff bitumen material

**Figure 34.36** An oxidised fretting pavement surface

**Figure 34.37** Fatigue cracking of a pavement surface

#### *34.9.7.3 Pavement materials and quality control*

Bituminous materials are quality controlled using focused testing procedures at the following points in manufacture.

- At the batching plant; before mixing
  - (i) aggregate grading and quality
  - (ii) bitumen pen and softening point
- At the point of loading into a lorry; coated products after mixing
  - (i) completed bitumen content
  - (ii) mixing temperature
  - (iii) aggregate grading
- At the point of laying
  - (i) delivery temperature
  - (ii) complete bituminous mat temperature immediately after compaction
- From the completed pavements
  - (i) void content to control compaction
  - (ii) recovered bitumen, aggregate quality and dynamic stiffness of extracted cores.

Each material testing step is intended to ensure that the pavement materials are correctly constructed.

### 34.9.8 Failures and construction problems

Bituminous materials have a limited pavement life; regular maintenance is needed to surface course materials, to limit degradation of the pavement surface. Even the most precisely specified and well-built pavement will have a limited service life

**Figure 34.38** 0/4 Dense Bitumen Macadam showing a worn pavement surface to a footway

without regular maintenance. The surface course material will crack, oxidise and fret as the bitumen degrades with time and the action of trafficking. Various techniques can be used to restore; the surface can be surface dressed, overlaid or planed out and replaced. The most cost-effective techniques are:

1. surface dressing to minor roads at approximately 15 year intervals
2. planning out and replacing the surface course and occasionally the binder and upper base course to major motorways at 10 to 15 year intervals.

Pavement defects and maintenance problems may be classified as either construction defects or general deterioration problems.

#### *34.9.8.1 Construction defects*

The following are descriptions of typical construction problems.

1. *Poor compaction.* A general lack of pavement durability can result from inadequate compaction, giving an uneven fretting pavement surface. A well-compacted material must achieve a void content of between 3% and 6%, depending on the specification, to achieve a durable, long-lasting material.
2. *Segregation and binder drainage.* Poor mixing, laying and transport techniques can induce material segregation. The mixed material can become separated into coarse and fine material. A segregated material will have on open, patchy surface finish.
3. *Cold-laying.* Thin, low pen materials can be susceptible to cold-laying – the effect obtained when a pavement material cannot be correctly compacted as the material has fallen below the minimum rolling temperature. A cold-laid material lacks durability and will disintegrate easily. Cold-laying commonly occurs at joints. The problem is a function of temperature and wind speed; high winds will induce premature cooling of the paving material, preventing efficient compaction.
4. *Poor level control.* Inadequate level control can produce an uneven, poor ride. Poor site quality control can occur, producing layers that are either too thick or too thin. This problem is of greater significance than might be expected, especially on large, high-pressure motorway projects.
5. *Poor bond between pavement layers.* Low pen bituminous materials require stiff construction platforms, bond coats and good construction weather conditions. In some cases a bond cannot develop between pavement layers. An unbonded pavement is significantly weaker than a bonded structure.

*34.9.8.2 Durability problems*

The following list is a brief summary of pavement deterioration problems – typical maintenance problems found on UK roads.

1 *Rutting.* Two types of pavement rutting can be induced within a pavement structure. Surface course material rutting can occur in soft materials when heavy vehicles travel in a canalised pattern over the pavement structure. Adjusting the material mix design reduces the problem. Subgrade rutting occurs in weak thin pavements where the pavement structure is unable to protect the subgrade.
2 *Cracking.* Stiff bituminous materials are susceptible to environmentally induced cracking problems. General, long, regular cracks can be caused by over stiff, low bitumen content, mixes that are unable to creep in response to thermal strains.
3 *Reflective cracking.* Thin bituminous materials laid over concrete slab system will produce reflective cracking problems. The underlying concrete slabs move thermally producing strains which then induce cracking in the overlaying bituminous material. Increasing the thickness of bituminous overlay material prevents the problem.
4 *Oxidisation problems.* Open textured bituminous materials will oxidise and degrade producing an unstable pavement surface. The problem is typically found in open texture surface course materials.
5 *Fatigue cracking.* Pavement surface course materials will eventually fatigue producing a cracked pavement surface known as 'alligator' cracking. The problem is most obvious in old thin materials but will eventually occur in all surface course materials.
6 *Aggregate polishing problems.* Limestone aggregates can become polished by heavy truck wheel abrasion. A polished surface becomes slippery producing a safety hazard.
7 *Lack of surface texture.* A worn pavement surface can lack sufficient texture to prevent skidding. Skidding resistance is a function of surface texture, low texture depths produce problems in wet weather conditions; high texture depths can maintain skidding resistance when heavy rain occurs.

The above is not an exhaustive list.

### 34.9.9 Pavement management systems (PMS)

No text on bituminous pavement would be complete without some comment on pavement management systems (PMS). Two PMS systems are used in the UK: the Highway Agency uses a system known as HAPMS; county councils use a second system, UKPMS.

The HAPMS system is an inventory of trunk roads and motorways. The system is a centrally managed inventory of pavement constructions complete with machine-gathered records of the pavement condition. The system monitors the following indicators: (a) cracking, (b) surface texture, (c) rutting, (d) friction condition, and (e) the ride quality using the road profile.

The UKPMS system is a similar management tool but focused to describe the general condition of county roads; the system also includes comments on footway condition, road signs and other highway features.

Pavement management systems are an area of pavement technology that will develop to assist in focusing maintenance spending, improving highway efficiency.

## 34.10 Further information and references


- Shell. 1995. *The Shell bitumen industrial handbook.* London, UK: Thomas Telford.
- The Mastic Asphalt Council. www.masticasphaltcouncil.co.uk
- Refined Bitumen Association. www.bitumenuk.com
- Eurobitume. www.eurobitume.eu

1 Powell W.D., Potter J.F., Mayhew H.C. and Nunn M.E. TRRL Laboratory Report 1132, *The structural design of bituminous roads*, TRRL, Crowthorne, UK, 1984.
2 Nunn M.E, Brown A., Weston D. and Nicholls J. TRL Report 250, *Design of long-life flexible pavements for heavy traffic*, TRRL, Crowthorne, UK, 1997.
3 Nunn M., TRL Report TRL615, *Development of a more versatile approach to flexible and flexible composite pavement design*, TRRL, Crowthorne, UK, 2004.
4 *Manual of Contract Documents for Highway Works* Volume 1 Specification for Highway Works, Series 900 Road Pavements – Bituminous Bound Materials, November 2004.
5 *Manual of Contract Documents for Highway Works* Volume 2 Notes for Guidance on the Specification for Highway Works, Series NG900 Road Pavements – Bituminous Bound Materials, November 2004.
6 *Design Manual for Roads and Bridges*, Volume 7 Section 2 Part 3 HD 26/01 Pavement Design August 2001.
7 *Design Manual for Roads and Bridges*, Volume 7 Section 2 Part 3 HD 39/01 Footway Design May 2001.
8 British Standard BS EN 13108 Part 1 Asphalt Concrete, formerly BS4987-1:2005 Coated macadam (asphalt concrete) for roads and other paved areas – Part 1: Specification for constituent materials and for mixtures.
9 British Standard BS4987-2:2005 Coated macadam (asphalt concrete) for roads and other paved areas – Part 2: Specification for transportation laying and compaction.
10 British Standard BS13108 Part 4 Hot Rolled Asphalt, formerly BS 594-1:2005 Hot rolled asphalt for roads and other paved areas – Part 1: Specification for constituent materials and asphalt mixtures.
11 British Standard BS594987 Asphalt for roads and other paved areas – specification for transport, laying, compaction and type test protocols, formerly BS594-2:2005 Hot rolled asphalt for roads and other paved areas – Part 2: Specification for transportation, laying and compaction of hot rolled asphalt.
12 British Standard BS594987 Asphalt for roads and other paved areas – specification for transport, laying, compaction and type test protocols, formerly BS598:2004 Sampling and examination of bituminous mixtures for roads and other paved areas.
13 Nicholls J.C., Road Note 39, *Design guide for road surface dressing*, 4th Edition.
14 BS EN 12591: 2000 Bitumen and bituminous binders: Specification for paving grade bitumens.
15 PD 6691:2007 Guidance on the use of BS EN 13108 Bituminous mixtures – Materials specifications.
16 BS EN 13108: Part 2 Asphalt Concrete for very thin layers.
17 BS EN 13108: Part 5 Stone Mastic Asphalt.
18 BS EN 13108: Part 7 Porous Asphalt.
19 BS EN 12697: Part 26 ITSM test.
20 BS EN 12697: Part 34 Marshall Test.
21 BS8218 1998: Code of practice for mastic asphalt roofing.
22 BS6925 1988: Specification for building and civil engineering (limestone aggregate).
23 BS EN 13108-6 2006: Bituminous mixtures. Material specifications for mastic asphalt.
24 BS EN 13707 2004: Flexible sheets for waterproofing – reinforced bitumen sheets for waterproofing definitions and characteristics.
25 BS8217 2005: Code of practice for reinforced bitumen membranes for roofing.
26 BS EN 13304: Framework for specification of oxidised bitumens.
27 BS EN14023: Specification framework for polymer modified bitumens.
28 BS5284 1993: Methods of sampling and testing of mastic asphalt used in building and civil engineering.
29 BS8000-4 1989: Workmanship on building sites. Code of practice for waterproofing.
30 Doran D.K. (Ed.) 2009. Mastic asphalt. In: *Site engineers manual* (2nd Edition]. Whittles Publishing, Caithness, UK.

35

# Geosynthetics in Civil and Environmental Engineering

**Geoffrey B. Card** Eur Ing BSc(Eng) PhD FICE
GB Card & Partners

## Contents

## 35.1 Introduction

This chapter describes the use of geosynthetics in civil and environmental engineering and the more recent advances in ground applications to control the migration of ground pollutants. Plastics are used in many aspects of civil and environmental engineering, particularly in issues related to the ground. Historically plastics have been used to replace more traditional materials such as ductile iron and clay in buried piped systems for water and gas supply, surface and foul drainage. Plastic sheet membranes have also been developed for landfill liners as a substitute for natural low-permeability clay materials. With greater emphasis on sustainability, the use of recycled plastics is an increasingly attractive option rather than consuming natural materials. Plastics can also be synthesised with chemical and physical properties to provide bespoke products for particular applications. In this respect a particular group of plastics, known as geosynthetics, has been developed over recent years and now has a major application to civil and environmental engineering.

Geosynthetics are generally regarded to encompass four main products: geotextiles, geogrids (known also as geonets), geomembranes and geocomposites. The synthetic nature of the products makes them suitable for use in the ground, where high levels of durability are required because of chemical and microbial activity. Geosynthetics are available in a wide range of forms and materials, each to suit a slightly different end use.

Applications include highways, airfields, railways, embankments, retaining structures, reservoirs, canals, dams, slope protection and coastal engineering.

The early development of geosynthetics has been slow, mainly due to the limitations of the materials used. With advances in polymer science the development of geosynthetics has moved along swiftly over the past 25 years or so. A broad clasification and the suitablity of geosynthetics for a particular application is summarised in Table 35.1. Figure 35.1 shows some typical applications.

## 35.2 Characteristics

### 35.2.1 Geotextiles

The prefix of geotextile, *geo*, means earth and the *textile* means fabric. Therefore, according to the definition in ASTM 4439, a geotextile is "A permeable geosynthetic comprised solely of textiles. Geotextiles are used with foundation, soil, rock, earth, or any other geotechnical engineering-related material as an integral part of human-made project, structure, or system."

The ASAE (American Society of agricultural & Biological Engineers) defines a geotextile as a "fabric or synthetic material placed between the soil and a pipe, gabion, or retaining wall to enhance water movement and retard soil movement, and as a blanket to add reinforcement and separation". A geotextile should

**Table 35.1** Suitable applications

| | *Geotextiles* | *Geogrids/Geonets* | *Geomembranes* | *Geocomposites* |
|---|---|---|---|---|
| Filtration | 1 | | | 1 |
| Drainage | 2 | | | 1 |
| Separation | 1 | | | |
| Reinforcement | 1 | 1 | | |
| Barrier | 3 | | 1 | |

1 = suitable; 2 = partly suitable (in certain landfill applications); 3 = partly suitable (when appropriately coated).

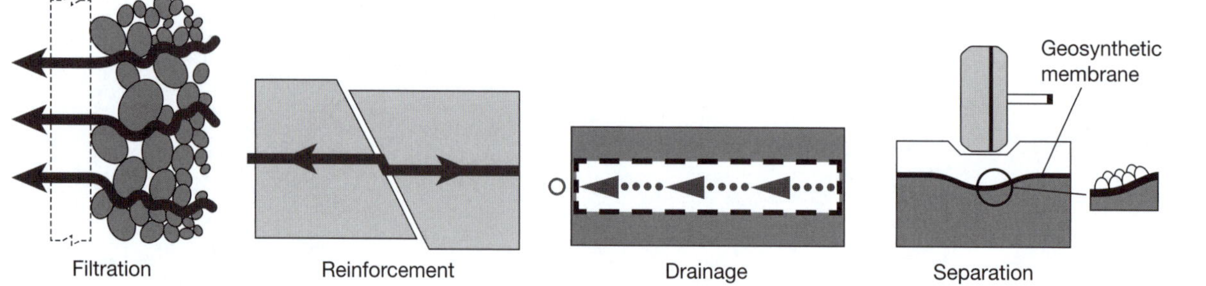


**Figure 35.1** Typical applications

consist of a stable network that retains its relative structure during handling, placement, and long-term service.

*35.2.1.1 Types of geotextile*

These are usually produced as either woven or non-woven geotextiles and are manufactured from polymeric yarns and fibres, consisting primarily of polypropylene, polyester, polyethylene and polyamide. The oldest of these is polyethylene, which was discovered in 1933 in the research laboratories of ICI. Another group of polymers with a long production history is the polyamide family, the first of which was discovered in 1935. The next oldest of the four main polymer families relevant to geotextile manufacture is polyester, which was first announced in 1941. The most recent polymer family relevant to geotextiles is polypropylene, which was discovered in 1954. The comparative properties of these four polymer plastics are given in Table 35.2.

The properties of polymer material are affected by its average molecular weight (MW) and its statistical distribution. The effects are summarised in Table 35.3.

Woven geotextiles are produced by the interlacing of yarns of polymer plastic to leave a finished material that has a discernible warp and weft. Non-woven geotextiles are produced by various methods other than weaving, mainly heat bonding, needle punching and chemical bonding of plastic filaments. There are a small group of geotextiles that are produced from fibrous materials, used mainly in erosion control. Their degradable characteristics are beneficial in some applications. In general, the vast majority of geotextiles are made from polypropylene or polyester formed into fabrics as follows:

- woven mono- and multi-filament
- woven slit-film mono- and multi-filament
- nonwoven continuous filament heat bonded
- nonwoven continuous filament needle punched
- nonwoven staple needle punched
- nonwoven resin bonded
- other woven and nonwoven combinations
- knitted.

*35.2.1.2 Woven geotextiles*

Woven geotextiles are manufactured using traditional weaving looms in which transverse elements (the weft) are woven between longitudinal elements (the warp). The structure of the finished fabric and, consequently, the properties of the geotextile will be governed by the particular form of the warp and weft and the polymer type used in the manufacture of these elements. By far the most common structure is formed by the weaving of flat tapes which may be extruded directly as tapes or slit from a wide sheet of film. In very heavy-weight fabrics the flat tape may be twisted into a tape yarn. Most commonly, polypropylene is used in the manufacture of tape, and the corresponding structure is termed "tape-on-tape". This particular structure has a small open area ratio and, consequently, would be expected to have a comparatively low capacity for conducting a flow of water normal to the plane of the fabric. In contrast, if the tape elements are replaced by circular cross-section monofilaments, the resulting structure, termed "monofilament-on-monofilament", has a much larger open area ratio and an extremely high capacity to transmit water normal to the plane of the fabric. Consequently, this type of structure, usually employing polyethylene and/or polypropylene monofilaments, is appropriate in drainage applications where high flow rates are to be accommodated. Where very high tensile strength is required, yet another structure is employed in which the warp and weft elements comprise low twist yarn manufactured from fine polyester (pes or pet) filaments. The resulting structure is termed "multifilament-on-multifilament". Depending on the requirements to be met, these different structures can be combined. So, for example, the requirement for a geotextile with

**Table 35.2** Property comparison

| *Plastic* | *Polyester* | *Polyamide* | *Polypropylene* | *Polyethylene* |
|---|---|---|---|---|
| Strength | H | M | L | L |
| Elastic modulus | H | M | L | L |
| Strain at failure | M | M | H | H |
| Creep | L | M | H | H |
| Specific gravity | H | M | L | L |
| Cost | H | M | L | L |
| UV resistance | H | M | H | H |
| Alkali resistance | L | H | H | H |
| Fungus vermin resistance | M | M | L | L |
| Hydrocarbon resistance | M | M | L | L |
| Detergent resistance | H | H | H | H |

H = high, M = moderate, L = low.

**Table 35.3** Effect of molecular weight

| *Action* | *Effect* |
|---|---|
| Increase in molecular weight of polymer | Increase in tensile strength, elongation, impact strength, stress crack resistance, heat resistance |
| Narrowing the molecular weight distribution of polymer | Increase in impact strength; decreased stress crack resistance and processability |
| Increase in crystallinity of polymer | Increase in stiffness or hardness, heat resistance, tensile strength, elastic modulus, chemical resistance. Decrease in permeability, elongagtion, flexibility, impact strength, stress crack resistance |

very high unidirectional strength and a very high water flow capacity may be met using the multifilament-on-monofilament structure. This is the essence of woven geotextiles where the fabric structure, and polymer type, may be chosen to provide the required properties defined in the design process.

*35.2.1.3 Non-woven geotextiles*

The basic element in the manufacture of non-woven geotextiles is the continuous filament which is extruded from overhead "spinnerets" onto a conveyor belt to form a loose web. The general term "spunbonded" is used to describe this method of forming the web. An alternative method of web formation makes use of staple fibre which is essentially short lengths of filament as opposed to continuous filament. The loose web so formed is subsequently given strength and coherence by one or sometimes a combination of three processes.

1 *Heat bonding* involves heating and rolling the loose web of continuous filaments. This causes the filaments to melt and subsequently fuse together at their crossover points. This process is also known as thermal bonding, melt bonding or melding.
2 *Chemical or resin bonding* involves pre-treatment of the web with a chemical adhesive or synthetic resin which upon rolling causes the filaments to bond together at their points of contact. This process is used in a small minority of non-woven geotextiles.
3 *Mechanical bonding* is achieved using a battery of thousands of fine barbed needles which reciprocate rapidly. The motion of the needles is normal to the plane of the loose web, with the needles penetrating the full thickness of the web. The actual mechanism of bonding is simply entanglement of the filaments. Following "needle punching" the entangled web is rolled to give a finish. Needle punching is also frequently used to bond staple fibres, which are short lengths, typically 100 mm of cut continuous filaments.

As would be expected, the structure of the geotextile and polymer type employed has a radical effect on the mechanical and hydraulic properties. Some aspects of these variations are addressed in Sections 35.3 and 35.4, which consider geotextile properties in the context of design functions.

## 35.2.2 Geogrids

Geogrids are grid-like plastic materials (hence the name) with relatively large open spaces called apertures between plastic polymer ribs. Just as with other types of geosynthetics such as geotextiles, geogrids can be made of a number of different polymers. They are generally stiffer than geotextiles and have relatively large voids within the material. Methods of production vary but include extrusion, bonding and interlacing. They can be produced from a range of plastic polymeric materials similar to those used for geotextiles (see Section 35.2.1.1). Geogrids are used to strengthen fill materials in geotechnical applications. They provide increased shear strength between soil strata interfaces. Their tensile strength can prevent or decrease the degree of differential settlement in some applications by transmitting the load over a broader area of soil, thereby diminishing the vertical stress in the soil.

A geogrid is defined[1] as a "geosynthetic used for reinforcement which is formed by a regular network of tensile elements with apertures of sufficient size to allow strike-through of surrounding soil, rock or other geotechnical material".

Geogrids can be either uniaxial or biaxial. Uniaxial geogrids are designed to endure stress in one direction. The ribs of this variety of geotextile tend to be thicker and the apertures are long, narrow slits. This variety of grid tends to be stronger than biaxial, but can only be applied in situations where stresses occur in a single direction (unless two uniaxial grids are placed in opposite directions).

Biaxial geogrids can take stresses in two directions. The apertures are more evenly dimensioned. They are useful in situations where stresses are applied in two directions, but do not have as much tensile strength in either direction as a uniaxial geogrid does in its direction of application.

The high tensile strength that geogrids possess makes them ideal for soil reinforcement applications. Their large aperture size limits their effectiveness as a filtration layer unless used with a widely graded soil with the coarsest material (gravel or rocks) adjacent to the geogrid or in conjunction with a geotextile to act as a separator or filtration layer. Some main uses of geogrids are given in Table 35.4.

## 35.2.3 Geomembranes

*35.2.3.1 Characteristics*

Geomembranes are virtually impermeable, thin synthetic barriers used to control fluid migration. As a result of their low permeability the principal driving mechanism for contaminant transport will be chemical concentration gradients giving rise to diffusion, rather than hydraulic gradients giving rise to advection flow. The vast majority of geomembranes are made up of thin sheets of polymeric materials which are also sometimes referred to as flexible membrane liners (FMLs). Geomembranes can also be made from the impregnation of geotextiles with asphalt or elastomer sprays.

Geomembranes are manufactured in several ways, excluding woven methods as they would leave unacceptably large voids in the material. The polymers used to make geomembranes are synthetic chemical compounds of high molecular weight and are generally thermoplastic. Typical materials include polyvinyl chloride (PVC), polypropylene, low-density polyethylene (LDPE) and high-density polyethylene (HDPE). Table 35.5 lists a number of common geomembranes.

The individual properties of polymeric geomembranes have been categorised and described in detail by Koerner (1990).[2] He subdivided the properties into the following categories, which are often used to characterise individual geomembranes.

**Table 35.4** Typical geogrid applications

| *Type of work* | *Application* |
|---|---|
| Highway pavements, railways and runways | Increase in sub-grade strength, as asphalt reinforcement in surface wearing course |
| Earthworks, embankment fills | Beneath surcharge fills, reinforcement of embankment fills, construction of stiff granular mattresses over soft and compressible ground |
| Retaining structures | As soil-reinforced retaining walls, gabions for wall construction and erosion control, sheet anchors for retaining wall facing panels |
| Slopes | Repairing slope failures and landslides |

**Table 35.5** Common types of geomembranes (after Koerner, 1990[2])

| *Thermoplastic polymers* | *Thermoset polymers* | *Combinations* |
|---|---|---|
| • Polyvinyl chloride (PVC)<br>• Polyethylene (VLDPE, LDPE, LLDPE, MDPE, HDPE, referring to very low, low, linear low, medium, and high density)<br>• Chlorinated polyethylene (CPE)<br>• Elasticised polyolefin (3110)<br>• Ethylene interpolymer alloy (EIA, also called XR-5)<br>• Polyamide (PA) | • Isoprene-isobutylene (UR), or butyl<br>• Epichlorohydrin rubber<br>• Ethylene propylene diene monomer (EPDM)<br>• Polychloroprene (neoprene)<br>• Ethylene propylene terpolymer (EPT)<br>• Ethylene vinyl acetate (EY A) | • PVC-nitrile rubber<br>• PE-EPDM<br>• PVC-ethyl vinyl acetate<br>• Cross-linked CPE<br>• Chlorosulphonated polyethylene (CSPE, also called "Hypalon") |

1 *Physical properties:* Thickness, density, mass per unit area, permeability to water vapour transmission.
2 *Mechanical properties:* Tensile strength, tear and impact resistance, puncture resistance, stress fatigue cracking, flexibility, membrane friction.
3 *Chemical properties:* Chemical resistance, ozone and ultraviolet light resistance.
4 *Thermal properties:* Expansion and contraction properties, affect mechanical properties.

*35.2.3.2 Design criteria*

The choices open to a designer with respect to geomembranes include the choice of material, the thickness and the joining or seaming requirements.

*35.2.3.2.1 Choice of geomembrane material* Some advantages and disadvantages of the basic polymers of geomembranes are given by Koerner (1990)[2] and reproduced in Table 35.6. One of the important advantages is resistance to chemical attack. HDPE has the best all-round chemical resistance properties of the geomembrane types currently available. It also can be seamed effectively and is relatively cheap compared with other material types. For these reasons, HDPE is by far the most common geomembrane material used in contamination containment barriers. HDPE does, however, have certain disadvantages such as low-friction surfaces, which can lead to stability problems on side slopes and it is relatively easily punctured. Certain polyethylene geomembranes are available with a textured surface, in an attempt to combat the low-friction surface problem. However, this can lead to a reduction in the tensile strength and to difficulties in seaming, and may be less effective if the surface becomes muddy. Good seaming is easier to achieve with medium-density polyethylene (MDPE) than with HDPE; however, the lower the density, the less durable the material is considered. This presents a degree of compromise to the designer, who may need to consider the trade-off

**Table 35.6** Some advantages and disadvantages of the basic polymers of geomembranes (after Koerner, 1990[2])

| *Advantages* | *Disadvantages* |
|---|---|
| *Polyvinyl chloride (PVC) thermoplastic* | |
| • low cost<br>• tough without reinforcement<br>• lightweight as single ply<br>• good seams – dielectric, solvent, and heat<br>• large variation in thickness | • plasticised for flexibility<br>• poor weathering, backfill cover required<br>• plasticiser leaches over time<br>• poor cold crack resistance<br>• poor high-temperature performance<br>• blocking possible |
| *Chlorinated polyethylene (CPC) thermoplastic* | |
| • good weathering resistance<br>• easy seams – dielectric and solvent<br>• cold crack resistance is good<br>• chemical resistance is good | • moderate cost<br>• plasticised with PVC<br>• seam reliability<br>• delamination possible |
| *Chlorosulphonated polyethylene (CSPE) thermoplastic rubber* | |
| • excellent weathering resistance<br>• cold crack resistance is good<br>• chemical resistance is good<br>• good seams – heat and adhesive | • moderate cost<br>• fair performance in high remperatures<br>• field repairs are difficult<br>• blocking possible |
| *Elasticised polyolefin (3110), thermoplastic EPDM-cured rubbers* | |
| • good weathering resistance<br>• lightweight as single ply<br>• cold crack resistance is good<br>• chemical resistance is good | • unsupported only<br>• poor high-temperature performance<br>• field repairs are difficult |

*(Continued)*

**Table 35.6** *(Continued)*

| *Advantages* | *Disadvantages* |
|---|---|
| *EPDM (4060) thermoplastic rubber* | |
| • good weathering resistance | • moderate cost |
| • cold crack resistance | • fair in high temperatures |
| • good seams – heat bonded | • blocking possible |
| • no adhesives required | • fair chemical resistance |
| *Butyl, butyl, EPDM, EPDM-cured rubbers* | |
| • fair to good weathering resistance | • moderate to high cost |
| • low permeability to gases | • poor field seams |
| • high temperature resistance is good | • small panels |
| • nonblocking | • fair chemical resistance |
| *Chloroprene (neoprene) cured rubber* | |
| • good weathering resistance | • high cost |
| • good high temperature resistance | • fair field seams – solvent and tape |
| • good chemical resistance | • fair seams to foreign surfaces |
| *High density polyethylene (HDPE) semicrystalline thermoplastic* | |
| • chemical resistance is excellent | • low-friction surfaces |
| • good seams – thermal and extrusion | • stress crack sensitive |
| • large variation in thickness | • seam workmanship critical |
| • low cost | • high thermal expansion/contraction |
| *Medium, low, very low density polyethylene (MDPE, LDPE, VLDPE) semicrystalline thermoplastic* | |
| • chemical resistance is good | • moderate thermal expansion/contraction |
| • good seams – thermal and extrusion | • low-friction surfaces |
| • large variation in thickness | • seam workmanship – critical |
| • low cost | |
| • no stress crack | |
| *Linear low density polyethylene (LLDPE) semicrystalline thermoplastic* | |
| • chemical resistance is very good | • moderate cost |
| • good seams – thermal and extrusion | |
| • large variation in thickness | |
| • high-friction surface | |
| • no stress crack | |

between flexibility and ease of seaming on one hand and long-term resistance to chemical attack on the other. Notwithstanding this, very good seams are made regularly with HDPE but do require greater skill and a high level of site quality assurance.

Very low-density polyethylene (VLDPE) is becoming more widely used in cover systems, particularly where differential settlement is expected, since VLDPE is very flexible and is capable of more elongation than HDPE.

If plasticity of the geomembrane is a key design consideration, it must be understood that in the long term the material will lose plasticiser and become less flexible. This may create durability problems in the future and should be taken into account in the design.

Geomembranes have traditionally been used in construction in ground or sub-structures as barriers against water and water vapour ingress. For many years they have also been developed in landfill engineering where geomembranes are used as impermeable liners to waste containment cells preventing landfill leachate and landfill gas migration into the surrounding ground.

The corrosive environment of a landfill has resulted in much research and development of high-resistant plastic membranes that can survive in a harsh chemically aggressive environment and sustain high elongation and deformation without rupture. For this reason an extensive range of geomembranes are widely available to suit all many aspects of civil and environmental engineering such as barrier systems to groundwater, effluent and leachate, and landfill gas (see Sections 35.3 and 35.4).

### 35.2.4 Geocomposites

Geocomposites can be defined as a combination of any two or more geosynthetic materials to form a product for a particular application. The materials and manufacturing methods vary with the constituent geosynthetics used. They are usually bespoke products for particular applications, such as prefabricated vertical band drains and gas vent wells (see Sections 35.3 and 35.4).

Geocomposites combine the characteristics of the individual geosynthetic components. For example, geocomposites that can act as drainage layers when combined with a geomembrane are particularly useful in the construction of landfill waste containment cells and environmental remediation barriers. The thickness of the geocomposite blanket is of the order of 100 mm while a conventional granular drainage layer and clay impermeable layer can be of the order of 1000 mm, i.e. 10-fold greater in thickness. For this reason geocomposite blankets can result in significant savings of volumes of material as well as on the additional excavation required to accommodate an engineered drainage and impermeable layer constructed of natural materials.

### 35.2.5 Geosynthetic clay liners

Geosynthetic clay liners (GCLs) is a generic name given to a group of prefabricated liner materials generally consisting of a thin layer of dry clay powder sandwiched between two geotextiles or glued to a geomembrane. They have also been known as clay mat, bentonite matting, bentonite clay liner, prefabricated clay blanket, or prefabricated clay liner. All the currently manufactured GCLs use approximately 5 $kg/m^2$ of bentonite as the powdered clay material.

Since their introduction, their potential as a component within a contamination barrier has been considered. Comprehensive overviews of the four types of GCLs available are given by Eith, Boschuk and Koerner (1991)[3] and Daniel (1993).[4]

GCLs, like geomembranes, are fabricated and placed on rolls in a factory, stored, transported to the site and unrolled in their final location. GCLs are generally self-sealing at overlaps between panels, where water hydrates the clay which swells and automatically seals the overlap. After the GCL is placed, it must be covered immediately. If the GCL becomes wet, e.g. from rainwater, the clay swells in an uneven manner and the material will not self-seam properly. Also, unconfined swelling of the bentonite will greatly increase its hydraulic conductivity and this may compromise the design requirements. A GCL should not be installed in standing water or in rain; it must be dry when installed and dry when covered. Products are being developed that incorporate two geotextiles to help prevent unrestrained swelling. Hydration of the bentonite should only be allowed once sufficient load has been applied to counter the effects of uncontrolled swelling. The magnitude of this load is not defined absolutely but is often stated to be equivalent to the weight of the overlying drainage layer, typically a minimum of 150 mm to 200 mm thick. Where a GCL is used immediately below an FML as a composite liner it often remains unhydrated for the most part, only becoming hydrated locally by liquid contaminants passing through a physical defect in the overlying geomembrane. In such situations the first hydration of the bentonite will not be by water but by some contaminated water (or even a non-aqueous liquid). It is important to understand whether or not such a liquid would be able to hydrate the bentonite and effect a low hydraulic conductivity seal. This is unlikely to be the case, for example, where oily contaminants with very low water content are to be contained.

GCLs' primary use with respect to contamination barriers is as the low permeability barrier component within a cover system. They have been used for this purpose in the UK, the USA, Germany and the Netherlands. They have also been used in the USA as the clay component of a primary composite liner for double-liner systems for landfill sites.

## 35.3 Applications in civil engineering

### 35.3.1 Separation and filtration

Ingold[5] defines the function of a separator as a geotextile placed between two different soil types to prevent intermixing. An example of this is a granular fill placed over a cohesive soil formation with a geotextile at the fill–formation interface. Another application is the use of geotextile separators to act simply as a physical marker layer between differing soil layers, such as in landscaping to identify sub-soil and topsoil growing media.

More usually the separation function is performed in parallel with one of the other prime functions. When a geotextile is laid and fitted over a cohesive clay formation it is unusual for the formation soil to be extruded through the geotextile. However. if the formation is waterlogged, then water would drain through the geotextile during construction. If the granular fill is not to be contaminated by fines from the formation soil, this introduces the need for some filtration criterion. This situation becomes more demanding for separators in paved or unpaved roads where formation pumping is to be controlled, since for the geotextile to separate the formation soil and the granular layers of the pavement it must function as a dynamic filter and drain.

Where a geotextile is required to act purely as a separator then design considerations extend only to consideration of serviceability, where use can be made of past experience and specifications can be drawn up using simple index test properties such as puncture, tear and burst resistance. Where the geotextile is required to perform additional functions, these must be designed for accordingly. It cannot be emphasised too strongly that it is vital to give full consideration to geotextile serviceability in this application.

Filtration can significantly enhance the performance of a geotechnical structure, and geosynthetics can be used to produce an effective filtration system. The function of a filter is to allow water to pass through the plane of the filter while retaining particles of the filtered soil, and thus control erosion of the structure. When fines get washed out of a soil it can reduce the cohesion of the matrix and thus the strength of the soil, referred to as piping. Mitigating erosion also improves the durability of a structure. Geotextile filters can improve the reliability and performance of traditional graded soil filters and require less work to construct.

There are two broad criteria to be met for a geotextile to perform adequately as a filter.

1 The pore sizes of the geotextile must be small enough to prevent significant loss of soil particles from the base soil.
2 The geotextile must have a sufficiently high permeability to water, normal to its plane, to transmit water flowing from the base soil.

These are the same basic criteria that apply to the design of a granular filter medium.

#### *35.3.1.1 Pore size*

In designing geotextile filters to prevent piping or erosion under steady-state unidirectional flow conditions, it is usual to relate the pore size of the geotextile to the particle size of the base soil being protected. A typical criterion would be

$$0_{90} \leq \lambda d_{90}$$

where $d_{90}$ is the 90% passing particle size of the base soil and $\lambda$ is a coefficient that normally has a value between 1 and 2. There are a wide range of design rules originating from numerous workers. Very often the design rules are related to a particular definition of the characteristic pore size determined by a particular test method. Consequently, some care must be exercised not to mix a particular design method with the manufacturer's product data determined using an incompatible test method. For example, some methods of filtration design rely on soil grading curves and determination of the uniformity coefficient of the base soil and flow conditions (see Table 35.7).

In woven geotextiles there is a comparatively small variation in pore sizes for a given fabric structure and these pores are regularly spaced over the fabric. For non-woven fabrics there is a somewhat wider variation in pore sizes and their distribution for a given fabric. For both types of fabric the range of pore sizes is

**Table 35.7** Geotextile filtration design criteria

| | |
|---|---|
| *Non-cohesive soils* | *Laminar flow* |
| Determine uniformity coefficient $U = d_{60}/d_{10}$ | If $U \geq 5$ then $O_{90} < 10\, d_{50}$ & $O_{90} < d_{90}$<br>If $U < 5$ then $O_{90} < 2.5\, d_{50}$ & $O_{90} \leq d_{90}$ |
| | *Turbulent flow* |
| | $O_{90} < d_{50}$ |
| *Cohesive soils* | *For all flow conditions* |
| | $O_{90} < 10\, d_{50}$ & $O_{90} < d_{90}$ & $O_{90} < d_{50}$ |

determined by sieving soil, or glass beads, of a known grading through the geotextile. The pore sizes will take up a distribution similar in form to the distribution of particle sizes for a soil. To simplify design procedures the filtration characteristics of a given geotextile are represented by a single pore size measured in micrometres. Common use is made of the $0_{90}$ pore size. This means that 90% of the pores in the geotextile are this size or smaller. In comparing pore sizes of different geotextiles, care should be exercised on several accounts. Firstly the $0_{90}$ pore size is not universally accepted as the standard pore size characterising the geotextile.

Pore size, say $0_{90}$, will vary with the structure of the geotextile and its mass per unit area. For non-woven fabrics, pore size tends to decrease with increasing mass per unit area. This trend is more pronounced in heat-bonded than mechanically bonded fabrics. The $0_{90}$ dry sieving pore size for non-woven fabrics varies from a minimum of approximately 30 μm to a maximum of some 300 μm. With woven-tape fabrics, pore size tends to increase with mass per unit area and at heavy weights the structure may change from flat tape to tape yarn. The $0_{90}$ pore size for such woven fabrics varies from approximately 100 to 500 μm. Much larger pore sizes are obtained for woven monofilament-and-monofilament fibres and these vary typically from 200 μm to values in excess of 1000 μm. The smallest pore size for a woven fabric is obtained using a multifilament-on-multifilament structure where $0_{90}$ values of less than 100 μm can be obtained.

#### *35.3.1.2 Permeability*

The other aspect of behaviour that must be taken into account in designing a filter is compatibility of permeability between geotextile and base soil. As an index test the water permeability of geotextiles is measured without the geotextile being in contact with soil. Since it is difficult to measure accurately the hydraulic gradient for flow normal to the plane of a geotextile, the permeability is defined in terms of the flow rate per unit area under a known driving head. In the UK and Europe this is commonly taken as a constant head of 100 mm of water. The resulting water flow, in litres per second per square metre (l/s $m^2$), will be governed, among other things, by the structure of the geotextile. The lowest water permeabilities are returned by the woven-tape fabrics at typically 10 to 30 l/s $m^2$. Heat-bonded non-woven fabrics have typically a water permeability of 30 to 250 l/s $m^2$. Mechanically bonded non-woven fabrics typically return to 50 to 300 l/s $m^2$, while the highest flow rates, in some cases exceeding 1000 l/s $m^2$, are obtained with high open ratio woven monofilament fabrics. Initially for a dry fabric, however, a driving head of greater than 100 mm of water may be required to initiate water flow.

In translating these index test values of water permeability into the design, several aspects of behaviour must be taken into account. For example, with the thicker non-woven fabrics, water flow normal to the plane of the fabric decreases as the level of normal stress increases. In addition, when a geotextile is in contact with base soil there will be some clogging and blocking of the geotextile pores by soil fines transported from the base soil. This will result in a reduced flow rate through the geotextile. To compensate for this in non-critical applications it is common practice to select a geotextile with a higher permeability than that of the base soil.

One reason why the capacity of a geotextile to transmit water normal to its plane is expressed in terms of "water permeability" is that this parameter does not involve quantification of the hydraulic gradient operating across the geotextile. This leads to complications in matching the hydraulic properties of geotextiles to the soils that they filter in terms of the coefficient of permeability, $k$ (m/s).

Groundwater flow is generally laminar and, therefore, the flow rate is proportional to hydraulic gradient as defined by Darcy's law: $q = kiA$, where $q$ is the volumetric flow rate ($m^3$/s), $k$ is the coefficient of permeability (m/s), $i$ is a dimensionless hydraulic gradient and $A$ is the cross-sectional area of flow ($m^2$). Flow through most thick non-woven fabrics is close to laminar and, therefore, if the thickness of the geotextile, the driving head and the flow rate are known, it is reasonable to calculate a coefficient of permeability for the geotextile, $k_g$, using Darcy's law. However, for thin non-woven fabrics and most woven fabrics, water flow under test conditions is non-laminar. In this connection Darcy's law is modified to take the form: $q^n = kiA$. The index, $n$, varies between 1 for laminar flow and 2 for purely turbulent flow.

Flow conditions for a particular geotextile can only be assessed by studying a plot of flow rate against hydraulic gradient. Where this is linear, Darcy's law may be applied to determine $k_g$. Where the plot is non-linear, $k_g$ can only be assessed using the modified version of Darcy's law. For non-critical applications the permeability of the geotextile, $k_g$, measured at an appropriate normal stress level is generally selected to be at least 10 times greater than the permeability of the soil being filtered.[6] In more critical applications the time-dependent flow characteristics of the soil–geotextile system should be investigated in the laboratory.

### 35.3.2 Drainage

Geosynthetics are widely used in drainage situations such as to remove surface water infiltration from land or to reduce hydraulic pressures behind retaining structures. Traditionally granular materials have been used as drainage materials. However, with increasing costs and scarcity of natural raw materials there is an increasing use of geosynthetic substitute products. Geocomposites constructed of a high-permeability geotextile core sandwiched between geotextile filter layers are used extensively as drains for water and/or soil gas as discussed below.

Drainage, as a geotextile function, is taken as the ability to transmit fluid in the plane of the geotextile. Consequently, it is only the thicker non-woven fabrics, particularly mechanically bonded structures, that exhibit any useful flow capacity. This in-plane capacity, which is the product of the geotextile thickness and coefficient of permeability, is called the transmissivity and is usually measured in units of litres per second per metre width of fabric (l/s m) for a given hydraulic gradient. Since the thickness of mechanically bonded non-woven fabrics reduces as the normal stress level increases, so then does the transmissivity.

In general, geotextile fabrics are not now used for drainage of water because much higher permeable geocomposites are available. These are laminates comprising a geotextile filter and a water-conducting core. The transmissivity of such composites can be several orders of magnitude higher than a simple geotextile fabric, and flow rates of up to 10 l/s m at normal stress levels in excess of 150 kN/$m^2$ can be attained. In this connection geocomposites are particularly useful for structural drainage, for example behind retaining walls where they replace sand or gravel or no-fines concrete drains, and for leakage detection and interception beneath geomembrane landfill liners.

The transmissivity of geotextile or geocomposite drainage fabrics to soil gas is much higher than to water and, consequently, such products can be used to disperse soil gas (see Section 35.4). In this regard geocomposite drainage products are used beneath geomembrane liners to landfill waste sites, where they function as both a gas and leachate drainage layer.

### 35.3.3 Barriers

In some geotechnical applications it is necessary to separate or protect one section of the works from another. This could be for a multitude of reasons, including stopping leachate seepage, protecting a structure from moisture and protecting a geotechnical structure from erosion. Geosynthetics such as geomembranes have been widely used over the past 30 years or so in such situations as landfill lining for leachate and landfill gas containment, vertical hydraulic barriers to groundwater or effluent flow.

A development from the geomembrane is stiff plastic sheet panels as in-ground vertical barriers. An example is the use of plastic sheet panels installed to some 5 m deep along a 500 m section of coastline to protect the existing residential area and proposed future building developments behind the beach from seawater intrusion. The solution is both cost-effective and based on sustainable engineering principles.

The geomembrane vertical barrier comprises recycled polyvinyl chloride sheets that slide together along a male and female clutch. The clutch is sealed using a hydrophilic strip that swells in contact with water, forming an impermeable water-tight seal.

The system has major advantages over more traditional materials such as steel sheet piles or bentonite–cement slurry walls. These include:

- the relatively low material cost of recycled PVC
- high chemical resistance in the soil environment
- the fact that PVC sheets are lightweight and portable.

### 35.3.4 Structural walls

A natural progression from plastic sheet cut-off barriers is the development of plastic sheet piling as structural retaining walls. The product has already been extensively used in North America and Europe, particularly for waterway and marine applications. Plastic piling has been employed in the UK for a variety of projects including:

- earth retaining structures
- slope stabilisation
- noise barriers
- flood control and erosion walls.

Plastic piling has many apparent advantages over conventional steel piling. The advantages include:

- ease of handling and transportation of the lightweight piles
- improved corrosion resistance to both fresh and salt water
- better resistance to chemicals
- the opportunity to use recycled materials in manufacture
- better appearance.

Plastic sheet piling is also competitively priced and can potentially be extruded and moulded to meet design requirements.

There are however some disadvantages regarding the use of plastic piling in retaining wall applications. The most obvious concerns are:

- structural performance in terms of driveability
- a lower bending moment resistance than steel products of similar cross-section
- creep under sustained load
- Long-term weathering and reduction in durability of exposed faces, from ultraviolet light.

Carder et al. (2002)[7] indicate the potential use of plastic sheet piles in highways schemes, including deep trenching, construction of road/rail interchanges where grade separations are required, improving slope stability, noise barriers, and the retention of land for the construction of footpaths and cycle-ways.

Recycled plastic (unplasticised PVC) piling currently manufactured in the UK uses window-grade material that has failed to meet the manufacturer's quality control standard for dimensional requirements and would otherwise be scrapped. The material for recycling is granulated and re-extruded without the addition of further additives. Because the material used to make plastic PVC piles in the UK was originally manufactured for another purpose, it would seem possible that its formulation is not the optimal composition. However, with the increasing use of plastic sheet piles to form structural walls, specific formulation of the PVC with additives and stabilisers is possible to create stronger and more rigid materials.

A summary of the typical properties of plastic sheet piles used in the UK is given in Table 35.8. As can be seen, the typical tensile strength of PVC sheet piles is in the range 40 to 46 MN/m$^2$. This can be compared with a minimum tensile strength of 410 MN/m$^2$ for steel sheet piles of grade S270GP supplied to ENI0248.[8] Other mechanical properties for plastic piles, such as tensile modulus, flexural strength and modulus, are also significantly below those expected when using steel. For these reasons, the allowable bending moments specified by the manufacturers of plastic piling are significantly lower than those of an equivalent steel section. The magnitude of the allowable moment can be increased by the addition of ribs to the plastic piles during their manufacture.

Because of the limitation on allowable moment, the safe retained height of a cantilever wall constructed using plastic piles will be less than for an equivalent steel section. This reduced ability to resist bending forces can in part be overcome by the use of a suitable anchoring or tieback system that will reduce the wall bending moments. Such anchoring systems have already been used successfully in the US to tie back plastic sheet pile walls for marine and waterway applications. A variety of tie-back types have been employed including concrete anchor slabs, timber anchor piles, helical tie-backs, injection boring, and pre-stressed anchors. Clearly there is a cost implication in providing additional support in this way but, as plastic piles are cheaper than steel piles of the same dimensions, an economical design can still be achieved.

Alternative techniques to reduce the applied moment on the plastic sheet pile wall include the use of a load-relieving platform constructed on the retained side of the wall, possibly incorporated as part of the capping beam.

While plastic piles are lighter, cheaper and more environmentally friendly than steel, there is uncertainty regarding their long-term durability. In particular there are also few exposure data to establish whether the mechanical properties of plastics are stable over the design life of a retaining structure – typically 120 years – required for highway structures.

The ability of plastic piling to resist impact damage and the likelihood of an impact occurring need to be ascertained at the design stage for any potential use of plastic piling. It must also be borne in mind that this ability can be seriously reduced by the influence of ultraviolet light.

Carder et al. (2002)[9] indicates that generally a sensible limit for the maximum bending moment in serviceability limit state design would be about 4 kN m per m length of wall. However, it should be noted that there are a number of options that would increase the scope for using plastic sheet piling, including:

- an increase in the section modulus and/or thickness of the pile
- the use of pile stiffeners (either steel or plastic) to improve the section modulus
- propping the sheet piled wall near the top using plastic walings and ground anchors or soil nails.

As with both steel and plastic piles, the ease of driving depends on the type of strata encountered, the required penetration, and the flexural rigidity of the pile. In hard driving conditions it may be necessary to increase the section size to prevent deformation of the pile toe and the pile going off line. This problem is likely to be more acute with the more flexible plastic piles; however, because of their lower bending moment capacity, the required penetrations are often less.

Some manufacturers of plastic piles claim that standard piling rigs can be used for driving their particular piles. Although this may be the case in some ground conditions, special rigs have been

**Table 35.8** Summary of the pile dimensions and mechanical properties of commercially available products

| *Product name* | *Z-Ribbed* | *Box-Ribbed* | *GeoGuard 150 to 500* | *GeoGuard 550 to 700* | *C-LOC CL 4500* | *C-LOC CL 9000* | *Series 1900 to 9400* | *Geoflex* | *Superloc™ 1540/1560* |
|---|---|---|---|---|---|---|---|---|---|
| *Manufacturer, country* | HL Plastics,UK | | Materials International Inc, USA | | Crane Products Limited, USA | | Northstar Vinyl Products, LLC, USA | Geotechnics Holland B V, Holland | Lee Composites Inc, USA |
| *Material* | Recycled uPVC | | Vinyl sheet | | Vinyl sheet | | Vinyl sheet | Vinyl sheet | Pultraded fibre reinforced polymer |
| *Configuration* | Panels interlock on their sides | Panels interlock front and back | Panels have strong interlock front and back | | Panels interlock on their sides | Panels interlock at back | Panels interlock front and back | Panels interlock front and back | Panels interlock at back |
| | | | No ribs | Reinforcing ribs | | | | | |
| Available length (m) | 0.5–6 | 0.5–6 | 1.8–5.5 | 4.3–6.7 | Variable | Variable | Variable | 0.5–8 | Variable |
| Material thickness (mm) | 5 | 5 | 5.1–10.2 | 10.2–11.4 | 5.7 | 7.1 | 5.1–20.3 | 5/6 | 3.2/5.1 |
| Section depth (mm) | 75 | 150 | 127–203 | 203–254 | 114 | 229 | 163–249 | 180 | 102/152 |
| Section coverage (mm) | 305 | 270 | 267–305 | 305 | 305 | 610 | 305–457 | 250 | 457 |
| Weight (kg/m) | 2.8 | 2.8 | | | | | 3.9–11.2 | 3 | |
| Weight (kg/m$^2$) | 9.2 | 10.4 | 12.7–24.4 | 26.4–39.1 | | | | 12 | 8.8/15.7 |
| Density (kg/m$^3$) | 1450 | 1450 | | | 1400 | 1420 | | 1450 | 1700 |
| Section modulus (cm$^3$/m) | 80 | 350 | 320–1020 | 1210–2150 | 322 | 878 | 480–2350 | 300 | 160/430 |
| Allowable bending moment (kNm/m) | 1.25<br>3.75 (max) | 5.25<br>137 (max) | 5.9–18.8 | 22.2–39.5 | 4.4 | 16.4 | 8.7–41.8 | 8<br>24 (max) | 24.5/50.1 |
| Flexural strength (MN/m$^2$) | | | | | 73 | 73 | | 80 | 124 (crosswise)<br>228 (lengthwise) |
| Flexural modulus (MN/m$^2$) | 2550–2150 | 2550–2150 | 2620 | 2620 | 2758 | 2758 | | 2300 | 6895 (crosswise)<br>11032 (lengthwise) |
| Tensile strength (MN/m$^2$) | 40 | 40 | 43 | 43 | 46 | 43 | | 41 | 69 (crosswise)<br>228 (lengthwise) |
| Tensile modulus (MN/m$^2$) | | | | | 2586 | 2551 | | | 10342 (crosswise)<br>24132 (lengthwise) |
| Impact strength (MNm/m$^2$) | | | 1.9–2.6 | 2.6 | | | | 0.032 | |

Further product development means that the above properties should be checked with the manufacturer at the time of usage.

developed by other manufacturers to minimise the stresses on the plastic piles, which are inherently less robust than steel piles.

### 35.3.5 Reinforced soils

*35.3.5.1 Soil walls and slopes*

Geosynthethics can be used to reinforce a soil mass in order to increase the effective angle of shear and increase the stability of an earth structure. In the reinforcement function, the geosynthetic is subjected to a sustained tensile force. Soil and rock materials are able to withstand compressive forces but have relatively low capacity to sustain tensile forces. In much the same way that tensile forces are taken up by steel in a reinforced concrete beam, the geosynthetic supports tensile forces that cannot be carried by the soil in a soil/geosynthetic system. Geogrids and some geotextiles are best suited to this function.

When reinforced, significant tensile stresses can be carried by the geogrid reinforcement, resulting in a composite structure that possesses wider margins of strength. This extra strength means that steeper slopes can be built. Geogrids (and some stronger geotextiles) have been utilised in the construction of reinforced soil walls since the early 1970s. In summary, geogrid or geotextile sheets are used to wrap compacted soil in layers, producing a stable composite structure. Reinforced soil walls somewhat resemble the popular sandbag walls that have been used for some decades. However, geogrid-reinforced soil walls can be constructed to significant height because of the higher strength and a simple mechanized construction procedure.

Table 35.9 provides a list of advantages and disadvantages of geogrid reinforced soil walls.

Geogrid-reinforced walls can be substantially more economical to construct than conventional walls. However, since geogrid application to walls is relatively new, long-term effects such as creep, ageing, and durability on design life are not known based on actual experience. Therefore, significant consequences of failure, or high repair or replacement costs could offset a lower first cost. Serious consideration should be given before utilisation in critical structures. Applications of geogrid-reinforced walls range from construction of temporary road embankments to permanent structures remedying slope instability issues and widening highways effectively. Such walls can be constructed as noise barriers or even as abutments for bridges. Because of their flexibility, soil-reinforced walls can be constructed in areas where poor foundation material exists or areas susceptible to earthquake activity.

*35.3.5.2 Design considerations*

Soil-reinforced walls or slopes can be vertical or inclined. This can be for structural reasons (internal stability), ease of construction, or architectural purposes. Generally all geogrids are equally spaced so that construction is simplified, and extend to the same vertical plane (see Figure 35.2).

Geogrids exposed to ultraviolet light may degrade and for this reason they should not be left exposed at the surface of the structure or slope. Generally the geogrid is protected by a layer of soil and/or vegetation that is allowed to establish over the surface. In some instances it may be appropriate to apply a concrete gunite or sprayed layer to prevent exposure and erosion.

When aesthetic appearance is important, a low-cost solution is to use plants to create a "green wall". Plants and their root network will increase the stability of the near surface soils as well as minimising soil erosion.

The soil wrapped by the geogrid sheets is termed "retained soil". This soil must be free-draining and non-plastic. Typically the amount of fines in the soil should be limited to 12 per cent passing sieve No. 200. This restriction is imposed because of possible migration of fines being washed away by seeping water. The fines may be trapped by geotextile sheets, thus eventually creating low-permeability layers that may retain water. Typically, the permeability of the retained soil[10] should be more than $1 \times 10^{-5}$ m/s. A minimum of 95 per cent of the maximum dry unit weight should be attained during construction.

The soil supported by the reinforced wall is termed "backfill soil". This soil has a direct effect on the external stability of the wall. Therefore, it should be carefully selected. Generally, backfill specifications used for conventional retaining walls should be employed here as well. Clay, silt, or any other material with low permeability should be avoided next to a permanent wall. If low-quality materials are used, then a drainage layer needs to be provided between the backfill and the retained soil. These drains can themselves be geosynthetic products, usually geocomposite materials specifically designed for applications with soil reinforcement as described in Section 35.3.1.

*35.3.5.3 Design methodology*

In the UK the design method recommended for retaining walls reinforced with geotextiles or geogrids is set out in BS8006:1995[11], to which detailed reference should be made for specific design situations and the appropriate factors of safety to be applied to loadings and materials. In addition a number of computer programs are available that allow the designer to carry out the

**Table 35.9** Reinforced walls – advantages and disadvantages

| *Advantages* | *Disadvantages* |
|---|---|
| They are economical compared to conventional structural walls. | High repair or replacement costs can offset a lower initial cost for installation. |
| Relatively easy and rapid to construct. Do not require highly skilled labour. | Some decrease in geogrid strength may occur because of possible damage during construction. |
| Regardless of the height or length of the wall, support of the structure is not required during construction as for conventional retaining walls. | Some decrease in strength may occur with time at constant load and soil temperature. |
| They are relatively flexible and can tolerate large lateral deformations and large differential vertical settlements. | The construction of geogrid-reinforced walls in cut slopes requires a wider excavation than conventional retaining walls. |
| They are potentially better suited for earthquake loading because of the flexibility and inherent energy absorption capacity of the coherent earth mass. | Excavation behind the geogrid-reinforced wall is restricted. |

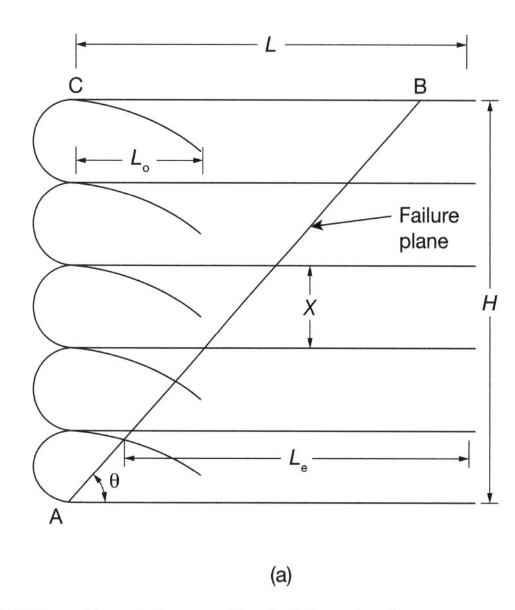


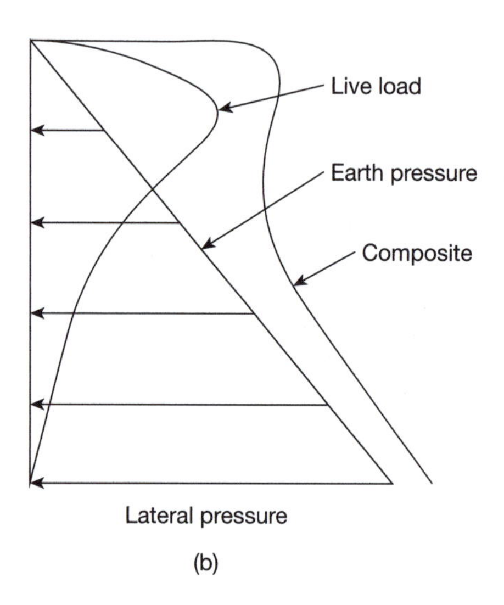


**Figure 35.2** General layout of a geogrid soil reinforced wall

design calculations rapidly. The fundamental design approach is to consider the earth pressure and the applied live load pressure and calculate the applied tension in the geogrid. This tension is then resisted by the pull-out resistance of the geogrid over a determined length of soil/geogrid interface. The primary equations are as follows.

*35.3.5.3.1 Earth pressure,* $\sigma_{ho}$ Lateral earth pressure at any depth below the top of the wall (Figure 35.2(a)) is given by:

$$\sigma_{ho} = K_a \gamma d$$

where $\sigma_{ho}$ = lateral earth pressure acting on the wall, $K_a$ = active earth pressure coefficient, $\gamma$ = bulk density of backfill, $d$ = depth below the top of the wall.

A typical earth pressure distribution is shown in Figure 35.2(b). It should be noted that in some circumstances the "at rest" pressure coefficient, $K_o$, may be more appropriate for calculation of the lateral pressure than $K_a$.

*35.3.5.3.2 Live load pressure,* $\sigma_{hl}$ Lateral pressures from live loads are calculated for a point load acting on the surface of the backfill using:

$$\sigma_{hl} = Px^2z/R^5$$

where $P$ = vertical design live load, $x$ = horizontal distance from load to wall and perpendicular to the wall, $z$ = vertical distance from load to point where stress is being calculated, $y$ = horizontal distance from load to wall, and parallel to the wall, $R = \sqrt{(x^2 + y^2 + z^2)}$.

*35.3.5.3.3 Tension in geogrid* Tension in any geogrid layer is equal to the lateral stress at the depth of the layer times the face area that the fabric must support. For a vertical fabric spacing of $X$, a unit width of fabric at depth $d$ must support a force of $\sigma_h X$, where $\sigma_h$ is the average total lateral pressure (composite of factored dead plus live design loads) over the vertical interval $X$.

*35.3.5.3.4 Pull-out resistance in geogrid* A sufficient length of geogrid must be embedded behind the failure plane to resist pullout. Thus, in Figure 35.2(a) only the length, $Le$, of fabric behind the failure plan AB would be used to resist pullout. Pull-out resistance can be calculated from:

$$P_a = 2d\gamma\, Le \tan(2/3\ \phi')$$

where $P_a$ = pull-out resistance, $d$ = depth of retained soil below top of retaining wall, $\gamma$ = bulk density of retained soil, $\phi'$ = angle of internal friction of retained soil, $Le$ = length of embedment behind the failure plane.

It can be seen from this expression that pull-out resistance is the product of overburden pressure, $\gamma d$, and the coefficient of friction between retained soil and fabric which is assumed to be $\tan(2/3\ \phi')$. This resistance is multiplied by the surface area of $2Le$ for a unit width. Where different soils are used above and below the geogrid layer, the expression can be modified to account for different coefficients of friction for each soil:

$$P_a = d\gamma\, Le[\tan(2/3\ \phi_1') + \tan(2/3\ \phi_2')]$$

The calculations for the geogrid dimensions for overlap, embedment length and vertical spacing should include a factor of safety ($fs$). BS8006 provides useful information in this regard.

The length of geotextile required at depth, $d$, for equilibrium of forces can be determined from:

$$\sigma_h X \leq P_a/fs$$

Thus the expression for length of geogrid, $Le$, behind failure plane at depth, $d$ to resist pullout is

$$Le \geq \sigma_h X\, fs/2d\gamma \tan(2/3\ \phi')$$

Length of fabric overlap for the folded portion of fabric at the face, $L_o$, must be long enough to transfer the stress from the lower section of geotextile to the longer layer above. The pullout resistance of the geotextile, $P_f$, is given by:

$$P_f = 2d_F \gamma \tan(2/3\ \phi')Lo$$

where $dF$ = depth to overlap. Tension in the geotextile is:

$$T = \sigma_h(X/2)$$

This can be solved for the length of overlap required:

$$Lo = \phi_h X fs/2d_F\gamma\tan(2/3\ \phi')$$

*35.3.5.3.5 External wall stability* Once the internal stability of the structure is satisfied, the external stability against overturning, sliding and foundation-bearing capacity should be checked. This is accomplished in the same manner as for a retaining wall without any soil reinforcement. Overturning loads are developed from the lateral pressure diagram for the back of the wall. This may be different from the lateral pressure diagram used in checking internal stability, particularly due to placement of live loads. Overturning is checked by summing moments of external forces about the bottom at the face of the wall. Sliding along the base is checked by summing external horizontal forces. Bearing capacity is checked using standard foundation bearing capacity analysis. Theoretically, the fabric layers at the base could be shorter than at the top. However, because of external stability considerations, particularly sliding and bearing capacity, all fabric layers are normally of uniform width.

### *35.3.5.4 Reinforced embankments*

*35.3.5.4.1 General considerations* Quite often, conventional techniques cannot be used to construct embankments such as on very soft ground because it may not be cost-effective, operationally practical, or technically feasible. Nevertheless, geogrid-reinforced embankments have been designed and constructed by being made to float on very soft foundations or constructed with a geogrid-reinforced granular mattress supported on piled foundations. The design and construction of a geogrid reinforced embankment on soft ground requires the following potential failure modes to be considered:

1 horizontal sliding, and spreading of the embankment and foundation
2 rotational slope and/or foundation failure
3 excessive vertical and differential foundation displacement
4 horizontal sliding and spreading of the embankment fill
5 rotational failure through the embankment and underlying foundation.

The geogrid must resist the unbalanced forces necessary for embankment stability and must develop moderate to high tensile forces at relatively low to moderate strains. It should exhibit enough soil fabric resistance to prevent pull-out of the geogrid. The geogrid tensile forces resist the unbalanced forces, and its tensile elastic modulus controls the vertical and horizontal displacement of the embankment and foundation. Adequate development of soil–geogrid friction allows the transfer of embankment load to the geogrid. Developing geogrid tensile stresses during construction at small material elongations or strains is essential.

Horizontal sliding and spreading of the embankment can result from excessive lateral earth pressure. The forces are determined from the embankment height, slope angle, and fill material properties. During conventional construction the embankment will resist these modes of failure through shear forces developed along the embankment/foundation interface. Where geogrids are used between the soft foundation and the embankment, the geogrid will increase the resisting forces of the foundation. Geogrid reinforcement may fail by fill material sliding off the geogrid surface, geogrid tensile failure, or excessive elongation. These failures can be prevented by specifying the geogrids that meet the required tensile strength, tensile modulus, and soil–geogrid internal friction properties.

Geogrid reinforced embankments constructed to a given height and side slope will resist classic rotational failure if the foundation and embankment fill shear strengths plus the geogrid tensile strength are adequate. The rotational failure mode of the embankment fill can only occur through the foundation layer and geogrid. For cohesionless fill materials, the embankment side slopes are less than the internal angle of friction. Since the geogrid does not have flexural strength, it must be placed such that the critical arc determined from a conventional slope stability analysis intercepts the horizontal layer. Embankments constructed on very soft foundations will require a high tensile strength geogrid to control the large unbalanced rotational moments.

Consolidation settlements of embankment foundations, whether geogrid-reinforced or not, will be similar. Consolidation of a geogrid-reinforced embankment usually results in more uniform settlements than for non-reinforced embankments. Classic consolidation analysis is a well-known theory, and foundation consolidation analysis for geotextile-reinforced embankments seems to agree with predicted classical consolidation values. Soft foundations may fail partially or totally in bearing capacity before classic foundation consolidation can occur. One purpose of geogrid reinforcement is to hold the embankment together until foundation consolidation and strength increase can occur. Generally, only two types of foundation bearing capacity failures may occur:

1 partial or centre-section foundation failure
2 rotational slope and foundation stability.

Partial bearing failure, or "centre sag" along the embankment alignment may be caused by improper construction procedure, such as working in the centre of the embankment before the geogrid edges are covered with fill materials to provide anchorage. If this procedure is used, geogrid tensile forces are not developed and no benefit is gained from the geogrid. A foundation bearing capacity failure may occur as in conventional embankment construction.

Centre sag failure may also occur when low-tensile strength or low-modulus geogrids are used, and embankment spreading occurs before adequate geogrid tensile stresses can be developed to carry the embankment fill weight and reduce the stresses on the foundation. If the foundation capacity is exceeded, then the geogrid must elongate to develop the required geogrid stress to support the embankment weight. Foundation bearing-capacity deformation will occur until either the geogrid fails in tension or carries the excess load. Low-modulus geogrids generally fail because of excessive foundation displacement that causes these low tensile strength geogrids to elongate beyond their ultimate strength. High-modulus geogrids may also fail if their strength is insufficient. This type of failure may occur where very steep embankments are constructed, and where outside edge anchorage is insufficient.

Earthwork failures due to inadequate drainage at the zone of the geogrid and foundation have been reported in the UK (Finlayson et al., 1984).[12] For this reason it is good practice to provide a granular free-draining mattress typically 600 mm to 1000 mm thick with single or multiple layers of geogrid "sandwiched" and compacted within the mattress to avoid creating a potential incipient failure plane. The mattress acts as a stiff member relative to the embankment and will also assist in "spreading" load/settlement beneath the embankment so that it is more uniform. The granular mattress also acts as a drainage blanket and prevents the build-up of water pressure at the zone between the formation and embankment base as a result of the embankment consolidating the underlying soils and squeezing water out of the soil matrix.

*35.3.5.4.2 Basis for design* The limit state equilibrium analysis is commonly adopted for design of geogrid-reinforced embankments. These design procedures are quite similar to conventional bearing capacity or slope stability analysis. Even though the rotational stability analysis assumes that ultimate tensile strength will occur instantly to resist the active moment, some geogrid strain, and consequently embankment displacement, will be necessary to develop tensile stress in the geotextile. The amount of movement within the embankment may be limited by the use of high tensile modulus geogrids that exhibit good soil–geogrid frictional properties. Conventional slope stability analysis assumes that the geogrid reinforcement acts as a horizontal force to increase the resisting moment. The following analytical procedures should be conducted for the design of a geogrid-reinforced embankment:

1 overall bearing capacity
2 edge bearing capacity or slope stability
3 sliding wedge analysis for embankment spreading/splitting
4 analysis to limit geogrid deformation, and determine geogrid strength in a direction transverse to the longitudinal axis of the embankment or the longitudinal direction of the geogrid
5 embankment settlements and creep must also be considered in the overall analysis.

Design guidance for soil-reinforced embankments on soft ground is given in BS8006:1995[13] together with recommended partial factors of safety to be applied to applied loads, soil materials, geogrid reinforcement and soil/reinforcement interaction.

Piled embankments are sometimes used where high embankments are required on soft ground. In these cases a granular mattress reinforced with geogrids is constructed at the base of the embankment to provide a load distribution or load transfer platform between the embankment and the underlying formation/pile caps. Useful design case studies include Padfield and Sharrock (1983),[14] Reid and Buchanan (1983)[15] and Card and Carter (1995).[16] BS8006:1995 provides useful guidance on the design of piled embankments with geogrid-reinforced granular mattress acting as a load distribution platform.

## 35.4 Applications in environmental engineering

### 35.4.1 Cover systems

Redevelopment of brownfield land often requires ground remediation measures to overcome ground and groundwater contamination to reduce environmental and human risks. Cover systems comprising geosynthetic materials can play an important role in providing separation, drainage and barrier functions to prevent the migration or exposure of soil and groundwater contaminants to potential receptors, such as humans, flora and the aquatic environment. Many geosynthetic materials can be used in these applications, often with multiple functions. An important criterion is that the materials themselves are chemically and biologically inert such that their durability and physical properties such as strength, elasticity and impermeability are not compromised.

Geomembranes have been used for many years in the waste management industry for landfill lining purposes. In this connection much research has taken place to develop geosynthetic membranes that are durable in contact with chemical landfill leachate for landfill containment lining systems.[17] As with liner systems, many of the design principles of cover systems come from the waste management industry. The materials used and testing required are described in more detail in Section 35.2.

Cover systems are by far the most common form of contamination barrier used in the UK, where they have been used as a "low-tech" alternative to the "do nothing" option, with little regard to considered design principles. However, covers can be very effective in specific circumstances, particularly where a site is not grossly contaminated. In other countries, such as the USA and the Netherlands, cover systems are not accepted as a long-term solution to contaminated land, except in the context of landfill covers. They may be used in these cases, however, as a temporary measure to provide immediate protection against a hazardous situation, or until a "clean-up" operation can be funded.

Geosynthetics can be integrated within a cover system. Figure 35.3 is a schematic cross-section showing the variety of component layers that could be used within a cover system (after Privett et al.[18]) Not all of these components will be necessary for every contaminated site and one particular layer may perform more than one function. Figure 35.4 shows how a number of geosynthetic products may be combined in a particular site-specific instance. The functions of the various layers are outlined below.

*35.4.1.1 Surface layer*

The surface layer of a cover system is likely to consist of hard surfacing or topsoil. The surface layer can help to reduce water infiltration. Hard surfaced areas are typically of low permeability and adequately drained to avoid surface flooding. If prevention of water infiltration is a prime objective of the cover system, surface water collected in the drainage system should be disposed of to a surface water drainage system and not into a soakaway.

Land not covered by hard surfacing is best vegetated, typically with shallow rooted grass species, to protect it from erosion, to improve soil stability and aesthetics, and to remove surface water through evapotranspiration. Vegetated areas should be adequately sloped and drained to avoid surface water ponding and increased infiltration.

*35.4.1.2 Protection layer*

The primary function of a protection layer is to protect the underlying layers from intrusion by plants, animals and humans, and any underlying mineral barrier layer from the effects of desiccation and freeze/thaw. The layer, therefore, should be of sufficient depth to reduce sufficiently any of these likely risks.

A protection layer may have to perform a secondary function such as having to contain services and support shallow foundations or act as a supply of soil moisture for vegetation. When it has to support a structural load, all materials selected in the cover system (except topsoil) should be suitable for proper compaction.

Additional protection can be afforded by the inclusion of a "biobarrier", a physical barrier layer consisting of cobbles, hardcore or synthetic materials in order to prevent penetration by animals and vegetation and to attempt to prevent human intrusion. However, in most UK situations, compacting the protection layer to a sufficiently high density should prevent animal and root intrusion.

*35.4.1.3 Barrier layer*

The primary function of a barrier layer is to minimise infiltration of surface water (mainly rainwater), and the escape of any gases. A barrier layer is likely to consist of a compacted low hydraulic conductivity mineral layer or a geosynthetic layer such as a geomembrane similar in nature to those used as landfill liners.

*35.4.1.4 Drainage layers*

Drainage layers can be used on both sides of the barrier layer (see Figure 35.4). The upper layer's function is to drain away infiltrating

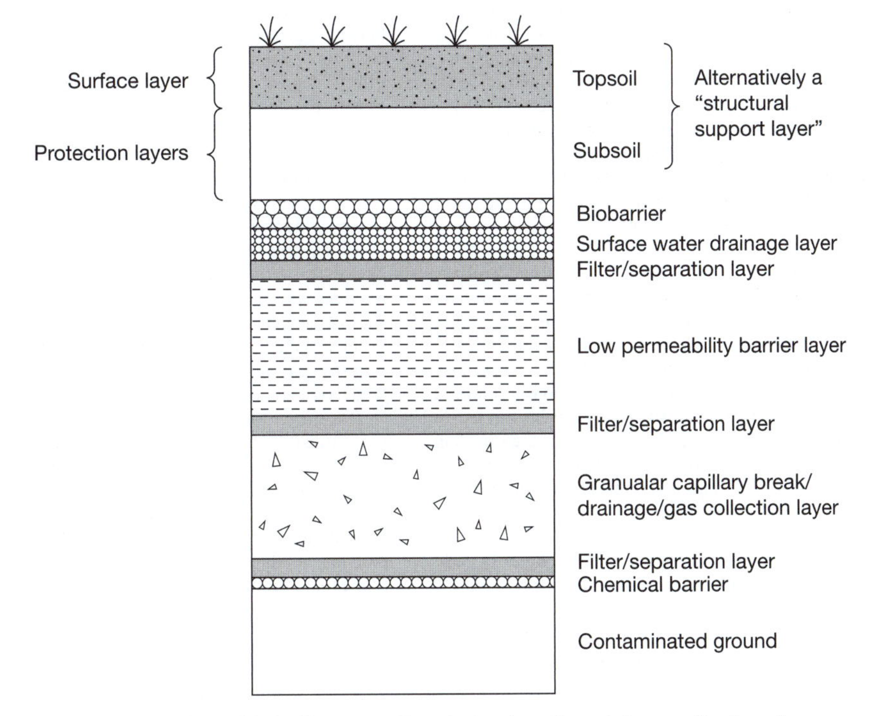


**Figure 35.3** Schematic cross-section illustrating the variety of geosynthetic component layers (after Privett et al., 1996).

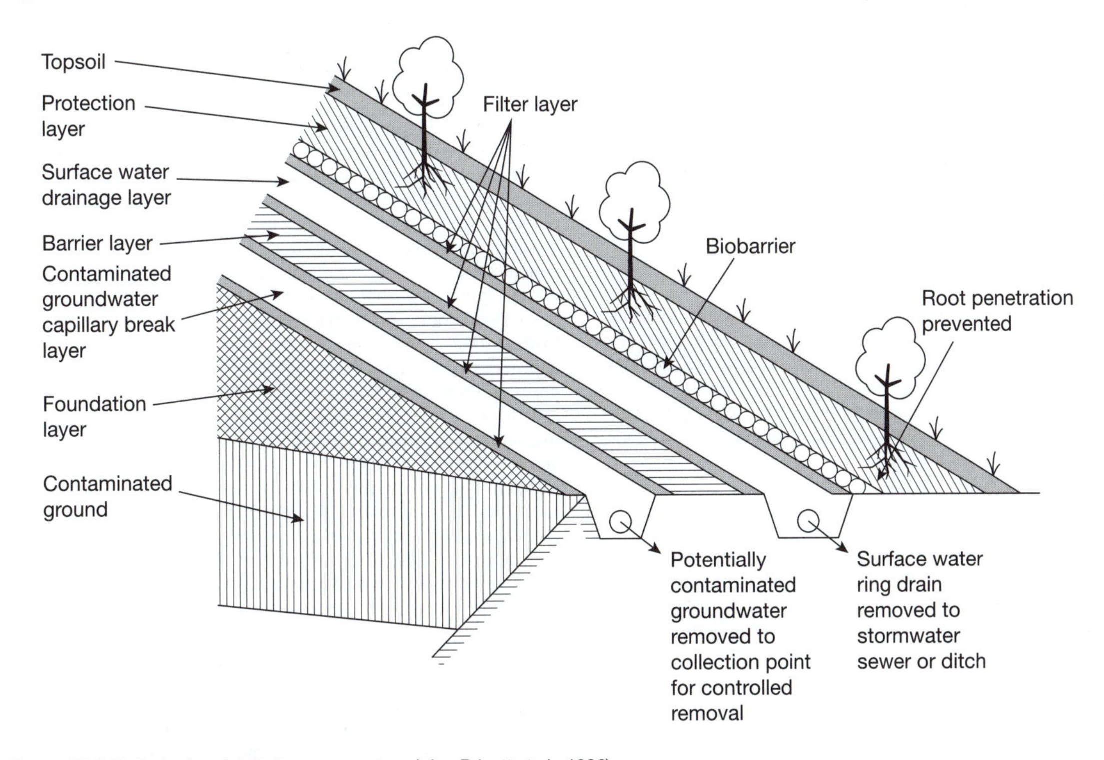


**Figure 35.4** Typical edge detail of a cover system (after Privett et al., 1996)

water so as to minimise barrier layer contact and to dissipate seepage forces. The lower layer's function is to limit the bulk movement of contaminants within, or floating on, the groundwater, resulting from any rise in the groundwater table. Each of these drainage layers is likely to consist of a geosynthetic (geocomposite) drainage medium comprising a high permeable geotextile core sandwiched between geotextile filter fabrics. Water collected in the upper layers should be disposed of to surface water drains, whereas the potentially contaminated water collected in the lower drainage layer is likely to require treatment.

*35.4.1.4.1 Capillary break layer* The function of a capillary break layer is to minimise soil-suction-induced contaminant migration during drought periods. It is likely to consist of granular material and the thickness will be dependent on the material properties and the depth to the groundwater table. A capillary break layer designed specifically against soil suction is required only for water-soluble contaminants (which could be carried to the surface), and only in cases where sensitive targets are less than about 3 m above the highest possible contaminated water table in fine-grained materials. A capillary break layer may usually be combined with the lower drainage layer.

*35.4.1.4.2 Gas collection layer* A gas collection layer is only required if upward movement of gases is likely. It should comprise a permeable "gas-drainage" layer of poorly graded gravel or high transmissivity geosynthetic which is vented at suitable locations. A gas collection layer will also require an overlying barrier layer as part of a gas control system.

#### *35.4.1.5 Insulation barrier layer*

Subsurface combustible contaminated materials must be prevented from being heated by any above-ground thermal source. A 1 m thickness of material is usually sufficient insulation. Since the thickness of a typical cover system is likely to exceed 1 m, there is generally no need to incorporate a special layer for this purpose.

#### *35.4.1.6 Chemical barrier layer*

In some cases, chemical barriers designed to attenuate certain contaminants have been used. Geomembrane lining materials that are suitable for landfill containment purposes can be used to satisfy this function.

#### *35.4.1.7 Filter and geomembrane protection layers*

Filter layers will be required at the boundaries of coarse granular layers or geosynthetic drainage layers in order to prevent the ingress of fines and subsequent reduction in drainage/capillary break potential.

If a coarse drainage material is to be placed onto a geomembrane, a protection layer is required to protect the geomembrane from puncture and overstressing.

#### *35.4.1.8 Foundation layer*

A foundation layer may be necessary directly on top of the contaminated land if a particular gradient is required of the cover or simply to act as a formation to take the construction of a drainage layer if the contaminated ground cannot fulfil this role directly.

### 35.4.2 Gas and vapour membranes

Geosynthtics have an important role in the control of soil gas and vapours into buildings, particularly in the protection of landfill gas. In the first instance, however, the floor slab itself is the first barrier against gas ingress. Several types of floor slab are commonly seen on developments in the UK, and each type will offer a degree of protection against gas ingress into the finished structure.

Reinforced concrete cast in situ suspended floor slabs can give reasonable gas resistance, if they are detailed correctly with anti-crack reinforcement. They also need to be sealed at joints (possibly with water bars – a design measure to ensure joints are water-tight) and sealed at all service penetrations with water bars.

Block and beam slabs used in many housing developments have many joints between the blocks and beams which means they provide minimal resistance to ground gas ingress. However, a well-constructed block and beam floor should have sand–cement grout brushed into the surface to fill all the joints, which can improve gas resistance by a limited amount. This also gives a better surface into which to lay a gas-resistant membrane.

If a ground-bearing slab is used on a site affected by ground gas it must be designed to minimise the risk of differential movement between it and the walls of the building. Such differential movement has been one of the main causes of damage to gas-resistant membranes (the movement has torn the membrane). The membrane must therefore be specified and detailed to accommodate any likely movement (e.g. by providing a fold in the membrane that allows movement, by limiting the movement of the slab or by specifying membranes with a high elongation before failure).

#### *35.4.2.1 Membrane types*

Polyethylene and polypropylene are the principal materials used in the manufacture of low-permeability gas membranes. Polyethylene is available in a variety of thicknesses and types:

- *HDPE* – high-density polyethylene
- *LDPE* – low-density polyethylene (this can be low-grade DPM or higher quality reinforced membrane)
- *MDPE* – medium density polyethylene, which is not commonly used as a gas-resistant membrane material
- *LLDPE* – linear low-density polyethylene.

The membranes can also be reinforced to improve the durability of the material and prevent over-elongation of the membrane. There are other types of materials and membranes available and a specific assessment should be made where these materials are proposed (e.g. liquid applied membranes and geocomposite membranes formed of aluminium foil bonded to geotextiles or geomembranes). The disadvantages and points to consider for various types of membrane are provided in Table 35.10.

Some membranes are marketed as carbon dioxide, methane or hydrocarbon membranes. This distinction is misleading as it appears to be based on the assumption that carbon dioxide always poses a lower risk than methane and that somehow a lower specification membrane can be used (it is also probably led by marketing rather than any technical justification). This is not true, and it is up to consultants and other designers to assess the risk and construction conditions and propose a suitable membrane. It is also untrue that different membranes to those for methane and carbon dioxide are required to prevent vapour migration into buildings from hydrocarbons. Any of the geosynthetic materials listed in Table 35.8 will be sufficiently resistant to all the most commonly encountered gases/vapours, and quality of installation is the key issue (Wilson et al. 2007).[23] The use of geosynthetic clay liners (a thin layer of dry bentonite clay powder sandwiched between two geotextiles) as gas-resistant membranes below buildings is not usually recommended. This is because the liner relies on the bentonite material being permanently wet to allow the clay particles to swell and form an impermeable gel. This cannot be guaranteed over the design life of the building structure.

**Table 35.10** Gas-resistant membranes

| *Advantages* | *Disadvantages* | *Points to consider* |
|---|---|---|
| *Polypropylene* | | |
| • Excellent chemical resistance<br>• Good elongation before failure<br>• Flexible<br>• Does not suffer stress cracking | • Relatively high methane permeability<br>• Higher unit cost<br>• Reduced adhesion with self-adhesive membranes | • Requires specialist welding and should not be tape jointed. Usually supplied in large rolls and therefore is more suitable to large floor areas and also requires plant for handling on site. Welds can be pressure tested as part of QA to ascertain joint integrity.<br>• Flexible, so more suitable to complex details and high elongation before rupture, so suitable where settlement may occur. |
| *High-density polyethylene* | | |
| • Excellent chemical resistance<br>• Good puncture resistance<br>• Large roll sizes, therefore less jointing<br>• Low unit cost | • Can be prone to stress cracking<br>• Rigid<br>• Difficult to achieve complex detailing<br>• Heavy and difficult to handle<br>• Reduced adhesion with self adhesive membranes | • Requires specialist welding and should not be tape jointed. Usually supplied in large rolls and therefore is more suitable to large floor areas and also requires plant for handling on site.<br>• Welds can be pressure tested as part of QA to ascertain joint integrity.<br>• Complex detailing can be hard to achieve due to the rigidity of the membrane |
| *Low-density polyethylene DPM* | | |
| • Good chemical resistance<br>• Flexible<br>• Low unit cost<br>• Easy handling<br>• Widely available | • Low-grade recycled raw materials<br>• Easily damaged<br>• Relatively high methane permeability<br>• Low UV resistance | • Most standard polyethylene DPM is manufactured from recycled LDPE building film which is considered to be a low-grade LDPE.<br>• The tolerances for building film thickness can be as much as ±10%. The minimum thickness of this type of membrane should be 0.5 mm or 2000 g.[a] |
| *Reinforced LDPE or HDPE with an aluminium core* | | |
| • Best resistance to ground gas or vapour migration if correctly installed<br>• Flexible<br>• Low cost<br>• Increased tear resistance<br>• Good chemical resistance | • Aluminium core can corrode on thinner polyethylene versions from prolonged contact with alkaline substances<br>• Low UV resistance<br>• Foils susceptible to rupture where settlement occurs, unless allowed for in design and installation | • Reinforced LDPE membrane is normally manufactured from high-quality virgin polymer laminating film as the thickness tolerance is critical for the heat bonding process (this should be confirmed). If this is not the case the reinforced membrane will be prone to delamination.<br>• Minimum recommended thickness 0.4 mm to provide sufficient protection to aluminium from alkaline substances.<br>• Polypropylene (PP) type membranes bonded to aluminium are not acceptable due to the poor adhesion of PP to aluminium. |

[a]Mallett, H., Taf fell-Andureau, L., Wilson, S. and Corban M. (2013). *Good practice on verification and testing of ground gas protection systems for buildings*. CIRIA Report 960, CIRIA, London, UK

Even if it is pre-wetted the clay can dry out and crack, which will allow gas to migrate through it.

Where membranes are used in the ground as a vertical barrier they can be subject to high stresses and puncture forces from the surrounding ground, settlement of backfill causing drag-down, etc. Therefore very robust materials and protection must be provided (see Card, 1995[19] and Privett et al., 1996[20]).

*35.4.2.2 Basis of design and use*

The gas resistance of floor slabs is improved by providing gas-resistant membranes. The purpose of the gas-resistant membrane is to provide a low-permeability barrier against the ingress of gas from the ground into the building fabric.

There are many different types of membrane on the market, each with advantages and disadvantages and with varying degrees of methane permeability. No membrane is however completely impermeable to ground gas or vapours. Based on the work by Wilson et al. (2008),[21] the permeability to methane of some typical proprietary membranes is given in Table 35.11, together with the time for gas to leak through and fill the overlying room with methane to 5% v/v. The results show that preventing holes due to damage and incorrect installation is the critical issue, as once any of the membranes have even the slightest puncture the rate of gas flow into an overlying room will increase by a factor of over 12 million!

It is recommended that gas membranes are specified on the basis of the need to survive construction, i.e. use one that is puncture- and tear-resistant. The joints should be sealed in accordance with the manufacturer's instructions. Thickness should not be the main method of determining the suitability of a membrane for a particular site, although it can indirectly influence some of the key performance parameters.

For guidance it should be clear that proprietary gas-resistant membranes are manufactured using robust materials that are often reinforced and have thicknesses from 1600 g (0.4 mm) upwards. Damp-proof membranes are usually made from lower quality materials (typically lower grade LDPE) than gas-resistant membranes and thus require a greater thickness. A 1200 g polyethylene DPM material (0.3 mm), as shown in previous BRE guidance, is

**Table 35.11** Methane permeability of membranes

| *Membrane* | *Methane permeability (ml/m²/24 h/atm)* | *Time to fill room to 5% v/v methane through intact membrane (years)* | *Time to fill room to 5% v/v methane through punctured membrane – 1 mm diameter puncture (hours)* |
|---|---|---|---|
| Polypropylene 1 mm thick (4000 gauge) | 0.15 | 3.7 million | 12 |
| Polyethylene 0.4 mm thick (1600 gauge) reinforced | 31 | 17,675 | 12 |
| LDPE sandwich with aluminium foil core. 0.4 mm thick (1600 gauge) reinforced | <0.001 | 547 million | 12 |

Room assumed to be 2 m high by 5 m by 4 m. Driving pressure of gas through membrane is 50 Pa. Divide gauge by 4 to establish equivalent thickness in micrometres.

unlikely to be installed correctly (it will usually be installed by general labourers) and is easily damaged during installation and by follow-on trades. Therefore the minimum thickness for a DPM (as opposed to a proprietary gas-resistant membrane) used as a gas protection should be 2000 g. In any event the BRE guidance suggests that the use of 1200 g polyethylene is only suitable for sites where methane concentrations are less than 1%. In very low-risk situations (Characteristic Situation 2 with an in situ reinforced concrete floor slab) a 1200 g DPM may be acceptable provided it is provided with adequate protection and is verified as adequate by extensive independent inspection.

Comparison of the performance characteristics of different membranes can be difficult because sometimes different test methods are used to measure the same property. In the UK, membrane testing should be undertaken following the methods in British or European standards (BS or EN). It is common to see test data using American standards (American Society for Testing and Materials – ASTM), but these results are not directly comparable with data from British or European standard tests. The only exception to this is where a standard is an ISO (International Standards Organisation) standard, which is recognised internationally.

When specifying a membrane, only those criteria that have an impact on the performance should be stated. The quoted limits should reflect the level of performance that is required in a particular application. The critical characteristics that should be specified for a gas-resistant membrane are summarised in Table 35.12 and focus on the short-term survivability during installation.

Membranes should ideally be resistant to puncture, abrasion and tearing. They should also be resistant to ultraviolet light, shrinkage, water and organic solvents and bacteriological action. In general terms, the thinner the material, the more susceptible the membrane will be to mechanical or chemical damage.

In practice because of the nature of construction sites the durability, survivability and robustness of membranes are more significant properties than methane (or any other gas) permeability. For this reason the need for high-quality workmanship during the installation of a gas membrane should not be underestimated. The

**Table 35.12** Suggested membrane testing (after Wilson et al., 2007)

| *Characteristic* | *Test method* | *Unit* | *Considerations* |
|---|---|---|---|
| Tensile properties | BS EN ISO 10319:1996 Geotextiles. Wide-width tensile test | kN/m | The greater the tensile strength, the greater the resistance to tearing and puncture. The amount of elongation before failure is important where movement of the membrane is anticipated. |
| CBR puncture resistance | BS EN ISO 12236:2006 Geosynthetics – Static puncture test (CBR test) | N | The acceptable puncture resistance is dependent on factors such as the nature of the surface onto which the membrane is laid (e.g. is it smooth concrete or a jagged aggregate?), the type of work that will be carried out on it after installation (e.g. will concrete trucks need to run on it?), the type of protection provided (boards, geotextile fleece) and the level of independent inspection. The higher the risk of damage occurring, the greater the CBR puncture resistance that should be specified. Example values for HDPE membrane are 160 N for 0.5 m thick sheet and 320 N for a 1 mm thick sheet. |
| Resistance to tearing (nail shank) | BS EN 12310-2:2000 Flexible sheets for waterproofing. Determination of resistance to tearing (nail shank). Plastic and rubber sheets for roof waterproofing | N | The acceptable level of resistance to tearing is dependent on the same factors as CBR puncture resistance. |
| Methane permeability | BS903: Part A30 (modified) | ml/m²/day | The minimum value should be based on a risk assessment and the time for gas to migrate through the membrane and accumulate to a hazardous concentration. Typically a minimum value of 35 (ml/m²/24 h/atm) is more than adequate. |

health and safety of the occupants of a building is dependent on its satisfactory performance. If installed incorrectly or damaged during the construction process, the gas membrane is rendered ineffective and therefore fails to provide adequate protection against the ingress of ground gas or vapours. This is why the installation of a gas membrane should be independently validated.

A lower specification membrane (e.g. 2000 g DPM) that is well installed and not damaged by follow-on trades is far better than a poorly installed high-quality thick membrane that is extensively damaged or not correctly sealed. However, it should be noted that it is usually more difficult to seal a thin lower quality membrane (for example, they cannot be welded easily) and they are more prone to damage and so require greater on-site protection.

*35.4.2.3 Sealing*

Sealing and jointing membranes can be achieved using welding or specialist adhesive tapes. For both operations the two most critical factors are that:

1 the membrane must be clean and dry to allow welding or taping
2 the temperature must not be too low to allow welding or taping.

In low temperatures it may not be possible to joint membranes or pre-warming using hot air may be required (typically at temperatures less than 5°C). It is also essential that there is sufficient membrane material to form a weld. More information on the methods of jointing is given in CIRIA Special Publication 125 (Privett et al., 1996).

For taped joints the tape used should be that recommended by the manufacturer. Some of the issues that need to be considered when using taped joints are as follows.

- Compatibility – numerous tapes are not actually adhesive to HDPE and polypropylene.
- Age hardening – many polymerised bitumen tapes (the majority of adhesive tapes) age-harden and become less adhesive, stiffen and can eventually de-bond. Although this may be less of a problem in horizontal installations with a weight of concrete on them, it does mean that vertical joints require mechanical support in the form of battens; otherwise the joints can peel open.
- Adhesive tapes usually fail at point of installation due to dust contamination – dust on the membrane, dust from the substrate or dust on the adhesive tape.
- Effectiveness of tape joints relies on a firm pressure being applied to the overlaps and for this reason (and the previous dust concern) reputable manufacturers of adhesive membranes and tapes all recommend a concrete blinding to provide a supporting substrate
- Tape-jointed systems are more susceptible to moisture at time of installation than welded joints.
- Tensile strength and adhesion on taped joints are far lower than on welded joints and are prone to de-bonding during installation as a result of thermal expansion if the membrane is subject to wide temperature gradients.
- Tape joints cannot be tested as comprehensively as welded joints, relying on the “screwdriver test” and visual inspection.

*35.4.2.4 Detailing*

Of equal importance to the performance of a membrane is good detailing of joints etc., especially around complex structural forms. Good detailing should ensure that installation and jointing of the membrane is as easy as possible. This should be considered at the design stage to prevent problems during installation. An experienced engineer on site during the installation can also be beneficial in dealing with detailing problems.

Note that if there is likely to be any differential movement between the floor slab and the wall foundations, this must be allowed for in the design and specification of the gas membrane.

*35.4.2.5 Durability*

Membranes will degenerate over time due to processes such as oxidation, biological degradation, chemical attack and degradation caused by ultraviolet (UV) light. It is necessary to understand how long a given membrane will maintain its performance over the design life. As a general rule membranes used in gas protection are not exposed to UV light after installation. The membranes should only be exposed to UV for a short period of time during installation and in addition membranes usually have UV stabilisers added to protect against attack by UV. Where a membrane is exposed to UV light its life will be reduced.

Another common issue is the resistance of membranes to attack by hydrocarbon vapours. This is unlikely on most sites as they are usually above a venting layer or above the floor slab and are not in direct contact with ground contamination. The vapours passing through the system will be at very low concentrations that are unlikely to cause damage to the materials used to make gas membranes (most testing that shows damage by hydrocarbons to membranes is from using 100% concentration liquids).

Oxidation and biological degradation will occur, but for most membranes chemical stabilisers are added to the materials and this will retard degradation. The main effect of the aging processes is to reduce the strength of the membrane. Where the liner is subjected to long-term stresses, stress cracking will also lead to the development of holes. As gas membranes do not usually have to carry any great load after installation, the adverse effects associated with these ageing processes should therefore not be significant. Gas-resistant membranes should in most cases therefore have a very long service life (at least 100 years) at an ambient temperature of about 20°C. The service life will be reduced as temperature increases and where membranes are subject to long-term stress.

### 35.4.3 Soil gas dispersion

*35.4.3.1 Band drains*

An approach necessary on landfill gassing sites is to remove/reduce the gassing regime prior to development. This can be achieved using in-ground venting techniques to dilute and disperse gas. In this way the overall scope and level of gas protection measures required to buildings and structures can be reduced. A recent new technique is the application of prefabricated vertical band drains to de-gas, in a controlled manner, active or closed landfill sites or natural gassing soils such organic silts and peats.

Band drains have long been used in civil engineering to accelerate consolidation of soft soil deposits by allowing free drainage of pore water. The determination of the length and spacing of the prefabricated band drain for gas dispersion is similar to the theory for conventional pore pressure dissipation and soil consolidation (Holtz et al., 1991[22]). In this regard the permeability and consolidation characteristics of the gassing soil are expressed in terms of intrinsic permeability and the specific gravity and viscosity of the gas or gases required to be dispersed by the band drains. Normally the pressure of the gas in the pore space of the soil is sufficiently high to provide a driving pressure and advective flow through the band drain. In some circumstances this can be enhanced either by applying a fill surcharge to the soil or applying a vacuum pressure to the band drains.

The technique has been used in the UK and Europe to release trapped landfill gas in waste materials and organic silts by the same physical mechanism. The process involves the installation into the waste or gassing soil prefabricated vertical band drains comprising a geocomposite strip made up of free-draining elemental core wrapped in a porous filter geotextile. The drains are installed on a closely spaced regular grid, typically 1 m to 2 m spacing, and penetrating the full depth of the soil gassing zone. The puncture effect of the drains causes a rapid release of pressure of the trapped soil gas and this, in turn, results in ground consolidation. By incorporating a sealed horizontal granular geocomposite drainage layer over the grid of prefabricated band drains, it is possible to control the dispersion of gas and vent at discrete points. The use of prefabricated band drains for gas dispersion can rapidly reduce the gassing regime on a site such that building development may proceed safely.

### *35.4.3.2 Vent wells and barriers*

*35.4.3.2.1 Description* Wilson and Shuttleworth (2002) describe a system for constructing a combined passive venting barrier using geocomposite materials to de-gas soils and waste materials.[23] The concept of the *passive dilution barrier* is to form a low-pressure area relative to the surrounding gassing ground, to encourage gas to flow towards the barrier. This is achieved by driving discrete vent nodes into the ground which are connected to a collection/dilution duct running along the top of the strips. The nodes comprise highly efficient geocomposite strips. The duct has a high flow of fresh air through it by means of passive ventilation. This is one of the key advantages of the system, as it:

- dilutes gas emissions to tolerable levels
- causes a suction effect in the geocomposite vents, which enhances gas flow from the ground towards the vents.

Ventilation of the duct can be achieved using a combination of vent stacks, bollards or ground level boxes, depending on the gas regime and wind conditions at a particular site. A schematic layout for the barrier is shown in Figure 35.5.

The system is particularly effective where gas migration is occurring through shallow layers of sand and gravel up to 5 m depth, underlain by an impermeable layer. This is typical of many situations encountered in the UK. The nodes can be installed to a maximum depth of 5 m below starting level. The starting level can be in trenches up to 3 m depth, giving a maximum effective depth of 8 m. As the depth of the migration pathway increases below the toe of the nodes, the barrier becomes less effective.

The passive dilution barrier is installed using a unique no-dig method in which a steel mandrel is vibrated up to 5 m into the ground, using a vibrating piling hammer supported by a 360° excavator. Once the hollow mandrel is in the ground the central cutting shoe can be removed and a geocomposite strip inserted. The mandrel is then withdrawn, leaving the vent in the ground.

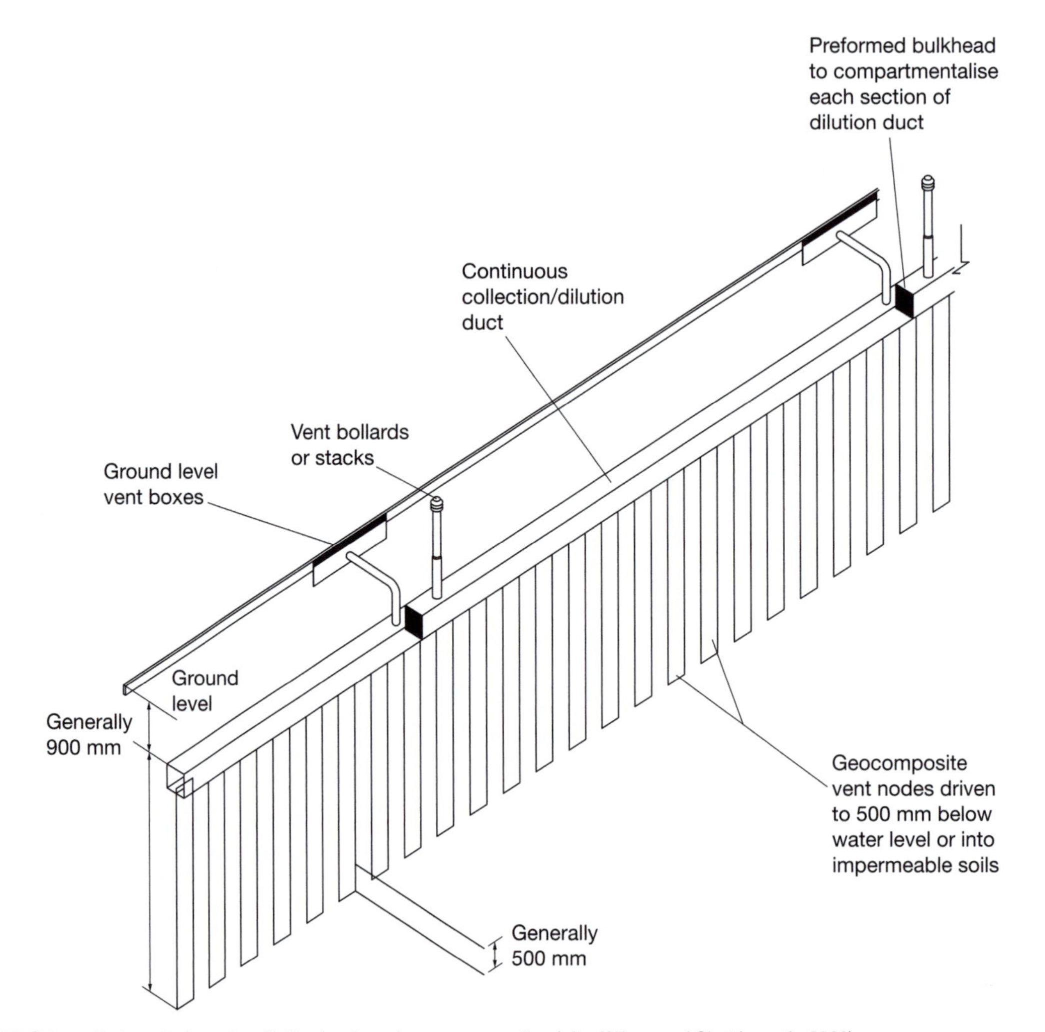


**Figure 35.5** Schematic layout of passive dilution barrier using geocomposites (after Wilson and Shuttleworth, 2002)

The key advantages of this method of installation are:

- *speed* – up to 30 vents per day can be installed
- *cost* – there is a reduction in excavation costs and disposal of spoil that is frequently contaminated
- *safety* – contact with contaminated materials by the installers is minimised.

A further advantage is that walls can be constructed very close to site boundaries and in areas where access is restricted and conventional barriers could not be constructed. In common with all passive ventilation measures that comprise a series of vents, ducts or pipework, there is uncertainty of the fluid dynamics of the gas flow within such systems. An assumption is usually made that Darcy's formula for advective fluid flow can be applied to gas flow and the design of passive ventilation. While this assumption provides a mathematical solution, laminar flow is certainly undesirable in practice as experience suggests that this can lead to conditions where undiluted landfill gas is transported through the ducts of the system to external vents as stream filaments or a stream-tube, resulting in potentially hazardous environments. This phenomenon was initially demonstrated by Reynolds in the 1880s in his early experiments on the flow of fluids in pipes. In order to prevent this condition it is necessary to ensure that turbulent or intermittent turbulent conditions exist within the ducts so that landfill gas entering the system from the waste mass is mixed, diluted and discharged safely at the vent. Turbulence can be induced simply by including baffles to provide irregular flow pattern or abrupt changes in length and cross-sectional area of the ducts.

*35.4.3.2.2 Design methodology* Generally the flow of gas of gas in the ground towards a well or barrier can be modelled using the equations for planar flow of fluids based on Darcy's law. One of the most common situations is shown in the conceptual model in Figure 35.6, where a permeable layer is overlain by an impermeable barrier (a capping layer or hard cover). A relatively shallow groundwater table or impermeable clay layer typically provides a lower confinement to gaseous flow. In this situation we may use the equation for flow of fluids in a confined aquifer towards a horizontal slot based on work by Chapman (1959)[24] relating to groundwater flow and the United States Environmental Protection Agency (1975)[25] which looked at the performance of vent wells in landfills. This approach is also proposed by Oweis and Khera (1990).[26] When analysing gas flows in the ground the conceptual model of the migration pathways must be correct, otherwise the analytical results will be unrealistic. Other models may require the use of different flow equations, for example if the impermeable cover layer is absent.

Based on Darcy's law, flow to a slot in a confined aquifer is given by

$$Q = K_i \gamma T L \Delta P_g / \mu L_o$$

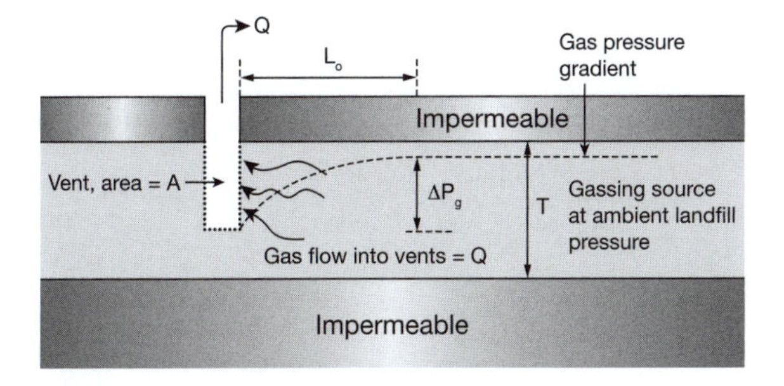


**Figure 35.6** Flow to a slot penetrating a confined aquifer

where $Q$ = flow of ground gas in $m^3/s$ from one side of barrier, over length of barrier, $L$, $K_j$ = intrinsic permeability of the ground in $m^2$, $\gamma$ = unit weight of landfill gas in $N/m^3$, $\mu$ = viscosity of gas being considered in N $s/m^2$, $T$ = thickness of confined aquifer or migration layer in m, $L$ = length of section of barrier being considered in m, $\Delta P_g$ = driving pressure of gas from ground as head of water in metres (gas pressure/unit weight), $L_o$ = distance of influence of barrier in m. This estimated gas flow from the ground is the volume that requires dilution in the duct.

The calculated flow of gas towards the barrier is very sensitive to the chosen value of the distance of influence of the barrier and the permeability of the ground. Based on experimental trials[27] and case studies,[28] the radius of influence for passive vent wells is most likely to vary between 1 m and 10 m depending on ground conditions. It seems reasonable to use similar values for $L_o$ in this case.

However, because the calculated value of gas flow is sensitive to any variation in $L_o$ and permeability, a sensitivity analysis should usually be carried out. It is also important to note that the actual flows are affected by many other factors, some of which are dynamic (e.g. changes in atmospheric pressure, temperature, gas compression), and therefore difficult to allow for except in complex mathematical models. For this reason the results from the simplified analysis method proposed should not be considered as absolute, but rather as an aid to assess the effects of varying layouts and to support engineering judgement.

Using these equations and the measured pressures from monitoring wells, the flow of gas to the line of vent wells can be estimated. When a gas monitoring borehole is left in the ground with the tap closed, the atmosphere within it comes into equilibrium with the surrounding ground, and on opening a peak value of pressure is recorded as the well comes into equilibrium with atmospheric pressure. This represents the ambient pressure in the surrounding ground, which is considered as the driving pressure for gas in the ground towards the vent nodes. At the top of the vent node the differential pressure is approximated to atmospheric pressure less the reduction due to flow of air through the dilution duct.

*35.4.3.2.3 Flow capacity* The flow capacity of a single geocomposite vent can be calculated directly using Darcy's law, and the value of intrinsic permeability, $K_j$, for the particular geocomposite used.

The gas flow in the vents is given by:

$$Q_v = [K_i \gamma A i / \mu] N$$

where $K_j$ = intrinsic permeability of geocomposite in $m^2$, $A$ = surface area of vents in $m^2$, $N$ = number of geocomposite vents, $i$ = pressure gradient along vent node = $\Delta P_v$ (in m head of water)/length of vent node.

The sum of the flows from all the geocomposite vents must be greater than the flow into the system from the surrounding ground. An alternative check is to assume that the driving pressure in the ground reduces to zero at the top of the node. The sum of the permeability and the flow area of the vents per metre length of barrier can then be simply compared with that of the ground, such that the following condition must apply to ensure the vents can provide a preferential flow path for gas:

$$K_j \text{ vent} \cdot A_{vent} > K_j \text{ ground} \cdot A_{ground}$$

The flow of fresh air through the collection/dilution duct (see Figure 35.7), required to dilute the methane to less than 1%, can be calculated using the gas flow, $Q_v$, and the guidance in Card (1995)[29] and BS 5925.[30]

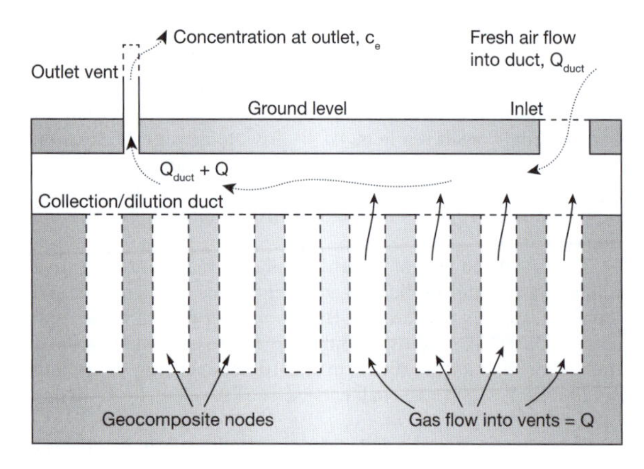


**Figure 35.7** Ground gas and air flow through the geocomposite dilution barrier and ducting

Using the formula as described in Card (1995),

$$Q_{\text{duct}} = Q(100 - C_e)/C_e$$

where $C_e$ = design equilibrium gas concentration in duct and at outlet vents, typically $C_e < 1\%$, $Q_{duct}$ = fresh air flow through each duct length, $L$, in m$^3$/h, $Q$ = flow of ground gas into the system for each duct length, $L$, in m$^3$/h.

The ventilation area required to provide the fresh air flow equivalent to $Q_{duct}$ can be calculated using the guidance for designing natural ventilation provided in BS5925. Using these design criteria the arrangement and type of ventilation can be determined.

There will be some friction losses in the duct, but considering the very low flow rates that occur in passive ventilation systems, these are considered to have negligible effect. In any event they can be allowed for by increasing the amount of ventilation provided.

The calculations require a factor of safety to be incorporated to allow for the effects of:

- uncertainty in the gas regime
- friction losses due to constrictions to flow in the system
- blocking of vents or other breakdowns of the system.

Wilson et al. (2007)[31] provide useful information on the design ventilation systems for ground gas and appropriate factors of safety.

## 35.5 References

1 ASTM Committee 0-35.
2 Koerner, R.M. (1990). *Designing with geosynthetics*. 2nd edition. Prentice Hall (Englewood Cliffs, NJ).
3 Eith, A.W., Boschuk, J. and Koerner, R.M. (1991). Prefabricated bentonite clay liners. *Geotextiles and Geomembranes* 10, 575–599.
4 Daniel, D.E. (1993). Clay liners. In: *Geotechnical practice for waste disposal*, Ed. D.E. Daniel, 137–163. Chapman & Hall (London).
5 Ingold, T.S. (1990). Geotextile properties related to structure and design. *Highways and Transportation*, Vol. 36, No. 8.
6 John, N.W.M. (1987). *Geotextiles*. Blackie (Glasgow, UK).
7 Carder, D.R., Darley, P. and Baker, K.J. (2002). *Guidance on the structural use of plastic sheet piling in highway applications*. TRL Report TRL533, Crowthorne, UK.
8 BS EN 10248:1996. Parts 1 and 2. Hot rolled sheet piling of non-alloy steels. British Standards Institution (London, UK).
9 Carder, D.R., Darley, P. and Baker, K.J. (2002). *Guidance on the structural use of plastic sheet piling in highway applications*. TRL Report TRL533 (Crowthorne, UK).
10 DAAF (1995). *Technical Manual – Engineering use of geotextiles*. Ref: Army TM 5-818-8 Air Force AFJMAN 32-1030. Department of the Army and the Air Force, United States, July 1995.
11 BS8006:1995. *Code of practice for strengthened/reinforced soils and other fills*. British Standards Institution (London, UK).
12 Finlayson, D.M., Greenwood, J.R., Cooper, G.C. and Simmons, N.E. (1984). Lessons to be learnt from an embankment failure. In: *Proc. Inst. Civil Engrs*, Part 1, 76, 207–220.
13 BS8006:1995. *Code of practice for strengthened/reinforced soils and other fills*. British Standards Institution (London, UK).
14 Padfield, C.J. and Sharrock, M.J. (1983). *Settlement of structures on clay soils*. CIRIA Special Publication 27 (London, UK).
15 Reid, W.M. and Buchanan, N.W. (1983). Bridge approach support piling. In: *Proceedings of the Conference on Piling and Ground Treatment*. Institution of Civil Engineers, London, UK, 267–274.
16 Card, G.B. and Carter, G.R. (1995). Case history of a piled embankment in London's Docklands. In: *Engineering geology of construction*, Eds. Eddleston, M., Walthall, S., Cripps, J.C. and Culshaw, M.G. The Geological Society (London, UK).
17 Bagchi, A. (1994). *Design, construction and monitoring of landfills*. 2nd edition. Wiley (Chichester, UK).
18 Privett, K.D., Matthews, S.C. and Hodges, R A. (1996). *Barriers, liners and cover systems for containment and control of land contamination*. CIRIA Special Publication 124 (London, UK).
19 Card, G.B. (1995). *Protecting development from methane*. CIRIA Report 149 (London, UK).
20 Privett et al., op. cit.
21 Wilson, S.A., Card, G.B. and Haines, S. (2008). *The ground gas handbook for local authority officers*. Chartered Institute of Environmental Health (London, UK).
22 Holtz, R.D., Jamiolkowski, M.B., Lancellotta, R. and Pedroni, R. (1991). *Prefabricated vertical drains – design and performance*. CIRIA–Butterworth-Heinemann (London, UK).
23 Wilson, S. and Shuttleworth A. (2002). Design and performance of a passive dilution gas migration barrier, *Ground Engineering* 35(1): 34–37.
24 Chapman, T.G. (1959). Groundwater flow to trenches and wellpoints. *Journal of the Institution of Civil Engineers*, Australia, October–November, 275–280.
25 USEPA (1975). *An evaluation of landfill gas migration and a prototype gas migration barrier*. Prepared by Winston-Salem NC and Enviro Engineers Inc. Grant No. S-801519.
26 Oweis, I.S. and Khera, R.P. (1990). *Geotechnology of waste management*. Butterworths (London, UK).
27 Pecksen, G.N. (1985). *Methane and the development of derelict land*. London Environmental Supplement, No. 13, Summer 1985. Greater London Council.
28 Harries, C.R., Witherington, P.J. and McEntee, J.M. (1995). *Interpreting measurements of gas in the ground*. CIRIA Report 151, Construction Industry Research and Information Association, London.
29 Card, G.B. (1995). *Protecting development from methane*. CIRIA Report 149. Construction Industry Research and Information Association, London.
30 BS: 5925:1991. *Code of practice for ventilation principles and designing natural ventilation*. British Standards Institution, London.
31 Wilson, S., Oliver, S., Mallett, H., Hutchings, H. and Card, G. (2007). *Assessing risks posed by hazardous ground gases to buildings*. CIRIA C665. Construction Industry Research and Information Association, London.

# 36

# Structural Fabrics and Foils

**Ian Liddell** CBE FREng MA DIC CEng MICE
Consultant, formerly Buro Happold

## Contents

## 36.1 Introduction

The distinctive property of both fabrics and foil materials is that they are flexible and can only carry loads by tension. Fabrics are made from woven yarns made historically from vegetable or animal fibres and more recently from manufactured fibres. Woven fabrics are not just bendable like paper or sheet steel, but because the angle of the weave is free to change they can be formed into doubly curved surfaces without stretching and can be folded up into a bundle. This property is used for clothes, which can flex and curve with movements of the body. If one imagines clothes made of paper they would be extremely uncomfortable

To contain air or water and to protect the fibres, modern woven fibres are often coated with thermoplastic polymers or elastomers. This restricts the angle changing of the weave and hence changes the mechanical properties of the material. The coating also makes the product more difficult to recycle as it has to be removed for recycling separately from the fibres.

Foils are thermoplastic films manufactured by an extrusion process sometimes followed by blowing to make finer films. Foils are made from a range of different materials that have different properties for a variety of end uses. Large extension by yielding is useful for the ubiquitous polyethylene bag since it warns of impending failure by overloading, and very low gas permeability is important for wrapping fresh meat.

## 36.2 Applications

Technical fabrics are used for a wide range of applications. These include agricultural uses, e.g. water storage tanks, silage covers; temporary buildings such as marquees and frame tents; geotechnical fabrics; leisure products including sails, clothing, sport kites and parafoils, microlight aircraft, artificial turf; general uses such as large containers for grain or sand, pillow tanks for water storage; and insulative and protective clothing. Fabrics are also used in transport for vehicle airbags, spacecraft landing bags, and the commonly seen lorry side curtains.

This section covers mainly architectural applications where these materials are used for canopies or translucent roofing. The particular requirements include durability, reliability, fire resistance, and usually translucency or sometimes transparency. This restricts the range of fibres and coatings that are generally used.

Foils also have a wide range of uses in construction such as damp-proof membranes, pond linings, poly-tunnels and greenhouse glazing. Most recently there has been an upsurge of the architectural use of ethylene tetrafluoroethylene (ETFE) foil as lightweight transparent roofing for wide-span structures. See Figures 36.1–36.5 for examples of fabric and foil projects.

## 36.3 Woven fabrics

The earliest fibres for yarns and ropes came from vegetable fibres such as papyrus fibres and flax. There is evidence of these being manufactured in 3000 BC for ropes and sails and used in boats in the Mediterranean. They would also have been used for clothing and containers such as bags for grain. Machine spinning of yarns was introduced in the 17th century, leading to an improvement in the mechanical properties of both ropes and canvas.

The development of balloons and airships at the end of the 19th century led to the scientific investigation of the complex

**Figure 36.1** Millennium Dome, London

**Figure 36.2** Fabric canopy, Portobello Market, London

**Figure 36.3** Butlins, UK

**Figure 36.4** The Water Cube, Beijing

mechanical properties of woven cloth. At that time they were using silk for high-strength applications and animal membranes for gas retention. The envelope of the ill-fated *Hindenburg* airship was clad with a linen fabric painted with oil-based paint filled with aluminium powder for heat reflection and to protect the fibres from UV light degradation. It is thought that the high flammability of the painted fabric was a major factor in the disastrous fire at Akron, NJ.

In the 1940s man-made fibres, initially Nylon, were introduced. Since then other manmade fibres with higher stiffness and strength have been introduced, particularly polyester, which was developed in 1953.

## 36.4 Current fibres

The fibres listed in Table 36.1 are currently available. Natural fires are little used now, having been replaced by synthetics which have continuous filaments and more predictable properties; even uncoated canvas is now made from acrylic fibre. Those that are commonly used for industrial fabrics are polyester, polyamide, S glass and Beta (3.5 micron) glass.

High-performance fibres such as aramids (Nomex, Kevlar, Twaron), Vectran and especially PBO offer remarkably high strengths for their weights. However, there are limits to the stress levels to which these materials can safely be used. Very often stiffness and low weight are the important factors in their selection when their high cost can be justified.

## 36.5 Yarns

Fabrics are made from woven or knitted yarns that form a two-dimensional cloth. The yarns are bundles of fibres held together by twist. Natural fibres such has cotton, flax or wool have a relatively short staple length and require twist to generate the friction necessary to get the fibres to work together. Artificial fibres are continuous filament and generally have only sufficient twist to hold the fibres as a yarn. High-twist yarns have lower strength but greater flexibility and are used for particular applications such as automobile tyres and hovercraft skirts where flexibility is essential.

Cord or rope is formed from twisting a number of yarns or strands together. Modern ropes are also made by plaiting or braiding. Often with ropes made from high-performance fibres the core is plaited with a very long lay and then covered with a braid of lower cost material to protect the core from abrasion and UV light.

## 36.6 Weaves

A range of different weaves are used for industrial fabrics. The warp is the yarns that run in long lengths through the weaving machine while the weft or fill is the yarns that go across the width of the cloth. Plain weave used for traditional canvas consists of single yarns woven over and under each other. Panama weave is similar but with pairs or sometimes triple yarns. In satin weave the

**Figure 36.5** Foil roofing development: (top) C & W hospital; (centre) 'City in the Arctic' proposal; (bottom) tennis centre

**Table 36.1** Typical properties of fibres used for flexible tensile elements

| | *Density g/cm³* | *Modulus GPa* | *Tenacity N/tex** | *Strength GPa* | *Elongation %* |
|---|---|---|---|---|---|
| *Natural fibres* | | | | | |
| Cotton | 1.4 | 5 | 0.21 | 0.3 | 6 |
| Flax | 1.6 | 24 | 0.39 | 0.6 | 2.5 |
| *Metal fibres* | | | | | |
| Drawn carbon steel wire | 7.8 | 200 | 0.26 | 2 | 2 |
| Stainless steel wire | 7.8 | 200 | 0.19 | 1.45 | |
| *Mineral fibres* | | | | | |
| E glass 20 μm | 2.5 | 73 | 1.15 | 2.8 | 3.8 |
| S glass 6–9 μm | 2.5 | 85 | 1.84 | 4.6 | 5.4 |
| Beta glass 3.5 μm | 2.5 | 85 | | | |
| *Synthetic fibres* | | | | | |
| Polyethylene tape PE | 0.95 | | 0.39–0.53 | 0.38–0.50 | 12 |
| Polypropylene PP | 0.91 | | 0.75 | 0.68 | |
| Polyester PET | 1.38 | 7.6 | 0.66 | 0.91 | 12 |
| Polyamide PA | 1.14 | 5.5 | 0.75 | 0.85 | 18 |
| Aramid AR | 1.44 | 62 | 1.93 | 2.8 | 4 |
| High-mod aramid HMAR | 1.44 | 131 | 1.93 | 2.8 | 2.5 |
| LCP “Vectran” | 1.41 | 65 | 2 | 2.8 | 3.3 |
| High-mod polyethylene HMPE | 0.97 | 110 | 2.64 | 3.5 | 3.5+ |
| Carbon | 1.8 | 230 | 1.94 | 3.5 | 1.5 |
| PBO “Zylon” | 1.54 | 280 | 3.7 | 5.8 | 3.5 |

*Tex is the weight in grams of 1000 m of the yarn, hence tenacity is a measure of the strength to weight ratio.

**Figure 36.6** Weave types: plain weave is shown; basket or Panama weave is similar but with yarns in pairs; in malimo weave straight yarns are overlaid and stitch-bonded so there is no crimp

fill yarns go over and under three or more warp yarns, but the bundle of warp yarns moves by one yarn at each row. The aim is to produce a cloth with a smoother finish. In any of the above weaves the yarns undulate under and over those of the cross set. The amount of the undulations is called crimp and causes changes to the load extension properties of the fabric.

At least one fabric producer uses malimo or stitch-bonded cloth, a material construction more often used for fibre-reinforced polymer (FRP) products. In this the fill yarns are laid over the warp and stitched together, thus avoiding the effects of crimp resulting in a stiffer fabric.

## 36.7 Coated fabrics

Canvas from natural fibres relies on the tightness of the weave and proofing agents that increase the surface tension to make it waterproof. Artificial fibres do not work so well with this technique, and coatings are used. As well as keeping water out, coatings protect the fibres from external abrasion and UV light and provide flame resistance. Coating materials can be thermoplastics such as PVC, Hypalon (chlorosulfonated polyethylene) and PU; or other cured rubbers such as Neoprene, EPDM or silicone. PTFE is a sintered polymer with a very high melting/gasifying point of about 370°C that can only be used on glass-fibre base cloth.

The coating materials and processes are considerably more complicated than it would seem from the above. The most popular coating PVC is a blend of typically PVC powder from two different suppliers, plasticisers, fillers, pigments and flame-proofing additives and fungicides. The mixture is blended and allowed to stand for a week or so before use. After application onto the fabric by a spreader blade or calender rolls, it is heated to about 200°C to polymerise the PVC. Unsurprisingly the coaters are reluctant to give out their secrets, so it can be difficult to compare the performance over time of different manufacturers' products.

With PTFE-coated glass the PTFE powder is mixed with a solvent carrier and sometimes glass micro-spheres as a filler. The coating process involves passing the glass cloth through a bath of this mixture and heating it in a tower at over 300°C to sinter the PTFE to the cloth. Heavy coatings may involve six or seven passes. Only glass fibres can be used as the base cloth because of the high processing temperatures. Silanes are used as a bonding agent to get the PTFE to adhere to the glass. The rigidity of the coating and the ageing of the bonding agents may affect the tear strength of the cloth over time, but this has not been systematically investigated.

The most used rubber coating is thermoplastic polyurethane, which can be seamed by heat sealing. With this, RFL (resorcinol formaldehyde latex) is used as a bonding agent. The coating is bonded only to the surface fibres of the bundles and when the cloth fails the inner unbonded fibres pull out. Rubber coatings are generally insufficiently flame-resistant for use in buildings. Silicone coated glass can be made flame-resistant with a heavy

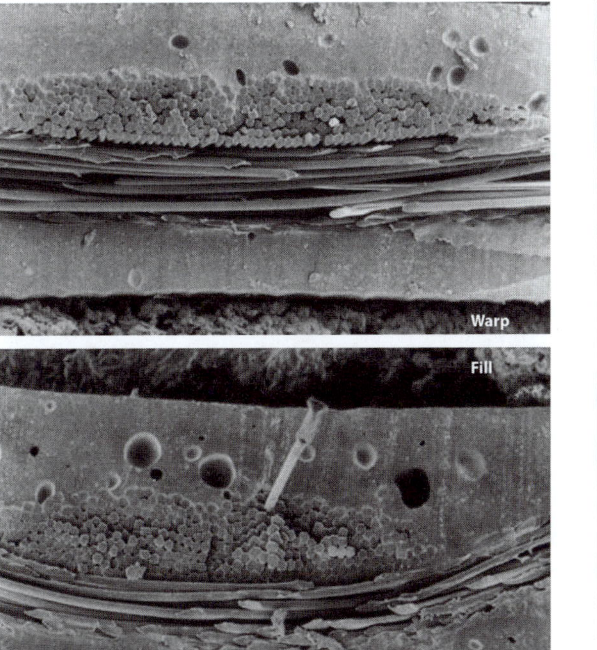


Figure 36.7a SEM pictures of sections of PVC-coated polyester cloth, 15μm fibres.

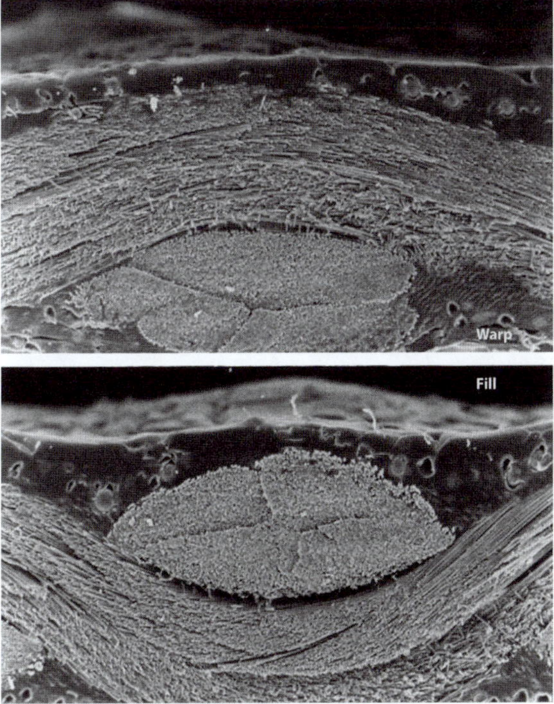


Figure 36.8a SEM pictures of fabric sections of PTFE-coated glass cloth, 3.5 μm fibres.

content of flame-inhibiting fillers and offers useful properties, but it collects surface dirt and has to be seamed by gluing, a slow process that makes the fabrication expensive.

The coatings keep water from penetrating through the cloth but they do not keep water from getting into the yarns. This is especially so for PVC-coated polyester cloth; the polyester fibres are about 30 μm in diameter, leaving appreciable spaces through which water can migrate by capillary action from cut edges or surface scratches. With translucent fabrics this water allows a fungus to grow fed by nutrients from the plasticisers that turn purple and then black. Manufacturers have made attempts to cure this using non-wicking treatment to the yarns, but with limited success.

Some of these coating processes are incompatible with the fibre materials. For example, PTFE and silicone coating processes use temperatures that will damage thermoplastic fibres. The commonly used architectural fabric combinations are PVC-coated polyester and PTFE-coated glass. The glass fibres used are 3.5 μm, which have the best strength and flexibility properties. Six micrometre glass is sometimes used with caution, and 9 μm should never be used.

W. L. Gore has for some years produced a fabric using woven PTFE fibres called Tenara. This has amazing properties of long life, no dirt retention, and mechanical flexibility but its use was limited by the high cost and the difficulty of repairing damage on site. Its main use has been foldable canopies, where it has withstood several thousand operations. The company now produces a new PTFE fabric coated with a thermoplastic fluoropolymer. This can be welded at 200°C and has excellent translucency and long-life properties. It can also be welded on site with a hot bar welder using an interlayer tape.

## 36.8 Fabric properties

The amount of fibre in a yarn is defined by the tex number. This is the mass in grams in 1000 m. Denier is the mass in grams in 9000 m. The fabric industry uses decitex (dtx), the mass in grams in 10,000 m. The amount of fibre in woven cloth in grams per square metre is defined by the number of ends/picks per cm multiplied by the yarn count in dtex/100. Ends refer to the warp and picks to the fill. The strength of the cloth can be calculated using the strength/tex and appropriate factors to allow for the twist and crimp. In practice the strength is measured by strip tensile tests, usually on 50 mm wide strips.

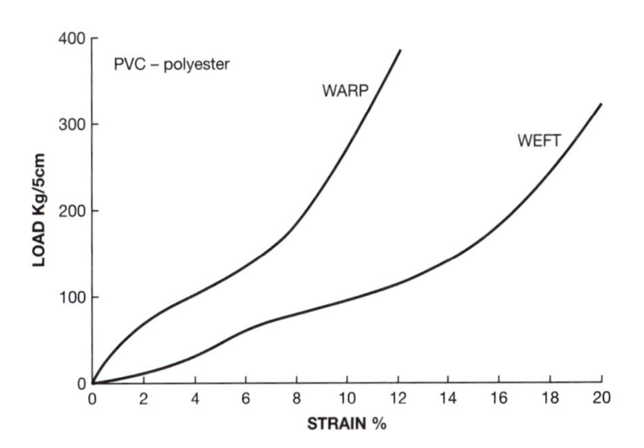


Figure 36.7b Typical uniaxial tensile test characteristics for PVC-coated polyester cloth, 15μm fibres.

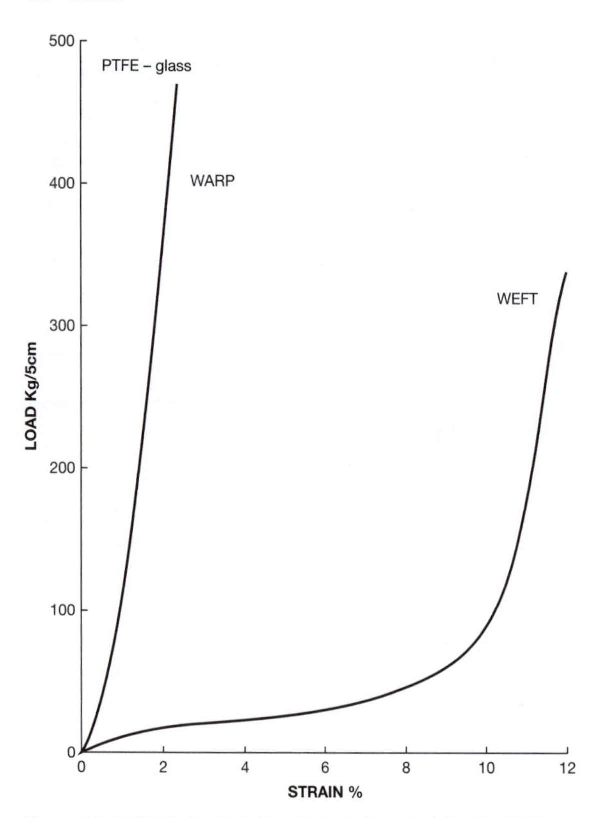


**Figure 36.8b** Typical uniaxial tensile test characteristics for PTFE-coated glass cloth, 3.5 μm fibres.

Standard testing includes the following:

- *strip tensile tests:* usually the breaking stress of a 50 mm strip
- *trapezoidal tear tests:* a strip with a cut in one edge and with the clamps placed at an angle of 30° so that the yarns at the end of the cut are loaded one by one (Figure 36.9)
- *peel tests:* to test the adhesion of the coating (Figure 36.10).

Tear strength is the most important property. The trapezoidal tear test loads a cut in a test strip at the edge so only a few yarns are loaded. The results depend on how the yarns can slip to allow them to share the force. The results of this test are not a quantitative guide to the effect of small cuts on the strength, but are a useful comparative test.

## 36.9 Crack propagation stress

For a true understanding of the effects of damage to fabric it is necessary to carry out wide panel tear tests. Ideally the sample would be at least 800 mm × 400 mm with up to a 40 mm cut in the centre. Such tests demonstrate that the crack propagation stress falls off very quickly to about 25% of the strip tensile strength with a 40 mm cut. With larger samples it continues to fall to about 10–15% when the crack reaches 500 mm or so.

All large fabric structures have flaws from fabrication imperfections or damage during installation or afterwards. Because of this engineers prefer to keep the peak stresses below 25% of the strip tensile. If the stress is above 40% then the fabric may suddenly split from small flaws without warning (see Figures 36.11–36.13).

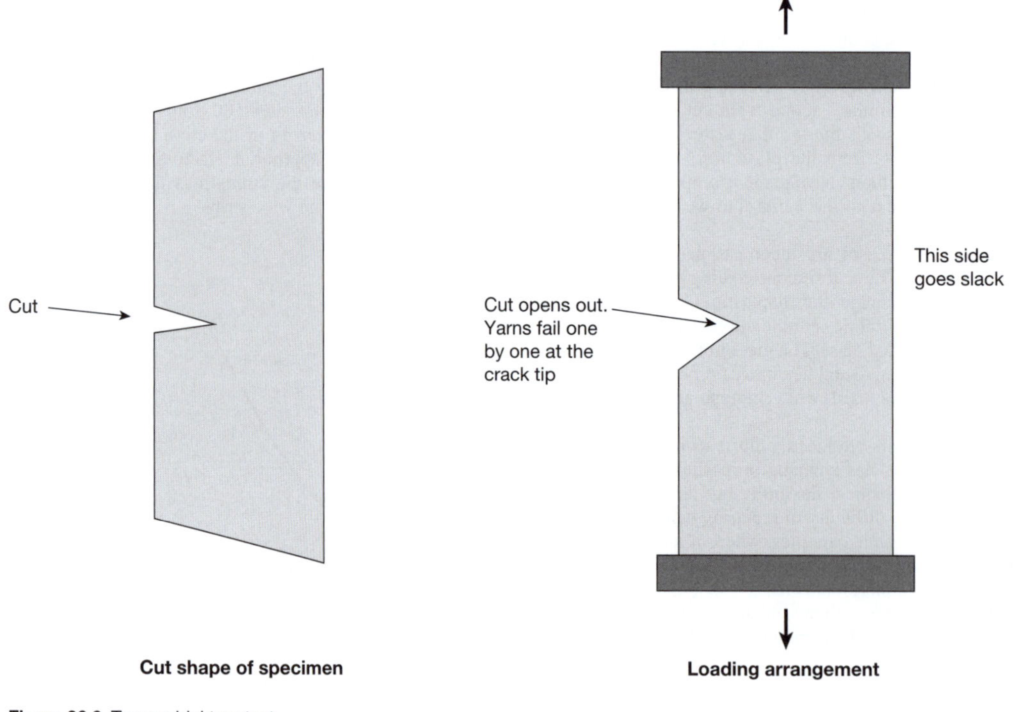


**Figure 36.9** Trapezoidal tear test

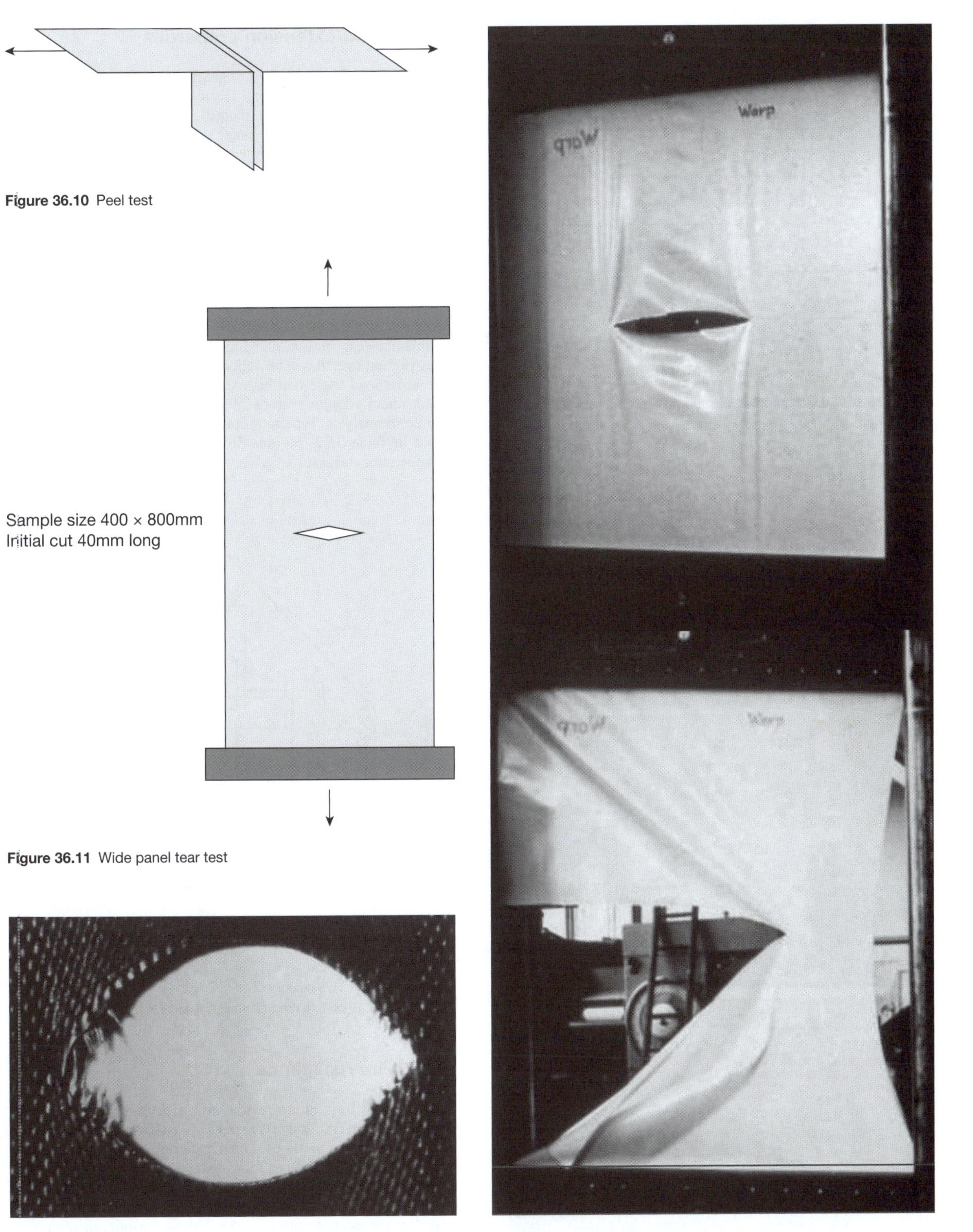


**Figure 36.10** Peel test

**Figure 36.11** Wide panel tear test

**Figure 36.12** Results of wide panel tear tests

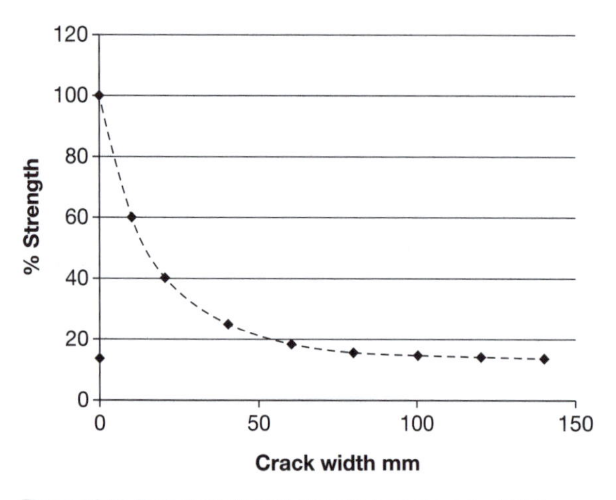


Figure 36.13 Typical graph of % strength against crack width for coated fabrics

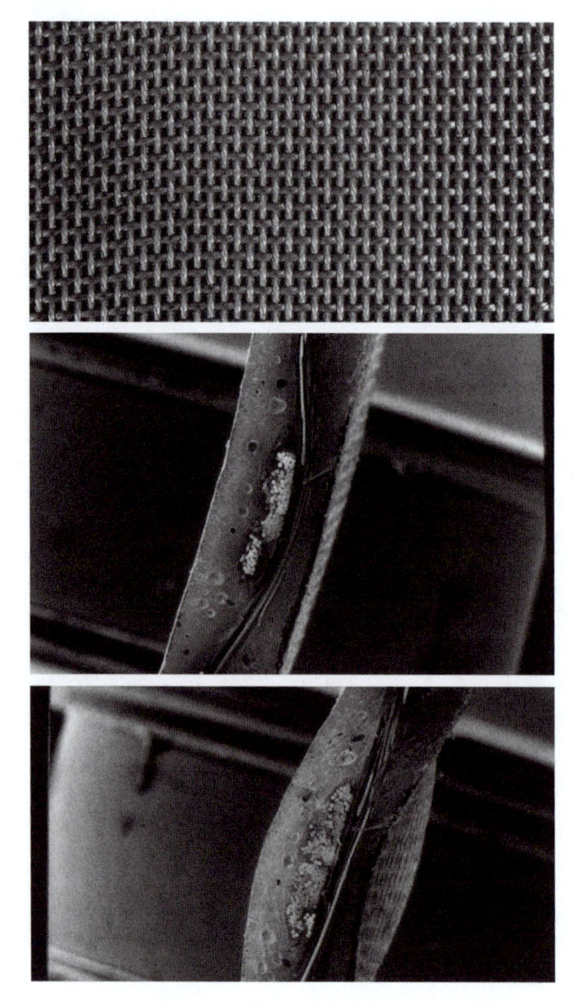

Figure 36.14 Crimp interchange in PVC/polyester fabric: fill yarns straighten out under tension and warp shortens; effect can be represented by Poisson's ratio for nearly equal tension ration, say 1:2–2:1

## 36.10 Load/extension properties

Biaxial testing has to be used to get the load extension properties most relevant for architectural fabrics. These are greatly affected by construction stretch, i.e. inelastic extension from the weave and crimp compaction and under tension. The elastic properties are also affected by crimp interchange between the warp and fill directions, which causes an effect that is similar to Poisson's ratio but non-linear.

In coated fabrics the warp yarns are nearly straight with the crimp pulled out while the fill yarns have more crimp. In traditional fabric, which is tightly woven to make it waterproof, the fill yarns are tensioned by the beaters while the warp has more crimp (Figure 36.14).

With glass fibre based cloth the yarn stretch to failure is of the order of 4% while the construction stretch plus crimp extension in a uniaxial test in the fill direction may be 8 or 10%. The construction stretch will be pulled out when the fabric is tensioned but under load there will be a strong residual crimp interchange effect under changing ratios of stress which is very difficult to handle in analysis. Typical properties of architectural fabrics are given in Table 36.2. Figures 36.15 and 36.16 show biaxial test arrangement and results.

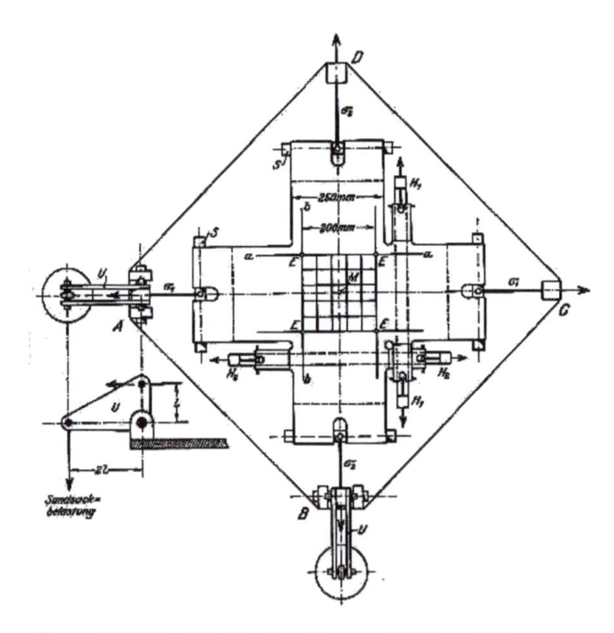


Figure 36.15 Biaxial testing (from Haas and Dietzius, 1917)

## 36.11 Fire resistance

To comply with building regulations, architectural fabrics must pass the tests in BS476 Parts 6 and 7. Glass-base cloth fabrics can be tested by these methods. Other materials with polymer fibres may not be robust enough to withstand the test, which was not designed with these materials in mind. It does not mean that these materials are an unacceptable fire hazard. The fabrics can be tested in accordance with BS7157. This is a standardisation of an ad hoc test in which about 1 $m^2$ of fabric is held at 45° over a small fire and the amount of burning noted.

Most fabrics that are used in the UK come from Germany or France. Both these countries have standard test procedures that enable them to grade the fire resistance of the fabric. The grading

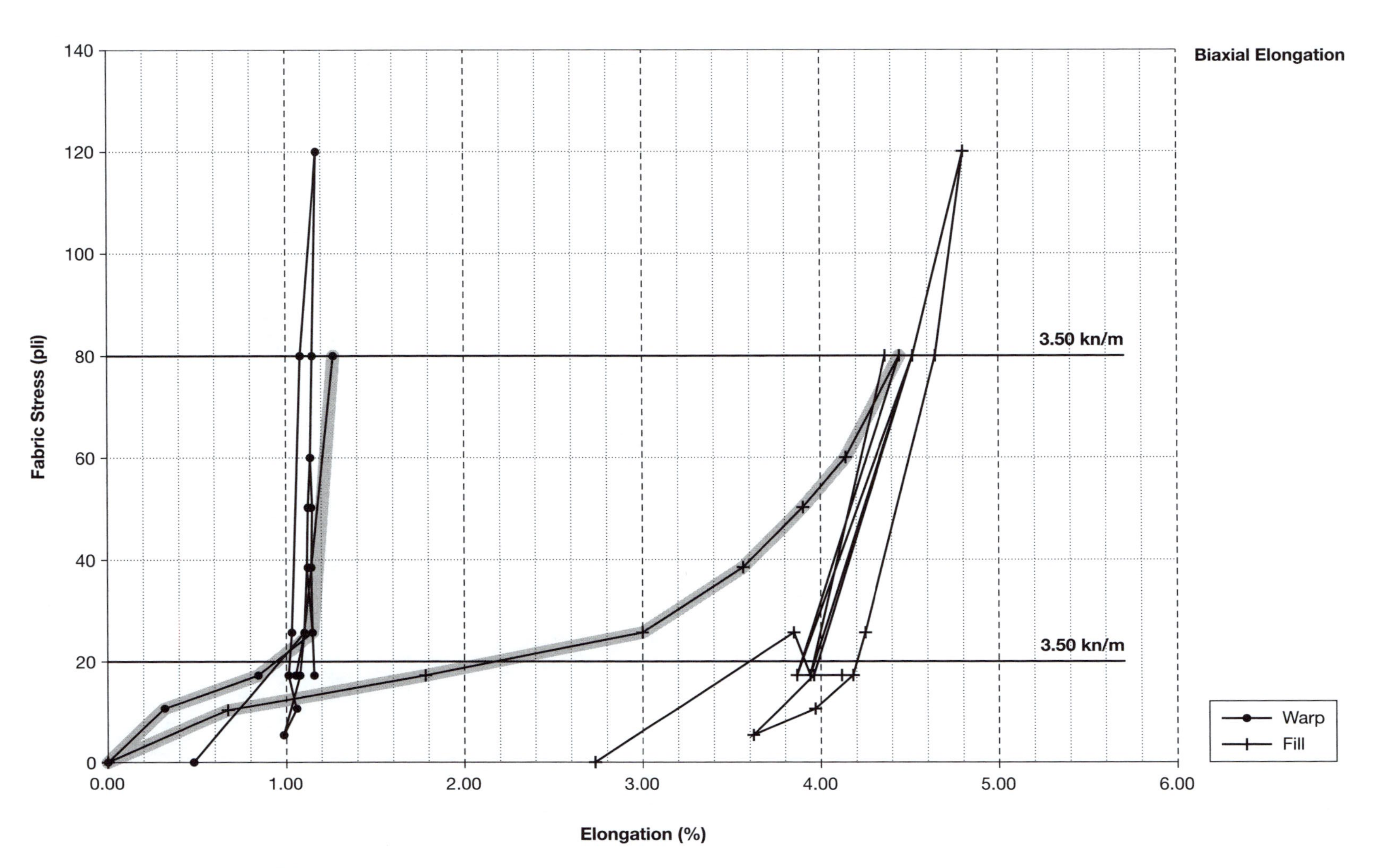

**Figure 36.16** Biaxial testing on PTFE/glass cloth at 1:1 stress ratio: compensation at 20 PLI = 1.1% Warp and 4% Fill

**Table 36.2** Typical properties of architectural fabrics

| | *Weight g/m²* | *Tensile strength warp/ fill kN/m* | *Trap tear strength N warp/fill* | *Translucency %* |
|---|---|---|---|---|
| PVC/Polyester type 2 | 900 | 88/79 | 520/580 | 12 to 15 |
| PVC/Polyester type 5 | 1450 | 196/166 | 1600/1800 | 7 to 9 |
| PTFE/Glass G5 | 1200 | 124/100 | 400/400 | 16 |
| PTFE/Glass G7 | 1600 | 170/158 | 450/450 | 8 |
| Tenara type 2 | 830 | 84/80 | 925/925 | 38 or 19 |

systems are of course different. PTFE/glass is inherently fire-resistant with very low spread of flame and considered virtually incombustible, since it has a limiting oxygen index of 90%, greater than the oxygen content of air. Tenara also has inherently zero spread of flame. PVC-coated fabrics require the addition of fire-inhibiting compounds to pass the fire tests.

## 36.12 Seaming

For a technical fabric to be used structurally it must be possible to make satisfactory seams that can transfer the full strength of the fabric. In traditional canvas made in a plain weave, the fill yarns are returned at each edge, resulting in a selvedge. The width of such cloth is quite small, of the order of one metre. To make large sheets for sails and tents the cloths can be stitched together with a small overlap. Because of the return of the fill yarns at the selvedge, a simple overlap seam with two rows of stitching will suffice. If the edge is cut this simple seam will slip, so a double folded "French" seam will be required.

Coated fabrics are seamed by heat welding an overlap; the principle being to heat up the thermoplastic to melting point and allow it to cool under pressure. Those with low-temperature thermoplastic coatings (e.g. PVC and PU) are often heated using a radio-frequency (RF) heating bar. This is similar to microwave heating and has the benefit of heating the plastic while the bar, with added cooling, remains cool enough to re-set the plastic when the power is switched off (Figure 36.17).

Alternative welding methods are hot air and hot wedge. Both of these can be used like a sewing machine where the fabric is moved past the tool. The seam is closed by a roller that is just behind the heat source. There are hot wedge machines available that crawl along the fabric that is placed on the floor. With both these methods there is a difficulty of closing the seam at the start. This can be done with a hand-held hot air gun after the rest of the seam has been made.

Site welds can be made with a hand-held hot air gun but these are difficult to do to get a full-strength weld. As PVC fabric ages and the plasticiser migrates, it becomes increasingly difficult to make satisfactory site welds.

**Figure 36.17** Welding fabric

PTFE/glass fabrics have to be heat sealed at about 350°C. This is done with a heated bar running at about 380°C. Cooling under pressure is achieved by switching the hot bar for a cold bar and reapplying the pressure. This welding cycle is slower than the RF heating method.

Site welds can be done satisfactorily with a hand-held hot bar, usually about 300 mm long, and a roller to close the seam. The task is very slow and it is difficult for workers to maintain concentration, so faulty welds are fairly common. It is also difficult to do in cold weather, and impossible in rain or snow.

## 36.13 Ageing and degradation of architectural fabrics

PVC coatings start beautifully clean but they are porous and with time the plasticiser migrates and dirt sticks to the surface. Oily dirt will impregnate the PVC. It is also prone to mould growth within the fibre bundles that show as black streaks. These processes cause the fabric to become dirty with time, and manufacturers have developed a range of surface treatments to improve the situation.

The effect of surface dirt is reduced by the application of surface lacquers. The standard coating is acrylic lacquer, which helps. Improved coatings are based on fluoropolymers. These can be various combinations of PVDF with acrylic applied as a liquid or PVF (Tedlar) applied as a laminate. These materials provide progressively better non-stick surfaces and the higher performance ones have to be removed by grinding before the panels can be welded. The performance of the fluoropolymer surface coatings relies on the integrity of the surface lacquer being maintained and it remaining adhered to the surface. If it comes off the resulting appearance is worse than with a standard acrylic coating, and it would seem that over time the fluoropolymer coatings will eventually de-laminate. Any surface cuts through the sealing coating should be repaired with a permanent tape to avoid water and dirt penetrating to the fibres and the accompanying mould growth. Potential users should view existing structures made with the proposed PVC-coated fabric to verify that the material can perform to the desired standard.

Another problem is that PVC is prone to mould growth within the fibres. If moisture gets into the fabric from cut edges or surface scratches, purple or black mould will develop. This mould spreads along the yarns being supported by nutrients in the PVC. These problems are extremely disfiguring but do not affect the structural properties significantly. Manufacturers have attempted to reduce this problem by using yarns treated to prevent wicking. This helps but does not entirely eliminate the problem. PVC can be made with a black core and different colours on the two faces. This makes it non-translucent but the mould growth will not be a problem and the life will be extended.

With old PVC-coated fabrics the PVC hardens with loss of plasticiser and the tear strength reduces because the yarns cannot bunch up at the crack tips. PVC fabric can be repaired with hot air welding but this will be much less successful with old, hardened material. The loss of plasticiser causes noticeable hardening after 12 years. The oldest PVC structures have been kept in place for about 22 years, but at this age the PVC surface breaks down under the UV light and is washed off as powder, so from the outside it looks cleaner. When viewed from the inside the mould growth can clearly be seen; these structures are very dirty and are probably much less safe.

PTFE/glass fabrics have a better track record for durability and dirt retention; the oldest ones have lasted over 30 years and are still in place. Observations on two structures over 20 years old revealed surface craters that trapped dirt and gave it a grey appearance. The fibres are not affected by UV light but do suffer loss of strength from wetting and flexing causing fibre breaks. Tests on 20 year old material indicated that the strip tensile strength dropped by about 20–25% while the trapezoidal tear strength dropped by about 40–50%. This last could be caused by hardening of the silane coupling agent but it is worrying from the point of view of the safety of persons walking on the roof.

## 36.14 Engineering standards

Tension cable and fabric structures are non-linear in that the deflections and tensions are not proportional to the load and the deflections change the load-carrying behaviour. This means that every load case has to be the subject of a separate analysis. Against this they are excellent load-averaging structures, smoothing out the effects of local high loads. The analytical model needs to have the correct warp and fill directions of the fabric. Correct representation of the biaxial load extension properties is difficult, especially with glass cloth where the stiffness is very high in uniaxial loading but a lot lower where the stress ratios are such as to induce crimp interchange. It is best to keep the stiffness high when calculating the peak stresses, which will give higher estimates for them, but of course the estimated deflections will be less than reality.

Normally the structures are analysed for extreme, 1 in 50 year, service loads that the engineer judges will give the highest forces without additional load factors. When assessing the safety of the fabric, factors of safety on the strip tensile strength of the cloth are used to obtain limiting stresses. These factors are typically 4 for wind and 5 for snow loading. If one adopted a fracture mechanics approach based on tear propagation then these factors would be 6 and 8 respectively, especially for glass cloth, which has relatively poor tear strength and a higher economic loss.

The factors used by engineers vary depending on the importance of the structure, how long it is required to last, the view the engineer takes of the likelihood of occurrence of the maximum design loads, and so on. Clearly some degradation takes place on all fabric materials and the levels of safety will reduce. Since there is no clear end-point to the life and owners will generally want to keep them in place for as long as they look OK, it has to be accepted that at some time they might suffer disastrous tearing in a severe storm initiated by an initial cut. The supporting structural elements have to be designed with this in mind so that they remain stable and reusable.

Snow loading is a problem with tensioned fabric structures because of the risk of ponding. This occurs when the snow causes local accumulations on flat or near-flat areas that cause the fabric to deflect. The local deflections can initiate areas where snow continues to accumulate and where water no longer drains away. When the snow melts water will continue to run into the pond, potentially until the fabric ruptures. To avoid this, such areas should be avoided in the design stage by having sufficient slope or by introducing additional local drains.

## 36.15 Foils

Polymer foils have been used to provide flexible transparent coverings, particularly PVC, polyester (known as Mylar) and polyethylene. The last is widely used for poly-tunnels in agriculture. These foils cannot be both flame-resistant and transparent. PVF foil, known as "Tedlar", has good UV resistance and a relatively high yield strength and has been used for single-layer greenhouse glazing but it too has insufficient flame resistance to be used in buildings. There has been a recent introduction of ETFE foil, which has inherent flame resistance because of its high limiting oxygen index and is also long lasting. It has an unusual stress/strain behaviour with a low yield point but very large extension to failure. Because of this it is used as inflated

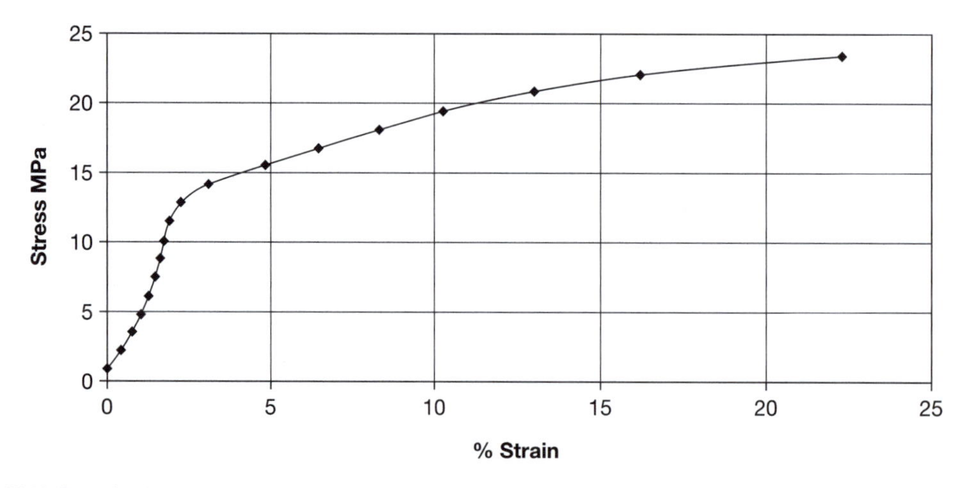


**Figure 36.18** Stress/strain diagram for 0.15 mm thick ETFE foil 50 mm wide loaded at 10 s intervals: the load was increased to 28.7 MPa stress and the sample extended to 210% strain after 30 min without failure when the test was stopped

cushions where the air pressure compensates for the creep extension and keeps the foil under tension. Use of this material has increased rapidly and it has been used for the cladding of the Chelsea and Westminster Hospital roof (1991), the Eden biodomes (2000), Munich Allianz stadium and the Beijing Olympic pool and stadium structures (see Figure 36.4).

ETFE foil is made in thicknesses from 50 to 250 μm. It has strange mechanical properties. It has a small elastic range up to 13 mPa followed by a very long plastic stage where it can stretch to 200% or more. This makes the material very resistant to tearing. It is also resistant to short-term overloads. It would be normal to allow the structure to be loaded to 14 mPa under maximum service loads. Under long-term loads the stresses should be kept below half of this (Figure 36.18).

As a cladding the material is generally used in the form of inflated cushions up to 3.5 m wide (see Figure 36.5). These are made with a minimum of patterning and rely on the plastic extension to form a smooth surface. As the cushions enlarge under creep, the tension under the inflation pressure reduces until an equilibrium point is reached and the creeping stops. Larger cushions can be made by patterning the foil to allow a greater rise and hence lower stress. The largest hexagonal cushions for the Eden project were 9 m across. These were patterned to allow more rise and the largest panels had two layers of foil to handle the wind pressures. With two layers a pinhole had to be made in the outer layer to prevent the lower layer from falling away.

If the cushions are punctured, the large ductility of the material makes it very resistant to tear propagation. Cuts can be repaired with ETFE tape coated with a pressure-sensitive adhesive. The repairs are virtually invisible and seem to last a long time. If they lose pressure they will flap noisily in the wind, but provided the edges are secured they do not tear. Under snow loading the air will be squeezed out until the foils are pressed together and they will deflect. If the slope is too low they can then fill up with melt water and rain. This risk needs to be addressed during the initial design stage.

## 36.16 Bibliography

Haas, R. and Dietzius, A. (1917). The stretching of the fabric and the deformation of the envelope in nonrigid balloons. In: *Annual report 3*. National Advisory Committee for Aeronautics, pp. 149–271 (originally published in German in 1912).

# 37 Cork

**Peter Olley**
Director, C. Olley & Sons Ltd

**Bob Cather** BSc CEng FIMMM
Consultant, formerly Director of Arup Materials Consulting

**David Doran** FCGI BSc(Eng) DIC CEng FICE FIStructE
Consultant, formerly Chief Engineer with Wimpey plc

## Contents

## 37.1 Introduction

Cork is a natural product obtained from the bark of the cork oak tree (*Quercus suber L.*) found primarily at the western end of the Mediterranean in Portugal, Spain and North Africa. The bark thickens and develops with age and can be harvested as a regular crop (every 9–12 years) from commercial plantations and natural forests. As a vegetable tissue, cork bark is composed of cells that are five-sided, impermeable and contain a large proportion of air. The cellular structure allows cork to be compressed and to recover its original dimensions without extrusion – a property yet to be equalled by synthetic products used in this field. Cork is, therefore, a perfect natural insulant.

Natural cork is harvested from the tree in sheets and they are steam treated to flatten them. It may then be laminated to any thickness. Cork is a natural material that regenerates itself without depletion of its source. It is ubiquitous and has a place in every phase of building.

Cork is unique in both its cellular form and its ability to combine functions, and has the following characteristics:

- compressibility
- recovery after compression
- sound-deadening
- anti-vibration
- thermal insulation properties
- non-carcinogenicity
- impervious to liquids
- resistant to most chemicals
- no insects, rodents, etc. live off it
- it is biodegradable without resort to processing other than mixing with soil in an outdoor environment
- low energy use during manufacture
- non-abrasive
- ease of removal
- 'grip' on surrounding materials (clutch action)
- ease of drilling and cutting
- softness
- environmentally friendly.

Material containing natural cork may be used for:

- pads for isolating columns
- wall and roof insulation
- anti-vibration products (see Figure 37.1)
- acoustic applications
- decorative purposes.

The material can combine low weight with great mechanical and structural strength. Natural cork is also granulated and this becomes the basis for a wide variety of cork-based materials. Many types and forms are used in construction.

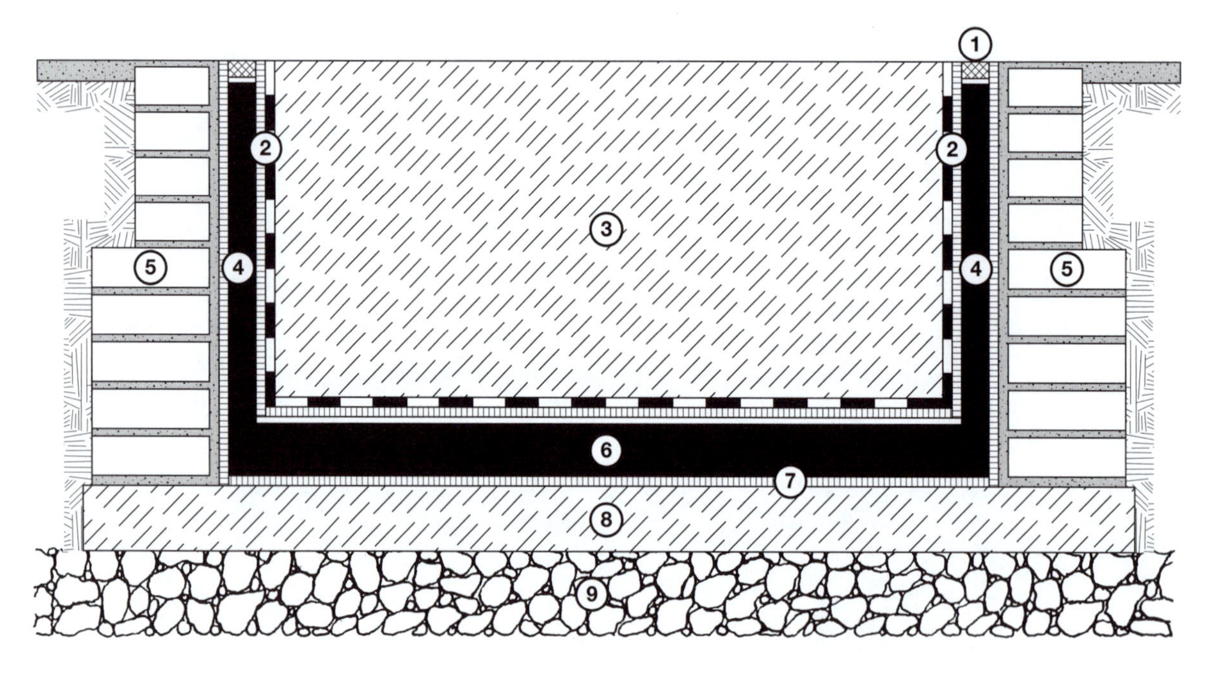

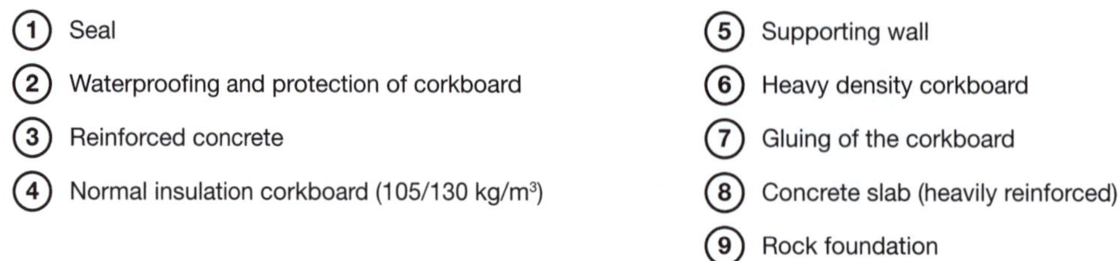


**Figure 37.1** Typical cork acoustic vibration insulation detail

## 37.2 Agglomerated corkboard

Corkboard is manufactured by baking granules (usually with superheated steam) at a temperature of 350°C and a pressure of 30,000 kg/m$^2$. There is no contact with air during this process and the cork's natural resins are released from the cells causing the granules to bond together, becoming dark brown in colour. After an appropriate period of 'cooking', a block of corkboard usually 1000 × 500 × 300 mm is produced; higher density boards, typically 914 × 305 × 300 mm are also produced. After 'baking' the compressed board is cooled, rested, trimmed to size and slabs of any thickness from 12 mm to 300 mm produced. Thicker sheets can be produced by post-laminating. The baking process causes natural cork granules with a density of approximately 200 kg/m$^3$ to undergo a large expansion in volume with the release of volatile elements and a commensurate decrease in density. At the same time, the expanded cork granules are restrained, causing their agglomeration to form a cork block with a higher density than that of freely expanded granules. This density can vary from a minimum of 80 kg/m$^3$ to over 600 kg/m$^3$ depending on:

- degree of compression
- quality of cork
- granule size
- purity of granulation
- manufacturing process
- temperature.

The combination of these variable factors enables the manufacture of different types and densities for numerous purposes. Low and medium densities are best for insulation of walls and roofs. Higher densities should be used for rooftop walkways and for machine bases in order to change the frequencies of vibrating machinery. Typically such products should have a minimum density of 176 kg/m$^3$. It is recommended that extra care be taken to scrutinise the intended use of any material with density below 96 kg/m$^3$. The ideal quality will be achieved where the manufacturer selects the material and eliminates the heavier woody inner part of the bark, which increases the weight at the penalty of reduced performance. The supplier must ensure that this has been achieved. For a chosen application the aim should be to achieve minimum density with maximum mechanical strength.

Corkboard insulation may also be used internally under a pitched roof.

The recovery of material should be tested on the basis of a loading test of 4 s duration at a load of 0–140 kN/m$^2$ repeated 1200 times with a maximum loss of thickness of 0.12 mm of the 60 mm sample tested

The thermal diffusivity (a measurement of heat flow calibrated against time), used in assessing the overall performance of a material, shows cork to be a good insulant and having resistance to changes in temperature (Figure 37.2 shows thermal insulation values of INS-OLL board). It can be used in conjunction with oil-based plastics foams to improve performance further. These composite boards recover after loading significantly better than oil-derived single component insulation products such as polystyrene foam.

The dimensional stability of cork is indicated by an expansion of 0.03 mm/m for a temperature rise of 1°C which equates to a maximum expansion (in the UK climate) of 0.9 mm/m. The corresponding values for foam plastics are considerably greater. Corkboard can therefore be described as benign, making it the ideal companion to other components of a roof structure. Material such as polyurethane when loaded and unloaded does not recover as well.

Cork cannot be successfully specified by density alone; it requires the judgement of a specialist to assist in the

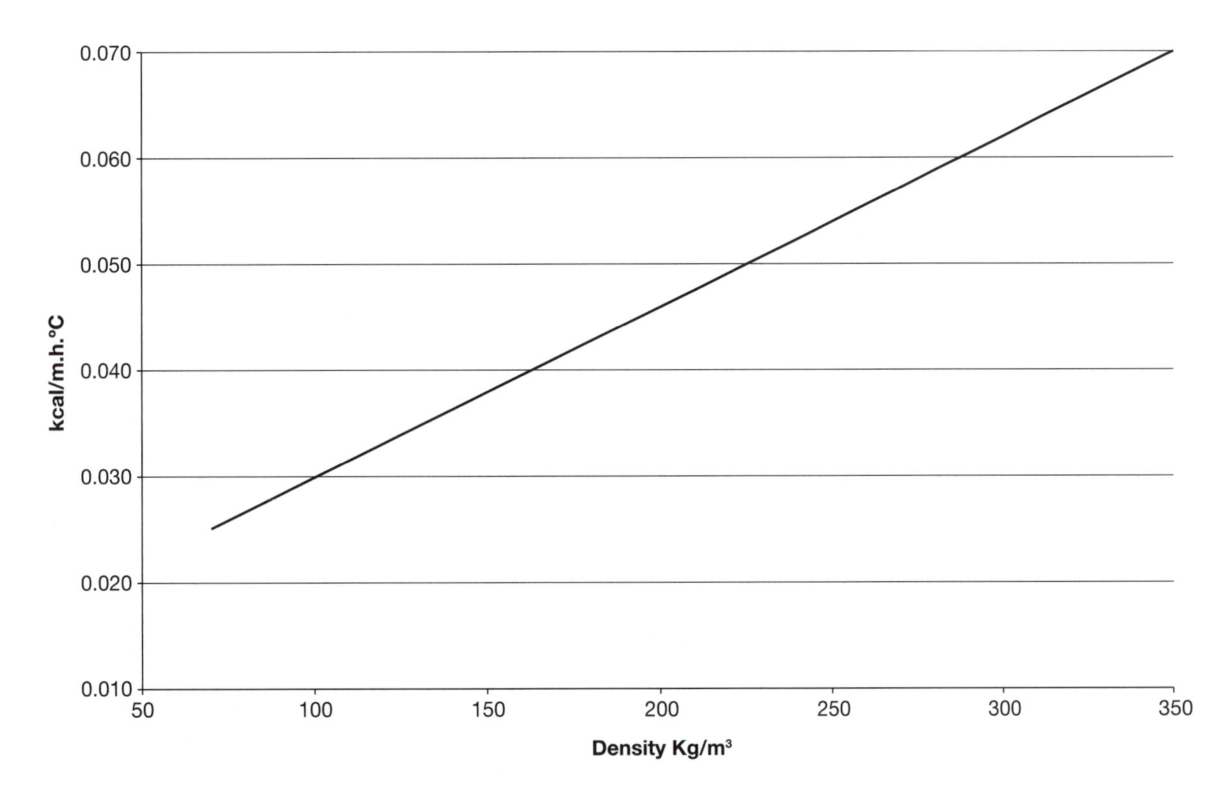


**Figure 37.2** Thermal insulation value of INS-OLL board at various densities

selection process to ensure that performance needs are properly met.

It is important when considering fire behaviour to take a holistic view of the performance of the whole construction, not just the cork being used as an insulant. However it can be stated that cork does not emit toxic fumes and, after a short period of smouldering, it will self-seal once a fire source has been extinguished.

## 37.3 Cork in flat roofing

In addition to its generally good properties discussed above, an attractive option for cork's use is the ability to cut the cork blocks to a taper, thereby creating a dry-construction roof to falls. Falls are essential for rapid shedding of rainwater from a roof.

## 37.4 Handling and storage

Material should be handled with care to avoid damage to board edges. Slabs are usually delivered to site shrink-wrapped in polythene or corrugated cardboard export cartons which provide some short-term protection. For long-term storage slabs should be stored indoors or off-ground under a waterproof cover.

For bitumen-based multi-layer 'built-up' roofing, corkboard is normally fully bonded to a roof with hot bonding grade bitumen using a staggered arrangement of joints. A vapour-resisting layer should always be installed below the insulation and the specification carefully chosen to suit the individual project. A vapour escape sheet fixed by partial bonding is not recommended with corkboard. All day-work joints should be fully sealed to prevent moisture ingress. In particularly exposed environmental conditions, additional mechanical fastenings may be required.

Corkboard is suitable for use with most single ply membranes, but it is recommended that the corkboard product is supplied with a sealed finish compatible with the adhesives to be used. The manufacturer's instructions for applying the membrane should be closely adhered to.

For roofs waterproofed by mastic asphalt there is often a requirement for a thermal protection layer between the insulation board and the hot applied asphalt. For corkboard insulation no thermal barrier layer is required. A loose laid standard mechanical isolation sheet membrane can be laid over the corkboard prior to application of asphalt to the recommendations of MACEF (Mastic Asphalt Council and Employers Federation)

Insulation requirement and design for buildings will normally be controlled by local or national Building Regulations and specific reference should be made to them. Cork insulation can be used to upgrade the performance of existing installations, and again reference to Regulations is advised.

## 37.5 Composition cork

Composition cork is made by the incorporation of a low volume (~0.02% by volume) of resin binder together with cork granules – dust to 6 mm – and subjecting them to heat and pressure. This approach permits the development of grades with tailored properties for particular applications, such as expansion joint filling or for decorative applications. It is usually produced in sheet sizes of 914 × 610 × 0.8 to 200 mm thick, although greater thicknesses can be achieved by laminating. Reels are also available up to a maximum thickness of 9 mm; grades vary from 110 kg/m$^3$ to 560 kg/m$^3$. Two types of composition cork are as follows.

Expand-O-Jointing is a composition cork with the ability to expand from 1 to 2 inches and is suited for road and tunnel construction. This material should be sealed until placed in situ. The appropriate standard is ASTM 1752.84 Parts 1 & 2.

Cork conforming to BS4332 is a high-density composition cork for use in construction. Only 29 to 35 lb material should be used for flooring, as this will resist foot traffic.

It is prudent to request a *certificate of conformity* of the supplier; in addition, for all industrial or domestic uses, a sample should be requested.

Some compressed cork will satisfy Class 2 of BS476:1971 as a surface with a very slow spread of flame and is therefore assessed as having a good resistance to fire. To achieve Class 1, cork panels and granules need to be treated by the COS-FP1 process.

Cork is also used as part of the construction of sky lights, where the material will absorb moisture but will then stabilise, forming a snug seal, without rotting. Natural and composition cork panels with a taper may be used to form a variable gap seal against, say, concrete or coatings.

A further use of cork is for the lining of containers used for the storage of sensitive materials such as nuclear waste. For typical material specification see Tables 37.1 and 37.2.

## 37.6 Cork rubber

Cork rubber provides another family of products that are made into sheets of cork with a rubber binder. These materials may, for example, contain a number of synthetic rubbers, e.g. nitrile. Sheets are usually 1000 × 1000 or 1200 × 1200. Reels are also available up to 1240 mm in width; lengths will depend on thickness and EU weight restrictions. Thicknesses vary from 0.8 mm to 150 mm for sheets and from 0.8 mm to 8 mm for reels. Applications include:

- gaskets
- seals
- anti-vibration pads
- flooring
- underlays
- electrical insulation mats
- machine mountings.

For typical properties of these products, see Table 37.3. The material will vary from soft for example for use as an underlay to hard for heavy duty use.

Cork-based product quality is important, as it is for all materials. Therefore manufacturer certification to quality assurance systems such as ISO9001:2008 should be sought. This will give recognition that the supplier and the manufacturer of these materials, of whatever specific type, have monitored the material sourcing and manufacture and can demonstrate it has the properties claimed.

From humble beginnings as ancient Egyptian fishing floats to the Apollo space programme – as part of the re-entry shield – cork has proved ideal for solving a range of problems.

## 37.7 Standards

- BS EN 233:1999. *Wall coverings in roll form – specification for finished wall-papers, wall vinyls and plastics wall-coverings*. BSI, London, UK.
- BS EN 1815:1998. *Resilient and textile floor coverings – assessment of static electrical propensity*. BSI, London, UK.
- BS4332:1989. *Specification for phenol formaldehyde resin bonded cork jointing*. BSI, London, UK.
- BS EN 12455:1999. *Specification for corkment underlay*. BSI, London, UK.

**Table 37.1** Example of cork material specification

| *Cork composition COS24* | |
|---|---|
| Grade defined as dense, resin-bonded, composition cork. The grain is pre- and post-mould treated to enhance performance consistency – including moisture stability. | Applications – high-performance anti-vibration pads, packings, polishing wheels, seals. |
| *Physical properties* | *Requirements* |
| Thickness: up to 2.50 mm | ASTM F104 Type 2 material +0.25 mm |
| Over 2.50 mm | +10% |
| Density | 380–460 kg/m$^3$ |
| Conditioning | ASTM F104 Type 2 |
| Compressibility | ASTM F36 procedure F<br>10–20% |
| Recovery | ASTM F procedure F<br>75% minimum |
| Tensile strength | ISO 7322<br>1700 kPa minimum |
| Flexibility factor | ASTM F147 Type 2 material. × 5 maximum (as received) |
| Thermal conductivity | CPTM No. 37 (Lees disc)<br>0.055 W/m K maximum |
| Boiling water resistance | No disintegration |
| Specification compliance | ASTM F104 F219000M9 |

The specification used for the above applications has passed the following tests: (i) fire, (ii) drop, (iii) density, (iv) moisture. Each batch must carry a certificate for full specification both at block production and at finished product stage. The certificate should be held by the product supplier. This material can be made 'fireproof' to BS476 Part 7 Class 1 and NES 804 Part 1 issue 2.

**Table 37.2** Example of cork material specification

| *Cork composition COS13F* | |
|---|---|
| Grade defined as general-purpose fine-grain bonded composition cork. The grain is pre- and post-mould treated to enhance performance consistency, including moisture stability. | Applications – multi-situations. |
| *Physical properties* | |
| Thickness: up to 2.50 mm | ASTM F104 Type 2 material +0.25 mm |
| Over 2.50 mm | +10% |
| Density | 170–240 kg/m$^3$ |
| Conditioning | ASTM F104 Type 2 |
| Compressibility | ASTM F36 procedure F<br>30–50% |
| Recovery | ASTM F procedure F<br>75% minimum |
| Tensile strength | ISO 7322<br>400 kPa minimum |
| Flexibility factor | ASTM F147 Type 2 material. × 5 maximum (as received) |
| Thermal conductivity | @ 25 C: 0.038–0.042 W/m K |
| Specification compliance | ISO 7322 |

**Table 37.3** Composition cork with rubber binder

| *Test method* | *Test type* | *Units* | *Grade reference* | | | | |
|---|---|---|---|---|---|---|---|
| | | | *Cor 66* | *Cor 65* | *Cor 58* | *Cor 658* | *Cor 14* |
| ASTM F1315 | Density | kg/m$^3$ | 700–800 | 850–950 | 600–700 | 550–650 | 850–1000 |
| ASTM D2240 | Hardness | Shore A | 65–75 | 65–85 | 50–70 | 40–55 | 65–75 |
| ASTM F36 | Compressibility @ 400 psi | % | 25–35 | 15–25 | 40–55 | 55–65 | 15–25 |
| ASTM F36 | Recovery (min) | % | 80 | 75 | 75 | 80 | 80 |
| ASTM F152 | Tensile strength (min) | MPa | 20 | 17.5 | 10.5 | 8.0 | 26.4 |
| ASTM F147 | Flexibility F5 | – | No cracks | No cracks | No cracks | No cracks | No cracks |

*(Continued)*

**Table 37.3** *(Continued)*

| *Test method* | *Test type* | *Units* | *Grade reference* | | | | |
|---|---|---|---|---|---|---|---|
| | | | *Cor 66* | *Cor 65* | *Cor 58* | *Cor 658* | *Cor 14* |
| ASTM F146 | Volume change after immersion in ASTM No. 1 oil 70 hours @ 100°C | % | –5 to + 10 | –5 to +10 | –10 to +10 | –15 max | –5 to +5 |
| ASTM F146 | IRM 903 oil 70 hours @ 100°C | % | –2 to +15 | –2 to +15 | –5 to +30 | –15 max | +10 to +38 |
| ASTM F146 | Fuel A, 22 hours @ 23°C | % | –2 to +10 | –2 to +10 | O–to +10 | –5 to +10 | Not applicable |
| | Compliance | – | BS 21F66 | – | – | – | – |
| | Binder | – | Nitrile | Nitrile | Nitrile | Chlorosulfonated polyethylene | Silicone |
| | Colour | – | Grey | Brown | Black | Various | Various |
| | Cork granule size | – | 0.5–1.0 mm | 0.5–1.0 mm | 2.0–3.0 mm | 0.5–1.0 mm | 0.5–1.0 mm |

Synthetic rubber binders are incorporated primarily to increase cork's natural resilience for sealing, flexibility and anti-vibration applications. The data in this table were resourced by C. Olley & Sons Ltd. The grade colour reference indicates the binder type used, helping to distinguish different grades with different properties. The ASTM oil and fuel figures are standard requirements and show the degree of alteration caused by the oils and fuels.


- BS EN 12105:1998. *Resilient floor coverings – determination of agglomerated composite cork.* BSI, London, UK.
- BS EN 12149: 1998. *Wall coverings in roll form – determination of migration of heavy metals and certain other elements of vinyl chloride monomer and of formaldehyde release.* BSI, London, UK.
- BS EN 12956:1999. *Wall coverings in roll form – determination of dimensions, straightness, spongeability and washability.* BSI, London, UK.
- BS EN 13187:1999. *Thermal performance of buildings – qualitative detection of thermal irregularities in building envelopes – infrared method.* BSI, London, UK.
- BS EN ISO 13786: 1999. *Thermal performance of building components – dynamic thermal characteristics – calculation methods.* BSI, London, UK.
- BS EN ISO 13789: 1999. *Thermal performance of buildings – transmission heat loss coefficient – calculation method.* BSI, London, UK.
- BS EN ISO 14683: 1999. *Thermal bridges in building construction – linear thermal transmittance – simplified methods and default values.* BSI, London, UK.
- ISO1997:1972. *Granulated cork and cork powder – specifications.* BSI, London, UK.
- ISO2030:1990. *Granulated cork – size analysis by mechanical sieving.* BSI, London, UK.
- ISO2031:1991. *Granulated cork – determination of bulk density.* BSI, London, UK.
- ISO2066:1986. *Expanded pure agglomerated cork – determination of moisture content.* BSI, London, UK.
- ISO2067:1988. *Granulated cork – sampling.* BSI, London, UK.
- ISO2077:1979. *Pure expanded cork board – determination of the modulus of rupture by bending.* BSI, London, UK.
- ISO2189:1986. *Expanded pure agglomerated cork – determination of bulk density.* BSI, London, UK.
- ISO2190:1988. *Granulated cork – determination of moisture content.* BSI, London, UK.
- ISO2191:1972. *Cork – expanded pure agglomerated – deformation under constant pressure.* BSI, London, UK.
- ISO2219:1989. *Expanded pure agglomerated cork for thermal insulation – characteristics, sampling and packaging.* BSI, London, UK.
- ISO2509:1989. *Sound-absorbing expanded pure agglomerated cork in tiles.* BSI, London, UK.
- ISO2510:1989. *Sound-reducing composition cork in tiles.* BSI, London, UK.
- ISO2569:1985. *Cork stoppers – types and general characteristics.* BSI, London, UK.
- ISO2582:1978. *Cork and cork products – determination of thermal conductivity – hot plate method.* BSI, London, UK.
- ISO3810:1987. *Floor tiles of agglomerated cork – methods of test.* BSI, London, UK.
- ISO3813:1987. *Floor tiles of agglomerated cork – characteristics, sampling and packing.* BSI, London, UK.
- ISO3863:1989. *Cylindrical cork stoppers – dimensional characteristics, sampling and packing.* BSI, London, UK.
- ISO3867:1982. *Agglomerated cork material of expansion joints for construction and building-test methods.* BSI London
- ISO3869:1981. *Agglomerated cork – filler material of expansion joints for construction and buildings-characteristics, sampling and packing.* BSI, London, UK.
- ISO4707:1981. *Cork – stoppers – sampling for inspection of dimensional characteristics.* BSI, London, UK.
- ISO4708:1985. *Cork – composition cork gasket material – test methods.* BSI, London, UK.
- ISO4709:2000. *Cork – composition gasket material – specifications.* BSI, London, UK.
- ISO4711:1987. *Agglomerated cork discs – specifications.* BSI, London, UK.
- ISO4714:1986. *Composition cork – test methods.* BSI, London, UK.
- ISO7322:1986. *Cork – composition cork – test methods.* BSI, London, UK.
- ISO8507:1985. *Agglomerated cork discs – methods of test.* BSI, London, UK.
- ISO8724:1989. *Cork decorative panels – specification.* BSI, London, UK.
- ISO9148:1987. *Composition cork in rolls for decoration – test methods.* BSI, London, UK.
- ISO9149:1987. *Composition cork in rolls for decoration – specifications.* BSI, London, UK.
- ASTM C640-83. *Standard specification for corkboard and cork pipe thermal insulation.* ASTM, West Conshohocken, PA.
- DIN 18 161 Pt 1: 1976 *German standard: Cork products as insulating building materials.* DIN, Berlin, Germany.
- ASLIB 6621:1998. *Guidance on the new European test method standards for thermal insulation materials.* ASLIB, Bingley, UK.
- ASTM C353-84. *Standard test method for adhesion of dried thermal insulating or finishing cement.* ASTM, West Conshohocken, PA

# 38 Asbestos

**Bob Cather** BSc CEng FIMMM
Consultant, formerly Director of Arup Materials Consulting

**David Doran** FCGI BSc(Eng) DIC CEng FICE FIStructE
Consultant, formerly Chief Engineer with Wimpey plc

Based on material from the first edition by

**R. Harris**
Wimpey Environmental

## Contents

## 38.1 Introduction

Over the centuries, asbestos became progressively more utilised as its many physical and chemical properties became known. The range of products in use that contained asbestos became very large. The common perception of asbestos as the corrugated sheeting on garage and factory roofs was merely the tip of the iceberg. However, the chemistry and physical structure that made it so useful were also the cause of the serious health problems that have now excluded asbestos from use. Although not used in new construction or products, asbestos-containing materials (ACMs) may well be present in existing buildings and come into exposure during refurbishment and demolition projects. Regulations impose a duty on a building occupier to survey, assess and manage possible ACMs in the building. The background detail on asbestos is presented here to foster awareness of the material.

This chapter describes the uses and identification of some ACMs known to have been used. The coverage of examples should not be taken as definitive or complete, as other, less common combinations may have been used. The Regulations relevant to asbestos have evolved and some historical background is given later in the chapter. To ensure currency, reference should be made to Health & Safety Executive sources for guidance and required actions (www.hse.gov.uk).

## 38.2 Structure and physical properties

Asbestos is the general term used to describe a number of naturally occurring crystalline metal silicates which possess an array of physical and chemical properties, such as resistance to chemical attack, excellent thermal stability and good tensile strength. The minerals, when crystallised, form narrow parallel rods in tight groups or bundles. These groups or bundles, if physically stressed, not only will break into shorter lengths, but will split into progressively finer fibrils. This particular facet is the major reason for the mineral's combination of physical properties.

Six main types of asbestos have been used, namely chrysotile (commonly known as white asbestos), amosite (known as brown asbestos), crocidolite (blue asbestos), anthophyllite, actinolite and tremolite. Several other minerals are classed as asbestos but have not been used commercially for various reasons. Asbestos can be classified into two groups, the serpentines and the amphiboles, labels that refer primarily to aspects of their crystalline structure (see Table 38.1).

The serpentine group is distinct from the amphibole group in that its crystalline structures are formed in layer lattices. Chrysotile appears white to the naked eye, hence the name 'white asbestos'.

Amphiboles, which include the remainder of the asbestos types mentioned, are chain silicates, which means that, in contrast to the layered crystalline formation of the serpentines, they form preferentially along the lengthwise direction of the fibres, giving them a strong lengthwise tensile strength. Crocidolite and amosite were the only amphibole minerals to have been mined in significant quantities. Visually, crocidolite appears deep blue to turquoise in colour and amosite a distinctive grey-brown.

Chrysotile, because of its crystalline structure, has a very soft nature when broken in fibre tufts, thus it was ideal for weaving. Amphiboles, and in particular amosite, are very rigid and were therefore unsuitable for weaving, but have excellent thermal properties which were utilised in a variety of insulation materials.

## 38.3 Historical origins of asbestos

Asbestos is an abundant naturally occurring material that has been used for over 4000 years. The earliest known use was as a binding and reinforcing additive to pottery, examples of which have been found in archaeological sites in Finland and are estimated as dating to before 2000 BC.

Reference has been made to the use of asbestos in ancient Egypt, China and Greece. In fact the ancient Greeks were so impressed by the material's amazing fire resistance and its hard-wearing nature that they gave it the name by which we know it today, derived from the Greek for 'indestructible' – asbestos. The Romans are known to have used asbestos as lamp wicks and, when woven, as heat-resistant cloth. Applications such as these, though, were very rare and the material was treated as something of a novelty. There is very little evidence of asbestos being used in the Middle Ages outside the occasional instances of use as lamp wicks in parts of Europe and Asia.

Asbestos was not fully exploited as a major commercial commodity until the 19th century. At this time the mineral caught the imagination of the textile industry in Europe, where textile manufacturing had boomed, primarily in France and the north of Britain. Mill-owners were quick to spot the market-place potential of a material whose fibres could be woven like wool or cotton to produce a virtually indestructible fire-resistant fabric.

As we have seen, chrysotile (white) asbestos was the most amenable to milling and manufacture, while the amosite and crocidolite varieties were too rigid and brittle to be woven. In 1877, the first large-scale commercial mining operations started in Canada: mines were sunk in Thetford and Coleraine, but could not keep pace with demand. Further mining operations started in other parts of Canada, to be followed shortly in Russia and many other parts of the world. From these small beginnings, production grew from 10,000 t year$^{-1}$ in 1880 to 95,000 t year$^{-1}$ in 1925.

In the 1920s amosite and crocidolite began to be mined in significant quantities. Asbestos-board products were manufactured on a large scale for the first time. Asbestos cement and later asbestos insulation boarding were produced in vast quantities in purpose-built factories.

In the 1930s the first serious doubts about the health effects of working with, or being exposed to, asbestos were confirmed and legislation, the first of a series of progressively more stringent codes of practice for work with asbestos, was issued. This had no effect whatsoever on the demand for the material, and production continued to rise at ever faster rates.

**Table 38.1** Asbestos types, chemical formulae and physical properties

| *Type* | *Approx. chemical composition* | *Group* | *Colour* |
|---|---|---|---|
| Chrysotile | $3MgO, 2SiO_2, 2H_2O$ | Serpentine | White/pale green |
| Amosite | $5.5Fe, 1.5MgO, 8SiO_2, H_2O$ | Amphibole | Pale brown |
| Crocidolite | $Na_2O, Fe_2O_3, 3FeO, 8SiO_2, H_2O$ | Amphibole | Blue/turquoise |
| Anthophyllite | $7MgO, 8SiO_2, H_2O$ | Amphibole | White/pale brown |
| Actinolite | $2CaO, 4MgO, FeO, 8SiO_2, H_2O$ | Amphibole | Pale green |
| Tremolite | $2CaO, 5MgO, 8SiO_2$ | Amphibole | White |

The advent of the Second World War, rather than slowing the production of asbestos, actually increased it to record levels. The virtual destruction of European cities led to an even greater need for asbestos – large numbers of families were homeless, and to provide immediate shelter estates of prefabricated dwellings were constructed all over Europe, these dwellings being almost entirely clad with asbestos cement sheeting. In the UK, some still existed in East London until the late 1990s.

Asbestos production increased up to the 1970s, when stringent controls were introduced. Eventually, in 1985, the importation of amosite and crocidolite was banned along with the installations of any product containing them. In the minds of the public and Government, asbestos was too dangerous to be considered for future use. The installation of asbestos is now to all practical extent at an end. Its safe management, removal and disposal is the huge task that remains.

## 38.4 Extraction

The most common extraction method was by open-cast mining, though occasionally asbestos was mined from deep pits. If the surrounding bedrock was of a hard formulation it was not uncommon for drilling or dynamiting to be used in order to expose the asbestos. The mining operation was often quite indiscriminate and extracted a high percentage of non-asbestos material enmeshed with the mineral.

## 38.5 Asbestos in construction and industry

Asbestos has been incorporated in a vast array of products that have been used by the construction industry, from the earliest chrysotile rope used as fire-proof gasketing to boilers and fires, through to asbestos cement sheeting as a roofing material, thermal insulation to hot-water services and as a fibre reinforcement to bitumen waterproofing, paper backing to decorative sheet vinyl flooring and a fibre reinforcement/filler to resin-based products.

Recognition of possible asbestos-containing products in existing buildings is not straightforward, as outwardly they may not seem obvious. Table 38.2 lists some of the more common products that may have been manufactured using asbestos.

**Table 38.2** Asbestos-containing products that might be found in existing construction

- Textured paint
- Bituminous roofing felt
- Thermoplastic vinyl asbestos floor tiles
- Asbestos cement sheeting
- Asbestos cement pipes
- Asbestos cement roofing 'slates'
- Chimney/boiler flues
- Insulation board
- Asbestos mill board
- Gaskets and jointing materials
- Brake linings
- Thermal insulation
- Electrical insulation
- Asbestos textiles
- Asbestos paper-backed sheet vinyl flooring
- Spray coatings

Since the implementation of strict controls on the use of asbestos, products of similar outward appearance may not contain asbestos. Specialist advice and practice are required to determine its presence.

### 38.5.1 Asbestos cement products

Asbestos cement products were by far the most common asbestos-containing materials in use, and it is estimated that approximately 70% of world asbestos production was used in this way. The range of products made with asbestos cement was diverse (Table 38.2). Its origins lie in the late 19th century, when crude cement products were first made; mechanised production processes were realised by 1910.

Asbestos cement was manufactured using chrysotile asbestos (occasionally crocidolite was used in addition to the chrysotile) mixed with Portland cement. The process involved mixing finely graded tufts of chrysotile with Portland cement and water, usually with chrysotile making up 10% of the overall bulk.

These raw materials were thoroughly mixed to form a paste or slurry which was then poured into flat trays.

Suspended in the slurry was a rotating arrangement of cylinders whose surfaces were meshed so as to sieve the slurry, picking up the cement-covered asbestos fibres on its surface. As the cylinders rotated, they came into contact with a moving conveyor belt, the speed of rotation of the cylinder matched that of the conveyor belt and, as the two moving surfaces came close to contact, strong suction pressure was applied through the conveyor belt, effectively sucking the slurry off the rotating cylinder and onto the belt.

At the far end of the conveyor belt the material was cut into lengths which were transferred onto another series of rotating cylinders. This process continued with more layers being added to the roller until the coating reached the required thickness.

The material at this stage was quite malleable but retained its form when cut and stripped from the cylinder. In this state it could be cut and formed to provide the raw material for a variety of asbestos cement products.

The material is extremely hard wearing but does undergo sight changes when first installed in an external environment. The surface will initially be affected by rainfall, which is slightly acidic. This reacts with the Portland cement bonding to dissolve the cement, exposing the asbestos tufts, but this effect is slight and quickly stabilises.

Generally, because of the hardness of cement products and the relatively small amount of asbestos present, it is very difficult to liberate fibres from the material. Elevated airborne fibre levels may be created if the material is roughly handled or machined.

#### *38.5.1.1 Sheeting (corrugated and flat)*

Asbestos cement sheeting was manufactured in a variety of forms, shapes and sizes. Boards were shaped by pressing the still-wet chrysotile cement sheet between two flat formers to produce flat asbestos cement panels, or into corrugated asbestos cement sheeting. The main use for these sheets was as roofing panels and exterior cladding to buildings. Indeed, as we have seen, after the Second World War asbestos cement sheeting was used in vast quantities to build prefabricated dwellings.

#### *38.5.1.2 Rainwater goods*

Asbestos cement was used to make a range of roofing accessories and rainwater goods including ridging tiles, guttering and drainpipes. The products were manufactured in the same way as asbestos cement sheeting, the only difference being in the final moulding operation. Similarly, the piping products, including rainwater pipes and larger diameter drain products, were formed by wrapping the cement material around a cylindrical former. This technique was widely used in the manufacture of asbestos cement water tanks and cisterns, which were installed in large numbers in post-war housing stock.

Non-asbestos replacements for these products are available, typically of plastics, aluminium and steel.

*38.5.1.3 Slates*

Asbestos cement slate-like roofing tiles were used extensively for many years and were a cheap alternative to other forms of roofing tiles, being relatively light and hence requiring a reduced level of roofing support.

The tiles were usually 6 mm thick, of very hard formulations and manufactured with various colouring additives or coatings to give the distinctive slate-grey appearance. They usually contained between 10% and 30% chrysotile asbestos.

They were all identical in size, shape and thickness; the external surface was very smooth and the obverse face had a characteristic stippled effect. In the 1980s several asbestos-free fibrereinforced cement slates came on the market. Some asbestos-free slates are virtually identical to their asbestos counterparts and they can only be told apart under microscopic examination.

*38.5.1.4 Decking, decorative goods and high-temperature flues*

Most asbestos cement production was taken up by the manufacture of asbestos cement sheeting and rainwater goods. A small proportion was used for various diverse applications.

Certain asbestos cement products are coloured and textured to provide highly decorative, hard-wearing products. One boarding was a high-gloss black surface finish and contained approximately 30% chrysotile. This was mainly used to make roof tiles, but has also been used as window ledging, for worktops and as battening to window surrounds. The material was produced in a range of shades and was popular in schools, where it was sometimes used as a chemical-resistant worktop for laboratories.

Flue piping, though generally constructed with chrysotile, was occasionally made with amosite asbestos. This was used for flues where high temperatures were likely and for which amosite, with its excellent thermal qualities and high tensile strength, proved durable.

Asbestos cement was used in the manufacture of walkway and decking tiles, which were a common feature on flat asphalt roofs. The tiles are normally 12 inches square and 0.5 inches thick and are held in place with bitumen adhesive to form hard-wearing pathways across the flat asphalt. The material was installed predominantly in the 1950s and 1960s. More recently similar products, but using glass fibres, have been used.

### 38.5.2 Asbestos insulating board

Asbestos insulating board products are generally softer and less dense than their asbestos cement counterparts, having a density of approximately 500 kg $m^{-3}$ and a greater percentage content of asbestos. Chrysotile, amosite and crocidolite were all used in the manufacture of insulating boards, but by far the most commonly used constituent was amosite.

The percentage content of asbestos varies, typically being 20–40%, but greater proportions of asbestos are not uncommon, especially in the millboard varieties.

The method of manufacture of the boards was intrinsically the same as for asbestos cement sheeting. However, there was one main difference with some product ranges, which employed autoclaved calcium silicate as the binder.

Asbestos insulation boarding was usually manufactured in 2400 mm × 1200 mm sheets and in a range of thicknesses, typically 6, 9 and 12 mm. Other standard products were also made, including ordinary and drilled acoustic ceiling tiles.

Typical installations of asbestos insulation boarding included fire-retardant panels to heater cupboards, ducts, doors and ceilings, partition-wall panels, internal thermal insulation to curtain walling, acoustic and general-purpose ceiling tiles and soffit boarding to stairs and eaves. Insulation boarding will be found, predominantly, in post-war buildings.

The production of asbestos insulation boarding ceased in the UK in 1980 and the import of amosite, the main fibrous additive to the boarding, was banned in 1985. It is possible that insulation boards manufactured before 1980 may have been installed some considerable time after this date and it is therefore not safe to assume, purely because of the date of installation of a board, that it cannot contain asbestos.

Non-asbestos insulating boards have been used to replace their asbestos counterparts as they were removed for maintenance or health-and-safety reasons. These boards are produced in a similar manner to asbestos boards, but contain a mixture of man-made or vegetable fibre generally mixed with mica.

### 38.5.3 Asbestos insulation materials (thermal)

The next most popular application of the material after asbestos boarding products was asbestos insulation. All types of asbestos were used in thermal insulation, but although amosite and crocidolite exhibited the best thermal characteristics, chrysotile was the most frequently used component in these products.

The main application of asbestos in this role was as insulation to hot-water and steam pipes and cladding to boilers and calorifiers, though occasionally the raw material was used in a variety of roles such as loft and cavity-wall insulation.

In its most common form the material was applied by hand as a thick paste to pipes, boilers and other hot surfaces where heat retention was required. This thermal insulation material was made with three main constituents: a bonding mortar, a filling material, and a fibrous binding agent.

The fibrous binding agent was usually one or more varieties of asbestos and could well have been mixed with some other non-asbestos fibre such as glass fibre or vegetable fibre. The bonding agent would normally be a lime or mortar compound, and the mixture was usually filled with diatomaceous earths, silica and a variety of other ingredients that were at hand when the batch was mixed. There were virtually no specified or proprietary brand mixtures for insulation materials and hence the make-up of the material was very haphazard.

A distinctive feature of asbestos insulation to pipe work was the rounded effect at the end of each length of insulation.

The installation method, however, was very slow, labour-intensive and, consequently, expensive. Therefore it was not surprising that eventually a manufacturing process was created to make preformed sections of asbestos pipe and boiler insulation. These were usually manufactured using chrysotile or amosite and were formed by compressing the material into shapes suitable for direct application onto pipes and boilers. The preformed sections were available in a full range of sizes to fit all commonly used pipes, and in various thicknesses designed to provide a range of thermal conductivities.

Asbestos insulation materials were installed primarily in large buildings such as schools, hospitals and factories and also on ships, all of which make use of large boiler-fired heating and hot-liquid systems.

### 38.5.4 Asbestos insulation materials (acoustic)

The adoption of asbestos as an acoustic insulation material dates back to just before the Second World War. Several product types were applied, namely sprayed asbestos, asbestos boarding, loose raw asbestos and composite materials containing asbestos.

The application of sprayed asbestos to steelwork was not used generally until after the Second World War. The first documented evidence on the use of this method in the UK was in 1935, when it was used as sprayed fire and thermal insulation to railway carriages.

Sprayed asbestos was one of the most popular noise-reducing agents in use, for theatres, where the material was applied to walls and ceilings, and for the ceilings of plant and boiler rooms and basement car-parks where there was occupied accommodation above.

Asbestos insulation board was popular in offices and public areas where general noise reduction was required. The material was itself not particularly noise-reductive, the noise-attenuation coming primarily from the shaping and marking of board surfaces.

Two main types of acoustic-insulation board were in common use: drilled board, which has holes (approximately 5 mm in diameter) drilled at regular intervals across it; and textured boards, which have a sculpted surface with many air gaps that act as deflective acoustic traps. Both types of insulation board were found as tiling to suspended ceilings, designed to cut down reflected noise.

The material was installed on a large scale in post-war Britain, but gave rise to maintenance problems. As the tiles were mainly attached to suspended ceilings, above which electrical and water services often run, they were regularly removed to facilitate maintenance. The tiles were normally screwed onto the suspended ceiling frame, which meant that inconsiderate maintenance removal led to regular breaking of the boarding due to its soft nature.

Modern acoustic tiling is manufactured with textured surfacing using either glass fibre or other man-made material. These tiles have the advantage of being light and are usually unfixed, just resting on the suspended ceiling frame, giving quick, uncomplicated access to the ceiling void.

Other acoustic insulation uses of asbestos have included the raw asbestos mineral in its loose fibrous form being used as acoustic padding to ceilings and walls. However, this type of installation was rare and is not thought to have been a standard use.

Asbestos has been used in partnership with other materials to provide acoustic insulation, such as insulation boarding with slabs of man-made mineral fibre, usually in a sandwich construction. These were added to walls and ceilings to reduce noise transmission and to increase thermal insulation. This was found in converted buildings for multiple occupation, where more stringent controls on noise and fire/thermal insulation are required.

Chrysotile mixed with coarse felt was occasionally used to provide an acoustic matting for use in auditoria. This was mainly installed in theatres and cinemas from 1920 onwards; it was stuck to the walls with heavy adhesive and covered with canvas.

### 38.5.5 Felts, plastics, textured paints and other asbestos containing materials

Asbestos, primarily chrysotile, has been included in many relatively minor applications in construction. Some of these are described below.

#### *38.5.5.1 Thermoplastics and reinforced plastics*

From the 1950s, asbestos was used in conjunction with plastics, particularly PVC (or vinyl), to produce a series of products. The best known use is as thermoplastic floor tiling commonly found in kitchens and bathrooms. Thermoplastic floor tiles were made with a small proportion (typically 5–10%) of chrysotile asbestos, which was finely opened before mixing with the thermoplastic resins and colouring additives. This means that the material is very well bonded and it is virtually impossible to extract asbestos from the compounded material.

There were several variants of this material, including linoleum products that contained chrysotile. This linoleum was relatively heavy duty and is most likely to be found in public buildings, offices and factories. Modern thermoplastic floor tiles do not contain asbestos.

#### *38.5.5.2 Textured paints*

Many textured paints contained a small proportion of chrysotile asbestos as a binding and texturing agent. The content was usually less than 5% and was often less than 1% in later brands. The paints were intended for both internal and external applications. They may pose a hazard during removal as the material bonds well to correctly prepared surfaces and is difficult to remove.

#### *38.5.5.3 Jointing materials and gaskets*

Chrysotile was used in the manufacture of gasket and pointing products for many years. It was used in the pure woven form in boiler and heavy plant applications but also in conjunction with other materials.

Chrysotile cloth was used with a rubberised coating for liquid gasket and has been used extensively in engines as a heat-resisting seal. This particular form of gasket is made of resinous material mixed with finely graded chrysotile, usually as much as 60%, evenly distributed throughout the matrix. In applications such as cylinder head gaskets in automobiles, the gasket compound is usually sandwiched in a metal casing. However, as ordinary gasket, it was used on its own.

#### *38.5.5.4 Asbestos paper*

Asbestos paper products were made for many years and were used in several applications. The manufacturing process for asbestos paper was much the same as for ordinary paper. Finely opened asbestos fibres were suspended in water to form a slurry, which was evenly sieved and spread onto a fine mesh felt forming surface, compressed to expel most of the water, and heated to dry the material, which was then rolled. Amosite was the most widely used type of asbestos in the manufacture of asbestos paper, though chrysotile and crocidolite were also used.

The paper, in this basic form, was used for fire-retardant purposes in locations such as linings to electrical fuse boxes, though its other main use is in a thermal insulation role. In this case the paper is made in various thicknesses and formed into a corrugated 'cardboard' effect sheeting. This was used as insulation to the outer casings of boilers and flue pipes, and to manufacture preformed sectional pipe.

In appearance, the material is light grey and looks like ordinary paper. It is easily abraded and can generate significant levels of asbestos fibres if damaged.

Another form of asbestos paper, with a rubber binding constituent, was used as a temperature-stable substrate for decorative sheet vinyl flooring for domestic and commercial use. Production of such flooring material was widespread in the 1960s and 1970s.

#### *38.5.5.5 Bituminous felts*

Felts produced with bituminous impregnation and coating were frequently manufactured with a small chrysotile content to increase the material's bonding strength and to prevent cracking due to temperature fluctuations. The main usage of this material was as a waterproof roof covering, but it has also been used as damp-proof coursing and as flashing.

Other minor uses of asbestos included its additions to sealants and mastics, where a very small proportion of chrysotile was added to provide bonding and to prevent slumping.

## 38.6 Health effects of exposure to asbestos

Since the first years of the large-scale commercial exploitation of asbestos minerals, there has been a growing awareness that people involved in working with the material were developing a series of disabling and fatal illnesses. The first signs of the effects of asbestos were noted very early, but were put down to various unassociated ailments. The early years of the 20th century saw the first serious investigations into lung disorders in asbestos workers, and several instances of pleural fibrosis were noted. In the 1930s the number of cases of fibrosis of the lung in asbestos workers was seen to be rising, and a cause–effect relationship became undeniable. The condition became known as 'asbestosis' and its discovery, largely as a result of epidemiological evidence, hastened the introduction of legislation to reduce the level of exposure of people working in the asbestos industry.

As the uses and applications of asbestos grew, the incidence of related illness grew also. Asbestos was being used on a large scale as an insulation material and, in addition to chrysotile, new types of asbestos – amosite and crocidolite – were being installed. Cases of asbestosis outside the asbestos-manufacturing industry were on the increase as well.

After the Second World War, more research was carried out into the links between asbestos and a range of illnesses. Epidemiological evidence was emerging which showed a high incidence of lung cancer in people occupationally exposed to asbestos. This was compounded in the 1950s by further evidence of asbestos workers and laggers having a greater incidence of a rare malignant tumour called mesothelioma, which invariably proved fatal.

Since then, mounting evidence of causal relationships of asbestos to these and other diseases has led to the virtual ban on the use of this material and extremely strict restrictions on the disturbance of installed asbestos.

## 38.7 Legislation

### 38.7.1 Historical development

At first the correlation between disease and exposure to asbestos was not really noticed, but by 1910 a series of papers had been published following years of research which linked cases of what we now call 'asbestosis' to people who had worked in the asbestos industry. The disease was of the lungs and was obviously related to inhaled airborne asbestos dust. In the 1920s many more cases of asbestosis were diagnosed among asbestos workers, and eventually the government was forced to act.

In 1931 the Asbestos Industry Regulations were drafted; they eventually came into force in 1933. These were aimed solely at the asbestos manufacturing industry and were designed to improve the working conditions of asbestos workers. Great emphasis was placed on lowering dust levels at the workplace. This meant the increasing introduction of air-extraction systems to industrial processes alongside tighter requirements on the standards of shopfloor cleanliness, plus compulsory medical surveillance of all asbestos workers.

In the 1920s the asbestos industry had expanded rapidly and into many new areas. Asbestos work was no longer confined to the factories and, although chrysotile was by far the most popular form of asbestos used, amosite and crocidolite were being installed in significant quantities. The installation of asbestos as insulation in particular caused many people to be exposed to very high levels of asbestos fibres.

Gradually, as the long-term effects of asbestos exposure of laggers and associated workers began to appear with even greater frequency, it became clear that the Asbestos Industry Regulations were inadequate. They were not diverse enough to cover all types of asbestos-related works. To compound the problem, new types of asbestos-related disease were being diagnosed, including lung cancer and mesothelioma, incidences of which appeared to be associated with exposure to crocidolite. Mounting pressure eventually resulted in the introduction of the Asbestos Regulations (1969). These applied to all factories including construction sites and anywhere else where asbestos works were undertaken. The only asbestos works exempt were those that did not give off asbestos dust.

The regulations again placed great emphasis on the ventilation of work processes, good housekeeping and the safeguarding of personnel by personal respiratory protection and protective clothing.

Strict codes of practice for all work involving asbestos were introduced, including the regular servicing of equipment and the requirement to notify Her Majesty's Inspectorate of Factories at least 28 days before works, including removal, on any asbestos material that contained crocidolite.

However, the one area not covered by legislation was non-occupational exposure to asbestos – for those not directly working with asbestos. This shortcoming was eventually covered by the Health and Safety at Work, etc. Act (1974)

### 38.7.2 Current legislation and control

Legislation and other controls on the use, handling and disposal of asbestos-containing materials have developed considerably. To ensure that the appropriate actions on asbestos are taken, current legislation at the time of publication is not summarised in this book as it may continue to develop and change. Anyone involved with identifying, or actions that might involve, asbestos-containing materials must consult current regulations to determine their actions and responsibilities. UK guidance and regulatory requirements for safe actions are given by the Health & Safety Executive (www.hse.gov.uk).

# Index

Note: Page numbers in **bold** refer to pages which include figures. Page numbers in *italic* refer to pages which include tables